Handbook of Experimental Pharmacology

Volume 151/I

Editorial Board

G.V.R. Born, London
D. Ganten, Berlin
H. Herken, Berlin
K. Starke, Freiburg i. Br.
P. Taylor, La Jolla, CA

Springer

*Berlin
Heidelberg
New York
Barcelona
Hong Kong
London
Milan
Paris
Singapore
Tokyo*

Purinergic and Pyrimidinergic Signalling I

Molecular, Nervous and Urogenitary System Function

Contributors

M.P. Abbracchio, M. Boarder, G. Burnstock, I.P. Chessell,
G.P. Connolly, T.V. Dunwiddie, G.D. Housley, P.P.A. Humphrey,
K.A. Jacobson, C. Kennedy, L.J.S. Knutsen, A. Lorenzen,
S.A. Masino, A.D. Michel, J.T. Neary, M. Salter, U. Schwabe,
A. Sollevi, B. Sperlágh, E.S. Vizi, T.E. Webb, M. Williams,
H. Zimmermann

Editors

M.P. Abbracchio and M. Williams

 Springer

Professor
MARIA PIA ABBRACCHIO
Institute of Pharmacological Sciences
University of Milan
Via Balzaretti, 9
20133 Milan
ITALY
e-mail: mariapia.abbracchio@unimi.it

Dr. MICHAEL WILLIAMS
Department of Molecular Pharmacology and Biological Chemistry
Northwestern University Medical School
Searle 8-477 (S-215)
303 East Chicago Avenue
Chicago, IL 60611
e-mail: ruerivoli@aol.com

With 64 Figures and 21 Tables

ISBN 3-540-67849-2 Springer-Verlag Berlin Heidelberg New York

Library of Congress Cataloging-in-Publication Data
Purinergic and pyrimidinergic signalling / contributors, M.P. Abbracchio . . . [et al.];
 editors, M.P. Abbracchio and M. Williams.
  p. cm. – (Handbook of experimental pharmacology; v. 151)
 Includes bibliographical references and index.
 Contents: 1. Muscular, nervous, and urogenitary system function – 2. Cardiovascular, respiratory, immune, metabolic, and gastrointestinal tract function.
 ISBN 3540678492 (v. 1: alk. paper) – ISBN 3540678484 (v. 2: alk. paper)
 1. Purines – Receptors. 2. Pyrimidines – Receptors. 3. Purines – Physiological effect.
4. Pyrimidines – Physiological effect. I. Abbracchio, M.P. (Maria P.), 1956–
II. Williams, Michael, 1947 Jan. 3. III. Series.
QP905.H3 vol. 151
[QP801.P8]
615′.1s – dc21
[615′.7]   00-048277

This work is subject to copyright. All rights reserved, whether the whole or part of the material is concerned, specifically the rights of translation, reprinting, re-use of illustrations, recitation, broadcasting, reproduction on microfilms or in any other way, and storage in data banks. Duplication of this publication or parts thereof is permitted only under the provisions of the German Copyright Law of September 9, 1965, in its current version, and permission for use must always be obtained from Springer-Verlag. Violations are liable for Prosecution under the German Copyright Law.

Springer-Verlag Berlin Heidelberg New York
a member of BertelsmannSpringer Science+Business Media GmbH

© Springer-Verlag Berlin Heidelberg 2001
Printed in Germany

The use of general descriptive names, registered names, etc. in this publication does not imply, even in the absence of a specific statement, that such names are exempt from the relevant protective laws and regulations and free for general use.

Product liability: The publishers cannot guarantee the accuracy of any information about dosage and application contained in this book. In every individual case the user must check such information by consulting the relevant literature.

Cover design: design & production GmbH, Heidelberg
Typesetting: Best-set Typesetter Ltd., Hong Kong

SPIN: 10705319   27/3020xv – 5 4 3 2 1 0 – printed on acid-free paper

Preface

The field of purinergic receptor research has gradually emerged from that of a controversial "left-field" intellectual curiosity of the 1970s to a mainstream biomedical research activity for the twenty-first century that promises – with the appropriate degree of rigor in the selection of disease targets in which purines play a key role in affecting tissue pathophysiology – to deliver novel medications within the next decade or two.

While many gifted individuals, both biologists and chemists, have contributed to the body of knowledge related to purinergic receptor function generated over the past 30 years, the genesis of the field is irrevocably associated with the vision, tenacity, passion, and proselytizing of Professor Geoffrey Burnstock. In two seminal reviews, BURNSTOCK (1972) and BURNSTOCK (1978), Geoff described the conceptual nucleus that provided the framework for the creation of the present body of knowledge that is contained in this volume. The spectacular advances in the molecular biology and functional pharmacology/physiology of P1 (adenosine) and P2 (ATP, ADP, UTP, UDP) receptors over the past decade are a direct result of: (a) Geoff's instance, in the face of considerable and highly vocal skepticism, that ATP and adenosine were key mediators of extracellular communication, acting via discrete cell surface receptors; and (b) the many collaborators, now numbering more than eight hundred (KING and NORTH 2000), who have worked with Geoff over the past three decades.

The present two-volume *Handbook of Experimental Pharmacology*, on *Purine and Pyrimidinergic Signalling* (HEP 151/I and HEP 151/II), presents a "state of the art" overview of the many aspects of P1 and P2 receptor function. Volume I covers *Molecular, Nervous and Urogenitary System Function* and Vol. II *Cardiovascular, Respiratory, Immune, Metabolic and Gastrointestinal System Function*.

Volume I starts with a brief historical overview of the field provided by the volume editors, MARIA P. ABBRACCHIO and MICHAEL WILLIAMS (Chap. 1). Among other things, the events and discoveries that lead to adopting the currently used P1 and P2 receptor nomenclature are reviewed.

Molecular Aspects of Purinergic Receptors

The task of reviewing P1 (adenosine) receptors, their structure and signal transduction properties is covered by ANNA LORENZEN and ULRICH SCHWABE in Chap. 2. In Chap. 3, IAN CHESSELL, ANTON D. MICHEL and PAT HUMPHREY review the molecular properties of the ionotropic P2X receptor family and in Chap. 4, MIKE BOARDER and TANIA WEBB review the structure and function of the P2Y G-protein coupled, metabotropic receptor family.

In the first of three chapters contributed to this monograph, GEOFF BURNSTOCK discusses developmental aspects of P2 receptor function in Chap. 5.

Medicinal Chemistry

In a key chapter that provides much of the structural information for other chapters in both parts of the monograph, KEN JACOBSON and LARS KNUTSEN review progress in the characterization and optimization of P1 and P2 purine and pyrimidine receptor ligands (Chap. 6).

Purine Release

In Chap. 7, BEATA SPERLAGH and SYLVESTER VIZI discuss the various physiological and pathophysiological stimuli involved in the release of purines into the extracellular space, while the expanding knowledge of the diversity of the ectonucleoside (E-NTPase) family of enzymes involved in the metabolism of purines and pyrimidines is covered in detail by HERBERT ZIMMERMAN (Chap. 8), highlighting the crucial role of these enzymes in regulating the functional effects of nucleotides and nucleosides.

Purines in the Nervous System

SUSAN MASINO and TOM DUNWIDDIE review the role of P1 and P2 receptors and their interactions in central nervous system function (Chap. 9), while CHARLES KENNEDY extends this topic to consider the role of purinergic receptors in the peripheral nervous system (Chap. 10).

In Chapter 11, JOE NEARY and MARIA ABBRACCHIO discuss the involvement of P1 and P2 receptor mechanisms in maintaining nervous system viability via modulation of the production, release and effects of trophic factors (both neurotrophins and pleiotrophins). GARY HOUSLEY (Chap. 12) reviews data on the distribution and function of P1 and P2 receptors (the latter in the form of unique splice variants) in auditory and ocular function, while MIKE SALTER and ALF SOLLEVI discuss the role(s) of adenosine and ATP in nociception (Chap. 13), covering in detail recent advances on the role(s) of these purines in modulating pain perception at the preclinical level and in the clinical setting. In Chap. 14, GERALD CONNOLLY focuses on the emerging functional evidence for distinct pyrimidine receptors sensitive to UTP and UDP, complimenting

BOARDER and WEBB's chapter on the molecular biology of P2Y receptors (Chap. 3).

In the final chapter in Vol. I of this monograph, GEOFF BURNSTOCK discusses the role of purines in urogenital function (Chap. 15), building on the "tube and sac" hypothesis (BURNSTOCK 1999). The latter focuses on increases in organ lumen volume, e.g., gut and gall bladder distension as a trigger for ATP release from endothelial sources which can then mediate peristalsis (physiological stimulus) and pain (pathophysiological stimulus).

Cardiovascular Function

In the first chapter (Chap. 16) of the second volume of this *Handbook*, MICHAEL BROAD and JOEL LINDEN provide an update on the well documented role of adenosine in modulating cardiovascular system function, a data base that has been instrumental in helping develop the purinergic nerve hypothesis. EDWIN JACKSON (Chap. 17) extends this update to a detailed overview of the role of purines in renal system physiology and its contribution to cardiovascular function.

AMIR PELLEG and GUY VASSORT (Chap. 18) review the role of ATP and P2 receptor activation in cardiovascular system function and VERA RALEVIC (Chap. 19) discusses the roles of purines and pyrimidines in endothelial function.

SUSANNA HOURANI (Chap. 20) provides information on the role of purines in the modulation of platelet function via ADP-sensitive receptor mechanisms, a highly topical subject given (a) the advancement of novel ADP antagonists to the clinic as improved antithrombotic agents and (b) the recent cloning of the highly elusive P_{2T}/$P2Y_T$ receptor as the $P2Y_{12}$ receptor (HOLLOPETER et al. 2000).

Gastrointestinal Tract Function

In his final contribution to this monograph, GEOFF BURNSTOCK provides a comprehensive discourse on the role of purines in gut function (Chap. 21), highlighting potential opportunities for drug development.

Respiratory Tract Function

ITALO BIAGGIONI and IGOR FEOKTISTOV (Chap. 22) review the involvement of adenosine mechanisms in pulmonary function with CRAIG WEGNER (Chap. 23), complimenting this chapter with an overview of the effects of ATP in respiratory tract function.

Immune Function

CARMEN MONTESINOS and BRUCE CRONSTEIN discuss the role of P1 receptors in inflammation and in the action of known antiinflammatory agents

(Chap. 24) and GEORGE DUBYAK (Chap. 25) discusses the dynamics of P2 receptors in immune system function. This is complimented by a chapter from FRANCESCO DI VIRGILIO, VENKATERAMAN VISHAWANATH and DAVIDE FERRARI (Chap. 26) on the emerging role of the cytolytic, pore forming $P2X_7$ receptor in immune cell apoptotic and necrotic processes.

Metabolic Function

PIERRE PETIT and his colleagues, D. HILLAIRE-BUYS, M. M. LOUBATIÈRES-MARIANI and J. CHAPAL, discuss the role of purinergic receptors in the pathophysiology of type 2 diabetes. (Chap. 27), and JACK FERRIER (Chap. 28) discusses similar roles for P1 and P2 receptors in the regulation of bone formation and function.

In the final chapter (Chap. 29), MICHAEL WILLIAMS provides an update on advances in purinergic based therapeutics in a variety of disease areas and also highlights the challenges in taking the vast body of basic research in the area of purinergic receptor function into the clinic arena.

The Editors are indebted to all of the authors for their hard work and patience in providing the excellent contributions in this monograph, most of which arrived in the time frame requested, and would also like to thank our principal Editor, Doris Walker at Springer-Verlag in Heidelberg, not only for her support and advice, but also for her gentle but constant prodding via e-mail to get this volume completed, one of the last projects she worked on before retiring. We would also like to thank Anja McKellar, who seamlessly took over the production of this two-volume set from Doris and brought it to fruition.

The editors are also deeply indebted to Angelo, Alessandro, Holly and Heather for their patience and continuous support as their spouse/parent spent additional evenings and weekends beyond the usual science-based activities that do not fit into the regular workday, editing the contributions of the authors.

Finally, they would like to thank Geoff Burnstock for his friendship, counsel and very active support in helping bring this project to completion. Without Geoff's vision, it is highly unlikely that the field would have reached a stage that the *Handbook of Experimental Pharmacology* would need to devote more than 1000 pages to cover the topic of *Purine and Pyrimidinergic Signalling*.

MARIA P. ABBRACCHIO and MICHAEL WILLIAMS
Milan and Lake Forest, August 2000

References and Additional Reading

Burnstock G (1972) Purinergic nerves. Pharmacol Rev 24:509–581
Burnstock G (1978) A basis for distinguishing two types of purinergic receptor. In: Straub RW, Bolis L (eds) Cell membrane receptors for drugs and hormones: a multidisciplinary approach. Raven, New York, pp 107–118

Burnstock G (1999) Release of vasoactive substances from endothelial cells by shear stress and purinergic mechano-sensory transduction. J Anat 194:335–342

Hollopeter G, Jantzen HM, Vincent D, Li G, England L, Ramakrishnan V, Yang RB, Nurden P, Nurden A, Julius D, Couley PB (2000) Identification of the Platelet ADP Receptor Targeted by Antithrombotic Drugs. Nature, in press

Jacobson KA, Jarvis MF (1997) Purinergic approaches in experimental therapeutics. Wiley-Liss, New York, pp 55–84

King BF, North RA (2000) Purines and the autonomic nervous system: from controversy to the clinic. J Auton Nerv Syst 81:1–300

NB: Because of the various changes in P1 receptor nomenclature over the years, many early papers described A_2 receptors. Since this receptor actually exists in two subforms, A_{2A} and A_{2B}, the terminology A_2-like is used in the present volumes to describe receptors that have been designated as A_2 receptor.

In addition, to achieve the best use of space and avoid duplication, the structures for many of the compounds referred to throughtout the text are collated in Chap. 6.

List of Contributors

ABBRACCHIO, M.P., Institute of Pharmacological Sciences, University of Milan, Via Balzaretti 9, 20133 Milano, Italy
e-mail: mariapia.abbracchio@ unimi.it

BOARDER, M., Cell Signalling Laboratory, Department of Biological Sciences, De Montfort University, The Gateway, Leicester LE1 9BH, UK
e-mail: mboarder@dmu.ac.uk

BURNSTOCK, G., Autonomic Neuroscience Institute, Royal Free and University College Medical School, Royal Free Campus, Rowland Hill Street, London NW3 2PF, UK
e-mail: g.burnstock@ucl.ac.uk

CHESSELL, I.P., Glaxo Institute of Applied Pharmacology, University of Cambridge, Department of Pharmacology, Tennis Court Road, Cambridge CB2 1QJ, UK
e-mail: ic44126@glaxowellcome.co.uk

CONNOLLY, G.P., Department of Chemical Pathology, Guy's, King's, Thomas' Medical School, London, UK
e-mail: gerald.connolly@kcl.ac.uk

DUNWIDDIE, T.V., Neuroscience Program and Department of Pharmacology, University of Colorado Health Sciences Center, 4200 E. 9th Avenue, Denver, CO 80262, USA
e-mail: tom.dunwiddie@uchsc.edu

HOUSLEY, G.D., Molecular Physiology Laboratory, Department of Physiology, University of Auckland, Private Bag 92019, Auckland, New Zealand
e-mail: g.housley@auckland.ac.nz

HUMPHREY, P.P.A., Glaxo Institute of Applied Pharmacology, University of Cambridge, Department of Pharmacology, Tennis Court Road, Cambridge CB2 1QJ, UK
e-mail: ppah2@cam.ac.uk

JACOBSON, K.A., Molecular Recognition Section, Laboratory of Bioorganic
  Chemistry, NIDDK, National Institutes of Health, Bethesda, MD, USA
  e-mail: kajacobs@helix.nih.gov

KENNEDY, C., Department of Physiology and Pharmacology,
  Strathclyde Institute of Biomedical Research, University of Strathclyde,
  27 Taylor Street, Glasgow G4 0NR, UK
  e-mail: c.kennedy@strath.ac.uk

KNUTSEN, L.J.S., Vernalis Limited, Oakdene Court, 613 Reading Road,
  Winnersh, Wokingham, Berkshire, RG41 5UA, UK
  e-mail: l.knutsen@cerebrus.ltd.uk

LORENZEN, A., Institute of Pharmacology, University of Heidelberg,
  D-69120 Heidelberg, Germany
  e-mail: anna.lorenzen@urz.uni-heidelberg.de

MASINO, S.A., Neuroscience Program and Department of Pharmacology,
  University of Colorado Health Sciences Center, 4200 E. 9th Avenue,
  Denver, CO 80262, USA
  e-mail: Susan.Masino@UCHSC.edu

MICHEL, A.D., Glaxo Institute of Applied Pharmacology, University of
  Cambridge, Department of Pharmacology, Tennis Court Road,
  Cambridge CB2 1QJ, UK
  e-mail: adm7393@glaxowellcome.co.uk

NEARY, J.T., Research Service, Veterans Affairs Medical Center, and
  Departments of Pathology and Biochemistry and Molecular Biology,
  University of Miami School of Medicine, Miami, Florida 33125, USA
  e-mail: jneary@med.miami.edu

SALTER, M., Programme in Brain and Behaviour, Hospital for Sick Children,
  Department of Physiology, University of Toronto, Toronto, Ontario,
  Canada
  e-mail: mike.salter@utoronto.ca

SCHWABE, U., Institute of Pharmacology, University of Heidelberg,
  D-69120 Heidelberg, Germany
  e-mail: ulrich.schwabe@urz.uni-heidelberg.de

SOLLEVI, A., Department of Anaesthesiology and Intensive Care,
  Huddinge University Hospital, S-141 86 Huddinge, Sweden
  e-mail: alf.sollevi@anaesth.hs.sll.se

SPERLÁGH, B., Department of Pharmacology, Institute of Experimental
Medicine, Hungarian Academy of Sciences, H-1450 Budapest,
PO Box 67, Hungary
e-mail: sperlagh@koki.hu

VIZI, E.S., Department of Pharmacology, Institute of Experimental
Medicine, Hungarian Academy of Sciences, H-1450 Budapest,
PO Box 67, Hungary
e-mail: esvizi@koki.hu

WEBB, T.E., Department of Cell Physiology and Pharmacology, University of
Leicester, PO Box 138, University Road, Leicester LE2 9HN, UK
e-mail: tewebb@dmu.ac.uk

WILLIAMS, M., Department of Molecular Pharmacology and Biological
Chemistry, Northwestern University Medical School, Searle 8-477
(S-215), 303 East Chicago Avenue, Chicago, IL 60611, USA
e-mail: ruerivoli@aol.com

ZIMMERMANN, H., AK Neurochemie, Biozentrum der J.W. Goethe-
Universität, Marie-Curie-Str. 9, D-60439 Frankfurt am Main, Germany
e-mail: H.Zimmermann@zoology.uni-frankfurt.de

Contents

CHAPTER 1

Purinergic Neurotransmission: An Historical Background
M.P. Abbracchio and M. Williams. With 1 Figure 1

A. Introduction 1
B. Purine Availability 3
C. P1 and P2 Receptor Nomenclature 6
    I. Adenosine (P1) Receptors 6
D. P2 (ATP and UTP) Receptors 8
    I. P2X Receptors 8
    II. P2Y Receptors 9
E. Mitochondrial Purinergic Receptors? 9
F. Transgenic Models of P1 and P2 Receptor Function 10
G. Exponential Diversity? 10
H. Conclusions 11
References 11

Section I: Receptor Classification

CHAPTER 2

P1 Receptors
A. Lorenzen and U. Schwabe. With 2 Figures 19

A. Introduction 19
B. Localization and Function of P1 Receptors 19
    I. A_1 Adenosine Receptors 20
    II. A_{2A} Adenosine Receptors 21
    III. A_{2B} Adenosine Receptors 22
    IV. A_3 Adenosine Receptors 23
C. Structure of Adenosine Receptors 23
D. Signal Transduction by Adenosine Receptors 26

E. Adenosine Effects with Unknown Mechanisms and
   Adenosine Binding Proteins 32
List of Abbreviations 35
References 35

CHAPTER 3

P2X Receptors
I.P. CHESSELL, A.D. MICHEL and P.P.A. HUMPHREY. With 2 Figures 47

A. Introduction 47
B. P2X Homomeric Subunits and Their Properties 48
   I. $P2X_1$ 48
   II. $P2X_2$ 50
   III. $P2X_3$ 50
   IV. $P2X_4$ 51
   V. $P2X_5$ and $P2X_6$ 53
   VI. $P2X_7$ 53
C. Effects of Ions on Homomeric P2X Receptor Assemblies 56
D. Receptor Binding Studies 57
E. Heteropolymeric Combinations of P2X Subunits 57
F. Conclusions 59
References 59

CHAPTER 4

P2Y Receptors: Structure and Function
M.R. BOARDER and T.E. WEBB. With 3 Figures 65

A. Introduction 65
B. P2Y Receptors 66
   I. G Protein-Coupled Receptors 66
   II. Diversity of P2Y Receptors: Cloning and Nomenclature ... 67
   III. Diversity of P2Y Receptors: Pharmacology 68
      1. ATP and $P2Y_1$ Receptors 69
      2. ATP and $P2Y_4$ Receptors: Species Differences Between
         Rat and Human 70
      3. Agonists at P2Y Receptors 70
      4. Antagonists at P2Y Receptors 71
C. Correlating Structure with Function 72
   I. Sequence Analysis 72
   II. Structure and Function 75
      1. Agonist/Receptor Interactions 75
      2. Events Downstream of Receptor Activation 77
   III. Post-Translational Modifications 77
      1. Asparagine-Linked Glycosylation 77
      2. Phosphorylation Sites 77

IV. Non-G Protein Associations	77
1. Integrins	77
2. Na^+/H^+ Exchanger Regulatory Factor	77
D. Desensitisation	79
E. Cell Signalling from P2Y Receptors	80
I. Coupling to Phospholipase C	80
II. Phospholipase C Independent Ca^{2+} Mobilisation	80
III. Regulation of Cyclic AMP	80
IV. Regulation of Ion Channels	81
V. Stimulation of Tyrosine Kinases and Mitogen Activated Protein Kinases	81
References	82

CHAPTER 5

Purinergic Signalling in Development
G. BURNSTOCK. With 8 Figures 89

A. Introduction	89
B. Embryological Development	89
I. Frog Embryos	89
II. Chick Embryos	91
1. Retina	91
2. Glial Cells	94
3. Skeletal Muscle	95
4. Cardiovascular System	95
5. Ganglia	97
6. Chondrocytes	98
III. Purinergic Receptors in Mammalian Embryos	98
1. Early Embryos	98
2. Central Nervous System (CNS)	101
3. Cardiorespiratory System	102
4. Skeletal Muscle	103
5. Gastrointestinal Tract	104
6. Skin	104
7. Bone	104
8. Lung	105
9. Ectoenzymes	105
IV. Human Embryos	106
C. Postnatal Development	106
I. Central Nervous System	106
II. Cardiovascular System	109
1. Heart	109
2. Vascular System	109
III. Airways	111

		IV. Gastrointestinal Tract	111
		V. Vas Deferens	114
		VI. Urinary Bladder	115
		VII. Other Organs	115
D. Summary			116
References			116

CHAPTER 6

P1 and P2 Purine and Pyrimidine Receptor Ligands
K.A. JACOBSON and L.J.S. KNUTSEN. With 13 Figures 129

A. Introduction					129
B. P1 Receptor Ligands					130
	I. P1 Receptor Agonists				130
		1. Non-Selective Adenosine Receptor Agonists			130
			a) NECA		130
			b) Metrifudil		130
			c) I-AB-MECA		133
		2. A_1 Receptor Agonists			133
			a) CPA		133
			b) R-PIA		133
			c) ADAC		134
			d) GR 79236		135
			e) SDZ WAG 994		135
			f) NNC 21–0136		136
		3. A_{2A} Receptor Agonists			136
			a) CGS 21680		136
			b) DPMA		137
			c) APEC		137
			d) HENECA		138
		4. A_3 Receptor Agonists			138
			a) IB-MECA		138
			b) Cl-IB-MECA		139
			c) NNC 21–0238		140
	II. P1 Receptor Antagonists				140
		1. A_1 Receptor Antagonists			140
			a) DPCPX		140
			b) KFM 19		142
			c) BG 9719		142
			d) WRC-0571		142
		2. A_{2A} Receptor Antagonists			142
			a) SCH 58261		142
			b) CSC		143
			c) KW 6002		144

d) ZM 241385	144
3. A_{2B} Receptor Antagonists	145
a) MRS 1754	145
4. A_3 Receptor Antagonists	145
a) L-249313	145
b) MRS 1191	146
c) MRS 1220	146
C. P1 Receptor Modulators	147
D. P2 Receptor Ligands	147
I. P2 Receptor Agonists	149
1. 2-MethylthioATP and 2-MethylthioADP	152
2. 2-(7-Cyanohexylthio)-ATP	153
3. α,β-MeATP	153
4. β,γ-Methylene-D-ATP and β,γ-Methylene-L-ATP	154
5. 3'-Benzylamino-3'-deoxyATP	154
6. Adenosine 5'-O-(3-thiotriphosphate) (ATPγS)	155
7. 2'- and 3'-O-(4-Benzoylbenzoyl)-ATP 2'- and 3'-O-(4-Benzoylbenzoyl)-ATP (BzATP)	155
8. UTP and UDP	156
9. UTPγS	156
10. HT-AMP and Other AMP-2-Thioethers	156
II. P2 Receptor Antagonists	157
1. Suramin, NF023, and NF279	157
2. Reactive Blue 2	157
3. Pyridoxal Phosphate Derivatives (e.g., PPADS)	160
4. Nucleotide Derivatives	162
5. MRS 2179 and Other Bisphosphate Analogues	162
6. KN-62	162
E. Conclusions	163
References	163

Section II: Neurotransmission

CHAPTER 7

Regulation of Purine Release
B. SPERLÁGH and E.S. VIZI. With 7 Figures 179

A. Introduction	179
B. Releasable Purine Stores in Neuronal and Non-Neuronal Cells	179
I. Releasable ATP Stores	179
II. Releasable Adenosine Stores	181
C. Release of Purines During Physiological Neuronal Activity Simulated by Electrical or Chemical Depolarization	182

	I. Release of Adenosine	182
	II. Release of ATP	183
	1. Source of ATP Release in the Central Nervous System (Presynaptic Origin)	184
	2. Source of ATP Release in the Peripheral Nervous System: Cascade Transmission (Postsynaptic Origin)	190
D.	Release of Purines by the Stimulation of Pre- or Post-Synaptic Receptors	194
E.	Release of Purines by Pathological Stimuli	195
	I. Release of Purines by Hypoxia/Hypoglycemia/Energy Deprivation	195
	II. Release of Purines by Inflammatory Stimuli	196
	III. Release of Purines by Cellular Hypotonia	198
	IV. Purine Release by Cell Death Apoptosis	199
F.	Concluding Remarks	199
List of Abbreviations		200
References		200

CHAPTER 8

Ecto-Nucleotidases

H. ZIMMERMANN. With 2 Figures 209

A.	Introduction	209
B.	The E-NTPDase Family	211
	I. From Yeast to Vertebrates	211
	II. Vertebrate Isoforms	211
	III. General Structural Properties	212
	IV. Catalytic Properties	214
	1. Hydrolysis of Purine and Pyrimidine Nucleotides and Cation Dependence	214
	2. K_m-Values	215
	V. Apyrase Conserved Regions and the β- and γ-Phosphate Binding Motif	216
	VI. Inhibitors of Ecto-Nucleotidases	217
	1. Search for Potency and Specificity	217
	2. Nucleotide Analogues	218
	3. Inhibitors of P2-Purinergic Receptors as Ecto-Nucleotidase Inhibitors	218
	4. Others	219
	VII. Wide and Overlapping Tissue Distribution	219
	1. Overall Distribution	219
	2. The Cardiovascular System	220
	3. The Nervous System	222
	VIII. Regulation of Expression	223
	IX. Physiological and Pathological Implications	223

	1.	Role in Thrombosis and Vascular Reperfusion	223
	2.	Upregulation Following Global Forebrain Ischemia	224
	3.	Alterations Following Plastic Changes in the Nervous System	225
C.	The E-NPP Family		225
	I.	Three Types and their Structural Properties	225
	II.	Catalytic Properties	226
		1. Substrates and Inhibitors	226
		2. Cation Dependence and K_m-Values	227
		3. The Catalytic Cycle	227
	III.	EF-Hand, Somatomedin B-Like Domain and RGD-Tripeptide	227
	IV.	Cell and Tissue Distribution	228
	V.	Functional Roles in Physiology and Pathology	229
D.	Extracellular Hydrolysis of Diadenosine Polyphosphates		230
E.	Ecto-5′-Nucleotidase		231
F.	Alkaline Phosphatase		232
G.	Ecto-Nucleoside Diphosphokinase and Ecto-Myokinase		233
H.	Extracellular Hydrolysis of NAD^+		233
I.	Ecto-Phosphorylation and Ecto-ADP Ribosylation		234
J.	Outlook		235
References			236

CHAPTER 9

Role of Purines and Pyrimidines in the Central Nervous System
S.A. MASINO and T.V. DUNWIDDIE 251

A.	Introduction	251
B.	P1 Adenosine Receptors in Nervous Tissue	251
	I. Physiological Responses	252
	1. Electrophysiological Actions	252
	a) Modulation of Transmission	253
	b) Modulation of Plasticity	254
	c) Atypical A_{2A} Receptors	254
	d) Interaction between Receptors	255
	e) Adenosine–Nitric Oxide Interactions	257
	2. Behavioral Actions	257
	a) Pain	258
	b) Sleep/Arousal	258
	c) Thermoregulation	260
	d) Addiction	260
	e) Ethanol/Adenosine	261
	f) Interactions with Dopamine Systems	263
	α) Role of Adenosine in Pathological Conditions	263

				β) Locomotor Effects	264
			3.	Functional Role(s) of Adenosine	264
				a) Neuroprotection	265
				b) Epilepsy	267
	II.	Regulation of Extracellular Adenosine Concentrations in Brain			268
			1.	Adenosine Transporters	268
			2.	Activity-Dependent Adenosine Release	269
			3.	cAMP Efflux	270
			4.	Extracellular Nucleotide Metabolism	270
			5.	Adenosine Phosphorylation and Degradation	271
	III.	General Conclusions and New Directions			271
C.	P2 Nucleotide Receptors				272
	I.	P2 Receptor Distribution			272
	II.	Physiological Responses to Application of ATP and Other Nucleotides			273
	III.	Synaptic Responses Mediated via ATP			275
	IV.	Physiological Effects of Diadenosine Polyphosphates			276
	V.	Responses Mediated by Pyrimidine Receptors			276
D.	Conclusions				277
References					277

CHAPTER 10

The Role of Purines in the Peripheral Nervous System
C. KENNEDY. With 2 Figures 289

A.	Introduction			289
B.	ATP and NA as Sympathetic Neurotransmitters			290
	I.	Storage in and Release of ATP Release from Sympathetic Nerves		290
	II.	Functional Effects of ATP Released from Sympathetic Nerves		292
		1. Electrophysiological Effects		292
			a) Excitatory Junction Potentials	292
			b) P2X Receptors	293
		2. Smooth Muscle Contraction		293
			a) Calcium Influx	293
			b) Contraction Profile	294
	III.	Termination of the Neurotransmitter Actions of ATP and NA		294
	IV.	Issues to be Resolved		294
		1. Source of ATP Released During Sympathetic Nerve Stimulation		295
		2. Transient or Maintained Release of ATP		295

C. ATP and ACh as Parasympathetic Neurotransmitters 296
  I. Functional Effects of ATP Released from
     Parasympathetic Nerves 296
     1. Excitatory Junctional Potentials 296
     2. Smooth Muscle Contraction 296
     3. Calcium Channels Involved in Neurotransmitter
        Release 297
     4. Termination of the Neurotransmitter Actions of
        ATP and ACh 298
  II. Parasympathetic Neurotransmission in the Human Urinary
      Bladder 298
      1. The Healthy Bladder 298
      2. Pathological Conditions 298
D. Concluding remarks.. 299
List of Abbreviations 299
References 300

CHAPTER 11

Trophic Roles of Purines and Pyrimidines
J.T. NEARY and M.P. ABBRACCHIO. With 1 Figure 305

A. Introduction 305
B. Effects on Muscle Cells 306
   I. Effects on Vascular Smooth Muscle and
      Endothelial Cells 306
   II. Effects on Myocardial Cells 308
   III. Effects on Skeletal Muscle Cells 309
C. Effects on Brain Cells 310
   I. Effects on Astrocytes 310
   II. Effects on Microglia 312
   III. Effects on Other Glial Cells 313
   IV. Effects on Neuronal Cells 313
D. Effects on Immune Cells 314
E. Effects on Secretory Cells 316
   I. Effects on Exocrine Cells 316
      1. Salivary Cells 316
      2. Kidney Cells 316
   II. Effects on Endocrine Cells 317
F. Effects on Fibroblasts 318
G. Effects on Bone Cells 319
H. Effects on Reproduction 320
I. Mechanisms 320
J. Conclusions and Future Perspectives 326
References 327

CHAPTER 12

Nucleoside and Nucleotide Transmission in Sensory Systems
G.D. HOUSLEY. With 2 Figures 339

A. Introduction 339
   I. Sensory Transmission Mediated by Nucleotides
      and Nucleosides 339
B. Olfactory and Gustatory Systems 341
C. Visual System 341
   I. Adenosine Actions on Retinal Neurotransmission 342
   II. Nucleosides and Nucleotides in Retinal Ontogeny 344
   III. Nucleotide Signaling in the Retina 344
   IV. Nucleoside and Nucleotide Signaling in Ciliary Ganglion,
      Ciliary Body, and Non-Pigmented Epithelium 346
      1. Regulation of Intraocular Pressure (IOP): Actions on
         the Ciliary Body 346
      2. Neural Signaling in the Ciliary Body and Iris 348
   V. Adenosine Regulation of Ocular Blood Flow:
      Protection During Ischemia and Hypoglycemia 349
   VI. Nucleotide Signaling in the Lens 349
   VII. Trophic Actions of Adenosine 350
      1. Vascular Endothelial Growth Factor 350
      2. Glucose Transport 350
D. Cochlear and Vestibular Systems 351
   I. Modulation of Hair Cell Sound Transduction and
      Cochlear Micromechanics 351
      1. Hair Cell Membrane Conductance 351
      2. Nucleotide Regulation of Cellular Mechanics 352
      3. Regulation of the Electrochemical Gradient across the
         Cochlear Partition 354
   II. Regulation of the K^+ Flux into Endolymph 355
   III. Putative Role in Auditory Neurotransmission 356
   IV. Localization of P2X and P2Y Receptor Expression:
      A Role for Nucleotides in Cochlear Ontogeny 356
   V. Significance of Alternative Splicing of the $P2X_2$ Receptor
      mRNA in the Inner Ear 357
   VI. Adenosine Actions on Afferent Neurotransmission, Free
      Radical Scavenging, Altered Cochlear Blood Flow, and Hair
      Cell Cation Channel Downregulation 358
E. Conclusion 359
References 359

CHAPTER 13

Roles of Purines in Nociception and Pain
M.W. Salter and A. Sollevi. With 1 Figure 371

A. Introduction 371
B. ATP and Adenosine in Somatosensory Neurotransmission 371
   I. Release of ATP 371
      1. Release of ATP from Primary Afferent Neurons 371
      2. Co-Release of ATP with GABA from
         Dorsal Horn Neurons 372
      3. Release of ATP from Other Cell Types in the
         Periphery 372
   II. P1 and P2 Purinergic Receptors 373
      1. Purinergic Receptor Expression by Primary Afferent
         Neurons 373
      2. Purinergic Receptor Expression by Dorsal Horn
         Neurons 373
   III. Actions of ATP and Adenosine 374
      1. Actions in Periphery 374
         a) ATP 374
         b) Adenosine 375
      2. Actions in the Dorsal Horn 376
         a) ATP 376
         b) Adenosine 377
         c) Extracellular Conversion of ATP to Adenosine
            Mediates an Inhibitory Postsynaptic Response
            to Low Threshold Primary Afferent Input 379
   IV. Potential Therapeutic Implications from Basic Studies 380
      1. Targeting P2 Purinoceptors 380
      2. Targeting Adenosine Receptors 382
      3. Other Potential Therapeutic Implications 382
C. Clinical Aspects of Nociception/Antinociception 383
   I. General Considerations 383
   II. Administration of Adenosine and Adenosine
     Analogues 384
      1. Safety and Adverse Effects 384
         a) IV Administration 384
         b) Intrathecal Administration 386
         c) Pharmacokinetics in CSF 387
      2. Effects of Adenosine in Healthy Volunteers 387
      3. Effects of Adenosine in Acute Surgical Pain 389
         a) Intraoperative Pain 389
         b) Postoperative Pain – Analgesic Requirements 389
      4. Effects of Adenosine in Chronic Neuropathic
         Pain Patients 390

		5. Possible Sites of Action	391
	III.	Adenosine as a Mediator of Analgesia Produced by TENS	392
D.	Summary and Conclusions		393
References			393

CHAPTER 14

Uridine and Pyrimidine Nucleotides in Cell Function
G.P. CONNOLLY 403

A.	Introduction	403
B.	Synthesis and Salvage of Pyrimidines	403
C.	Plasma and CSF Uridine	405
D.	Liver and Kidney Pathology	405
E.	Uridine Nucleotides and Cystic Fibrosis	406
F.	Pyrimidines and Reproductive Function	408
G.	Cancer and Antiviral Therapy	408
H.	Peripheral Nervous System Modulation	409
I.	CNS Modulation	410
	I. Clinical Studies	411
	II. Animal Studies	412
	1. Neuropeptide Interactions	412
	2. Dopamine Interactions	412
	3. Anticonvulsant and Anxiolytic Effects	415
	4. Sleep and Thermoregulation	415
	5. Additional Actions of Uridine and its Derivatives	416
References		417

CHAPTER 15

Purinergic Signalling in Lower Urinary Tract
G. BURNSTOCK. With 21 Figures 423

A.	Introduction	423
B.	Parasympathetic Purinergic Cotransmission	426
	I. Parallel Block of Responses to ATP and NANC Nerve Stimulation	428
	II. Purinergic Excitatory Junction Potentials in the Bladder	430
	III. Purinergic/Cholinergic Cotransmission	431
	IV. Extracellular Calcium, Calcium Channel Blockers and Potassium Channel Openers	434
	V. Involvement of Prostaglandins in Purinergic Transmission	436

	VI. Ectoenzymatic Breakdown of ATP	437
	VII. In Vitro and In Vivo Studies	439
C. Receptors to Purines and Pyrimidines		442
	I. P2X Receptors Mediating Contraction of the Bladder	442
	1. Agonist Potencies	442
	2. Localisation of P2X Receptors	443
	3. Structure-Activity Studies	446
	II. P2Y Receptors Mediating Relaxation of the Bladder	447
	III. P1 Receptors Mediating Relaxation and Contraction of the Bladder	450
D. Sympathetic Cotransmission		450
E. Intramural Bladder Neurones and Pelvic Ganglia		452
F. Neuromodulation in the Bladder		456
G. Afferent Pathways in Bladder and Purinergic Mechanosensory Transduction		458
	I. Evidence for a Suburothelial Sensory Nerve Plexus with $P2X_3$ Receptors	459
	II. Afferent Nerve Activity During Bladder Distension and Evidence of Involvement of Purines	460
	III. Evidence for ATP Release from Urothelial Cells During Distension	462
	IV. Evidence for Purinergic Involvement in Bladder Nociception	463
H. Perinatal Development of Purinergic Signalling in Urinary Bladder		465
I. Urethra		467
J. Ureter		469
K. Plasticity of Purinergic Signalling in Bladder, Urethra and Ureter		470
	I. Changes Occurring During Pregnancy or Hormone Therapies	470
	II. Changes Due to Selective Denervation	472
	III. Bladder Grafts	474
	IV. Hibernation	474
L. Purinergic Signalling in the Human Bladder in Health and Disease		474
	I. Healthy Bladder	474
	II. Bladder Outflow Obstruction: Unstable, Hypertrophic Bladders	478
	III. Neurogenic Hyperreflexive Bladder	479
	IV. Multiple Sclerosis (MS)	481
	V. Post-Irradiation Bladder Dysfunction	482
	VI. Ischaemic Bladder	482
	VII. Chronic Alcohol Consumption and Bladder Function	482
	VIII. Vitamin E Deficiency	483

IX.	Interstitial Cystitis	483
X.	Diabetes	485
M.	Concluding Comments	487
	References	488

Subject Index 517

Contents of Companion Volume 151/II

Purinergic and Pyrimidinergic Signalling II:
Cardiovascular, Respiratory, Immune,
Metabolic and Gastrointestinal Tract Function

Section III: Cardiovascular and Renal Systems

CHAPTER 16
P1 Receptors in the Cardiovascular System
R.M. BROAD and J. LINDEN 3

CHAPTER 17
P1 and P2 Receptors in the Renal System
E.K. JACKSON 33

CHAPTER 18
P2 Receptors in the Cardiovascular System
A. PELLEG and G. VASSORT 73

CHAPTER 19
Roles of Purines and Pyrimidines in Endothelium
V. RALEVIC 101

CHAPTER 20
P1 and P2 Receptors in Platelets
S.M.O. HOURANI 121

CHAPTER 21
Purinergic Signalling in Gut
G. BURNSTOCK 141

CHAPTER 22
P1 Receptors in the Respiratory System
I. BIAGGIONI and I. FEOKTISTOV 239

CHAPTER 23
P2 Receptors in the Respiratory System
C.D. WEGNER 281

Section IV: Immune System

CHAPTER 24
Role of P1 Receptors in Inflammation
M.C. MONTESINOS and B.N. CRONSTEIN 303

CHAPTER 25
Role of P2 Receptors in the Immune System
G.R. DUBYAK 323

CHAPTER 26
On the Role of the $P2X_7$ Receptor in the Immune System
F. DI VIRGILIO, V. VISHWANATH and D. FERRARI 355

Section V: Endocrine System

CHAPTER 27
Purinergic Receptors and the Pharmacology of Type 2 Diabetes
P. PETIT, D. HILLAIRE-BUYS, M.M. LOUBATIÈRES-MARIANI and
J. CHAPAL 377

CHAPTER 28
Purinergic and Pyrimidinergic Receptor Signaling in Bone Cells
J. FERRIER 393

CHAPTER 29
Clinical Opportunities in Purinergic Neuromodulation
M. WILLIAMS 407

Subject Index 435

CHAPTER 1
Purinergic Neurotransmission: An Historical Background

M.P. ABBRACCHIO and M. WILLIAMS

A. Introduction

It was not until the second decade of the last century that basic research findings on purines and their effects on tissue function began to appear sporadically in the literature. BASS (1914) reported on the presence of adenine (most probably as AMP) in the blood, while THANNHAUSER and BOMMES (1914) administered the purine nucleoside adenosine subcutaneously to humans, and found it, unlike adenine, to be reasonably non-toxic. Administration of AMP had depressor activity (FRUEND 1920) which led to the now seminal study of DRURY and SZENT-GYORGI (1929) on the cardiovascular actions of adenosine and AMP. These included the demonstration of sinus bradycardia, heart block, reduction in atrial conduction, and at higher doses, "apprehension" and somnolence. In that same year, the purine nucleotide, ATP was discovered (LOHMANN 1929; FISKE and SUBBAROW 1929).

Studies on the ability of adenosine to regulate coronary blood flow (LINDLER and RIGLER 1931) and the studies of Drury and Szent-Gyorgi led to the evaluation of the therapeutic potential of adenosine as an hypotensive agent in the 1930s (HONEY et al. 1930; JEZER et al. 1933) but the clinical efficacy of the nucleoside as an antihypertensive agent was limited by its short plasma half life (3–6s; PATERSON et al. 1987), a pharmacokinetic property that made the purine an ideal candidate for the treatment of supraventricular tachycardia and as a cardiac imaging agent over half a century later (BELARDINELLI et al. 1995).

In the early 1960s, the hypothesis that the purine nucleoside was a physiological regulator of coronary blood flow received additional support with the demonstration that adenosine released from hypoxic and ischemic heart muscle functioned to regulate coronary blood flow during reactive hyperemia (BERNE 1963; GERLACH et al. 1963) acting as a signal linking energy demand, cardiac oxygen usage, to increased coronary blood flow in order to provide the energy required for normal physiological function (BERNE 1980; OLSSON and PATTERSON 1976). The endogenous production of adenosine during hypoxic and ischemic conditions led to the concept of the nucleoside being termed a "retaliatory metabolite" (NEWBY 1984) or "homeostatic modulator"

(WILLIAMS 1989), functioning as an autocrine regulatory agent to protect traumatized/damaged tissues. Local adenosine levels in the brain are markedly increased following stroke (RUDOLPHI et al. 1992) and during epileptic episodes in humans (DURING and SPENCER 1992). Adenosine has accordingly been referred to as the "signal of life" (ENGLER 1991) while ATP is well known as the key energy source within the cell (ATKINSON 1977).

Two seminal review articles published in the 1970s by Burnstock (BURNSTOCK 1972, 1978) summarized the considerable evidence for the purines, adenosine and ATP, acting as neurotransmitter/neuromodulators in nearly all tissue systems with ATP playing a key role in non-adrenergic, non-cholinergic (NANC) transmission processes. While biochemical and pharmacological studies established the existence of multiple adenosine and ATP receptor-mediated tissue responses (BURNSTOCK and KENNEDY 1985; GORDON 1986; DUBYAK and EL MOATASSIM 1993), it was another quarter century before definitive molecular evidence for the existence of adenosine (P1) and ATP (P2) receptors became available (HARDEN et al. 1995; OLAH and STILES 1995; RALEVIC and BURNSTOCK 1998; NORTH and SURPRENANT 2000; CHESSELL et al., Chap. 3, this volume; BOARDER and WEBB, Chap. 4, this volume).

The purinergic nerve hypothesis was based on studies showing the effects of adenosine and ATP on various aspects of mammalian tissue function, from the already quoted work of Berne and Gerlach on the role of adenosine in cardiac function and the key demonstration by HOLTON (1959) that ATP was released from rabbit ear artery following antidromic sensory nerve stimulation (incidentally, it was another three decades before ATP was unequivocally demonstrated to act as a fast excitatory transmitter at central synapses; EDWARDS et al. 1992). Additional studies on the systemic administration of adenosine and the stable analog, 2-chloroadenosine (2-CADO) into the human brachial artery resulted in a forearm vasodilation that was associated with pain (BORN et al. 1965). This phenomenon was reinvestigated some two decades later (SYLVEN et al. 1988) and shown to be the result of acute bolus administration of the nucleoside since, when infused, adenosine has pronounced and long-lasting analgesic activity (KARLSTEN and GORDH 1996). Similarly, ATP injection can elicit pain responses (DRIESSAN et al. 1994) when injected into the skin and paw (BLAND-WARD and HUMPHREY 1998), being an example of the opposing physiological effects of ATP and adenosine (WILLIAMS and JARVIS 2000). The sedative actions of endogenous adenosine in the CNS, reflected in the stimulant activities of caffeinated beverages, caffeine acting as a CNS stimulant via its ability to block the effects of endogenous adenosine (FREDHOLM et al. 1999) and thus representing one of the most widely consumed psychoactive drugs in the world (DALY 1982), similarly contrast with the actions of ATP, which acts as an excitatory transmitter in the CNS (EDWARDS 1996).

B. Purine Availability

The factors governing the extracellular availability of ATP, UTP, and adenosine have been the topic of considerable debate ever since the purine hypothesis was formulated (BURNSTOCK and BARNARD 1996). Why a molecule that is integral to energy generation within the cell would be deliberately lost from the cell in order to communicate with other cells has presented an enigma to many scientists. However, it is pertinent to remember that all molecules involved in cell-to-cell communication involve the use of cellular energy in their synthesis. The obviousness of ATP as an energy-rich molecule can be contrasted with the less obvious energy-dependent steps in producing a peptide neuromodulator like corticotrophin-releasing hormone (CRH), where at least 39 peptide bonds need to be formed to produce a biologically active substance. It should also be noted that ATP can be produced on demand and that the body can produce its own weight in ATP in a single day (NOJI 1998). Extracellular ATP levels reach millimolar concentrations in the extracellular local environment following release or cellular perturbation. However, these concentrations are relatively miniscule when compared to the overall steady state nucleotide content of the cell (BURNSTOCK and BARNARD 1996).

ATP is co-released with other neurotransmitters including acetylcholine, norepinephrine, glutamate, GABA, CGRP, vasoactive intestinal peptide, and neuropeptide Y (SPERLAGH and VIZI 1996; BURNSTOCK 1999), and its dephosphorylation leads sequentially to adenosine formation via the actions of a family of approximately 11 ectonucleoside triphosphate diphosphohydrolyases (E-NTPDases) that metabolize ATP, ADP, and AMP, the diADO polyphosphates Ap4A and Ap5A and NAD. ATP is converted to AMP via ecto-ATPase action. Ecto-apyrases convert both ATP and ADP to AMP and ecto-5'-nucleotidase converts AMP to adenosine (ZIMMERMAN and BRAUN 1999). Ecto-apyrase and ecto-5'-nucleotidase activity change with cellular dynamics (CLIFFORD et al. 1997) and soluble nucleotidases are released together with ATP and norepinephrine from guinea-pig vas deferens (TODOROV et al. 1997), the latter representing a potential mechanism to limit the actions of extracellular ATP via enhancement of its inactivation.

The metabolic pathways linking ATP, ADP, AMP, and adenosine and the potential for each of these purines to elicit distinct receptor-mediated effects on cell function form the basis of a complex, physiologically relevant, purinergic cascade comparable to those involved in blood clotting and complement activation (Fig. 1; WILLIAMS and JARVIS 2000). This concept, originally proposed by Linda Slakey and Jeremy Pearson in the 1970s, provides a comprehensive, integrated scheme for the actions of ATP, its related nucleotides, and adenosine not only at P1 and P2 purinergic receptors but also on a variety of other cell surface and intracellular (e.g., mitochondrial) receptors whose activity can be regulated by ATP.

The ADP, AMP, and adenosine formed from ATP by E-NTPDase action have their own receptor-mediated functional activities that can be antagonistic

ATP: A Pluripotent Modulator of Cellular Communication

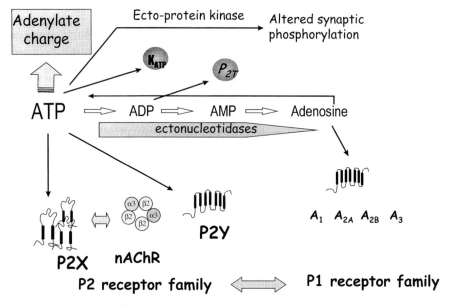

Fig. 1. ATP: a pluripotent modulator of cellular communication. ATP is released into the extracellular milieu from nerves or cells where it can produce direct effects on P2 and KATP channels, participates in the transfer of adenylate charge and acts as a substrate for ecto- and other protein kinases. Via the ectonucleotidase driven purinergic cascade, ATP is degraded to ADP and AMP and ADO. ADP interacts with P_{2T} ($P2Y_{12}$) receptors. ADO interacts with the various P1 receptors (A_1, A_{2A}, A_{2B}, A_3). ADO can also be formed by intracellular 5′-nucleotidase activity. P2X receptor function can be modulated by neuronal nicotinic receptor (nAChR) activation (shown as $\alpha 3\beta 2$) while P1 and P2 receptors modulate the effects of one another. See text for further discussion

to one another. ATP antagonizes ADP actions on platelet aggregation (HOURANI, Chap. 20, second volume) while, as already noted, adenosine-mediated CNS sedation contrasts with the excitatory actions of ATP on nerve cells. In the broader framework of ATP-modulated proteins [or ATP-binding cassette (ABC) proteins], ATP-sensitive potassium channels (K_{ATP}) are activated when intracellular ATP levels are reduced (COOPER and JAN 1999). As P2 receptor-mediated effects decrease with the hydrolysis of ATP to adenosine, P1 mediated responses and K_{ATP}-mediated responses become enhanced. Activation of P1 receptors can inhibit ATP release (ROBERTSON and EDWARDS 1998) and activation of hippocampal A_{2A} and A_3 receptors can desensitize A_1 receptor-mediated inhibition of excitatory neurotransmission (MASINO and DUNWIDDIE, Chap. 9, this volume). An additional facet of the role of purines in neurotransmission is their transfer from one cell to another in the context of cellular adenylate charge (Fig. 1; ATKINSON 1977) acting to transfer information and alter the target environment. Similarly, ATP is a substrate for synaptic ecto-protein kinases, which dynamically modulate the phosphorylation state of the synaptic membrane (EHRLICH and KORNECKI 1999) and, consequently, the intrinsic properties of the synapse including receptor, ion channel, transporter, integrin, adhesion molecule, and growth factor function. Thus, once in the extracellular space, ATP functions as a pluripotent modulator of synaptic function through a potential diversity of molecular targets (Fig. 1).

A corresponding role for UTP in synaptic signaling is less well defined (CONNOLLY and DIULEY 1999; CONNOLLY, Chap. 14, this volume) and while high concentrations of exogenous uracil can modulate dopaminergic systems in the CNS (AGNATI et al. 1989), there is only limited data on the existence of a "uridine receptor" equivalent to the P1 receptor (KARDOS et al. 1999). Recent evidence suggests that there may also be a receptor for guanosine (TRAVERSA et al. 2000).

At rest, extracellular adenosine levels are in the 30–300 nmol/l range (PHILLIS 1989; DUNWIDDIE and DIAO 1994; PAZZAGLI et al. 1995; LATINI et al. 1999) and are tightly regulated by ongoing metabolic activity, bidirectional nucleoside transporters and the enzymes, adenosine deaminase (ADA) and adenosine kinase (AK), regulating adenosine removal from the extracellular space (GEIGER et al. 1997). Extracellular adenosine subserves a physiological role in tissue homeostasis to regulate the tissue energy supply/demand balance in response to changes in blood flow and energy availability (NEWBY 1984), thus conserving tissue function under adverse conditions. Reduced oxygen or glucose availability due to tissue trauma, e.g., during stroke, epileptogenic activity, and reduced cerebral blood flow leads to ATP breakdown and the formation of ADP, AMP, and adenosine.

Several studies (ZHANG et al. 1993; WEISNER et al. 1999) have shown that AK inhibition is physiologically more relevant in increasing extracellular adenosine levels than inhibition of either ADA or adenosine transport. AK inhibitors are also more effective in enhancing the neuroprotective actions of endogenous adenosine than inhibitors of ADA or adenosine transport (GEIGER et al. 1997; IJZERMANN and VAN DER WENDEN 1997). Selective AK inhibitors like GP 3269 and ABT-702, as well as ADA inhibitors, have been described as "site and event specific" agents (ENGLER 1991) that can locally enhance levels of adenosine in areas of tissue trauma. By avoiding the global effects of adenosine on cardiovascular and/or CNS function, this approach has the potential to provide a more selective homeostatic action (WILLIAMS and JARVIS 2000). In vivo administration of AK inhibitors like GP 3269 and related compounds (ERION et al. 2000) has shown that acutely, at doses where efficacy is seen in animal models of epilepsy and pain, these agents produce microhemorrhages in the brain that have the potential to lead to cerebral infarcts and cognitive impairment. Based on this finding, and its independent confirmation (M.F. Jarvis, personal communication), the AK inhibitor approach to selectively modulating endogenous adenosine function does not appear to have a sufficiently wide therapeutic window in CNS tissues to be a viable drug discovery target. It remains to be seen whether ADA inhibitors have similar effects.

Adenosine has both pre and postsynaptic effects on neurotransmission processes (MASINO and DUNWIDDIE, Chap. 9, this volume) while ATP has excitatory actions in a variety of neuronal systems including rat trigeminal nucleus, nucleus tractus solitarius, dorsal horn, medial habenula, and locus ceruleus (EDWARDS 1996; NIEBER et al. 1997; EDWARDS and ROBERTSON 1999).

C. P1 and P2 Receptor Nomenclature

In the field of nomenclature, the P1/adenosine and the P2/ATP system originally proposed by Burnstock provided a cornerstone for receptor nomenclature (BURNSTOCK 1978). For the P1 receptor class, nomenclature has proved to be quite easy, with the recognition of four distinct G-protein-coupled receptors (GPRC)s (FREDHOLM et al. 1994, 1997; see also below). For P2 receptors, an initial further sub-classification into $P2_X$ and $P2_Y$ was proposed by BURNSTOCK and KENNEDY (1985) to differentiate excitant and relaxant effects mediated by these receptors. By further delineating ATP effects in a variety of tissue systems, Gordon defined the P_{2t} and P_{2z} receptors (GORDON 1986). P_{2u} and P_{2n} receptors were then introduced to represent a newer class of P2 receptor sensitive to the pyrimidine nucleotide UTP (for review, see ABBRACCHIO et al. 1992; ABBRACCHIO and BURNSTOCK 1994; FREDHOLM et al. 1994). Concerns regarding the *"apparent random walk through the alphabet"* of the naming of the P2 receptors (ABBRACCHIO et al. 1992), stimulated the IUPHAR Subcommittee for Purinoceptor Nomenclature and Classification to resolve the issue. In 1994, based on extensive analysis of the pre-existing literature, Abbracchio and Burnstock proposed a new means of classifying all the P2 receptors into two main families, the ionotropic P2X receptors, and the metabotropic GPC-P2Y receptors (ABBRACCHIO and BURNSTOCK 1994). This proposal has been accepted by the IUPHAR Subcommittee (FREDHOLM et al. 1994) and has represented the basis for the currently used $P2X_{1-n}$ and $P2Y_{1-n}$ classification.

Four distinct P1 receptors sensitive to adenosine and twelve P2 receptors sensitive to ADP, ATP, and UTP have been cloned and characterized so far (for a recent review, see RALEVIC and BURNSTOCK 1998), providing a diversity of discrete cellular targets through which adenosine, ADP, and ATP as well as UTP can modulate tissue function. The four adenosine-sensitive P1 receptors have been designated as A_1, A_{2A}, A_{2B}, and A_3 subtypes. Functional P2X receptors represent a family of seven ligand-gated ion channels (LGICs) ($P2X_1$–$P2X_7$), whereas the P2Y family consists of the $P2Y_1$, $P2Y_2$, $P2Y_4$, $P2Y_6$, and $P2Y_{11}$ GPCRs. The missing numbers in the P2Y family sequence are proposed receptors that have subsequently been found to lack functional responses, are species variants, or have been inadvertently assigned to the P2 receptor family (KING et al. 2000).

I. Adenosine (P1) Receptors

With the discovery of the intracellular signaling molecule cyclic AMP, a diversity of agents modulating the activity of adenylate cyclase was extensively studied. The pivotal biochemical studies of SATTIN and RALL (1970) showing that theophylline and caffeine, used in the adenylate cyclase assay as phosphodiesterase inhibitors to prevent cAMP degradation, could block the increase in adenylate cyclase activity produced by adenosine in guinea pig brain slices led to these two methylxanthines being classified as the first P1 (adenosine) antagonists and provided early evidence for the existence of discrete adenosine receptors. Additional work by the groups of McIlwain (KURODA and MCILWAIN 1974; PULL and MCILWAIN 1977) and Daly (HUANG et al. 1972; DALY et al. 1983) substantiated and expanded these original findings with further characterization of the putative adenosine receptors and studies on adenosine release (KURODA 1981). Efforts were also focused on improving the efficacy and bioavailability of the initial xanthine antagonist leads by substituting a diversity of groups in the 8-position of the pharmacophore (DALY 1982; JACOBSON et al. 1992), a medicinal chemistry activity that remains unsurpassed today (JACOBSON and KNUTSEN, Chap. 6, this volume).

Adenosine recognition sites were further delineated by LONDOS and WOLFF (1977), who demonstrated inhibitory and stimulatory effects of adenosine on adenylate cyclase activity that were mediated, respectively, by a P site on the enzyme and extracellular R site. The P site required an intact purine ring for activity while the R site required an intact ribose ring for efficacy. Subsequent, concurrent studies by VAN CALKER et al. (1979) and LONDOS et al. (1980) identified two extracellular G-protein coupled receptors sensitive to adenosine, the A_1, linked to inhibition of adenylate cyclase activity, and the A_2, linked to activation of adenylate cyclase (VAN CALKER et al. 1979). These same two receptors were designated as Ri and Ra by LONDOS et al. (1980) but the A designation has become the nomenclature of choice (FREDHOLM et al. 1994). The A_2 receptor in turn was found to exist in high and low affinity forms designated, respectively, as A_{2A} and A_{2B} receptor (DALY et al. 1983). All three receptors have been cloned, expressed, and shown to be discrete molecular entities (OLAH and STILES 1995; LORENZEN and SCHWABE, Chap. 2, this volume).

These biochemical studies also led to the development of a number of selective agonists and antagonists, some of which were developed for use in radioligand binding assays (JACOBSON and KNUTSEN, Chap. 6, this volume). Agonist radioligands for the A_1 receptor include the N^6-cyclohexyl (CHA) and N^6-cyclopentyl (CPA) analogs of adenosine while antagonists include the 8-substituted xanthines, cyclopentylxanthine (CPX) and cyclopentyltheophylline (CPT). The adenosine analog, CGS 21680, is an agonist radioligand for the A_{2A} receptor with the non-xanthine, SCH 58261 being an antagonist ligand. There are currently no radioligands that provide a robust analysis of the A_{2B} receptor. Selective ligands for the A_3 receptor include IB-MECA and its 2-chloro analog Cl-IB-MECA and the non-xanthine antagonist, MRE-3008F20, has been recently been described as a selective A_3 radioligand (VARANI et al. 2000). Selectivity of compounds for the different P1 receptor subtypes can also be defined in null cell lines transfected with the cDNA for each receptor provided that there is an iteration with the native receptor to avoid expression artifacts (KENAKIN 1996).

D. P2 (ATP and UTP) Receptors

P2 receptors were originally classified on the basis of the rank order potency of agonists structurally related to ATP (ABBRACCHIO and BURNSTOCK 1994; RALEVIC and BURNSTOCK 1998). The majority of these putative receptors (including the elusive *P2T* P2Y$_{12}$ receptor; HOURANI, Chap. 20, second volume) have been subsequently cloned and functionally characterized in various heterologous expression system.

Their functional characterization in native tissues and in animals has been limited by a paucity of potent, selective, and bioavailable ligands, both agonists and antagonists. All known P2 agonist ligands are analogs of ATP, UTP, and ADP and, irrespective of their degree of chemical modification, show varying degrees of susceptibility to extracellular degradation and differences in intrinsic activity (WILLIAMS and JARVIS 2000; JACOBSON and KNUTSEN, Chap. 6, this volume). Their functional selectivity and potency is thus very much tissue and species dependent. There have been few systematic evaluations of the P2 receptor selectivity of P2 receptor agonists. However, examination of BzATP, a widely used selective agonist for the P2X$_7$ receptor (EC$_{50}$ = 18 μmol/l), in recombinant cell lines in functional FLIPR assays showed that this ATP analog is far more potent at transfected rat and human P2X$_1$ (EC$_{50}$ = 1.9 nmol/l) and P2X$_3$ (EC$_{50}$ = 98 nmol/l) receptors (BIANCHI et al. 1999) and thus cannot be used, *a priori*, to define P2X$_7$ receptor-mediated cellular responses.

Radioligand binding assays for P2 receptors using ligands like [^{35}S]-ATPγS, [^{35}S]-dATPαS and [^{35}S]-ADPβS have been controversial, with no clear relationship between binding and function (SCHACHTER and HARDEN 1997) and binding pharmacology inconsistent with that derived from more traditional assays with the potential for interactions with ATP target proteins beyond P2 receptors. [^{35}S]-ATPγS does not discriminate between the P2X$_{1-4}$ receptors (MICHEL et al. 1997) while [^{35}S]-ATPγS and [^{35}S]-ADPβS showed high specific binding to HEK293 and 1321N1 astrocytoma cells that were devoid of detectable P2X and P2Y receptors (YU et al. 1999). At the present time, newer ligands, preferably antagonists, will be needed to assess whether binding assays for the P2 receptor family can be developed.

I. P2X Receptors

P2X receptors are ATP-gated LGICs formed from various P2X subunits that share a common motif of two transmembrane spanning regions (2TM; NORTH and SURPRENANT 2000; CHESSELL et al., Chap. 3, this volume), with an associated large extracellular domain with both the N and C termini located intracellularly, similar to that seen with the amiloride-sensitive epithelial Na$^+$

channel (NORTH and BARNARD 1997). Functional P2 receptors consist of multimeric, probably trimeric (NICKE et al. 1998) combinations of the various P2X subunits forming a non-selective pore permeable to Ca^{2+}, K^+, and Na^+ that mediates rapid (~10ms) neurotransmission events. In addition to putative $P2X_1$–$P2X_7$ homomers, $P2X_{1/5}$, $P2X_{2/3}$, and $P2X_{4/6}$ functional hetereomers have been identified (NORTH and SURPRENANT 2000). $P2X_5$ and $P2X_6$ receptors do not appear to exist in homomeric form, but rather as heteromers with other P2X receptor subtypes. Unlike other LGICs, e.g., nAChRs, the $5HT_3$ receptor, very little is known regarding nature of the agonist (ATP) binding site on P2X receptor constructs or of ancillary sites (KN-62, TNP-ATP, avermectin) that modulate receptor function. Many of the antagonists used to define P2 receptor function, e.g., suramin, NF-023, Ip5I, have an unusual symmetric structure while the ability of dinucleotide polyphosphates (Ap4A, Ap5A) to activate P2 receptors may be indicative of the need for interaction with two agonist (ATP) recognition sites for functional activity.

II. P2Y Receptors

P2Y receptors are traditional GPCRs activated by purine and/or pyrimidine nucleotides (COMMUNI and BOEYNAEMS 1997; BOARDER and WEBB, Chap. 4, this volume). The five mammalian functional subtypes, $P2Y_1$, $P2Y_2$, $P2Y_4$, $P2Y_6$, and $P2Y_{11}$ have been cloned and are coupled to Gq_{11} with receptor activation resulting in stimulation of PLC and IP_3 activation and the subsequent release of calcium from intracellular stores. The P_{2T} receptor present in platelets (HOURANI, Chap. 20, second volume, has recently been cloned as the $P2Y_{12}$ receptor (BOARDER and WEBB, Chap. 4, this volume). UTP is the preferred agonist for the $P2Y_4$ receptor with ATP and the nucleotide diphosphates being inactive. Diphosphates are more active at the $P2Y_6$ receptor than triphosphates leading to the classification of the $P2Y_6$ receptor as a UDP-preferring receptor. The $P2Y_{11}$ receptor is unique in regard to other P2Y receptors in that only ATP serves as a agonist for this receptor.

The diadenosine polyphosphates, Ap4A and Ap5A, represent another group of purine signaling molecules that modulate cell function via activation of cell surface P2 receptors. While release of Ap4A has been shown, and this molecule modulates neurotransmitter release, it is unclear whether diadenosine polyphosphate actions involve distinct receptor subtypes or reflect activation of known P2 receptors. The relationship of receptors activated by dinucleotide polyphosphates to the P2 receptor family has however been questioned (PINTOR and MIRAS-PORTUGAL 2000). Receptors for the diadenosine polyphosphates have yet to be cloned.

E. Mitochondrial Purinergic Receptors?

In addition to functioning as the key source of ATP within the cell, mitochondria play a key role in the apoptotic cascade via the release of cytochrome

c that results from alterations in mitochondrial transition pore function elicited by members of the *bcl-2* family of cell death proteins (SIMON and JOHNS 1999; HONIG and ROSENBERG 2000). Given the key role of adenosine and ATP in mitochondrial function, and the precedent for identifying bona fide neurotransmitter receptors on mitochondrial members (e.g., peripheral benzodiazepine receptor; GAVISH et al. 1999), it remains a possibility that P2 receptors may be present on mitochondria and play a role in ATP-mediated cytotoxicity, both apoptotic (DI VIRGILIO et al. 1998) and necrotic (SIKORA et al. 1999).

F. Transgenic Models of P1 and P2 Receptor Function

For both P1 and P2 receptors, the use of mice either deficient in (knockouts) or overexpressing (knockins) a targeted receptor provides a unique way to assess the functional role of the receptor. While this approach is not always straightforward since some phenotypes can be fatal, others show no discernable change and, in others, a compensation(s) in associated receptor systems leading to atypical phenotypes (GINGRICH and HEN 2000), knockouts can be helpful in the absence of selective antagonists or antisense probes.

P1 receptor knockouts show altered cardiovascular function (A_1) and reduced exploratory activity, aggressiveness, hypoalgesia, and high blood pressure (A_{2A}; LEDENT et al. 1997). P2 knockouts are associated with decreased male fertility (P2X$_1$; MULRYAN et al. 2000), decreased nociception and bladder hyporeflexia (P2X$_3$; COCKAYNE et al. 2000), decreased platelet aggregation and bleeding time (P2Y$_1$; FABRE et al. 1999), reduced chloride secretion (P2Y$_2$; CRESSMAN et al. 1999). A preliminary report on a P2X$_7$ knockout mouse has appeared (SIKORA et al. 1999).

G. Exponential Diversity?

Electrophysiological studies on P2X and neuronal nicotinic receptor (nAChR) -mediated responses indicate that these two LGICs can interact with one another resulting in a functional cross talk between the systems (SEARL et al. 1998). Recombination of the neuronal nicotinic cholinergic receptor construct, $\alpha 3\beta 4$, with P2X$_2$ receptor channels in *Xenopus* oocytes results in cross inhibition with co-activation using ATP and ACh, resulting in inhibition of both channel types (KHAKH et al. 2000). While this may occur via structurally related interactions involving the phenomenon known as conformational spread, the possibility that hetero-oligimerization can occur between these two different classes of LGICs cannot be discounted and may have precedence in GPCR heterodimerization (MILLIGAN 2000). This would provide an additional layer of complexity to an already complex situation in understanding the precise subunit composition of functional P2X receptors.

H. Conclusions

Building on early studies on purines at the beginning of the last century and an increased interest in the physiological role(s) of adenosine and ATP in the two decades after World War II, Burnstock's hypothesis of purinergic neurotransmission has provided the framework for an explosion in research in the area over the past quarter century that has resulted in the cloning and expression of a multitude of molecular targets that ATP, ADP, AMP, and adenosine (and UTP) interact with to produce their effects on mammalian tissue function. Preliminary studies on receptor knockouts are beginning to delineate the physiological and pathophysiological function of the various P1 and P2 receptors. However, the challenge ahead (WILLIAMS, Chap. 29, second volume) is to identify novel ligands that can be used both as experimental tools to elucidate the complex pathophysiological roles of these receptors and as lead molecules to develop novel therapeutic agents. Hopefully, with the tools of modern drug discovery, high throughput screening, (KENNY et al. 1998) and diversity (TRIGGLE 1999), this may be anticipated to occur within the next decade.

At the basic research level, it is also apparent that ATP, via the purinergic cascade and its interactions with other ATP-recognizing molecular targets either as a ligand or an energy-imparting substrate, occupies a central role in cellular interactions and disease pathophysiology, especially that involving cellular trauma and/or aging. In developing potential therapeutics acting via P1 and P2 receptors, it will be important to recognize such disease states as being dynamically different from "normal" tissues or transfected cell lines, the milieu in which the majority of biomedical research is currently conducted. To be overly reductionistic at a molecular level in studying the role of ATP and adenosine would be an injustice to the pivotal role of purines in cell function.

References

Abbracchio MP, Burnstock G (1994) Purinoceptors: are there families of P2X and P2Y purinoceptors? Pharm Ther 64:445–475

Abbracchio MP, Cattabeni F, Fredholm BB, Williams M (1992) Purinoceptor nomenclature: a status report. Drug Dev Res 28:207–213

Agnati LR, Fuxe K, Ruggeri M, Merlo-Pich E, Benefenati F, Volterra V, Ungerstedt U, Zini I (1989) Effects of chronic uridine on striatal dopamine release and dopamine-related behaviours in the absence or presence of chronic treatment with haloperidol. Neurochem Inter 15:107–113

Atkinson DE (1977) Cellular energy metabolism and its regulation. Academic, San Diego

Bass R (1914) Über die Purinkörper des menschlichen Blutes und den Wirkungsmodus der 2-phenyl-cholincarbosäure (Atohan) Arch Exp Pathol Phramakol 76:40

Belardinelli L, Shyrock JC, Song Y, Wang D, Shinivas M (1995) Ionic basis of the electrophysiological actions of adenosine on cardiomyocytes. FASEB J 9:359–365

Berne RM (1963) Cardiac nucleotides in hypoxia: possible role in the regulation of coronary blood flow. Am J Physiol 204:317–322

Berne RM (1980) The role of adenosine in the regulation of coronary blood flow. Circ Res 47:807–813

Bianchi B, Lynch KJ, Touma E, Niforatos W, Burgard EC, Alexander KM, Park HS, Yu H, Metzger R, Kowaluk E, Jarvis MF, van Biesen T (1999) Pharmacological characterization of recombinant human and rat P2X receptor subtypes. Eur J Pharmacol 376:127–138

Bland-Ward PA, Humphrey PPA (1997) Acute nociception mediated by hindpaw P2X receptor activation in the rat. Br J Pharmacol 122:365–371

Born GVR, Haslam RJ, Goldman N, Rowe RD (1965) Comparative effects of adenosine analogues on the inhibition of blood platelet aggregation as vasodilators in man. Nature 205:678–680

Burnstock G (1972) Purinergic nerves. Pharmacol Rev 24:509–581

Burnstock G (1978) A basis for distinguishing two types of purinergic receptor. In: Straub RW and Bolis L (eds) Cell Membrane Receptors for Drugs and Hormones A Multidisciplinary Approach. Raven, New York, pp 107–118

Burnstock G (1999) Purinergic cotransmission. Brain Res Bull 50:355–357

Burnstock G, Barnard EA (1996) ATP as a neurotransmitter. P2 purinoceptors: localization function and transduction mechanisms. CIBA Foundation Symp 198:262–265

Burnstock G, Kennedy C (1985) Is there a basis for distinguishing two types of P_2-purinoceptor? Gen Pharmacol 16:433–440

Clifford EE, Martin KA, Dalal P, Thomas R, Dubyak GR (1997) Stage specific expression of P2Y receptors, ecto-apyrase and ecto-5'-nucleotidase in myeloid leukocytes. Am. J. Physiol 273 (Cell Physiol 42):C973–C987

Cockayne DA, Zhu Q-M, Hamilton S, Dunn PM, Zhong Y, Berson A, Kassotakis L, Bardini M, Muraski J, Novakovic S, Lachnit W G, Burnstock G, McMahon SB, Ford APDW (2000) $P2X_3$-deficient mice display urinary bladder hyporeflexia and reduced nocifensive behavior. Drug Dev Res 50 (in press)

Communi O, Boeynaems JM (1997) Receptors responsive to extracellular pyrimidine nucleotides. Trends Pharmacol Sci 18:83–86

Connolly GP, Diuley JA (1999) Uridine and its nucleotides: biological actions therapeutic potential. Trends Pharmacol Sci 20:218–226

Cooper EC, Jan LY (1999) Ion channel genes and human neurological disease: recent progress prospects and challenges. Proc Natl Acad Sci USA 96:4759–4764

Cressman VL, Lazorowski E, Homolya L, Boucher RC, Koller BH, Grubb BR (1999) Effect of loss of $P2Y_2$ receptor gene expression on nucleotide regulation of murine epithelial Cl⁻ transport. J Biol Chem 274:26461–26468

Daly JW (1982) Adenosine receptors: targets for future drugs. J Med Chem 25:197–207

Daly JW, Butts-Lamb P, Padgett WL (1983) Subclasses of adenosine receptors in the central nervous system: interaction with caffeine and related methylxanthines Cel Mol Neurobiol 3:69–80

Di Virgilio F, Ferrari D, Chiozzi P, Falzoni S, Sanz JM, dal Susion M, Mutini C, Hanau S, Baricordi OR (1998) Purinoceptor function in the immune system. Drug Dev Res 43:319–329

Driessen B, Reimann W, Selve N, Friderichs E, Bultmann R (1994) Antinociceptive activity of intrathecally administered P2 purinoceptor antagonists in rats. Brain Res 666:182–188

Drury AN, Szent-Gyorgi A (1929) The physiological activity of adenine compounds with special reference to their effects upon the mammalian heart. J Physiol 68, 213–237

Dubyak GR, El Moatassim C (1993) Signal transduction via P2-purinergic receptors for extracellular ATP and other nucleotides. Am J Physiol 265:C577–C606

Dunwiddie TV, Diao L (1994) Extracellular adenosine concentrations in hippocampal brain slices and the tonic inhibitory modulation of evoked excitatory responses. J Pharm Exp Ther 268:537–545

During MJ, Spencer DD (1992) Adenosine: a potential mediator of seizure arrest and postictal refractoriness. Ann Neurol 32:618–624

Edwards FA (1996) Features of P2X receptor-mediated synapses in the rat brain: why doesn't ATP kill the postsynaptic cell. P2 purinoceptors: localization, function and transduction mechanisms. CIBA Foundation Symp 198:278–289

Edwards FA, Roberston SJ (1999) The function of A2 adenosine receptors in the mammalian brain: evidence for inhibition vs enhancement of voltage gated calcium channels an neurotransmitter release. Prog Brain Res 120:265–273

Edwards FA, Gibb AJ, Colquhoun D (1992) ATP receptor-mediated synaptic currents in the central nervous system. Nature 359:144–147

Ehrlich YH, Kornecki E (1999) Ecto-protein kinases as mediators for the action of secreted ATP in the brain. Prog Brain Res 120:411–426

Engler R (1991) Adenosine: the signal of life? Circ 84:951–954

Erion MD, Wiesner JB, Rosengren S, Ugarkar B, Boyer SH, Tsuchiya M, Nakane M, Pettersen BA, Nagahisa A (2000) Therapeutic potential of adenosine kinase inhibitors as analgesic agents. Drug Dev Res 50 (in press)

Fabre JE, Nguyen M, Latour A, Keifer JA, Audoly LP, Coffman TM, Koller BH (1999) Decreased platelet aggregation, increased bleeding time and resistance to thromboembolism in $P2Y_1$-deficient mice. Nature Med 5:1199–1202

Fiske CH, Subbarow Y (1929) Phosphorus compounds of muscle and liver. Science 70:381–382

Fredholm BB, Abbracchio MP, Burnstock G, Daly JW, Harden KT, Jacobson KA, Leff P, Williams M (1994) Nomenclature and classification of purinoceptors: a report from the IUPHAR subcommittee. Pharmacol Rev 46:143–156

Fredholm BB, Abbracchio MP, Burnstock G, Dubyak GR, Harden KT, Jacobson KA, Schwabe U, Spedding M, Williams M (1997) Towards a revised nomenclature for P1 and P2 receptors. Trends Pharm Sci 18:79–82

Fredholm BB, Bättig K, Holmén J, Nehlig A, Zvartau EE (1999) Actions of caffeine in the brain with special reference to factors that contribute to its widespread use. Pharmacol Rev 51:83–133

Fruend H (1920) Über die pharmkologischen Wirkungen des defibriniertin Blutes. Arch Exp Pathol Pharmakol 86:267–268

Gavish M, Bachman I, Shoukrun R, Katz Y, Veenman L, Weisinger G, Weizman A (1999) Enigma of the peripheral benzodiazepine receptor. Pharmacological Rev 51:629–650

Geiger JD, Parkinson FE, Kowaluk E (1997) Regulators of endogenous adenosine levels as therapeutic agents. In: Jacobson KA, Jarvis MF (eds) Purinergic approaches in experimental therapeutics. Wiley-Liss, New York, pp 55–84

Gerlach E, Deuticke B, Driesbach RH (1963) Der Nucleotid-Abbau in Herzmuskel bei Sauerstoffmangel und seine mögliche Bedeutung für die Coronar-Durchblutung. Naturwissenschaft 50:228–229

Gingrich JA, Hen R (2000) Commentary: The broken mouse: the role of development, plasticity and environment in the interpretation of phenotypic changes in knockout mice. Curr Opin Neurobiol 10:146–152

Gordon J (1986) Extracellular ATP: effects, sources and fate. Biochem J 233:309–319

Harden TK, Boyer JL, Nicholas RA (1995) P_2-purinergic receptors: subtype-associated signaling responses and structure. Ann Rev Pharmacol Toxicol 35:541–579

Holton P (1959) The liberation of adenosine triphosphate on antidromic stimulation of sensory nerves. J Physiol (Lond) 145:494–504

Honig LS, Rosenberg RN (2000) Apoptosis and neurologic disease. Am J Med 108:317–330

Honey RM, Ritchie WT, Thompson WAR (1930) The action of adenosine upon the human heart. Quart J Med 23:485–490

Huang M, Shimizu H, Daly JW (1972) Accumulation of cyclic adenosine monophosphate in incubated slices of brain tissue. 2. Effects of depolarizing agents, membrane stabilizers, phosphodiesterase inhibitors, and adenosine analogs. J Med Chem 15:462–466

Ijzermann AP, van der Wenden NM (1997) Modulators of adenosine uptake, release, and inactivation. In: Jacobson KA, Jarvis MF (eds) Purinergic approaches in experimental therapeutics. Wiley-Liss, New York, pp 129–148

Jacobson KA, van Galen PM, Williams M (1992) Perspective: Adenosine receptors: pharmacology, structure activity relationships, and therapeutic potential. J Med Chem 35:407–422

Jezer A, Oppenheimer BS, Schwartz SP (1933) The effect of adenosine on cardiac irregularities in man. Am Heart J 9:252–258

Kardos J, Kovács I, Szárics E, Kovács R, Skuban N, Nyitrai G, Dobolyi A, Juhász G (1999) Uridine activates fast transmembrane Ca^{2+} ion fluxes in rat brain homogenates. NeuroReport 10:1577–1582

Karlsten R, Gordh TJ (1996) An A_1-selective adenosine agonist abolishes allodynia elicited by vibration and touch after intrathecal injection. Anesth Analg 80:844–847

Kenakin TP (1996) The classification of seven transmembrane receptors in recombinant expression systems. Pharmacol Rev 48:413–465

Kenny BA, Bushfield M, Parry-Smith DJ, Fogarty S, Treherne JM (1998) The application of high throughput screening to novel lead discovery. Prog Drug Res 41: 246–269

King BF, Burnstock G, Boyer JL, Boeynaems JM, Weisman GA, Kennedy C, Jacobson KA, Humphries RG, Abbracchio MP, Gachet C, Miras-Portugal MT (2000) The P2Y receptors. The IUPHAR Compendium of Receptor Characterization and Classification. IUPHAR Media (in press)

Khakh BS, Zhou X, Sydes J, Galligan JJ, Lester HA (2000) State-dependent cross inhibition between transmitter gated ion channels. Nature 406:405–410

Kuroda Y (1981) Regulation of neurotransmission by adenosine and cyclic AMP in mammalian central nervous system. In: Rodnight R, Bachelard HS, Stahl WL (eds) Chemisms of the brain, basic and applied neurochemistry. Churchill-Livingstone, Edinburgh, pp 108–118

Kuroda Y, McIlwain H (1994) Uptake and release of ^{14}C adenine derivatives at beds of mammalian cortical synaptosomes in superfusion system. J Neurochem 22:691–700

Latini S, Bordoni F, Pedata F, Corradetti R (1999) Extracellular adenosine concentrations during in vitro ischaemia in rat hippocampal slices. Br J Pharmacol 127:729–739

Ledent C, Vaugeois JM, Schiffmann SN, Pedrazzini T, El Yacoubi M, Vanderhaeghen JJ, Costentin J, Heath JK, Vassart G, Parmentier M (1997) Aggressiveness, hypoalgesia and high blood pressure in mice lacking the adenosine A_{2a} receptor. Nature 388:674–678

Lindler F, Rigler R (1931) Über die Beeinflussung der Weite der Herzteranzgefässe durch Produkte des Zellkern-Stoffwechsels. Pflugers Archiv 226:697–707

Lohmann K (1929) Über die Pyrophosphatfraktion in Muskel. Naturwissenschaften 17:624–625

Londos C, Wolff J (1977) Two distinct adenosine-sensitive sites on adenylate cyclase. Proc Natl Acad Sci USA 74:5482–5486

Londos C, Cooper DMF, Woolf J (1980) Subclasses of external adenosine receptors. Proc Natl Acad Sci USA 77:2551–2554

Michel AD, Miller KJ, Lundtstrom K, Buell GN, Humphrey PPA (1997) Radiolabeling of the rat $P2X_4$ purinoceptor: evidence for allosteric interactions of purinoceptor antagonists and monovalent cations with P2X purinoceptors. Mol Pharmacol 51:524–532

Milligan G (2000) Receptors as kissing cousins. Science 288:65–67

Mulryan K, Gitterman DP, Lewis CH, Vial C, Leckie BJ, Cobb AL, Brown JE, Conley EC, Buell G, Pritchard CA, Evans RJ (2000) Reduced vas deferens contraction and male infertility in mice lacking $P2X_1$ receptors. Nature 403:86–89

Newby AC (1984) Adenosine and the concept of a retaliatory metabolite. Trends Biochem Sci 9:42–44

Nicke A, Baumert HG, Rettinger J, Eichele A, Lambrecht G, Mutschler E, Schmaizing F (1998) $P2X_1$ and $P2X_3$ receptors form stable trimers: a novel structural motif of ligand-gate ion channels. EMBO J 17:3016–3028

Noji H (1998) The rotary enzyme of the cell: the rotation of F1-ATPase. Science 282:1844–1845

North RA, Barnard E A (1997) Nucleotide receptors. Curr Opin Neurobiol 7:346–357

North RA, Surprenant A (2000) Pharmacology of cloned P2X receptors. Ann Rev Pharmacol Toxicol 40:563–580

Olah ME, Stiles GL (1995) Adenosine receptor subtypes: characterization and therapeutic regulation. Ann Rev Pharmacol Toxicol 35:581–606

Olsson RA, Patterson RE (1976) Adenosine as a physiological regulator of coronary blood flow. Prog Mol Subcell Biol 4:227–248

Paterson ARP, Jakobs ES, Ng CYC, Odegard RD, Adjei AA (1987) Nucleoside transport inhibition in vitro and in vivo. In: Gerlach, E, Becker, B (eds) Topics and perspectives in adenosine research. Springer, Berlin Heidelberg New York, pp 89–101

Pazzagli M, Corsi C, Fratti S, Pedata F, Pepeu G (1995) Regulation of extracellular adenosine levels in the striatum of aging rats. Brain Res 684:103–106

Phillis JW (1989) Adenosine in the control of the cerebral circulation Cerebrovasc Brain Metab Rev 1:26–54

Pintor J, Miras-Portugal MT (2000) Receptors for diadenosine polyphosphates P_{2D}, $P2Y_{ApnA}$, P4 and dinucleotide receptors: are there too many? Trends Pharmacol Sci 21:135

Pull IP, McIlwain H (1977) Adenine mononucleotides and their metabolites liberated from and applied to isolated tissues of the mammalian brain. Neurochem Res 2:203–216

Ralevic V, Burnstock G (1998) Receptors for purines and pyrimidines. Pharmacol Rev 60:413–492

Robertson SJ, Edwards FA (1998) ATP and glutamate are released from separate neurones in the rat medial habenula nucleus: frequency dependence and adenosine-mediated inhibition of release. J Physiol (Lond) 508:691–701

Rudolphi KA, Schubert P, Parkinson FE, Fredholm BB (1992) Neuroprotective role of adenosine in cerebral ischaemia. Trends Pharmacol Sci 13:439–445

Sattin A, Rall TW (1970) The effect of adenosine and adenine nucleotides on the cyclic adenosine 3',5'-monophosphate content of guinea pig cerebral cortex slices. Mol Pharmacol 6:13–23

Schacter JB, Harden TK (1997) An examination of deoxyadenosine 5'(alpha-thio) triphosphate as a ligand to define P2Y receptors and its selectivity as a low potency partial agonist of the $P2Y_1$ receptor. Br J Pharmacol 121:228–344

Searl TJ, Redman RS, Silinsky EM (1998) Mutual occlusion of P2X ATP receptors and nicotinic receptors on sympathetic neurons of the guinea-pig. J Physiol (Lond) 510:783–791

Sikora A, Liu J, Brosnan C, Buell G, Chessel I, Bloom BR (1999) Cutting edge: purinergic signaling regulates radical-mediated bacterial killing mechanism in macrophages through a $P2X_7$-independent mechanism. J Immunol 163:558–561

Simon DK, Johns DR (1999) Mitochondrial disorders: clinical and genetic features. Ann Rev Med 50:111–127

Sylven C, Jonzon B, Fredholm BB, Kaijser L (1988) Adenosine injection into the brachial artery produces ischemia like pain or discomfort in the forearm. Cardiovasc Res 22:674–678

Sperlagh B, Vizi ES, (1996) Neuronal synthesis, storage and release of ATP. Seminar Neurosci 8:175–186

Thannhauser SJ, Bommes A (1914) Experimentelle Studien über den Nucleinstoffwechsel. 2. Mitteilung. Stoffwechselversuche mit Adenosin und Guanosin. Hoppe-Seyler's Z Physiol Chem 91:336–344

Traverso V, Florio T, Virgilio A, Caciagli F, Rathbone MP (2000) Are neuroprotective effects of guanosine mediated by guanosine receptors? Soc Neurosci Abstr 26, 148.15

Triggle DJ (1998) Chemical diversity, Current Protocols Pharmacol. Wiley, New York, pp 901–918

Tordorov LD, Mihaylova-Todorova S, Westfall TD, Sneddon P, Kennedy C, Bjur RA, Westfall DP (1997) Neuronal release of soluble nucleotidases and their role in neurotransmitter inactivation. Nature 387:76–79

Van Calker D, Muller M, Hamprecht, B (1979) Adenosine regulates via two different receptors, the accumulation of cyclic AMP in cultured brain cells. J Neurochem 33:999–1005

Vranai K, Merighi S, Gessi S, Klotz K-N, Leung E, Baraldi PG, Cacciari B, Romagnoli R, Spalluto G, Borea, PA (2000) [3H]MRE 3008F20: a novel antagonist radioligand for the pharmacological and biochemical characterization of human A_3 adenosine receptors. Mol Pharmacol 57:968–975

Wiesner JB, Ugarkar BG, Castellino AJ, Barankiewicz J, Dumas DP, Gruber HE, Foster AC, Erion MD (1999) Adenosine kinase inhibitors as a novel approach to anticonvulsant therapy J Pharmacol Exp Ther 289:1669–1677

Williams M (1989) Adenosine: the prototypic neuromodulator. Neurochem Int 14: 249–264

Williams M, Jarvis MF (2000) Purinergic and pyrimidinergic receptors as potential drug targets. Biochem Pharmacol, 59:1173–1185

Yu H, Bianchi B, Metzger R, Lynch KJ, Kowaluk EA, Jarvis MF, van Biesen T (1999) Lack of specificity of [^{35}S]-ATPγS and [^{35}S]-ADPβS as radioligands for inotropic and metabotropic P2 receptor binding. Drug Dev Res 48:84–93

Zhang G, Franklin PH, Murray TF (1993) Manipulation of endogenous adenosine in the rat piriform cortex modulates seizure susceptibility. J Pharmacol Exp Ther 264:1415–1424

Zimmerman H, Braun N (1999) Ecto-nucleotidases – molecular structures, catalytic properties, and functional roles in the nervous system. Prog Brain Res 120:371–385

Section I
Receptor Classification

CHAPTER 2
P1 Receptors

A. LORENZEN and U. SCHWABE

A. Introduction

Adenosine is an endogenous nucleoside, present in all cells and body fluids, that modulates a wide variety of physiological processes. In their seminal paper, DRURY and SZENT-GYÖRGYI (1929) showed that adenosine and other adenine nucleotides could modulate cardiovascular system function. Over the following 70 years, specific receptors for adenosine were pharmacologically characterized and eventually cloned and a number of selective agonists and antagonists identified (RALEVIC and BURNSTOCK 1998; ABBRACCHIO and WILLIAMS, Chap. 1, this volume; JACOBSON and KNUTSEN, Chap. 6, this volume). Four G-protein-coupled heptahelical receptors (GPCRs) sensitive to adenosine constitute the P1 receptor family and these are designated A_1, A_{2A}, A_{2B}, and A_3 (FREDHOLM et al. 1994) (Fig. 1).

B. Localization and Function of P1 Receptors

In general, the physiological role of adenosine is that of a feedback regulator. The release of adenosine, formed through catabolism of ATP (ZIMMERMANN, Chap. 8, this volume), can restore the balance between energy supply and demand. For example, in hypoxia, adenosine is released from the heart and improves the ratio between oxygen supply and demand by relaxing the vasculature and thus increasing blood flow (BERNE 1963). In the brain, adenosine is released by nerve stimulation (SCHUBERT et al. 1976) and in epileptic seizures, adenosine inhibits neuronal activity and aids in seizure termination (DRAGUNOW et al. 1985; DURING and SPENCER 1992).

The identification of the four adenosine receptor subtypes has greatly facilitated understanding of the many physiological and pharmacological effects of adenosine in almost every cell type. Thus, the localization of the different P1 receptor subtypes in target tissues is an important further step in the characterization of adenosine actions. The elucidation of the molecular structure of adenosine receptors has also made possible the study of the four receptors by Northern blot analysis, in situ hybridization, reverse

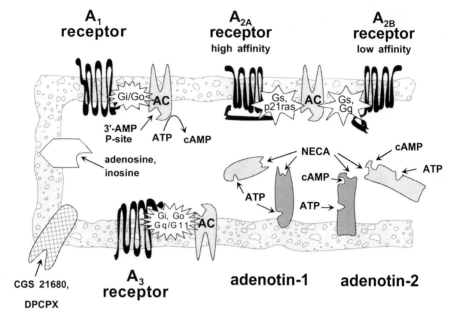

Fig. 1. Adenosine receptors and adenosine binding proteins

transcription-polymerase chain reaction (RT-PCR), and immunostaining with receptor antibodies.

I. A_1 Adenosine Receptors

The tissue distribution of A_1 adenosine receptors was first demonstrated using binding with agonist and antagonist radioligands (Schwabe 1981). Autoradiographic techniques have also been used to study the localization of A_1 receptors in the brain of different species and in peripheral tissues with lower receptor density such as heart, kidney, and adrenals (Weber et al. 1988). The high density of A_1 receptors in brain tissue has been confirmed by mRNA analysis with high levels being expressed in the cerebral cortex, hippocampus, cerebellum, thalamus, brain stem, and spinal cord (Reppert et al. 1991; Dixon et al. 1996). A_1 receptor RNA is also ubiquitous in peripheral tissues with considerable levels being present in vas deferens, testis, white adipose tissue, stomach, spleen, pituitary, adrenal, heart, aorta, liver, eye, and urinary bladder, with only low levels being found in lung, kidney, and small intestine (Reppert et al. 1991; Stehle et al. 1992; Dixon et al. 1996). The specific distribution of A_1 receptors in the brain has also been confirmed by immunohistochemistry (Rivkees et al. 1995b).

The inhibitory feedback regulator role of adenosine is mediated by the A_1 receptor activation, a concept supported by the effects of cardiac A_1 receptor activation that results in negative chronotropic, inotropic, and dromotropic effects (DRURY and SZENT-GYÖRGYI 1929; OLSSON and PEARSON 1990). Furthermore, cardiac A_1 receptors are involved in the protective effect of adenosine in preconditioning and during ischemia and reperfusion injury in the heart (LASLEY et al. 1990). In the kidney, A_1 receptors are implicated in the vasoconstriction and inhibition of renin release (OSSWALD et al. 1978) and in the CNS, adenosine actions at A_1 receptors have been associated with sedative, anxiolytic, anticonvulsant, and locomotor depressant effects (SNYDER et al. 1981). A_1 receptors also mediate the prejunctional inhibition of neurotransmission (FREDHOLM and HEDQVIST 1980; MASINO and DUNWIDDIE, Chap. 9, this volume). A_1 receptors are also involved in bronchoconstrictive (MANN and HOLGATE 1985), antilipolytic (SCHWABE et al. 1973), and antinociceptive responses (REEVE and DICKENSON 1995; SALTER and SOLLEVI, Chap. 13, this volume) to adenosine.

II. A_{2A} Adenosine Receptors

Adenosine A_{2A} receptors were originally defined on the basis of their high affinity for adenosine in stimulating adenylate cyclase activity (DALY et al. 1983). Like the A_1 receptor, the A_{2A} subtype was first detected in the brain, but with a completely different pattern of distribution being mainly restricted to the striatum, nucleus accumbeus and olfactory tubercle (JARVIS and WILLIAMS 1989). RT-PCR revealed a much wider distribution of the A_{2A} receptor transcript product in all brain areas (DIXON et al. 1996) with high levels of the amplification product being observed in the striatum, nucleus accumbeus, and thalamus. A_{2A} receptor mRNA was also found to be widely distributed in many peripheral tissues (STEHLE et al. 1992; DIXON et al. 1996).

A_{2A} receptor agonists inhibit spontaneous locomotor activity and induce a hypomotility resembling that induced by typical neuroleptics (BRIDGES et al. 1987). In addition, A_{2A} receptor stimulation reduces dopamine D_2 agonist affinity, which may contribute to these neuroleptic-like effects of A_{2A} receptor agonists (FUXE et al. 1998). Transgenic mice lacking the A_{2A} receptor exhibit aggressive behavior and lack the stimulant effects of caffeine, suggesting that A_{2A} receptors exert a tonic central depressant action (LEDENT et al. 1997).

In peripheral tissues, the most pronounced A_{2A}-selective agonist effect appears to be coronary vasodilatation (HUTCHISON et al. 1989), the mechanism via which adenosine regulates coronary blood flow during myocardial hypoxia (BERNE 1980). Adenosine and A_{2A}-selective agonists inhibit human platelet aggregation via increased formation of cAMP (HASLAM and ROSSON 1975). Adenosine A_{2A} receptors also mediate the adenosine-induced inhibition of neutrophil activation (FREDHOLM et al. 1996), the inhibition of human T-cell activation (KOSHIBA et al. 1999), and the mitogenic action of adenosine on human endothelial cells (SEXL et al. 1997).

III. A_{2B} Adenosine Receptors

A_{2B} adenosine receptors were originally described as the low affinity component of A_2-receptor mediated responses in the brain (DALY et al. 1983). The major cellular population in the brain containing A_{2B} receptors is represented by glial cells that were responsible for the marked increase of cAMP observed in brain slices in response to adenosine and adenine nucleotides (SATTIN and RALL 1970). After the A_{2B} receptor had been cloned, expression of the receptor mRNA was detected in almost every tissue. Highest mRNA levels were found in the cecum, large intestine, and urinary bladder and lower levels in brain, spinal cord, lung, epididymis, vas deferens, and pituitary (STEHLE et al. 1992; DIXON et al. 1996). Although A_{2B} receptors are widely distributed, their receptor density is relatively low and micromolar concentrations of adenosine are generally needed to induce a functional response. However, such high concentrations may only be obtained under extreme conditions of severe hypoxia.

Functional A_{2B} receptors are present in fibroblasts (BRUNS 1981), vascular beds (DUBEY et al. 1996), hematopoetic cells (FEOKTISTOV and BIAGGIONI 1993), mast cells (MARQUARDT et al. 1994), intestinal epithelial cells (STROHMEIER et al. 1995), endothelium (IWAMOTO et al. 1994), and neurosecretory cells (MATEO et al. 1995). The most important functional role of A_{2B} receptors may be the involvement in adenosine-induced vasodilation in some vascular beds (WEBB et al. 1992). However, the relative contribution of the A_{2B} receptor component in relation to A_{2A} receptors as well as the additional role of endothelial NO generation to the vascular effects of adenosine is still difficult to evaluate at present. A_{2B} receptors modulate secretion of neurotransmitters in the brain and in the periphery. In rat cortex, A_{2B} receptors increase the release of the excitatory transmitters aspartate and acetylcholine and decrease the release of the inhibitory amino acid GABA (PHILLIS et al. 1993a,b). Acetylcholine release in electrically stimulated rat bronchial smooth muscle is potentiated via presynaptic A_{2B} receptors (WALDAY and AAS 1991). A_{2B} receptors may elicit relaxation of intestinal longitudinal smooth muscle (MURTHY et al. 1995; NICHOLLS et al. 1996; PRENTICE and HOURANI 1997) as well as contraction of the intestinal muscularis mucosae (NICHOLLS et al. 1996). The adenosine-induced stimulation of chloride secretion (STROHMEIER et al. 1995) and inhibition of fluid secretion (HANCOCK and COUPAR 1995) may be of therapeutic interest. A_{2B} receptors, depending on the experimental model investigated, either directly activate (FEOKTISTOV and BIAGGIONI 1995; AUCHAMPACH et al. 1997) or potentiate (MARQUARDT and WALKER 1990; PEACHELL et al. 1991) activation of mast cells from different origins. The adenosine-induced bronchoconstriction in human asthmatics is best explained by an A_{2B} receptor. This view is supported by the potent antiasthmatic effects of enprofylline, which is presently the most selective A_{2B} antagonist (ROBEVA et al. 1996). Since A_{2B} receptors potentiate both bronchoconstriction by an increase of acetylcholine release from cholinergic neurons (WALDEY and AAS 1991) and mast cell activation, the development of selective and potent A_{2B} antagonists is a promising therapeutic target. The lamin-related secretory protein netrin-1 may produce its effects on axon outgrowth via A_{2B}-receptor activation (CORSET et al. 2000).

IV. A_3 Adenosine Receptors

In contrast to A_1, A_{2A}, and A_{2B} adenosine receptors, the A_3 receptor was not anticipated on the basis of physiological and pharmacological criteria but rather was identified by homology cloning. The rat A_3 receptor was originally detected only in the testis and with low levels in the brain (ZHOU et al. 1992). Subsequent studies in sheep, humans and rat revealed a wide mRNA expression in many tissues including testis, lung, liver, aorta, kidney, placenta, heart, brain, spleen, uterus, bladder, jejunum, proximal colon, and eye (LINDEN et al. 1993; SALVATORE et al. 1993; DIXON et al. 1996). Expression was also observed in discrete loci of the central nervous system (corpus callosum, substantia nigra, thalamus, subthalamic nucleus, spinal cord, hippocampus) as well as in adrenal cortex and medulla (ATKINSON et al. 1997). Cloning and expression of A_3 receptors from different species has revealed considerable differences in their pharmacological characteristics, especially concerning the sensitivity to xanthine-based antagonists. Rat, rabbit, and gerbil brain A_3 receptors bind xanthine derivatives only weakly, whereas A_3 receptors from sheep and human display high affinity (LINDEN 1994; JI et al. 1994).

Important functional responses mediated by the A_3 receptor are the stimulation of histamine release from a rat mast cell line (RAMKUMAR et al. 1993), depression of locomotor activity in mice (JACOBSON et al. 1993), and the inhibition of chemotaxis of human eosinophils (WALKER et al. 1997). A_3 receptors may also be involved in the cardioprotective and neuroprotective actions af adenosine (VON LUBITZ et al. 1994; LIANG and JACOBSON 1998).

C. Structure of Adenosine Receptors

Human A_1 (LIBERT et al. 1992; TOWNSEND-NICHOLSON and SHINE 1992), A_{2A} (FURLONG et al. 1992) and A_{2B} (PIERCE et al. 1992) receptors were cloned from hippocampus and A_3 receptors from striatum (SALVATORE et al. 1993) and heart (SAJJADI et al. 1993).

The percentage of identities in amino acid sequences of adenosine receptor subtypes ranges from 36% to 56% (Table 1). The genes coding for the human receptors have been located to chromosomes 1q32.1, 22q11.2, 17p11.2, and 1p13.3 for A_1 (TOWNSEND-NICHOLSON et al. 1995a), A_{2A} (LE et al. 1996; PETERFREUND et al. 1996), A_{2B} (TOWNSEND-NICHOLSON et al. 1995b; JACOBSON et al. 1995), and A_3 (ATKINSON et al. 1997) receptors, respectively (Table 2). The genes for adenosine receptors are interrupted within the coding region by a single intron located in the second intracellular loop. Alternative splicing occurs for the A_1 receptor subtype (REN and STILES 1994).

Table 1. Amino acid identities between human adenosine receptor subtypes

	A_1	A_{2A}	A_{2B}	A_3
A_1	100			
A_{2A}	47	100		
A_{2B}	44	56	100	
A_3	47	38	36	100

Table 2. Structure and coupling of human adenosine receptors

	A_1	A_{2A}	A_{2B}	A_3
Molecular mass	36.5 kDa	44.7 kDa	36.3 kDa	36.2 kDa
Amino acids	326	412	332	318
Sequence information	P30542	P29274	P29275	P33765
Gene	1q32.1	22q11.2	17p11.2	1p13.3
G proteins	G_i, G_o	G_s, G_{olf}, $p21^{ras}$	G_s, G_q	G_i, G_q
Effectors	↓AC	↑AC	↑AC	↓AC
	↑PLC		↑PLC	↑PLC
	↑K^+			
	↓Ca^{2+}			

AC, adenylate cyclase; PLC, phospholipase C; K^+, potassium channels; Ca^{2+}, calcium channels. ↑ and ↓ denote stimulation or inhibition, respectively. Sequence information is given as the SwissProt accession number for human adenosine receptors.

The sequence alignment of the four human P1 receptor subtypes is shown in Fig. 2. Sequence identity between the different adenosine receptors is highest in the transmembrane domains (TMs). Potential attachment sites for N-linked glycosylation are located on the second extracellular loop, which are not required for ligand binding to A_{2A} receptors (PIERSEN et al. 1994). Potential sites for palmitoylation on the C terminal portion are present in A_1, A_{2B}, and A_3 receptor proteins, and potential sites for formation of intramolecular disulfide bridges are found in all four P1 receptor subtypes.

Based on formation of chimeric receptors and site-directed mutagenesis studies, the agonist and antagonist binding regions of adenosine receptors have been localized mainly to a binding pocket formed by the transmembrane regions, with some contribution of the second extracellular loop to ligand binding (OLAH et al. 1994a; KIM et al. 1996). Chemical modification of histidine residues with diethylpyrocarbonate pointed to the presence of one histidine which is required for both agonist and antagonist binding, and a second histidine residue required only for antagonist binding (KLOTZ et al. 1988; OLAH et al. 1992). His-278 in TM 7 is crucial for ligand binding and is conserved in all four human adenosine receptor subtypes, whereas His-251 in TM 6 has a role only in antagonist recognition and is present in A_1, A_{2A}, and A_{2B} receptors (Fig. 2). Species differences in ligand binding to A_1 receptors are due to

```
                              TM 1
AA1R_HUMAN    -MPPSISAFQAA---YIGIEVLIALVSVPGNVLVIWAVKVNQALRDATFC 46
AA3R_HUMAN    -MPNNSTALSLANVTYITMEIFIGLCAIVGNVLVICVVKLNPSLQTTTFY 49
AA2A_HUMAN    -MPIMGSSV------YITVELAIAVLAILGNVLVCWAVWLNSNLQNVTNY 43
AA2B_HUMAN    MLLETQDAL------YVALELVIAALSVAGNVLVCAAVGTANTLQTPTNY 44
                 :    :.        *: :*: *.   :: *****  .*    *:   *

                     TM 2                      TM 3
AA1R_HUMAN    FIVSLAVADVAVGALVIPLAILINIGPQTYFHTCLMVACPVLILTQSSIL 96
AA3R_HUMAN    FIVSLALADIAVGVLVMPLAIVVSLGITIHFYSCLFMTCLLLIFTHASIM 99
AA2A_HUMAN    FVVSLAAADIAVGVLAIPFAITISTGFCAACHGCLFIACFVLVLTQSSIF 93
AA2B_HUMAN    FLVSLAAADVAVGLFAIPFAITISLGFCTDFYGCLFLACFVLVLTQSSIF 94
              *:****.**:***  ::*:**  :. *         : **::*  *:*:.*::**:

                                      TM 4
AA1R_HUMAN    ALLAIAVDRYLRVKIPLRYKMVVTPRRAAVAIAGCWILSFVVGLTPMFGW 146
AA3R_HUMAN    SLLAIAVDRYLRVKLTVRYKRVTTHRRIWLALGLCWLVSFLVGLTPMFGW 149
AA2A_HUMAN    SLLAIAIDRYIAIRIPLRYNGLVTGTRAKGIIAICWVLSFAIGLTPMLGW 143
AA2B_HUMAN    SLLAVAVDRYLAICVPLRYKSLVTGTRARGVIAVLWVLAFGIGLTPFLGW 144
              :***.*:***:   :  .:**  :.*    *      *:::* :****::**

                                          TM 5
AA1R_HUMAN    NN----LSAVERAWAANGSMGEP---VIKCEFEKVISMEYMVYFNFFVWV 189
AA3R_HUMAN    N------MKLTSEYHRN----VT---FLSCQFVSVMRMDYMVYFSFLTWI 186
AA2A_HUMAN    N-------NCGQPKEGKNHSQGCGEGQVACLFEDVVPMNYMVYFNFFACV 186
AA2B_HUMAN    NSKDSATNNCTEPWDGTTNESCC---LVKCLFENVVPMSYMVYFNFFGCV 191
               *                .         :  * * .*: *.*****.*:  :

AA1R_HUMAN    LPPLLLMVLIYLEVFYLIRKQLNKKVSASSG--DPQKYYGKELKIAKSLA 237
AA3R_HUMAN    FIPLVVMCAIYLDIFYIIRNKLSLNLSNSK---ETGAFYGREFKTAKSLF 233
AA2A_HUMAN    LVPLLLMLGVYLRIFLAARRQLKQMESQPLPGERARSTLQKEVHAAKSLA 236
AA2B_HUMAN    LPPLLIMLVIYIKIFLVACRQLQRTELM----DHSRTTLQREIHAAKSLA 237
              : **::*  :*: :*       .:*.         .    :*.: ****

                     TM 6                       TM 7
AA1R_HUMAN    LILFLFALSWLPLHILNCITLFCPSC--HKPSILTYIAIFLTHGNSAMNP 285
AA3R_HUMAN    LVLFLFALSWLPLSIINCIIYFNG----EVPQLVLYMGILLSHANSMMNP 279
AA2A_HUMAN    IIVGLFALCWLPLHIINCFTFFCPDCS-HAPLWLMYLAIVLSHTNSVVNP 285
AA2B_HUMAN    MIVGIFALCWLPVHAVNCVTLFQPAQGKNKPKWAMNMAILLSHANSVVNP 287
              :::  :***.***: :**. * :**      * .   .  :.*.*:* **  :**

AA1R_HUMAN    IVYAFRIQKFRVTFLKIWNDHFRCQPA--PPIDEDLPEERPDD------- 326
AA3R_HUMAN    IVYAYKIKKFKETYLLILKACVVCHPS--DSLDTSIEKNSE--------- 318
AA2A_HUMAN    FIYAYRIREFRQTFRKIIRSHVLRQQEPFKAAGTSARVLAAHGSDGEQVS 335
AA2B_HUMAN    IVYAYRNRDFRYTFHKIISRYLLCQAD-VKSGNGQAGVQPALG-----VG 331
              ::***::   :..*:  *:   *              

AA1R_HUMAN    -------------------------------------
AA3R_HUMAN    -------------------------------------
AA2A_HUMAN    LRLNGHPPGVWANGSAPHPERRPNGYALGLVSGGSAQESQGNTGLPDVEL 385
AA2B_HUMAN    L------------------------------------------------- 332

AA1R_HUMAN    -------------------------
AA3R_HUMAN    -------------------------
AA2A_HUMAN    LSHELKGVCPEPPGLDDPLAQDGAGVS 412
AA2B_HUMAN    -------------------------
```

Fig. 2. Aligned amino acid sequences of human adenosine receptors

distinct differences in amino acids in positions 270 and 277 in TM 7. Amino acid 270 (Ile in bovine, Met in human A_1 receptors) interacts with the N^6-substituted region of agonists and C8-substituents of xanthine antagonists, whereas a threonine residue in position 277 mediates the interaction of the receptor with the 5'-substituent in NECA (TOWNSEND-NICHOLSON and SCHOFIELD 1994; TUCKER et al. 1994). In the human A_{2A} receptor, the corresponding residue, Ser-277, is required for high affinity binding of agonists, but not antagonists (KIM et al. 1995). Binding of 5'-substituted agonists in the rat A_3 receptor is also favored by a stretch of six amino acids (FFSFIT) in TM 5 neighboring the second extracellular loop (OLAH et al. 1994b). TMs 1–4 are sufficient to confer agonist and antagonist binding specificity to the human A_1 adenosine receptor (RIVKEES et al. 1995a). More specifically, a serine residue in TM 3 (Ser-94) is required for agonist as well as antagonist binding. Glu-16 in TM 1 and Asp-55 in TM 2 are critical to agonist rather than antagonist binding, and Asp-55 mediates the effect of sodium ions on ligand binding to the A_1 receptor (BARBHAIYA et al. 1996). A glutamic acid residue (Glu-13) in the human A_{2A} receptor, which corresponds to Glu-16 in A_1 receptors, has also been shown to be crucial for agonist, but not antagonist, binding to A_{2A} receptors (IJZERMAN et al. 1996). In the human A_1 receptor, Thr-91 and Gln-92 in TM 3 interact with the adenine moiety of adenosine derivatives (RIVKEES et al. 1999). In human A_{2A} receptors, the equivalent residues are also involved in ligand binding, Thr-88 mediating high-affinity binding of agonists and Gln-89 being involved in agonist and antagonist recognition (JIANG et al. 1996). For A_1-selective high-affinity recognition of N^6-substituted agonists or adenine-based antagonists, amino acids Pro-86 and Leu-88 are essential in the human A_1 receptor (RIVKEES et al. 1999). Replacement in the A_{2A} receptor sequence of Asp-181 in TM 5 selectively diminishes the affinities of N^6- or C2-substituted agonists, but has almost no effect on binding of 5'-substituted derivatives (KIM et al. 1995). In conclusion, amino acid residues relevant for agonist and antagonist binding to A_1 and A_{2A} receptors have been identified. Evidence is emerging which clarifies the relationships between ligand structure and recognition by the receptor protein of specific structural characteristics of ligands. Due to the lack of suitable ligands, mutational studies addressing ligand binding to A_{2B} and A_3 adenosine receptors have not been performed. Future investigations might also address more functional aspects of receptor activation by full and partial agonists and selective G protein activation by adenosine receptor subtypes.

D. Signal Transduction by Adenosine Receptors

Signal transduction by all adenosine receptor subtypes proceeds through activation of specific G protein subsets (Table 2). The modulation by adenosine receptors of guanine nucleotide interactions with G proteins has been studied by investigation of GDP release or GTP binding, which represent the activation steps of the G protein, and by measurement of GTPase activity, which represents the off-switch step of G protein activation subsequent to the guanine nucleotide exchange reaction. MARALA and MUSTAFA (1993) have demonstrated the release of $[\alpha\text{-}^{32}P]GDP$ from G_s via A_{2A} receptor stimulation by CGS 21680 in rat striatum after prelabeling of G proteins with $[\alpha\text{-}^{32}P]GTP$. A_1 receptor agonists induce a release of $[^{35}S]GDP\beta S$ from rat brain membranes (LORENZEN et al. 1996a). Determination of $[^{35}S]GTP\gamma S$ binding serves as a measure for ligand potencies and intrinsic activities at A_1 and A_3 receptors (LORENZEN et al. 1993a, 1996a; SHRYOCK et al. 1998a; VAN TILBURG et al. 1999). More specifically, the subtypes of G protein α subunits activated by A_3 receptors have been identified by agonist-induced covalent incorporation of the photoaffinity label 4-azidoanilido-$[\alpha\text{-}^{32}P]$ GTP into α subunits, which were consecutively immunoprecipitated with specific antisera (PALMER et al. 1995).

The A_1 adenosine receptor is coupled to pertussis toxin-sensitive G proteins of the G_i/G_o family (MUNSHI and LINDEN 1991; MUNSHI et al. 1991; FREISSMUTH et al. 1991; JOCKERS et al. 1994). Species differences in this coupling may direct signaling differently depending on the species. Bovine A_1 receptors display an approximately tenfold preference for $G_{i\alpha 3}$ compared to $G_{i\alpha 1}$, $G_{i\alpha 2}$, and $G_o\alpha$ (FREISSMUTH et al. 1991), whereas human and rat A_1 receptors preferentially interact with G_i isoforms (JOCKERS et al. 1994; LORENZEN et al. 1998a) rather than with G_o. A further species difference in receptor-G protein interactions is the apparent inability of the coupling cofactor to stabilize the ternary complex (agonist-receptor-G protein) in human brain membranes, whereas A_1 receptor-G_i protein complexes in membranes from rat and bovine brain are highly resistant to the uncoupling effect of GTP due to coupling cofactor activity (NANOFF et al. 1995, 1997). Although the protein has been partially purified, the structure of the coupling cofactor is presently unknown (NANOFF et al. 1997).

Mutational studies to locate specific regions in the A_1 receptor which interact with the G protein have not been performed. The contact site between receptors and G proteins must be distinct even between receptors which couple to identical or similar G proteins. These differences in receptor-G protein interaction may facilitate the development of drugs which block signal transduction selectively at a site distinct from agonist or antagonist binding. High-affinity agonist binding to the A_1 receptor and signal transduction can be blocked by peptides corresponding to the C-terminal eleven residues of the α subunits of $G_{i1/2}$ or G_o; some α-subunit-related peptides are able to distinguish between A_1 and $GABA_B$ receptors, although both receptors activate identical G proteins (GILCHRIST et al. 1998). Suramin analogues acting on the receptor-G protein interaction sites discriminate between A_1 adenosine and D_2 dopamine receptors, although both receptors couple to identical G proteins (BEINDL et al. 1996; WALDHOER et al. 1998).

The relevance of the contribution of G protein $\beta\gamma$ subunits to signal transduction via adenosine receptors has been less well studied. Evidence for their importance comes from studies which demonstrate that the composition of $\beta\gamma$

subunits modulates agonist binding and the interaction of the A_1 receptor with G protein α subunits (FIGLER et al. 1996, 1997). The prenyl group on the G protein γ subunit influences high-affinity agonist binding to A_1 receptors and agonist-induced guanine nucleotide exchange on the α subunit (YASUDA et al. 1996).

The relative roles of the different G protein α and $\beta\gamma$ subunits in A_1-mediated signal transduction are becoming clearer. A_1 adenosine receptors inhibit the activity of adenylate cyclase through α subunits of G_i, which results in antiadrenergic effects in the heart (PELLEG and BELARDINELLI 1993) and inhibition of lipolysis in adipocytes (MOXHAM et al. 1993). The role of adenylate cyclase inhibition in the brain is less clear, because one of the main actions of adenosine through A_1 receptors, the inhibition of neurotransmitter release in the brain, proceeds independently from cAMP (DUNWIDDIE and FREDHOLM 1985; FREDHOLM and LINDGREN 1987).

A_1 adenosine receptor agonists stimulate potassium currents through several types of K^+ channels (BELARDINELLI and ISENBERG 1983; TRUSSELL and JACKSON 1985, 1987). G protein $\beta\gamma$ subunits, by occupying binding domains lacking specificity for various combinations of $\beta\gamma$ subtypes, mediate the activation of inwardly rectifying K^+ channels in the atria of the heart following activation of A_1 adenosine or M_2 muscarinic receptors (ITO et al. 1992; YAMADA et al. 1998). As shown by targeted inactivation of G protein α subunit genes in embryonic stem cell-derived cardiocytes, for this activation of inwardly rectifying K^+ channels the presence of α_{i2} or α_{i3} is required, but not of α_o or α_q (SOWELL et al. 1997). The negative chronotropic effects of adenosine, however, are preserved even in the absence of α_{i2} and α_{i3} subunits and without activation of inwardly rectifying K^+ channels (SOWELL et al. 1997), suggesting that other G proteins and ion currents can serve as substitutes. Activation of A_1 receptors also increases the activity of ATP-sensitive K^+ channels, most probably via G_i protein α subunits (KIRSCH et al. 1990; ITO et al. 1992, 1994). The functional relevance of A_1 receptor regulation of K_{ATP} channels in cerebral ischemic preconditioning has been underscored by the protective effects of the selective A_1 receptor agonist, N^6-cyclopentyladenosine, and of the K_{ATP} channel opener, levcromakalim (HEURTEAUX et al. 1995).

Inhibition of Ca^{2+} currents through L-, N-, and P-type channels by A_1 adenosine receptors has been characterized in a variety of cell types but mainly in neurons (e.g., DOLPHIN et al. 1986; SCHOLZ and MILLER 1991a,b; UMEMIYA and BERGER 1994; MEI et al. 1996). In the brain, A_1 receptors reduce L-type Ca^{2+} currents via the α subunit of G_o (SWEENEY and DOLPHIN 1995). In the heart, the functional relevance of A_1 receptor coupling to G_o is not understood (ASANO et al. 1995).

A_1 adenosine receptors affect inositol phosphate production by phospholipase C in an either inhibitory (DELAHUNTY et al. 1988) or stimulatory manner. Stimulation of phospholipase C by A_1 receptors may be permissive, such that responses to other PLC-coupled receptors, e.g., to muscarinic M_3 acetylcholine receptors, are augmented (AKBAR et al. 1994), or may be a consequence of activation of A_1 receptors alone (WHITE et al. 1992; FREUND et al. 1994). In the interaction between H_1 histamine receptors and A_1 adenosine receptors, adenosine receptor agonists may either inhibit or potentiate PLC activation induced by histamine receptor agonists (for review, see DICKENSON and HILL 1994). A_1 receptors have been reported to act synergistically with bradykinin (GERWINS and FREDHOLM 1992a) and ATP (GERWINS and FREDHOLM 1992b) in DDT_1 MF-2 smooth muscle cells, with α_1 adrenergic receptors (OKAJIMA et al. 1989), and with cholecystokinin CCK_1 receptors (DICKENSON and HILL 1996). The activation of PLC by G protein depends on the enzyme subtype which is to be activated, and on the receptor which activates PLC. PLC β isoenzymes are target enzymes for α subunits of the G_q family, but also for G protein $\beta\gamma$ subunits (EXTON 1996). For example, in smooth muscle cells from rabbit intestine, the PLC β_1-isoform is activated by cholecystokinin involving the G protein α_q subunit. In contrast, A_1 adenosine receptors activate PLC β_3 in these cells, and activation via A_1 receptors can be inhibited by antibodies against G protein α_{i3} and β subunits (MURTHY and MAKHLOUF 1995). $\beta\gamma$ Subunits mediate the A_1-induced increase in inositolphosphate production in COS-7 cells in response to thyrotropin (TOMURA et al. 1997). $\beta\gamma$ Subunits are also the active G protein subunits in the direct activation of PLC by A_1 receptors expressed in CHO cells and in the synergistic activation of PLC by A_1 receptors and $P2Y_2$ receptors, whereas the stimulatory effect of $P2Y_2$ receptors does not involve $\beta\gamma$ subunits (DICKENSON and HILL 1998). Since higher concentrations of $\beta\gamma$ subunits than of α subunits are required to interact effectively with effectors, the receptor densities or the proportion of agonist-occupied receptors for induction of responses through $\beta\gamma$ subunits is generally higher than for α subunit-mediated effector activation (BIRNBAUMER 1992; STERNWEIS 1994). Therefore, activation of PLC by A_1 adenosine receptors in cultured astrocytes from rat brain critically depends on the level of A_1 receptor expression (BIBER et al. 1997). A_1 adenosine receptors activate, probably also via $\beta\gamma$ subunits, the mitogen-activated protein kinase cascade in a pertussis toxin sensitive and protein tyrosin kinase-dependent manner (FAURE et al. 1997; DICKENSON et al. 1998).

Allosteric modulators specific for the A_1 adenosine receptor, e.g., PD 81723, increase agonist binding by slowing the agonist dissociation rate from the receptor, and by stabilization of receptor-G protein interactions in the presence of agonists (BRUNS and FERGUS 1990; KOLLIAS-BAKER et al. 1994). This enhancing effect of PD 81723 seems to be tissue-specific, since it was detected in brain and cardiac membranes, but was absent in adipocyte membranes (JARVIS et al. 1999). Some actions of this allosteric enhancer on adenylate cyclase activity are not mediated via A_1 adenosine receptors, but are due to direct inhibition of the enzyme (MUSSER et al. 1999). The putative mechanism of action of PD 81723 on A_1-mediated signal transduction is most probably based on an allosterically mediated conformational isomerization of the receptor from the inactive to a spontaneously active state. This assumption is substantiated by the finding that PD 81723, even in the absence of agonists, increases G protein activation and inhibits adenylate cyclase through human A_1 receptors stably expressed in CHO cells (KOLLIAS-BAKER et al. 1997). Agonist-independent actions of PD 81723 are probably indicative of spontaneously active receptors, since the reduction of cAMP content by PD 81723 is antagonized by the neutral antagonist WRC-0342 in CHO cells overexpressing the human A_1 receptor (SHRYOCK et al. 1998a). Further evidence for constitutively active A_1 adenosine receptors comes from the recent subclassification of A_1 receptor antagonists into neutral antagonists, which do not alter the level of basal activity, and inverse agonists, which inhibit the basal G protein activation and increase cAMP production in the absence of agonists (SHRYOCK et al. 1998a).

The most thoroughly investigated signal transduction mechanism of A_2 adenosine receptors is the stimulation of adenylate cyclase via G_s protein, leading to activation of protein kinase A (PKA). The coupling of A_2 receptors to their signal transduction cascade is unusually tight (BRAUN and LEVITZKI 1979). The selective activation of the protein G_s by A_{2A} adenosine receptors is dependent on the aminoterminal region of the third intracellular loop, especially Lys-209 and Glu-212, with some additional modulation by Gly-118 and Thr-119 (OLAH 1997). Other signaling pathways independent of increased cAMP have been described, but are less well understood. A_{2A} adenosine receptors on striatal nerve terminals inhibit the release of GABA through a mechanism which involves N-type Ca^{2+} channels and protein kinase C, but is independent of PKA or PKG (KIRK and RICHARDSON 1995). In the A_{2A} receptor-induced increase of acetylcholine release from striatal nerve terminals, two distinct signaling pathways are activated: a cholera-toxin sensitive mechanism involving G_s, cAMP increase, PKA, and P-type Ca^{2+} channels, and, secondly, a cholera-toxin-insensitive pathway in which PKC and N-type channels contribute to the signal (GUBITZ et al. 1996). In neutrophils, superoxide anion generation is inhibited by A_{2A} receptors in a cAMP-independent manner by activation of a serine/threonine protein phosphatase (REVAN et al. 1996). Another cAMP-independent mechanism, which is also independent from generation of inositol phosphates and does not affect Na^+/Ca^{2+} exchange, underlies the A_{2A} receptor-induced Ca^{2+} entry via L-type Ca^{2+} channels into embryonic chick ventricular cells (LIANG and MORLEY 1996). The A_{2A} receptor-induced activation of mitogen-activated protein kinase (MAP kinase) is not mediated by G_s, does not require pertussis toxin-sensitive G proteins, PKC or epidermal growth factor signaling, but involves activation of $p21^{ras}$ and mitogen activated kinase kinase 1 (SEXL et al. 1997). In contrast, the partial inhibition by A_{2A} receptors of thrombin-induced activation of MAP kinase is mediated by an increase of cAMP through G_s and adenylate cyclase (HIRANO et al. 1996).

A_{2B} adenosine receptors are coupled to adenylate cyclase in a stimulatory manner (DALY et al. 1983; BRACKETT and DALY 1994; FEOKTISTOV and BIAGGIONI 1995; AUCHAMPACH et al. 1997) and to activation of phospholipase C (PLC) via a pertussis- and cholera-toxin-insensitive G protein, probably G_q (YAKEL et al.

1993; FEOKTISTOV and BIAGGIONI 1995; AUCHAMPACH et al. 1997). In a probably cAMP- and PLC-independent manner, A_{2B} adenosine receptors reduce nicotinic agonist-stimulated catecholamine release from bovine adrenal chromaffin cells, possibly by activation of a protein phosphatase in the cytosol (MATEO et al. 1995). Mitogenic signaling through activation of $p21^{ras}$ secondary to G_s or G_q activation is stimulated by A_{2B} receptors in human mast cells (FEOKTISTOV et al. 1999) and HEK-293 cells (GAO et al. 1999).

A_3 adenosine receptors interact with G_i (α_{i2} and α_{i3}) and G_q-like G proteins (PALMER et al. 1995) inhibiting adenylate cyclase (ZHOU et al. 1992) via a pertussis toxin sensitive G protein and stimulating PLC, leading to inositol phosphate generation and increased levels of intracellular Ca^{2+} (RAMKUMAR et al. 1993; ABBRACCHIO et al. 1995). In the rat RBL-2H3 mast cell line, pertussis toxin treatment and microinjection of neutralizing antibodies to G protein α subunits indicate that adenosine-mediated Ca^{2+} mobilization involves the α subunits, α_{i3} and α_q (HOFFMAN et al. 1997).

In the light of the great variety of adenosine receptor-mediated signal transduction pathways and the almost ubiquitous presence of these receptors in mammalian tissues, it would probably be advantageous in developing potential therapeutic applications of receptor agonists to activate distinct signal transduction pathways selectively. Signal trafficking by structurally distinct agonists of GPCRs (KENAKIN 1996) has been shown for many receptors coupled to more than a single G protein. Although the A_1 receptor is coupled to at least four distinct G protein subtypes (G_{i1}, G_{i2}, G_{i3}, and G_o), none of these G proteins appear to be activated preferentially by structurally distinct full agonists (WISE et al. 1999) or partial agonists (LORENZEN et al. 1998a). Another way of selectively targeting tissues may be to take advantage of different receptor densities, since the intrinsic activity of a ligand depends on the level of receptor expression (for review, see KENAKIN 1996). Indeed, 8-alkylamino-substituted derivatives of N^6-cyclopentyladenosine, that are partial agonists at A_1 receptors, display full agonist intrinsic activity as anti-lipolytic agents, whereas their intrinsic activities in the reduction of heart rate are only 30%–80% of the maximum effects induced by full agonists (VAN SCHAICK et al. 1998). In rat adipocytes, agonist occupation of only 6% of A_1 receptors is sufficient to produce a half-maximum effect in inhibition of cAMP accumulation (LOHSE et al. 1986). Adenosine A_{2A} receptor-induced half-maximal coronary vasodilation was observed at agonist concentrations that occupied between 1.3% and 9% of the receptors (SHRYOCK et al. 1998b). Even in a single tissue, the receptor reserve may vary depending on the signal. Whereas half-maximal inhibition of isoproterenol-stimulated L-type Ca^{2+} currents is achieved at only 4% A_1 receptor occupancy, adenosine activation of inwardly rectifying K^+ currents in guinea pig atria required 40% A_1 receptor occupancy (SRINIVAS et al. 1997). Since only a single subtype of A_1 receptor has been cloned, these differences in receptor reserve in a single tissue cannot be attributed to more than one receptor subtype. They may, however, be related to the types of G protein subunits, α or $\beta\gamma$, that are involved in transducing the signal, because higher agonist concentrations, and, therefore, higher degrees of receptor occupancy are required to activate G protein $\beta\gamma$-transduced signals. Therefore, it seems conceivable that selectivity in the application of adenosine receptor-related drugs may be achieved by taking into account distinct receptor densities in tissues and by taking advantage of different G protein subunits involved in the desired and unwanted side effects. In this context, the possible role of rapid receptor desensitization remains to be elucidated.

E. Adenosine Effects with Unknown Mechanisms and Adenosine Binding Proteins

Additional adenosine binding sites may complicate the pharmacological characterization of P1 receptors. A schematic presentation is given in Fig. 1. The physiological function of adenosine binding proteins distinct from the four cloned receptors is presently unknown. They may, however, mediate some actions of purines that cannot be attributed to purinergic receptors. Adenosine and inosine protect astrocytes in hypoxia and glucose deprivation by acting on an intracellular xanthine-insensitive site (HAUN et al. 1996). Adenosine and inosine stimulate neurite outgrowth and catecholamine synthesis in rat PC12 pheochromocytoma cells and chick sympathetic neurons (BRAUMANN et al. 1986; ZURN and DO 1988). These effects are also mediated by an intracellular site and are not antagonized by xanthines. Adenosine and 5′-substituted, but not N^6-substituted adenosine derivatives or 2-CADO inhibit phosphorylation of myosin light chain, troponin I and phosphatidylinositol (DOCTROW and LOWENSTEIN 1985, 1987; ROSENTHAL and LOWENSTEIN 1991). Adenosine has mitogenic effects on endothelial cells (MEININGER et al. 1988). Several studies indicate that extracellular adenosine receptors mediate this effect (MEININGER and GRANGER 1990; ETHIER et al. 1993), specifically the A_{2A} receptor, which stimulates proliferation of primary human endothelial cells via activation of $p21^{ras}$ and mitogen-activated protein kinase (SEXL et al. 1995, 1997). There is also evidence that additional mechanisms are probably involved in the growth-promoting effects of nucleosides. The proliferative effects on aortic endothelial cells of adenosine, inosine, and hypoxanthine cannot be reconciled with activation of a known adenosine receptor subtype, because these effect are not mimicked by either R-PIA or NECA and are not antagonized by 8-phenyltheophylline (van DAELE et al. 1992). According to ETHIER and DOBSON (1997), Na^+/H^+ exchange and phospholipase A_2, but not classical adenosine receptors, are involved in the stimulation of DNA synthesis by adenosine in human umbilical vein endothelial cells. The inhibition of adenosine- and inosine-induced stimulation of cellular [methyl-^{3}H]thymidine uptake by 5-(N,N-dimethyl)amiloride also points to a potential role of the Na^+/H^+ exchanger in mediating the proliferative effects of nucleosides (LEMMENS et al. 1996). Inosine dilates smooth muscles in kidney arteries (SINCLAIR et al. 1985) and aorta (COLLIS et al. 1986) in an adenosine-independent and xanthine-insensitive manner. The site of action in the aorta is located in the intracellular space (COLLIS et al. 1986). Adenosine and some of its derivatives induce a xanthine-insensitive relaxation of aorta (COLLIS and BROWN 1983; COLLIS et al. 1986; PRENTICE and HOURANI 1996) and taenia caeci (PRENTICE et al. 1995; PRENTICE and HOURANI 1997), that is most probably not mediated by the A_3 adenosine receptor subtype. The mechanism of these adenosine actions described above has not been clarified to date.

On the other hand, adenosine binding sites of unknown function have been identified that may mediate some of the effects of purines that are not attributable to P1 or P2 receptors. In binding studies, adenosine binding proteins share with all adenosine receptors a high affinity for the non-selective agonist, NECA. Multiple binding sites have been detected by the use of the A_{2A}-selective agonist [^{3}H]CGS 21680 (WAN et al. 1990). In addition to A_{2A} receptors, this radioligand labels a sodium- and guanine-nucleotide-sensitive site in rat striatum that is the predominant binding site for CGS 21680 in rat brain cortex and hippocampus (CUNHA et al. 1996). The agonist and antagonist potency ratios at this site demonstrate that it is distinct from known adenosine receptors (JOHANSSON et al. 1993; JOHANSSON and FREDHOLM 1995; CUNHA et al. 1996). Thus A_{2A} adenosine receptors are more selectively labeled using the selective antagonist, SCH 58261, which does not interact with the non-A_{2A}-binding site for CGS 21680 (LINDSTRÖM et al. 1996). The nature of the non-A_{2A} binding site for CGS 21680 remains to be clarified.

Three additional adenosine binding proteins have been characterized using [^{3}H]NECA. Their pharmacological profiles are compared in Table 3. Adenotin-1 is a ubiquitous adenosine-binding protein that is found in distinct membrane-associated and cytosolic forms in all rat tissues investigated and is enriched in the endoplasmic reticulum, particularly in the liver (LORENZEN et al. 1998b). It has been characterized in many cells and tissues. Adenotin-1 has

Table 3. Pharmacological profiles of adenosine binding proteins determined in [^{3}H]NECA binding studies. Affinities are reported for adenotin-1 purified from human platelets[a] (FEIN et al. 1994) or partially purified from bovine striatum (LORENZEN et al. 1992). Affinities for adenotin-2 were determined in rat brain membranes (LORENZEN et al. 1996b). Ligand affinities for the adenosine/inosine site were measured in partially purified fractions from bovine striatum (LORENZEN et al. 1992)

	Adenotin-1	Adenotin-2	Adenosine/inosine site
Adenosine	14 300[a]	7 060	15
NECA	323	57	23
Inosine	>100 000	>100 000	71
cAMP	11 200[a]	196	80
2-CADO	1 540	>100 000	1 300
R-PIA	>100 000	>100 000	3 600
Adenine	>100 000[a]	5 910	>100 000
XAC	>100 000	>100 000	>100 000

been purified from human placenta (HUTCHISON et al. 1990) and human platelets (FEIN et al. 1994) and exhibits aminoterminal sequence homology to stress proteins. The human proteins tumor rejection antigen 1, gp96 homolog, endoplasmin precursor, and the glucose-regulated protein grp94 (accession P14625) are identical proteins and share with adenotin-1 a sequence identity over 14 amino acids at or near the N termini of these proteins (DDEVDVDGTVEEDL). Since adenotin-1 has not been cloned, it is not clear if it is identical with or distinct from the known human homologs. Likewise, the relevance of adenosine binding to adenotin-1 needs to be clarified. Adenotin-1 binds adenosine with low affinity (~10 μmol/l) (Table 3), suggesting that adenosine binding to adenotin-1 can only occur when adenosine levels are substantially elevated, e.g., in severe hypoxia.

Adenotin-2 (LORENZEN et al. 1996b) shares with adenotin-1 a micromolar affinity for adenosine, and the lack of affinity for N^6-substituted adenosine derivatives and xanthines (Table 3). Substitution in C-2 position of the purine ring also leads to loss of binding activity to adenotin-2. However, NECA, cyclic AMP, and adenine are much more potent ligands for adenotin-2 than for adenotin-1. The complex interactions of adenotin-2 with cyclic AMP, ATP, and protein kinase activators and inhibitors may indicate that adenotin-2 is involved in phosphorylation reactions (LORENZEN et al. 1996b). cAMP inhibits radioligand binding to adenotin-2 with an apparent K_i of <200nmol/l (Table 3). However, this is due to an allosteric mechanism. The metabolically stable ATP derivative ATPγS increases [^{3}H]NECA binding to adenotin-2 in particulate preparations. The protein kinase inhibitors H-9, which is selective for cyclic nucleotide-dependent kinases, and the nonselective inhibitor K-252a, both of which act at the ATP binding sites of kinases, inhibit [^{3}H]NECA binding to adenotin-2. However, selective inhibitors of PKA (KT 5720) and PKG (KT 5723) were ineffective. Therefore, adenotin-2 cannot be identified as a known protein kinase.

Another [^{3}H]NECA binding site has been characterized in bovine striatum (LORENZEN et al. 1992) and cortex (LORENZEN et al. 1993b). It shows high affinities for adenosine, NECA, inosine, and 5'-substituted adenosine derivatives like S-adenosylhomocysteine or adenine nucleotides, intermediate affinities for most adenosine receptor agonists, but it does not bind adenosine receptor antagonists (Table 3) (LORENZEN et al. 1992). It has been distinguished from S-adenosylhomocysteine hydrolase by the low affinities of eritadenine and adenosine-2',3'-dialdehyde, which are potent inhibitors of the enzyme. This protein, which binds adenosine and inosine with nanomolar affinities (Table 3) may, based on its pharmacological characteristics, be involved in xanthine-insensitive actions of adenosine, inosine and adenine nucleotides.

In summary, adenosine-binding proteins are interesting candidates which might be targets of actions of adenosine and related compounds which are not mediated by the known purinoceptors. Purification, cloning, and expression of these proteins are required to gain insight into their physiological roles.

List of Abbreviations

ATPγS	adenosine-5'-(γ-thio)-triphosphate
CGS 21680	2-[p-(2-carboxyethylphenethylamino]-5'-N-ethylcarboxami-doadenosine
2-CADO	2-chloroadenosine
GDPβS	guanosine-5'-(β-thio)-diphosphate
GTPγS	guanosine-5'-(γ-thio)-triphosphate
NECA	5'-N-ethylcarboxamidoadenosine
PD 81723	(2-amino-4,5-dimethyl-3-thienyl)-[3-(trifluoromethyl)-phenyl]methanone
PIA	N^6-phenylisopropyladenosine
PLC	phospholipase C
SCH 58261	5-amino-7-(2-phenylethyl)-2-(2-furyl)-pyrazolo-(4,3-e)1,2,4-triazolo-(1,5-c)pyrimidine
XAC	xanthine amine congener

References

Abbracchio MP, Brambilla R, Ceruti S, Kim HO, von Lubitz DKJE, Jacobson KA, Cattabeni F (1995) G Protein-dependent activation of phospholipase C by adenosine A_3 receptors in rat brain. Mol Pharmacol 48:1038–1045

Akbar M, Okajima F, Tomura H, Shimegi S, Kondo Y (1994) A single species of A_1 adenosine receptor expressed in Chinese hamster ovary cells not only inhibits cAMP accumulation but also stimulates phospholipase C and arachidonate release. Mol Pharmacol 45:1036–1042

Asano T, Shinohara H, Norota I, Kato K, Endoh M (1995) The G-protein G_o in mammalian cardiac muscle: Localization and coupling to A_1 adenosine receptors. J Biochem 117:183–189

Atkinson MR, Townsend-Nicholson A, Nicholl JK, Sutherland GR, Schofield PR (1997) Cloning, characterisation and chromosomal assignment of the human adenosine A3 receptor (ADORA3) gene. Neurosci Res 29:73–79

Auchampach JA, Jin X, Wan TC, Caughey GH, Linden J (1997) Canine mast cell adenosine receptors: Cloning and expression of the A_3 receptor and evidence that degranulation is mediated by the A_{2B} receptor. Mol Pharmacol 52:846–860

Barbhaiya H, McClain R, IJzerman A, Rivkees SA (1996) Site-directed mutagenesis of the human A_1 adenosine receptor: Influences of acidic and hydroxy residues in the first four transmembrane domains on ligand binding. Mol Pharmacol 50:1635–1642

Beindl W, Mitterauer T, Hohenegger M, IJzerman AP, Nanoff C, Freissmuth M (1996) Inhibition of receptor/G protein coupling by suramin analogues. Mol Pharmacol 50:415–423

Belardinelli L, Isenberg G (1983) Isolated atrial myocytes: Adenosine and acetylcholine increase potassium conductance. Am J Physiol 244:H734–H737

Berne RM (1963) Cardiac nucleotides in hypoxia: Possible role in regulation of coronary blood flow. Am J Physiol 204:317–322

Berne RM (1980) The role of adenosine in the regulation of coronary blood flow. Circ Res 47:807–813

Biber K, Klotz K-N, Berger M, Gebicke-Härter PJ (1997) Adenosine A_1 receptor-mediated activation of phospholipase C in cultured astrocytes depends on the level of receptor expression. J Neurosci 17:4956–4964

Birnbaumer L (1992) Receptor-to-effector signaling through G proteins: Roles for $\beta\gamma$ dimers as well as α subunits. Cell 71:1069–1072

Brackett LE, Daly JW (1994) Functional characterization of the A_{2b} adenosine receptor in NIH 3T3 fibroblasts. Biochem Pharmacol 47:801–814

Braumann T, Jastorff B, Richter-Landsberg C (1986) Fate of cyclic nucleotides in PC12 cell cultures: Uptake, metabolism, and effects of metabolites on nerve-growth factor-induced neurite outgrowth. J Neurochem 47:912–919

Braun S, Levitzki A (1979) Adenosine receptor permanently coupled to turkey erythrocyte adenylate cyclase. Biochemistry 18:2134–2138

Bridges AJ, Moos WH, Szotek DL, Trivedi BK, Bristol JA, Heffner TG, Bruns RF, Downs DA (1987) N^6-(2,2-Diphenylethyl)adenosine, a novel adenosine receptor agonist with antipsychotic-like activity. J Med Chem 30:1709–1711

Bruns RF (1981) Adenosine receptor activation in human fibroblasts: nucleoside agonists and antagonists. Can J Physiol Pharmacol 55:673–691

Bruns RF, Fergus JH (1990) Allosteric enhancement of adenosine A_1 receptor binding and function by 2-amino-3-benzoylthiophenes. Mol Pharmacol 38:939–949

Collis MG, Brown CM (1983) Adenosine relaxes the aorta by interacting with an A_2 receptor and an intracellular site. Eur J Pharmacol 96:61–69

Collis MG, Palmer DB, Baxter GS (1986) Evidence that the intracellular effects of adenosine in the guinea-pig aorta are mediated by inosine. Eur J Pharmacol 121:141–145

Corset V, Nguyen-Ba-Charvet KT, Forcet C, Moyse E, Chedotal A, Mehlen P (2000) Netrin-1-mediated axon outgrowth and cAMP production requires interaction with adenosine A_{2B} receptor. Nature 290:131–134

Cunha RA, Johansson B, Constantino MD, Sebastião AM, Fredholm BB (1996) Evidence for high-affinity binding sites for the adenosine A_{2A} receptor agonist [^{3}H]CGS 21680 in the rat hippocampus and cerebral cortex that are different from striatal A_{2A} receptors. Naunyn-Schmiedebergs Arch Pharmacol 353:261–271

Daly JW, Butts-Lamb P, Padgett W (1983) Subclasses of adenosine receptors in the central nervous system: Interactions with caffeine and related methylxanthines. Cell Mol Neurobiol 3:69–80

Delahunty TM, Cronin MJ, Linden J (1988) Regulation of GH_3-cell function via adenosine A_1 receptors. Biochem J 255:69–77

Dickenson JM, Hill SJ (1994) Interactions between adenosine A_1- and histamine H_1-receptors. Int J Biochem 26:959–969

Dickenson JM, Hill SJ (1996) Synergistic interactions between human transfected adenosine A_1 receptors and endogenous cholecystokinin receptors in CHO cells. Eur J Pharmacol 302:141–151

Dickenson JM, Hill SJ (1998) Involvement of G-protein $\beta\gamma$ subunits in coupling the adenosine A_1 receptor to phospholipase C in transfected cells. Eur J Pharmacol 355:85–93

Dickenson JM, Blank JL, Hill SJ (1998) Human adenosine A_1 receptor and $P2Y_2$-purinoceptor-mediated activation of the mitogen-activated protein kinase cascade in transfected CHO cells. Br J Pharmacol 124:1491–1499

Dixon AK, Gubitz AK, Sirinathsinghji DJS, Richardson PJ, Freeman TC (1996) Tissue distribution of adenosine receptor mRNAs in the rat. Br J Pharmacol 118: 1461–1468

Doctrow SR, Lowenstein JM (1985) Adenosine and 5′-chloro-5′-deoxyadenosine inhibit the phosphorylation of phosphatidylinositol and myosin light chain in calf aorta smooth muscle. J Biol Chem 260:3469–3476

Doctrow SR, Lowenstein JM (1987) Inhibition of phosphatidylinositol kinase in vascular smooth muscle membranes by adenosine and related compounds. Biochem Pharmacol 36:2255–2262

Dolphin AC, Forda SR, Scott RH (1986) Calcium-dependent currents in cultured rat dorsal root ganglion neurones are inhibited by an adenosine analogue. J Physiol 373:47–61

Dragunow M, Goddard GV, Laverty R (1985) Is adenosine an endogenous anticonvulsant? Epilepsia 26:480–487

Drury AN, Szent-Györgyi A (1929) The physiological activity of adenine compounds with especial reference to their action upon the mammalian heart. J Physiol (Lond) 68:213–237

Dubey RK, Gillespie DG, Osaka K, Suzuki F, Jackson EK (1996) Adenosine inhibits growth of rat aortic smooth muscle cells: possible role of A_{2b} receptor. Hypertension 27:786–793

Dunwiddie TV, Fredholm BB (1985) Adenosine modulation of synaptic responses in rat hippocampus: Possible role of inhibition or activation of adenylate cyclase. Adv Cyclic Nucleotide Protein Phosphorylation Res 19:259–272

During MJ, Spencer DD (1992) Adenosine: a potential mediator of seizure arrest and postictal refractoriness. Ann Neurol 32:618–624

Ethier MF, Dobson JG Jr (1997) Adenosine stimulation of DNA synthesis in human endothelial cells. Am J Physiol 272:H1470–H479

Ethier MF, Chander V, Dobson Jr JG (1993) Adenosine stimulates proliferation of human endothelial cells in culture. Am J Physiol 265:H131–H138

Exton JH (1996) Regulation of phosphoinositide phospholipases by hormones, neurotransmitters, and other agonists linked to G proteins. Annu Rev Pharmacol Toxicol 36:481–509

Faure M, Voyno-Yasenetskaya TA, Bourne HR (1997) cAMP and $\beta\gamma$ subunits of heterotrimeric G proteins stimulate the mitogen-activated protein kinase pathway in COS-7 cells. J Biol Chem 269:7851–7854

Fein T, Schulze E, Bär J, Schwabe U (1994) Purification and characterization of an adenotin-like adenosine binding protein from human platelets. Naunyn-Schmiedebergs Arch Pharmacol 349:374–380

Feoktistov I, Biaggioni I (1993) Characterization of adenosine receptors in human erythroleukemia cells: further evidence for heterogeneity of adenosine A_2 receptors. Mol Pharmacol 43:909–914

Feoktistov I, Biaggioni I (1995) Adenosine A_{2b} receptors evoke interleukin-8 secretion in human mast cells. An enprofylline-sensitive mechanism with implications for asthma. J Clin Invest 96:1979–1986

Feoktistov I, Goldstein AE, Biaggioni I (1999) Role of p38 mitogen-activated protein kinase and extracellular signal-regulated protein kinase kinase in adenosine A_{2B} receptor-mediated interleukin-8 production in human mast cells. Mol Pharmacol 55:726–735

Figler RA, Graber SG, Lindorfer MA, Yasuda H, Linden J, Garrison JC (1996) Reconstitution of recombinant bovine A_1 adenosine receptors in Sf9 cell membranes with recombinant G proteins of defined composition. Mol Pharmacol 50:1587–1595

Figler RA, Lindorfer MA, Graber SG, Garrison JC, Linden J (1997) Reconstitution of bovine A_1 adenosine receptors and G proteins in phospholipid vesicles: $\beta\gamma$-subunit composition influences guanine nucleotide exchange and agonist binding. Biochemistry 36:16288–16299

Fredholm BB, Hedqvist P (1980) Modulation of neurotransmission by purine nucleotides and nucleosides. Biochem Pharmacol 29:1635–1643

Fredholm BB, Lindgren E (1987) Effects of N-ethylmaleimide and forskolin on noradrenaline release from rat hippocampal slices: Evidence that prejunctional adenosine and α-receptors are linked to N-proteins but not to adenylate cyclase. Acta Physiol Scand 130:95–105

Fredholm BB, Abbracchio MP, Burnstock G, Daly JW, Harden TK, Jacobson KA, Leff P, Williams M (1994) Nomenclature and classification of purinoceptors. Pharmacol Rev 46:143–156

Fredholm BB, Zhang Y, van der Ploeg I (1996) Adenosine A_{2A} receptors mediate the inhibitory effect of adenosine on formyl-Met-Leu-Phe-stimulated respiratory burst in neutrophil leucocytes. Naunyn-Schmiedebergs Arch Pharmacol 354:262–267

Freissmuth M, Schütz W, Linder ME (1991) Interactions of the bovine brain A_1-adenosine receptor with recombinant G protein α-subunits. Selectivity for $rG_{i\alpha-3}$. J Biol Chem 266:17778–17783

Freund S, Ungerer M, Lohse MJ (1994) A_1 Adenosine receptors expressed in CHO-cells couple to adenylyl cyclase and to phospholipase C. Naunyn-Schmiedebergs Arch Pharmacol 350:49–56

Furlong TJ, Pierce KD, Selbie LA, Shine J (1992) Molecular characterization of a human brain adenosine A_2 receptor. Brain Res Mol Brain Res 15:62–66

Fuxe K, Ferré S, Zoli M, Agnati LF (1998) Integrated events in central dopamine transmission as analyzed at multiple levels. Evidence for intramembrane adenosine A_{2A}/dopamine D_2 and adenosine A_1/dopamine D_1 receptor interactions in the basal ganglia. Brain Res Rev 26:258–273

Gao Z, Chen T, Weber MJ, Linden J (1999) A_{2B} Adenosine and $P2Y_2$ receptors stimulate mitogen-activated protein kinase in human embryonic kidney-293 cells. Crosstalk between cyclic AMP and protein kinase C pathways. J Biol Chem 274:5972–5980

Gerwins P, Fredholm BB (1992a) Stimulation of adenosine A_1 receptors and bradykinin receptors, which act via different G proteins, synergistically raises inositol 1,4,5-triphosphate and intracellular free calcium in DDT_1 MF-2 smooth muscle cells. Proc Natl Acad Sci USA 89:7330–7334

Gerwins P, Fredholm BB (1992b) ATP and its metabolite adenosine act synergistically to mobilize intracellular calcium via the formation of inositol 1,4,5-triphosphate in a smooth muscle cell line. J Biol Chem 267:16081–16087

Gilchrist A, Mazzoni MR, Dineen B, Dice A, Linden J, Proctor WR, Lupica CR, Dunwiddie TV, Hamm HE (1998) Antagonists of the receptor-G protein interface block G_i-coupled signal transduction. J Biol Chem 273:14912–14919

Gubitz AK, Widdowson L, Kurokawa M, Kirkpatrick KA, Richardson PJ (1996) Dual signalling by the adenosine A_{2a} receptor involves activation of both N- and P-type calcium channels by different G proteins and protein kinases in the same striatal nerve terminals. J Neurochem 67:374–381

Hancock DL, Coupar IM (1995) Functional characterization of the adenosine receptor mediating inhibition of intestinal secretion. Br J Pharmacol 114:152–156

Haslam RJ, Rosson GM (1975) Effects of adenosine on levels of adenosine cyclic 3′,5′-monophosphate in human blood platelets in relation to adenosine incorporation and platelet aggregation. Mol Pharmacol 11:528–544

Haun SE, Segeleon JE, Trapp VL, Clotz MA, Horrocks LA (1996) Inosine mediates the protective effect of adenosine in rat astrocyte cultures subjected to combined glucose-oxygen deprivation. J Neurochem 67:2051–2059

Heurteaux C, Lauritzen I, Widmann C, Lazdunski M (1995) Essential role of adenosine, adenosine A1 receptors, and ATP-sensitive K^+ channels in cerebral ischemic preconditioning. Proc Natl Acad Sci USA 92:4666–4670

Hirano D, Aoki Y, Ogasawara H, Kodama H, Waga I, Sakanaka C, Shimizu T, Nakamura M (1996) Functional coupling of adenosine A_{2a} receptor to inhibition of the mitogen-activated protein kinase cascade in Chinese hamster ovary cells. Biochem J 316:81–86

Hoffman HM, Walker LL, Marquardt DL (1997) Mast cell adenosine induced calcium mobilization via G_{i3} and G_q proteins. Inflammation 21:55–68

Hutchison AJ, Webb RL, Oei HH, Ghai HH, Zimmerman MB, Williams M (1989) CGS 21680C, an A_2 selective adenosine receptor agonist with preferential hypotensive activity. J Pharmacol Exp Ther 251:47–55

Hutchison KA, Nevins B, Perini F, Fox IH (1990) Soluble and membrane-associated human low-affinity adenosine binding protein (adenotin): Properties and homology with mammalian and avian stress proteins. Biochemistry 29:5138–5144

IJzerman AP, von Frijtag Drabbe Künzel JK, Kim J, Jiang Q, Jacobson K (1996) Site-directed mutagenesis of the human adenosine A_{2A} receptor. Critical involvement of Glu[13] in agonist recognition. Eur J Pharmacol 310:269–272

Ito H, Tung RT, Sugimoto T, Kobayashi I, Takahashi K, Ui M, Kurachi Y (1992) On the mechanism of G protein $\beta\gamma$ subunit activation of the muscarinic K^+ channel in guinea pig atrial cell membrane. Comparison with the ATP-sensitive K^+ channel. J Gen Physiol 99:961–983

Ito H, Vereecke J, Carmeliet E (1994) Mode of regulation by G protein of the ATP-sensitive K^+ channel in guinea-pig ventricular cell membrane. J Physiol (Lond) 478:101–107

Iwamoto T, Umemura S, Toya Y, Uchibori T, Kogi K, Takagi N, Ishii M (1994) Identification of adenosine A_2 receptor-cAMP system in human aortic endothelial cells. Biochem Biophys Res Commun 199:905–910

Jacobson KA, Nikodijevic O, Shi D, Gallo-Rodriguez C, Olah ME, Stiles GL, Daly JW (1993) A role for central A_3-adenosine receptors. Mediation of behavioral depressant effects. FEBS Lett 336:57–60

Jacobson MA, Johnson RG, Luneau CJ, Salvatore CA (1995) Cloning and chromosomal localization of the human A_{2b} adenosine receptor gene (ADORA2B) and its pseudogene. Genomics 27:374–376

Jarvis MF, Williams M (1989) Direct autoradiographic localization of adenosine A_2 receptors in the rat brain using the A_2-selective agonist, [^{3}H]CGS 21680. Eur J Pharmacol 168:243–246

Jarvis MF, Gessner G, Shapiro G, Merkel L, Myers M, Cox BF, Martin G (1999) Differential effects of the adenosine A_1 receptor allosteric enhancer PD 81,723 on agonist binding to brain and adipocyte membranes. Brain Res 840:75–83

Ji XD, von Lubitz DKJE, Olah, ME, Stiles, GL, Jacobson KA (1994) Species differences in ligand affinity at central A_3-adenosine receptors. Drug Dev Res 33:51–59

Jiang Q, van Rhee M, Kim J, Yehle S, Wess J, Jacobson KA (1996) Hydrophilic side chains in the third and seventh transmembrane helical domains of human A_{2A} adenosine receptors are required for ligand recognition. Mol Pharmacol 50:512–521

Jockers R, Linder ME, Hohenegger M, Nanoff C, Bertin B, Strosberg AD, Marullo S, Freissmuth M (1994) Species differences in the G protein selectivity of the human and bovine A_1-adenosine receptor. J Biol Chem 269:32077–32084

Johansson B, Fredholm BB (1995) Further characterization of the binding of the adenosine receptor agonist [^{3}H]CGS 21680 to rat brain using autoradiography. Neuropharmacology 34:393–403

Johansson B, Georgiev V, Parkinson FE, Fredholm BB (1993) The binding of the adenosine A_2 receptor selective agonist [^{3}H]CGS 21680 to rat cortex differs from its binding to rat striatum. Eur J Pharmacol 247:103–110

Kenakin T (1996) The classification of seven transmembrane receptors in recombinant expression systems. Pharmacol Rev 48:413–463

Kim J, Wess J, van Rhee M, Schöneberg T, Jacobson KA (1995) Site-directed mutagenesis identifies residues involved in ligand recognition in the human A_{2a} adenosine receptor. J Biol Chem 270:13987–13997

Kim J, Jiang Q, Glashofer M, Yehle S, Wess J, Jacobson KA (1996) Glutamate residues in the second extracellular loop of the human A_{2a} adenosine receptor are required for ligand recognition. Mol Pharmacol 49:683–691

Kirk IP, Richardson PJ (1995) Inhibition of striatal GABA release by the adenosine A_{2a} receptor is not mediated by increases in cyclic AMP. J Neurochem 64:2801–2809

Kirsch GE, Codina J, Birnbaumer L, Brown AM (1990) Coupling of ATP-sensitive K^+ channels to A_1 receptors by G proteins in rat ventricular myocytes. Am J Physiol 259:H820–H826

Klotz K-N, Lohse MJ, Schwabe U (1988) Chemical modification of A_1 adenosine receptors in rat brain membranes. J Biol Chem 263:17522–17526

Kollias-Baker C, Ruble J, Dennis D, Bruns RF, Linden J, Belardinelli L (1994) Allosteric enhancer PD 81,723 acts by novel mechanism to potentiate cardiac actions of adenosine. Circ Res 75:961–971

Kollias-Baker CA, Ruble J, Jacobson M, Harrison JK, Ozeck M, Shryock JC, Belardinelli L (1997) Agonist-independent effect of an allosteric enhancer of the

A_1 adenosine receptor in CO cells stably expressing the recombinant human A_1 receptor. J Pharmacol Exp Ther 281:761–768

Koshiba M, Rosin DL, Hayashi N, Linden J, Sitkovsky MV (1999) Patterns of A_{2A} extracellular adenosine receptor expression in different functional subsets of human peripheral T cells. Flow cytometry studies with anti-A_{2A} receptor monoclonal antibodies. Mol Pharmacol 55:614–624

Lasley RD, Rhee JW, van Wylen DGL, Menzer RM (1990) Adenosine A_1-mediated protection of the globally ischemic rat heart. J Mol Cell Cardiol 22:39–47

Le F, Townsend-Nicholson A, Baker E, Sutherland GR, Schofield PR (1996) Characterization and chromosomal localization of the human A2a adenosine receptor gene: ADORA2A. Biochem Biophys Res Commun 223:461–467

Ledent C, Vaugeois J-M, Schiffmann SN, Pedrazzini T, Yacoubi ME, Vanderhaeghen J-J, Costentin J, Heath JK, Vassart G, Parmentier M (1997) Aggressiveness, hypoalgesia and high blood pressure in mice lacking the adenosine A_{2a} receptor. Nature 388:674–678

Lemmens R, Vanduffel L, Teuchy H, Culic O (1996) Regulation of proliferation of LLC-MK$_2$ cells by nucleosides and nucleotides: the role of ecto-enzymes. Biochem J 316:551–557

Liang BT, Jacobson KA (1998) A physiological role of the adenosine A_3 receptor: sustained cardioprotection. Proc Natl Acad Sci USA 95:6995–6999

Liang BT, Morley JF (1996) A new cyclic AMP-independent, G_s-mediated stimulatory mechanism via the adenosine A_{2a} receptor in the intact cardiac cell. J Biol Chem 271:18678–18685

Libert F, van Sande J, Lefort A, Czernilofsky A, Dumont JE, Vassart G, Ensinger HA, Mendla KD (1992) Cloning and functional characterization of a human A_1 adenosine receptor. Biochem Biophys Res Commun 187:919–926

Linden J (1994) Cloned adenosine A_3 receptors: pharmacological properties, species differences and receptor functions. Trends Pharmacol Sci 15:298–306

Linden J, Taylor HE, Robeva AS, Tucker AL, Stehle JH, Rivkees SA, Fink JS, Reppert SM (1993) Molecular cloning and functional expression of a sheep A_3 adenosine receptor with widespread tissue distribution. Mol Pharmacol 44:524–532

Lindström K, Ongini E, Fredholm BB (1996) The selective adenosine A_{2A} receptor antagonist SCH 58261 discriminates between two different binding sites for [^{3}H]-CGS 21680 in the rat brain. Naunyn-Schmiedebergs Arch Pharmacol 354:539–541

Lohse MJ, Klotz K-N, Schwabe U (1986) Agonist photoaffinity labeling of A_1 adenosine receptors: Persistent activation reveals spare receptors. Mol Pharmacol 30:403–409

Lorenzen A, Grün S, Vogt H, Schwabe U (1992) Identification of a novel high affinity adenosine binding protein from bovine striatum. Naunyn-Schmiedebergs Arch Pharmacol 346:63–68

Lorenzen A, Fuss M, Vogt H, Schwabe U (1993a) Measurement of guanine nucleotide-binding protein activation by A_1 adenosine receptor agonists in bovine brain membranes: Stimulation of guanosine-5'-O-(3-[^{35}S]thio)triphosphate binding. Mol Pharmacol 44:115–123

Lorenzen A, Nitsch-Kirsch M, Vogt H, Schwabe U (1993b) Characterization of membrane-bound and solubilized high-affinity binding sites for 5'-N-ethylcarboxamido-[^{3}H]adenosine from bovine cerebral cortex. J Neurochem 60:745–751

Lorenzen A, Guerra L, Vogt H, Schwabe U (1996a) Interaction of full and partial agonists of the A_1 adenosine receptor with receptor/G protein complexes in rat brain membranes. Mol Pharmacol 49:915–926

Lorenzen A, Großekatthöfer B, Kerst B, Vogt H, Fein T, Schwabe U (1996b) Characterization of a novel adenosine binding protein sensitive to cyclic AMP in rat brain cytosolic and particulate fractions. Biochem Pharmacol 52:1375–1385

Lorenzen A, Lang H, Schwabe U (1998a) Activation of various subtypes of G-protein α subunits by partial agonists of the adenosine A_1 receptor. Biochem Pharmacol 56:1287–1293

Lorenzen A, Engelhardt J, Kerst B, Schwabe U (1998b) Heterogeneous forms of adenotin-1 of different subcellular localization. Biochem Pharmacol 55:455–464

Mann JS, Holgate ST (1985) Specific antagonism of adenosine-induced bronchoconstriction in asthma by oral theophylline. Br J Clin Pharmac 19:685–692

Marala RB, Mustafa SJ (1993) Direct evidence for the coupling of A_2-adenosine receptor to stimulatory guanine nucleotide-binding-protein in bovine brain striatum. J Pharmacol Exp Ther 266:294–300

Marquardt DL, Walker LL (1990) Modulation of mast cell responses to adenosine by agents that alter protein kinase C activity. Biochem Pharmacol 39:1929–1934

Marquardt DL, Walker LL, Heinemann S (1994) Cloning of two adenosine receptor subtypes from mouse bone marrow-derived mast cells. J Immunol 152:4508–4515

Mateo J, Castro E, Zwiller J, Aunis D, Miras-Portugal MT (1995) 5'-(N-ethylcarboxamido)-adenosine inhibits Ca^{2+} influx and activates a protein phosphatase in bovine adrenal chromaffin cortex. J Neurochem 64:77–84

Mei YA, Le Foll F, Vaudry H, Cazin L (1996) Adenosine inhibits L- and N-type calcium channels in pituitary melanotrophs. Evidence for the involvement of a G protein in calcium channel gating. J Neuroendocrinol 8:85–91

Meininger CJ, Granger HJ (1990) Mechanisms leading to adenosine-stimulated proliferation of microvascular endothelial cells. Am J Physiol 258:H198–H206

Meininger CJ, Schelling ME, Granger HJ (1988) Adenosine and hypoxia stimulate proliferation and migration of endothelial cells. Am J Physiol 255:H554–H562

Moxham CM, Hod Y, Malbon CM (1993) $G_i\alpha_2$ mediates the inhibitory regulation of adenylylcyclase in vivo: Analysis in transgenic mice with $G_i\alpha_2$ suppressed by inducible antisense RNA. Dev Genet 14:266–273

Munshi R, Linden J (1991) Co-purification of A_1 adenosine receptors and guanine-nucleotide-binding proteins from bovine brain. J Biol Chem 264:14853–14859

Munshi R, Pang I-H, Sternweis PC, Linden J (1991) A_1 adenosine receptors of bovine brain couple to guanine nucleotide-binding proteins G_{i1}, G_{i2}, and G_o. J Biol Chem 266:22285–22289

Murthy KS, Makhlouf GM (1995) Adenosine A_1 receptor-mediated activation of phospholipase C-β_3 in intestinal muscle: Dual requirement for α and $\beta\gamma$ subunits of G_{i3}. Mol Pharmacol 47:1172–1179

Murthy KS, McHenry L, Grider JR, Makhlouf GM (1995) Adenosine A_1 and A_{2b} receptors coupled to distinct interactive signaling pathways in intestinal muscle cells. J Pharmacol Exp Ther 274:300–306

Musser B, Mudumbi RV, Liu J, Olson RD, Vestal RE (1999) Adenosine A_1 receptor-dependent and -independent effects of the allosteric enhancer PD 81,723. J Pharmacol Exp Ther 288:446–454

Nanoff C, Mitterauer T, Roka F, Hohenegger M, Freissmuth M (1995) Species differences in A_1 adenosine receptor/G protein coupling: Identification of a membrane protein that stabilizes the association of the receptor/G protein complex. Mol Pharmacol 48:806–817

Nanoff C, Waldhoer M, Roka F, Freissmuth M (1997) G Protein coupling of the rat A_1-adenosine receptor – Partial purification of a protein which stabilizes the receptor-G protein association. Neuropharmacology 36:1211–1219

Nicholls J, Brownhill VR, Hourani SM (1996) Characterization of P_1-purinoceptors on rat isolated duodenum longitudinal muscle and muscularis mucosae. Br J Pharmacol 117:170–174

Okajima F, Sato K, Sho K, Kondo Y (1989) Stimulation of adenosine receptor enhances α_1-adrenergic receptor-mediated activation of phospholipase C and Ca^{2+} mobilization in a pertussis-toxin-sensitive manner in FRTL-5 thyroid cells. FEBS Lett 248:145–149

Olah ME (1997) Identification of A_{2a} adenosine receptor domains involved in selective coupling to G_s. Analysis of chimeric A_1/A_{2a} adenosine receptors. J Biol Chem 272:337–344

Olah ME, Ren H, Ostrowski J, Jacobson KA, Stiles GL (1992) Cloning, expression, and characterization of the unique bovine A_1 adenosine receptor. Studies on the ligand binding site by site-directed mutagenesis. J Biol Chem 267:10764–10770

Olah ME, Jacobson KA, Stiles GL (1994a) Role of the second extracellular loop of adenosine receptors in agonist and antagonist binding. Analysis of chimeric A_1/A_3 adenosine receptors. J Biol Chem 269:24692–24698

Olah ME, Jacobson KA, Stiles GL (1994b) Identification of an adenosine receptor domain specifically involved in binding of 5'-substituted adenosine agonists. J Biol Chem 269:18016–18020

Olsson RA, Pearson JD (1990) Cardiovascular purinoceptors. Physiol Rev 70:761–845

Osswald H, Schmitz H-J, Kemper R (1978) Renal actions of adenosine: effect on renin secretion in the rat. Naunyn-Schmiedebergs Arch Pharmacol 303:95–99

Palmer TM, Gettys TW, Stiles GL (1995) Differential interaction with and regulation of multiple G-proteins by the rat A_3 adenosine receptor. J Biol Chem 270:16895–16902

Peachell PT, Lichtenstein LM, Schleimer RP (1991) Differential regulation of human basophil and lung mast cell function by adenosine. J Pharmacol Exp Ther 256:717–726

Pelleg A, Belardinelli L (1993) Cardiac electrophysiology and pharmacology of adenosine: Basic and clinical aspects. Cardiovasc Res 27:54–61

Peterfreund RA, MacCollin M, Gusella J, Fink JS (1996) Characterization and expression of the human A_{2a} adenosine receptor gene. J Neurochem 66:362–368

Phillis JW, O'Regan MH, Perkins LM (1993a) Effect of adenosine receptor agonist on spontaneous and K^+-evoked acetylcholine release from the in vivo rat cerebral cortex. Brain Res 605:293–297

Phillis JW, Perkins LM, O'Regan MH (1993b) Potassium-evoked efflux of transmitter amino acids and purines from rat cerebral cortex. Brain Res Bull 31:547–552

Pierce KD, Furlong TJ, Selbie LA, Shine J (1992) Molecular cloning and expression of an adenosine A_{2b} receptor from human brain. Biochem Biophys Res Commun 187:86–93

Piersen CE, True CD, Wells JN (1994) A carboxyl-terminally truncated mutant and nonglycosylated A_{2a} adenosine receptors retain ligand binding. Mol Pharmacol 45:861–870

Prentice DJ, Hourani SMO (1996) Activation of multiple sites by adenosine analogues in the rat isolated aorta. Br J Pharmacol 118:1509–1517

Prentice DJ, Hourani SMO (1997) Adenosine analogues relax guinea pig taenia caeci via an A_{2B} receptor and a xanthine-resistant site. Eur J Pharmacol 323:103–106

Prentice DJ, Shankley NP, Black JW (1995) Pharmacological analysis of the interaction between purinoceptor agonists and antagonists in the guinea-pig taenia caecum. Br J Pharmacol 115:549–556

Ralevic V, Burnstock G (1998) Receptors for purines and pyrimidines. Pharmacol Rev 50:413–492

Ramkumar V, Stiles GL, Beaven MA, Ali H (1993) The A_3 adenosine receptor is the unique adenosine receptor which facilitates release of allergic mediators in mast cells. J Biol Chem 268:16887–16890

Reeve AJ, Dickenson AH (1995) The roles of spinal adenosine receptors in the control of acute and more persistent nociceptive responses of dorsal horn neurons in the anaesthetized rat. Br J Pharmacol 116:2221–2228

Ren H, Stiles GL (1994) Characterization of the human A_1 adenosine receptor gene. Evidence for alternative splicing. J Biol Chem 269:3104–3110

Reppert SM, Weaver DR, Stehle JH, Rivkees SA (1991) Molecular cloning and characterization of a rat A_1-receptor that is widely expressed in brain and spinal cord. Mol Endocrinol 5:1037–1048

Revan S, Montesinos MC, Naime D, Landau S, Cronstein BN (1996) Adenosine A_2 receptor occupancy regulates stimulated neutrophil function via activation of a serine/threonine protein phosphatase. J Biol Chem 271:17114–17118

Rivkees SA, Lasbury ME, Barbhaiya H (1995a) Identification of domains of the human A_1 adenosine receptor that are important for binding receptor subtype-selective ligands using chimeric A_1/A_{2a} adenosine receptors. J Biol Chem 270:20485–20490

Rivkees SA, Price SL, Zhou FC (1995b) Immunohistochemical detection of A_1 adenosine receptors in rat brain with emphasis on localization in the hippocampal formation, cerebral cortex, cerebellum, and basal ganglia. Brain Res 677:193–203

Rivkees SA, Barbhaiya H, IJzerman AP (1999) Identification of the adenine binding site of the human A_1 adenosine receptor. J Biol Chem 274:3617–3621

Robeva AS, Woodard RL, Jin X, Gao Z, Bhattacharya S, Taylor, HE, Rosin DL, Linden J (1996) Molecular characterization of recombinant human adenosine receptors. Drug Dev Res 39:243–252

Rosenthal RA, Lowenstein JM (1991) Inhibition of phosphorylation of troponin I in rat heart by adenosine and 5'-chloro-5'-deoxyadenosine. Biochem Pharmacol 42:685–692

Sajjadi FG, Firestein GS (1993) cDNA cloning and sequence analysis of the human A_3 adenosine receptor. Biochim Biopys Acta 1179:105–107

Salvatore CA, Jacobson MA, Taylor HE, Linden J, Johnson RG (1993) Molecular cloning and characterization of the human A_3 adenosine receptor. Proc Natl Acad Sci USA 90:10365–10369

Sattin A, Rall TW (1970) The effect of adenosine and adenine nucleotides on the cyclic adenosine 3',5'-monophosphate content of guinea pig cerebral cortex slices. Mol Pharmacol 6:13–23

Scholz KP, Miller RJ (1991a) Analysis of adenosine actions on Ca^{2+} currents and synaptic transmission in cultured rat hippocampal pyramidal neurones. J Physiol (Lond) 435:373–393

Scholz KP, Miller RJ (1991b) Presynaptic inhibition at excitatory hippocampal synapses: Development and role of presynaptic Ca^{2+} channels. J Neurophysiol 76:39–46

Schubert P, Lee K, West M, Deadwyler S, Lynch G (1976) Stimulation-dependent release of [^{3}H]adenosine derivatives from central axon terminals to target neurons. Nature 260:541–542

Schwabe U, Ebert R, Erbler HC (1973) Adenosine release from isolated fat cells and its significance for the effects of hormones on cyclic 3',5'-AMP levels and lipolysis. Naunyn-Schmiedebergs Arch Pharmacol 276:133–148

Schwabe U (1981) Direct binding studies of adenosine receptors. Trends Pharmacol Sci 2:299–303

Sexl V, Mancusi G, Baumgartner-Parzer S, Schütz W, Freissmuth M (1995) Stimulation of human umbilical vein endothelial cell proliferation by A_2-adenosine and β_2-adrenoceptors. Br J Pharmacol 114:1577–1586

Sexl V, Mancusi G, Höller C, Gloria-Maercker E, Schütz W, Freissmuth M (1997) Stimulation of the mitogen-activated protein kinase via the A_{2A}-adenosine receptor in primary human endothelial cells. J Biol Chem 272:5792–5799

Shryock JC, Ozeck MJ, Belardinelli L (1998a) Inverse agonists and neutral antagonists of recombinant human A_1 adenosine receptors stably expressed in Chinese hamster ovary cells. Mol Pharmacol 53:886–893

Shryock JC, Snowdy S, Baraldi PG, Cacciari B, Spalluto G, Monopoli A, Ongini E, Baker SP, Belardinelli L (1998b) A_{2A}-adenosine receptor reserve for coronary vasodilation. Circulation 98:711–718

Sinclair RJ, Randall JR, Wise GE, Jones CE (1985) Response of isolated renal artery rings to adenosine and inosine. Drug Dev Res 6:391–396

Snyder SH, Katims IJ, Annau Z, Bruns RF, Daly JW (1981) Adenosine receptors and behavioral actions of methylxanthines. Proc Natl Acad Sci USA 78:3260–3264

Sowell MO, Ye C, Ricupero DA, Hansen S, Quinn SJ, Vassilev PM, Mortensen RM (1997) Targeted inactivation of α_{i2} or α_{i3} disrupts activation of the cardiac muscarinic K^+ channel, $I_{K^+_{ACh}}$, in intact cells. Proc Natl Acad Sci USA 94:7921–7926

Srinivas M, Shryock JC, Dennis DM, Baker SP, Belardinelli L (1997) Differential A_1 adenosine receptor reserve for two actions of adenosine on guinea pig atrial myocytes. Mol Pharmacol 52:683–691

Stehle JH, Rivkees SA, Lee JJ, Weaver DR, Deeds JD, Reppert SM (1992) Molecular cloning and expression of the cDNA for a novel A_2-adenosine receptor subtype. Mol Endocrinol 6:384–393

Sternweis PC (1994) The active role of $\beta\gamma$ in signal transduction. Curr Opin Cell Biol 6:198–203

Strohmeier GR, Reppert SM, Lencer WI, Madara JL (1995) The A_{2b} adenosine receptor mediates cAMP responses to adenosine receptor agonists in human intestinal epithelia. J Biol Chem 270:2387–2394

Sweeney MI, Dolphin AC (1995) Adenosine A_1 agonists and the Ca^{2+} channel agonist Bay K 8644 produce a synergistic stimulation of the GTPase activity of G_o in rat frontal cortical membranes. J Neurochem 64:2034–2042

Tomura H, Itoh H, Sho K, Sato K, Nagao M, Ui M, Kondo Y, Okajima F (1997) $\beta\gamma$ Subunits of pertussis toxin-sensitive G proteins mediate A_1 adenosine receptor agonist-induced activation of phospholipase C in collaboration with thyrotropin. A novel stimulatory mechanism through the cross-talk of two types of receptors. J Biol Chem 272:23130–23137

Townsend-Nicholson A, Shine J (1992) Molecular cloning and characterisation of a human brain A_1 adenosine receptor cDNA. Mol Brain Res 16:365–370

Townsend-Nicholson A, Schofield PR (1994) A threonine residue in the seventh transmembrane domain of the human A_1 adenosine receptor mediates specific agonist binding. J Biol Chem 269:2373–2376

Townsend-Nicholson A, Baker E, Schofield PR, Sutherland GR (1995a) Localization of the adenosine A_1 receptor subtype gene (ADORA1) to chromosome 1q32.1. Genomics 26:423–425

Townsend-Nicholson A, Baker E, Sutherland GR, Schofield PR (1995b) Localization of the adenosine A_{2b} receptor subtype gene (ADORA2B) to chromosome 17p11.2–p12 by FISH and PCR screening of somatic cell hybrids. Genomics 25:605–607

Trussell LO, Jackson MB (1985) Adenosine-activated potassium conductance in cultured striatal neurons. Proc Natl Acad Sci USA 82:4857–4861

Trussell LO, Jackson MB (1987) Dependence of an adenosine-activated potassium current on a GTP-binding protein in mammalian central neurons. J Neurosci 7:3306–3316

Tucker AL, Robeva AS, Taylor HE, Holeton D, Bockner M, Lynch KR, Linden J (1994) A_1 adenosine receptors. Two amino acids are responsible for species differences in ligand recognition. J Biol Chem 269:27900–27906

Umemiya M, Berger AJ (1994) Activation of adenosine A_1 and A_2 receptors differentially modulates calcium channels and glycinergic synaptic transmission in rat brainstem. Neuron 13:1439–1446

Van Daele P, van Coevorden A, Roger PP, Boeynaems J-M (1992) Effects of adenine nucleotides on the proliferation of aortic endothelial cells. Circ Res 70: 82–90

Van Schaick EA, Tukker HE, Roelen HCPF, IJzerman AP, Danhof M (1998) Selectivity of action of 8-alkylamino analogues of N^6-cyclopentyladenosine in vivo: Haemodynamic versus anti-lipolytic responses in rats. Br J Pharmacol 124:607–618

Van Tilburg EW, von Frijtag Drabbe Künzel J, de Groote M, Vollinga RC, Lorenzen A, IJzerman AP (1999) N^6,5′-Disubstituted adenosine derivatives as partial agonists for the human adenosine A_3 receptor. J Med Chem 42:1393–1400

Von Lubitz DKJE, Lin RC-S, Popik P, Carter MF, Jacobson KA (1994) Adenosine A_3 receptor stimulation and cerebral ischemia. Eur J Pharmacol 263:59–67

Walday P, Aas P (1991) Prejunctional stimulation of cholinergic nerves in rat airway smooth muscle by an adenosine analogue. Pulm Pharmacol 4:114–119

Waldhoer M, Bofill-Cardona E, Milligan G, Freissmuth M, Nanoff C (1998) Differential uncoupling of A_1 adenosine and D_2 dopamine receptors by suramin and didemethylated suramin (NF037). Mol Pharmacol 53:808–818

Walker BA, Jacobson MA, Knight DA, Salvatore CA, Weir T, Zhou D, Bai TR (1997) Adenosine A_3 receptor expression and function in eosinophils. Am J Resp Cell Mol Biol 16:531–537

Wan W, Sutherland GR, Geiger JD (1990) Binding of the adenosine A_2 receptor ligand [^{3}H]CGS 21680 to human and rat brain: Evidence for multiple affinity sites. J Neurochem 55:1763–1771

Webb RL, Sills MA, Chovan JP, Balwierczak JL, Francis JE (1992) CGS 21680: a potent selective adenosine A_2 receptor agonist. Cardiovasc Drug Rev 10:26–53

Weber RG, Jones CR, Palacios JM, Lohse MJ (1988) Autoradiographic visualization of A_1-adenosine receptors in brain and peripheral tissues of rat and guinea pig using ^{125}I-HPIA. Neurosci Lett 87:215–220

White TE, Dickenson JM, Alexander SPH, Hill SJ (1992) Adenosine A_1-receptor stimulation of inositol phospholipid hydrolysis and calcium mobilisation in DDT_1 MF-2 cells. Br J Pharmacol 106:215–221

Wise A, Sheehan M, Rees S, Lee M, Milligan G (1999) Comparative analysis of the efficacy of A_1 adenosine receptor activation of $G_{i/o}\alpha$ G proteins following coexpression of receptor and G protein and expression of A_1 adenosine receptor-$G_{i/o}\alpha$ fusion proteins. Biochemistry 38:2272–2278

Yakel JL, Warren RA, Reppert SM, North RA (1993) Functional expression of adenosine A_{2b} receptor in *Xenopus* oocytes. Mol Pharmacol 43:277–280

Yamada M, Inanobe A, Kurachi Y (1998) G Protein regulation of potassium ion channels. Pharmacol Rev 50:723–757

Yasuda H, Lindorfer MA, Woodfork KA, Fletcher JE, Garrison JC (1996) Role of the prenyl group on the G protein γ subunit in coupling trimeric G proteins to A_1 adenosine receptors. J Biol Chem 271:18588–18595

Zhou QY, Li C, Olah ME, Johnson RA, Stiles GL, Civelli O (1992) Molecular cloning and characterization of an adenosine receptor: the A3 adenosine receptor. Proc Natl Acad Sci USA 89:7432–7436

Zurn AD, Do KQ (1988) Purine metabolite inosine is an adrenergic neurotrophic substance for cultured chicken sympathetic neurons. Proc Natl Acad Sci USA 85:8301–8305

CHAPTER 3
P2X Receptors

I.P. CHESSELL, A.D. MICHEL and P.P.A. HUMPHREY

A. Introduction

P2X receptors are a new sub-class of ligand-gated cation channels (BARNARD and HUMPHREY 1998). Of the seven subunits cloned to date, all have a similar predicted topology, and are able to form functional cation channels when expressed as homomeric proteins. Each subunit is around 400 amino acids in length (NORTH and BARNARD 1997), except for the $P2X_7$ subunit (SURPRENANT et al. 1996; RASSENDREN et al. 1997b) which has 595 amino acids, and all have intracellular N and C termini, two transmembrane domains and a long extracellular loop. Thus, the molecular architecture of the P2X receptor is quite different to that of other ligand-gated ion channels. P2X subunits, with the amiloride-sensitive FMRFamide peptide-gated sodium channels (FnaC; NORTH 1996), comprise two distinct but related sub-classes of ligand-gated ion channels. While the overall topology of the P2X receptors (Fig. 1) is shared with various channel types, including amiloride-sensitive epithelial Na^+ channels (EnaC), FnaC, the inward rectifier K^+ channel, and the mechanosensitive channel of *Eschericha coli*, they do not share primary sequence homology. Interestingly, the sequence homology between the P2X subunits themselves is rather low; subunit identity is shown in Table 1. Not only are P2X receptors structurally distinct from other ion channels, but they also possess unique phenotypic properties, in that some members (particularly $P2X_7$, see below) are able to undergo a transition whereby the ion channel pore dilates with prolonged agonist application (SURPRENANT et al. 1996; KHAKH et al. 1999a; VIRGINIO et al. 1999).

While the stoichiometry of subunit assembly into functional channels is not yet clear, a recent study suggests that $P2X_1$ and $P2X_3$ receptors may form stable, and functional homo-trimers, and it seems likely that these trimers can aggregate to form larger complexes (NICKE et al. 1998). The ability of P2X subunits to polymerise with other types of P2X subunits is an important consideration when comparing the functional properties of heterologously expressed single subunits with those of P2X receptors expressed in native tissues. Indeed, functional studies, particularly within the CNS, often reveal apparent phenotypes which are quite different from those of homomeric recombinant P2X

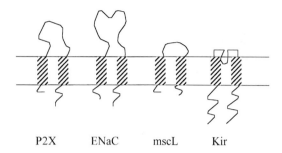

Fig. 1. Schematic showing structural motif of P2X receptors, epithelial amiloride-sensitive Na^+ channels (*EnaC*), *Eschericha coli* mechanosensitive channels (*mscL*) and inward rectifier K^+ channels (*Kir*)

Table 1. Percentage primary sequence identities between the seven known rat P2X receptors (HUMPHREY et al. 1998)

	$P2X_1$	$P2X_2$	$P2X_3$	$P2X_4$	$P2X_5$	$P2X_6$	$P2X_7$
$P2X_1$		38	41	50	41	43	35
$P2X_2$			44	46	38	39	28
$P2X_3$				45	43	38	36
$P2X_4$					52	45	44
$P2X_5$						48	27
$P2X_6$							32
$P2X_7$							

receptors. It is now apparent that at least three distinct heteropolymeric receptors can be formed from different subunits, namely $P2X_2+P2X_3$ (LEWIS et al. 1995), $P2X_4+P2X_6$ (LÊ et al. 1998) and $P2X_1+P2X_5$ (TORRES et al. 1998). The existence of these heteromers at least begins to explain the major discrepancies between operational characteristics of native and recombinant P2X receptor phenotypes (see below). With knowledge of the properties of the P2X homo- and heteromeric combinations, it is now finally possible, at least partially, to rationalise some previous pharmacological classifications of these receptors to their respective component subunits (Tables 2 and 5).

B. P2X Homomeric Subunits and Their Properties

I. $P2X_1$

The isolation of the first two subunit members of the P2X family, $P2X_1$ and $P2X_2$, was reported in rapid succession. The cDNA encoding the $P2X_1$ protein was isolated from rat vas deferens (VALERA et al. 1994), and represented the "archetypal" P2X receptor in that it is sensitive to the agonist, previously though to be diagnostic for P2X receptors, α,β-methylene-ATP ($\alpha\beta$meATP),

Table 2. Phenotypic features of heterologously expressed homomeric P2X subunits

Group 1	Group 2	Group 3	Group 4
$\alpha\beta$meATP sensitive	$\alpha\beta$meATP insensitive	$\alpha\beta$meATP insensitive	$\alpha\beta$meATP insensitive
Rapidly desensitising	Slowly desensitising	Slowly desensitising	Non-desensitising
Suramin and PPADS sensitive	Suramin and PPADS sensitive	Suramin and PPADS insensitive	Suramin and PPADS sensitive (moderate)
$P2X_1$, $P2X_3$	$P2X_2$, $P2X_5$	$P2X_4$, $P2X_6$	$P2X_7$

Table 3. Chromosomal localisation of P2X receptor genes

Receptor	Chromosomal location	References
Human $P2X_1$	17p13.3	VALERA et al. (1995); LONGHURST et al. (1996)
Human $P2X_3$	11q12	GARCIA-GUZMAN et al. (1997b)
Mouse $P2X_3$	2p	SOUSLOVA et al. (1997)
Human $P2X_4$	12q24.32	GARCIA-GUZMAN et al. (1997a)
Human $P2X_5$	17p13	BUELL et al. (1999)
Human $P2X_7$	12q24.2	BUELL et al. (1999)

and to the P2 antagonist, suramin (HUMPHREY et al. 1995). This P2X receptor, found in myocytes from vas deferens, bladder and vascular smooth muscle, desensitises very rapidly at concentrations of ATP > 1 μmol/l, and this has proved to be a problem when studying the properties of this receptor, as often intervals of 15–20 min are required between agonist applications to obtain reproducible responses. Since identification of the cDNA encoding the rat $P2X_1$ receptor, both human (VALERA et al. 1995) and mouse orthologues have been cloned, and the gene encoding the human $P2X_1$ receptor has been localised to chromosome 17 (VALERA et al. 1995; Table 3).

The pharmacological profile of the $P2X_1$ receptor is similar across species (unlike, for example $P2X_7$; see below); of the commonly used agonists, ATP is the most potent, followed by 2-methylthio-ATP (2MeSATP) and $\alpha\beta$meATP. Interestingly, an agonist considered selective for $P2X_7$, benzoyl-benzoylATP (BzATP), was found in a recent study to be the most potent agonist at $P2X_1$ (BIANCHI et al. 1999). The $P2X_1$ receptor is blocked by suramin, and pyridoxal-phosphate-6-azophenyl-2',4'-disulphonic acid (PPADS). A recent report describes potent antagonism of the $P2X_1$ receptor by trinitrophenyl-substituted nucleotides including TNP-ATP (VIRGINIO et al. 1998). While these compounds also block the $P2X_3$ receptor, TNP-ATP nevertheless represents a useful tool for delineating the first phenotypic group of P2X receptors (Table 2) from the remaining members of this family. Better antagonists that are able to differentiate between $P2X_1$ and $P2X_3$ are eagerly awaited.

II. P2X$_2$

The cDNA for the rat P2X$_2$ subunit was first isolated from PC12 cells (BRAKE et al. 1994) encoding a protein of 472 amino acids in length. Homomeric P2X$_2$ receptors desensitise slowly and, together with homomeric P2X$_5$ receptors, form the second phenotypic group of P2X receptors, being insensitive to $\alpha\beta$meATP but sensitive to the antagonists suramin and PPADS. The P2X$_2$ receptor has a wide distribution, being found in sensory ganglia, adrenal chromaffin cells and a number of brain nuclei (VULCHANOVA et al. 1996; FUNK et al. 1997).

Since the P2X$_2$ receptor expresses robust, non-desensitising currents when expressed heterologously, it has been used to investigate some of the structural characteristics of P2X receptors. Thus, RASSENDREN et al. (1997a) used the substituted cysteine accessibility method to demonstrate that part of the pore of the P2X receptor is formed by the second hydrophobic domain, and later the same group demonstrated that concatenated cDNAs encoding proteins where the C terminus of one P2X$_2$ subunit was joined to the N terminus of another gave rise to fully functional channels, supporting the original proposal that P2X receptors have a topology of intracellular N and C termini, with a large extracellular loop. More recent studies by our own and other groups have identified amino acid residues that contribute to the rate of desensitisation of the P2X$_2$ receptor (KOSHIMIZU et al. 1998; SMITH et al. 1999). Both studies involved mutation of the C-terminal amino acids, and concluded that the proximal C-terminal region between TM2 and Lys374, and residues 373–376, play an important role in controlling the rate of desensitisation of this receptor.

The work examining the residues controlling desensitisation of the P2X$_2$ receptor was initiated after functional splice variants, with differing desensitisation rates, had been identified (BRANDLE et al. 1997; SIMON et al. 1997). The functional P2X$_{2(b)}$ receptor has a serine/proline-rich segment deleted from the intracellularly-located C-terminal domain, and desensitises more rapidly than the originally described larger isoform (P2X$_{2(a)}$). Nevertheless, it appears that the P2X$_{2(b)}$ splice variant is physiologically relevant, as it is expressed on a variety of sensory and sympathetic ganglia, as well as on a number of structures in the neonatal rat brain (SIMON et al. 1997).

III. P2X$_3$

The P2X$_3$ receptor is the second member of the first phenotypic group of P2X receptors. In common with homomeric P2X$_1$ receptors, homomeric P2X$_3$ receptors are activated by $\alpha\beta$meATP, and desensitise rapidly in the continued presence of agonist. As with P2X$_1$, BzATP is also a potent agonist (BIANCHI et al. 1999). The P2X$_3$ receptor is also sensitive to the antagonists suramin and PPADS (Table 2). Genes encoding the human, rat and mouse orthologues of P2X$_3$ have been identified (LEWIS et al. 1995; CHEN et al. 1995; GARCIA-

GUZMAN et al. 1997b; SOUSLOVA et al. 1997), and each encodes a 397 amino acid receptor subunit. A recent study has shown the gene encoding the human P2X$_3$ subunit is localised to chromosome 11 (Table 3).

The P2X$_3$ receptor has generated intense interest as a potential drug target, which is mainly attributable to its apparently localised cellular distribution in a subset of sensory neurones. Thus, P2X$_3$ is found on a subset of nociceptive dorsal root ganglia (DRG) neurones (CHEN et al. 1995), and is co-localised with the vanilloid-1 (VR-1) receptors and isolectin B4 (IB4) binding sites (GUO et al. 1999). This corresponds well with previous reports that ATP is able to depolarise DRG neurones (KRISHTAL et al. 1988a), and it has long been known that ATP causes pain when applied to the human blister base preparation (BLEEHAN and KEELE 1977). The hypothesis that P2X receptors may play a role in nociception has been extended by studies showing that P2 agonists activate sensory neurones in the rat tail preparation (TREZISE and HUMPHREY 1997), and cause nociceptive behaviours when injected into the rat hindpaw (BLAND-WARD and HUMPHREY 1997; HAMILTON et al. 1999). More recent studies have demonstrated that ATP is able to activate directly nociceptive units of rat tooth pulp afferents (COOK et al. 1997), and medial articular nerve terminals in rat knee joints (DOWD et al. 1998).

An often-cited critique of the putative involvement of P2X$_3$ receptors in nociception is the desensitisation profile of the receptor. Reconciliation of this property of the receptor and its involvement in activation of pain pathways has been proposed in a variety of ways. First, it is now known that P2X$_3$ subunits can heteropolymerise with P2X$_2$ subunits to yield a novel phenotype which desensitises slowly (see below). Second, recent reports suggest that the desensitisation profile of P2X$_3$ can be dramatically altered by the action of calcium, such that elevated extracellular calcium can speed P2X$_3$ receptor recovery from desensitisation (COOK and McCLESKEY 1997; COOK et al. 1998). As it has been proposed that extracellular Ca^{2+} concentrations may be altered by tissue trauma (VALERA et al. 1995), it is possible that changes in calcium concentrations also contribute to the nociceptive role of P2X$_3$ receptors.

IV. P2X$_4$

As with some of the other P2X receptors, several separate reports describing cloning of the cDNA encoding the rat P2X$_4$ receptor were published almost simultaneously (Bo et al. 1995; SÉGUÉLA et al. 1996; BUELL et al. 1996). Subsequently, the gene for the human orthologue of the P2X$_4$ receptor was cloned (GARCIA-GUZMAN et al. 1997a), and its chromosomal location described (Table 3). More recently, a cDNA encoding the mouse P2X$_4$ receptor has been identified (SIMON et al. 1999; TOWNSEND-NICHOLSON et al. 1999). In all cases, the cDNAs encode a 388 amino acid subunit. Homomeric P2X$_4$ receptors are slowly desensitising, insensitive to the agonist, $\alpha\beta$meATP, and are relatively insensitive to the antagonists, suramin and PPADS. It has been proposed that PPADS acts as an antagonist at P2X receptors by forming a Schiff's base with the lysine at position 249. BUELL et al. (1996) demonstrated that modification of the glutamate residue at this position in $P2X_4$ to lysine increased the sensitivity to PPADS. In addition, GARCIA-GUZMAN and co-workers (1996) have identified a region in the $P2X_4$ receptor (amino acids 81–183) which contribute to the increased sensitivity to PPADS of the human $P2X_4$ receptor compared to the rat orthologue. Finally, the same group identified position 78 as being important for the higher potency of suramin at the human compared to the rat receptor (GARCIA-GUZMAN et al. 1997a).

The interaction of antagonists with the $P2X_4$ receptor appears complex, and antagonist effects reported vary considerably between studies. Thus, even though most studies demonstrate that P2 antagonists have low potency at the $P2X_4$ receptor compared to other P2X receptor types, some studies have shown that antagonists can actually potentiate agonist-mediated responses (Bo et al. 1995; WANG et al. 1996). In the case of cibacron blue, we have shown potentiation of responses which are attributable to an allosterically mediated increase in the apparent affinity of ATP for the rat $P2X_4$ receptor (MICHEL et al. 1997b; MILLER et al. 1998), and similar potentiation has been noted at the mouse $P2X_4$ receptor (TOWNSEND-NICHOLSON et al. 1999), although in our own studies we noted only inhibition of responses by antagonists at the murine orthologue. We have recently completed a full comparison of agonist and antagonist sensitivities at rat, human and mouse $P2X_4$ receptors, and these are summarised in Table 4 (JONES et al. 1999).

The $P2X_4$ receptor has a wide distribution, being found in cerebellum, hippocampus, piriform cortex, thalamus and a number of brainstem nuclei (SÉGUÉLA et al. 1996), as well as in epithelia and salivary glands (BUELL et al. 1996). A detailed summary of the localisation of $P2X_4$ mRNA in the central nervous system has been described by COLLO et al. (1996).

The human $P2X_4$ receptor gene is alternatively spliced (DHULIPALA et al. 1998). In the shorter isoform, the 5′-untranslated region and the first 90 amino acids in the coding region of full-length human $P2X_4$ are replaced by a 35 amino acid coding sequence that is highly homologous with a region of chaparonin proteins in the hsp-90 family. Injection of the alternatively spliced

Table 4. Agonist and antagonist potencies at heterologously expressed recombinant $P2X_4$ receptor species orthologues (JONES et al. 1999)

$P2X_4$ species	Agonist EC_{50} (μmol/l) ATP	Antagonist IC_{50} (μM)	
		PPADS	Suramin
Mouse	1.8	33.0	>300
Human	1.6	11.7	>100
Rat	4.1	124	>300

RNA into *Xenopus* oocytes does not result in functional channels, which is not surprising since the 5′ splicing results in the deletion of the predicted TM1 region. However, the use of RT-PCR has enabled identification of the alternatively spliced mRNAs in both smooth muscle and brain (our own unpublished observations).

V. $P2X_5$ and $P2X_6$

As compared to the other P2X receptor subunits, comparatively little is known about $P2X_5$ and $P2X_6$ subunits. For $P2X_5$, cDNAs for human (Lê et al. 1997) and rat (COLLO et al. 1996) orthologues have been identified, encoding proteins of 422 and 455 amino acids, respectively. However, the human $P2X_5$ subunit is truncated and non-functional, with analysis predicting only one transmembrane domain. The rat $P2X_5$ subunit, when expressed heterologously, produces a homomeric receptor which desensitises slowly, is $\alpha\beta$meATP insensitive, but is sensitive to the antagonists, suramin and PPADS. In situ hybridisation provided no evidence for $P2X_5$ receptor RNA in the brain, with the exception of the mesencephalic nucleus of the trigeminal nerve. However, this subunit is expressed in the heart, sensory ganglia, and spinal cord (TORRES et al. 1998).

The rat $P2X_6$ cDNA encodes a 379 amino acid protein, which can form functional channels when expressed heterologously (COLLO et al. 1996), although one study failed to obtain functional responses from *Xenopus* oocytes injected with cRNA for $P2X_6$ (SOTO et al. 1996b). Functionally, $P2X_6$ appears to be similar to $P2X_4$, and even shares a similar pattern of distribution. Thus, $P2X_6$ subunits are insensitive to $\alpha\beta$meATP, desensitise slowly, and appear to be insensitive to the P2 antagonists, suramin and PPADS. In a similar manner to $P2X_4$, alteration of the leucine residue at position 251 (equivalent to position 246 of the $P2X_4$ receptor) to lysine confers sensitivity to PPADS, presumably by acting as a Schiff base substrate (COLLO et al. 1996).

VI. $P2X_7$

It was noted twenty years ago that ATP can cause receptor-mediated permeabilisation of membranes in certain cell types, including mast cells (COCKCROFT and GOMPERTS 1979), macrophages (STEINBERG et al. 1987) and microglia (FERRARI et al. 1996). These effects were believed to be mediated by a receptor, originally termed P2Z, which when activated induced membrane permeabilisation by formation of large aqueous pores, with consequent cytolysis upon persistent receptor activation. This receptor was originally considered to be distinct from P2X receptors on the basis of this unique transductional characteristic. However, the P2Z receptor has now been identified as belonging to the family of P2X receptors (SURPRENANT et al. 1996), and it is accepted that the $P2X_7$ receptor is, or at least forms a critical part of, the P2Z receptor. Rat (SURPRENANT et al. 1996), human (RASSENDREN et al. 1997b) and mouse

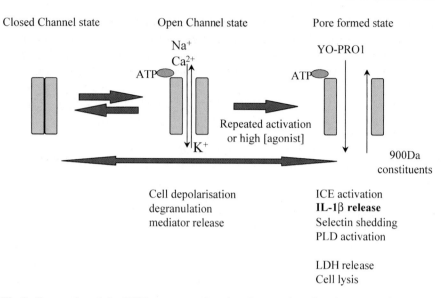

Fig. 2. Properties of the P2X$_7$ receptor, showing the two functional states and some of the consequences of receptor activation. The fluorescent DNA binding dye, *YO-PRO1*, has been shown to pass through the "large pore" of the P2X$_7$ receptor, to which the channel form is impermeable. In some immune cells, P2X$_7$ activation leads to IL-1β release via activation of interleukin converting enzyme (*ICE*), and also activates phospholipase D (*PLD*). Later events with continued receptor activation include release of lactate dehydrogenase (*LDH*; release upon cell lysis)

(CHESSELL et al. 1998b) subunit cDNAs have been identified, and all encode proteins of 595 amino acids, making P2X$_7$ considerably larger than the chromosome P2X receptor subunits. The human P2X$_7$ gene has been localised to 12q24.2 (Table 3). In all cases, when expressed as homomers, these P2X$_7$ subunits act as typical non-selective cationic channels with brief agonist application (DI VIRGILIO et al., Chap. 26, second volume). However, repeated, or prolonged application of higher agonist concentrations, especially in solutions of low divalent cation concentration, causes conversion of the channel to a much larger aqueous pore (Fig. 2). Formation of this "large pore" allows entry of molecules with molecular weights of up to around 800 Da, such as the DNA binding dye, YO-PRO-1, and thus these dyes are often used to monitor large pore formation.

At the present time it is unknown how channel dilation into the large pore form occurs. Recent studies using a monoclonal antibody directed to the extracellular loop of the receptor suggest that a conformational change at the level of the receptor must occur for large pore formation to proceed (BUELL et al. 1998; CHESSELL et al. 1999). It is thought that the long C-terminal tail of the receptor contributes to large pore formation, as evidenced by the finding that a truncated receptor, lacking a portion of the C terminal tail, does not form the large pore (SURPRENANT et al. 1996). However, it is not clear how the C terminal tail contributes to large pore formation. Moreover, recent studies have demonstrated that other P2X subunits, lacking this long C-terminal tail, are able to undergo pore dilation, so it may be that pore formation is intrinsic to P2X subunits (KHAKH et al. 1999a; VIRGINIO et al. 1999). For $P2X_7$ at least, transition to the large pore state of the receptor also appears to differ in its phenotype between species. For the rat $P2X_7$ receptor, large pore formation seems to be associated with a change in the deactivation rate of the receptor, such that repeated agonist application leads to progressively longer open receptor times (SURPRENANT et al. 1996). Conversely, with repeated agonist applications, the mouse $P2X_7$ receptor produces progressively larger inward currents, with no change in deactivation rate (CHESSELL et al. 1997b), while the human receptor appears to display an intermediate phenotype, with changes both in deactivation rate and inward current magnitude (CHESSELL et al. 1998a, 1999). Recent studies in our own laboratory suggest that the kinetics of large pore formation are dependent on the receptor density, and we hypothesise that one mechanism by which large pore formation may occur is by individual subunits coming together in the membrane (CHESSELL et al. 1999).

The pharmacology of the recombinant $P2X_7$ receptor is unlike that of the other recombinant P2X receptor types. Of the commonly used agonists, only benzoyl-benzoyl-ATP (BzATP) and ATP itself activate the receptor. Whilst BzATP is not selective for $P2X_7$, as it has activity at other subunits (notably $P2X_1$ and $P2X_3$; BIANCHI et al. 1999), this agonist remains useful for delineating $P2X_7$-mediated responses. $\alpha\beta$meATP has very little effect at the $P2X_7$ receptor (Table 2; CHESSELL et al. 1997b). When the receptor gene was first cloned, it was reported to be relatively insensitive to PPADS and suramin (SURPRENANT et al. 1996; RASSENDREN et al. 1997b). However, more recent studies suggest that PPADS is a potent antagonist, so long as appropriate incubation times are used (CHESSELL et al. 1998a). Unique to the human $P2X_7$ receptor is the action of the isoquinoline derivative, KN-62; this compound is a potent antagonist at this subunit, but has little effect at the rat or mouse orthologues (GARGETT and WILEY 1997; HUMPHREYS et al. 1998; CHESSELL et al. 1998a). Recently, we have provided evidence that KN-62 acts as an allosteric modulator at the $P2X_7$ receptor, binding to a different site to PPADS or oxidised-ATP (MICHEL et al. 1998). To date, $P2X_7$ receptor mRNA or protein has not been localised to neurones. Most of the literature supports the localisation of $P2X_7$ to immune cells, and these include macrophages, monocytes, mast cells, microglia and some lymphocytes. This limited distribution has generated much interest in the role of $P2X_7$ in the immune system. There are a number of reports describing $P2X_7$ mediated release of the cytokine, IL-1β (FERRARI et al. 1997a,b; GRAHAMES et al. 1999). The use of a monoclonal antibody specific for human $P2X_7$ has unequivocally demonstrated that it is indeed activation of this receptor that is responsible for ATP-mediated externalisation of mature IL-1β (BUELL et al. 1998). The $P2X_7$ receptor is discussed in further detail in this volume (DI VIRGILIO et al., Chap. 26, second volume).

C. Effects of Ions on Homomeric P2X Receptor Assemblies

P2X receptor function can be dramatically modified by extracellular ions; protons, divalent cations and, more recently, monovalent ions have been shown to affect agonist potency and magnitude of responses. In a number of instances these ionic effects can be attributed to a direct action on the receptor. However, protons and cations can bind to ATP and result in the formation of a large number of different molecular species of ATP and, since it is not always clear which ionic species activate the various P2X receptors, it is often difficult to dissociate direct ionic effects upon the receptor from those effects due to ions binding to ATP.

Protons affect responses at P2X receptors in rat nodose ganglion (LI et al. 1996), at rat recombinant $P2X_2$ receptors (KING et al. 1996) and at the recombinant $P2X_1$-$P2X_4$ receptors (STOOP et al. 1997). In the case of the rat nodose P2X receptor, as well as the recombinant $P2X_2$ subunit and heteromultimers of recombinant $P2X_2$ and $P2X_3$ subunits, protons increased the potency of ATP by approximately tenfold whereas at the $P2X_1$, $P2X_3$ and $P2X_4$ receptors, protons decreased responses to ATP. These various actions of protons are thought to result from a direct effect upon the receptor.

Divalent cations can also produce dramatic effects on P2X receptor function. Calcium and magnesium ions invariably inhibit ATP responses at P2X receptors. The inhibitory effects of magnesium on $P2X_{2/3}$ receptors (LI et al. 1997) and of calcium and magnesium ions on $P2X_1$, $P2X_2$ and $P2X_7$ receptors (EVANS et al. 1996; VIRGINIO et al. 1997) are thought to be mediated through a direct action upon the receptor. However, it should be noted that the effects of calcium and magnesium on the $P2X_7$ receptor have also been attributed to chelation of ATP (STEINBERG et al. 1987). There are no reports of calcium or magnesium increasing responses to ATP, although on the $P2X_3$ receptor raised calcium levels can speed recovery from desensitisation (COOK et al. 1998).

Heavy metals such as zinc and copper may also dramatically affect P2X receptor-mediated responses. Zinc was originally shown to increase responses at native P2X receptors in sensory ganglia (LI et al. 1993; CLOUES et al. 1993) and subsequently this has been confirmed on the recombinant $P2X_2$ and $P2X_4$ receptors (BRAKE et al. 1994; SOTO et al. 1996a; SÉGUÉLA et al. 1996). In contrast, heavy metals reduce responses at $P2X_7$ receptors, with copper, in particular, exhibiting high affinity as an antagonist (VIRGINIO et al. 1997; MICHEL et al. 1997a).

Finally, at the $P2X_7$ receptor, monovalent cations and anions can affect agonist potency with 20- to 30-fold increases in agonist potency being reported when sodium chloride is replaced with sucrose or other sodium and chloride free buffers (KAIHO et al. 1997; MICHEL et al. 1999). The changes in agonist potency are due to both anionic and cationic influences (MICHEL et al. 1999).

It is not known at present whether protons, heavy metals, divalent cations and monovalent ions act at the same or distinct regulatory sites on the P2X receptor subunits.

D. Receptor Binding Studies

There are no selective and potent P2X antagonists and so most receptor binding studies have utilised radiolabelled ATP or its structural analogues and this has resulted in a number of complications in interpreting the results obtained. Thus, because ATP and its analogues can be subject to metabolism, binding studies often need to be conducted at reduced temperature or in the absence of divalent cations (Bo et al. 1992; Michel and Humphrey 1996), so making comparisons with functional studies complicated. In addition, endogenous non-P2X subunit related nucleotide binding sites are ubiquitous and we have found that it is only possible to detect specific binding to P2X subunits when they are expressed at very high density (>2 pmol/mg protein).

Despite these complications, it has proved possible to label the recombinant $P2X_1$-$P2X_4$ subunits using $[^{35}S]ATP\gamma S$ (Michel et al. 1996b, 1997b) and both recombinant and native $P2X_1$ subunits using $[^3H]\alpha,\beta$me-ATP (Bo et al. 1992; Michel et al. 1996a). The binding characteristics of these subunits are complex, with ATP and its analogues possessing very high affinity (nanomolar or sub-nanomolar) compared to their potency determined in functional studies (micromolar). The reason for this discrepancy has never been fully explained, although it has been suggested, by analogy with studies on nicotinic receptors, that binding studies identify a desensitised high affinity state of the P2X receptor (Michel et al. 1997b). In contrast to the data obtained with agonist ligands, antagonist affinity estimates at the recombinant receptors are similar in both binding and functional studies (Khakh et al. 1994; Michel et al. 1997b).

There is also evidence for allosteric regulation of $P2X_4$ receptor ligand binding by monovalent cations and cibacron blue which may well relate to the ability of monovalent cations to affect the function of the P2X receptor and cibacron blue to potentiate responses at $P2X_4$ receptors in functional studies (Michel et al. 1997b; Miller et al. 1998). Furthermore, ivermectin has recently been shown to function as an allosteric regulator of the $P2X_4$ receptor (Khakh et al. 1999b).

E. Heteropolymeric Combinations of P2X Subunits

One difficulty already discussed relating to P2X receptor research is the reconciliation of the properties of recombinant receptors with those present in whole tissues. As an example, both ATP and $\alpha\beta$meATP evoke slowly-desensitising currents in nodose and dorsal root ganglia (Bean 1990; Krishtal et al. 1988b), but none of the appropriate homomeric P2X receptors possess this phenotype. The explanation for this was proposed when it was

Table 5. Operational properties of P2X receptor heteromultimers

P2X$_{2/3}$ (CHEN et al. 1995; LEWIS et al. 1995)	P2X$_{4/6}$ (LÊ et al. 1998)	P2X$_{1/5}$ (TORRES et al. 1998)
αβmeATP sensitive Slowly desensitising Antagonist sensitive	αβmeATP sensitive Moderately desensitising Suramin sensitive PPADS insensitive	αβmeATP sensitive Biphasic desensitisation Unknown antagonist sensitivity

noted that P2X$_2$ and P2X$_3$ receptors, when co-expressed, are able to form a novel heteropolymer, with a phenotype similar to that observed in dorsal root ganglia cells and nodose ganglia (LEWIS et al. 1995). Thus, the properties of P2X$_{2/3}$ appear to be a combination of P2X$_2$ and P2X$_3$, in that this receptor is slowly desensitising, and is also sensitive to αβmeATP (Table 5). This receptor combination is also sensitive, like P2X$_3$, to blockade by TNP-ATP. Subsequent co-immunoprecipitation studies have shown that P2X$_2$ and P2X$_3$ receptors form stable heteromeric channels (RADFORD et al. 1997), and studies by COOK et al. (1997) also suggest that these heteromultimeric receptors exist on pain-sensing neurons from the trigeminal ganglia.

Another apparent discrepancy between recombinant and native P2X receptor phenotypes arises when experiments are performed on neurones from the CNS. Neurones from both the locus coeruleus and the medial vestibular nucleus can be depolarised with αβmeATP (SHEN and NORTH 1993; SANSUM et al. 1998; CHESSELL et al. 1997a), and these responses are slowly desensitising. Only P2X$_1$ and P2X$_3$ homomeric receptors are sensitive to αβmeATP, and responses at these receptors desensitise very rapidly. In addition, there is little evidence for transcripts or protein for P2X$_1$ or P2X$_3$ subunits in the adult CNS, although these receptors are found in neonate brain (KIDD et al. 1995, 1998). A recent report (LÊ et al. 1998) describes coassembly of P2X$_4$ and P2X$_6$ subunits to produce a heteromer with a novel phenotype which is αβmeATP-sensitive, and slowly desensitising (see Table 5). Given that both P2X$_4$ and P2X$_6$ are strongly expressed in the CNS, it seems likely that this heteromeric combination of subunits could account for at least some of the αβmeATP-sensitive responses observed in the CNS.

The most recent heteropolymeric combination of subunits to be described is that of P2X$_1$ and P2X$_5$ (TORRES et al. 1998). Confirmation of heteromeric assembly was provided by co-immunoprecipitation, and this heteromer again has an αβmeATP sensitive phenotype, with a biphasic desensitisation, the first phase of which resembles that of P2X$_1$, and the second that of P2X$_5$. Confirmation of its existence in vivo is awaited, and the same applies to other potential heteromeric combinations recently identified in vitro (TORRES et al. 1999).

F. Conclusions

Knowledge about the structure, function and localisation of the P2X receptors is rapidly increasing. ATP is now firmly established as a neurotransmitter in the central and peripheral nervous systems and, as described above, P2X receptors are found throughout the body. In addition, there is now overwhelming evidence that P2X receptors can mediate a plethora of effects as diverse as muscle contraction and the release of inflammatory cytokines. Further understanding of the stoichiometry, polymerisation and functional properties of the known receptor subunits will aid better understanding of the endogenous receptor types involved, but it should be noted that, even with the large volume of ongoing research into P2X receptors, this area is still severely hampered by the lack of selective agonists and antagonists (particularly for use in vivo). It remains to be seen whether their identification will lead subsequently to development of new therapeutic agents.

References

Barnard EA, Humphrey PPA (1998) International Union of Pharmacology. XIX. The IUPHAR receptor code: a proposal for an alphanumeric classification system. Pharmacol Rev 50:271–277

Bean BP (1990) ATP-activated channels in rat and bullfrog sensory neurons: concentration dependence and kinetics. J Neurosci 10:1–10

Bianchi BR, Lynch KJ, Touma E, Niforatos W, Burgard EC, Alexander KM, Park HS, Yu H, Metzger R, Kowaluk E, Jarvis MF, van Biesen T (1999) Pharmacological characterisation of recombinant human and rat P2X receptor subtypes. Eur J Pharmacol 376:127–138

Bland-Ward PA, Humphrey PPA (1997) Acute nociception mediated by hindpaw P2X receptor activation in the rat. Br J Pharmacol 122:365–371

Bleehan T, Keele CA (1977) Observations on the algogenic actions of adenosine compounds on the human blister base preparation. Pain 3:367–377

Bo X, Simon J, Burnstock G, Barnard EA (1992) Solubilization and molecular size determination of the P2x purinoceptor from rat vas deferens. J Biol Chem 267:17581–17587

Bo X, Zhang Y, Nassar M, Burnstock G, Schoepfer R (1995) A P2X purinoceptor cDNA conferring a novel pharmacological profile. FEBS Lett 375:129–133

Brake AJ, Wagenbach MJ, Julius D (1994) New structural motif for ligand-gated ion channels defined by an ionotropic ATP receptor. Nature 371:519–523

Brandle U, Spielmanns P, Osteroth R, Sim J, Surprenant A, Buell G, Ruppersberg JP, Plinkert PK, Zenner H, Glowinski E (1997) Desensitization of the $P2X_2$ receptor controlled by alternative splicing. FEBS Lett 404:294–298

Buell G, Lewis C, Collo G, North RA, Surprenant A (1996) An antagonist-insensitive P_{2X} receptor expressed in epithelia and brain. EMBO J 15:55–62

Buell G, Chessell IP, Michel AD, Collo G, Salazzo M, Herren S, Gretener D, Grahames CBA, Kaur R, Kosco-Vilbois M, Humphrey PPA (1998) Blockade of human $P2X_7$ receptor function with a monoclonal antibody. Blood 92:3521–3528

Buell G, Talabot F, Gos A, Lorenz J, Lai E, Morris MA, Antonarakis SE (1999) Gene structure and chromosomal localisation of the human $P2X_7$ receptor. Receptors Channels (in press)

Chen C, Akopian AN, Sivilotti L, Colquhoun D, Burnstock G, Wood JN (1995) A P2X purinoceptor expressed by a subset of sensory neurons. Nature 377:428–431

Chessell IP, Michel AD, Humphrey PPA (1997a) Functional evidence for multiple purinoceptor subtypes in the rat medial vestibular nucleus. Neuroscience 77: 783–791

Chessell IP, Michel AD, Humphrey PPA (1997b) Properties of the pore-forming $P2X_7$ purinoceptor in mouse NTW8 microglial cells. Br J Pharmacol 121:1429–1437

Chessell IP, Michel AD, Humphrey PPA (1998a) Effects of antagonists at the human recombinant $P2X_7$ receptor. Br J Pharmacol 124:1314–1320

Chessell IP, Simon J, Hibell AD, Michel AD, Barnard EA, Humphrey PPA (1998b) Cloning and functional characterisation of the mouse $P2X_7$ receptor. FEBS Lett 439:26–30

Chessell IP, Michel AD, Humphrey PPA (1999) Determinants of human $P2X_7$ receptor large pore formation. Br J Pharmacol 126:19P

Cloues R, Jones S, Brown DA (1993) Zn^{2+} potentiates ATP-activated currents in rat sympathetic neurones. Pflügers Arch 424:152–158

Cockcroft S, Gomperts BD (1979) ATP induces nucleotide permeability in rat mast cells. Nature 279:541–542

Collo G, North RA, Kawashima E, Merlo-Pich E, Neidhart S, Surprenant A, Buell G (1996) Cloning of P2X5 and P2X6 receptors, and the distribution and properties of an extended family of ATP-gated ion channels. J Neurosci 16:2495–2507

Cook SP, Vulchanova L, Hargreaves KM, Elde R, McCleskey EW (1997) Distinct ATP receptors on pain-sensing and stretch-sensing neurons. Nature 387:505–508

Cook SP, Rodland KD, McCleskey EW (1998) A memory for extracellular Ca^{2+} by speeding recovery of P2X receptors from desensitization. J Neurosci 18:9238–9244

Cook SP, McCleskey EW (1997) Desensitization, recovery and Ca^{2+}-dependent modulation of ATP-gated P2X receptors in nociceptors. Neuropharmacology 36: 1303–1308

Dhulipala PDK, Wang Y, Kotlikoff MI (1998) The human $P2X_4$ receptor gene is alternatively spliced. Gene 207:259–266

Dowd E, McQueen DS, Chessell IP, Humphrey PPA (1998) P2X receptor-mediated excitation of nociceptive afferents in the normal and arthritic rat knee joint. Br J Pharmacol 125:341–346

Evans RJ, Lewis C, Virginio C, Lundstrom K, Buell G, Surprenant A, North RA (1996) Ionic permeability of, and divalent cation effect on, two ATP-gated cation channels (P2X receptors) expressed in mammalian cells. J Physiol 497.2:413–422

Ferrari D, Villalba M, Chiozzi P, Falzoni S, Ricciardi-Castagnoli P, Di Virgilio F (1996) Mouse microglial cells express a plasma membrane pore gated by extracellular ATP. J Immunol 156:1531–1539

Ferrari D, Chiozzi P, Falzoni S, Dal Susino M, Melchiorri L, Baricordi OR, Di Virgilio F (1997a) Extracellular ATP triggers IL-1b release by activating the purinergic P2Z receptor of human macrophages. J Immunol 159:1451–1458

Ferrari D, Chiozzi P, Falzoni S, Hanau S, Di Virgilio F (1997b) Purinergic modulation of interleukin-1-β release from microglial cells stimulated with bacterial endotoxin. J Exp Med 185:579–582

Funk GD, Kanjan R, Walsh C, Lipski J, Comer AM, Parkis MA, Housley GD (1997) P2 receptor excitation of rodent hypoglossal motoneuron activity in vitro and in vivo. J Neurosci 17:6325–6337

Garcia-Guzman M, Soto F, Laube B, Stuhmer W (1996) Molecular cloning and functional expression of a novel rat heart P2X purinoceptor. FEBS Lett 388:123–127

Garcia-Guzman M, Soto F, Gomez-Hernandez JM, Lund P, Stumer W (1997a) Characterization of recombinant human $P2X_4$ receptor reveals pharmacological differences to the rat homologue. Mol Pharmacol 51:109–118

Garcia-Guzman M, Stuhmer W, Soto F (1997b) Molecular characterization and pharmacological properties of the human $P2X_3$ purinoceptor. Mol Brain Res 47:59–66

Gargett CE, Wiley JS (1997) The isoquinoline derivative KN-62 a potent antagonist of the P2Z-receptor of human lymphocytes. Br J Pharmacol 120:1483–1490

Grahames CBA, Michel AD, Chessell IP, Humphrey PPA (1999) Pharmacological characterization of ATP- and LPS-induced IL-1beta release in human monocytes. Br J Pharmacol 127:1915–1921

Guo A, Vulchanova L, Wang J, Li X, Elde R (1999) Immunocytochemical localization of the vanilloid receptor 1 (VR1): relationship to neuropeptides, the $P2X_3$ purinoceptor and IB4 binding sites. Eur J Neurosci 11:946–958

Hamilton SG, Wade A, McMahon SB (1999) The effects of inflammation and inflammatory mediators on nociceptive behaviour induced by ATP analogues in the rat. Br J Pharmacol 126:326–332

Humphrey PPA, Buell G, Kennedy I, Khakh B, Michel AD, Surprenant A, Trezise DJ (1995) New insights on P_{2X} receptors. Naunyn-Schmiedebergs Arch Pharmacol 352:585–596

Humphrey PPA, Khakh BS, Kennedy C, King BF, Burnstock G (1998) P2X receptors. IUPHAR compendium of receptor characterization and classification 197–207

Humphreys BD, Virginio C, Surprenant A, Rice J, Dubyak GR (1998) Isoquinolines as antagonists of the $P2X_7$ nucleotide receptor: high selectivity for the human versus rat receptor homologues. Mol Pharmacol 54:22–32

Jones CA, Chessell IP, Simon J, Humphrey PPA (1999) Operational properties of mouse $P2X_4$ receptors: a species comparison. Br J Pharmacol (in press)

Kaiho H, Kimura J, Matsuoka I, Nakanishi H (1997) Effects of anions on ATP-activated non-selective cation current in NG108–15 cells. J Neurophysiol 77:2717–2722

Khakh BS, Michel AD, Humphrey PPA (1994) Estimates of antagonist affinities at P_{2X} purinoceptors in rat vas deferens. Eur J Pharmacol 263:301–309

Khakh BS, Bao XR, Labarca C, Lester HA (1999a) Neuronal P2X transmitter-gated cation channels change their ion selectivity in seconds. Nature Neurosci 2:322–330

Khakh BS, Proctor WR, Dunwiddie TV, Labarca C, Lester HA (1999b) Allosteric control of gating and kinetics at P2X4 receptor channels. J Neurosci 19:7289–7299

Kidd EJ, Grahames CBA, Simon J, Michel AD, Barnard EA, Humphrey PPA (1995) Localization of P2X purinoceptor transcripts in the rat nervous system. Mol Pharmacol 48:569–573

Kidd EJ, Miller KJ, Sansum AJ, Humphrey PPA (1998) Evidence for $P2X_3$ receptors in the developing rat brain. Neuroscience 87:533–539

King BF, Ziganshina LE, Pintor J, Burnstock G (1996) Full sensitivity of $P2X_2$ purinoceptor to ATP revealed by changing extracellular pH. Br J Pharmacol 117:1371–1373

Koshimizu T, Tomic M, Koshimizu M, Stojilkovic SS (1998) Identification of amino acid residues contributing to desensitization of the $P2X_2$ receptor channel. J Biol Chem 273:12853–12857

Krishtal OA, Marchenko SM, Obukhov AG (1988a) Cationic channels activated by extracellular ATP in rat sensory neurons. Neuroscience 27:995–1000

Krishtal OA, Marchenko SM, Obukhov AG, Volkova TM (1988b) Receptors for ATP in rat sensory neurones: the structure-function relationship for ligands. Br J Pharmacol 95:1057–1062

Lê K, Paquet M, Nouel D, Babinski K, Séguéla P (1997) Primary structure and expression of a naturally truncated human P2X ATP receptor subunit from brain and immune system. FEBS Lett 418:195–199

Lê K, Babinski K, Séguéla P (1998) Central $P2X_4$ and $P2X_6$ channel subunits coassemble into a novel heteromeric ATP receptor. J Neurosci 18:7152–7159

Lewis C, Neidhart S, Holy C, North RA, Buell G, Surprenant A (1995) Coexpression of $P2X_2$ and $P2X_3$ receptor subunits can account for ATP-gated currents in sensory neurons. Nature 377:432–435

Li C, Peoples RW, Li Z, Weight FF (1993) Zn^{2+} potentiates excitatory action of ATP on mammalian neurons. Proc Natl Acad Sci USA 90:8264–8267

Li C, Peoples RW, Weight FF (1997) Mg^{2+} inhibition of ATP-activated current in rat nodose ganglion neurons: Evidence that Mg^{2+} decreases the agonist affinity of the receptor. J Neurophysiol 77:3391–3395

Li CY, Peoples RW, Weight FF (1996) Acid pH augments excitatory action of ATP on a dissociated mammalian sensory neuron. Neuroreport 7:2151–2154

Longhurst PA, Schwegel T, Folander K, Swanson R (1996) The human P2X$_1$ receptor: molecular cloning, tissue distribution, and localization to chromosome 17. Biochim Biophys Acta 1308:185–188

Michel AD, Lundstrom K, Buell GN, Surprenant A, Valera S, Humphrey PPA (1996a) The binding characteristics of a human bladder recombinant P2X purinoceptor, labeled with [^{3}H]-α-β-meATP, [^{35}S]-ATP-gamma-S or [^{33}P]-ATP. Br J Pharmacol 117:1254–1260

Michel AD, Lundstrom K, Buell GN, Surprenant A, Valera S, Humphrey PPA (1996b) A comparison of the binding characteristics of recombinant P2X$_1$ and P2X$_2$ purinoceptors. Br J Pharmacol 118:1806–1812

Michel AD, Chessell IP, Humphrey PPA (1997a) Potent inhibition of P2Z (P2X$_7$) receptor-mediated effects by copper, zinc and nickel ions. Br J Pharmacol 122:13P

Michel AD, Miller KJ, Lundstrom K, Buell GN, Humphrey PP (1997b) Radiolabeling of the rat P2X4 purinoceptor: evidence for allosteric interactions of purinoceptor antagonists and monovalent cations with P2X purinoceptors. Mol Pharmacol 51:524–532

Michel AD, Chessell IP, Humphrey PPA (1998) Inhibition of human P2X$_7$ receptor-mediated YO-PRO-1 influx by PPADS and KN-62. Br J Pharmacol 123:103P

Michel AD, Chessell IP, Humphrey PPA (1999) Ionic effects on human recombinant P2X$_7$ receptor function. Naunyn-Schmiedebergs Arch Pharmacol 359:102–109

Michel AD, Humphrey PPA (1996) High affinity P2X-purinoceptor binding sites for [^{35}P]-adenosine 5′-O-[3-thiotriphosphate] in rat vas deferens membranes. Br J Pharmacol 117:63–70

Miller KJ, Michel AD, Chessell IP, Humphrey PPA (1998) Cibacron blue allosterically modulates the rat P2X$_4$ receptor. Neuropharmacology 37:1579–1586

Nicke A, Baumert HG, Rettinger J, Eichele K, Lambrecht G, Mutschler E, Schmalzing G (1998) P2X$_1$ and P2X$_3$ receptors form stable trimers: a novel structural motif of ligand-gated ion channels. EMBO J 17:3016–3028

North RA (1996) P2X purinoceptor plethora. Sem Neurosci 8:187–194

North RA, Barnard EA (1997) Nucleotide receptors. Curr Opin Neurobiol 7:346–357

Radford K, Virginio C, Surprenant A, North RA, Kawashima E (1997) Baculovirus expression provides direct evidence for heteromeric assembly of P2X$_2$ and P2X$_3$ receptors. J Neurosci 17:6529–6533

Rassendren F, Buell G, Newbolt A, North RA, Surprenant A (1997a) Identification of amino acid residues contributing to the pore of a P2X receptor. EMBO J 16:3446–3454

Rassendren F, Buell G, Virginio C, North RA, Surprenant A (1997b) The permeabilizing ATP receptor (P2X$_7$): Cloning and expression of a human cDNA. J Biol Chem 272:5482–5486

Sansum AJ, Chessell IP, Hicks GA, Trezise DJ, Humphrey PPA (1998) Evidence that P2X purinoceptors mediate the excitatory effects of α,β-methylene-ADP in rat locus coeruleus neurones. Neuropharmacology 37:875–885

Séguéla P, Haghighi A, Soghomonian J, Cooper E (1996) A novel P$_{2X}$ ATP receptor ion channel with widespread distribution in the brain. J Neurosci 16:448–455

Shen KZ, North RA (1993) Excitation of rat locus coeruleus neurons by adenosine 5′-triphosphate: ionic mechanism and receptor characterisation. J Neurosci 13:894–899

Simon J, Kidd EJ, Smith FM, Chessell IP, Murrell-Lagnado R, Humphrey PPA, Barnard EA (1997) Localization and functional expression of splice variants of the P2X$_2$ receptor. Mol Pharmacol 52:237–248

Simon J, Chessell IP, Jones CA, Michel AD, Barnard EA, Humphrey PPA (1999) Molecular cloning and characterisation of splice variants of the mouse P2X$_4$ receptor. Br J Pharmacol 126:127P

Smith FM, Humphrey PPA, Murrell-Lagnado R (1999) Identification of amino acids within the $P2X_2$ receptor C-terminus that regulate desensitization. J Physiol 520:91–99

Soto F, Garcia-Guzman M, Gomez-Hernandez JM, Hollmann M, Karschin C, Stuhmer W (1996a) P2X4: an ATP-activated ionotropic receptor cloned from rat brain. Proc Natl Acad Sci USA 93:3684–3686

Soto F, Garcia-Guzman M, Karschin C, Stuhmer W (1996b) Cloning and tissue distribution of a novel P2X receptor from rat brain. Biochem Biophys Res Comm 223:456–460

Souslova V, Ravenall S, Fox M, Wells D, Wood JN, Akopian AN (1997) Structure and chromosomal mapping of the mouse $P2X_3$ gene. Gene 195:101–111

Steinberg TH, Newman AS, Swanson JA, Silverstein SC (1987) ATP^{4-} permeabilizes the plasma membrane of mouse macrophages to fluorescent dyes. J Biol Chem 262:8884–8888

Stoop R, Surprenant A, North RA (1997) Different sensitivities to pH of ATP-induced currents at four cloned P2X receptors. J Neurophysiol 78:1837–1840

Surprenant A, Rassendren F, Kawashima E, North RA, Buell G (1996) The cytolytic P_{2Z} receptor for extracellular ATP identified as a P_{2X} receptor ($P2X_7$). Science 272:735–738

Torres GE, Haines WR, Egan TM, Voigt MM (1998) Co-expression of $P2X_1$ and $P2X_5$ receptor subunits reveals a novel ATP-gated ion channel. Mol Pharmacol 54:989–993

Torres GE, Egan TM, Voigt MM (1999) Hetero-oligomeric assembly of P2X receptor subunits. Specificities exist with regard to possible partners. J Biol Chem 274:6653–6659

Townsend-Nicholson A, King BF, Wildman SS, Burnstock G (1999) Molecular cloning, functional characterization and possible cooperativity between the murine $P2X_4$ and $P2X_{4a}$ receptors. Mol Brain Res 64:246–254

Trezise DJ, Humphrey PPA (1997) Activation of cutaneous afferent neurons by adenosine triphosphate in the neonatal rat tail-spinal cord preparation in vitro. In: Olesen J, Edvinsson L (eds) Headache pathogenesis: monoamines, neuropeptides, purines and nitric oxide. Lippincott-Raven, New York, pp 111–116

Valera S, Hussy N, Evans RJ, Adami N, North RA, Surprenant A, Buell G (1994) A new class of ligand-gated ion channel defined by P_{2X} receptor for extracellular ATP. Nature 371:516–519

Valera S, Talabot F, Evans RJ, Gos A, Antonarakis SE, Morris MA, Buell GN (1995) Characterization and chromosomal localization of a human P2X receptor from the urinary bladder. Receptors Channels 3:283–289

Virginio C, Church D, North RA, Surprenant A (1997) Effects of divalent cations, protons and calmidazolium at the rat $P2X_7$ receptor. Neuropharmacology 36:1285–1294

Virginio C, Robertson G, Surprenant A, North RA (1998) Trinitrophenyl-substituted nucleotides are potent antagonists selective for $P2X_1$, $P2X_3$ and heteromeric $P2X_{2/3}$ receptors. Mol Pharmacol 53:969–973

Virginio C, MacKenzie A, Rassendren FA, North RA, Surprenant A (1999) Pore dilation of neuronal P2X receptor channels. Nature Neurosci 2:315–321

Vulchanova L, Arvidsson U, Riedl MS, Wang J, Buell G, Surprenant A, North RA, Elde R (1996) Differential distribution of two ATP-gated channels (P2X receptors) determined by immunohistochemistry. Proc Natl Acad Sci USA 93:8063–8067

Wang CZ, Namba N, Gonoi T, Inagaki N, Seino S (1996) Cloning and pharmacological characterization of a fourth P2X receptor subtype widely expressed in brain and peripheral tissues including various endocrine tissues. Biochem Biophys Res Commun 220:196–202

CHAPTER 4
P2Y Receptors: Structure and Function

M.R. BOARDER and T.E. WEBB

A. Introduction

Following the finding that nucleotides act as extracellular signalling molecules by interaction with cell surface transmembrane P2 receptors, it has now been recognised that this mode of regulation is widespread, exerting a ubiquitous influence on physiological function. Indeed, P2 receptors are present on most cells in the body (RALEVIC and BURNSTOCK 1998) P2 receptor regulation via disparate cell types is mirrored by the diversity of cellular responses that these receptors elicit on activation by endogenous nucleotides. This diversity in turn comes, in part from multiple P2 receptors, and in part from different responses to activation of the same receptors in different cell types. P2 receptors are subdivided into P2X receptors with intrinsic ion channels and G protein-coupled P2Y receptors (ABBRACCHIO and BURNSTOCK 1994; FREDHOLM et al. 1997). The wide spread distribution of multiple P2Y receptor subtypes represents an important challenge in characterising and classifying this subfamily of receptors in order to understand the manner in which their activation by native extracellular nucleotides gives rise to cognate cellular responses. It has been evident for some time that different P2Y receptor-subtypes can give rise to distinct cellular responses, even when two subtypes are located on the same cell, and even when both are coupled to a similar level of stimulation of phospholipase C. In this chapter, the relationships between the molecular and pharmacological properties of these P2Y receptors is discussed, as well as the contribution of structural diversity to the control of intracellular signalling pathways to understand at the cellular level how the different receptors of the P2Y subfamily generate appropriate functional responses. To pursue this objective, an overview of some current issues concerning the P2Y receptors is provided.

B. P2Y Receptors

I. G Protein-Coupled Receptors

P2Y receptors are members of the G protein-coupled (GPCR) superfamily of heptahelical receptors. These have in common an extracellular amino-terminus, followed by seven transmembrane domains, and an intracellular carboxy-terminus (Fig. 1). The native nucleotide agonists are highly charged molecules which dock with aspects of the extracellular and transmembrane domains, causing a change in the interface between the receptor and G proteins located at the intracellular face of the cell membrane. The α subunit of the heterotrimeric G protein sheds GDP and takes up GTP, and then dissociates to generate α_{GTP} and $\beta\gamma$ subunits, which have the potential to control effector proteins. In P2Y signalling, these effector proteins are often assumed to be the enzyme phospholipase C and adenylate cyclase, but may also include ion channels and other signalling proteins (see Sect. E). These primary signalling events are initially determined in part by the structure of intracellular domains of the receptor (intracellular loops and carboxy-terminus). The dissociation-association cycle of α and $\beta\gamma$ subunits, and consequences for signalling, have been the subject of many recent reviews (e.g. LINDER and GILMAN 1992; GUDERMAN et al. 1996), and will not be further discussed here.

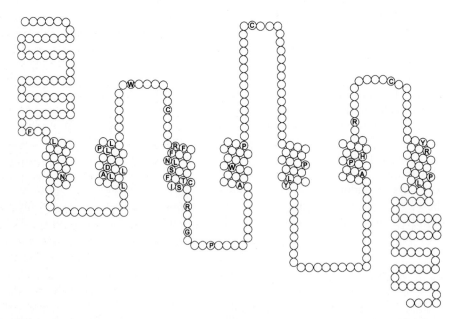

Fig. 1. Serpentine representation of the human $P2Y_1$ receptor. Residues conserved in all five human receptors are indicated by the *one letter amino acid code*

II. Diversity of P2Y Receptors: Cloning and Nomenclature

The most secure foundation for the discussion of diversity and classification of P2Y receptors lies in the structural aspects revealed by cloning studies. Structure and function issues are developed in detail in Sect. C below. In summary, the cloning work has revealed a number of DNA sequences encoding putative P2Y receptors. It is important that clear criteria for the acceptance that these do indeed represent functionally significant P2Y receptors exist; two that may be usefully employed are:

1. That the encoding sequence, when heterologously expressed in a suitable host system, does lead to nucleotide receptor expression
2. That the relevant messenger RNA sequence is expressed in cells which demonstrate appropriate functional responses

There are five mammalian P2Y receptors that fulfil these criteria. These are $P2Y_1$, $P2Y_2$, $P2Y_4$, $P2Y_6$ and $P2Y_{11}$. The general pharmacology of these receptors has been described in various recent reviews (BOARDER and HOURANI 1998; KING et al. 1998). A prefix letter is used to indicate species of origin, as in h-$P2Y_1$ for human $P2Y_1$ receptors or r-$P2Y_4$ for rat $P2Y_4$. Each of the five h-P2Y subtypes meets the criteria for acceptance as a true P2Y receptor, whereas r-$P2Y_{11}$ has not yet been cloned. Various chicken, turkey, bovine and mouse sequences have also been set against these criteria. These, with the relevant cloning literature, are indicated in Table 1.

Some of the missing numbers in the P2Y sequence were those assigned as P2Y receptors but later shown not to meet these criteria. The sequence assigned to $P2Y_7$ has since been shown to be a leukotriene receptor (YOKOMIZO et al. 1997). $P2Y_3$ was assigned to a chick P2Y receptor and is now known to be an orthologue of the mammalian $P2Y_6$ (LI et al. 1998), while numbers have been assigned to other sequences which await confirmation by criteria such as those indicated above.

There are clear indications that other P2Y sequences remain to be cloned, so this numbering will be extended in the future. For example, G_i coupled $P2Y_1$-like receptors, which directly inhibit adenylate cyclase, have not been cloned despite widespread reports of their existence, notably in both platelets and C6 glioma cells (GORDON 1986; BOYER et al. 1995).

The inclusion of expression studies at the messenger RNA level in the criteria set out above is contentious in that the most common method for demonstrating this, the reverse transcriptase polymerase chain reaction (RT-PCR), is capable of detecting specific RNA species which are present a such a low level that they are not indicative of expression of a functional receptor protein. The introduction of truly quantitative RT-PCR procedures, giving a measure of absolute copy numbers per microgram of RNA for the transcripts encoding each P2Y subtype (ERLINGE et al. 1998), is therefore of considerable importance. At the time of writing there are no widely applicable radioligand binding

Table 1. P2Y receptor subtypes sequences in the EMBL database

Receptor	Species	EMBL AC	SWISSPROT/ TREMBL AC	Citation
P2Y$_1$	Chicken	X73268	P34996	WEBB et al. 1993
	Mouse	U22830	P49650	TOKUYAMA et al. 1995
	Rat	U22829	P49651	TOKUYAMA et al. 1995
	Turkey	U09842	P49652	FILTZ et al. 1994
	Bovine	X87628 U34041	P48042	HENDERSON et al. 1995
	Human	AJ006945, Z49205 U42029, U42030 S81950	P47900	DENG et al. 1998 LEON et al. 1996 AYYANATHAN et al. 1996 JANSSENS et al. 1996
P2Y$_2$	Mouse	L14751 S83099	P35383	LUSTIG et al. 1993 ENMOTO et al. 1996
	Human	U07225	P41231	PARR et al. 1994
	Rat	U09402 U56839 U56839	P41232	RICE et al. 1995 SEYE et al. 1997 CHEN et al. 1996
p2y$_3$[a]	Chicken	X98283	Q98907	WEBB et al. 1996
	Turkey	AF031897	O93361	LI et al. 1998
P2Y$_4$	Human	X91852 U40223 X96597	P51582	COMMUNI et al. 1995 NGUYEN et al. 1995 STAM et al. 1996
	Rat	Y11433 Y14705	O35811	WEBB et al. 1998 BOGDANOV et al. 1998
P2Y$_6$	Rat	D63665	Q63371	CHANG et al. 1995
	Human	X97058 U52464 AF007891, AF007892	Q15077	COMMUNI et al. 1996 SOMERS et al. 1997 MAIER et al. 1996
p2y$_8$	Xenopus	X99953	P79928	BOGDANOV et al. 1997
tp2y	Turkey	AF031897	O57466	BOYER et al,. 1997
P2Y$_{11}$	Human	AF030335	O43190	COMMUNI et al. 1997

[a] Non-mammalian receptors are in lowercase in accordance with IUPHAR recommendations.

procedures, or antibodies, which would enable expression studies at the protein level for the P2Y receptors.

III. Diversity of P2Y Receptors: Pharmacology

The original division by BURNSTOCK and KENNEDY (1985) of P2 receptors was by rank order of agonist potency. In the absence of robust antagonists and binding assays, this has remained the basis of the pharmacological characterisation of P2Y receptors, and the rank order of agonist profiles for selected agonists for the five h-P2Y receptors are shown in Table 2. Obviously, a pharmacology based on agonist responses is intrinsically insecure. This is particu-

Table 2. Principle nucleotide selectivity of cloned P2Y receptors

Subclass	Receptor	Rank order of agonist potency	Inactive nucleotides
Adenine selective	hP2Y$_1$	2-MeSADP > ADP > 2-MeSATP > ATP	UTP, UDP
	hP2Y$_{11}$	ATP > 2-MeSATP >> ADP, 2-MeSADP	UTP, UDP
Uridine preferring	hP2Y$_4$	UTP	ADP, ATP, 2-MeSADP, 2-MeSATP
	hP2Y$_6$	UDP > ADP > 2-MeSATP	UTP, ADP, 2-MeSADP
No adenine/ uridine preference	hP2Y$_2$	UTP = ATP	ADP, 2-MeSADP, 2-MeSATP
	rP2Y$_4$	UTP = ATP	ADP, 2-MeSADP, 2-MeSATP

larly true with responses to nucleotides due to metabolism in the extracellular compartment, in which interconversion between di- and tri-phosphates may occur (KENNEDY and LEFF 1995; HARDEN et al. 1997). The agonist-based pharmacology has been sustained in this case by association with the cloning studies. The relationship between the native and the recombinant P2Y receptors has been explored recently (KING et al. 1998). Some important issues in P2Y receptor pharmacology are discussed below.

1. ATP and P2Y$_1$ Receptors

There is an apparently conflicting literature concerning the status of ATP as an agonist at the P2Y$_1$ receptor. Numerous studies, on both native and transfected P2Y$_1$ receptors, reveal responses on application of ATP (e.g. WILKINSON et al. 1994; CHARLTON et al. 1996a) and yet recent reports have suggested that ATP is an antagonist, not an agonist, at P2Y$_1$ receptors (LEON et al. 1997). This is a particularly important issue, since it has long been known that platelet activation in response to ADP is antagonised by ATP.

This issue has its origins in the earliest single cell study on co-existing P2Y subtypes, which indicated there are two nucleotide receptors on hepatocytes, one selective for ATP over ADP (DIXON et al. 1990) and another subsequently shown to be selective for ADP (DIXON et al. 1993, 1995), providing the first evidence for a receptor which is selective for ADP over ATP in cells other than platelets. This receptor, which we now know is probably P2Y$_1$, was originally suggested by DIXON et al. (1990) to be the same as the phospholipase C coupled ADP-selective receptor described in platelets (GORDON et al. 1986). This suggestion has been confirmed by recent studies which show the presence in platelets of a phospholipase C coupled P2Y$_1$ receptor at which ATP is an antagonist (LEON et al. 1997; HECHLER et al. 1998; JANTZEN et al. 1999).

A recent study by PALMER et al. (1998) investigates two explanations for responses to ATP seen at P2Y$_1$ receptors in many studies. First, conversion of ATP to ADP in the extracellular compartment, and second, that ATP is an agonist with low intrinsic efficacy at P2Y$_1$ receptors, eliciting responses only when there is a high receptor reserve. Under conditions in which ATP metabolism was controlled, they report that ATP can be a full agonist when receptor reserve is high. However, a reduction of receptor reserve, which shifts the response to ADP to the right while maintaining its maximal response, results in an abolition of response to ATP. This provides direct evidence that ATP has a lower efficacy than ADP at the P2Y$_1$ receptor and consequently can act as a full agonist, a partial agonist or an antagonist dependent on the receptor reserve. Evidence suggests that either situation, that is ATP acting as an agonist or an antagonist, may exist in native receptors, presumably depending on the receptor reserve. For example, it seems likely that ATP acts as an agonist at P2Y$_1$ receptors on endothelial cells (WILKINSON et al. 1993, 1994) but not at P2Y$_1$ receptors on hepatocytes or platelets (DIXON et al. 1993, 1995; LEON et al. 1997; HECHLER et al. 1998).

2. ATP and P2Y$_4$ Receptors: Species Differences Between Rat and Human

Subsequent to the original subdivision of the P$_2$ receptors into P$_{2X}$ and P$_{2Y}$ receptors by BURNSTOCK and KENNEDY (1985), evidence emerged for other receptors with extended pharmacology. The salient example was an equal response to both ATP and UTP, leading to the designation of the P$_{2U}$ receptor (O'CONNOR et al. 1991). When P2 receptors were divided into the ion channel P2X and the G protein-coupled P2Y receptors (ABBRACCHIO and BURNSTOCK 1994), the responses previously designated as P$_{2Y}$ were widely assumed to be at the P2Y$_1$ receptor, and those designated as P$_{2U}$ were assumed to be due to the presence of P2Y$_2$ receptors. As indicated in Table 2, P2Y$_2$ responds equally to both UTP and ATP, the common criteria for P$_{2U}$ receptors in the earlier nomenclature. This is true for heterologously expressed r-P2Y$_2$ as well as h-P2Y$_2$. When h-P2Y$_4$ was originally cloned and the expressed receptor characterised, UTP was a full agonist but ATP was either a partial agonist (COMMUNI et al. 1995) or inactive (NGUYEN et al. 1995). This difference presumably reflects the low efficacy of ATP at this receptor, and the resultant influence of receptor reserve on the responses generated (see above). However, subsequent study of r-P2Y$_4$ (BOGDANOV et al. 1998; WEBB et al. 1998) showed that ATP and UTP are agonists with apparently equal efficacy and potency. This means that in studies done with the rat, those earlier designated as P$_{2U}$ may be at either r-P2Y$_2$ or r-P2Y$_4$ receptors.

3. Agonists at P2Y Receptors

Despite the problems caused by uncontrolled receptor reserves, when extracellular agonist interconversion is prevented, it is possible to assign unequivocal rank orders of agonist potencies and produce a characteristic profile for each of the five receptor subtypes. This is shown in Table 2 for each of the principle native agonists, with 2-methylthioATP (2MeSATP) and 2-methylthio-ADP (2MeSADP) also included as the most widely used synthetic analogues showing useful discrimination. It can be seen that there are two human P2Y receptors for ATP ($P2Y_2$ and $P2Y_{11}$). In rat there are three ATP preferring P2Y receptors, including r-$P2Y_4$. $P2Y_1$ is the only ADP preferring receptor currently cloned. There are two receptors for UTP ($P2Y_2$ and $P2Y_4$) and one for UDP ($P2Y_6$). In addition we know there remains a further as yet uncloned receptor for ADP which is expressed in platelets and is negatively coupled to adenylate cyclase.

In addition to these native agonists, synthetic nucleotide derivatives include adenosine 5'-O-(3-thiotrisphosphate), which is a relatively ectonucleotidase resistant derivative of ATP which shares with it many of the pharmacological characteristics. It is frequently supplied contaminated with other nucleotides, and should be purified by high-pressure liquid chromatography prior to use. The 2-methylthiolated derivatives of ATP and ADP (Table 2) are useful as discriminating agonists as indicated, but are not resistant to metabolism in the extracellular compartment, and again are often used in the impure state. 2-Methylthio ATP is significantly contaminated by 2-methylthioADP, and this contamination increases on storage. This may be overcome by some combination of purification by chromatography and the use of ATP regenerating system (e.g. see HECHLER et al. 1998). It is also worth noting that 2-meSATP, but not 2-meSADP, is an agonist at the P2X receptors. Methylene derivates of ATP on the phosphate chain (notably α,β-methyleneATP and β,γ-methyleneATP) show significant resistance to nucleotidase breakdown and are useful as potent agonists at some P2X receptors but are inactive at P2Y receptors. Beyond this, series of modified nucleotide analogues, some members of which have high potency at P2Y receptors (FISCHER et al. 1993; BOYER et al. 1995) are explored elsewhere (JACOBSON and KNUTSEN, Chap. 6, this volume).

4. Antagonists at P2Y Receptors

Comment here will be limited to the degree of selectivity which currently available antagonists have between different P2Y receptors. Drugs such as cibacron 2/reactive blue are widely active at P2 receptors, and their selectivity between P2Y receptor subtypes has not been effectively investigated. Suramin also shows a wide spectrum of activity at both P2 receptors and other sites of action. This includes activity as an inhibitor of ectonucleotidase activity (HOURANI and CHOWN 1989) which may confound interpretation of experiments with nucleotide agonists. Despite this, suramin has been usefully employed as a non-selective P2 antagonist (e.g. DUNN and BLAKELY 1988; LEFF et al. 1990; HOYLE et al. 1990). Further work on recombinant receptors established that suramin is as effective at $P2Y_1$ receptors as at P2X, with some selectivity among P2Y responses, with a pA_2 of 6 for $P2Y_1$, a pA_2 of 4.3 for $P2Y_2$, and ineffective at $P2Y_4$ receptors (CHARLTON et al. 1996a,b). This provides an explanation for the earlier observation of suramin sensitive and insensitive P_{2U} responses, which can be interpreted as at rat $P2Y_2$ and $P2Y_4$ respectively (MURRIN and BOARDER 1992; DAINTY et al. 1994). Pyridoxalphosphate-6-azophenyl-2',4'-disulphonic acid (PPADS) was introduced as an antagonist at P2X receptors by LAMBRECHT et al. (1992), but was later shown to be effective at some, but not all, native P2Y receptors (BOYER et al. 1994; WINDSCHEIF et al. 1994; BROWN et al. 1995). Again, studies with transfected recombinant receptors clarified this, showing that PPADS is equally potent at $P2Y_1$ as at P2X receptors, but ineffective at $P2Y_2$ and $P2Y_4$ receptors (CHARLTON et al. 1996a,b). The activity at P2 receptors of a series of PPADS derivatives has been reported (KIM et al. 1998), although it is not clear whether these will generate compounds distinguishing between P2Y subtypes. A useful series of bisphosphate analogues has been reported which show competitive and selective antagonism at $P2Y_1$ receptors (BOYER et al. 1996, 1998; BULTMANN et al. 1998). Summarising these various antagonists, they each provide the potential to distinguish between $P2Y_1$ responses and those at other P2Y receptors, and antagonism by suramin at $P2Y_2$ enables discrimination from $P2Y_4$ receptors.

C. Correlating Structure with Function

I. Sequence Analysis

In context of the GPCR superfamily, P2Y receptors fall within class A (rhodopsin- and β-adrenergic receptor-like) and phylogenetic studies further assign them to group V of this class along with receptors such as those for angiotensin II, opioids and chemokines. There are currently 18 full-length sequences in the EMBL database which have been proven to encode functional P2Y receptors (Table 1). None of these sequences share a significant sequence homology with the adenosine receptors that reside in group I of class A of this receptor superfamily.

Despite the diversity of species that the P2Y receptors have been isolated from, when aligned their sequences fall into discreet subtypes that can be correlated broadly with their agonist preferences (Table 2). This sequence relatedness is illustrated in Fig. 2 in the form of a dendrogram. There are five branch points, with each branch containing a human receptor. In terms of the primary sequence of these receptors, $P2Y_{11}$ is clearly the most divergent, with the $P2Y_1$ receptors also forming a distinct grouping, while the remainder of the sequences are more closely related. From the dendrogram of the multiple sequence alignment of their entire sequences, $P2Y_6$ receptors (including chick $P2Y_3$, the avian orthologue of mammalian $P2Y_6$) align more closely with $P2Y_1$ than do $P2Y_2$ and $P2Y_4$ sequences. However, dendrograms similarly derived from human P2Y transmembrane or extracellular sequences (Fig. 3A and Fig. 3B respectively) show that the branch points for the $P2Y_2/P2Y_4$ grouping and the $P2Y_6$ receptor are altered, i.e. the $P2Y_4$ and $P2Y_2$ receptors sequences

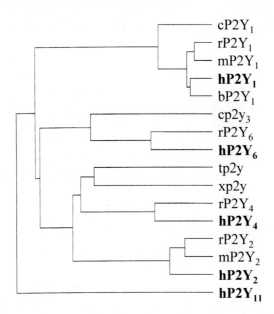

Fig. 2. Dendogram demonstrating the amino acid sequence similarity shared within the P2Y receptor family. The turkey P2Y$_1$ and P2Y$_3$ receptors were not included in the alignment as they have minimal amino acid sequence difference from the chicken sequence. *h*, human; *b*, bovine; *r*, rat; *m*, mouse; *t*, turkey; *x*, xenopus. Note sequence relatedness is inversely proportional to the length of the branches. The alignment was generated using CLUSTAL

share greater sequence similarity in these regions with the P2Y$_1$ receptor than the P2Y$_1$ receptor does with the P2Y$_6$ subtype.

It is of interest to relate this dendrogram analysis to the agonist pharmacology in Table 2. Using the dendrogram derived from alignments of the entire coding sequence, the adenine nucleotide preferring receptors (P2Y$_1$ and P2Y$_{11}$) are clearly distinct from those which respond to uridine nucleotides (P2Y$_{2, 4}$ and $_6$) which are clustered together. It is of particular interest to note that rat and human P2Y$_4$ align together (Fig. 2) even though r-P2Y$_4$ agonist pharmacology is closer to that of P2Y$_2$ from either species. This emphasises the need for structural analysis at a higher resolution, as described below.

As discussed above, the P2Y$_{11}$ receptor shares the lowest degree of sequence identity with the rest of the h-P2Y receptors, and has the largest second and third extracellular loops (Table 3). In spite of this, there are 44 residues that are totally conserved in position, relative to the predicted transmembrane domains, in all five receptors, and these residues are indicated on the serpentine representation of the h-P2Y$_1$ receptor presented in Fig. 1. However, it should be noted that if the P2Y$_{11}$ receptor sequence is excluded from such an analysis the number of conserved residues in the remaining four sequences increases to 70. As can be seen from Fig. 1, the majority of the con-

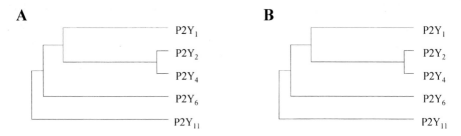

Fig. 3. Sequence relatedness of the human P2Y receptor amino acid sequences derived from the predicted transmembrane regions (**A**) and the extracellular domains (**B**)

Table 3. Lengths of extracellular and intracellular domains of the human P2Y receptors

Domain[a]	$P2Y_1$	$P2Y_2$	$P2Y_4$	$P2Y_6$	$P2Y_{11}$
Amino terminal	53	32	35	26	26
First intracellular loop	9	12	12	13	13
First extracellular loop	17	13	13	17	17
Second intracellular loop	19	23	19	23	20
Second extracellular loop	23	23	24	23	36
Third intracellular loop	20	23	22	23	17
Third extracellular loop	23	12	14	13	42
Carboxy terminal	47	70	55	35	44

[a] Domains were predicted using TMHMM V0.1 (SONNHAMMER et al. 1998).

served residues are located within the predicted transmembrane domains with just 9 residues located outside these domains. A subset of the 44 totally conserved residues in the h-P2Y receptors are present in either all group V receptors or a significant proportion of them. Using amino acid numbering from the human $P2Y_1$ receptor: N^{69} in TM1; L^{93}, D^{97} in TM2; C^{124} in the first extracellular loop, R^{149} in the second intracellular loop, W^{176} in TM4; P^{275} and P^{321} in TM6 and TM7 respectively are conserved in all of the group V receptors. Residues conserved in the majority of the group V sequences are A^{94} in TM2; W^{117} in the first extracellular loop; N^{135}, S^{139} in TM3; P^{156}, P^{195} and P^{229} in the second intracellular loop, TM4 and TM5 respectively and Y^{237} in TM5. The importance of some of these residues to the functional properties of these receptors is discussed in the next section.

II. Structure and Function

1. Agonist/Receptor Interactions

As nucleotides are small molecules their receptors might be expected to have the agonist binding site within the transmembrane domains as is the case for the receptors for biogenic amines (STRADER et al. 1995). However, as the P2Y receptors are more related in sequence to peptide receptors their extracellular regions might be expected to contribute to the binding process as well (STRADER et al. 1995). Both of these hypotheses have been explored for the P2Y receptors with the delineation of the specific residues involved in agonist binding concentrated on the $P2Y_1$ and $P2Y_2$ subtypes. This will be described below and extrapolated to the other members of this receptor family where possible.

A number of studies have indicated that the fully ionised form of nucleotides are agonists of the P2Y receptors (MOTTE et al. 1993; SUH et al. 1997). Using this rationale, ERB et al. (1995) used site directed mutagenesis to change positive amino acids within the transmembrane domains of the m-$P2Y_2$ receptor to probe for regions of agonist-receptor interaction. Four positively charged residues, H^{262} and R^{265} in TM 6 and K^{289} and R^{292} in TM 7 (m-$P2Y_2$ notation) were deduced to be important for the agonist potency and specificity at the $P2Y_2$ receptor. The mutations H262L, R265L and R292L all caused a decrease in the potency but not efficacy of ATP and UTP, while the mutant K289I did not differ significantly from the wildtype with regard to the response to ATP and UTP. However, the mutation K289R led to an increase in the activity of ADP and UDP at this receptor (ERB et al. 1995). The analogous residues of the human $P2Y_1$ receptor have also been mutated, along with S^{314} and S^{317} in TM7 (JIANG et al. 1997). 2MeSATP had reduced potency at the two TM6 mutants H277A and K280A, of 17- and 950-fold respectively. The potency was also reduced at the TM7 Q307A mutant (207-fold). The mutations R310A and R310S both abolished agonist activity while mutation to lysine, present in the $P2Y_2$ receptor led to a decrease in potency (680-fold). S^{314}, but not S^{317}, was also found to be of importance to the binding site as were residues in TM3 and TM5 (JIANG et al. 1997).

Of the four residues mutated in both receptors, the TM6 histidine and TM7 arginine are totally conserved in all five human receptors (Fig. 1), and indeed all P2Y receptor sequences. The former residue is highly conserved amongst G protein-coupled receptors in general and is thought to be important for the general architecture of the receptors, supported by the reduction but not abolition of activity of agonists at both the $P2Y_1$ and P2Y receptors. The latter residue is predicted to interact with the phosphate moiety of nucleotides, and the reduced potency of agonists at both receptors may indicate that this residue plays the same role in all P2Y receptors. The TM6 arginine is conserved in the $h-P2Y_2$, $P2Y_4$ and $P2Y_{11}$ receptors (i.e. those with a preference for triphosphate nucleotides), but replaced by lysine in the $P2Y_1$ and $P2Y_6$ receptors (dinucleotide preferring) and again this trend is followed for all 18 P2Y receptor sequences. Modelling predicts that this residue interacts directly with the phosphate moiety (ERB et al. 1995; MORO et al. 1999). The TM7 lysine is conserved in the $P2Y_2$, $P2Y_4$ and $P2Y_6$ receptors (pyrimidine tolerating) and is thought to be close to the binding pocket of the $P2Y_2$ receptor (ERB et al. 1995). In the $P2Y_1$ and $P2Y_{11}$ receptors (adenine specific) this residue is replaced with a neutral glutamine and has been predicted to interact with the adenine group in the $P2Y_1$ receptor (MORO et al. 1999).

A more recent study has concentrated on the possible contribution of the extracellular regions of the $P2Y_1$ receptor to agonist binding and a refinement of the $P2Y_1$ binding site model (HOFFMANN et al. 1999; MORO et al. 1999). Single point mutations of all charged residues in the extracellular domains, and along with cysteine residues, to alanine were performed and revealed that the $P2Y_1$ receptor has two disulphide bridges and that a number of residues in the extracellular regions are also involved in agonist recognition. Mutagenesis studies of other receptors of group V suggest that the presence of a second disulphide bridge may be present in a number of receptors of this class (OHYAMA et al. 1995), the presence of which, between the amino-terminal region and third extracellular loop, would place further steric constraints on the receptor structure. Three of these four cysteine residues are totally conserved in the five human receptors while the amino terminal cysteine is conserved in position in the $P2Y_1$, $P2Y_2$, $P2Y_4$ and $P2Y_6$ receptors but is closer to the amino-terminus methionine of the $P2Y_{11}$ receptor. A model for the $P2Y_1$ receptor has been proposed, based on this mutagenesis data, that places the first and second extracellular loops at the surface of the receptor and in proximity to TM3, TM5, TM6 and TM7 previously predicted to form the binding pocket of this receptor (MORO et al. 1998, 1999).

It is perhaps not unreasonable to assume that the extracellular regions of other members of this receptor family will also contribute to the agonist binding process. Thus the difference in agonist selectivity of the $h-P2Y_4$ and $r-P2Y_4$ receptors might be contributed to by one or more of the five residues in the extracellular region of the $r-P2Y_4$ receptor which are also present in the $r-P2Y_2$ and $h-P2Y_2$ receptors but which are not conserved in the $h-P2Y_4$ receptor. However, the six non-conservative substitutions within the transmembrane domains may also contribute to the inactivity of ATP at the human receptor, in particular the species substitutions P209C and Q266L (rat to human), which are also conserved as phenylalanine and glutamine in the $P2Y_2$ receptor from both species.

2. Events Downstream of Receptor Activation

Nucleotide binding to a P2Y receptor must alter the conformation of the receptor allowing/altering the interaction of the receptor with a G protein and initiating downstream events. The exact nature of this conformation change is unknown, as are the regions of the P2Y receptors that interact with G proteins. Obviously, interactions must involve regions of the intracellular loops and/or the cytoplasmic tail; however for the P2Y receptor family these regions await definition. By analogy with other receptors, regions in the second and third cytoplasmic loops and C-terminal tail are likely to play a major role (STRADER et al. 1995; GUDERMAN et al. 1996).

III. Post-Translational Modifications

1. Asparagine-Linked Glycosylation

Of the human receptors, $P2Y_4$ is the only subtype devoid of potential asparagine-linked glycosylation sites. The $P2Y_1$ receptor has two sites in the amino-terminal region with single sites present in the first and second extracellular loops. The $P2Y_2$ receptor has two potential sites in the amino-terminal region while the $P2Y_6$ receptor has one site each in the amino-terminal region and second extracellular loop. Finally the $P2Y_{11}$ receptor has a single potential site for glycosylation in the third extracellular loop.

2. Phosphorylation Sites

All of the human P2Y receptors contain consensus sequences for phosphorylation by serine/threonine protein kinases (Table 4). These sites are concentrated in the third intracellular loop and the cytoplasmic tail. It is highly likely that P2Y receptors are also substrates for G protein coupled receptor kinases. This will be a major area for future research. The significance of phosphorylations is explored in Sects D and E.

IV. Non-G Protein Associations

1. Integrins

The mouse and human $P2Y_2$ receptors possess the tri-peptide sequence R-G-D in the first extracellular loop. This sequence is predicted to act as an integrin binding domain and has been shown to facilitate the binding of ab_3 and ab_5 integrins to this receptor subtype (ERB et al. 1997; WEISMAN et al. 1998). The same motif is also present in the rat $P2Y_6$ receptor at the same relative position and might therefore be expected to act in an analogous manner.

2. Na⁺/H⁺ Exchanger Regulatory Factor

The $P2Y_1$ has a species conserved motif D-T-S-L at the end of its cytoplasmic tail. This sequence has been predicted to be the optimal sequence for binding

Table 4. Potential phosphorylation sites of intracelluar domains of the human P2Y receptors

Domain	Site	P2Y$_1$	P2Y$_2$	P2Y$_4$	P2Y$_6$	P2Y$_{11}$
i1	CaMII[a]		Thr^{64}		Thr^{59}	
	CKI[b]	Ser^{87}	Ser^{68}		Thr^{53}	
	CKII[c]					
	PKA[d]		Thr^{64}		Thr^{59}	Ser^{51}
	PKG[e]					
	PKC[f]				Thr^{53}, Ser^{54}	Ser^{51}
i2	CaMII[a]					
	CKI[b]	Ser^{146}	Ser^{128}, Ser^{246}	Ser^{130}	Ser^{121}	Ser^{121}
	CKII[c]					
	PKA[d]	Ser^{151}				
	PKG[e]					
	PKC[f]		Ser^{141}			
i3	CaMII[a]	Ser^{258}	Ser^{246}	Ser^{243}		
	CKI[b]	Ser^{258}	Ser^{243}	Ser^{238}, Ser^{243}		Thr^{236}
	CKII[c]					Thr^{236}
	PKA[d]	Ser^{258}	Ser^{243}	Ser^{243}		
	PKG[e]	Ser^{258}	Ser^{243}			
	PKC[f]		Ser^{243}	Ser^{238}, Ser^{243}		
	CKI[b]	Thr^{339}, Ser^{346}, Thr^{358}	Ser^{327}, Thr^{344}, Thr^{365}, Thr^{372}	Ser^{334}		
				Ser^{355}, Ser^{359}		Ser^{366}
	CKII[c]	Ser^{352}	Ser^{355}	Ser^{339}, Thr^{351}		
	PKA[d]	Ser^{336}, Ser^{339}, Ser^{343}, Ser^{346}	Ser^{341}, Thr^{262}			Ser^{338}
	PKG[e]	Ser^{336}, Ser^{343}				
	PKC[f]	Thr^{330}, Thr^{339}, Ser^{343}	Ser^{341}	Ser^{345}, Thr^{359}	Thr^{304}, Thr^{320}	

Receptor amino acid sequences were screened for potential phosphorylation sites using PhosphoBase V2.0 (KREEGIPUU et al. 1999).
[a] CaMII, calcium calmodulin kinase II. Consensus sequence R-X-X-S/T-X (KENNELLY and KREBS 1991; KEMP and PEARSON 1990).
[b] CKI, caesin kinase I. Consensus sequence S/T-X$_{2-3}$-S/T-X, note that the N-terminal S/T must be pre-phosphorylated (PINNA and RUZZENE 1996).
[c] CKII, caesin kinase II. Consensus sequence X-S/T-X-X-D/E (HOFFMANN et al. 1999).
[d] PKA, cAMP-dependant protein kinase. Consensus sequence R-X$_{1-2}$-S/T-X (KENNELLY and KREBS 1991).
[e] PKG, cGMP-dependant protein kinase. Consensus sequence R/K$_{2-3}$-X-S/T-X (KENNELLY and KREBS 1991).
[f] PKC, calcium-dependant protein kinase. Consensus sequence X-S/T-X-R/K (HOFFMANN et al. 1999).

of the first PDZ domain of the Na$^+$/H$^+$ exchanger regulatory factors NHERF and NHERF-2 and binding between the P2Y$_1$ receptor and NHERF has been demonstrated in vitro (HALL et al. 1998a). Thus the P2Y$_1$ receptor is potentially able to modulate Na$^+$/H$^+$ exchange in a manner independent of its interaction with G proteins as in the case of the β$_2$-adrenergic receptor (HALL et al. 1998b). The P2Y$_2$ receptor also has a species conserved motif at the end of its cytoplasmic tail (D-I-R-L) which, although similar to the optimal C-terminal motif for binding to the first PDZ domain of NHERF (D-S/T-X-L), has been shown not to facilitate a significant level of binding (HALL et al. 1998a). It is interesting to note that in the β_2-adrenergic receptor serine at position 2 in this motif is a site for phosphorylation by G protein receptor kinase 5 (FREDRICKS et al. 1996).

D. Desensitisation

Agonists have been shown to induce the desensitisation (WILKINSON et al. 1994) and internalisation of the $P2Y_2$ receptor (SROMEK and HARDEN 1998; GARRAD et al. 1998). However, these two events are not strictly associated as desensitisation still occurs when internalisation is prevented (SROMEK and HARDEN 1998). The pattern of desensitisation seen is dependent on the experimental protocol used, and in the case of native co-existing $P2Y_1$ and $P2Y_2$ receptors has been shown to involve both homologous and heterologous desensitisation, and both protein kinase C dependent and independent mechanisms. The carboxy-terminal tail of the $P2Y_2$ receptor has been shown to be involved in both desensitisation and internalisation of the receptor. Truncations of this region had no effect on the agonist activated calcium response but increased the IC_{50} for agonist-mediated desensitisation and internalisation (GARRAD et al. 1998). It is anticipated that phosphorylations are involved in desensitisation, both by protein kinase C and other kinases. The consensus phosphorylation sites present in the intracellular domains of the P2Y receptors for some of these kinases are shown in Table 4. Some common themes emerge. For example, $P2Y_{11}$ has no sites for phosphorylation by Ca^{2+}/calmodulin kinase II, and is the only h-P2Y sequence with no phosphorylation sites for protein kinase C in the C-terminal tail. $P2Y_2$ has a consensus site for protein kinase C in the third intracellular loop while $P2Y_1$ does not, correlating with the observation that $P2Y_2$ and not $P2Y_1$ receptors are subject to protein kinase C-mediated feedback inhibition (WILKINSON et al. 1993). Also of particular interest, evidence has been presented for a role for casein kinase I phosphorylation of the third intracellular loop in desensitisation of muscarinic receptors (TOBIN et al. 1997); unlike the other h-P2Y receptors, $P2Y_6$ has no consensus sites for casein kinase I phosphorylation in the third intracellular loop. In fact $P2Y_6$ stands out in that it has no sites in this loop for phosphorylation by any of the kinases in Table 4 and has been reported to undergo slow desensitisation (ROBAYA et al. 1997). Other kinases are likely to be involved in the phosphorylation and desensitisation of P2Y receptors, notably the G protein-receptor kinases, a family of kinases known to phosphorylate, and downregulate, a number of G protein-coupled receptors.

E. Cell Signalling from P2Y Receptors

I. Coupling to Phospholipase C

The five cloned human P2Y receptors have all been shown to couple to phospholipase C, activating the inositol (1,4,5)trisphosphate – Ca^{2+} mobilisation cascade (PALMER et al. 1998; PARR et al. 1994; COMMUNI et al. 1995, 1997; NGUYEN et al. 1995; NICHOLAS et al. 1996), with the potential downstream recruitment of various protein kinases. Some selectivity between P2Y receptor subtypes in coupling to phospholipase C has been reported, both with respect to protein kinase C feedback (see above) and pertussis toxin sensitivity (BOEYNAEMS et al. 1998). The latter indicates the likelihood that the P2Y receptors are not coupled to phospholipase C in a pertussis toxin sensitive manner, with the exception of $P2Y_2$, which often displays partial sensitivity. This leads to the conclusion that P2Y receptors are likely to be coupled to phopholipase C via G_q, with the exception of $P2Y_2$, which is at least in part coupled via $G_{i/o}$.

II. Phospholipase C Independent Ca^{2+} Mobilisation

Observations of responses to nucleotides in cultured brain capillary endothelial cells has shown that these cells respond to UTP with a conventional phospholipase C/Ca^{2+} response, but that the same cells respond to 2-methylthioATP or ADP with a Ca^{2+} mobilisation which is in the absence of detectable inositol phosphate formation, including formation of inositol 1,4,5-trisphosphate measured within a few seconds of stimulation (FRELIN et al. 1993; ALBERT et al. 1997). Two alternative explanations are that the Ca^{2+} response is dependent on levels of inositol 1,4,5-trisphosphate which do not rise in a detectable manner, or that the response is truly independent of inositol phosphates and is mediated by an alternative mechanism, such as the formation of sphingosine phosphate (MEYER ZU HERINGDORF et al. 1998). This response can be assumed to generate a mobilisation of Ca^{2+} in the absence of diacylglycerol formation and therefore of protein kinase C activation. This is of interest with respect to downstream signalling pathways.

III. Regulation of Cyclic AMP

The various signalling repertoires which can lead to the regulation of cyclic AMP levels by P2 receptors have been recently discussed (BOARDER 1998). Essentially these comprise either direct regulation of adenylate cyclase via G_s or G_i, and indirect regulation of adenylate cyclase or phosphodiesterases by events downstream of phospholipase C activation. When considering the fidelity of responses to a given P2Y subtype it is worth noting that these indirect responses are likely to vary considerably with respect to the cell in which the receptors are located. At the cloned $P2Y_1$ receptor, expressed in 1321N1 cells, small or non-existent effects have been reported (FILTZ et al. 1994;

SCHACHTER et al. 1996), $P2Y_4$ receptors in 1321N1 cells and $P2Y_6$ receptors in C6 glioma cells (NGUYEN et al. 1995; CHANG et al. 1995) are not coupled to changes in cyclic AMP levels, and only $P2Y_{11}$ receptors show a clear stimulation of cyclic AMP, interpreted as due to direct Gs coupling to adenylyl cyclase (COMMUNI et al. 1997). Native P2Y receptors have been shown variously to be coupled to either elevation (e.g. JOHNSON et al. 1991; COTE et al. 1993; MATSUOKA et al. 1995; POST et al. 1996; ALBERT et al. 1997) or reduction (OKAJIMA et al. 1987; CRYSTALLI and MILLS 1993; SCHACHTER et al. 1996) of cyclic AMP levels. CONIGRAVE et al. (1998) have described a P2Y receptor coupled to increases in cyclic AMP in HL-60 leukaemia cells which is likely to be $P2Y_{11}$. ANWAR et al. (1999) report a forskolin-dependent cyclic AMP stimulation in response to nucleotides in brain capillary endothelial cells which is probably due to an augmentation of a G_s-coupled response by forsklolin, similar to that described by COTE et al. (1993).

IV. Regulation of Ion Channels

As with control of cyclic AMP, the regulation of ion channels by P2Y receptors is either direct, via G proteins, or indirect downstream of another effector cascade. In many studies it is unclear which is the case, and it is often not clear whether the effect is of ATP or of a metabolic product. For example, STROEBAEK et al. (1996) describe activation by ATP of K^+ and Cl^- channels downstream of Ca^{2+} in human coronary artery vascular smooth muscle cells. G protein involvement (G_i) for the P2Y activation of the muscarinic K^+ channel of guinea-pig myocardial cells has been described (MATSUURA et al. 1996). UTP has also been shown to have an effect on myocardial ion channels (MURAKI et al. 1998). MOSBACHER et al. (1998) pursued uridine nucleotide responses with a different approach: when expressed in oocytes $P2Y_2$ and $P2Y_6$ receptors both coupled to Ca^{2+} dependent Cl^- channels, but only $P2Y_2$ receptors coupled to inwardly rectifying K^+ channels. This K^+ channel link was by a pertussis toxin sensitive G protein, while the Cl^- channel effect was not. In addition to indicating differential coupling of transfected P2Y receptors to ion channels, they also show that the $P2Y_2$ receptor may couple to two different responses via different G proteins. This is also seen in another series of experiments in this area describing the heterologous expression of $P2Y_2$ or $P2Y_6$ receptors, this time in rat sympathetic neurons. This showed that these receptors are each coupled to both N-type Ca^{2+} and M-type K^+ channels via PTX-sensitive and insensitive G proteins respectively (FILIPPOV et al. 1998, 1999).

V. Stimulation of Tyrosine Kinases and Mitogen Activated Protein Kinases

Tyrosine kinase/mitogen activated protein kinase (MAPK) pathways were originally studied as responses to growth factors with respect to regulation of cell proliferation, but are now known to be widely activated by G protein-coupled receptors and to control diverse short term and long term cellular responses (BOARDER 1998).

There are three MAPK cascades: the only one widely studied as a response to P2Y receptor activation is the ERK (extracellular signal related kinase) p42/p44 MAPK cascade. ISHIKAWA et al. (1994) and HUWILER and PFEILSCHIFTER (1994), working on renal mesangial cells, provided the first report of activation of MAPK by P2Y receptor stimulation, followed by the demonstration by NEARY et al. (1995) that co-existing $P2Y_1$ and $P2Y_2$ receptors on astrocytes are coupled to MAPK and proliferation. At the same time BOWDEN et al. (1995) showed that $P2Y_1$ and $P2Y_2$ receptors on endothelial cells were both coupled to tyrosine phosphorylation, and furthermore that this was required for a short term response previously ascribed to raised Ca^2, the production of prostacyclin. Following from this, PATEL et al. (1996) showed that p42/44 MAPK is phosphorylated and activated on stimulation of these two endothelial receptors, and that it is this which is required for activation of phospholipase A_2 and subsequent prostacyclin production. P2Y receptors are also coupled to p42/44 MAPK activation in both rat and human vascular smooth muscle cells (HARPER et al. 1998; WHITE et al. 1999). Here, the functional response of the cells is stimulation of DNA synthesis, but in the human cells there was no clear correlation between DNA synthesis and p42/p44 MAPK activation. This was also seen in 1321N1 cells transfected with different P2Y subtypes: $P2Y_1$ and $P2Y_2$ stimulation gave substantial phosphorylation and activation of p42/44 MAPK, while $P2Y_4$ and $P2Y_6$ were not coupled to MAPK, and yet it was only in response to activation of $P2Y_1$ receptors that a mitogenic response was seen (ROBERTS et al. 1999). These studies provide examples of the significance of the study of P2Y responses, since they relate to the role of P2Y receptors in human vascular proliferative disease.

Note added in proof: While this book was in press, the cloning of the highly elusive $P_{2T}/P2Y_T$ platelet ADP receptor as the $P2Y_{12}$ receptor was reported (HOLLOPETER et al. 2000).

References

Abbracchio MP, Burnstock G (1994) Purinoceptors – are there families of P2X and P2Y purinoceptors? Pharmacol Therap 64:445–475

Albert JL, Boyle JP, Roberts JA, Challiss RAJ, Gubby SE, Boarder MR (1997) Regulation of brain capillary endothelial cells by P2Y receptors coupled to Ca^{2+}, phospholipase C and mitogen-activated protein kinase. Br J Pharmacol 122:935–941

Anwar Z, Albert JL, Gubby SE, Boyle JP, Roberts JA, Webb TE, Boarder MR (1999) Regulation of cyclic AMP by extracellular ATP in cultured brain capillary endothelial cells. Br J Pharmacol 128:465–471

Ayyanathan K, Webb TE, Sandhu AK, Athwal RS, Barnard EA, Kunapuli SP (1996) Cloning and chromosomal localization of the human $P2Y_1$ purinoceptor. Biochem Biophys Res Commun 218:783–788

Boarder MR (1998) Cyclic AMP and tyrosine kinase cascades in the regulation of cellular function by P2Y nucleotide receptors. In: Turner JT, Weisman GA, Fedan JS (eds) The P2 nucoleotide receptors. Humana Press, Totowa NJ, pp 185–209

Boarder MR, Hourani SMO (1998) The regulation of vascular function by P_2 receptors: multiple sites and multiple receptors. Trends Pharmacol Sci 19:99–107

Boeynaems JM, Communi D, Janssens R, Motte S, Robaye B, Pirotton S (1998) Nucleotide receptors coupling to the phospholipase C signaling pathway. In: Turner JT, Weisman GA, Fedan JS (eds) The P2 nucoleotide receptors. Humana Press, Totowa NJ, pp 169–183

Bogdanov YD, Dale L, King BF, Whittock N, Burnstock G (1997) Early expression of a novel nucleotide receptor in the neural plate of Xenopus embryos. J Biol Chem 272:12583–12590

Bogdanov YD, Wildman SS, Clements MP, King BF, Burnstock G (1998) Molecular cloning and characterisation of rat $P2Y_4$ nucleotide receptor. Br J Pharmacol 124:428–430

Bowden A, Patel V, Brown C, Boarder MR (1995) Evidence for requirement of tyrosine phosphorylation in endothelial P_{2Y} purinoceptor and P_{2U} purinoceptor stimulation of prostacyclin release. Br J Pharmacol 116:2563–2568

Boyer JL, Zohn IE, Jacobson KA, Harden TK (1994) Differential-effects of P_2 purinoceptor antagonists on phospholipase C-coupled and adenylyl cyclase-coupled P2Y- purinoceptors. Br J Pharm 113:614–620

Boyer JL, O'Tuel JW, Fischer B, Jacobsen KA, Harden TK (1995) 2-Thioether derivatives of adenosine nucleotides are exceptionally potent agonists at adenylyl cyclase-linked P_{2Y}-purinergic receptors. Br J Pharmacol 116:2611–2616

Boyer JL, Romero-Avila T, Schachter JB, Harden TK (1996) Identification of competitive antagonists of the $P2Y_1$ receptor. Mol Pharmacol 50:1323–1329

Boyer JL, Waldo GL, Harden TK (1997) Molecular cloning and expression of an avian G protein-coupled p2y receptor. Mol Pharmacol 52:928–934

Boyer JL, Mohanram A, Camaioni E, Jacobsen KA, Harden TK (1998) Competitive and selective antagonism of $P2Y_1$ receptors by N^6-methyl 2′-deoxyadenosine 3′,5′-bisphosphate. Br J Pharmacol 124:1–3

Brown C, Tanna B, Boarder MR (1995) PPADS – an antagonist at endothelial P_{2Y}-purinoceptors but not P_{2U}-purinoceptors. Br J Pharmacol 116:2413–2416

Bultmann R, Trendelenburg M, Tuluc F, Wittenburg H, Starke K (1999) Concomitant blockade of P2X-receptors and ecto-nucleotidases by P_2-receptor antagonists: functional consequences in rat vas deferens. Naunyn-Schmiedebergs Archiv Pharmacol 359:339–344

Burnstock G, Kennedy C (1985) Is there a basis for distinguishing 2 types of P2-purinoceptor. Gen Pharmacol 16:433–440

Chang KG, Hanoka K, Kumada M, Takuwa Y (1995) Molecular cloning and functional analysis of a novel P_2 nucleotide receptor. J Biol Chem 270:26152–26158

Charlton SJ, Brown CA, Weisman GA, Turner JT, Erb L, Boarder MR (1996a) PPADS and suramin as antagonists at cloned P_{2Y}- and P_{2U}-purinoceptors. Br J Pharmacol 118:704–710

Charlton SJ, Brown CA, Weisman GA, Turner JT, Erb L, Boarder MR (1996b) Cloned and transfected $P2Y_4$ receptors: characterisation of a suramin and PPADS-insensitive response to UTP, Br J Pharmacol 119:1301–1303

Chen ZP, Krull N, Xu S, Levy A, Lightman SL (1996) Molecular cloning and functional characterisation of a rat pituitary G protein-coupled adenosine triphosphate (ATP) receptor. Endocrinology 137:1833–1840

Communi D, Pirotton S, Parmentier M, Boeynaems JM. (1995) Cloning and functional expression of a human uridine nucleotide receptor. J Biol Chem 270:30849–30852

Communi D, Govaerts C, Parmentier M, Boeynaems JM (1997) Cloning of a human purinergic P_{2Y} receptor coupled to phospholipase C and adenylyl cyclase. J Biol Chem 272:31969–31973

Conigrave AD, Lee JY, Van der Leyden L, Jiang L, Ward P, Tasevski V, Luttrell BM, Morris MB (1998) Pharmacological profile of a novel cyclic AMP-linked P2 receptor on undifferentiated HL-60 leukaemia cells. Br J Pharmacol 124:1580–1585

Cote S, van Sande J, Boeynaems JM (1993) Enhancement of endothelial cAMP accumulation by adenine nucleotides: role of methylxanthine-sensitive sites. Am J Physiol 264:H1498–H1503

Crystalli G, Mills CB (1993) Identification of a receptor for ADP in blood platelets by photoaffinity labelling. Biochem J 291:875–881

Dainty IA, Pollard CE, Roberts SM, Franklin M, McKechnie KCW, Leff P (1994) Evidence for subdivisions of P_{2U}-receptors based on suramin sensitivity. Br J Pharmacol 112:578P

Deng GM, Matute C, Kumar CK, Fogarty DJ, Miledi R (1998) Cloning and expression of a P_{2Y} purinoceptor from the adult bovine corpus callosum. Neurobiol of Disease 5:259–270

Dixon CJ, Woods NM, Cuthbertson KSR, Cobbold PH (1990) Evidence for 2 Ca^{2+}-mobilizing purinoceptors on rat hepatocytes. Biochem J 269:499–502

Dixon CJ, Cobbold PH, Green AK (1993) Adenosine 5'-[α-β-methylene]triphosphate potentiates the oscillatory cytosolic Ca^{2+} responses of hepatocytes to ATP, but not to ADP. Biochem J 293:757–760

Dixon CJ, Cobbold PH, Green AK (1995) Actions of ADP, but not ATP, on cytosolic free Ca^{2+} in single rat hepatocytes mimicked by 2-methylthioATP. Br J Pharmacol 116:1979–1984

Dunn PM, Blakeley AGH (1988) Suramin – a reversible P_2-purinoceptor antagonist in the mouse vas deferens. Br J Pharmacol 93:243–245

Enomoto K, Furuya K, Moore RC, Yamagishi S, Oka T, Maeno T (1996) Expression cloning and signal transduction pathway of P2U receptor in mammary tumor cells. Biol Signals 5:9–21

Erb L, Garrad R, Wang YJ, Quinn T, Turner JT, Weisman GA (1995) Site-directed mutagenesis of P_{2U} purinoceptors – positively charged amino-acids in transmembrane helix-6 and helix-7 affect agonist potency and specificity. J Biol Chem 270:4185–4188

Erb L, Ockerhausen J, Garrad R, Gresham H, Weisman GA, Turner JT (1997) Specific binding between the a33 integrin and an RGD domain of the $P2Y_2$ nucleotide receptor. FASEB J 11:3357

Erlinge D, Hou M, Webb TE, Barnard EA, Moller S (1998) Phenotype changes of the vascular smooth muscle cell regulate P2 receptor expression as measured by quantitative RT-PCR. Biochem Biophys Res Commun 248:864–870

Filippov AK, Webb TE, Barnard EA, Brown DA (1998) $P2Y_2$ nucleotide receptors expressed heterologously in sympathetic neurons inhibit both N-type Ca^{2+} and M-type K^+ currents. J Neurosci 18:5170–5179

Filippov AK, Webb TE, Barnard EA, Brown DA (1999) Dual coupling of heterologously-expressed rat $P2Y_6$ nucleotide receptors to N-type Ca^{2+} and M-type K^+ currents in rat sympathetic neurones. Br J Pharmacol 126:1009–1017

Filtz TM, Li Q, Boyer JL, Nicholas RA, Harden TK (1994) Expression of a cloned p2y purinergic receptor that couples to phospholipase C. Mol Pharmacol 46:8–14

Fischer B, Boyer JL, Hoyle CHV, Ziganshin AU, Bizzolara AL, Knight GE, Zimmet J, Burnstock G, Harden TK, Jacobsen KA (1993) Identification of potent, selective P_{2Y}-purinoceptor agonists: structure activity relationships for 2-thioether derivatives of adenosine-5'-triphosphate. J Med Chem 36:3937–3946

Fredericks ZL, Pitcher JA, Lefkowitz RJ (1996) Identification of the G protein-coupled receptor kinase phosphorylation sites in the human b_2-adrenergic receptor. J Biol Chem 271:13796–13803

Frelin C, Breittmayer JP, Vigne P (1993) ADP induces inositol phosphate independent intracellular Ca^{2+} mobilisation in brain capillary endothelial cells. J Biol Chem 268:8787–8792

Fredholm BB, Abbrachio MP, Burnstock G, Dubyak GR, Harden TK, Jacobson KA, Schwabe U, Williams M (1997) Towards a revised nomenclature for P1 and P2 receptors. Trends Pharmacol Sci 18:79–82

Garrad RC, Otero MA, Erb L, Theiss PM, Clarke LL, Gonzalez FA, Turner JT, Weisman GA (1998) Structural basis of agonist-induced desensitisation and sequestration of the $P2Y_2$ nucleotide receptor – consequences of truncation of the C terminus. J Biol Chem 273:29437–29444

Gordon JL (1986) Extracellular ATP: effects, sources and fate. Biochem J 233:309–319

Guderman T, Kalkbrenner F, Schultz G (1996) Diversity and selectivity of receptor-G protein interaction. Annu Rev Pharmacol 36:429–459

Hall RA, Ostedgaard LS, Premont RT, Blitzer JT, Rahman N, Welsh MJ, Lefkowitz RJ (1998a) A C-terminal motif found in the b_2-adrenergic receptor, $P2Y_1$ receptor and cystic fibrosis transmembrane conductance regulator determines binding to the Na^+/H^+ exchanger regulatory factor family of PDZ proteins. Proc Natl Acad Sci USA 95:8496–8501

Hall RA, Premont RT, Chow CW, Blitzer JT, Pitcher JA, Claing A, Stoffel RH, Barak LS, Shenolikar S, Weinman EJ, Grinstein S, Lefkowitz RJ (1998b) The b_2-adrenergic receptor interacts with the Na^+/H^+-exchanger regulatory factor to control Na^+/H^+ exchange. Nature 392:626–630

Harden TK, Lazarowski ER, Boucher RC (1997) Release, metabolism and interconversion of adenine and uridine nulceotides: implications for G protein-coupled P2 receptor activation. Trends Pharmacol Sci 18:43–46

Harper S, Webb TE, Charlton SJ, Ng LL, Boarder MR (1998) Evidence that $P2Y_4$ nucleotide receptors are involved in the regulation of rat aortic smooth muscle cells by UTP and ATP. Br J Pharmacol 124:703–710

Hechler B, Vigne P, Leon C, Breittmayer JP, Gachet C, Frelin C (1998) ATP derivatives are antagonists of the $P2Y_1$ receptor: similarities to the platelet ADP receptor. Mol Pharmacol 53:727–733

Henderson DJ, Elliot DG, Smith GM, Webb TE, Dainty IA. Cloning and characterisation of a bovine P2Y receptor. Biochem Biophys Res Commun 212:648–656

Hoffman C, Moro S, Jacobson KA (1999) J Biol Chem (in press)

Hoffmann K, Bucher P, Falquet L, Bairoch A (1999) The PROSITE database, its status in 1999. Nucleic Acids Res 27:215–219

Hollopeter G, Jantzen HM, Vincent D, Li G, England L, Ramakrishnan V, Yang RB, Nurden P, Nurden A, Julius D, Conley PB (2000) Identification of the Platelet ADP Receptor Targeted by Antithrombotic Drugs. Nature, in press

Hourani SMO, Chown JA (1989) The effects of some possible inhibitors of ectonucleotidases on the breakdown and pharmacological effects of ATP in the guinea pig urinary bladder. Gen Pharmacol 20:413–416

Hoyle CHV, Knight GE, Burnstock G (1990) Suramin antagonises responses to P_2-purinoceptor agonists and purinergic nerve stimulation in the guinea pig urinary bladder and taenia coli. Br J Pharmacol 99:617–621

Huwiler A, Pfeilschifter J (1994) Stimulation by extracellular ATP and UTP of the mitogen-activated protein-kinase cascade and proliferation of rat renal mesangial cells. Br J Pharmacol 113:1455–1463

Ishikawa S, Kawasumi M, Kusaka I, Komatsu N, Iwao N, Saito T (1994) Extracellular ATP promotes cellular growth of glomerular mesangial cells mediated via phospholipase C. Biochem Biophys Res Commun 202:234–240

Janssens R, Communi D, Pirotton S, Samson M, Parmentier M, Boeynaems JM (1995) Cloning and tissue distribution of the human $P2Y_1$ receptor. Biochem Biophys Res Commun 221:588–593

Jantzen HM, Gousset L, Bhaskar V, Vincent D, Tai A, Reynolds EE, Conley PB (1999) Evidence for two distinct G protein-coupled ADP receptors mediating platelet activation. Thromb Haemost 81:111–117

Jiang Q, Guo D, Lee BX, Van Rhee AM, Kim YC, Nicholas RA, Schachter JB, Harden TK, Jacobson KA (1997) A mutational analysis of residues essential for ligand recognition at the human $P2Y_1$ receptor. Mol Pharmacol 52:499–507

Johnson JA, Friedman J, Halligan RD, Birnbaumer M, Clark RB (1991) Sensitisation of adenylyl cyclase by P2 purinergic and M5 muscarinic receptor agonists in L cells. Mol Pharmacol 39:539–546

Kennedy C, Leff P (1995) How should P2X receptors be classified pharmacologically? Trends Pharmacol. Sci. 16:168–174

Kim YC, Camaioni E, Ziganshin AU, Ji XD, King BF, Wildman SS, Rychkov A, Yoburn J, Kim H, Mohanram A, Harden TK, Boyer JL, Burnstock G (1998) Synthesis and structure-activity relationships of pyridoxal-6-arylazo-5′-phosphate and phosphonate derivatives as P2 receptor antagonists. Drug Dev Res 45:52–66

King BF, Townsend-Nicholson A, Burnstock G (1998) Metabotropic receptors for ATP and UTP: exploring the correspondence between native and recombinant nucleotide receptors. Trends Pharmacol Sci 19:506–514

Kreegipuu A, Blom N, Brunak S (1999) PhosphoBase, a database of phosphorylation sites: release 2.0. Nucleic Acids Res 27:237–239

Lambrecht G, Friebe T, Grimm U, Windscheif U, Bungardt E, Hildebrandt C, Baumert HG, Spatzkumbel G, Mutschler E (1992) PPADS, a novel functionally selective antagonist of P2 purinoceptor-mediated responses. Eur J Pharmacol 217:217–219

Leff P, Wood BE, O'Connor SE (1990) Suramin is a slowly-equilibrating but competitive antagonist at P2X receptors in the rabbit isolated ear artery. Br J Pharm 101:645–649

Leon C, Vial C, Cazenave JP, Gachet C (1996) Cloning and sequencing of a human cDNA encoding endothelial $P2Y_1$ purinoceptor. Gene 171:295–297

Leon C, Hechler B, Vial C, Leray C, Cazenave JP, Gachet C (1997) The $P2Y_1$ receptor is an ADP receptor antagonized by ATP and expressed in platelets and megakaryoblastic cells. FEBS Letts 403:26–30

Li Q, Olesky M, Palmer RK, Harden TK, Nicholas RA (1998) Evidence that the $p2y_3$ receptor is the avian homologue of the mammalian $P2Y_6$ receptor. Mol Pharmacol 54:541–546

Linder ME, Gilman AG (1992) G proteins. Sci Am 267:56–61, 64, 65

Lustig KD, Shiau AK, Brake AJ, Julius D (1993) Expression cloning of an ATP receptor from mouse neuroblastoma-cells. Proc Natl Acad Sci USA 90:5113–5117

Maier R, Glatz A, Mosbacher J, Bilbe G (1997) Cloning of $P2Y_6$ cDNAs and identification of a pseudogene: comparison of P2Y receptor subtype expression in bone and brain tissues. Biochem Biophys Res Commun 237:297–302

Matsuoka I, Zhou Q, Ishimoto H, Nakinishi H (1995) Extracellular ATP stimulates adenylyl cyclases and phospholipase C through distinct purinoceptors in NG108–15 cells. Mol Pharmacol 47:855–862

Matsuura H, Sakaguchi M, Tsuruha Y, Ehara T (1996) Activation of the muscarinic K^+ channel by P_2-purinoceptors via pertussis toxin-sensitive G proteins in guinea pig atrial cells. J Physiol 490:659–671

Meyer Zu Heringdorf D, Lass H, Alemany R, Laser KT, Neumann E, Zhang CY, Schmidt M, Rauen U, Jakobs KH, vanKoppen CJ (1998) Sphingosine kinase-mediated Ca^{2+} signalling by G-protein-coupled receptors. EMBO J 17:2830–2837

Moro S, Guo D, Camaioni E, Boyer JL, Harden TK, Jacobson KA (1998) Human $P2Y_1$ receptor: molecular modelling and site-directed mutagenesis as tools to identify agonist and antagonist recognition sites. J Medicinal Chem 41:1456–1466

Moro S, Hoffmann C, Jacobson KA (1999) Role of the extracellular loops of G protein-coupled receptors in ligand recognition: a molecular modelling study of the human $P2Y_1$ receptor. Biochemistry 38:3498–3507

Mosbacher J, Maier R, Fakler B, Glatz A, Crespo J, Bilbe G (1998) P2Y receptor subtypes differentially couple to inwardly rectifying potassium channels. FEBS Lett 436:104–110

Motte S, Pirotton S, Boeynaems JM (1993) Evidence that a form of ATP uncomplexed with divalent cations is the ligand of P2Y and nucleotide/P2U receptors on aortic endothelial cells. Br J. Pharmacol. 109:967–971

Muraki K, Imaizumi Y, Watanabe M (1998) Effects of UTP on membrane current and potential in rat aortic myocytes. Eur J Pharmacol 360:239–247

Murrin RJA, Boarder MR (1992) Neuronal nucleotide receptor linked to phospholipase-C and phospholipase-D – stimulation of PC12 cells by ATP analogs and UTP. Mol Pharmacol 41:561–568

Neary JT, Zhu Q, Bruce JM, Moore AN, Dash PK (1995) Signalling from P2 purinoceptors to MAP kinase in astrocytes involves protein kinase C. Soc Neurosci Abst 21:581

Nguyen T, Erb L, Weisman GA, Marchese A, Heng HHQ, Garrad RC, George SR, Turner JT, O'Dowd BF (1995) Cloning, expression, and chromosomal localisation of the human uridine nucleotide receptor gene. J Biol Chem 270:30845–30848

Nicholas RA, Watt WC, Lazarowski ER, Li Q, Harden TK (1996) Uridine nucleotide selectivity of three phospholipase C-activating P2 receptors: identification of a UDP-selective, a UTP-selective, and an ATP- and UTP-specific receptor. Mol Pharmacol 50:224–229

O'Connor SE, Dainty IA, Leff P (1991) Further subclassification of ATP receptors based on agonist studies. Trends Pharmacol Sci 12:137–141

Ohyama K, Yamano Y, Sano T, Nakagomi Y, Hamakubo T, Morishima I, Inagami T (1995) Disulphide bridges in extracellular domains of angiotensin-II receptor-type I-A, Regulatory Peptides 57:141–147

Okajima F, Tokumitsu Y, Kondo Y, Ui M (1987) P_2-purinergic receptors are coupled to two signal transduction systems leading to inhibition of cAMP generation and to production of inositol trisphosphate in rat hepatocytes. J Biol Chem 262:13483–13490

Palmer RK, Boyer JL, Schachter JB, Nicholas RA, Harden TK (1998) Agonist action of adenosine triphosphates at the human $P2Y_1$ receptor. Mol Pharmacol 54:1118–1123

Parr CE, Sullivan DM, Paradiso AM, Lazarowski ER, Burch LH, Olsen JC, Erb L, Weisman GA, Boucher RC, Turner JT (1994) Cloning and expression of a human P2U nucleotide receptor. a target for cystic fibrosis pharmacotherapy. Proc Natl Acad Sci USA 91:3275–3279

Patel V, Brown C, Goodwin A, Wilkie N, Boarder MR (1996) Phosphorylation and activation of p42 and p44 mitogen-activated protein kinase are required for the P2 purinoceptor stimulation of endothelial prostacyclin production. Biochem J 320:221–226

Post SR, Jacobson JP, Insel PA (1996) P_2 purinergic receptor agonists enhance cAMP production in Madin-Darby canine kidney epithelial cells via an autocrine/paracrine mechanism. J Biol Chem 271, 2029–2023

Rice WR, Burton FM, Fiedeldey DT (1995) Cloning and expression of the alveolar type II cell P_{2U}-purinergic receptor. Am J Respir Cell Mol Biol 12:27–32

Robaye B, Boeynaems JM, Communi D (1997) Slow desensitisation of the human $P2Y_6$ receptor. Eur J Pharmacol 329:231–236

Roberts JA, Boarder MR (1999) Differential regulation of mitogenesis by transfected P2Y receptors. Br J Pharmacol 126:82P

Schachter JB, Li Q, Boyer JL, Nicholas RA, Harden TK (1996) Second messenger cascade specificity and pharmacological selectivity of the human $P2Y_1$-purinoceptor. Br J Pharmacol 118:167–173

Savarese TM, Fraser CM (1992) In vitro mutagenesis and the search for structure-function- relationships among G protein-coupled receptors. Biochem J 283:1–19

Seye CI, Gadeau AP, Daret D, Dupuch F, Alzieu P, Capron L, Desgranges C (1997) Overexpression of the $P2Y_2$ purinoceptor in intimal lesions of the rat aorta, Arteriosclerosis Thrombosis & Vasc Biol 17:3602–3610

Somers GR, Hammet F, Woollatt E, Richards RI, Southey MC, Venter DJ (1997) Chromosomal localisation of the human $P2Y_6$ purinoceptor gene and phylogenetic analysis of the P2Y purinoceptor family. Genomics 44:127–130

Sonnhammer ELL, von Heijne G, Krogh A (1998) A hidden Markov model for predicting TM helices in protein sequences. In: Glasgow EJ, Littlejohn T, Major F, Lathrop R, Sankoff D, Sensen D (eds) Proceedings of sixth international conference on intelligent systems for molecular biology. AAAI Press, Menlo Park, CA, pp 175–182

Sromek SM, Harden TK (1998) Agonist-induced internalisation of the $P2Y_2$ receptor. Mol Pharmacol 54:485–494

Stam NJ, Klomp J, Van Der Heuvel M, Olijve W (1996) Molecular cloning and characterisation of a novel orphan receptor (P2P) expressed in human pancreas that shows high structural homology to the P2U purinoceptor. FEBS LETT. 384: 260–264

Strader CD, Fong TM, Graziano MP, Tota MR (1995) The family of G protein-coupled receptors. FASEB J 9:745–754

Stroebaek D, Dissing CP, Olesen SP (1996) ATP activates K^+ and Cl^- channels via purinoceptor-mediated release of Ca^{2+} in human coronary artery smooth muscle. Am J Physiol 271:C1463–C1471

Suh BC, Son JH, Joh TH, Kim KT (1997) Two distinct P2 purinergic receptors, P2Y and P2U, are coupled to phospholipase C in mouse pineal gland tumor cells. J Neurochem 68 1622–1632

Tobin AB, Totty NF, Sterlin AE, Nahorski SR (1997) Stimulus-dependent phosphorylation of G-protein-coupled receptors by casein kinase 1α. J Biol Chem 272:20844–20849

Tokuyama Y, Hara M, Jones EMC, Fan Z, Bell GI (1995) Cloning of rat and mouse P2Y purinoceptors. Biochem Biophys Res Commun 211:211–218

Webb TE, Simon J, Krishek BJ, Bateson AN, Smart TG, King BF, Burnstock G, Barnard EA (1993) Cloning and functional expression of a brain G protein-coupled ATP receptor. FEBS Letters 324:219–225

Webb TE, Henderson D, King BF, Wang S, Simon J, Bateson AN, Burnstock G, Barnard EA (1996) A novel G protein-coupled P2 purinoceptor $P2Y_3$ activated preferentially by nucleoside diphosphates. Mol Pharmacol 50:258–265

Webb TE, Henderson DJ, Roberts JA, Barnard EA (1998) Molecular cloning and characterisation of the rat $P2Y_4$ receptor. J Neurochem 71:1348–1357

Weisman GA, Erb L, Garrad RC, Theiss PM, Santiago-Perez LI, Flores RV, Santos-Berrios C, Mendez Y, Gonzalez FA (1998) P2Y nucleotide receptors in the immune system: signalling by a $P2Y_2$ receptor in U937 monocytes. Drug Dev Res 45:222–228

White PJ, Kumari R, Porter KE, London NJM, Boarder MR (1999) ATP and PDGF stimulate proliferation differentially in vascular smooth muscle cells from the human saphenous vein and internal mammary artery. Br J Pharm 127:4P

Wilkinson GF, Purkiss JR, Boarder MR (1993) The regulation of aortic endothelial-cells by purines and pyrimidines involves coexisting P2Y-purinoceptors and nucleotide receptors linked to phospholipase-C. Br J Pharm 108:689–693

Wilkinson GF, Purkiss JR, Boarder MR (1994) Differential heterologous and homologous desensitisation of 2 receptors for ATP (P_{2Y} purinoceptors and nucleotide receptors) coexisting on endothelial cells. Mol Pharmacol 45:731–736

Windscheif U, Ralevic V, Baumert HG, Mutschler E, Lambrecht G, Burnstock G (1994) Vasoconstrictor and vasodilator responses to various agonists in the rat perfused mesenteric arterial bed – selective-inhibition by PPADS of contractions mediated via P2X-purinoceptors. Br J Pharmacol 113:1015–1021

Yokomizo T, Izumi T, Chang K, Takuwa Y, Shimizu T (1997) A G-protein-coupled receptor for leukotriene B-4 that mediates chemotaxis. Nature 387:620–624

CHAPTER 5
Purinergic Signalling in Development

G. BURNSTOCK

A. Introduction

Most early studies of the roles of nucleotides in development have been discussed in terms of their intracellular roles and as a source of energy. However, since it is now generally accepted that purines and pyrimidines have potent extracellular actions mediated by the activation of specific membrane receptors (see BURNSTOCK 1997), many of these previous studies can now be reinterpreted. ATP and adenosine play key roles from the very beginnings of life, i.e. the moment of conception. ATP is obligatory for sperm movement (YEUNG 1986) and is a trigger for capacitation, the acrosome reaction necessary to fertilise the egg (FORESTA et al. 1992). Extracellular ATP also promotes a rapid increase in Na^+ permeability of the fertilised egg membrane through the activation of a specific ATP receptor (KUPITZ and ATLAS 1993). Mg^{2+}-ATPase activity has been localised on the entire surface of unfertilised eggs and in pre- and post-implantation embryos (SMITH et al. 1983; ISHIKAWA and SEGUCHI 1985). Together with the demonstration that ATP-activated spermatozoa show very high success rates in fertilisation tests (FORESTA et al. 1992), this strongly suggests that ATP is a key sperm-to-egg signal in the process of fertilisation.

In this article, the extracellular roles of purines and pyrimidines will be considered as signalling molecules in both embryological and postnatal development in a wide variety of systems in amphibians, birds and mammals, including humans.

B. Embryological Development

I. Frog Embryos

A novel P2Y purinoceptor (X1P2Y or $P2Y_8$) has been cloned and sequenced in my laboratory that is expressed (as seen by Northern blots and in situ hybridisation) in the neural plate of *Xenopus* embryos from stages 13 to 18 and again at stage 28 when secondary neurulation occurs in the tail bud (BOGDANOV et al. 1997). It differs from other members of the P2Y purinoceptor family in that it has an intracellular C terminus with 216 amino acid residues (compared to 16 to 67 in $P2Y_{1-7}$). When expressed as a recombinant receptor in *Xenopus* oocytes, it shows equipotent responses to the triphosphates ATP, UTP, ITP, CTP and GTP and smaller responses to diphosphates and tetraphosphates, but is not responsive to inorganic phosphates. Responses to activation of the X1P2Y receptor have a long duration (40–60min). These data suggest that this novel P2Y receptor may be involved in the early formation of the nervous system.

Suramin and trypan blue, both substances that are known to be antagonists at P2 receptors (see BURNSTOCK 1996a) as well as having other actions, have been shown to interfere with gastrulation (GERHART et al. 1989). If injected early when the dorsal lip is first invaginating, the *Xenopus* embryo develops no head or trunk and sometimes no tail; somites and notocord are also missing. If they are injected midway in gastrulation, embryos develop without heads, but with trunks and tails, while if injected at the end of gastrulation, the embryo is completely unaffected.

The nicotinic channels in myotomal muscle cells cultured from *Xenopus* embryos at stages 19–22 were shown to be opened by micromolar concentrations of exogenous ATP (IGUSA 1988), following the earlier demonstration that ATP increases the sensitivity of receptors in adult frog skeletal muscles without increasing the affinity of acetylcholine (ACh) for the receptor or inhibiting acetylcholinesterase (AKASU et al. 1981). Since then, there have been a number of studies of the actions of ATP in developing *Xenopus* neuromuscular synapses (see FU 1995). Extracellular applications of ATP to developing *Xenopus* neuromuscular synapses in culture potentiate ACh responses of developing muscle cells during the early phase of synaptogenesis (FU and Poo 1991; FU 1994; FU and HUANG 1994). The possibility that extracellular ATP, coreleased with ACh, may serve as a positive trophic factor at developing neuromuscular synapses has also been raised (FU and Poo 1991; FU 1995). It is further suggested that calcitonin gene-related peptide (CGRP) and ATP coreleased with ACh from the nerve terminal may act together to potentiate postsynaptic ACh channel activity during the early phase of synaptogenesis (LU and FU 1995); it is claimed that CGRP actions are mediated by cyclic adenosine monophosphate- (cAMP-) dependent protein kinase A, while ATP exerts its effects via protein kinase. In the most recent report from this group (FU et al. 1997), they present results that suggest that endogenously released ATP, acting in concert with various protein kinases, is involved in the maintenance and/or development of the quantum size of synaptic vesicles at embryonic neuromuscular synapses.

Regulation of rhythmic movements by purinergic transmitters in frog embryos has been described (DALE and GILDRAY 1996). It was shown that ATP is released during swimming that activates P2Y receptors to reduce voltage-gated K^+ currents and cause an increase in the excitability of the spinal motor circuits. It was also shown that adenosine, resulting from the breakdown of ATP, acts on P1 receptors to reduce the voltage-gated Ca^{2+} currents to lower excitability of the motor circuits, thereby opposing the actions of ATP. The authors suggest that a gradually changing balance between ATP and adenosine underlies the run-down of the motor pattern for swimming in *Xenopus*. In a later study, DALE (1998) presented evidence to suggest that delayed production of adenosine underlies temporal modulation of swimming in the frog embryo and is likely to result from feed-forward inhibition of the 5'-ectonucleotidase in the spinal cord.

II. Chick Embryos

Together with muscarinic cholinergic receptors, extracellular receptors to ATP were shown to be the first functionally active membrane receptors in chick embryo cells at the time of germ layer formation (LAASBERG 1990). In gastrulating chick embryo, ATP causes rapid accumulation of inositol-phosphate and Ca^{2+} mobilisation in a similar way and to the same extent as ACh, whereas other neuroendocrine substances such as insulin and noradrenaline (NA) have much weaker effects (LAASBERG 1990). This suggests that, alongside ACh, other phylogenetically old and universal regulators of cell metabolism such as ATP (and perhaps nitric oxide) might play a leading role in the functional regulation of gastrulation via the activation of specific receptors triggering Ca^{2+} mobilisation.

ATP has been shown to induce precocious evagination of the embryonic chick thyroid, an event which has been hypothesised to be involved in the formation of the thyroid gland from the thyroid primordium (HILFER et al. 1977). The requirement for ATP was very precise, since it could not be replaced by pyrophosphate, AMP or ADP, nor by GTP, suggesting a high degree of specificity of the ATP-induced effect.

ATP acts on embryonic and developing cells of both nervous and non-nervous systems by increasing intracellular Ca^{2+} concentrations. Release of Ca^{2+} from intracellular stores is evoked in the otocyst epithelium of the early embryonic chick, incubated for 3 days (stage 18 to 19) (NAKAOKA and YAMASHITA 1995) (Fig. 1), in developing chick myotubes (HÄGGBLAD and HEILBRONN 1988) and in dissociated cells from whole early embryonic chicks (LAASBERG 1990; LOHMANN et al. 1991).

In a recent study, expression of the G protein-coupled $P2Y_1$ receptor during embryonic development of the chick has been described (MEYER et al. 1999). During the first 10 days of embryonic development, the $P2Y_1$ receptor is expressed in a developmentally regulated manner in the limb buds, mesonephros, brain, somites and facial primordia (Fig. 2), suggesting that there may be a role for ATP and $P2Y_1$ receptors in the development of these systems.

1. Retina

A study of embryonic chick neural retina (SUGIOKA et al. 1996) has shown that the ATP-induced rise in intracellular Ca^{2+} is mediated by P_{2U} (= $P2Y_2$)

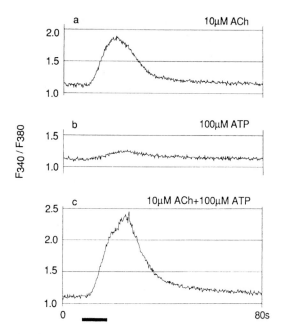

Fig. 1a–c. Interaction between acetylcholine and ATP recorded in an otocyst from chick embryo. **a** The Ca^{2+} response to 10 μmol/l acetylcholine. **b** The response to 100 μmol/l ATP. **c** The response to the co-application of 10 μmol/l acetylcholine and 100 μmol/l ATP. The records in **a**–**c** were taken in this order at 5 min intervals. The bath solutions contained 25 mM Ca^{2+}. (Reproduced with permission from NAKAOKA and YAMASHITA 1995)

purinergic receptors and that there is a dramatic decline of the ATP-induced rise in intracellular Ca^{2+} just before synaptogenesis. Suramin and Reactive blue 2 almost completely block these responses (Fig. 3). These authors also reported unpublished data that injection of Reactive blue 2 into early embryonic chicks produced severe effects in embryogenesis. A later study showed that while both the muscarinic and purinergic Ca^{2+}-mobilisations utilise inositol trisphosphate- (IP_3-) sensitive Ca^{2+} stores, different signal transduction pathways were involved (SAKAI et al. 1996). P2 purinergic receptors activated by autocrine or paracrine release of ATP have been claimed to be involved in the regulation of DNA synthesis in the neural retina at early embryonic stages (SUGIOKA et al. 1999). ATP increased [^{3}H]thymidine incorporation in retinal cultures from embryonic day 3 (E3) and suramin and pyridoxalphosphate-6-azophenyl-2′,4′-disulphonic acid (PPADS) inhibited these activities in a dose-dependent manner; the concentration of ATP increased 25-fold in the medium of E3 retinal organ cultures within 1h of incubation and was maintained for at least 24h (3 μmol/l ATP at E3, declined to 0.15 μmol/l at E7). In a recent review it was suggested that the change in Ca^{2+} signalling mediated by P_{2U} receptors during development may underlie the differentiation of neuroepithelial cells

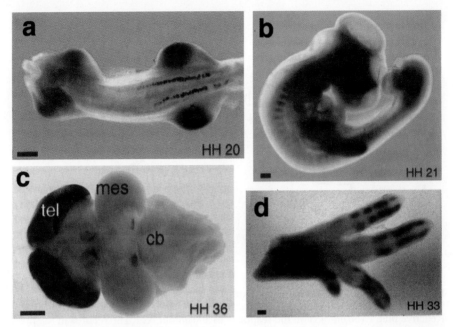

Fig. 2a–d. Expression of P2Y$_1$ receptors during embryonic development of the chick as visualized by whole-mount in situ hybridization. Stages of development are shown in *bottom right corner*. **a** Ventral view of stage 20 embryo showing P2Y$_1$ expression in mesonephros and limb buds (*scale bar* = 200 µm). **b** Lateral view of the chick somite at stage 21 showing P2Y$_1$ expression in the anterior region. The *dark area* in the head region is due to an artefact of photography (*scale bar* = 200 µm). **c** Dorsal view of stage 36 brain (*anterior to the left*), showing increased levels of expression in telecephlon (*tel*), dorsal diencephlon and posterior midbrain. *mes*, mesencephalon; *cb*, cerebellum (*scale bar* = 1 mm). **d** An anterior-uppermost view of a leg at embryonic stage 33. Expression of P2Y$_1$ in seen in the digits, but not in areas of joint formation. The same expression pattern is also seen in the wing (*scale bar* = 100 µm). (Reproduced with permission from MEYER et al. 1999)

or undifferentiated progenitor cells into neurons (YAMASHITA and SUGIOKA 1998).

Adenosine has also been implicated in chick retinal development. Adenosine induction of cyclic AMP increased strongly from the 14th to the 17th embryonic day, P1 (A$_1$) subtype receptors modulating D$_1$ dopamine receptor-mediated stimulation of adenylate cyclase activity (PAES DE CARVALHO and DE MELLO 1982, 1985; DE MELLO et al. 1982). In a later study it was suggested that A$_1$ receptors may have different functions in the embryonic retina as compared to mature chick retina (PAES DE CARVALHO 1990), and the localisation of A$_1$ receptors and uptake sites in the developing chick retina examined (PAES DE CARVALHO et al. 1992). A$_1$ receptors were localised predominantly in plexiform regions by embryonic day 12 (E12). They were absent in the retina at E8, but were detected at E12 in the ganglion cell layer, as well as cells in the nuclear cell layer and photoreceptors.

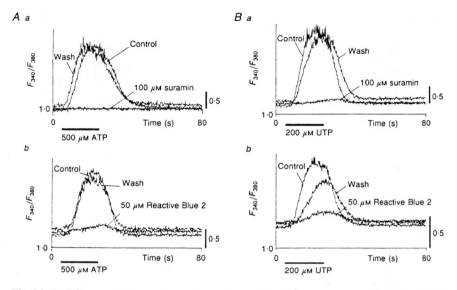

Fig. 3A,B. Effects of P2 receptor antagonists on the Ca^{2+} responses to ATP and UTP in embryonic (E3) chick neural retinas. **A** The effects of: **a** suramin (100μmol/l), **b** Reactive blue 2 (50μmol/l) on the response to 500μmol/l ATP. The records in the presence of suramin or Reactive blue 2 were taken 7 min after changing the bath solutions to the antagonist-containing medium. The recovery controls (*Wash*) were taken after washing suramin for 7 min or Reactive blue 2 for 25 min. The duration of ATP application (20s) is indicated by the *bars*. All records were taken in the bath solutions containing 2.5 mmol/l Ca^{2+}. **B** The effects of: **a** suramin (100μmol/l), **b** Reactive blue 2 (50μmol/l) on the response to 200μmol/l UTP. The records in the presence of suramin or Reactive blue 2 were taken 7 min after changing the bath solutions to the antagonist-containing medium. The recovery controls were taken after washing suramin for 7 min or Reactive blue 2 for 15 min. The duration of UTP application (20s) is indicated by the *bars*. All records were taken in the bath solutions containing 2.5 mmol/l Ca^{2+}. (Reproduced with permission from SUGIOKA et al. 1996)

2. Glial Cells

In a study of 48 h-old cultured ciliary ganglia and confluent peripheral and central nervous system glial cultures taken from 12- to 14-day-old embryonic chicks, MEGHJI et al. (1989) concluded that adenosine is formed intracellularly and exported out of the cell by the nucleoside transporter; the participation of *ecto-5'*-nucleotidase was excluded. Adenosine transport *into* primary cultures of neurons and glial cells from chick embryonic brain has also been demonstrated (THAMPY and BARNES 1983a,b). A_2-like adenosine receptors were postulated to be present in glial cell membranes of chick embryonic brains (BARNES and THAMPY 1982). Widespread programmed cell death has been demonstrated in proliferative regions of chick optic tectum during early development, particularly in the ventricular zone between stages E7.5 and E8 (ZHANG and GALILEO 1998). This is of particular interest since purinoceptors (particularly P1 and P2X$_7$) are apoptosis-signalling molecules (see DI VIRGILIO et al. 1995, 1998; WAKADE et al. 1995).

3. Skeletal Muscle

A transmitter-like action of ATP on patched membranes of myoblasts and myotubes cultured from 12-day-old chicken embryos was first demonstrated by KOLB and WAKELAM (1983). Using biochemical methods, ATP-induced cation influx was later demonstrated in myotubes prepared from 11-day-old chick embryos and shown to be additive to cholinergic agonist action (HÄGGBLAD et al. 1985). Later papers from this group claimed that the myotube P2 purinoceptor triggers phosphoinositide turnover (HÄGGBLAD and HEILBRONN 1987, 1988) and alters Ca^{2+} influx through dihydropyridine-sensitive channels (ERIKSSON and HEILBRONN 1989). ATP has a potent depolarizing action on myotubes derived from pectoral muscle cultured from 11-day chick embryos (HUME and HÖNIG 1986) and its physiological and pharmacological properties have been described in a series of papers (HUME and THOMAS 1988; THOMAS and HUME 1990a,b, 1993; THOMAS et al. 1991). The myotube P2 purinoceptor is not activated by ADP, AMP, adenosine or the non-hydrolysable ATP analogues α,β-methylene ATP (α,β-meATP) or β,γ-methylene ATP (HUME and HÖNIG 1986). A single class of ATP-activated ion channel conducts both cations and anions in the myotube (THOMAS and HUME 1990a) and the P2 purinoceptors involved showed marked desensitisation (THOMAS and HUME 1990b). The sensitivity to extracellular ATP has been tested at various stages of development of different muscles (WELLS et al. 1995). At embryonic day 6 (stage 30 of HAMBERGER and HAMILTON 1951) ATP (50–100 µmol/l) elicits vigorous contractions in all the muscles tested, but by embryonic day 17 (stage 43) none of the muscles contract in response to ATP (Fig. 4). However, denervation of muscles in newly hatched chicks leads to the reappearance of sensitivity to ATP, suggesting that the expression of ATP receptors is regulated by motor neurons. An immunohistochemical study of the distribution of 5'-nucleotidase during the development of chick striated muscle shows that the adult exhibits a more restricted distribution compared to the embryo (MEHUL et al. 1992).

4. Cardiovascular System

Studies of the development of pharmacological sensitivity to adenosine analogues in embryonic chick heart (BLAIR et al. 1989; HATAE et al. 1989) show that pharmacological sensitivity to A_1 agonists begins at embryonic day 7 and then increases continuously to day 12, when the atria became fully responsive. Ligand binding shows that A_1 receptors are present at days 5 and 6, but are not responsive to adenosine, and the author concluded that the development of sensitivity to A_1 adenosine receptor-mediated negative chronotropic responses was not paralleled by developmental changes in adenosine inhibition of adenylate cyclase, or by the development of sympathetic and para-

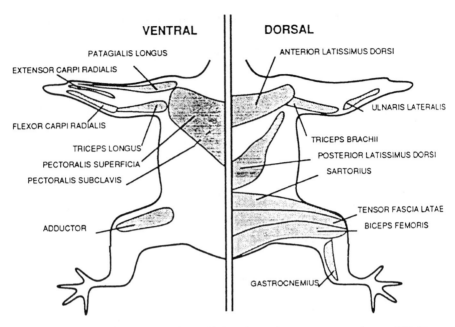

Fig. 4. Location of the muscles of the chick embryo that were responsive to ATP. Three chick embryos from stages 35–37 were sacrificed, and each muscle was identified and tested in at least two of the three embryos. All muscles tested in embryos of these ages contracted in response to ATP. By embryo day 17 (stage 43) none of the muscles contracted in response to ATP. (Reproduced with permission from WELLS et al. 1995)

sympathetic innervation. Chronic exposure of the embryonic chick heart (15–17 days old) to R-(–)-N^6-2-phenylisopropyladenosine (R-PIA) produces down-regulation of A_1 adenosine receptor and desensitisation of the negative inotropic response to adenosine (SHRYOCK et al. 1989). A study of ventricular cells cultured from chick embryos 14 days in ovo showed that a functional A_{2A} receptor is expressed and mediates augmentation of myocyte contractility (LIANG and HALTIWANGER 1995). The A_{2A} receptor coexists with an A_{2B} receptor, although it has a 50 times higher affinity, and the authors suggest that high affinity A_{2A} receptors play an important modulatory role in the presence of low levels of adenosine, while the low affinity A_{2B} receptor becomes functionally important when the adenosine level is high.

At a recent purine meeting in Leipzig (August 1998) a paper was presented where it was claimed that a new subunit of the P2X receptor family had been isolated from cardiomyocytes and brain from 14-day-old chick embryos; the primary sequence shares 75% identity with the rat and human $P2X_4$ receptor, suggesting that the cDNA isolated may be the corresponding chick isoform and the pharmacological properties of the receptor expressed in *Xenopus* oocytes was consistent with this view (RUPPELT et al. 1999).

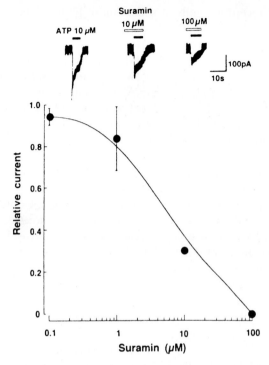

Fig. 5. Chick embryo (day 14) ciliary ganglion cells: the inhibition of ATP-induced inward current by suramin. The neurons were pretreated with suramin of various concentrations for 2 min. In the *upper panel* the *filled and open horizontal bars* indicate the periods of application of ATP and suramin, respectively. In the *lower panel*, the responses in the presence of suramin are normalised to the peak current amplitude induced by 10 μmol/l alone. Each point is the average of four neurons, and the *vertical bars* indicate standard error of the mean. (Reproduced with permission from ABE et al. 1995)

Adenosine has been implicated in growth regulation of the vascular system in the chick embryo (ADAIR et al. 1989), in common with a similar role claimed for experimental angiogenesis in the chorio-allantoic membrane (FRASER et al. 1979; TEUSHER and WEIDLICH 1985; DUSSEAU et al. 1986).

5. Ganglia

Responses to ATP have been described in ciliary neurons acutely dissociated from embryonic chick ciliary ganglia taken at day 14 (ABE et al. 1995). The relative potency of agonists in producing transient inward currents with patch recording is ATP > 2-methylthioATP (2meSATP) > ADP; neither adenosine, AMP nor α,β-meATP are effective, but suramin is an antagonist (Fig. 5). The authors suggest that the P2 receptor subtype involved might be P2Y, but in view of more recent knowledge about the functional properties of cloned subtypes of the P2 receptor family, it seems more likely to belong to the P2X receptor family.

Adenosine inhibits neurite outgrowth of chick sympathetic neurons taken from 11-day chick embryos and kills by apoptosis about 80% of sympathetic nerves supported by growth factor over the next 2 days in culture (WAKADE et al. 1995). Specific A_1 or A_2 agonists are not neurotoxic. The toxic effects of adenosine are not antagonised by aminophylline, but are prevented by an adenosine transporter or adenosine deaminase inhibitor, suggesting an intracellular site of action for the toxic effects of adenosine. The authors conclude that adenosine and its breakdown enzymes play an important role in the regulation of growth and development of sympathetic neurons. In follow-up experiments, the authors suggest that adenosine induces apoptosis by inhibiting mRNA and protein synthesis (KULKARNI et al. 1998). A study of P2 receptors modulating noradrenaline release from chick sympathetic neurons cultured from 12-day-old embryos suggested that two different P2 receptor subtypes were involved: a facilitatory receptor and an inhibitory receptor (ALLGAIER et al. 1995a). In a sister paper, the authors postulated that the ATP receptor involved in noradrenaline release acted via a subclass of the nicotinic cholinoceptor (ALLGAIER et al. 1995b).

6. Chondrocytes

Chondrocytes, isolated from the cephalic region of day 19 chick embryo sterna, release nucleotides into the extracellular milieu, although they are rapidly degraded; it is claimed that they are involved in both chondrocyte maturation and matrix mineralisation (HATORI et al. 1995). Extracellular ATP modulates $[Ca^{2+}]_i$ in retinoic acid-treated chondrocytes isolated from the cephalic portion of day 14 chick embryonic sterna (HUNG et al. 1997). They speculate that immature chondrocytes may generate adenine nucleotides that then act in a paracrinal manner on chondrocytes that are at a later stage of maturation.

III. Purinergic Receptors in Mammalian Embryos

1. Early Embryos

Puff-applied ATP has been shown to have two main effects on a mouse mesodermal stem cell line: an increase in intracellular Ca^{2+} concentrations and a subsequent hyperpolarisation due to Ca^{2+}-activated K^+ conductance (KUBO 1991a) (Fig. 6). The author speculates that the transient increase in intracellular Ca^{2+} may influence mesodermal cell differentiation, particularly in relation

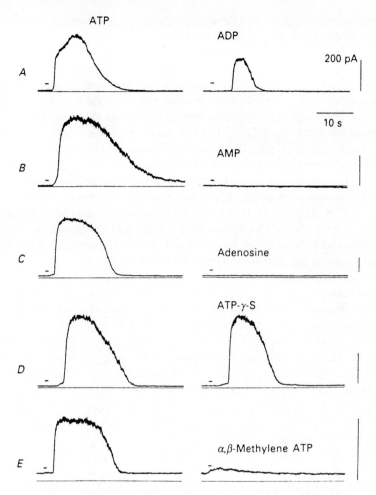

Fig. 6. K⁺ responses to ATP analogues of a mouse mesodermal cell line. Each of the two traces was obtained from the same cell (*A*, *B*, *C*, *D* and *E*). The responses induced by ATP (*left traces*) and ATP analogues (*right traces*) are shown. The names of the analogues are shown near the traces. Each drug was applied at 20 µmol/l, and the holding potentials were 0 mV. (Reproduced with permission from Kubo 1991a)

to muscle differentiation. In a later paper (Kubo 1991b), two myoblastic cell lines, one from rat, the other from mouse, showed similar properties to those of the myogenic clonal cells derived from the mouse mesodermal stem cell line described above.

ATP and ADP have been shown to enhance, reduce or have no effect (depending on the dose used) on the incidence of trypan blue-induced teratogenic malformations in the rat foetus at day 20 (Beaudoin 1976). Concomitant administration of ATP and cortisone in mice either decrease the teratogenic effect of cortisone (50μg ATP) or enhance its teratogenic effect (>100μg ATP) (GORDON et al. 1963).

Mouse heads of embryos from 14 to 24 pairs of body somites exposed to an ATP-containing medium have been demonstrated to undergo rapid epithelial thickening and invagination, a process that appears to take part in the shaping of nasal pits and formation of primary palate (SMUTS 1981).

Besides ATP, a number of reports implicate adenosine as one of the endogenous effectors that can selectively modulate cell growth during embryonic development. For example, adenosine is shown to potentiate the delaying effect of dibutyryl cyclic adenosine monophosphate (a membrane-permeable analogue of cyclic AMP) on meiosis resumption in denuded mouse oocytes (PETRUNGARO et al. 1986). The role of adenosine has been particularly well characterised in the morphogenetic outgrowth of vertebrate limb buds (KNUDSEN and ELMER 1987). Embryonic limb development in the mouse is driven by rapid mesenchymal cell proliferation induced by trophic substances secreted by the apical ectodermal ridge. This interaction can be restricted experimentally by pharmacological agents that elevate intracellular cAMP levels, or physiologically by the onset of "programmed cell death" triggered by naturally occurring negative regulators of growth. Mutations that affect the pattern of limb/bud outgrowth provide invaluable experimental means to investigate these growth-regulatory processes. KNUDSEN and ELMER (1987) studied the regulation of polydactylous outgrowth [an expression of the Hemimelia-extra toe ($Hm^{x/+}$) mutant phenotype] in hind-limb buds explanted into a serum-free in vitro system at stage 18 of gestation. Its expression was promoted by exposure to exogenous adenosine deaminase, the enzyme which catalyses the inactivation of endogenous adenosine, and conversely suppressed by co-exposure to hydrolysis-resistant adenosine analogues. Adenosine-induced effects were mediated by activation of specific extracellular receptors, since the P1 receptor antagonist, caffeine, could completely prevent suppression of polydactylous outgrowth. Measurement of both adenosine and adenosine deaminase levels in embryonal plasma and whole embryos argued against an endocrine mechanism of adenosine secretion, in favour of an autocrine (self-regulatory) or paracrine (proximate-regulatory mechanisms). These results suggest that the in vitro outgrowth of the prospective polydactylous region is induced upon escape from the local growth-inhibitory influence of extracellular adenosine.

Micromolar concentrations of adenosine, inosine and hypoxanthine, but not guanosine block the second or third cleavage of mouse embryos developing in vitro (NUREDDIN et al. 1990). Zygotes or early two-cell embryos, cultured in a purine-containing medium for 24 h, resume development following transfer to purine-free conditions. The precise mechanism of the purine-sensitive process is not known, but embryos conceived in vivo are sensitive until approximately 28–30 h after fertilisation and are no longer sensitive by 34h (LOUTRADIS et al. 1989). However, a later study by this group has shown that the purine-induced block can be reversed by compounds that elevate cAMP (FISSORE et al. 1992).

Taken together, these results point to a role for purines in both physiological fertilisation and normal development and also underline that alterations of the purinergic regulation of embryonal growth might be involved in the onset of morphological malformations. Depending upon the purine derivative, and probably upon the purinoceptor involved as well, ATP and adenosine can act as both positive and negative regulators of growth. This is also consistent with data obtained from in vitro cell lines which implicates purines in both cell proliferation and apoptosis. Further studies are needed to characterise better the receptor subtypes involved and also to identify more precisely the developmental events specifically controlled by purines.

2. Central Nervous System (CNS)

Radioligand binding studies have provided information about the development of A_1 receptors in guinea-pig and rat brain, in particular the forebrain and cerebellum (MORGAN et al. 1987, 1990; NICOLAS and DAVAL 1993). In guinea-pig forebrain it appears that A_1 receptors are present from embryonic day 19, with adult binding levels achieved about 25 days postpartum. In guinea-pig cerebellum, however, A_1 receptor binding is low until just prior to birth, when a dramatic increase in binding is observed which then continues to increase up to adulthood. A similar development is seen in rat forebrain and cerebellum with A_1 receptor binding changing very gradually in the forebrain, whereas binding in the cerebellum increases markedly after birth (MARANGOS et al. 1982; GEIGER et al. 1984). The developmental appearance of A_1 receptor gene expression was examined in rats by in situ hybridisation (WEAVER 1996). Expression of A_1 receptor mRNA in brain was first detected on gestation day 14, and was restricted to portions of neuroepithelium caudate putamen, piriform cortex, hypoglossal nucleus and ventral horn of spinal cord. A_1 receptor activation mediates ethanol-induced inhibition of stimulated glutamate release in the hippocampus of near-term foetal guinea-pig (REYNOLDS and BRIAN 1995). The ontogeny of adenosine uptake sites in the guinea-pig brain has been described (DECKERT et al. 1988).

There are a number of reports about changes in the distribution of the ectoenzymes involved in the breakdown of ATP and adenosine in the brain during foetal and neonatal development. 5'-Nucleotidase shows a marked redistribution during development of the cat visual cortex and is thought to be involved in the remodelling of ocular dominance columns (SCHOEN et al. 1990). A later electron microscopic study by the same group has suggested that synapse-bound 5'-nucleotidase activity plays a role in synaptic malleability during development; its later association with glial cell profiles may reflect other functions for this enzyme (SCHOEN et al. 1993). At 30 and 35 days of gestation of foetal guinea-pigs, 5'-nucleotidase levels were low, but increased rapidly during the 40 to 60 day period; in contrast, adenosine deaminase was present at 30 days of gestation and remained at the same level until 60 days (MISHRA et al. 1988). Complex changes in the activity of adenosine deaminase in the different regions of the developing rat brain suggest that there are important roles for purines in very early stages of development from 15–18 days of gestation in specific regions of the brain, namely the hypoglossal motor nucleus, cingulate, retrosplenial and visual cortex, posterior basal hypothalamus and in the facial motor nucleus (GEIGER and NAGY 1987; SENBA et al. 1987a,c). Adenosine deaminase-containing neurons were seen in the olfactory cortex of rat embryos as early as E15; this was suggested to indicate precocious development of purinergic neurotransmission within this system (SENBA et al. 1987b). A histochemical study of Ca^{2+}-ATPase in the rat spinal cord during embryonic development demonstrated intense activity in the roof and floor plates, rather than in the basal and lateral plates at embryonic day 12, indicating a possible role for Ca^{2+}-ATPase in early differentiation of neuroepithelial cells (YOSHIOKA et al. 1987; YOSHIOKA 1989). ATP induces rises in intracellular Ca^{2+} in embryonic spinal cord astrocytes (SALTER and HICKS 1995).

A recent immunohistochemical study has revealed intense labelling of $P2X_3$ receptors in the embryonic rat brain (KIDD et al. 1998) (Fig. 7). The staining was restricted to the hind brain, in particular the mesencephalic trigeminal nucleus, the superior and inferior olives, the intermediate reticular zone, the spinal trigeminal tract and the prepositus hypoglossal nucleus. Other areas labelled included the mandibular nerves to the teeth, the auditory nerve from the inner ear and the trigeminal ganglion. In the E19 rat embryo $P2X_7$ receptor mRNA was detected by in situ hybridisation in brain ependyma but not neurons (COLLO et al. 1997).

3. Cardiorespiratory System

In foetal sheep, centrally administered adenosine influences cardiac function (EGERMAN et al. 1993). The ontogeny of A_1 adenosine receptors was studied in rats with binding assays (using $[^3H]$1,3-dipropyl-8-cyclopentylxanthine, DPCPX), an A_1 antagonist, and by in situ hybridisation of mRNA (RIVKEES 1995). In a later study of mouse embryo cardiac function (HOFMAN et al. 1997), adenosine, via A_1 receptors, was shown to regulate potently heart rate via multiple effector systems at very early stages of prenatal development (9–12 days postconceptual). At gestational days 8–11, mRNA expression for A_1 receptor was detected in the atrium (one of the earliest G protein-coupled receptor genes to be expressed in the heart), but not in other foetal structures, while at gestational day 14, A_1 mRNA was present in the CNS (thalamus, ventral horn of spinal cord) as well as the atrium; by gestational age 17, patterns of A_1 receptor expression in the brain were similar to those observed in adults (WEBER et al. 1990; REPPERT et al. 1991). Determination of A_1 receptor density in developing rat heart using $[^3H]$DPCPX, showed that functional A_1 receptors are present in greater numbers in the immature perinatal heart than in the adult heart (COTHRAN et al. 1995).

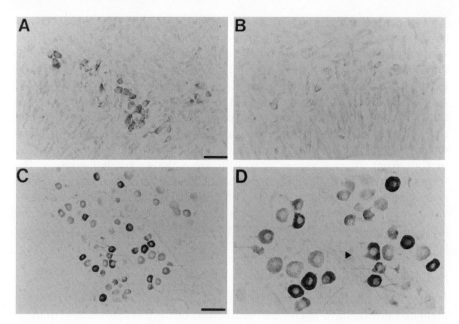

Fig. 7. Immunoreactivity for P2X$_3$ receptors in the trigeminal nucleus (Me5) of: **A,B** E16; **C,D** P7 rats. Immunohistochemistry was performed with a polyclonal antibody raised to a nine-amino acid peptide identical to the carboxy terminus of the rat P2X$_3$ receptor. The sections were counter-stained using Methyl green. Labelled cell bodies can be detected for the majority of the Me5 neurons in the E16 rat (**A, B**) while in the P7 rat (**C, D**) a diminishing subpopulation of cells was labelled. Fibres and processes can clearly be seen in the P7 animal (*arrowhead*, **D**). No staining was seen in control sections incubated with pre-immune serum in place of the P2X$_3$ antibody (**B**). *Scale bars* = 25 µm (**A, B, D**), 50 µm (**C**). (Reproduced with permission from KIDD et al. 1998)

Intravenous infusion of adenosine analogues into foetal lambs produced dose-dependent bradycardia and hypotension (YONEYAMA and POWER 1992; KONDURI et al. 1992a; Koos et al. 1993). In contrast, in the newborn, 5'-N-ethylcarboxamidoadenosine (NECA) produced dose-dependent tachycardia, while R-PIA and cyclohexyladenosine (CHA) produced dose-dependent bradycardia. Although adenosine causes cardiovascular changes in pregnant ewes, the effects are well tolerated and do not significantly affect the cardiorespiratory status of the foetus (MASON et al. 1993).

Foetal breathing movements were interrupted by all analogues, but they did not produce apnea in the newborn (TOUBAS et al. 1990). ATP and ADP are claimed to be important mediators of oxygen-induced pulmonary vasodilatation in foetal lambs (KONDURI et al. 1993, 1997).

4. Skeletal Muscle

Purinergic receptors have been characterised in mouse C2C12 myotubes (HENNING et al. 1992, 1993a,b, 1996). Adenosine-sensitive P1 receptors activating cyclic AMP formation were identified and a P2 receptor was also postulated, sensitive to ATP, ADP and adenosine-5'-O-3-thiotriphosphate (ATPγS), which also activates the formation of cAMP. This receptor was also sensitive to UTP, but not to α,β-meATP, 2meSATP, GTP or CTP, thus resembling the P2Y$_2$ (= P$_{2U}$) purinoceptor identified in other mammals. The response to ATP and UTP was biphasic, a transient hyperpolarisation being followed by a slowly declining depolarisation; the hyperpolarisation was blocked by apamin and suramin and abolished under Ca^{2+}-free conditions. Occupation of the receptor by ATP or UTP led to formation of inositol trisphosphate and release of Ca^{2+} from internal stores as well as from the extracellular space. The responses to ATP of myotubes prepared from E18 mouse embryos from normal and mutant mdg/mdg mice with muscular dysgenesis were studied by Tassin et al. (1990). Using Fura-2 as a probe, they showed that many of the mdg/mdg myotube preparations showed little or no increase in cytoplasmic Ca^{2+} levels.

5. Gastrointestinal Tract

In the gastrointestinal tract, non-adrenergic, non-cholinergic (NANC) nerve-mediated effects have been observed before birth in rat stomach (Ito et al. 1988) and in mouse and rabbit small intestine (Gershon and Thompson 1973). Also, quinacrine fluorescence, which indicates the presence of high levels of bound ATP, was observed before birth in enteric neurons of rabbit ileum and stomach, about 3 days before catecholamine fluorescence was detected in enteric nerves (Crowe and Burnstock 1981). NANC inhibitory and cholinergic excitatory innervation appear simultaneously in the rabbit at 17 days of gestation and both were present in the mouse by the 16th day of gestation; however, the development of adrenergic innervation lagged far behind the other two components, clearly establishing that the intrinsic innervation of the gut is not adrenergic (Gershon and Thompson 1973; Miyazaki et al. 1982).

6. Skin

Merkel cells appear in the epidermus of planum nasale or rat foetuses from the 16th day of intrauterine development and nerve fibres form close association with them by day 20 (Pác 1984). This is of interest since it is known that Merkel cells contain high levels of peptide-bound ATP and are in close association with sensory fibres expressing P2X$_3$ receptors (see Burnstock and Wood 1996).

7. Bone

Rapid deamination of adenosine in cultures of foetal mouse calvarial bones was shown and taken to account, at least in part, for the failure to observe effects of adenosine in bone metabolism in culture (Fredholm and Lerner 1982). Cells of osteogenic and chondrogenic lineage derived from foetal metatarsal bones were exposed to ATP^{4-}; cells of haemopoietic origin were permeabilised and killed, while cells of non-haemopoietic origin (e.g. osteoblasts, chondrocytes) were insensitive to ATP^{4-} and survived (MODDERMAN et al. 1994). This system allows the study of the properties and functions of osteogenic or chondrogenic cells without interference by the presence of cells of haemopoietic origin. ATP pyrophosphohydrolase has been purified and partially characterised from foetal bovine epiphyseal cartilage of patients with chondrocalcinosis (HSU 1983). The most prominent location of $P2X_7$ receptor mRNA in E19 rat embryos is bone marrow; but bone marrow cells from mouse femur also showed strong immunoreactivity for P2X receptors (COLLO et al. 1997).

8. Lung

ATP and UTP evoke $[Ca^{2+}]_i$ signals in rat foetal lung epithelial cells, but only if grown into functionally polarised epithelia on permeable supports; moreover, reverse transcriptase-polymerase chain reaction (RT-PCR) identification of the mRNA for the $P2Y_2$ receptor was clearly expressed in the polaroid cells (CLUNES et al. 1998). In another study of epithelia explanted from foetal rat lung, receptors to adenosine, ATP and UTP were present on apical membranes throughout the lung; basolateral receptors for these agonists in distal lung and ATP/UTP receptors in trachea function later in gestation (BARKER and GATZY 1998). In E19 rat embryos $P2X_7$ mRNA was detected by in situ hybridisation in bronchial epithelium (as well as salivary gland, liver bone marrow and brain endyma) (COLLO et al. 1997).

9. Ectoenzymes

Changes in expression of adenosine deaminase have been described in lymphoid tissues of rat embryos (CHECHIK et al. 1985). In the thymus, cortical lymphocytes begin to express significant amounts of the enzyme at 17 days of gestation; in the spleen and lymph node, adenosine deaminase was initially detected in T cell areas, but not primary follicles; in the duodenum, epithelial cells of villi and the neck of crypts showed positive staining; in the cartilage of 15-day foetuses, positive staining was seen in perichondrial and hypertrophic cells, while Kupffer cells in the liver and vascular endothelial cells showed positive staining for adenosine deaminase at every autogenetic stage studied. In mouse embryonic development adenosine deaminase increased 74-fold between days 7 and 9; deoxyadenosine kinase increased 5.4-fold during the same period; adenosine kinase, deoxyguanosine kinase and purine nucleoside phosphorylase exhibited less than 2-fold changes in activity between days 7 and 13 (JENUTH et al. 1996). The authors concluded that phosphorylation of adenosine was the principal route of metabolism up to day 9, after which there is a switch to deamination. The possible role of ectoadenosine deaminase in the development of the nervous system and the neurological abnormalities that occur in adenosine deaminase-deficient patients is discussed in a review by FRANCO et al. (1997).

IV. Human Embryos

A few studies have been made of receptors to purines and pyrimidines in human embryos.

Human embryonic kidney cells (HEK293) endogenously express $P2Y_1$ and $P2Y_2$ receptors (SCHACHTER et al. 1997). These embryonic cells have also been shown to express an endogenous A_{2B} adenosine receptor (COOPER et al. 1997).

ATP and ATPγS were shown to stimulate DNA synthesis in human foetal astrocyte cultures (NEARY et al. 1998). In addition, ATP stimulated a mitogen-activated protein kinase (MAPK) termed ERK (extracellular signal-regulated kinase), a key component of signal transduction pathways involved in cellular proliferation and differentiation. The activation of MAPK was mediated, at least in part, by P2 receptors since suramin produced 50% block.

There has been a study of plasma ATP levels in the foetus at the time of obstetrical delivery, samples being collected immediately after clamping of the cord (FUKUDA et al. 1990); the results showed that the plasma ATP was significantly higher in arterial compared to venous or maternal venous blood. It was suggested therefore that the ATP in arterial blood was of foetal origin and that the levels decrease in response to stress during vaginal delivery and correlate with the oxygen supplied from the placenta.

In a study of human fibroblasts, differential sensitivity to adenosine was demonstrated in donors of different ages (BYNUM 1980). Foetal fibroblasts were the most sensitive to adenosine, which produced inhibition of growth and RNA synthesis; in contrast, fibroblasts taken from 4-year-old donors showed growth stimulation to adenosine.

Activation of A_2-like receptors by adenosine stimulates L-arginine transport and nitric oxide synthase in human foetal endothelial cells (SOBREVIA et al. 1997).

C. Postnatal Development

A growing number of reports are available concerning purinergic signalling in postnatal development, particularly in brain, heart, blood vessels, lung, skeletal neuromuscular junctions and gut.

I. Central Nervous System

The emphasis in earlier studies of purines in the CNS was largely about adenosine (DUNWIDDIE 1985; FREDHOLM 1995), but since purinergic neurotransmission involving ATP as a transmitter was clearly demonstrated in the medial habenula of the rat brain (EDWARDS et al. 1992), many more papers have been appearing concerning P2 receptors in the brain and spinal cord (see GIBB and HALLIDAY 1996; ABBRACCHIO 1997).

Postnatal changes in expression of A_{2A} receptors, as indicated by ligand binding of the adenosine receptor antagonist [^{3}H]CGS 21680, have been described in various brain regions (JOHANSSON et al. 1997). The authors suggest that postnatal changes in these adenosine receptors may explain age-dependent differences in stimulatory effects of caffeine and endogenous protection against seizures. Further, since A_{2A} receptors show a co-distribution with D_2 receptors throughout development, they also speculate that caffeine may partly exert its actions via dopamine receptors. A recent study claimed wide distribution of $P2Y_1$ receptors in the 1-day-old chick brain, based on in vitro ligand autoradiography of [^{35}S]2'-deoxy 5'-O-(1-thio) ATP binding sites and in situ hybridisation histochemistry (WEBB et al. 1998).

Age-dependent changes in presynaptic neuromodulation via P1 (A_1) adenosine receptors have been studied in hippocampus of mouse (PAGONOPOULOU and ANGELATOU 1992) and rat (JIN et al. 1993; SPERLÁGH et al. 1997). There is a down-regulation and reduced responsiveness to pre-synaptic A_1 receptors modulating acetylcholine release during postnatal development and ageing.

A study of the distribution of adenosine-binding sites, using [^{3}H]CHA and [^{3}H]NECA in the cat visual cortex, revealed changes in the laminar binding pattern during postnatal development (SHAW et al. 1986). Adenosine receptors were detected in the rat forebrain in very young neonates, preceding N-protein coupling (MORGAN and MARANGOS 1987). From a study of the developing guinea-pig brain, it was claimed that in cerebellum, but not in forebrain, postnatal coupling of adenosine A_1 receptors to associated G-proteins is much more extensive than in the pre-natal period (MORGAN et al. 1990). Postnatal development of ATPase, ADPase and ATP diphosphohydrolase activity in the cerebral cortex have also been studied (MÜLLER et al. 1993; OLIVEIRA et al. 1994). The activities increased steadily from birth, reaching maximum values at 21 days of age. A marked increase in activity of 5'-nucleotidase was also seen in rat olfactory bulb during neonatal development (CLEMOW and BRUNJES 1996). During early postnatal life, the enzyme was found within synapses in the brain but a glial pattern of expression dominated in the adult, with the exception of the olfactory bulb where both glial and synaptic staining was present in mature adults.

Both adenosine (HERLENIUS et al. 1997) and ATP (FUNK et al. 1997) have been shown to modulate the activity of inspiratory neurons in the brain stem of neonatal rats. Adenosine depressed both the neurons in the rostral ventro-lateral medulla and the respiratory motor output, with a more pronounced decrease in respiratory activity in younger animals. ATP excitation of glutamate inspiratory drive to mouse hypoglossal neurons was demonstrated; it was shown to remain constant during the first 2 weeks of postnatal development. A secondary inhibitory response most likely results from enzymatic breakdown of extracellular ATP to adenosine and subsequent activation of A_1 adenosine receptors. ATP and adenosine mediate responses of sympathetic preganglionic neurons in a neonatal rat brain stem-spinal cord preparation

(DEUCHARS 1995). P2 purinoceptors have been demonstrated in rat locus coeruleus neurons; they first appear to be functional soon after birth, thereafter increasing to reach maturity in animals older than 18 days (WIRKNER et al. 1998).

The developmental properties of adenosine A_{2A} receptors differ from those of A_1 receptors during postnatal development of rat striatum. A_{2A} receptor binding sites were low at birth (about 3% of adult levels) and then increased, mostly between birth and 5 days, and then again from 15 days to adulthood (DORIAT et al. 1996). In contrast, A_1 receptors are widely distributed at birth (about 10% of adult levels) and then increase gradually until adulthood, with a peak during the second week of postnatal life (MARANGOS et al. 1982; GEIGER et al. 1984; DAVAL et al. 1991). The ratio of adenosine (A_{2A}) receptors to dopamine D_2 receptors in the rat striatum increases with age, involving both presynaptic and postsynaptic mechanisms (POPOLI et al. 1998). A decrease in striatal A_2 receptor mRNA expression has been demonstrated with in situ hybridisation histochemistry in rat striatum between 3 and 24 months, but it was suggested that this may be related to neuronal loss over the same period (SCHIFFMAN and VANDERHAEGHEN 1993).

In vitro studies of sensorimotor cortical neurons from 14-day-old (P14) and 30-day-old (P30) rats have shown that Ca^{2+} release could be evoked by ATP indicating the presence of P2Y receptors (LALO and KOSTYUK 1998). Almost all P14 neurons appeared to possess such receptors, whereas only about one third of neurons from P30 rats responded to ATP, suggesting that substantial changes in signalling mechanisms occur in neocortical neurons in the third-fourth week of postnatal development.

Ca^{2+}/Mg^{2+}-ATPase was shown to be present in the rat cerebellum (YOSHIOKA and TANAKA 1989). It first appeared in immature Purkinje cells at birth and increased throughout postnatal development; it was observed in migratory granular neurons during the critical period from 3 to 15 postnatal days. In contrast, peak alkaline phosphatase activity in the developing cerebellum did not appear on the granular neurons until day 7 postnatal. Under-nutrition during foetal and postnatal development impairs and delays the activity of the enzymes involved in purine nucleotide salvage pathways particularly in the cerebellum of rat brain (VANELLA et al. 1983). Adenosine receptors appeared earlier and reached higher adult levels in the brains (most notably in the cerebellum) of mice pups chronically exposed in utero to caffeine (MARANGOS et al. 1984).

Northern blotting of the brain of newborn rats showed the 6kb RNA transcript of the $P2X_7$ receptor, but there was no detectable reactivity in the adult brain (COLLO et al. 1997).

II. Cardiovascular System

1. Heart

A_1 receptors are present in greater numbers in the pre-innervated immature rat myocytes compared to the adult and they are functionally coupled to their effector sites (COTHRAN et al. 1995). An increase in density of A_1 receptors in rabbit heart in old age has been claimed (MUDUMBI et al. 1995) in contrast to the diminished β-adrenergic responsiveness in the senescent heart (VESTAL et al. 1979). In the immature (25-day-old) rat heart, adenosine was found to have little role in the modulation of contractile responsiveness via activation of either A_1 or A_2 receptors, but enhanced antiadrenergic and stimulatory functions of adenosine via A_1 and A_2 receptor activation were present in the heart of mature (79-day-old) rats (SAWMILLER et al. 1996). Reduced adenosine A_1 receptor and G_α protein-coupling in rat ventricular myocardium during ageing has been reported (CAI et al. 1997). Age-related decline in β-adrenergic and adenosine A_1 receptor function in the rat heart are attenuated by dietary restriction (GAO et al. 1998). Adenosine formation and release by neonatal rat ventricular myocytes in culture has been described (MEGHJI et al. 1985, 1993). Adenosine and hypoxia effects on the atrioventricular node of neonatal rabbit hearts have been reported (YOUNG et al. 1987).

Multiple P2Y receptor subtypes are expressed in rat heart and the expression in myocytes changes from neonate to the adult (WEBB et al. 1996). $P2Y_1$ is expressed at higher levels in comparison to $P2Y_2$, $P2Y_4$ and $P2Y_6$ in the neonatal myocyte, while $P2Y_4$ could not be detected in the adult myocytes. In the neonatal fibroblasts in the heart, $P2Y_1$ and $P2Y_6$ receptors appear to be expressed at higher levels than $P2Y_2$ and $P2Y_4$ receptors. In a later RT-PCR study of the foetal human heart, $P2X_1$, $P2X_3$ and $P2X_4$ subtypes were identified together with $P2Y_2$, $P2Y_4$ and $P2Y_6$ receptors (BOGDANOV et al. 1998).

Extracellular ATP has been shown to inhibit adrenergic agonist-induced hypertrophy of neonatal cardiac myocytes and alter differentially the changes in gene expression that accompany hypertrophy (ZHENG et al. 1996). ATP had been previously shown by this group to increase expression of the immediate-early genes c-fos and jun B in cultured neonatal cardiac myocytes, but by a different pathway from that produced by noradrenaline (ZHENG et al. 1994).

The adenosine agonist NECA increases cardiac output in developing *Xenopus* tadpoles through a combination of increased filling and accelerated growth of heart and vessels (TANG and ROVAINEN 1996).

2. Vascular System

Adenosine appears to play an important role in the regulation of coronary blood flow in the newborn lamb (MAINWARING et al. 1985). Adenosine-induced vasodilatation of the guinea-pig coronary vascular bed is greater in immature guinea-pig heart (5-days) compared to mature heart (1–2 months) (MATHERNE et al. 1990).

The endothelium-dependent relaxant response of the aorta to ATP, mediated by nitric oxide, was greatest in 4-week-old rats, but declined progressively at 45 and 105 weeks (UEDA and MORITOKI 1991).

Age-related changes have been described concerning the relative importance of noradrenaline and ATP as mediators of the contractile responses of the rat tail artery to sympathetic nerve stimulation; the ATP component is dominant in young (100–150g) rats, but declines with age (BAO et al. 1989). In some young rats, ATP appeared to be the sole mediator of the sympathetic contractile response in the tail artery. A reduction in spontaneous and α-adrenoceptor-induced release of ATP from endothelial cells of the rat tail artery occurs with advancing age (HASHIMOTO et al. 1995). Recordings in vitro of spinal L_3 and L_5 ventral root d.c. potential, made in 1- to 2-day-old rats, showed that ATP and α,β-meATP activate sensory, possibly nociceptive, P2X receptors on nerves in the rat tail (TREZISE and HUMPHRY 1996).

Oxygen-induced pulmonary vasodilatation is mediated by ATP in newborn lambs; it is attenuated by combined administration of P1 and P2Y receptor antagonists (CROWLEY 1997). However it appears to differ from the mechanism of oxygen-induced pulmonary vasodilatation in the late gestational foetal lamb (MORIN and EGAN 1992; KONDURI et al. 1992b, 1993). Infusion of ATP caused a significant decrease in pulmonary vascular resistance in normoxic and hypoxic newborn lambs (KONDURI and WOODWARD 1991). Adenosine is also a pulmonary vasodilator in newborn lambs (KONDURI et al. 1993).

Responses of isolated mesenteric arteries of the beagle to both β-adrenoceptors and P1 purinoceptors are less in old compared to young animals (SHIMIZU and TODA 1986). In the mesenteric artery of the rat, the sympathetic and sensory nerve fibre plexuses develop over the first three postnatal weeks, but functionally mature nerve-mediated contractile responses cannot be elicited before 14 days postnatal (SERIO et al. 1996), correlating with the appearance of adult-like excitatory junction potentials (EJPs) (HILL et al. 1983). Prior to this period, intracellular recordings from animals aged 4 to 9 days, showed slow depolarising potentials which were mediated by α-adrenoceptors. From day 9 onwards, EJPs, which were resistant to α-adrenergic antagonists, were recorded (HILL et al. 1983) and are likely to be mediated by ATP (see BURNSTOCK 1995).

Adenosine is claimed not to be as effective as a vasodilator of internal carotid arteries in the newborn as it is in the adult pig (LAUDIGNON et al. 1990). It is suggested that neonatal brain adenosine may play a role in regulating blood flow during hypoxia (PARK et al. 1987). Following denervation studies in the rat mesenteric vascular bed, electrical responses similar to those seen during the early stages of development were recorded (HILL et al. 1985), suggesting that a similar sequence of events occurs during regeneration as takes place during development. A developmental profile for P2X receptor subtype mRNA expression in rat mesenteric artery showed high expression for $P2X_1$

and P2X$_4$ at postnatal day 7 which was retained during development until day 360 (the oldest animals examined) (PHILLIPS 1998).

III. Airways

The effect of the adenosine agonist R-PIA on respiration was studied in rabbit pups (1–8 days old) (RUNOLD et al. 1986). R-PIA caused a decrease in ventilation which was blocked by theophylline indicating mediation by P1 receptors. R-PIA caused a considerably more pronounced effect in 1- to 3-day-old animals than in 8-day-old animals and was shown to bind with higher affinity in brains from newborn animals compared to older animals. The authors suggested that this might explain the potent therapeutic effect of the adenosine antagonist, theophylline, on recurrent apnoea in pre-term infants. In studies of anaesthetised newborn piglets, it was concluded that adenosine contributes to ventilatory depression caused by hypoxia (LAGERCRANTZ et al. 1984; DARNELL and BRUCE 1987). In another investigation of the role of adenosine in the hypoxic ventilatory response of the newborn piglet, the authors concluded that adenosine plays a central role in modulating ventilation in the newborn piglet and is involved in the diphasic ventilatory responses to hypoxia (LOPES et al. 1994). Postnatally, at 3 days, adenosine released from the CNS and within the kidney is a major contributor to the secondary fall in ventilation and renal vasoconstriction, whereas at 3 weeks adenosine makes little contribution to the ventilatory responses or renal vasoconstriction, although it is largely responsible for hypoxia-induced vasodilatation in skeletal muscle (ELNAZIR et al. 1996).

Adenosine has a regulatory role in lung surfactant secretion in the newborn rabbit (ROONEY and GOBRAN 1988). Ventilation-induced increase in secretion of lung surfactant was inhibited by P1 receptor antagonists. ATP stimulation of surfactant in type II cells in adult rats is mediated by both a P2Y$_2$ receptor coupled to phospholipase C and a receptor coupled to adenylate cyclase; UTP also activated the P2Y$_2$ receptor, but did not stimulate cAMP formation. In newborn rats, ATP increased secretion as early as day 1, but the effect of UTP did not become significant until 4 days after birth (GOBRAN and ROONEY 1997). ATP increased cAMP formation in newborns, but did not promote phospholipase D activation until day 4. Thus the adenylate cyclase-coupled ATP signalling mechanism is functional early in development, but the P2Y$_2$ receptor pathway is not.

IV. Gastrointestinal Tract

There have been several studies of developmental changes in purinergic signalling in the small intestine (see also HOURANI 1999) (Fig. 8). In rat duodenal segments, ATP and ADP produced *contractile* responses on postnatal day 1; this response increased with age, peaking on day 7, followed by a gradual decrease and was non-existent by day 21 (FURUKAWA and NOMOTO 1989). In

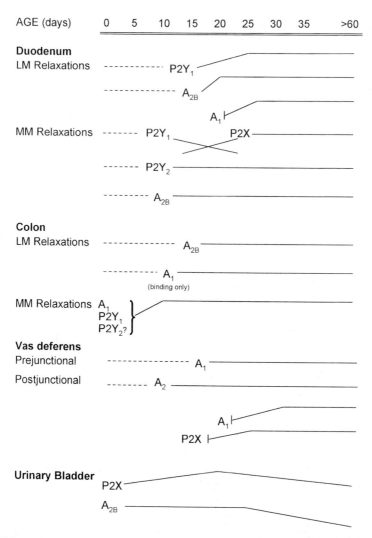

Fig. 8. Diagram summarizing the development of functional responses mediated by purine receptors in the rat duodenum, colon, urinary bladder and vas deferens. The *dashed lines* represent ages at which it was not possible to study functional responses, and the *solid lines* show when responses were observed, with the slope of the line indicating whether a response, in general, increased or decreased with age. (Reproduced with permission from HOURANI 1999)

contrast, the *relaxant* response to ATP and ADP was apparent at day 21 and continued to increase during the period examined, i.e. up to day 70. Adenosine or AMP did not elicit responses before day 14, which were then small relaxations that increased with age. In a later study of rat duodenum by NICHOLLS et al. (1990) it was reported that, if the tissues were precontracted with carbachol, *low* concentrations of ATP could be shown to produce relaxations from day 2 increasing with age, while higher concentrations of ATP (3 μmol/l and above) were excitatory, but only until day 15 and they postulated that P2Y receptors mediated both relaxations and contractions. They also reported weak responses to adenosine as early as day 2. The response of the rat duodenum to the ganglion stimulant, nicotine, was contraction in neonatal rats, but changed from contraction to relaxation around the third postnatal week (IRIE et al. 1994). The striking switch from contractile responses to purines to relaxant effects is probably associated with the major changes that take place in the gut at weaning, which occurs during the third postnatal week, when the food source and composition change from being liquid and rich in fat to being solid and rich in carbohydrate (HENNING 1981).

In a study of the ontology of P1 receptor signalling in the longitudinal muscle of the rat duodenum, it was shown that A_{2B} receptors are present at day 15, but A_1 receptors did not appear until after day 20; both receptor subtypes mediating relaxation (BROWNHILL et al. 1996). The authors also reported that A_{2B} receptors mediated *contraction* of the muscularis mucosa from day 10. In later experiments, the ontogeny of P2 purinoceptors in the duodenum was examined (BROWNHILL et al. 1997). It was concluded that P2Y receptors mediate *relaxation* of the longitudinal muscle at day 25, while in the muscularis mucosa, P2Y receptors mediate *contraction*, which is well developed before day 20, but that after this time P2X receptors mediate the contraction, as well as P2U receptors.

Responses to adenosine, ATP and α,β-meATP have examined in the rat *colon* longitudinal muscle and muscularis mucosa during postnatal development (HOURANI et al. 1993b). The longitudinal muscle *relaxes* via A_{2B} and P2Y receptors, while the muscularis mucosa *contracts* through A_1 and probably P2Y receptors, with UTP also causing contraction. The contractile responses of the muscularis mucosa to all three agonists were observed from the day after birth, but much lower than in the adult; the responses increased with time to reach a maximum at days 10–15, at which time they were greater than in the adult. There do not appear to be any major changes associated with weaning in the purinoceptor population in the colon, in contrast to the situation in the duodenum, although nitric oxide synthase activity increase during the first 20 postnatal days and then fall after weaning (BROWN and TEPPERMAN 1997).

An electrophysiological study of developmental changes in the innervation of the guinea-pig taenia coli has been carried out (ZAGORODNYUK et al. 1993). The non-adrenergic (largely purinergic) inhibitory system appeared before and matured faster than the cholinergic excitatory system. The NANC inhibitory system was present by 8 weeks of gestation, while cholinergic excitatory transmission was not seen until birth. Responses to α,β-meATP were also recorded in the foetal taenia coli.

The distribution and localisation of the breakdown enzyme for adenosine, adenosine deaminase, was studied during postnatal development of the mouse alimentary tract (CHINSKY et al. 1990). Adenosine deaminase was predominantly localised to the keratinised squamous epithelium that lines the mucosal layer of the oesophagus, fore-stomach and surface of the tongue and the simple columnar epithelium of proximal small intestine. The levels of adenosine deaminase in these tissues were low at birth, but achieved very high levels within the first two weeks of postnatal life. The authors suggested that adenosine deaminase is subject to strong cell-specific developmental regulation during functional differentiation of certain foregut derivatives in mice.

V. Vas Deferens

Changes in purinergic signalling in the vas deferens might be expected to occur later than in the gut, because rats are not sexually active until about 10 weeks, although the morphology of the vas deferens appears mature by day 35 (FRANCAVILLA et al. 1987). As far back as 1970, FURNESS et al. (1970) showed that EJPs, now known to be produced by ATP (see BURNSTOCK 1995), in response to nerve stimulation of the vas deferens were not observed for mice of less than 18 days postnatal. Another early study of postnatal development of functional neurotransmission in the rat vas deferens showed that at 3 weeks postnatal (the earliest time studied) the responses to field stimulation with single or trains of pulses lacked the adrenergic component, although the non-adrenergic component was present (MACDONALD and MCGRATH 1984). ATP and noradrenaline are now well established as cotransmitters in the sympathetic nerves supplying vas deferens (see BURNSTOCK 1990, 1995). Responses to ATP also first appeared at day 15 and increased with age (HOURANI et al. 1993a).

Examination of the ontogeny of P1 purinergic receptors, which mediate inhibition of neurotransmission by sympathetic nerves in the rat vas deferens, showed that adenosine, acting via prejunctional A_1 receptors, inhibited when nerve-mediated contractions were first seen at day 15, but its potency decreased with age. Inhibitory postjunctional A_2-like receptors were also identified in the rat vas deferens, although the selective agonists and antagonists available at that time do not make the observation decisive. In a later study, this group claimed that inhibitory postjunctional A_2-like receptors and prejunctional A_1 receptors were present from days 10 and 15, respectively. In contrast, they identified postjunctional *excitatory* A_1 receptors that did not appear until after day 20 (PEACHEY et al. 1996).

In 2-week-old guinea-pigs, stimulation of the hypogastric nerve produced monophasic contractions which were only partially blocked by the combination of prazosin and α,β-meATP, suggesting the involvement of an unknown transmitter; however, in 10- to 15-week animals, stimulation produced a biphasic contraction, which was almost completely inhibited by both blockers (NAGAO et al. 1994).

Studies of developmental changes on sympathetic nerve-evoked contractions of the *circular* muscle layer of the guinea-pig vas deferens showed that the contractions produced significant decrease with increasing age, apparently due to postjunctional rather than prejunctional mechanisms, responses to α,β-meATP decreasing in parallel (REN et al. 1996). An increase in $P2X_1$ receptor mRNA expression has been demonstrated between postnatal days 10 and 42 (LIANG et al. 1998).

VI. Urinary Bladder

ATP and acetylcholine are cotransmitters in parasympathetic nerves supplying the bladder (see BURNSTOCK et al. 1978; HOYLE and BURNSTOCK 1991; BURNSTOCK 1995).

In an early study of the responses of the rabbit urinary bladder to autonomic neurotransmitters, receptors to ATP and acetylcholine were recognised in the newborn animals, but adrenoceptors were poorly expressed until a later stage (LEVIN et al. 1981). In a later study of rabbit bladder, newborn bladders were shown to generate much greater tension in response to ATP than adult tissue and then decline, while the response to cholinergic agonists did not decline (KEATING et al. 1990; ZDERIC et al. 1990; SNEDDON and MCLEES 1992). The ontogeny of purinoceptors in *rat* urinary bladder was examined by NICHOLLS et al. (1990). Responses to adenosine (inhibitory) and ATP (excitatory) mediated by P1 and P2X purinoceptors, respectively, were present as early as postnatal day 2, the earliest day studied. Adenosine was more potent in the neonate than in the adult, while the potency of ATP initially increased with age, but then declined, being highest between postnatal days 10 and 25. In vivo evidence for the functional roles of cholinergic and purinergic components of parasympathetic cotransmission for micturition contractions in normal unanaesthetised rats has been presented (IGAWA et al. 1993).

The rate and pattern of breakdown of ATP and adenosine by ectoenzymes in the rat urinary bladder was shown to be identical in neonates and adults, indicating that the marked differences in potency to ATP and adenosine during development is likely to be due to changes in receptor number and/or agonist affinity or efficacy (NICHOLLS et al. 1992).

VII. Other Organs

The *hepatic* ATP/ADP ratio showed significant decrease during the first day of postnatal life in starved newborn rats, correlating directly to the gluconeogenic flux (CUEZVA et al. 1983).

Ectoenzymes for purines have been measured in the developing rat *testis*; it was concluded that full metabolic involvement in terms of Mg^{2+} ATPase and 5'-nucleotidase is not achieved until 45 days postnatal (GAZDZIK and KAMINSKI 1986).

Functional $P2Y_1$ receptors are expressed in immature *salivary glands* and the receptor activity decreases as the glands mature (PARK et al. 1997).

In rat *kidney*, intrarenal adenosine is a physiological regulator of glomerular filtration rate and renal blood flow and appears to play a key role in the hypoxaemia-induced renal insufficiency in newborn rats (GOUYON and GUINGNARD 1989).

Adenosine is important in regulating the action of insulin on rat fat cell metabolism during postnatal development and ageing (ROLBAND et al. 1990).

ATP and ADP and, to a much lesser extent AMP and adenosine, increase insulin secretion from the isolated, perfused newborn dog pancreas (CHAPAL et al. 1981).

RT-PCR studies of $P2X_7$ mRNA in postnatal rats (P23–P210) showed positive identification in the retina; in the adult retina immunolabelling for $P2X_7$ receptors was detected in amacrine and ganglion cells (BRÄNDLE et al. 1998).

D. Summary

Several interesting generalisations emerge from the analysis in this review

1. Purinergic signalling appears to play a major role in early development (before classical adrenergic signalling) and usually declines with maturation; sometimes this involves reduction in potency of purines and pyrimidines, sometimes particular purinoceptor subtypes are no longer expressed.
2. There are examples where a similar sequence of events occurs during regeneration after trauma to adult tissues as occurs during development, e.g. skeletal muscle, mesenteric arteries.
3. During postnatal development, responses to nucleosides and nucleotides increase or decrease with age depending on the physiological demands of the particular system involved.
4. Both P1 and P2 receptors are present early in development of many systems, but there are a few examples where the earliest effects of ATP are mediated via adenylate cyclase, and only later in receptor activation does the IP_3 second messenger system for P2Y receptor transduction come into operation (e.g. type III pneumocytes in lung). Fast P2X purinergic signalling may appear a little later in development compared to P1 and P2Y receptors, except perhaps in urinary bladder.

This is reminiscent of what appears to occur during phylogeny (see BURNSTOCK 1996b), thus being consistent with the general principle that "ontogeny repeats phylogeny".

References

Abbracchio MP (1997) ATP in brain function. In: Jacobson KA, Jarvis MF (eds) Purinergic approaches in experimental therapeutics. Wiley-Liss, New York, pp 383–404
Abe Y, Sorimachi M, Itoyama Y, Furukawa K, Akaike N (1995) ATP responses in the embryo chick ciliary ganglion cells. Neuroscience 64:547–551

Adair TH, Montani JP, Strick DM, Guyton AC (1989) Vascular development in chick embryos: a possible role for adenosine. Am J Physiol 256:H240–H246

Akasu T, Hirai K, Koketsu K (1981) Increase of acetylcholine-receptor sensitivity by adenosine triphosphate: a novel action of ATP on ACh-sensitivity. Br J Pharmacol 74:505–507

Allgaier C, Wellmann H, Schobert A, von Kügelgen I (1995a) Cultured chick sympathetic neurons: modulation of electrically evoked noradrenaline release by P_2-purinoceptors. Naunyn Schmiedebergs Arch Pharmacol 352:17–24

Allgaier C, Wellmann H, Schobert A, Kurz G, von Kügelgen I (1995b) Cultured chick sympathetic neurons: ATP-induced noradrenaline release and its blockade by nicotinic receptor antagonists. Naunyn Schmiedebergs Arch Pharmacol 352:25–30

Bao JX, Eriksson IE, Stjärne L (1989) Age-related variations in the relative importance of noradrenaline and ATP as mediators of the contractile response of rat tail artery to sympathetic nerve stimulation. Acta Physiol Scand 136:287–288

Barker PM, Gatzy JT (1998) Effects of adenosine, ATP, and UTP on chloride secretion by epithelia explanted from fetal rat lung. Pediatr Res 43:652–659

Barnes EM Jr, Thampy KG (1982) Subclasses of adenosine receptors in brain membranes from adult tissue and from primary cultures of chick embryo. J Neurochem 39:647–652

Beaudoin AR (1976) Effect of adenosine triphosphate and adenosine diphosphate on the teratogenic action of trypan blue in rats. Biol Neonate 28:133–139

Blair TA, Parenti M, Murray TF (1989) Development of pharmacological sensitivity to adenosine analogs in embryonic chick heart: role of A_1 adenosine receptors and adenylyl cyclase inhibition. Mol Pharmacol 35:661–670

Bogdanov YD, Dale L, King BF, Whittock N, Burnstock G (1997) Early expression of a novel nucleotide receptor in the neural plate of *Xenopus* embryos. J Biol Chem 272:12583–12590

Bogdanov Y, Rubino A, Burnstock G (1998) Characterisation of subtypes of the P2X and P2Y families of receptors in the foetal human heart. Life Sci 62:697–703

Brändle U, Kohler K, Wheeler-Schilling TH (1998) Expression of the $P2X_7$-receptor subunit in neurons of the rat retina. Brain Res Mol Brain Res 62:106–109

Brown JF, Tepperman BL (1997) Ontogeny of nitric oxide synthase activity and endotoxin-mediated damage in the neonatal rat colon. Pediatr Res 41:635–640

Brownhill VR, Hourani SMO, Kitchen I (1996) Differential ontogeny of adenosine receptors in the longitudinal muscle and muscularis mucosae of the rat isolated duodenum. Eur J Pharmacol 317:321–328

Brownhill VR, Hourani SMO, Kitchen I (1997) Ontogeny of P2-purinoceptors in the longitudinal muscle and muscularis mucosae of the rat isolated duodenum. Br J Pharmacol 122:225–232

Burnstock G (1990) Noradrenaline and ATP as cotransmitters in sympathetic nerves. Neurochem Int 17:357–368

Burnstock G (1995) Noradrenaline and ATP: cotransmitters and neuromodulators. J Physiol Pharmacol 46:365–384

Burnstock G (1996a) P2 Purinoceptors: historical perspective and classification. In: Chadwick DJ, Goode JA (eds) P2 purinoceptors: localization, function and transduction mechanisms. Ciba Foundation Symposium 198. John Wiley and Sons, Chichester, pp 1–29

Burnstock G (1996b) Purinoceptors: ontogeny and phylogeny. Drug Dev Res 39:204–242

Burnstock G (1997) The past, present and future of purine nucleotides as signalling molecules. Neuropharmacology 36:1127–1139

Burnstock G, Wood JN (1996) Purinergic receptors: their role in nociception and primary afferent neurotransmission. Curr Opin Neurobiol 6:526–532

Burnstock G, Cocks T, Crowe R, Kasakov L (1978) Purinergic innervation of the guinea-pig urinary bladder. Br J Pharmacol 63:125–138

Bynum JW (1980) Differential adenosine sensitivity in fibroblasts from different age donors. Exp Geront 15:217–225

Cai G, Wang HY, Gao E, Horwitz J, Snyder DL, Pelleg A, Roberts J, Friedman E (1997) Reduced adenosine A_1 receptor and G alpha protein coupling in rat ventricular myocardium during aging. Circ Res 81:1065–1071

Chapal J, Loubatières-Mariani MM, Roye M (1981) Effect of adenosine and phosphated derivatives on insulin release from the newborn dog pancreas. J Physiol (Paris) 77:873–875

Chechik BE, Sengupta S, Hibi T, Fernandes B (1985) Immunomorphological localization of adenosine deaminase in rat tissues during ontogeny. Histochem J 17:153–170

Chinsky JM, Ramamurthy V, Fanslow WC, Ingolia DE, Blackburn MR, Shaffer KT, Higley HR, Trentin JJ, Rudolph FB, Knudsen TB, Kellems RE (1990) Developmental expression of adenosine deaminase in the upper alimentary tract of mice. Differentiation 42:172–183

Clemow DB, Brunjes PC (1996) Development of 5′-nucleotidase staining in the olfactory bulbs of normal and naris-occluded rats. Int J Dev Neurosci 14:901–911

Clunes MT, Collett A, Baines DL, Bovell DL, Murphie H, Inglis SK, McAlroy HL, Olver RE, Wilson SM (1998) Culture substrate-specific expression of P2Y2 receptors in distal lung epithelial cells isolated from foetal rats. Br J Pharmacol 124:845–847

Collo G, Neidhart S, Kawashima E, Kosco-Vilbois M, North RA, Buell G (1997) Tissue distribution of the $P2X_7$ receptor. Neuropharmacology 36:1277–1283

Cooper J, Hill SJ, Alexander SP (1997) An endogenous A_{2B} adenosine receptor coupled to cyclic AMP generation in human embryonic kidney (HEK 293) cells. Br J Pharmacol 122:546–550

Cothran DL, Lloyd TR, Taylor H, Linden J, Matherne GP (1995) Ontogeny of rat myocardial A_1 adenosine receptors. Biol Neonate 68:111–118

Crowe R, Burnstock G (1981) Perinatal development of quinacrine-positive neurons in the rabbit gastrointestinal tract. J Auton Nerv Syst 4:217–230

Crowley MR (1997) Oxygen-induced pulmonary vasodilation is mediated by adenosine triphosphate in newborn lambs. J Cardiovasc Pharmacol 30:102–109

Cuezva JM, Fernandez E, Valcarce C, Medina JM (1983) The role of ATP/ADP ratio in the control of hepatic gluconeogenesis during the early neonatal period. Biochim Biophys Acta 759:292–295

Dale N (1998) Delayed production of adenosine underlies temporal modulation of swimming in frog embryo. J Physiol (Lond) 511:265–272

Dale N, Gilday D (1996) Regulation of rhythmic movements by purinergic neurotransmitters in frog embryos. Nature 383:259–263

Darnall RA, Bruce RD (1987) Effects of adenosine and xanthine derivatives on breathing during acute hypoxia in the anesthetized newborn piglet. Pediatr Pulmonol 3:110–116

Daval J-L, Werck MC, Nehlig A, Pereira de Vasconcelos A (1991) Quantitative autoradiographic study of the postnatal development of adenosine A_1 receptors and their coupling to G proteins in the rat brain. Neuroscience 40:841–851

de Mello MC, Ventura AL, Paes de Carvalho R, Klein WL, de Mello FG (1982) Regulation of dopamine- and adenosine-dependent adenylate cyclase systems of chicken embryo retina cells in culture. Proc Natl Acad Sci USA 79:5708–5712

Deckert J, Morgan PF, Daval JL, Nakajima T, Marangos PJ (1988) Ontogeny of adenosine uptake sites in guinea pig brain: differential profile of [^{3}H]nitrobenzylthioinosine and [^{3}H]dipyridamole binding sites. Brain Res 470:313–316

Deuchars S (1995) Effect of ATP and adenosine on sympathetic preganglionic neurons in a neonatal rat brainstem-spinal cord preparation. J Physiol (Lond) 489:154P

Di Virgilio F, Zanovello P, Zambon A, Bronte V, Pizzo P, Murgia M (1995) Cell membrane receptors for extracelluar ATP: a new family of apoptosis-signalling molecules. Fundam Clin Immunol 3:80–81

Di Virgilio F, Chiozzi P, Falzoni S, Ferrari D, Sanz JM, Venketaraman V, Baricordi OR (1998) Cytolytic P2X purinoceptors. Cell Death Diff 5:191–199

Doriat J-F, Humbert A-C, Daval J-L (1996) Brain maturation of high-affinity adenosine A_2 receptors and their coupling to G-proteins. Brain Res Dev Brain Res 93:1–9

Dunwiddie TV (1985) The physiological role of adenosine in the central nervous system. Int Rev Neurobiol 27:63–139

Dusseau JW, Hutchins PM, Malbasa DS (1986) Stimulation of angiogenesis by adenosine on the chick chorioallantoic membrane. Circ Res 59:163–170

Edwards FA, Gibb AJ, Colquhoun D (1992) ATP receptor-mediated synaptic currents in the central nervous system. Nature 359:144–147

Egerman RS, Bissonnette JM, Hohimer AR (1993) The effects of centrally administered adenosine on fetal sheep heart rate accelerations. Am J Obst Gynecol 169:866–869

Elnazir B, Marshall JM, Kumar P (1996) Postnatal development of the pattern of respiratory and cardiovascular response to systemic hypoxia in the piglet: the roles of adenosine. J Physiol (Lond) 492:573–585

Eriksson H, Heilbronn E (1989) Extracellularly applied ATP alters the calcium flux through dihydropyridine-sensitive channels in cultured chick myotubes. Biochem Biophys Res Commun 159:878–885

Fissore R, O'Keefe S, Kiessling AA (1992) Purine-induced block to mouse embryo cleavage is reversed by compounds that elevate cyclic adenosine monophosphate. Biol Reprod 47:1105–1112

Foresta C, Rossato M, Di Virgilio F (1992) Extracellular ATP is a trigger for the acrosome reaction in human spermatozoa. J Biol Chem 267:19443–19447

Francavilla S, Moscardelli S, Properzi G, De Matteis MA, Scorza Barcellona P, Natali PG, De Martino C (1987) Postnatal development of epididymis and ductus deferens in the rat. Cell Tissue Res 249:257–265

Franco R, Casado V, Ciruela F, Saura C, Mallol J, Canela EI, Lluis C (1997) Cell surface adenosine deaminase: much more than an ectoenzyme. Prog Neurobiol 52:283–294

Fraser RA, Ellis EM, Stalker AL (1979) Experimental angiogenesis in the chorioallantoic membrane. Bibl Anat 18:25–27

Fredholm BB (1995) Adenosine receptors in the central nervous system. News Physiol Sci 10:122–128

Fredholm BB, Lerner U (1982) Metabolism of adenosine and 2'-deoxy-adenosine by fetal mouse calvaria in culture. Med Biol 60:267–271

Fu W-M (1994) Potentiation by ATP of the postsynaptic acetylcholine response at developing neuromuscular synapses in *Xenopus* cell cultures. J Physiol (Lond) 477:449–458

Fu W-M (1995) Regulatory role of ATP at developing neuromuscular junctions. Prog Neurobiol 47:31–44

Fu W-M, Huang F-L (1994) Potentiation by endogenously released ATP of spontaneous transmitter secretion at developing neuromuscular synapses in *Xenopus* cell cultures. Br J Pharmacol 111:880–886

Fu W-M, Poo M-M (1991) ATP potentiates spontaneous transmitter release at developing neuromuscular synapses. Neuron 6:837–843

Fu WM, Chen YH, Lee KF, Liou JC (1997) Regulation of quantal transmitter secretion by ATP and protein kinases at developing neuromuscular synapses. Eur J Neurosci 9:676–685

Fukuda S, Katoh S, Yamamoto K, Hashimoto M, Kitao M (1990) Correlation between levels of plasma adenosine triphosphate and stress to the fetus at delivery. Biol Neonate 57:150–154

Funk GD, Parkis MA, Selvaratnam SR, Walsh C (1997) Developmental modulation of glutamatergic inspiratory drive to hypoglossal motoneurons. Respir Physiol 110:125–137

Furness JB, McLean JR, Burnstock G (1970) Distribution of adrenergic nerves and changes in neuromuscular transmission in the mouse vas deferens during postnatal development. Dev Biol 21:491–505

Furukawa K, Nomoto T (1989) Postnatal changes in response to adenosine and adenine nucleotides in rat duodenum. Br J Pharmacol 97:1111–1118

Gao E, Snyder DL, Roberts J, Friedman E, Cai G, Pelleg A, Horwitz J (1998) Age-related decline in β-adrenergic and adenosine A_1 receptor function in the heart are attenuated by dietary restriction. J Pharmacol Exp Ther 285:186–192

Gazdzik T, Kaminski M (1986) Evolution of localization of the reactions of adenosine triphosphatase (Mg^{++}-ATP-ase), 5'nucleotidase (5'nt), alkaline phosphatase (AP), and acid phosphatase (AcP) in developing rat testis. I. Physiological conditions. Acta Histochem 79:199–204

Geiger JD, Nagy JI (1987) Ontogenesis of adenosine deaminase activity in rat brain. J Neurochem 48:147–153

Geiger JD, LaBella FS, Nagy JI (1984) Ontogenesis of adenosine receptors in the central nervous system of the rat. Brain Res Dev Brain Res 13:97–104

Gerhart J, Danilchik M, Doniach T, Roberts S, Rowning B, Stewart R (1989) Cortical rotation of the *Xenopus* egg: consequences for the anteroposterior pattern of embryonic dorsal development. Development 107 [Suppl]:37–51

Gershon MD, Thompson EB (1973) The maturation of neuromuscular function in a multiply innervated structure: development of the longitudinal smooth muscle of the foetal mammalian gut and its cholinergic excitatory, adrenergic inhibitory, and non-adrenergic inhibitory innervation. J Physiol (Lond) 234:257–277

Gibb AJ, Halliday FC (1996) Fast purinergic transmission in the central nervous system. Semin Neurosci 8:225–232

Gobran LI, Rooney SA (1997) Adenylate cyclase-coupled ATP receptor and surfactant secretion in type II pneumocytes from newborn rats. Am J Physiol 272: L187–L196

Gordon HW, Tkaczyk W, Peer LA, Bernhard WG (1963) The effect of adenosine triphosphate and its decomposition products on cortisone induced teratology. J Embryol Exp Morph 11:475–482

Gouyon JB, Guignard JP (1989) Adenosine in the immature kidney. Dev Pharmacol Ther 13:113–119

Häggblad J, Heilbronn E (1987) Externally applied adenosine-5'-triphosphate causes inositol triphosphate accumulation in cultured chick myotubes. Neurosci Lett. 74:199–204

Häggblad J, Heilbronn E (1988) P_2-purinoceptor-stimulated phosphoinositide turnover in chick myotubes. Calcium mobilization and the role of guanyl nucleotide-binding proteins. FEBS Lett 235:133–136

Häggblad J, Eriksson H, Heilbronn E (1985) ATP-induced cation influx in myotubes is additive to cholinergic agonist action. Acta Physiol Scand 125:389–393

Hamberger V, Hamilton HL (1951) A series of normal stages in the development of the chick embryo. J Morphol 88:49–92

Hashimoto M, Shinozuka K, Bjur RA, Westfall DP, Hattori K, Masumura S (1995) The effects of age on the release of adenine nucleosides and nucleotides from rat caudal artery. J Physiol (Lond) 489:841–848

Hatae J, Sperelakis N, Wahler GM (1989) Development of the response to adenosine during organ culture of young embryonic chick hearts. J Dev Physiol 11:342–345

Hatori M, Teixeira CC, Debolt K, Pacifici M, Shapiro IM (1995) Adenine nucleotide metabolism by chondrocytes in vitro: role of ATP in chondrocyte maturation and matrix mineralization. J Cell Physiol 165:468–474

Henning SJ (1981) Postnatal development: coordination of feeding, digestion, and metabolism. Am J Physiol 241:G199–G214

Henning RH, Nelemans A, Van den Akker J, Den Hertog A (1992) The nucleotide receptors on mouse C2C12 myotubes. Br J Pharmacol 106:853–858

Henning RH, Duin M, Den Hertog A, Nelemans A (1993a) Activation of the phospholipase C pathway by ATP is mediated exclusively through nucleotide type P_2-purinoceptors in C2C12 myotubes. Br J Pharmacol 110:747–752

Henning RH, Duin M, Den Hertog A, Nelemans A (1993b) Characterization of P_2-purinoceptor mediated cyclic AMP formation in mouse C2C12 myotubes. Br J Pharmacol 110:133–138

Henning RH, Duin M, van Popta JP, Nelemans A, Den Hertog A (1996) Different mechanisms of Ca^{2+} handling following nicotinic acetylcholine receptor stimulation, P_{2U}-purinoceptor stimulation and K^+-induced depolarization in C2C12 myotubes. Br J Pharmacol 117:1785–1791

Herlenius E, Lagercrantz H, Yamamoto Y (1997) Adenosine modulates inspiratory neurons and the respiratory pattern in the brainstem of neonatal rats. Pediatr Res 42:46–53

Hilfer SR, Palmatier BY, Fithian EM (1977) Precocious evagination of the embryonic chick thyroid in ATP-containing medium. J Embryol Exp Morphol 42:163–175

Hill CE, Hirst GD, Van Helden DF (1983) Development of sympathetic innervation to proximal and distal arteries of the rat mesentery. J Physiol (Lond) 338:129–147

Hill CE, Hirst GD, Ngu MC, Van Helden DF (1985) Sympathetic postganglionic reinnervation of mesenteric arteries and enteric neurones of the ileum of the rat. J Auton Nerv Syst 14:317–334

Hofman PL, Hiatt K, Yoder MC, Rivkees SA (1997) A_1 adenosine receptors potently regulate heart rate in mammalian embryos. Am J Physiol 273:R1374–R1380

Hourani SMO (1999) Postnatal development of purinoceptors in rat visceral smooth muscle preparations. Gen Pharmacol 32:3–7

Hourani SMO, Nicholls J, Lee BS, Halfhide EJ, Kitchen I (1993a) Characterization and ontogeny of P_1-purinoceptors on rat vas deferens. Br J Pharmacol 108:754–758

Hourani SMO, Shaw DA, Kitchen I (1993b) Ontogeny of purinoceptors in the rat colon muscularis mucosae. Pharmacol Commun 2:317–322

Hoyle CHV, Burnstock G (1991) ATP receptors and their physiological roles. In: Stone TW (ed) Adenosine in the nervous system. Academic Press, London, pp 43–76

Hsu HH (1983) Purification and partial characterization of ATP pyrophosphohydrolase from fetal bovine epiphyseal cartilage. J Biol Chem 258:3463–3468

Hume RI, Hönig MG (1986) Excitatory action of ATP on embryonic chick muscle. J Neurosci 6:681–690

Hume RI, Thomas SA (1988) Multiple actions of adenosine 5'-triphosphate on chick skeletal muscle. J Physiol (Lond) 406:503–524

Hung CT, Allen FD, Mansfield KD, Shapiro IM (1997) Extracellular ATP modulates $[Ca^{2+}]_i$ in retinoic acid-treated embryonic chondrocytes. Am J Physiol 272:C1611–C1617

Igawa Y, Mattiasson A, Andersson KE (1993) Functional importance of cholinergic and purinergic neurotransmission for micturition contraction in the normal, unanaesthetized rat. Br J Pharmacol 109:473–479

Igusa Y (1988) Adenosine 5'-triphosphate activates acetylcholine receptor channels in cultured *Xenopus* myotomal muscle cells. J Physiol (Lond) 405:169–185

Irie K, Furukawa K, Nomoto T, Fujii E, Muraki T (1994) Developmental changes in the response of rat isolated duodenum to nicotine. Eur J Pharmacol 251:75–81

Ishikawa T, Seguchi H (1985) Localization of Mg^{++}-dependent adenosine triphosphatase and alkaline phosphatase activities in the postimplantation mouse embryos in day 5 and 6. Anat Embryol Berl 173:7–11

Ito S, Kimura A, Ohga A (1988) Development of non-cholinergic, non-adrenergic excitatory and inhibitory responses to intramural nerve stimulation in rat stomach. Br J Pharmacol 93:684–692

Jenuth JP, Mably ER, Snyder FF (1996) Modelling of purine nucleoside metabolism during mouse embryonic development: relative routes of adenosine, deoxyadenosine, and deoxyguanosine metabolism. Biochem Cell Biol 74:219–225

Jin ZL, Lee TF, Zhou SJ, Wang LC (1993) Age-dependent change in the inhibitory effect of an adenosine agonist on hippocampal acetylcholine release in rats. Brain Res Bull 30:149–152

Johansson B, Georgiev V, Fredholm BB (1997) Distribution and postnatal ontogeny of adenosine A_{2A} receptors in rat brain: comparison with dopamine receptors. Neuroscience 80:1187–1207

Keating MA, Duckett JW, Snyder HM, Wein AJ, Potter L, Levin RM (1990) Ontogeny of bladder function in the rabbit. J Urol 144:766–769

Kidd EJ, Miller KJ, Sansum AJ, Humphrey PPA (1998) Evidence for $P2X_3$ receptors in the developing rat brain. Neuroscience 87:533–539

Knudsen T, Elmer W-A (1987) Evidence for negative control of growth by adenosine in the mammalian embryo: induction of $Hm^{x/+}$ mutant limb outgrowth by adenosine deaminase. Differentiation 33:270–279

Kolb H-A, Wakelam MJO (1983) Transmitter-like action of ATP on patched membranes of cultured myoblasts and myotubes. Nature 303:621–623

Konduri GG, Woodard LL (1991) Selective pulmonary vasodilation by low-dose infusion of adenosine triphosphate in newborn lambs. J Pediatr 119:94–102

Konduri GG, Woodard LL, Mukhopadhyay A, Deshmukh DR (1992a) Adenosine is a pulmonary vasodilator in newborn lambs. Am Rev Respir Dis 146:670–676

Konduri GG, Theodorou AA, Mukhopadhyay A, Deshmukh DR (1992b) Adenosine triphosphate and adenosine increase the pulmonary blood flow to postnatal levels in fetal lambs. Pediat Res 31:451–457

Konduri GG, Gervasio CT, Theodorou AA (1993) Role of adenosine triphosphate and adenosine in oxygen-induced pulmonary vasodilation in fetal lambs. Pediatr Res 33:533–539

Konduri GG, Mital S, Gervasio CT, Rotta AT, Forman K (1997) Purine nucleotides contribute to pulmonary vasodilation caused by birth-related stimuli in the ovine fetus. Am J Physiol 272:H2377–H2384

Koos BJ, Mason BA, Ducsay CA (1993) Cardiovascular responses to adenosine in fetal sheep: autonomic blockade. Am J Physiol 264:H526–H532

Kubo Y (1991a) Properties of ionic currents induced by external ATP in a mouse mesodermal stem cell line. J Physiol (Lond) 442:691–710

Kubo Y (1991b) Electrophysiological and immunohistochemical analysis of muscle differentiation in a mouse mesodermal stem cell line. J Physiol (Lond) 442:711–741

Kulkarni JS, Prywara DA, Wakade TD (1998) Adenosine induces apoptosis by inhibiting mRNA and protein synthesis in chick embryonic sympathetic neurons. Neurosci Lett 248:187–190

Kupitz Y, Atlas D (1993) A putative ATP-activated Na^+ channel involved in sperm-induced fertilization. J Neurochem 63:S39

Laasberg T (1990) Ca^{2+}-mobilizing receptors of gastrulating chick embryo. Comp Biochem Physiol C Comp Pharmacol 97:1–12

Lagercrantz H, Yamamoto Y, Fredholm BB, Prabhakar NR, von Euler C (1984) Adenosine analogues depress ventilation in rabbit neonates. Theophylline stimulation of respiration via adenosine receptors? Pediatr Res 18:387–390

Lalo U, Kostyuk P (1998) Developmental changes in purinergic calcium signalling in rat neocortical neurones. Brain Res Dev Brain Res 11:43–50

Laudignon N, Aranda JV, Varma DR (1990) Effects of adenosine and its analogues on isolated internal carotid arteries from newborn and adult pigs. Biol Neonate 58:91–97

Levin RM, Malkowicz SB, Jacobowitz D, Wein AJ (1981) The ontogeny of the autonomic innervation and contractile response of the rabbit urinary bladder. J Pharmacol Exp Ther 219:250–257

Liang BT, Haltiwanger B (1995) Adenosine A_{2a} and A_{2b} receptors in cultured fetal chick heart cells. High- and low-affinity coupling to stimulation of myocyte contractility and cAMP accumulation. Circ Res 76:242–251

Liang SX, Phillips WD, Lavidis N (1998) Development of fast purinergic transmission in the mouse vas deferens. Proc Aust Neurosci Soc 9:119

Lohmann K (1929) Über die pyrophosphat fraction in Muskel. Naturwissenschaften 17:624–628

Lohmann F, Drews U, Donie F, Reiser G (1991) Chick embryo muscarinic and purinergic receptors activate cytosolic Ca^{2+} via phosphatidylinositol metabolism. Exp Cell Res 197:326–329

Lopes JM, Davis GM, Mullahoo K, Aranda JV (1994) Role of adenosine in the hypoxic ventilatory response of the newborn piglet. Pediatr Pulmonol 17:50–55

Loutradis D, John D, Kiessling AA (1989) Hypoxanthine causes a 2-cell block in random-bred mouse embryos. Biol Reprod 37:311–316

Lu B, Fu W-M (1995) Regulation of postsynaptic responses by calcitonin gene related peptide and ATP at developing neuromuscular junctions. Can J Physiol Pharmacol 73:1050–1056

MacDonald A, McGrath JC (1984) Post-natal development of functional neurotransmission in rat vas deferens. Br J Pharmacol 82:25–34

Mainwaring RD, Mentzer RM, Jr., Ely SW, Rubio R, Berne RM (1985) The role of adenosine in the regulation of coronary blood flow in newborn lambs. Surgery 98:540–546

Marangos PJ, Patel J, Stivers J (1982) Ontogeny of adenosine binding sites in rat forebrain and cerebellum. J Neurochem 39:267–270

Marangos PJ, Boulenger JP, Patel J (1984) Effects of chronic caffeine on brain adenosine receptors: regional and ontogenetic studies. Life Sci 34:899–907

Mason BA, Ogunyemi D, Punla O, Koos BJ (1993) Maternal and fetal cardiorespiratory responses to adenosine in sheep. Am J Obstet Gynecol 168:1558–1561

Matherne GP, Headrick JP, Berne RM (1990) Ontogeny of adenosine response in guinea pig heart and aorta. Am J Physiol 259:H1637–H1642

Meghji P, Holmquist CA, Newby AC (1985) Adenosine formation and release by neonatal-rat heart cells in culture. Biochem J 229:799–805

Meghji P, Tuttle JB, Rubio R (1989) Adenosine formation and release by embryonic chick neurons and glia in cell culture. J Neurochem 53:1852–1860

Meghji P, Skladanowski AC, Newby AC, Slakey LL, Pearson JD (1993) Effect of 5′-deoxy-5′-isobutylthioadenosine on formation and release of adenosine from neonatal and adult rat ventricular myocytes. Biochem J 291:833–839

Mehul B, Doyennette-Moyne M-E, Aubery M, Codogno P, Mannherz HG (1992) Enzymatic activity and in vivo distribution of 5′-nucleotidase, an extracellular matrix binding glycoprotein, during the development of chicken striated muscle. Exp Cell Res 203:62–71

Meyer MP, Clarke JDW, Patel K, Townsend-Nicholson A, Burnstock G (1999) Selective expression of purinoceptor $cP2Y_1$ suggests a role for nucleotide signalling in development of the chick embryo. Dev Dynam 214:152–158

Mishra OP, Wagerle LC, Delivoria Papadopoulos M (1988) 5′-Nucleotidase and adenosine deaminase in developing fetal guinea pig brain and the effect of maternal hypoxia. Neurochem Res 13:1055–1060

Miyazaki H, Ohga A, Saito K (1982) Development of motor response to intramural nerve stimulation and to drugs in rat small intestine. Br J Pharmacol 76:531–540

Modderman WE, Weidema AF, Vrijheid Lammers T, Wassenaar AM, Nijweide PJ (1994) Permeabilization of cells of hemopoietic origin by extracellular ATP^{4-}: elimination of osteoclasts, macrophages, and their precursors from isolated bone cell populations and fetal bone rudiments. Calcif Tissue Int 55:141–150

Morgan PF, Marangos PJ (1987) Ontogenetic appearance of the adenosine receptor precedes N-protein coupling in rat forebrain. Brain Res 432:269–274

Morgan PF, Montgomery P, Marangos PJ (1987) Ontogenetic profile of the adenosine uptake sites in rat forebrain. J Neurochem 49:852–855

Morgan PF, Deckert J, Nakajima T, Daval J-L, Marangos PJ (1990) Late ontogenetic development of adenosine A_1 receptor coupling to associated G-proteins in guinea pig cerebellum but not forebrain. Mol Cell Biochem 92:169–176

Morin FC, Egan EA (1992) Pulmonary hemodynamics in fetal lambs during development at normal and increased oxygen tension. J Appl Physiol 73:213–218

Mudumbi RV, Olson RD, Hubler BE, Montamat SC, Vestal RE (1995) Age-related effects in rabbit hearts of N^6-R-phenylisopropyladenosine, an adenosine A_1 receptor agonist. J Gerontol A Biol Sci Med Sci 50:B351–B357

Müller J, Rocha JB, Battastini AM, Sarkis JJ, Dias RD (1993) Postnatal development of ATPase-ADPase activities in synaptosomal fraction from cerebral cortex of rats. Neurochem Int 23:471–477

Nagao T, Fujita A, Takeuchi T, Hata F (1994) Changes in neuronal contribution to contractile responses of vas deferens of young and adult guinea pigs. J Auton Nerv Syst 50:87–92

Nakaoka Y, Yamashita M (1995) Ca^{2+} responses to acetylcholine and adenosine triphosphate in the otocyst of chick embryo. J Neurobiol 28:23–34

Neary JT, McCarthy M, Kang Y, Zuniga S (1998) Mitogenic signaling from P1 and P2 purinergic receptors to mitogen-activated protein kinase in human fetal astrocyte cultures. Neurosci Lett 242:159–162

Nicholls J, Hourani SM, Kitchen I (1990) The ontogeny of purinoceptors in rat urinary bladder and duodenum. Br J Pharmacol 100:874–878

Nicholls J, Hourani SMO, Kitchen I (1992) Degradation of extracellular adenosine and ATP by adult and neonate rat duodenum and urinary bladder. Pharmacol Commun 2:203–210

Nicolas F, Daval J-L (1993) Expression of adenosine A_1 receptors in cultured neurons from fetal rat brain. Synapse 14:96–99

Nureddin A, Esparo E, Kiessling AA (1990) Purines inhibit the development of mouse embryos in vitro. J Reprod Fertil 90:455–446

Oliveira EM, Rocha JB, Sarkis JJ (1994) In vitro and in vivo effects of $HgCl_2$ on synaptosomal ATP diphosphohydrolase (EC 3.6.1.5) from cerebral cortex of developing rats. Arch Int Physiol Biochim Biophys 102:251–254

Pác L (1984) Contribution to ontogenesis of Merkel cells. Z Mikrosk Anat Forsch 98:36–48

Paes de Carvalho R (1990) Development of A_1 adenosine receptors in the chick embryo retina. J Neurosci Res 25:236–242

Paes de Carvalho R, de Mello FG (1982) Adenosine-elicited accumulation of adenosine 3′, 5′-cyclic monophosphate in the chick embryo retina. J Neurochem 38: 493–500

Paes de Carvalho R, de Mello FG (1985) Expression of A_1 adenosine receptors modulating dopamine-dependent cyclic AMP accumulation in the chick embryo retina. J Neurochem 44:845–851

Paes de Carvalho R, Braas KM, Alder R, Snyder SH (1992) Developmental regulation of adenosine A_1 receptors, uptake sites and endogenous adenosine in the chick retina. Dev Brain Res 70:87–95

Pagonopoulou O, Angelatou F (1992) Reduction of A_1 adenosine receptors in cortex, hippocampus and cerebellum in ageing mouse brain. Neuroreport 3:735–737

Park TS, Van Wylen DG, Rubio R, Berne RM (1987) Increased brain interstitial fluid adenosine concentration during hypoxia in newborn piglet. J Cereb Blood Flow Metab 7:178–183

Park MK, Garrad RC, Weisman GA, Turner JT (1997) Changes in $P2Y_1$ nucleotide receptor activity during the development of rat salivary glands. Am J Physiol 272:C1388–C1393

Peachey JA, Brownhill VR, Hourani SM, Kitchen I (1996) The ontogenetic profiles of the pre- and postjunctional adenosine receptors in the rat vas deferens. Br J Pharmacol 117:1105–1110

Petrungaro S, Salustri A, Siracusa G (1986) Adenosine potentiates the delaying effect of dbcAMP on meiosis resumption in denuded mouse oocytes. Cell Biol Int Rep 10:993

Phillips JK (1998) Neuroreceptor mediated vascular control mechanisms in the rat. PhD Thesis. Australian National University, Canberra

Popoli P, Betto P, Rimondini R, Reggio R, Pezzola A, Ricciarello G, Fuxe K, Ferre S (1998) Age-related alteration of the adenosine/dopamine balance in the rat striatum. Brain Res 795:297–300

Ren LM, Hoyle CHV, Burnstock G (1996) Developmental changes in sympathetic contraction of the circular muscle layer in the guinea-pig vas deferens. Eur J Pharmacol 318:411–417

Reppert SM, Weaver DR, Stehle JH, Rivkees SA (1991) Molecular cloning and characterization of a rat A_1-adenosine receptor that is widely expressed in brain and spinal cord. Mol Endocrinol 5:1037–1048

Reynolds JD, Brien JF (1995) The role of adenosine A_1 receptor activation in ethanol-induced inhibition of stimulated glutamate release in the hippocampus of the fetal and adult guinea pig. Alcohol 12:151–157

Rivkees SA (1995) The ontogeny of cardiac and neural A_1 adenosine receptor expression in rats. Dev Brain Res 89:202–213

Rolband GC, Furth ED, Staddon JM, Rogus EM, Goldberg AP (1990) Effects of age and adenosine in the modulation of insulin action on rat adipocyte metabolism. J Gerontol 45:B174–B178

Rooney SA, Gobran LI (1988) Adenosine and leukotrienes have a regulatory role in lung surfactant secretion in the newborn rabbit. Biochim Biophys Acta 960:98–106

Runold M, Lagercrantz H, Fredholm BB (1986) Ventilatory effect of an adenosine analogue in unanesthetized rabbits during development. J Appl Physiol 61:255–259

Ruppelt A, Liang BT, Soto F (1999) Cloning, functional characterization and development expression of a P2X receptor from chick embryo. Prog Brain Res 120:81–92

Sakai Y, Fukuda Y, Yamashita M (1996) Muscarinic and purinergic Ca^{2+} mobilisation in the neural retina of early embryonic chick. Int J Dev Neurosci 14:691–699

Salter MW, Hicks JL (1995) ATP causes release of intracellular Ca^{2+} via the phospholipase $C\beta/IP_3$ pathway in astrocytes from dorsal spinal cord. J Neurosci 15:2961–2971

Sawmiller DR, Fenton RA, Dobson JG Jr (1998) Myocardial adenosine A_1-receptor sensitivity during juvenile and adult stages of maturation. Am J Physiol 274:H627–H635

Schachter JB, Sromek SM, Nicholas RA, Harden TK (1997) HEK293 human embryonic kidney cells endogenously express the $P2Y_1$ and $P2Y_2$ receptors. Neuropharmacology 36:1181–1187

Schiffmann SN, Vanderhaeghen JJ (1993) Age-related loss of mRNA encoding adenosine A_2 receptor in the rat striatum. Neurosci Lett 158:121–124

Schoen SW, Leutenecker B, Kreutzberg GW, Singer W (1990) Ocular dominance plasticity and developmental changes of 5′-nucleotidase distributions in the kitten visual cortex. J Comp Neurol 296:379–392

Schoen SW, Kreutzberg GW, Singer W (1993) Cytochemical redistribution of 5′-nucleotidase in the developing cat visual cortex. Eur J Neurosci. 5:210–222

Senba E, Daddona PE, Nagy JI (1987a) Transient expression of adenosine deaminase in facial and hypoglossal motoneurons of the rat during development. J Comp Neurol 255:217–230

Senba E, Daddona PE, Nagy JI (1987b) Adenosine deaminase-containing neurons in the olfactory system of the rat during development. Brain Res Bull 18:635–648

Senba E, Daddona PE, Nagy JI (1987c) Development of adenosine deaminase-immunoreactive neurons in the rat brain. Brain Res 428:59–71

Serio M, Montagnani M, Potenza MA, Mansi G, De Schaepdryver AF, Mitolo-Chieppa D (1996) Postnatal developmental changes of receptor responsiveness in rat mesenteric vascular bed. J Auton Pharmacol 16:63–68

Shaw C, Hall SE, Cynader M (1986) Characterization, distribution, and ontogenesis of adenosine binding sites in cat visual cortex. J Neurosci 6:3218–3228

Shimizu I, Toda N (1986) Alterations with age of the response to vasodilator agents in isolated mesenteric arteries of the beagle. Br J Pharmacol 89:769–778

Shryock J, Patel A, Belardinelli I, Linden J (1989) Downregulation and desensitization of A_1-adenosine receptors in embryonic chicken heart. Am J Physiol 25: H321–H327

Smith R, Köenig C, Pereda J (1983) Adenosinetriphosphatase (Mg-ATPase) activity in the plasma membrane of preimplantation mouse embryo as revealed by electron microscopy. Anat Embryol Berl 168:455–466

Smuts MS (1981) Rapid nasal pit formation in mouse embryos stimulates by ATP-containing medium. J Exp. Zool 216:409–414

Sneddon P, McLees A (1992) Purinergic and cholinergic contractions in adult and neonatal rabbit bladder. Eur J Pharmacol 214:7–12

Sobrevia L, Yudilevich DL, Mann GE (1997) Activation of A_2-purinoceptors by adenosine stimulates L-arginine transport (system y^+) and nitric oxide synthesis in human fetal endothelial cells. J Physiol (Lond) 499:135–140

Sperlágh B, Zsilla G, Baranyi M, Kékes-Szabó A, Vizi ES (1997) Age-dependent changes of presynaptic neuromodulation via A_1-adenosine receptors in rat hippocampal slices. Int J Dev Neurosci 15:739–747

Sugioka M, Fukuda Y, Yamashita M (1996) Ca^{2+} responses to ATP via purinoceptors in the early embryonic chick retina. J Physiol (Lond) 493:855–863

Sugioka M, Zhou WL, Hofmann HD, Yamashita M (1999) Involvement of P2 purinoceptors in the regulation of DNA synthesis in the neural retina of chick embryo. Int J Dev Neurosci 17:135–144

Tang Y-Y, Rovainen CM (1996) Cardiac output in *Xenopus laevis* tadpoles during development and in response to an adenosine agonist. Am J Physiol 270: R997–R1004

Tassin AM, Häggblad J, Heilbronn E (1990) Receptor-triggered polyphosphoinositide turnover produces less cytosolic free calcium in cultured dysgenic myotubes than in normal myotubes. Muscle Nerve 13:142–145

Teuscher E, Weidlich V (1985) Adenosine nucleotides, adenosine and adenine as angiogenesis factors. Biomed Biochim Acta 44:493–495

Thampy KG, Barnes EM Jr (1983a) Adenosine transport by primary cultures of neurons from chick embryo brain. J Neurochem 40:874–879

Thampy KG, Barnes EM Jr (1983b) Adenosine transport by cultured glial cells from chick embryo brain. Arch Biochem Biophys 220:340–346

Thomas SA, Hume RI (1990a) Permeation of both cations and anions through a single class of ATP-activated ion channels in developing chick skeletal muscle. J Gen Physiol 95:569–590

Thomas SA, Hume RI (1990b) Irreversible desensitization of ATP responses in developing chick skeletal muscle. J Physiol (Lond) 430:373–388

Thomas SA, Hume RI (1993) Single potassium channel currents activated by extracellular ATP in developing chick skeletal muscle: a role for second messengers. J Neurophysiol 69:1556–1566

Thomas SA, Zawisa MJ, Lin X, Hume RI (1991) A receptor that is highly specific for extracellular ATP in developing chick skeletal muscle in vitro. Br J Pharmacol 103:1963–1969

Toubas PL, Sekar KC, Sheldon RE, Seale TW (1990) Fetal and newborn lambs differ in their cardiopulmonary responsiveness to adenosine agonists. Dev Pharmacol Ther 15:68–81

Trezise DJ, Humphrey PPA (1996) Activation of peripheral sensory neurons in the neonatal rat tail by ATP. Br J Pharmacol 117:103P

Ueda H, Moritoki H (1991) Possible association of decrease of ATP-induced vascular relaxation with reduction of cyclic GMP during aging. Arch Int Pharmacodyn Ther 310:35–45

Vanella A, Barcellona ML, Serra I, Ragusa N, Avola R, Avitabile M, Giuffrida AM (1983) Effect of undernutrition on some enzymes involved in the salvage pathway of purine nucleotides in different regions of developing rat brain. Neurochem Res 8:151–158

Vestal RE, Wood AJ, Shand DG (1979) Reduced β-adrenoceptor sensitivity in the elderly. Clin Pharmacol Ther 26:181–186

Wakade TD, Palmer KC, McCauley R, Przywara DA, Wakade AR (1995) Adenosine-induced apoptosis in chick embryonic sympathetic neurons: a new physiological role for adenosine. J Physiol (Lond) 488:123–138

Weaver DR (1996) A_1-adenosine receptor gene expression in fetal rat brain. Brain Res Dev Brain Res 94:205–223

Webb TE, Boluyt MO, Barnard EA (1996) Molecular biology of P_{2Y} purinoceptors: expression in rat heart. J Auton Pharmacol 16:303–307

Webb TE, Simon J, Barnard EA (1998) Regional distribution of $[^{35}S]2'$-deoxy 5'-O-(1-thio) ATP binding sites and the $P2Y_1$ messenger RNA within the chick brain. Neuroscience 84:825–837

Weber RG, Jones CR, Lohse MJ, Palacios JM (1990) Autoradiographic visualization of A_1 adenosine receptors in rat brain with $[^3H]$8-cyclopentyl-1,3-dipropylxanthine. J Neurochem 54:1344–1353

Wells DG, Zawisa MJ, Hume RI (1995) Changes in responsiveness to extracellular ATP in chick skeletal muscle during development and upon denervation. Dev Biol 172:585–590

Wirkner K, Franke H, Inoue K, Illes P (1998) Differential age-dependent expression of $α_2$-adrenoceptor- and P2 purinoceptor-functions in rat locus coeruleus neurons. Naunyn-Schmiedebergs Arch Pharmacol 357:186–189

Yamashita M, Sugioka M (1998) Calcium mobilization systems during neurogenesis. News Physiol Sci 13:75–79

Yeung CH (1986) Temporary inhibition of the initiation of motility of demembranated hamster sperm by high concentrations of ATP. Int J Androl 9:359–370

Yoneyama Y, Power GG (1992) Plasma adenosine and cardiovascular responses to dipyridamole in fetal sheep. J Dev Physiol 18:203–209

Yoshioka T (1989) Histochemical examination of adenosine nucleotidases in the developing rat spinal cord: possible involvement in enzymatic chain of ATP degradation. Acta Histochem Cytochem 22:685–694

Yoshioka T, Tanaka O (1989) Histochemical localization of Ca^{2+}, Mg^{2+}-ATPase of the rat cerebellar cortex during postnatal development. Int J Dev Neurosci 7:181–193

Yoshioka T, Inomata K, Tanaka O (1987) Cytochemistry of Ca^{2+}-ATPase in the rat spinal cord during embryonic development. Acta Histochem Cytochem 20:511–526

Young ML, Ramza BM, Tan RC, Joyner RW (1987) Adenosine and hypoxia effects on atrioventricular node of adult and neonatal rabbit hearts. Am J Physiol 253:H1192–H1198

Zagorodnyuk V, Hoyle CHV, Burnstock G (1993) An electrophysiological study of developmental changes in the innervation of the guinea-pig taenia coli. Pflugers Arch 423:427–433

Zderic SA, Duckett JW, Wein AJ, Snyder HM 3rd, Levin RM (1990) Development factors in the contractile response of the rabbit bladder to both autonomic and non-autonomic agents. Pharmacology 41:119–123

Zhang Z, Galileo DS (1998) Widespread programme death in early developing chick optic tectum. Neuroreport 9:2797–2801

Zheng JS, Boluyt MO, O'Neill L, Crow MT, Lakatta EG (1994) Extracellular ATP induces immediate-early gene expression but not cellular hypertrophy in neonatal cardiac myocytes. Circ Res 74:1034–1041

Zheng JS, Boluyt MO, Long X, O'Neill L, Lakatta EG, Crow MT (1996) Extracellular ATP inhibits adrenergic agonist-induced hypertrophy of neonatal cardiac myocytes. Circ Res 78:525–535

CHAPTER 6
P1 and P2 Purine and Pyrimidine Receptor Ligands

K.A. JACOBSON and L.J.S. KNUTSEN

A. Introduction

Research focused on P1 and P2 purinergic receptor function is currently at a critical stage where considerable efforts in the disciplines of medicinal chemistry, molecular biology, and pharmacology have resulted in the identification of a number of novel ligands that have been used to enhance understanding of the roles of both P1 (adenosine) and P2 (ATP, ADP, UTP and UDP) receptors in human tissue function.

The present overview of P1 and P2 receptor ligands takes a somewhat different approach to that of previous reviews (JACOBSON et al. 1992, 1997a; JACOBSON and VAN RHEE 1997; POULSEN and QUINN 1998) that summarized the structure–activity relationships of a large array of ligands with only brief coverage of their biological properties. Instead, the authors will focus on what can be considered as those P1 and P2 receptor ligands that have contributed most significantly to advances in the field, providing a context for other chapters in this monograph. The focus herein is on pharmacologically active substances, many of which have been used as radioligands, as well as compounds for which human clinical trial data is available. Inevitably, because of the longer and more sustained effort in the area of P1 receptor chemistry (JACOBSON et al. 1992), there is a relative imbalance in informational content on P2 vs P1 receptor ligands, as focused efforts in the area of P2 medicinal chemistry and ligand design are relatively recent.

Chemical efforts to develop P1 receptor selective ligands preceded the cloning of the various P1 receptor subtypes and were facilitated by robust in vitro binding and functional assays. In many instances the ligands were then used to identify and characterize cloned receptors. The study of P1 receptor-mediated responses has also been facilitated by the ready availability of a number of potent and selective antagonists.

In contrast, for the P2 receptor family, information from cloning of receptors preceded the identification of potent and selective ligands while the characterization of P2 responses is hampered by a current lack of potent, selective, and bioavailable antagonists. Utilizing classical molecular modeling techniques in concert with high throughput screening approaches using in vitro functional assays with transfected rat and human P2 receptor constructs, this situation is anticipated to change.

B. P1 Receptor Ligands

Structure–activity relationships for P1 and P2 receptor agonists and P1 receptor antagonists are outlined in Fig. 1. A number of other structures are included in this chapter as a uniform cross reference to other chapters in this monograph.

I. P1 Receptor Agonists

1. Non-Selective Adenosine Receptor Agonists (Fig. 2)

a) NECA

NECA (*N*-ethylcarboxamidoadenosine) was one of the first potent P1 receptor agonists reported in the early 1970s. At the time of its discovery, its mechanism of action was not understood since the concept of P1 receptors had not been established. NECA had potent cardiovascular (CV) activity, being 20,000 times more potent than adenosine in promoting cardiac efficacy (flow increase integrated over the time of action) following intravenous (i.v.) dosing in dogs acting via A_1 (depression of cardiac activity) and A_{2A} receptors (vasorelaxation; RABERGER et al. 1977). NECA also had potent behavioral effects following i.c.v. dosing, reducing spontaneous locomotor activity (LMA) in rats, an effect blocked by the adenosine antagonist, caffeine.

In the presence of 50 nmol/l CPA to block binding to A_1 receptors, [^{3}H]-NECA labels both A_{2A} and A_{2B} receptors in rat striatal membranes (BRUNS et al. 1986). Originally regarded as a selective A_2 agonist, NECA non-selectively interacts with human recombinant A_1, A_{2A}, A_{2B} and A_3 receptors with affinities of 14, 18, 2400 and 4 nmol/l, respectively (KLOTZ et al. 1998). As a result of its unusual 5'-uronamide structure, NECA has been used as a template for other P1 ligands by a number of investigators (OLSSON et al. 1986) resulting in more selective adenosine agonists, e.g., CGS 21680, HENECA, and IB-MECA.

b) Metrifudil

Metrifudil (*N*-(2-methylphenyl)methyladenosine) is an agonist at rat brain A_1 and A_{2A} receptors (K_i = 25 nmol/l and 98 nmol/l, respectively; KLITGAARD et al. 1993; KNUTSEN et al. 1999) and also binds to recombinant human adenosine A_3 receptors (K_i = 360 nmol/l; SIDDIQI et al. 1995). Metrifudil was one of the first P1 agonists to be evaluated in humans (SCHAUMANN and KUTSCHA 1972). At doses of 0.03 mg/kg i.v. and 0.35 mg/kg p.o., the compound increased heart rate and cardiac output without impairing atrioventricular node (AVN) conduction or altering the electrocardiogram. At doses of 0.1 mg/kg and

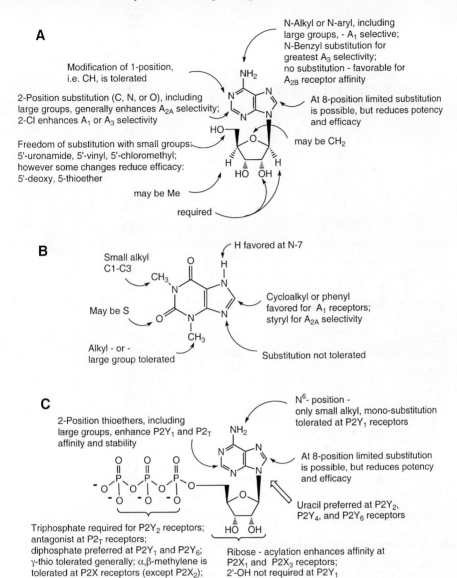

Fig. 1A–C. Summary of structure–activity relationships for: **A** adenosine receptor agonists; **B** xanthine antagonists; **C** P2 receptor agonists [Note to figures: Structures described in Chap. 6 and those mentioned elsewhere in this volume are shown. K_i values, expressed in nmol/l, determined in rat adenosine receptor binding assays are shown for adenosine receptor ligands at $A_1/A_{2A}/A_3$ receptors. K_a or IC_{50} values at turkey $P2Y_1$ receptors are shown for P2 receptor ligands]

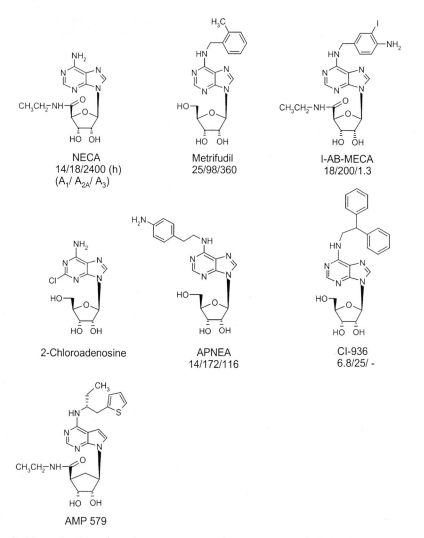

Fig. 2. Non-selective adenosine receptor agonists

0.47–0.53 mg/kg i.v. and p.o. respectively, "uneasiness" and other side effects were noted. Serum fatty acids were lowered by i.v. administration of metrifudil, most probably an A_1-mediated effect. In patients with coronary heart disease, 40 mg/kg i.v. metrifudil increased coronary-venous O_2 saturation following an increase in coronary blood flow (KUGLER and WESTERMANN 1974).

Metrifudil has potent anticonvulsant activity in rodents (KLITGAARD et al. 1993; KNUTSEN et al. 1995; KRAHL et al. 1995) and has neuroprotectant effects in a gerbil ischemia model (SHEARDOWN and KNUTSEN 1996; KNUTSEN et al.

1999). When given orally, metrifudil lowered blood pressure in spontaneously hypotensive rats SHRs (YAGIL and MIYAMOTO 1995).

c) I-AB-MECA

I-AB-MECA (N-(4-amino-3-iodophenyl)methyladenosine) was initially identified as an adenosine A_3 receptor agonist (GALLO-RODRIGUEZ et al. 1994) and used as a radioligand for this receptor (OLAH et al. 1994; VARANI et al. 1998). It was subsequently found to label rat brain A_1 and A_{2A} receptors (SHEARMAN and WEAVER 1997) and has been replaced as an A_3 agonist by the desamino analogues, IB-MECA and 2-chloro-IB-MECA (KIM et al. 1994).

2. A_1 Receptor Agonists (Fig. 3)

a) CPA

CPA (N-cyclopentyladenosine) is one of a series of N^6-cycloalkyladenosine derivatives that are potent A_1 receptor agonists (A_1 K_i = 2.3 nmol/l). Originally identified as an inhibitor of platelet aggregation (KIKUGAWA et al. 1973), CPA is 780-fold selective for A_1 vs A_{2A} receptors in binding (Moos et al. 1985) and functional assays (DALY and PADGETT 1992).

CPA has both cardioprotective and cerebroprotective activity (VON LUBITZ et al. 1994; KNUTSEN et al. 1999) and, like other adenosine agonists, lowers LMA (HEFFNER et al. 1989) and blood pressure in rats (KNUTSEN et al. 1999). Given acutely, CPA has anticonvulsant properties (KLITGAARD et al. 1993; KNUTSEN et al. 1995, 1997) and depresses synaptic transmission (DUNWIDDIE et al. 1986). Tolerance to the hemodynamic effects of the close analog of CPA, 2-chloro-N-cyclopentyladenosine (CCPA) in SHRs occurs on repeat dosing (CASATI et al. 1994).

b) R-PIA

R-PIA ((R)-N-phenylisopropyladenosine) was identified in the late 1960s as a cardiovascular agent. The first clinical trials on this compound were conducted prior to knowledge of its mechanism of action as a selective adenosine receptor agonist (A_1 K_i = 2 nmol/l) being discovered. [^{3}H]-R-PIA has been extensively used as an A_1 receptor ligand (SCHWABE et al. 1980).

Administration of PIA (presumably as the R- and S- mixture) at doses of 0.5 mg and 1 mg p.o. lowered serum free fatty acids without altering heart rate, blood pressure, or electrocardiogram (SCHAUMANN et al. 1972). R-PIA has anticonvulsant (DUNWIDDIE and WORTH 1982; KNUTSEN et al. 1995) and cerebroprotective activity in rats (MACGREGOR et al. 1998). While effective in rodent focal ischemia models (RUDOLPHI et al. 1992; KNUTSEN et al. 1999), R-PIA was without significant neuroprotective effect in a severe temporary forebrain ischemia model in gerbils (KNUTSEN et al. 1999). R-PIA has analgesic (VON HEIJNE et al. 1998), hypotensive, and bradycardic actions in conscious rats

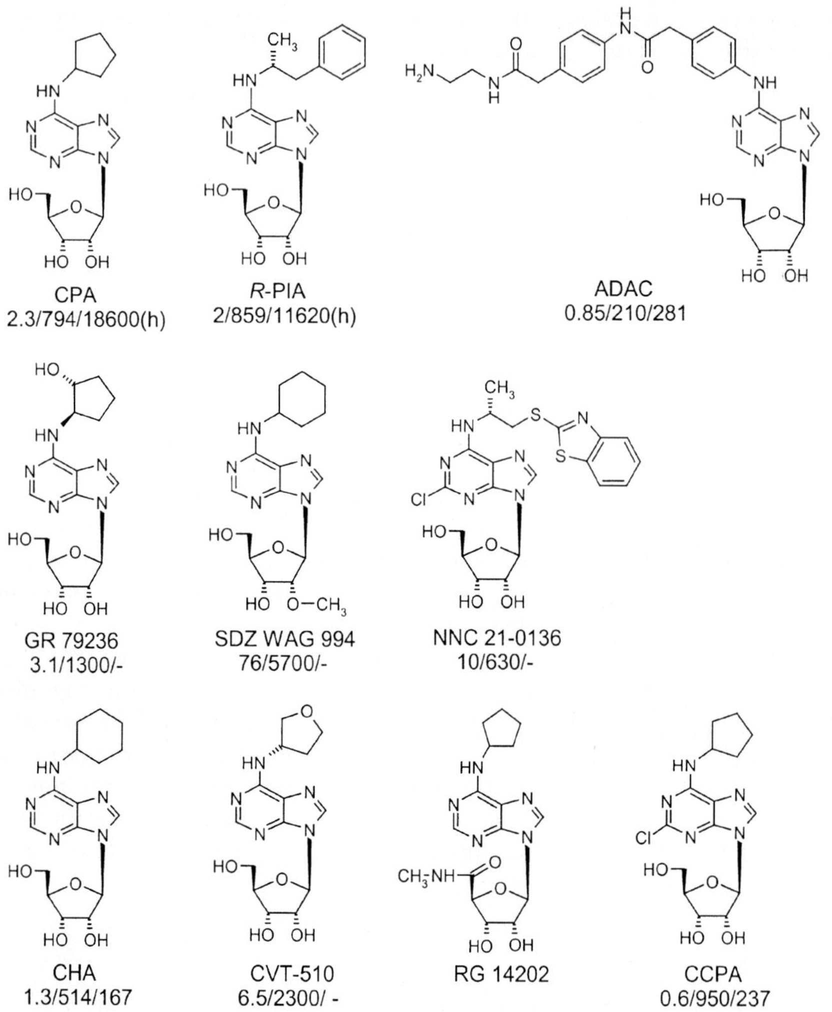

Fig. 3. A₁ selective agonists

(MATHOT et al. 1995) and monkeys (COFFIN and SPEALMAN 1987) suggesting that it may have potential as an antifibrillatory agent (WAINWRIGHT et al. 1997).

c) ADAC

ADAC [(N-[4-[2-[[4-[2-[(2-aminoethyl)amino]-2-oxoethyl]phenyl]amino]-2-oxoethyl]phenyl]adenosine; adenosine amine congener] is a selective A₁ receptor agonist that is a functionalized "congener" containing a primary amino group at the terminal position of a strategically placed elongated chain that allows conjugation to carriers without losing receptor binding (JACOBSON et al. 1985). When conjugated with bifunctional cross-linking reagents, e.g., p-phenylene diisothiocyanate, ADAC binds irreversibly to A_1 receptors (JACOBSON et al. 1989). [^{3}H]-ADAC is a useful high affinity ligand for the A_1 receptor with minimal non-specific binding, despite the presence of aromatic rings in the functionalized chain.

Administered acutely at doses of 25–100 μg/kg i.p. to gerbils prior to bilateral carotid artery occlusion, ADAC limited post-ischemic damage and also protected against ischemia-induced spatial memory loss assessed in the Morris water maze (VON LUBITZ et al. 1996).

d) GR 79236

GR 79236 (N-[(1$S,trans$)-2-hydroxycyclopentyl]adenosine), a CPA analog, is a selective adenosine A_1 agonist (GURDEN et al. 1993) with pronounced antilipolytic effects. K_i values for rat brain A_1 and A_{2A} receptors were 3.1 nmol/l and 1300 nmol/l, respectively (KNUTSEN et al. 1995, 1999).

GR 79236 inhibited catecholamine-induced lipolysis in isolated human, rat, and dog adipocytes. Given via i.v. infusion to fasted, pithed rats, or orally to fasted conscious rats and dogs, it produced time- and dose-dependent decreases in plasma non-esterified fatty acids. In fasted rats, hypotriglyceridemia and anti-ketotic effects were observed (STRONG et al. 1993). In conscious rats, GR 79236 elicited antilipolytic effects that were accompanied by bradycardia and decreased arterial blood pressure (MERKEL et al. 1995) and also modulated insulin actions in human adipocytes in vitro affecting lipolysis and glucose transport (HESELTINE et al. 1995). GR 79236 may also have utility in myocardial ischemia (LOUTTIT et al. 1999) and cerebral ischemia (KNUTSEN et al. 1999).

e) SDZ WAG 994

SDZ WAG 994 (N-cyclohexyl-2'-O-methyladenosine) is an orally active adenosine A_1 agonist that is noteworthy in that it is modified at the ribose 2-position. It is a full agonist at pig striatal A_1 receptors (K_D = 23 nmol/l) being approximately 1000-fold selective vs the A_{2A} receptor, although the compound was much weaker than expected in vitro given its striking in vivo activity (KNUTSEN et al. 1999). SDZ WAG 994 at doses of 0.1–0.3 mg/kg i.v. or 0.2–0.6 mg/kg p.o. in rodents and dogs and in squirrel monkeys, elicited dose-related bradycardia and suppression of plasma renin activity without altering blood pressure (BP). The compound also slowed the atrial rate and prolonged the atrioventricular node conduction time in spontaneously beating guinea pig hearts (BELARDINELLI et al. 1994). The compound may thus have potential as a hypoglycemic (ISHIKAWA et al. 1998) or anti-ischemic agent (KNUTSEN et al. 1999). In a group of 50 cardiac patients who received 1 mg, 2 mg or 5 mg SDZ WAG 994, no significant hemodynamic changes were observed at rest in a subset of these patients with left ventricular dysfunction, but increases in the

PR interval and atrial natriuretic peptide levels, consistent with adenosine A_1 receptor activation, were observed (BERTOLET et al. 1996).

f) NNC 21–0136

NNC 21–0136 (2-chloro-*N*-[*R*-(2-benzothiazolyl)thio-2-propyl]adenosine) is a representative of a new series of adenosine A_1 agonists with potent neuroprotective properties that have a decreased hypotensive effect compared to other P1 receptor agonists with a similar in vitro binding profile (KNUTSEN et al. 1999). NNC 21–0136 is effective in both focal and global ischemia models, in the latter paradigm with post-ischemic dosing (SHEARDOWN et al. 1995) and has been characterized as a tritiated radioligand (THOMSEN et al. 1997).

3. A_{2A} Receptor Agonists (Fig. 4)

a) CGS 21680

CGS 21680 (2-[*p*-(carboxyethyl)phenylethylamino]-5′-*N*-ethylcarboxamidoadenosine) was originally designed as a antihypertensive agent. It is 140-fold selective for rat A_{2A} vs A_1 receptors (K_i = 22 nmol/l at A_{2A} receptors; HUTCHISON et al. 1989), increased coronary flow in an isolated perfused

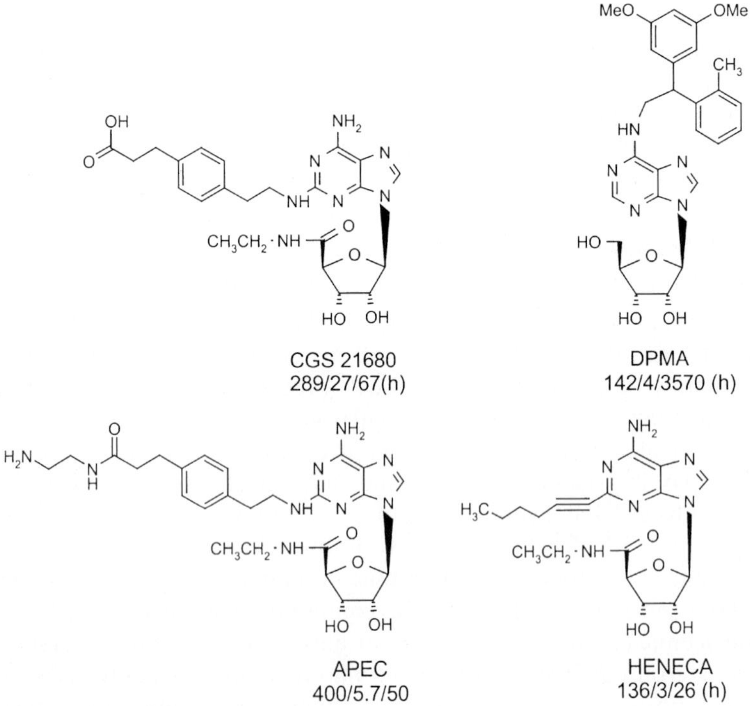

Fig. 4. A_{2A} selective agonists

working rat heart model and, given i.v., reduced blood pressure in an anesthetized normotensive rat model ($ED_{50} = 9\,mg/kg$). CGS 21680 is widely used as a selective A_{2A} receptor ligand and is selective for the A_{2A} vs. A_{2B} receptor (HIDE et al. 1992). At recombinant human adenosine A_3 receptors its affinity is 67 nmol/l (KLOTZ et al. 1998), representing only a threefold selectivity over human A_{2A} receptors.

[^{3}H]-CGS 21680 is a selective ligand for rodent A_{2A} receptors (JARVIS et al. 1989) binding to A_{2A} receptors in the caudate-putamen, accumbens, olfactory tubercle, and globus pallidus, cerebral and cerebellar cortex, hippocampus, thalamus, and brainstem nuclei in rat brain (JOHANSEN and FREDHOLM 1995). In cerebral cortex and hippocampus, [^{3}H]-CGS 21680 binds to a site distinct from the striatal A_{2A} receptor that may represent an "atypical" binding site. High-affinity binding sites for [^{3}H]-CGS 21680 have been identified in rat hippocampus and cerebral cortex that are different from striatal A_{2A} receptors (CUNHA et al. 1996). The A_{2A} antagonist, SCH 58261 can also discriminate between two different binding sites for [^{3}H]-CGS 21680 in rat brain (LINDSTROEM et al. 1996).

In contrast to other A_{2A} agonists, CGS 21680 is neuroprotective in the gerbil ischemia model (SHEARDOWN and KNUTSEN 1996) and in vitro (JONES et al. 1998b) and can inhibit LMA (JANUSZ and BERMAN 1993; BARRACO 1994; KNUTSEN et al. 1999) although the compound was originally designed not to cross the blood-brain barrier.

b) DPMA

DPMA [N-[2-(3,5-dimethoxyphenyl)-2-(2-methylphenyl)ethyl]adenosine); PD 125944] is one of the few lipophilic N-substituted adenosine agonists that have high A_{2A} affinity ($K_i = 4.4\,nmol/l$). The diastereoisomer mixture was 30-fold selective for the rat A_{2A} receptor (TRIVEDI and BRUNS 1989). In conscious SHRs DPMA produced dose-related reductions in mean arterial pressure (WEBB et al. 1991) and also reduced renal vascular resistance producing a concomitant increase in renal blood flow. Responses to DPMA as well as CGS 21680 indicate that systemic vasodilation, with resultant cardioexcitation and stimulation of renin release, are the predominant hemodynamic effects of selective A_{2A} agonists in the conscious SHR.

In hippocampal neurones, DMPA partially blocked kainate-induced excitotoxicity (JONES et al. 1998b) and facilitated synaptosomal Ca^{2+} uptake (GONCALVES and RIBEIRO 1996). Unlike CGS 21680, the compound was not neuroprotective in severe forebrain ischemia in gerbils (SHEARDOWN and KNUTSEN 1996).

c) APEC

APEC (2-[4-(2-(2-(4-aminophenyl)methylcarbonylamino)ethylaminocarbonyl)ethyl)phenyl) ethylamino]-5'-N-ethylcarboxamidoadenosine) is an amine-functionalized congener of CGS 21680 with additional substitution at the purine 2-position. The covalent conjugate of APEC with *p*-aminophenyl-acetic acid is a high affinity radioligand for A_{2A} receptors and was used for photoaffinity cross-linking to distinguish A_1 and A_{2A} receptors as distinct molecular species (JACOBSON et al. 1989). APEC is moderately selective for A_{2A} adenosine receptors (A_{2A} Ki = 5.7 nmol/l, A_3 = 50 nmol/l) readily crosses the blood-brain barrier, decreases LMA (NIKODIJEVIC et al. 1990), promotes sleep (SATOH et al. 1998), and has neuroprotectant properties (VON LUBITZ et al. 1995b).

d) HENECA

HENECA (1-[6-amino-2-(1-hexynyl)-9H-purin-9-yl]-1-deoxy-*N*-ethyl-*β*-D-ribofuranuronamide) is a NECA analog (CRISTALLI et al. 1992) that is a potent human A_{2A} receptor agonist (K_i = 2 nmol/l; KLOTZ et al. 1998) but also binds to the human adenosine A_3 receptor (K_i = 26 nmol/l), showing only a tenfold selectivity for the A_{2A} receptor. HENECA has vasodilator actions similar to NECA and CGS 21680 that correlated with its A_{2A} receptor affinity (CONTI et al. 1993). HENECA protected against pentylenetetrazole-induced seizures (ADAMI et al. 1995; KNUTSEN and MURRAY 1997) but did not significantly modify sleep patterns in rats although A_{2A} receptors mediate purine effects on sleep (BERTORELLI et al. 1996; SAITOH et al. 1997).

Since no truly selective adenosine A_{2B} receptor agonists have been identified at the time of writing, adenosine A_{2B} receptor ligands are covered under A_{2B} antagonists.

4. A_3 Receptor Agonists (Fig. 5)

a) IB-MECA

IB-MECA (1-[6-[[(3-iodophenyl)methyl]amino]-9H-purin-9-yl]-1-deoxy-*N*-methyl-*β*-D-ribofuranuronamide) was the first selective adenosine A_3 receptor agonist (K_i = 1.1 nmol/l) and is 50-fold selective for the A_3 receptor versus both A_1 or A_{2A} receptors in vitro (GALLO-RODRIGUEZ et al. 1994). IB-MECA stimulates apoptosis in human promyelocytic HL-60 cells (KOHNO et al. 1996) and in the human macrophage cell line, U937, alters the cytokine milieu by decreasing the proinflammatory cytokine tumor necrosis factor-α (TNF-α) (SAJADI et al. 1996). In collagen-induced arthritis model in mice, IB-MECA (0.5 mg/kg per day) reduced the severity of joint inflammation, inhibited the formation of macrophage inflammatory protein (MIP)-1α and IL-12 in the paws and suppressed neutrophil infiltration (SZABO et al. 1998).

In a gerbil ischemia model, IB-MECA improved post-occlusive cerebral blood flow and reduced neuronal damage and mortality (VON LUBITZ et al. 1999). Given acutely, IB-MECA protected against chemical but not electrically induced seizures. When dosed chronically, protection against the latter was observed (VON LUBITZ et al. 1995a). IB-MECA protects against injury

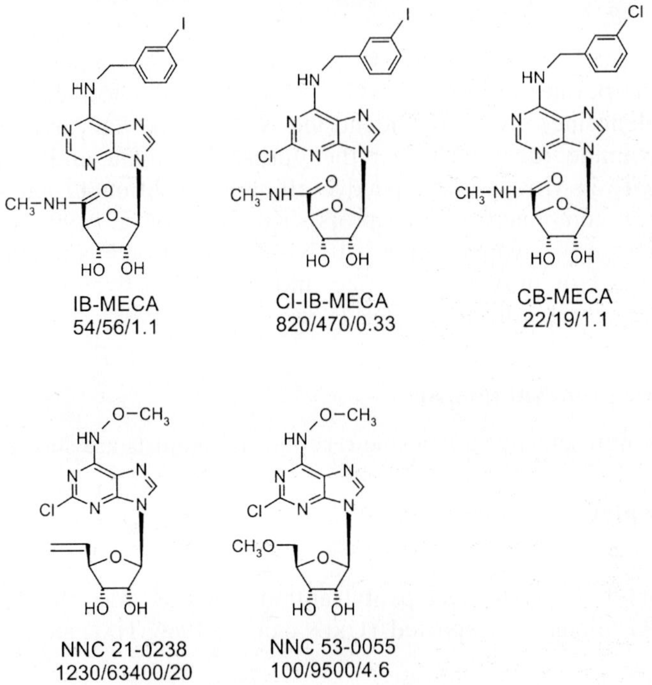

Fig. 5. A_3 selective agonists

during acute myocardial ischemia and reperfusion via a protein kinase C-mediated mechanism in conscious rabbits (AUCHAMPACH et al. 1997).

b) Cl-IB-MECA

Cl-IB-MECA (1-[2-chloro-6-[[(3-iodophenyl)methyl]amino]-9H-purin-9-yl]-1-deoxy-*N*-methyl-β-D-ribofuranuronamide) is an IB-MECA analog with enhanced A_3 receptor selectivity ($K_i = 0.33$ nmol/l; KIM et al. 1994) being 2500- and 1400-fold selective for the A_3 vs A_1 and A_{2A} receptors. Low concentrations (less than 100 nmol/l) of Cl-IB-MECA antagonized the A_1 receptor-mediated inhibition of excitatory post synaptic potentials (EPSPs) in hippocampal slices (DUNWIDDIE et al. 1997) but higher concentrations induce apoptosis (KOHNO et al. 1996). In human ADF astroglial cells, 100 nmol/l Cl-IB-MECA caused a marked reorganization of the cytoskeleton, with the appearance of stress fibers and numerous cell protrusions enriched in the anti-apoptotic protein Bcl-x_L (ABBRACCHIO et al. 1997). At higher concentrations (greater than 10 μmol/l), Cl-IB-MECA was lethal to cultured rat cerebellar granule neurons and potentiated the toxic effects of glutamate (SEI et al. 1997). In conscious rats, administration of 2-Cl-IB-MECA (200 mg/kg i.v.) resulted in a short-lasting hypotension, accompanied by a 50 to 100-fold increase in plasma histamine levels (VAN SCHAICK et al. 1996).

c) NNC 21–0238

NNC 21–0238 (2-chloro-9-(5,6-dideoxy-β-D-ribo-hex-5-enofuranosyl)-*N*-methoxy-9H-purine) and NNC 53–0055 (2-chloro-5′-*O*-methyl-*N*-methoxyadenosine) are newer adenosine A_3 agonists that have an unusual *N*-methoxyamino functionality at the purine 6-position and display high A_3 selectivity in vitro, with respective affinities of 20nmol/l and 4.6nmol/l at human recombinant A_3 receptors (KNUTSEN et al. 1998). These *N*-alkoxyadenosine derivatives inhibited TNF-α production in a rat whole blood assay, indicative of an A_3 agonist effect and were protective in rodent seizure and gerbil cerebral ischemia models (KNUTSEN et al. 1998).

II. P1 Receptor Antagonists

A series of non-selective adenosine receptor antagonists are shown in Fig. 6.

1. A_1 Receptor Antagonists (Fig. 7)

a) DPCPX

DPCPX [(1,3-dipropyl-8-cyclopentylxanthine); CPX] was the first highly selective A_1 antagonist reported (UKENA et al. 1986; HALEEN et al. 1987).

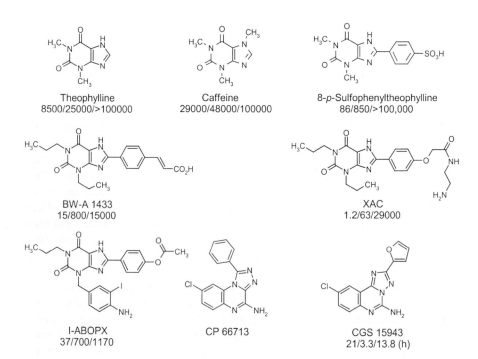

Fig. 6. Non-selective adenosine antagonists

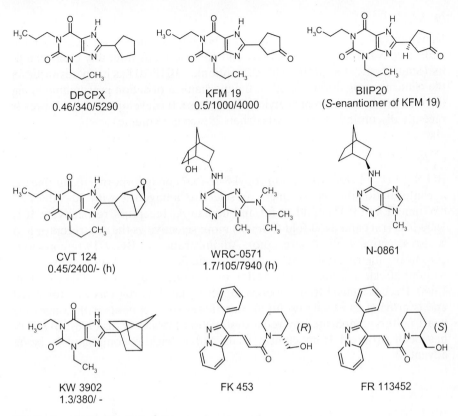

Fig. 7. A_1 selective antagonists

DPCPX had a K_i of 0.46 nmol/l for rat A_1 receptors vs 340 nmol/l for A_{2A} receptors. [^{3}H]-DPCPX has been widely used as a high affinity radioligand for the A_1 receptor (BRUNS et al. 1987). At human A_1, A_{2A}, A_{2B}, and A_3 receptors DPCPX has affinities of 4 nmol/l, 129 nmol/l, 56 nmol/l, and 3960 nmol/l, respectively (KLOTZ et al. 1998; JACOBSON et al. 1999) and is thus not overly selective for the A_1 receptor, limiting its use as a specific probe. In general, the A_3 receptor affinity of most xanthines is highly species dependent, with affinity at human receptors typically being greater that 100-fold that seen at rat receptors.

Adenosine A_1 receptor antagonists can reduce cardiac ischemia-reperfusion injury (NEELY et al. 1996). Despite potent proconvulsant-like effects in vitro, DPCPX does not induce seizures in vivo, in contrast to less selective adenosine antagonists (CHESI and STONE 1997; KNUTSEN and MURRAY 1997). DPCPX can promote chloride efflux in cystic fibrosis transmembrane conductance regulator (CFTR) cells (ARISPE et al. 1998), an action that is distinct from its A_1 receptor antagonist properties (JACOBSON et al. 1995).

b) KFM 19

KFM 19 [(±)8-(3-oxocyclopentyl)-1,3-dipropylxanthine] is a DPCPX analog with selective A_1 antagonist activity and good aqueous solubility bioavailability (SCHINGNITZ et al. 1991). The *S*-enantiomer, BIIP 20 has been advanced in the clinic as a cognition enhancer, the mechanism of action presumably being through the stimulation of acetylcholine release. Its development was however recently discontinued (WILLIAMS, Chap. 29, second volume).

c) BG 9719

BG 9719 (CVT 124, (*S*)-1,3-dipropyl-8-[2-(5,6-epoxynorbornyl)]xanthine) is a conformationallly constrained A_1 selective antagonist with K_i values of 0.67 nmol/l and 0.45 nmol/l at rat and human A_1 receptors, respectively. It is 1800-fold (rat) and 2400-fold (human) more selective vs the A_{2A} receptor and is also selective vs the A_{2B} receptor. Administration of BG 9719 to conscious chronically instrumented rats increased urine flow rate and sodium excretion without affecting potassium excretion or renal hemodynamics (GELLAI et al. 1998). Recent clinical trial evidence suggests that the drug was well tolerated and is proving especially useful in diuretic-resistant patients with congestive heart failure complicated by renal dysfunction (WOLFF et al. 1998). BG 9719 has been dropped from clinical trials and a backup compound is being advanced.

d) WRC-0571

WRC-0571 (8-(*N*-methylisopropyl)amino-*N*-(5′-endohydroxy-endonorbornyl)-9-methyladenine) is a potent non-xanthine A_1 antagonist (K_i = 1.1 nmol/l) with 200-fold selectivity vs A_{2A} receptors. At human A_1, A_{2A}, and A_3 receptors it had affinities of 1.7 nmol/l, 105 nmol/l, and 7940 nmol/l, respectively. In guinea pig isolated atria, WRC-0571 antagonized the A_1-mediated negative inotropic response to NECA with a K_B of 3.4 nmol/l, and was more than 2500-fold less potent at antagonizing NECA-induced A_{2B} mediated relaxation of guinea pig aorta. In anesthetized rats, WRC-0571 given orally antagonized adenosine-induced bradycardia at concentrations as low as 1 nmol/kg but failed to antagonize A_2-mediated hindquarter vasodilation at concentrations of ≤10 000 nmol/kg (MARTIN et al. 1996).

2. A_{2A} Receptor Antagonists (Fig. 8)

a) SCH 58261

SCH 58261 (5-amino-7-(2-phenylethyl)-2-(2-furyl)pyrazolo[4,3-*e*]-1,2,4-triazolo[1,5-*c*]pyrimidine) is a potent (K_i = 1–2 nmol/l) non-xanthine A_{2A} antagonist (BARALDI et al. 1994, 1996; ONGINI 1997). Its selectivity for A_{2A} vs A_1 receptors varies from 53- to 750-fold. [^{3}H]-SCH 58261 specifically binds to A_{2A} receptors in dopamine-rich regions in rat brain, e.g., caudate-putamen,

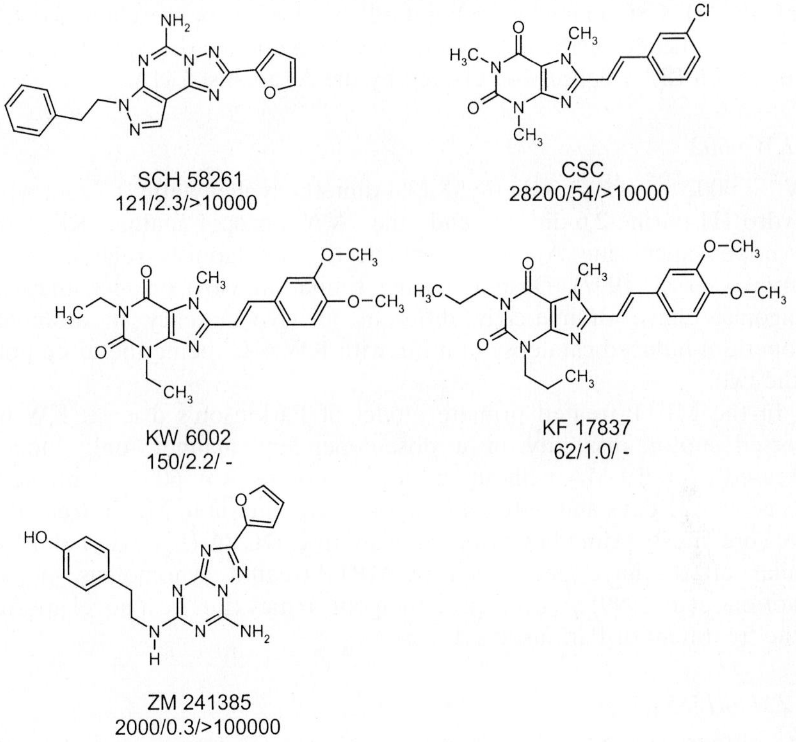

SCH 58261
121/2.3/>10000

CSC
28200/54/>10000

KW 6002
150/2.2/ -

KF 17837
62/1.0/ -

ZM 241385
2000/0.3/>100000

Fig. 8. A_{2A} selective antagonists

nucleus accumbens, and tuberculum olfactorium (FREDHOLM et al. 1998) and to porcine cardiac tissue, human neutrophils, lymphocytes, and human platelets (ONGINI 1987). In rats, SCH 58261 enhances LMA (POPOLI et al. 1998), increases wakefulness, and produces slight increases both blood pressure and heart rate. SCH 58261, although hampered by a lack of oral activity, potentiates the activity of L-DOPA or dopamine receptor agonists in a 6-hydroxydopamine-lesioned rat model of Parkinson's disease as well as reducing infarct size in a rat model of cerebral ischemia (BONA et al. 1997).

b) CSC

CSC (8-(3-chlorostyryl)caffeine) is a xanthine A_{2A} selective antagonist (JACOBSON et al. 1993b) that is 520-fold selective for rat brain A_{2A} adenosine vs A_1 receptors. Functionally, CSC was 22-fold selective for A_{2A} receptors in rat pheochromocytoma cells (K_b 60 nmol/l) vs A_1 receptors in rat adipocytes (K_b 1.3 mmol/l) in reversing agonist effects on adenylate cyclase. CSC (5 mg/kg) stimulated LMA by 22% and co-administration of CSC and DPCPX, both at non-stimulatory doses, increased activity by 37% over CSC alone, suggesting a behavioral synergy of A_1 and A_2 antagonist effects in the CNS (JACOBSON et al. 1993a). CSC (1 mg/kg i.p.) shifted the dose-response curve for LMA depression elicited by the A_{2A} agonist, APEC in mice to the right, but showed no effect on LMA depression elicited by the A_1 agonist CHA.

c) KW 6002

KW 6002 ((E)-1,3-diethyl-8-(3,4-dimethoxyphenylethyl)-7-methyl-3,7-dihydro-1H-purine-2,6-dione) and the N,N-dipropyl-analog, KF 17837, are newer adenosine A_{2A} receptor selective antagonists related to CSC (SHIMADA et al. 1997). Despite having similar in vitro profiles, these two antagonists have dramatically different in vivo potency in attenuating haloperidol-induced catalepsy in mice, with KW 6002 being the more potent of the two.

In the MPTP-treated primate model of Parkinson's disease, KW 6002 reversed motor disability in a dose-dependent manner, only modestly increased overall LMA without causing stereotypy. KW 6002 maintained its effects over 21 days and induced little or no dyskinesia in MPTP-treated primates previously primed by prior exposure to L-DOPA (KANDA et al. 1998a). Similar effects have been seen in MPTP-treated cynomolgus monkeys (GRONDIN et al. 1999). A_{2A} receptor antagonism may thus be a novel approach to the treatment of Parkinson's disease.

d) ZM 241385

ZM 241385 (4-(2-[7-amino-2-(2-furyl)[1,2,4]-triazolo[2,3-a][1,3,5]triazin-5-ylamino]ethyl) phenol) was one of the earliest A_{2A} selective antagonists identified (POUCHER et al. 1995). In rat pheochromocytoma cell membranes, ZM 241385 potently displaced [^{3}H]-NECA binding (pIC$_{50}$ = 9.53; ~0.3 nmol/l) but was far less effective (pIC$_{50}$ = 5.69; ~2 μmol/l) in displacing [^{3}H]-R-PIA (A_1 receptor) and I-AB-MECA (A_3 receptor; pIC$_{50}$ = 3.82; ~150 μmol/l) illustrating its selectivity for A_{2A} receptors. [^{125}I]-ZM 241385 is a high affinity antagonist radioligand (PALMER et al. 1995). ZM 241385 was equipotent in displacing [^{3}H]-CGS 21680 binding in hippocampal (K_i = 0.52 nmol/l) and striatal membranes (K_i = 0.35 nmol/l), in contrast to the A_{2A} agonist, HENECA (CUNHA et al. 1997). In anesthetized dogs, ZM 241385 (i.v.) was 140-fold more potent in attenuating the vasodilator effects of adenosine in a constant flow perfused hind limb, than the bradycardic effects of the purine. In conscious SHRs, ZM 241385 given p.o. attenuated the mean arterial blood pressure response produced by exogenous adenosine without altering the bradycardic effects (KEDDIE et al. 1996) In cats, the duration of action of ZM 241385 was ≤12 h after p.o. dosing (POUCHER et al. 1996). ZM 241385 exhibited protection against kainate-induced neuronal death when the neurotoxin was administered systemically 10 min after i.p. injection of the antagonist. ZM 241385 was effective when administered alone or in combination with CGS 21680 (JONES et al. 1998a) giving evidence of CNS penetration. Like KW 6002, ZM 241385 may be useful in the treatment of Parkinson's disease.

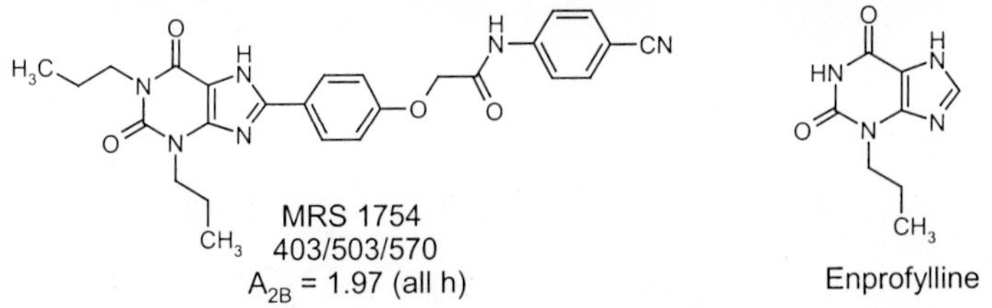

Fig. 9. A_{2B} antagonists

3. A_{2B} Receptor Antagonists (Fig. 9)

a) MRS 1754

MRS 1754 (*N*-(4-cyanophenyl)-2-[4-(2,6-dioxo-1,3-dipropyl-2,3,6,7-tetrahydro-1*H*-purin-8-yl)-phenoxy] acetamide). The division of brain A_2 receptors into two subtypes, the high affinity A_{2A} and the low affinity A_{2B}, was based on biochemical studies (DALY et al. 1983). The A_{2B} subtype was cloned (RIVKEES and REPPERT 1992) and its distribution and functional properties studied (FEOKTISTOV and BIAGGIONI 1997). Activation of A_{2B} receptors include potentiation of neurotransmitter release and activation of human mast cells with potential implications in asthma The distribution of A_{2B} receptors in the gastrointestinal tract may suggest utility in GI disorders.

Characterization of A_{2B} receptors has been hampered by a lack of selective ligands with the assessment of the receptor relying on functional data in CHO cells transfected with human A_{2B} receptor cDNA, in which inhibition of agonist-induced cAMP production is measured. A binding assay using the A_{2A} antagonist [^{3}H]-ZM 214385 in recombinant human A_{2B} adenosine receptors in HEK-293 cell membranes has been developed (JI and JACOBSON 1999) and selective antagonists have been reported (DE ZWART et al. 1999). MRS 1754, an 8-substituted xanthine, is the first truly selective A_{2B} antagonist (KIM et al. 2000) binding to A_{2B} receptors in HEK-293 cells (HEK-A_{2B}) with a Ki of 2 nmol/l, and having a selectivity of >200 to the other human adenosine receptors. Enprofylline has also been described as a selective A_{2B} receptor antagonist (FEOKTISTOV and BIAGGIONI 1997).

4. A_3 Receptor Antagonists (Fig. 10)

a) L-249313

L-249313 (5,9-dihydro-9-methyl-2-phenyl-[1,2,4]triazolo[5,1-a][2,7]naphthyridine-6-acetic acid) is a newer, potent (K_i = 5.4 nmol/l, hA_3 receptors), nonxanthine adenosine A_3 antagonist (JACOBSON 1998) that exhibits 1222- and 231-fold selectivity over human A_1 and A_{2A} receptors, respectively. L-249313 induces apoptotic cell death in human HL-60 leukemia and U-9377 lymphoma cell lines, an effect that can be blocked by the A_3 agonist, Cl-IB-MECA (YAO et al. 1997).

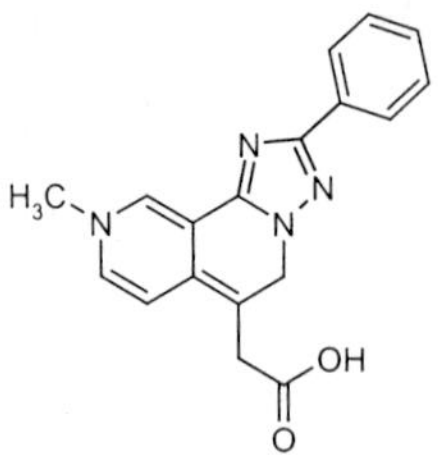

L-249313
6600/1200/5.4(h)

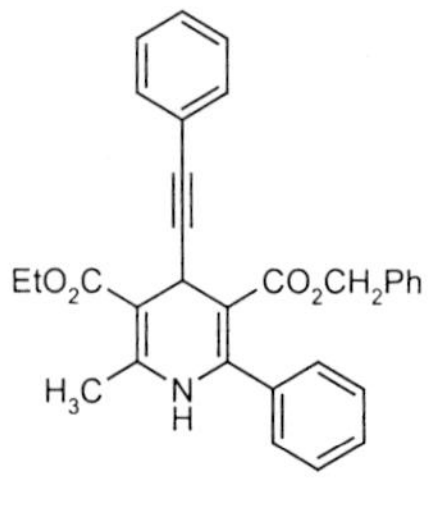

MRS-1191
40100/>10000/31.4(h)

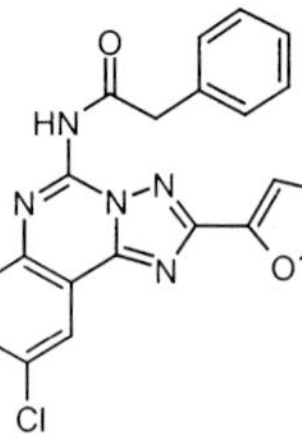

MRS-1220
305/52/0.65(h)

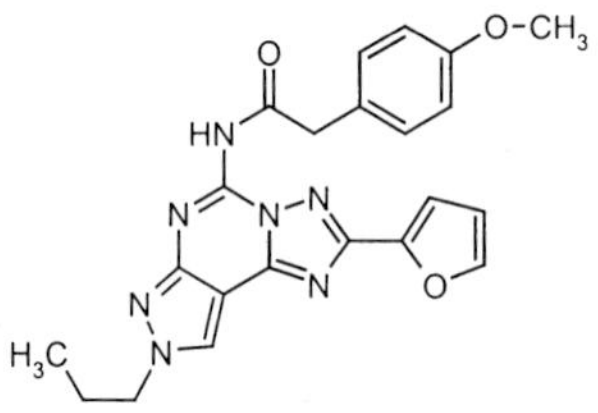

MRE 3008-F20
1200/141/0.29(h)

Fig. 10. A_3 selective antagonists

b) MRS 1191

MRS 1191 (1,4-dihydro-2-methyl-6-phenyl-4-(phenylethynyl)-3,5-pyridinedicarboxylic acid, 3-ethyl 5-(phenylmethyl) ester) is another novel selective A_3 antagonist derived from the dihydropyridine pharmacophore. MRS 1191 binds to the human A_3 receptor with a K_i value of 31 nmol/l, with minimal activity at A_1 and A_{2A} receptors (Jiang et al. 1996). MRS-1191 is a functional antagonist, competitively antagonizing the effects of IB-MECA on inhibition of adenylate cyclase in recombinant human A_3 receptor cell lines (Jacobson et al. 1997). Other dihydropyridine A_3 receptor antagonists have function activity, protecting in chick cardiac myocytes, where activation of A_3 receptors induces apoptosis (Shneyvays et al. 1998).

c) MRS 1220

MRS 1220 (*N*-[9-chloro-2-(2-furanyl)[1,2,4]triazolo[1,5-*c*]quinazolin-5-yl]benzeneacetamide) is an analog of CGS 15943 and is a potent ($K_i = 0.65$ nmol/l) inhibitor of binding to the human but not to the rat A_3 receptor (Kim et al. 1996). MRS 1220 reversed the effect of A_3 agonist-elicited inhibition of TNF-α formation in the human macrophage U-937 cell line ($IC_{50} = 0.3\,\mu$mol/l; Jacobson et al. 1997b), although in similar experiments MRS 1191 was ineffective suggesting the involvement of a non-A_3 receptor mechanism.

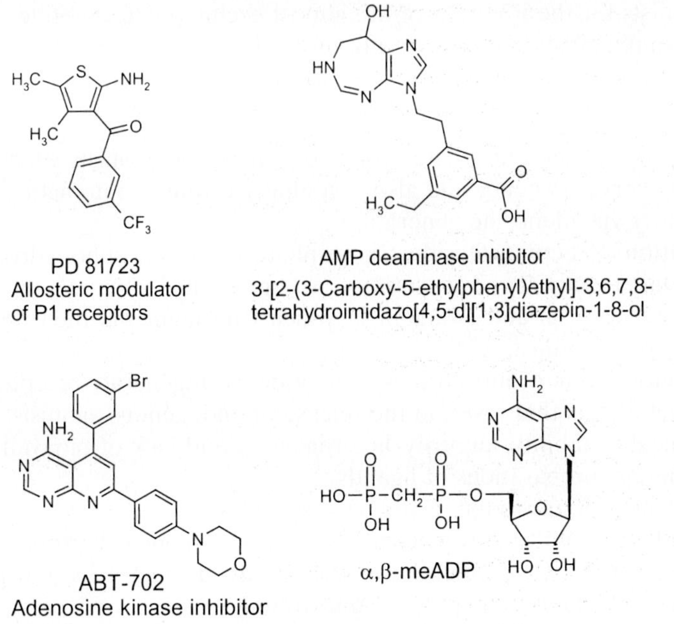

Fig. 11. Indirect modulators of purine receptor action

MRE 3008-F20 is a recently described A$_3$ receptor antagonist (VARANI et al. 2000).

C. P1 Receptor Modulators

P1 receptor activity can be altered by a number of compounds that do not interact directly with the active site of the receptor. These include the allosteric modulator PD 81,723 (2-amino-4,5-dimethylthien-3-yl)[3-(trifluoromethyl)phenyl]-methanone; BRUNS and FERGUS 1990; LINDEN 1997) as well as adenosine regulating agents (KOWALUK et al. 1998) that include inhibitors of adenosine kinase (ABT-702), adenosine deaminase (compound 1) and adenosine transport which may have potential applications for neuroprotection, seizure, and pain treatment (IJZERMAN et al. 1997; GEIGER et al. 1997). These agents are not covered in detail in the present chapter but their structures are shown in Fig. 11.

D. P2 Receptor Ligands

ATP modulates cell function via interaction with two superfamilies of membrane-bound P2 receptors: the P2X subtype, a family of seven ligand-gated ion channels that are activated by adenine nucleotides (RALEVIC and BURNSTOCK 1998; CHESSELL et al., Chap. 3, this volume) and P2Y, G protein-coupled (generally phospholipase C-activating) receptors that are activated by both adenine and uracil nucleotides (BOARDER and WEBB, Chap. 4, this volume).

Agonists for the P2 receptor are almost exclusively nucleotides, while P2 receptor antagonists are structurally more diverse. Pharmacological studies to characterize P2 receptor function can be confounded by the following considerations:

1. Hydrolysis of the phosphate groups of nucleotide agonists which reduces their observed potency and also potentially results in interactions with P1 receptors via adenosine generation
2. Inhibition of ectonucleoside triphosphate diphosphohydrolyase (E-NTPDase) activity (ZIMMERMANN, Chap. 8, this volume) by both P2 receptor agonists and antagonists, which can augments the effect of the endogenous agonist, ATP
3. Phosphorylation of nucleosides/nucleotides to regenerate receptor ligands
4. Indirect effects that result in the release of endogenous agonists
5. Chemical instability, impurity, heterogeneity and lack of bioavailability of commercial preparations of ligands
6. Differences in native and expressed receptors
7. Activation of multiple P2 receptor subtypes in single cell preparations
8. The presence of unique, pharmacologically distinct P2X receptor heteromers (WILLIAMS, Chap. 29, second volume)
9. Desensitization of P2X receptors – making agonists appearing to have antagonist activity

The resolution of these issues, as with other neurotransmitter receptor systems, will be dependent on the discovery of novel, potent, and selective pharmacophores from natural product, combinatorial, or medicinal chemical libraries.

The present limited knowledge regarding the molecular structure of the P2X receptor and a lack of robust radioligand binding assays (XU et al. 1999) has added to the complexity of identifying novel ligands. At the present time, the P2X receptor, a multimeric protein formed of subunits with a 2TM motif (NORTH and SURPRENANT 2000) is thought to form functional trimers (STOOP et al. 1997). These may be homomeric ($P2X_3$) or heteromeric ($P2X_{2/3}$) and there is also emerging evidence that P2X subunits can former multimers with subunits from other ligand gated ion channel families when transfected into cell lines, e.g., neuronal nicotinic cholinergic α-subunits (KHAKH et al. 2000); thus in defining the actual molecular target against which to screen compound libraries, knowledge is very rudimentary. It is also unknown what features define an activatable P2X receptor construct and also what additional ligand recognition sites may be present and what sites reflect the interactions with non-competitive antagonists like KN-62 (GARGETT and WILEY 1997). Given the unusual symmetric structure suggestive of a bidentate ligand of some of the classical P2 receptor antagonists, e.g., suramin, and the ability of dinucleotide polyphosphates to activate (Ap4A; MIRAS-PORTUGAL et al. 1999) and block (Ip5I; KING et al. 1999) P2 receptors, it is possible that functional P2 receptors, both P2X and P2Y, may require interactions with two ATP recognition sites for function.

I. P2 Receptor Agonists (Fig. 12)

The structure–activity relationships of the adenine, ribose, and triphosphate moieties of P2 receptor agonist have been previously reviewed (Cusack et al. 1990; Burnstock et al. 1994; Fischer et al. 1993; Wildman et al. 1999) and are summarized in Fig. 1C.

Currently, because of limitations in currently available antagonists, all P2 native receptors are characterized by the rank order potency of agonists. Depending on the species, the tissue, the assay conditions, and the cohort of P2 agonists used, very different pharmacological profiles may be obtained.

For the $P2Y_1$ receptor, the potency order for activation is 2-methylthioadenosine 5'-diphosphate (2-MeSADP) > 2-MeSATP > 2-(hexylthio)adenosine 5'-monophosphate (HT-AMP) > ADP > ATP, while AMP and UTP are inactive (Boyer et al. 1996). Whether the $P2Y_1$ receptor is activated or antagonized by ATP derivatives is a controversial issue. There is evidence that ATP is an antagonist or weak partial agonist at platelet $P2Y_1$ and at $P2Y_4$ receptors (Kennedy et al. 2000). Rat and human $P2Y_1$ receptors are ADP-specific

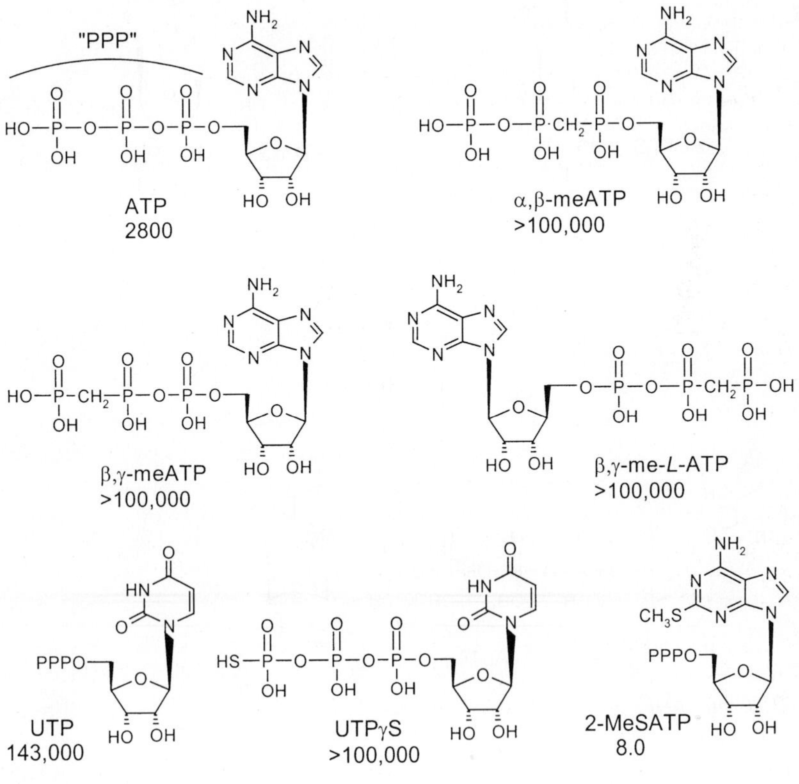

Fig. 12. P2 receptor agonists

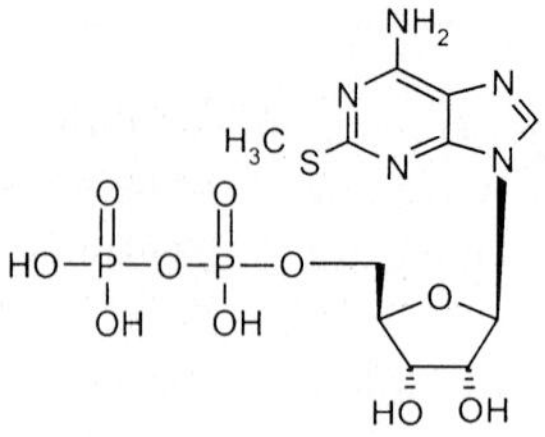

2-MeSADP

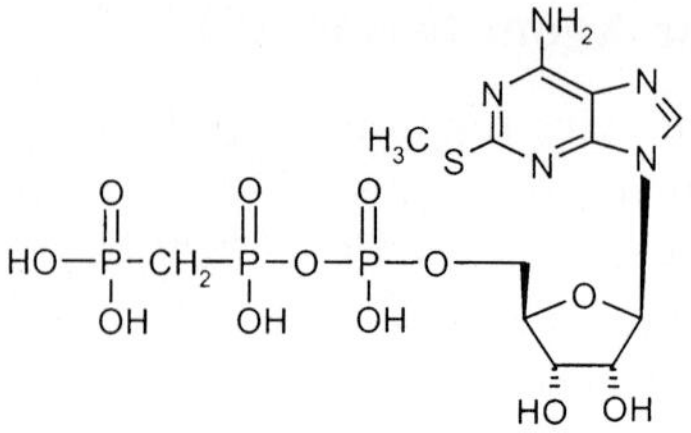

2-MeS-β,γ-meATP

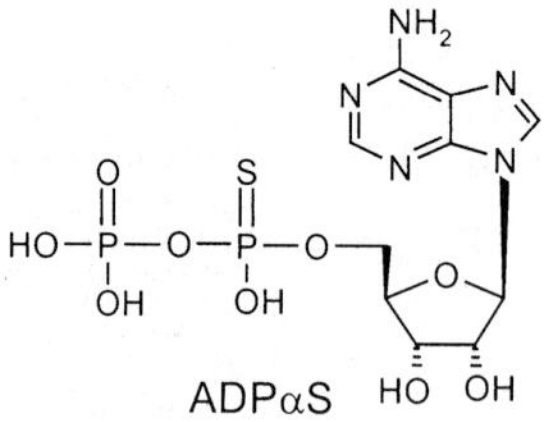

ADPαS

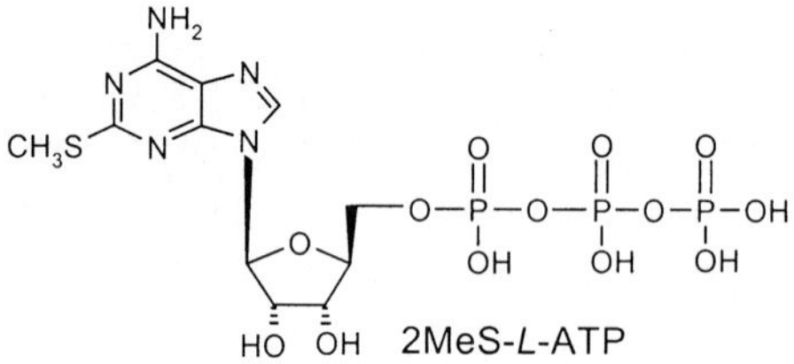

2MeS-L-ATP

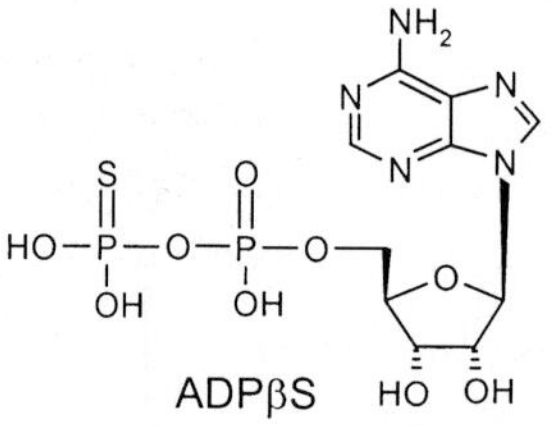

ADPβS

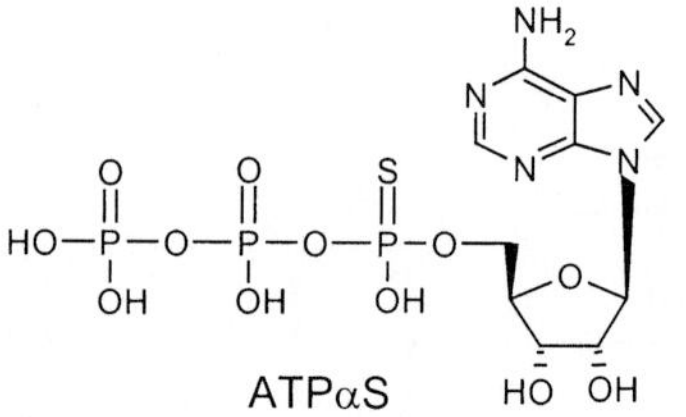

ATPαS

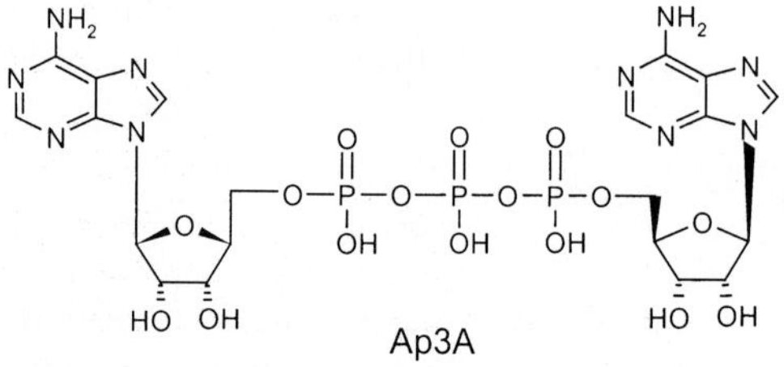

Ap3A

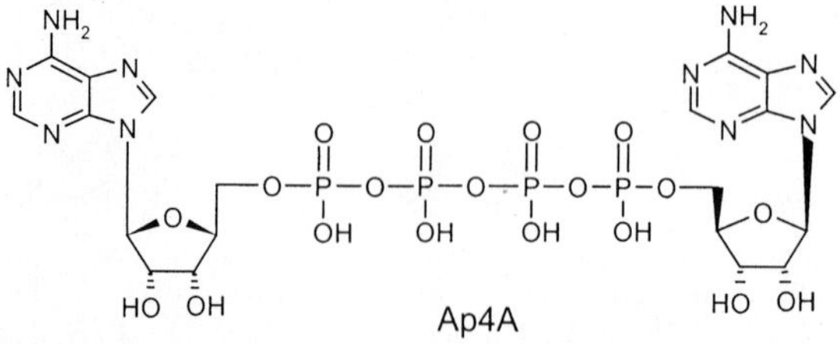

Ap4A

Fig. 12. *Continued*

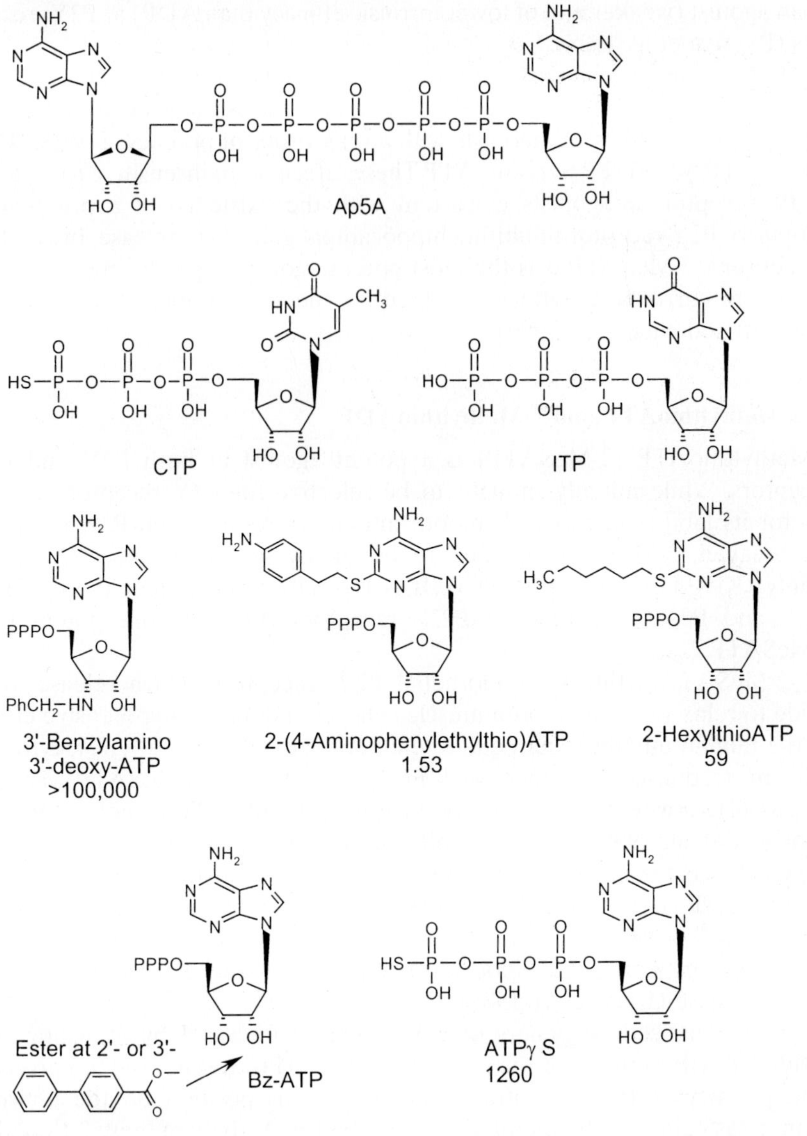

Fig. 12. *Continued*

receptors that recognize ADP and 2-meSADP (HECHLER et al. 1998), whereas ATP and 2-MeSATP are competitive antagonists, emphasizing the pharmacological similarity of the $P2Y_1$ receptor and the platelet ADP ($P2Y_{12}$, see BOARDER and WEBB, Chap. 4, this volume) receptor. In a study which minimized the breakdown of the triphosphate ATP, was found to be an agonist (weaker and of lower intrinsic efficacy than ADP) at $P2Y_1$ receptors (PALMER et al. 1998).

Intracellular calcium oscillations in rat cultured hippocampal neurons produced by postsynaptic responses to glutamate release (KOIZUMI and INOUE 1997) are inhibited by P2 agonists with a rank order of potency: 2-MeSATP > ATP > ATP-γ-S > UTP > α,β-meATP. These effects were insensitive to suramin or P1 receptor antagonists consistent with the existence of a presynaptic inhibitory P2Y-receptor inhibiting hippocampal glutamate release. In cerebellar neurons, 2-MeSATP was the most potent agonist in producing whole-cell potassium currents, an effect mediated by the $\beta\gamma$ subunits of a G protein (IKEUCHI and NISHIZAKI 1996).

1. 2-MethylthioATP and 2-MethylthioADP

2-MethylthioATP (2-MeSATP) is a potent agonist at both P2Y and P2X receptors. While initially thought to be selective for P2Y receptors, adjusting for its lability in classical smooth muscle assays in which P2X receptors are assayed, 2-MeSATP appears to be highly potent at the P2X superfamily (KENNEDY and LEFF 1995). It is typically more potent than ATP at $P2X_1$ and $P2X_3$ receptors. At $P2X_2$ receptors, ATP is more potent than 2-MeSATP.

2-MeSATP, acting via endothelial $P2Y_1$ receptors, which release nitric oxide to relax vascular smooth muscle cells, should have a hypotensive effect. In the human umbilical vein endothelial-derived cell line, ECV304, cytosolic calcium responses was increased most potently by 2-MeSATP (EC_{50} = 0.5μmol/l), however there was evidence for multiple P2Y receptor subtypes (CONANT et al. 1998). In the rabbit corpus cavernosum smooth muscle 2-MeSATP was the most potent agonist in inducing a nitric-oxide independent relaxation (FILIPPI et al. 1999). 2-MeSATP, likely acting at P2Y receptors on pancreatic β-cells in vivo, when infused directly in the pancreaticoduodenal artery in dogs to reach a blood level of approximately 15μmol/l, promoted insulin release (RIBES et al. 1988).

Functionalized congeners of 2-MeSATP synthesized by attaching substituted alkylthio chains at the purine 2-position (FISCHER et al. 1993) preserve high potency at P2Y receptors, establishing this position on the pharmacophore as being highly tolerated by the receptor. Activity of these 2-thioether ATP derivatives at P2Y receptors varied somewhat, depending on the distal structural features while activity at P2X receptors varied to an even greater degree. All eleven of the synthesized 2-thioethers of ATP (FISCHER et al. 1993) stimulated phospholipase C, increasing inositol phosphate production in turkey erythrocyte membranes with $pK_{0.5}$ values between 1.5nmol/l and 770nmol/l. In smooth muscle assay systems for P2Y receptor activity the, 2-thioethers had EC_{50} values between 10nmol/l and 1μmol/l. There was a significant correlation for the 2-thioether compounds between the $pK_{0.5}$ values for inositol phosphate production in turkey erythrocyte membranes and the pD$_2$ values for relaxation mediated via P2Y receptors in the guinea-pig taenia coli, but not for vascular P2Y receptors or for P2X receptors. At P2X receptors, no activity was observed in the rabbit saphenous artery, but variable degrees of activity were observed in the guinea pig vas deferens and bladder depending on distal substituents of the 2-thioether moiety.

2. 2-(7-Cyanohexylthio)-ATP

2-(7-Cyanohexylthio)-ATP is more potent than 2-MeSATP at the taenia coli P2Y receptor. 2-[2-(p-Aminophenyl)ethylthio]-ATP, intended as a substrate for radioiodination and potential cross-linking to the receptor, displayed the highest affinity of all the analogues at turkey erythrocyte P2Y$_1$ receptors. A p-nitrophenethylthio ether was relatively weak at P2Y$_1$ receptors, but provided selectivity for a subset (*vas deferens*) of the P2X receptor class. 7-Aminoheptylthio- and 7-thioheptylthio- ethers synthesized for the ease of the further derivatization by acylation or alkylation and to probe potential accessory binding sites on the receptor, displayed K$_{0.5}$ values of 73 nmol/l and 770 nmol/l, respectively, at erythrocyte P2Y receptors. A further benefit of the presence of a long chain thioether group at the 2-position was increased stability of the triphosphate group at the 5′-position (ZIMMET et al. 1993). It is likely that long chains, although at a site on the molecule distal to the triphosphate group, interfere with the binding site of E-NTPDases.

3. α,β-MeATP

α,β-MeATP (α,β-methylene ATP) is among the more potent P2X agonists and is selective for P2X$_1$ and P2X$_3$ subtypes (EC$_{50}$ = 1–10 μmol/l), at which it causes rapid desensitization. α,β-meATP also activates P2X$_{2/3}$ heteromers, which are much less susceptible to desensitization. α,β-meATP is much less potent than ATP at P2X$_2$ and P2X$_{4-7}$ subtypes. At P2Y receptors α,β-meATP is weak or inactive. Due to the presence of the non-hydrolyzable methylene bridge, α,β-meATP tends to be more stable than other agonists under most pharmacological assay conditions (BROWNHILL et al. 1997). For this reason, its potency in comparison to other, hydrolyzable nucleotide agonists can be overestimated leading to the description of α,β-meATP as the most potent P2X receptor agonist.

In isolated, constant-pressure perfused rat kidneys at basal vascular tone, systemic α,β-meATP was a potent vasoconstrictor acting via a P2X receptor (ELTZE and ULLRICH 1996). In kidneys with raised tone, there was an endothelium-dependent vasodilatation at low doses of 2meSATP. In isolated pancreatic islet cells and isolated perfused pancreas from rats, in addition to P2Y receptors, which potentiate glucose-induced insulin secretion, there are P2X receptors, at which α,β-meATP (5–50 μmol/l) transiently stimulated insulin release at low non-stimulating glucose concentrations (PETIT et al. 1998). In contrast, at a higher glucose concentration (3 mmol/l), the P2Y agonist ADP-β-S, acting at a P2Y receptor subject to desensitization, was 100-fold more potent than α,β-meATP in stimulating insulin release.

ATP is a fast neurotransmitter in the medial habenula (ROBERTSON et al. 1998) and α,β-meATP also acts potently at this site. In rat brain sections containing the locus ceruleus (SANSUM et al. 1998), α,β-meATP (1–300 µmol/l) and the related P2X-receptor selective agonist, α,β-methylene-ADP (α,β-meADP), elicited concentration-dependent increases in the spontaneous firing rate, and the effect was attenuated by suramin (100 µmol/l) and PPADS (30 µmol/l). In this system, both P2X and P2Y purinoceptor types appear to be present. In rat hippocampal slices, α,β-meATP (10 µmol/l), but not 2-meSATP or UTP, produced an increase in spontaneous frequency (Ross and STONE 1998) suggesting the presence of an excitatory P2X receptor.

4. β,γ-Methylene-D-ATP and β,γ-Methylene-L-ATP

β,γ-Methylene-D-ATP (β,γ-meATP) is a metabolically stable analogue of ATP due to the non-hydrolyzable methylene bridge. In rat juxtamedullary afferent arterioles (INSCHO et al. 1998), β,γ-meATP was equipotent to ATP in eliciting a concentration-dependent vasoconstriction; however, another P2X agonist, α,β-meATP, evoked a marked receptor desensitization. 2-MeSATP evoked only a modest vasoconstriction, whereas UTP and ATPγS (see below) reduced afferent diameter markedly, suggesting that both P2X- and P2Y$_2$-receptor subtypes are present in renal microvasculature.

In isolated rat liver plasma membranes, phospholipase D (PLD) is activated by purinergic agonists (MALCOMB et al. 1995), the rank order of efficacy for stimulation of PLD activity in the presence of 0.2 µmol/l GTPγS being β,γ-meATP > ATPγS = ATP = 2-MeSATP > α,β-meATP = UTP. The effect was independent of the activation of phospholipase C, upon which α,β-meATP was inactive.

The corresponding nucleotide analogue containing the unnatural adenosine enantiomer, β,γ-methylene-L-ATP (L-adenyl 5'-(β,γ-methylene)diphosphonate, L-AMP-PCP) is more potent than ATP at certain P2X receptor subtypes (CUSACK and HOURANI 1990). It was first reported to be a potent P2 receptor agonist in the guinea pig bladder (P2X), but inactive in the guinea pig taenia coli (P2Y). L-AMP-PCP, unlike ATP, did not relax taenia coli, and was not degraded by E-NTPDases. In the rat dorsal root ganglia, the inactivity of this agonist in stimulating inward currents suggests that ATP and other agonists that were effective were acting at the P2X$_3$-subtype rather than the P2X$_1$-subtype (RAE et al. 1998).

5. 3'-Benzylamino-3'-deoxyATP

3'-Benzylamino-3'-deoxyATP is a potent P2X receptor agonist in the guinea-pig vas deferens with slightly less potent in the urinary bladder (BURNSTOCK et al. 1994). It was inactive at rabbit saphenous artery P2X receptors and at all P2Y receptors. The P2X-potency observed was approximately an order of magnitude greater than that of α,β-meATP, which is used widely in studies of P2X receptors. 3'-Benzylamino-3'-deoxy-ATP was not tested for the ability to desensitize P2X$_1$ receptors, however it caused the same profile of contractile response as ATP or α,β-meATP, e.g., a transient twitch-like contraction suggesting that it would cause rapid desensitization.

6. Adenosine 5'-O-(3-thiotriphosphate) (ATPγS)

Thiophosphate groups tend to increase the stability of nucleotide analogues towards degradation by E-NTPDases, and in some cases improve potency at P2 receptors. Adenosine 5'-O-(3-thiotriphosphate) is a potent agonist at various P2Y and P2X subtypes. ATPγS (BEUKERS et al. 1995) also inhibited ecto-ATPase competitively with micromolar affinity.

ATPγS acts at phospholipase C-coupled P2Y (P$_{2u}$) receptors to regulate ion transport in epithelial cells, including Medin-Darby canine kidney (MDCK) cells (YANG et al. 1997). The order of agonist potency was UTP = ATPγS > ATP > α,β-meATP = 2-MeSATP. In undifferentiated HL-60 cells there is evidence of a novel cyclic AMP-linked P2Y receptor (CONIGRAVE et al. 1998) with the order of agonist potency in stimulating cAMP production being ATPγS > BzATP > ATP. The response to UTP was poor.

[^{35}S]-ATPγS has been used as a radioligand (MICHEL et al. 1996) for cloned P2X$_1$–P2X$_4$ receptors expressed in CHO cells. The affinities for antagonists determined by binding were similar to those obtained in functional assays. Among agonist competitors, 2-MeSATP was more potent than α,β-meATP at all four subtypes. The radioligand appears to label a desensitized high affinity state of the receptors, which is subject to positive allosteric modulation by the P2 antagonist, Cibacron blue and negative modulation by d-tubocurarine. Subsequent studies (XU et al. 1999) showed that [^{35}S]-ATPγS was able to bind to cells that did not express any P2 receptors.

7. 2'- and 3'-O-(4-Benzoylbenzoyl)-ATP 2'- and 3'-O-(4-Benzoylbenzoyl)-ATP (BzATP)

2'- and 3'-O-(4-Benzoylbenzoyl)-ATP 2'- and 3'-O-(4-benzoylbenzoyl)-ATP (BzATP), available commercially only as a mixture of isomers, is among the most potent agonists (high micromolar range) at P2X$_7$ receptors This ligand also had nanomolar potency at P2X$_1$ receptors (BIANCHI et al. 1999). The P2X$_7$ receptor in mast cells has been affinity labeled with [^{3}H]-BzATP.

In human lymphocytes (WILEY et al. 1998), BzATP elicited a twofold greater maximal cation influx and had a tenfold lower EC$_{50}$ value (8μmol/l) than ATP. Activation of P2X$_7$ receptors in the immune system can induce apoptosis (DIVIRGILIO et al. 1990). BzATP (5mmol/l) caused apoptosis in dendritic cells, which play a large role in T-cell activation (COUTINHO-SILVA et al. 1999). BzATP (1mmol/l) was very effective in activating the transcription factor NFAT in N9 microglial cells, suggesting purinergic modulation of early inflammatory gene expression in the nervous and immune systems (FERRARI et al. 1999).

8. UTP and UDP

Among metabotropic P2Y receptors, $P2Y_2$, $P2Y_4$, and $P2Y_6$ respond to uridine nucleotides with the a similar or even greater degree of potency than the corresponding adenine nucleotides (FILTZ et al. 1997; COMMUNI and BOYNAEMS 1997). Thus, the previously pharmacologically defined P_{2U} subtype responsive to UTP may refer to any of these three receptors. At the $P2Y_2$ receptor, UTP is equipotent to ATP. At $P2Y_4$ receptors UTP and UDP are equipotent with the potency of ATP being species dependent. The intrinsic activity of ATP can also vary depending on species. At the rat $P2Y_4$ receptor, ATP is an agonist but at the human $P2Y_4$ receptor acts as an antagonist (KENNEDY et al. 2000). At $P2Y_6$ receptors UDP is the preferred ligand and is more potent than UTP.

UTP and UDP elicit vasoconstriction in the rat pulmonary vascular bed with equal potency (RUBINO et al. 1996) and there is a uracil nucleotide receptor on rat sympathetic-ganglia (CONNOLLY 1995). In neuroblastoma-glioma hybrid NG108–15 cells (SONG et al. 1996), multiple P2Y subtypes mediate inhibition of cAMP accumulation with the rank order of agonist potency of UTP > 2-MeSATP > ATP > ATPγS.

Activation of $P2Y_2$ receptor stimulates chloride efflux in airway epithelia by a mechanism independent of the CFTR chloride transporter (DONALDSON and BOUCHER 1998).

9. UTPγS

Uridine 5'-O-(3-thiotriphosphate) (UTPγS) is a potent $P2Y_2$ receptor agonist that is not susceptible to degradation by E-NTPDases (LAZAROWSKI et al. 1996) and may be useful in the treatment of pulmonary diseases.

10. HT-AMP and Other AMP-2-Thioethers

The addition of a functionalized chain at the 2-position of the adenine ring in ATP functionalized congeners (see above) allowed for truncation of the triphosphate group with retention of affinity, thus circumventing one of the major complications in interpreting ATP pharmacological results, e.g., the action of E-NTPDases. While AMP was inactive at P2Y receptors, 2-thioether analogues were full agonists at erythrocyte P2Y receptors (FISCHER et al. 1993; BOYER et al. 1996), although generally several orders of magnitude less potent than the corresponding 2-thioether triphosphate analogue. For example, the 2-hexenylthio ether of AMP was 8-fold more potent than ATP itself in the stimulation of phospholipase C, but was 33-fold less potent than the corresponding triphosphate. Thus, the long chain may act as a distal anchor of the ligand at an accessory binding site on the receptor. Also, several adenosine diphosphate 2-thioether analogues proved equipotent to the corresponding

ATP analogues at erythrocyte P2Y$_1$ receptors. 2-HexylthioAMP was the most potent at P2Y$_1$ receptors among a series of homologous thioethers and a highly potent agonist at the yet uncloned P2Y receptor in C6 glioma cells that is coupled to inhibition of adenylate cyclase (BOYER et al. 1996).

II. P2 Receptor Antagonists

P2 receptor antagonists are generally limited in their potency and receptor selectivity both for P2 receptors and other classes of protein target (BHAGWAT and WILLIAMS 1997; VAN RHEE and JACOBSON 1997). A series of key compounds are shown in Fig. 13.

1. Suramin, NF023, and NF279

The symmetrical, polysulfonated, trypanocidal drug suramin (1,3,5-naphthalenetrisulfonic acid, 8,8′-[carbonylbis[imino-3,1-phenylenecarbonylimino (4-methyl-3,1-phenylene) carbonylimino]] bis-, hexasodium salt) is a weak P2 receptor antagonist but nonetheless has been extensively used to study P2 receptors. At P2Y receptors (CHARLTON, et al. 1996), the sensitivity to suramin occurs in the rank order of P2Y$_1$ > P2Y$_2$ > P2Y$_4$.

Unfortunately, suramin suffers from a lack of both selectivity and competitive antagonism at P2 receptors. The affinity of both suramin and its truncated analogue NF023, also a P2 antagonist, in inhibiting G protein function is in the submicromolar range, e.g. at concentrations lower than those normally used to blocking P2 receptors (FREISSMUTH et al. 1996). NF023 appears to be selective for P2X receptors (BUELTMANN et al. 1996) antagonising effects at these receptors in rat, hamster, and rabbit isolated blood vessels (ZIYAL et al. 1996).

Suramin is a non-competitive inhibitor of ecto-ATPase in human blood cells having micromolar affinity (BEUKERS et al. 1995). The truncated suramin derivatives (VAN RHEE et al. 1994) XAMR0721 (1,3,5-naphthalenetrisulfonic acid, 8-[(3,5-dinitrobenzoyl)amino]-trisodium salt) antagonized P2 receptors without inhibiting E-NTPDases. Another suramin analogue, NF279 (8,8′-(carbonylbis(imino-4,1-phenylenecarbonylimino-4,1-phenylenecarbonylimino))bis(1,3,5-naphthalenetrisulfonic acid; DAMER et al. 1998)) antagonized P2X receptor-mediated contractions in rat vas deferens, with a pIC$_{50}$ of 5.71 (IC$_{50}$ ~2 µmol/l) without affecting α-adrenergic, adenosine, histamine, muscarinic, or nicotinic receptor function. The low inhibitory potency of NF279 on At P2Y receptors in guinea-pig taenia coli NF270 had a pA$_2$ = 4.10 and at E-NTPDases in folliculated *Xenopus* oocytes the IC$_{50}$ value was greater than 100 µmol/l indicating that NF279 is a novel, selective P2X receptor antagonist.

2. Reactive Blue 2

The P2-receptor mediated effects of ATP in smooth muscle assays (BURNSTOCK and WARLAND 1987) are antagonized by a number of highly negatively-

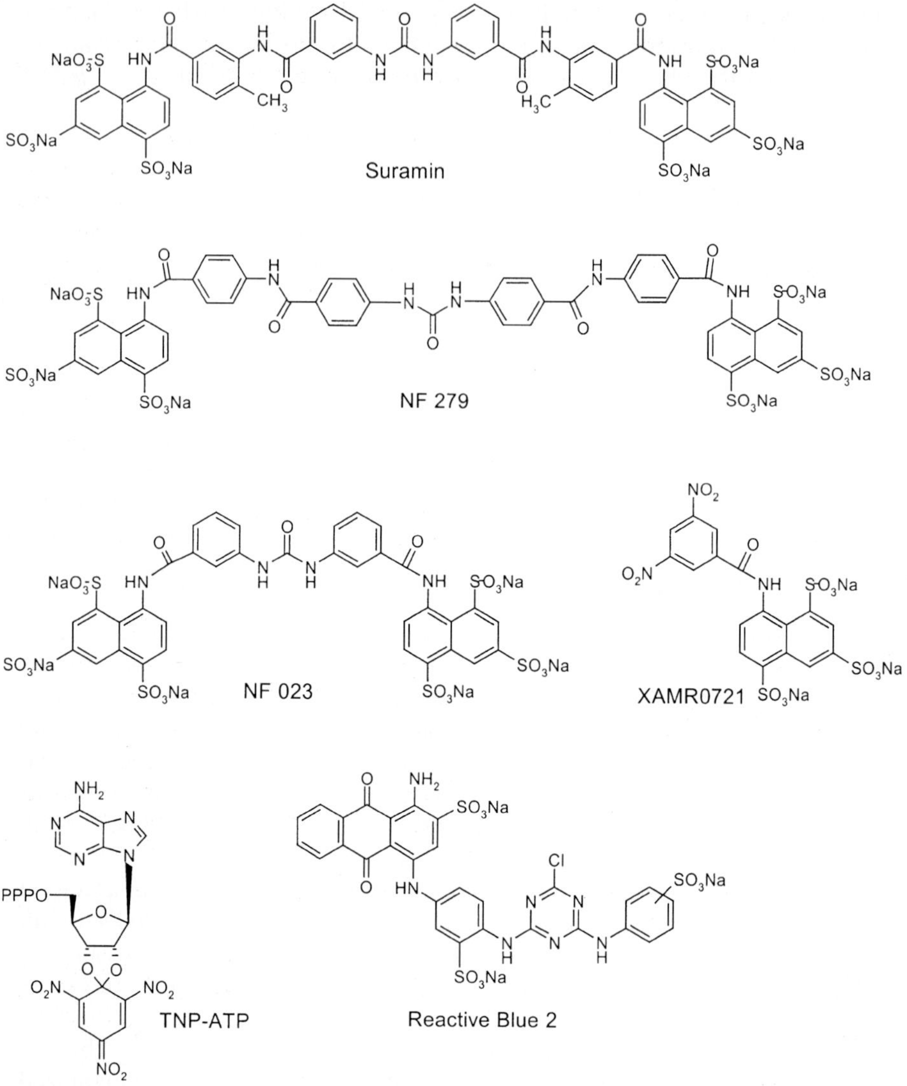

Fig. 13. P2 antagonists

charged, high molecular weight organic molecules. These include histochemical dyes, such as reactive blue 2 (RB2), which is not a chemically pure substance, but rather an isomeric mixture of Cibacron blue and basilen blue, each of which contain three aryl sulfonate groups. RB2 is more potent at $P2Y_6$ than at $P2Y_4$ receptors and is inactive at $P2Y_2$ receptors (JACOBSON et al. 1999). It antagonizes $P2Y_1$ receptors non-competitively (BOYER et al. 1994) and is considerably more potent at $P2X_2$ ($IC_{50} = 0.36 \mu mol/l$) than at $P2X_4$ receptors.

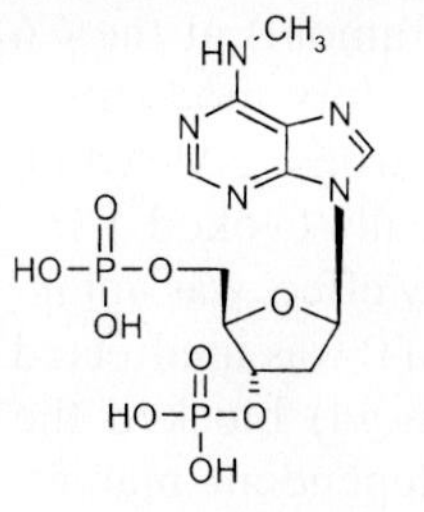

MRS 2179 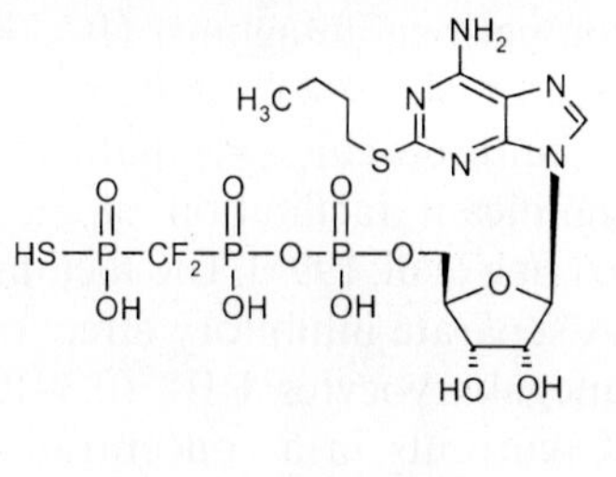
ARL 66,096 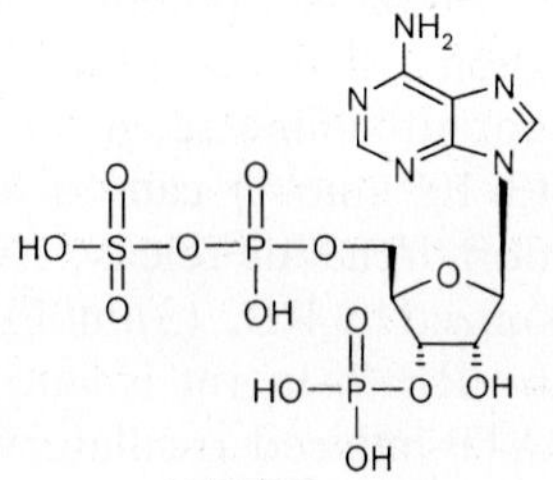
A3P5PS

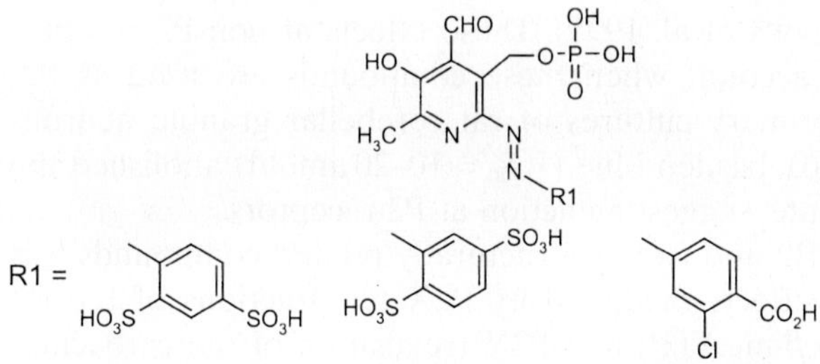

R1 =

PPADS
iso-PPADS
MRS 2160 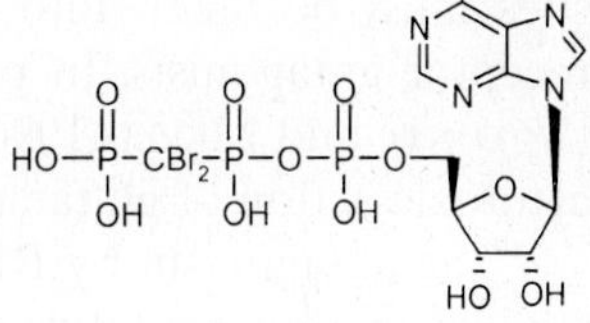
ARL-67156

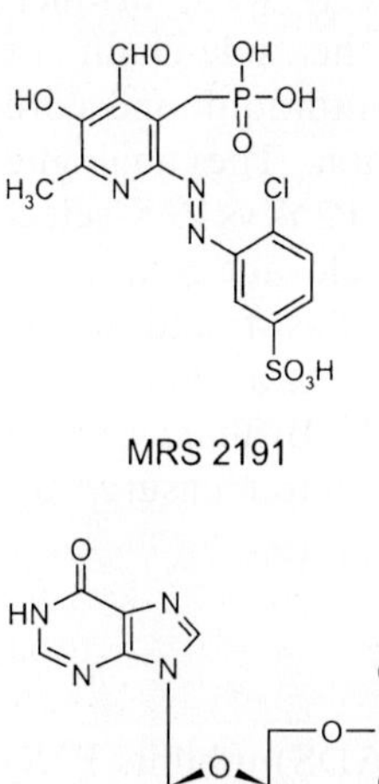

MRS 2191 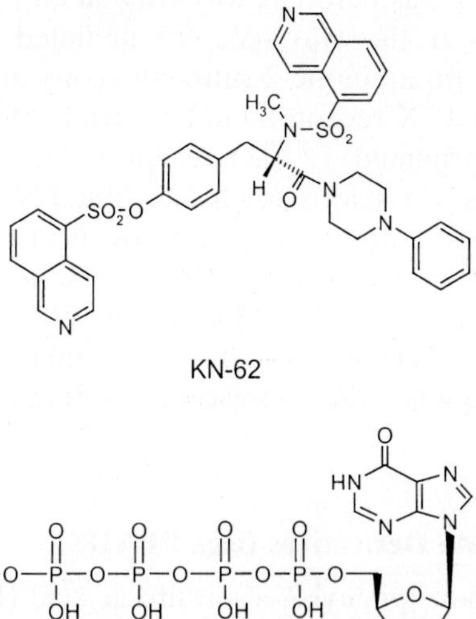
KN-62

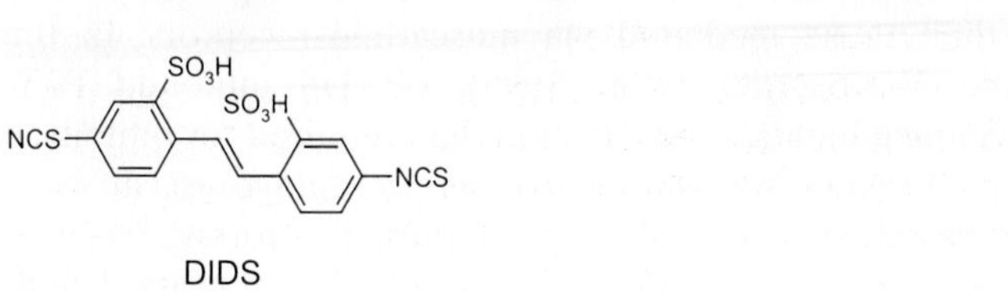

Ip5I

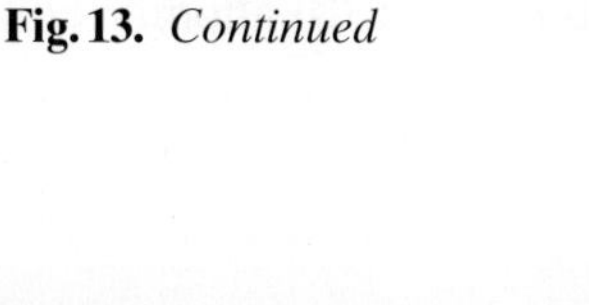

DIDS

Fig. 13. *Continued*

RB2 is a potent and competitive antagonist (IC_{50} = 30nmol/l) at the C6 glioma P2Y receptor (BOYER et al. 1994). At P2-purinoceptors modulating noradrenaline release from cultured chick sympathetic neurons, 2-MeSATP (3–100µmol/l) caused a significant facilitation of electrically-evoked [^{3}H]-noradrenaline release (ALLGAIER et al. 1995). The facilitatory effect was antagonized by RB2 (3µmol/l). A separate inhibitory effect of ATP was unaffected by RB 2. In rat isolated megakaryocytes RB2 (0.3–10µmol/l) blocked the ATP-induced oscillatory K^+-currents in a concentration-dependent manner (UNEYAMA et al. 1994).

Both RB2 and suramin (see below) can inhibit GABA and glutamate receptor channels (NAKAZAWA et al. 1995). These effects at non-P2 receptor sites must be taken into account when these compounds are used as P2-receptor antagonists. In primary cultures of rat cerebellar granule neurons (VOLONTE and MERLO 1996), basilen blue (IC_{50} = 10–20µmol/l) abolished the cytotoxic action of glutamate, suggesting action at P2 receptors.

The antagonism by RB2 and twelve structurally related compounds was studied in representative effects mediated by P2X (contractions of the rat vas deferens elicited by α,β-meATP) and P2Y (relaxation of the carbachol-precontracted guinea-pig taenia coli elicited by adenosine 5'-O-(2-thiodiphosphate) receptors (TULUC et al. 1998). The degradation of ATP in the medium by nucleotidases in the vas deferens was attenuated by RB2. The structure–activity relationships at the two receptor-mediated effects were distinct. Both the 1-amino-anthraquinone-2-sulfonate core and the 'side-chain' of RB2 are involved in P2X-receptor binding, while the anthraquinone core is the principal determinant of P2Y-receptor recognition. The analogue uniblue A was P2X- vs P2Y-selective, and acid blue 129 was P2Y- vs P2X-selective, with neither showing non-P2-receptor effects at the relevant concentrations. Another analogue, Coomassie Brilliant Blue G was found to be ~100-fold more potent than RB2 in blocking the effects of extracellular ATP on rat parotid acinar cells (SOLTOFF et al. 1989). Unlike RB2, Brilliant Blue G did not substantially quench the fluorescence of Fura 2, used to measure Ca^{2+} levels.

3. Pyridoxal Phosphate Derivatives (e.g., PPADS)

Pyridoxalphosphate-6-azophenyl-2',4'-disulfonic acid (PPADS) inhibits P2X-receptor-mediated responses in rabbit vas deferens and urinary bladder (ZIGANSHIN et al. 1993). It was selective for P2-purinoceptor-mediated contractions rather than those mediated via muscarinic receptors. In the rat mesenteric artery (WINDSCHEIF et al. 1995), PPADS inhibited P2Y-purinoceptors but at a much higher concentration than required for inhibition of P2X-purinoceptors. PPADS was ineffective as an antagonist at both vasoconstriction- and vasodilatation-mediating P_{2U}-purinoceptors.

PPADS antagonized ATP-agonist effects at several P2Y receptors, including $P2Y_1$ receptors and the cAMP-inhibitory receptor on rat C6 glioma cells

(BOYER et al. 1994). Initially, no distinction was made between PPADS and isoPPADS (pyridoxalphosphate-6-azophenyl-2',5'-disulphonic acid); however they have distinct pharmacological profiles (e.g., CONNOLLY 1995). PPADS is inactive or weakly active at most of the currently cloned P2Y subtypes other than $P2Y_1$ receptors. The $P2Y_2$ and $P2Y_4$ receptors are PPADS-insensitive (CHARLTON et al. 1996). At $P2X_1$, $P2X_2$, $P2X_3$, and $P2Y_1$ receptors, isoPPADS (JACOBSON et al. 1998) is more potent than PPADS (KIM et al. 1998).

The outer longitudinal muscles of guinea-pig urinary bladder contain P2X-purinoceptors, which are more sensitive to PPADS than suramin (USUNE et al. 1996). In the supraoptic nucleus, the increase in $[Ca^{2+}]_i$ elicited by ATP in a dose-dependent manner, was antagonized by PPADS (SHIBUYA et al. 1999). RT-PCR analysis P2X receptors mRNAs for $P2X_3$, $P2X_4$, and $P2X_7$ were predominant, suggesting functional activity of the $P2X_3$ subtype.

At the $P2X_7$ receptor stably expressed in HEK cells (CHESSELL et al. 1998), PPADS produced a concentration-dependent increase in maximal inward currents to BzATP, with an IC_{50} of 70 µmol/l.

The effect of ATP to depress the rate of basal, catecholamine-stimulated, or insulin-stimulated glucose transport by up to 60% in cardiac myocytes was suppressed by PPADS, but not by suramin or reactive blue 2 (FISCHER et al. 1999). Administration of PPADS in vivo reduced ischemia/reperfusion damage (NEELY 1995).

Modification of PPADS by functional group substitution on the sulfophenyl ring and at the phosphate moiety demonstrated that a phosphate linkage is not required for recognition by either P2X or turkey erythrocyte $P2Y_1$ receptors (KIM et al. 1998). Among the 6-phenylazo derivatives, a p-carboxyphenylazo phosphate derivative and its m-chloro analog, MRS 2160, which were selective for $P2X_1$ and $P2X_2$ vs $P2Y_1$ receptors. MRS 2160 was very potent at rat $P2X_2$ receptors expressed in *Xenopus* oocytes ($IC_{50} = 0.82$) while MRS 2159 was less potent ($IC_{50} = 11.9$ µmol/l). Among the phosphonate derivatives, [4-formyl-3-hydroxy-2-methyl-6-azo-(2'-chloro-5'-sulphonylphenyl)-5-pyridyl]-methylphosphonic acid (MRS 2192) had high potency at $P2Y_1$ receptors ($IC_{50} = 4.35$ µmol/l). The corresponding 2',5'-disulfonylphenyl derivative, MRS 2191, was nearly inactive at turkey erythrocyte $P2Y_1$ receptors. Thus a single ring substitution, sulpho instead of chloro, has a major effect on the selectivity of these methylphosphonates as P2Y receptor antagonists. MRS 2191 was relatively potent at recombinant $P2X_2$ receptors ($IC_{50} = 1.1$ µmol/l). The ethyl phosphonate, MRS 2142, while inactive at turkey $P2Y_1$ receptors, was particularly potent at recombinant $P2X_2$ receptors ($IC_{50} = 1.5$ µmol/l). A pyridoxine cyclic phosphate (cyclic pyridoxine-$\alpha^{4,5}$-monophosphate, MRS 2219) and its 6-azoaryl derivative (cyclic pyridoxine-a^{4,5}-monophosphate-6-azophenyl-2',5'-disulphonic acid, MRS 2220) selectively potentiate and antagonize, respectively, $P2X_1$ receptors expressed in *Xenopus* oocytes (JACOBSON et al. 1998).

4. Nucleotide Derivatives

Nucleotide derivatives are also antagonists at P2X and P2Y receptors. Trinitrophenyl-ATP (TNP-ATP), and the corresponding di- and mono-phosphate derivatives are nanomolar antagonists at $P2X_1$, $P2X_3$, and $P2X_{2/3}$ receptors expressed in *Xenopus* oocytes (VIRGINIO et al. 1998); however the potencies are much lower in other functional models. The inosine dinucleotide polyphosphate, Ip5I is a nanomolar antagonist (Ki = 1.5 nmol/l) at the $P2X_1$ receptor (KING et al. 1999).

At the ADP receptor (the as yet uncloned $P2_t$ receptor) in platelets, purine triphosphates are antagonists of aggregation (CUSACK et al. 1990). ARL-66,096 is an ADP receptor antagonist that inhibits platelet aggregation (HUMPHRIES et al. 1996). The corresponding β,γ-dichloromethylene derivative, ARL 67085, has been examined as a novel anti-thrombotic agent (INGALL et al. 1999) and a follow on compound, ARL 69931MX, the N^6-(2-methylthioethyl)-2-(3,3,3-trifluoropropylthio)-β,γ-dichloromethylene derivative, which is potent ADP receptor antagonist (IC_{50} = 0.4 nmol/l) is in clinical trials.

5. MRS 2179 and Other Bisphosphate Analogues

Adenosine 3',5'- and 2',5'-*bis*phosphates act as competitive antagonists at the $P2Y_1$ receptor (BOYER et al. 1996). Modifications at the 2-and 6-position of the adenine ring, on the ribose moiety, and on the phosphate groups in 2'- and 3'-deoxyadenosine *bis*phosphate analogues have resulted in more potent and selective $P2Y_1$ antagonists (CAMAIONI et al. 1998; NANDANAN et al. 1999). MRS 2179 (2'-deoxy-N^6-methyladenosine-3',5'-*bis*phosphate) competitively antagonizes the effects of 2-MeSADP at turkey erythrocyte $P2Y_1$ receptors (K_B = 104 nmol/l; BOYER et al. 1998). MRS 2179 had no agonist activity at either the turkey or human $P2Y_1$ receptor and is thus a pure antagonist. MRS 2179 was inactive at the P2Y receptor in C6 glioma cells which is coupled to inhibition of adenylate cyclase (J.L. Boyer et al., unpublished). MRS 2209 (2'-deoxy-N^6-methyladenosine-2',5'-*bis*phosphate) is a $P2Y_1$ receptor antagonist of similar potency.

Although MRS 2179 is selective for $P2Y_1$ from among the five different metabotropic P2 receptors, caution is required when using this compound since it can also antagonize, although less potently, the rat $P2X_1$ receptor expressed in *Xenopus* oocytes. Ion current induced by 3 µmol/l ATP acting at $P2X_1$ receptors was blocked by MRS 2179 with an IC_{50} value of 1.2 µmol/l. At the rat $P2X_3$ receptor, MRS 2179 is a weaker antagonist with an IC_{50} value of ~10 µmol/l. The compound at a concentration of 10 µmol/l was inactive at the rat $P2X_2$ receptor, while at the rat $P2X_4$ receptor, a potentiation of ion current by 25% was observed.

6. KN-62

The isoquinoline derivative 1-(N,O-bis[5-isoquinolinesulfonyl]-N-methyl-L-tyrosyl)-4-phenylpiperazine (KN-62), a selective antagonist of Ca^{2+}/calmodulin-dependent protein kinase II (CaMKII), is also a potent (Ki = 9–15 nmol/l) non-competitive antagonist at $P2X_7$ receptors, showing greater potency at the $P2X_7$ receptor than CaMKII. (GARGETT and WILEY 1997; CHESSELL et al. 1998). In human leukemic B lymphocytes (WILEY et al. 1998) KN-62 reduced the rate of permeability increase to larger permeant cations, like ethidium, induced by Bz-ATP with an IC_{50} of 13 nmol/l and complete inhibition of the flux at 500 nmol/l. KN-62 has no effect on responses mediated by the $P2Y_2$ receptor of neutrophils but does show species-dependent activity in its antagonist effects (HUMPHRIES and DUBYAK 1998).

E. Conclusions

Medicinal chemistry efforts in the purinergic receptor field have resulted in the identification of a large number of novel ligands over the past three decades. In the P1 area, a diversity of receptor subtype selective compounds have been identified. The structural diversity of these is greater for antagonists than agonists, the latter, as noted being almost exclusively analogs of adenosine. A number of P1 agonists and antagonists have been advanced to clinical trials but few of these compounds have advanced to the approval stage (WILLIAMS, Chap. 29, this volume).

Despite the tremendous advances in the molecular biology of P2 receptors (RALEVIC and BURNSTOCK 1998; CHESSELL et al., Chap. 3, this volume; BOARDER and WEBB, Chap. 4, this volume), progress in identifying novel, potent, selective, and bioavailable agonists and antagonists for these receptors has been slow. The majority of agonists are analogs of ADP, ATP, or UTP while antagonists, as discussed, represent a variety of novel structures with questionable potency and receptor selectivity. The development of robust radioligand binding assays and the use of these in conjunction with functional assays to screen diverse compound libraries, will – as illustrated from precedents with other receptor families (WILLIAMS 1996; TRIGGLE 1998) – improve this situation, providing the medicinal chemistry with pharmacophores to develop better compounds and pharmacologists the tools to characterize better the function and biomedical potential of P2 of receptor systems.

References

Abbracchio MP, Burnstock G (1998) Purinergic signalling: pathophysiological roles. Jap J Pharmacol 78:113–145

Abbracchio MP, Rainaldi G, Giammarioli AM, Ceruti S, Brambilla R, Cattabeni F, Barbieri D, Franceschi C, Jacobson KA, Malorni W (1997) The A_3 adenosine receptor mediates cell spreading, reorganization of actin cytoskeleton, and distribution of Bcl-x_L: studies in human astroglioma cells. Biochem Biophys Res Commun 241: 297–304

Adami M, Bertorelli R, Ferri N, Foddi MC, Ongini E (1995) Effects of repeated administration of selective adenosine A_1 and A_{2A} receptor agonists on pentylenetetrazole-induced convulsions in the rat. Eur J Pharmacol 294:383–389

Alberti C, Monopoli A, Casati C, Forlani A, Sala C, Nador B, Ongini E, Morganti A (1997) Mechanism and pressor relevance of the short-term cardiovascular and renin excitatory actions of the selective A_{2A}-adenosine receptor agonists. J Cardiovasc Pharmacol 30:320–324

Ali H, Choi OH, Fraundorfer PF, Yamada K, Gonzaga HMS, Beaven MA (1996) Sustained activation of phospholipase D via adenosine A_3 receptors is associated with enhancement of antigen- and Ca^{2+}-ionophore-induced secretion in a rat mast cell line. J Pharmacol Exp Ther 276:837–845

Allgaier C, Wellmann H, Schobert A, von Kuegelgen I (1995) Cultured chick sympathetic neurons: modulation of electrically evoked noradrenaline release by P2-purinoceptors. Naunyn-Schmiedebergs Arch Pharmacol 352:17–24

Arispe N, Ma J, Jacobson KA, Pollard HB (1998) Direct activation of cystic fibrosis transmembrane conductance regulator channels by 8-cyclopentyl-1,3-dipropylxanthine (CPX) and 1,3-diallyl-8-cyclohexylxanthine (DAX). J Biol Chem 273:5727–5734

Auchampach JA, Rizvi A, Qiu Y, Tang X-L, Maldonado C, Teschner S, Bolli R (1997) Selective activation of A_3 adenosine receptors with N^6-(3-iodobenzyl)adenosine-5'-N-methyluronamide protects against myocardial stunning and infarction without hemodynamic changes in conscious rabbits. Circ Res 80:800–809

Baraldi PG, Cacciari B, Spalluto G, Pineda de Villatoro MJ, Zocchi C, Dionisotti S, Ongini E (1996) Pyrazolo[4,3-e]-1,2,4-triazolo[1,5-c]pyrimidine Derivatives: Potent and Selective A_{2A} Adenosine Antagonists. J Med Chem 39:1164–1171

Baraldi PG, Manfredini S, Simoni D, Zappaterra L, Zocchi C, Dionisotti S, Ongini E (1994) Synthesis of new pyrazolo[4,3-e]1,2,4-triazolo[1,5-c]pyrimidine and 1,2,3-triazolo[4,5-e]1,2,4-triazolo[1,5-c]pyrimidine displaying potent and selective activity as A_{2A} adenosine receptor antagonists. Bioorg Med Chem Lett 4:2539–2544

Barraco RA, Aggarwal AK, Phillis JW, Moron MA, Wu PH (1984) Dissociation of the locomotor and hypotensive effects of adenosine analogs in the rat. Neurosci Lett 48:139–144

Barraco RA, Martens KA, Parizon M, Normile HJ (1994) Role of adenosine A_{2A} receptors in the nucleus accumbens. Prog Neuro-Psychopharmacol Biol Psychiatry 18:545–553

Belardinelli L, Lu J, Dennis D, Martens J, Shryock JC (1994) The cardiac effects of a novel A_1-adenosine receptor agonist in guinea pig isolated heart. J Pharmacol Exp Ther 271:1371–1382

Bertolet BD, Anand IS, Bryg RJ, Mohanty PK, Chatterjee K, Cohn JN, Khurmi NS, Pepine CJ (1996) Effects of A_1 adenosine receptor agonism using N^6-cyclohexyl-2'-O-methyladenosine in patients with left ventricular dysfunction. Circulation 94:1212–1215

Bertorelli R, Ferri N, Adami M, Ongini E (1996) Effects of selective agonists and antagonists for A_1 or A_{2A} adenosine receptors on sleep-waking patterns in rats Drug Dev Res 37:65–72

Beukers MW, Kerkhof CJM, van Rhee MA, Ardanuy U, Gurgel C, Widjaja H, Nickel P, IJzerman AP, Soudijn W (1995) Suramin analogs, divalent cations and ATPγS as inhibitors of ecto-ATPase. Naunyn-Schmiedebergs Arch Pharmacol 351:523–528

Bhagwat SS, Williams M (1997). P2 Purine and pyrimidine receptors: emerging superfamilies of G-protein coupled and ligand gated ion channel receptors. Eur J Med Chem 32:183–193

Bianchi BR, Lynch KJ, Touma E, Niforatos W, Burgard EC, Alexander KM, Park HS, Yu H, Metzger R, Kowaluk E, Jarvis M, van Biesen T (1999) Pharmacological characterization of recombinant human and rat P2X receptor subtypes. Eur J Pharmacol 376:127–138

Bona E, Aden U, Gilland E, Fredholm BB, Hagberg H (1997) Neonatal cerebral hypoxia-ischemia: the effect of adenosine receptor antagonists. Neuropharmacology 36:1327–1338

Boyer JL, Zohn IE, Jacobson KA, Harden TK (1994) Differential effects of P2-purinoceptor antagonists on phospholipase C- and adenylyl cyclase-coupled P2Y-purinoceptors. Br J Pharmacol 113:614–620

Boyer JL, Siddiqi S, Fischer B, Romera-Avila T, Jacobson KA, Harden TK (1996) Identification of potent P2Y purinoceptor agonists that are derivatives of adenosine 5′-monophosphate. Br J Pharmacol 118:1959–1964

Boyer JL, Mohanram A, Camaioni E, Jacobson KA, Harden TK (1998) Competitive and selective antagonism of P2Y$_1$ receptors by N^6-methyl 2′-deoxyadenosine 3′,5′-bisphosphate. Brit J Pharmacol 124:1–3

Bruns RF, Fergus JH, Badger EW, Bristol JA, Santay LA, Hartman JD, Hays SJ, Huang, CC (1987) Binding of the A$_1$-selective adenosine antagonist 8-cyclopentyl-1,3-dipropylxanthine to rat brain membranes. Naunyn-Schmiedebergs Arch Pharmacol 335:59–63

Bruns RF, Lu GH, Pugsley TA (1986) Characterization of the A$_2$ adenosine receptor labelled by [^{3}H]NECA in rat striatal membranes. Mol Pharmacol 29:331–346

Bueltmann R, Wittenburg H, Pause B, Kurz G, Nickel P, Starke K (1996) P2-purinoceptor antagonists. Part 3. Blockade of P2-purinoceptor subtypes and ecto-nucleotidases by compounds related to suramin. Naunyn-Schmiedebergs Arch Pharmacol 354:498–504

Burnstock G, Warland JJ (1987) P2-purinoceptors of two subtypes in the rabbit mesenteric artery: Reactive Blue 2 selectively inhibits responses mediated via the P2Y- but not the P2X-purinoceptor. Brit J Pharmacol 90:383–391

Burnstock G, Fischer B, Hoyle CHV, Maillard M, Ziganshin AU, Brizzolara AL, von Isakovics A, Boyer JL, Harden TK, Jacobson KA (1994) Structure–activity relationships for derivatives of adenosine-5′- triphosphate as agonists at P2 purinoceptors – heterogeneity within P2X and P2Y subtypes. Drug Dev Res 31:206–219

Camaioni E, Boyer JL, Mohanram A, Harden TK, Jacobson KA (1998) Deoxyadenosine-bisphosphate derivatives as potent antagonists at P2Y$_1$ receptors. J Med Chem 41:183–190

Casati C, Monopoli A, Dionisotti S, Zocchi C, Bonizzoni E, Ongini E (1994) Repeated administration of selective adenosine A$_1$ and A$_2$ receptor agonists in the spontaneously hypertensive rat: tolerance develops to A$_1$-mediated hemodynamic effects. J Pharmacol Exp Ther 268:1506–1511

Charlton S, Brown CA, Boarder MR (1996) Suramin and PPADS antagonists at transfected P2Y$_1$, P2Y$_2$, P2Y$_3$, and P2Y$_4$ receptors. Drug Dev Res 37:113

Chesi AJR, Stone TW (1997) Alkylxanthine adenosine antagonists and epileptiform activity in rat hippocampal slices *in vitro*. Exp Brain Res 113:303–310

Chessell IP, Simon J, Hibell AD, Michel AD, Barnard EA, Humphrey PPA (1998) Cloning and functional characterization of the mouse P2X$_7$ receptor. FEBS Lett 439:26–30

Coffin VL, Spealman RD (1987) Behavioral and cardiovascular effects of analogs of adenosine in cynomolgus monkeys. J Pharmacol Exp Ther 241:76–83

Communi O, Boeynaems JM (1997) Receptors responsive to extracellular pyrimidine nucleotides. Trends Pharmacol Sci 18:83 –86

Conant AR, Fisher MJ, McLennan AG, Simpson AWM (1998) Characterization of the P2 receptors on the human umbilical vein endothelial cell line ECV304. Br J Pharmacol 125:357–364

Connolly GP (1995) Differentiation by pyridoxal 5-phosphate, PPADS and isoPPADS between responses mediated by UTP and those evoked by α,β-methylene-ATP on rat sympathetic-ganglia. Br J Pharmacol 114:727–731

Conti A, Monopoli A, Gamba M, Borea PA, Ongini E (1993) Effects of selective A$_1$ and A$_2$ adenosine receptor agonists on cardiovascular tissues. Naunyn-Schmiedebergs Arch Pharmacol 348:108–112

Coutinho-Silva R, Persechini PM, Bisaggio RD, Perfettini JL, Neto AC, Kanellopoulos JM, Motta-Ly I, Dautry-Varsat A, OjciusDM (1999) P2Z/P2X$_7$ receptor-dependent apoptosis of dendritic cells. Am J Physiol 276:C1139–1147

Cristalli G, Eleuteri A, Vittori S, Volpini R, Lohse MJ, Klotz KN (1992) 2-Alkynyl derivatives of adenosine and adenosine-5'-N-ethyluronamide as selective agonists at A_2 adenosine receptors. J Med Chem 35:2363–2368

Cunha RA, Constantino MD, Ribeiro JA (1997) ZM 241385 is an antagonist of the facilitatory responses produced by the A_{2A} adenosine receptor agonists CGS 21680 and HENECA in the rat hippocampus. Br J Pharmacol 122:1279–1284

Cunha RA, Johansson B, Constantino MD, Sebastiao AM, Fredholm BB (1996) Evidence for high-affinity binding sites for the adenosine A_{2A} receptor agonist [^{3}H]CGS 21680 in the rat hippocampus and cerebral cortex that are different from striatal A_{2A} receptors. Naunyn-Schmiedebergs Arch Pharmacol 353:261–271

Cusack NJ, Hourani SMO (1990) Structure activity relationships for adenine nucleotide receptors on mast cells, human platelets, and smooth muscle, in Purines in Cellular Signalling: Targets for New Drugs, Jacobson, KA, Daly, JW, Manganiello, V, eds, Springer, New York, pp. 254–259

Daly JW, Padgett WL (1992) Agonist activity of 2- and 5'-substituted adenosine analogs and their N^6-cycloalkyl derivatives at A_1- and A_2-adenosine receptors coupled to adenylate cyclase. Biochem Pharmacol 43:1089–1093

Daly JW, Butts-Lamb P, Padgett WL (1983) Subclasses of adenosine receptors in the central nervous system: interaction with caffeine and related methylxanthines. Cel Mol Neurobiol 3:69–80

De Zwart M, Vollinga RC, Beukers MW, Sleegers DF, Von Frijtag Drabbe Kunzel J, De Groote M, IJzerman AP (1999) Potent antagonists for the human adenosine A_{2B} receptor. Derivatives of the triazolotriazine adenosine receptor antagonist ZM241385 with high affinity. Drug Dev Res 48:95–103

Di Virgilio F, Pizzo P, Zanovello P, Bronte V, Collavo D (1990) Extracellular ATP as a possible mediator of cell-mediated cytotoxicity. Immunol Today 11:274–277

Donaldson SH, Boucher RC (1998) Therapeutic applications for nucleotides in lung disease. Chapter 15 in The P2 Nucleotide Receptors, in the series "The Receptors", eds. John T. Turner, Gary Weisman, and Jeffrey Fedan, Humana Press, Clifton, NJ, pp. 413–424

Dunwiddie TV, Diao L, Kim HO, Jiang J-L, Jacobson KA (1997) Activation of hippocampal adenosine A_3 receptors produces a desensitization of A_1 receptor-mediated responses in rat hippocampus. J Neurosci 17:607–614

Dunwiddie TV, Worth T (1982) Sedative and anticonvulsant effects of adenosine analogs in mouse and rat. J Pharm Exp Ther 220:70–76

Dunwiddie TV, Worth TS, Olsson, RA (1986) Adenosine analogs mediating depressant effects on synaptic transmission in rat hippocampus: structure–activity relationships for the N^6 subregion. Naunyn-Schmiedebergs Arch Pharmacol 334:77–85

Eltze M, Ullrich B, (1996) Characterization of vascular P2 purinoceptors in the rat isolated perfused kidney. Eur J Pharmacol 306:139–152

Feoktistov I, Biaggioni I (1997) Adenosine A_{2B} receptors. Pharmacol Rev 49:381–402

Ferrari D, Stroh C, Schulze-Osthoff K (1999) P2X7/P2Z purinoreceptor-mediated activation of transcription factor NFAT in microglial cells. J Biol Chem 274:13205–13210

Filippi S, Amerini S, Maggi M, Natali A, Ledda F (1999) Studies on the mechanisms involved in the ATP-induced relaxation in human and rabbit corpus cavernosum. J Urol 161:326–331

Filtz TM, Harden TK, Nicholas RA (1997) Structure, pharmacological selectivity and second messenger properties of G protein coupled P2 purinergic receptors. In: Jacobson KA, Jarvis MF (eds) Purinergic Approaches in Experimental Therapeutics. Wiley-Liss, New York pp. 39–53

Fischer B, Boyer JL, Hoyle CHV, Ziganshin AU, Brizzolara AL, Knight GE, Zimmet J, Burnstock G, Harden TK, Jacobson KA (1993) Identification of potent, selective P2Y-purinoceptor agonists – structure–activity relationships for 2-thioether derivatives of adenosine 5'-triphosphate. J Med Chem 36:3937–3946

Fischer Y, Becker C, Loken C (1999) Purinergic inhibition of glucose transport in cardiomyocytes. J Biol Chem 274:755–761

Fredholm BB, Abbracchio MP, Burnstock G, Daly JW, Harden KT, Jacobson KA, Leff P, Williams M (1994) Nomenclature and classification of purinoceptors: a report from the IUPHAR subcommittee. Pharmacol Rev 46:143–156

Fredholm BB, Lindstrom K, Dionisotti S, Ongini E (1998) [^{3}H]SCH 58261, a selective adenosine A_{2A} receptor antagonist, is a useful ligand in autoradiographic studies. J Neurochem 70:1210–1216

Freissmuth M, Boehm S, Beindl W, Nickel P, IJzerman AP, Hohenegger M, Nanoff C, (1996) Suramin analogues as subtype selective G protein inhibitors. Mol Pharmacol 49:602–611

Froehlich R, Boehm S, Illes P (1996) Pharmacological characterization of P2 purinoceptor types in rat locus ceruleus neurons. Eur J Pharmacol 315:255–261

Gallo-Rodriguez C, Ji XD, Melman N, Siegman BD, Sanders LH, Orlina J, Pu QL, Olah ME, van Galen PJM, Stiles GL, Jacobson KA (1994) Structure–activity relationships of N^6-benzyladenosine-5'-uronamides as A_3-selective adenosine agonists. J Med Chem 37:636–646

Gargett CE, Wiley JS (1997) The isoquinoline derivative KN-62 a potent antagonist of the P2Z-receptor of human lymphocytes. Br J Pharmacol 120:1483–1490

Geiger JD, Parkinson FE, Kowaluk E (1997) Regulators of endogenous adenosine levels as therapeutic agents In: Jacobson KA, Jarvis MF (eds) Purinergic Approaches in Experimental Therapeutics. Wiley-Liss, New York pp. 55–84

Gellai M, Schreiner GF, Ruffolo RR Jr, Fletcher T, DeWolf R, Brooks DP (1998) CVT-124, a novel adenosine A_1 receptor antagonist with unique diuretic activity. J Pharmacol Exp Ther 286:1191–1196

Goncalves ML, Ribeiro JA (1996) Adenosine A_2 receptor activation facilitates $^{45}Ca^{2+}$ uptake by rat brain synaptosomes. Eur J Pharmacol 310:257–261

Grondin R, Bedard P, Tahar A, Hadj GL, Mori A, Kase H (1999) Antiparkinsonian effect of a new selective adenosine A_{2A} receptor antagonist in MPTP-treated monkeys. Neurology 52:1673–1677

Gurden MF, Coates J, Ellis F, Evans B, Foster M, Hornby E, Kennedy I, Martin DP, Strong P, Vardey CJ, Wheeldon A (1993) Functional characterization of three adenosine receptor types. Br J Pharmacol 109:693–698

Haleen SJ, Steffen RP, Hamilton HW (1987) PD 116,948, a highly selective A_1 adenosine receptor antagonist. Life Sci 40:555–561

He Z, Raman S, Guo Y, Reenstra WW (1998) Cystic fibrosis transmembrane conductance regulator activation by cAMP-independent mechanisms. Am J Physiol 275:C958–C966

Hechler B, Vigne P, Leon C, Breittmayer JP, Gachet C, Frelin C (1998) ATP derivatives are antagonists of the $P2Y_1$ receptor: similarities to the platelet ADP receptor. Mol Pharmacol 53:727–733

Heffner TG, Wiley JN, Williams AE, Bruns RF, Coughenour LL, Downs DA (1989) Comparison of the behavioral effects of adenosine agonists and dopamine antagonists in mice. Psychopharmacology 98:31–37

Heseltine L, Webster JM, Taylor R (1995) Adenosine effects upon insulin action on lipolysis and glucose transport in human adipocytes. Mol Cell Biochem 144:147–151

Hide I, Padgett WL, Jacobson KA, Daly JW (1992) A_{2A} Adenosine receptors from rat striatum and rat pheochromocytoma PC12 cells: Characterization with radioligand binding and by activation of adenylate cyclase. Mol Pharmacol 41:352–359

Humphries RG, Leff P, Robertson MJ (1996) P2T-purinoceptor antagonists: a novel class of anti-thrombotic agents. Drug Dev Res 37:175

Hutchison AJ, Webb RL, Oei HH, Ghai GR, Zimmerman MB, Williams M (1989) CGS 21680 C, an A_2 selective adenosine receptor agonist with preferential hypotensive activity. J Pharmacol Exp Ther 251:47–55

Hutchison AJ, Williams M, De Jesus R, Yokoyama R, Oei HH, Ghai GR, Webb RL, Zoganas HC, Stone GA, Jarvis MF (1990) 2-(Arylalkylamino)adenosin-5′-uronamides: a new class of highly selective adenosine A_2 receptor ligands. J Med Chem 33:1919–1924

IJzerman AP, van der Wenden N (1997) Modulators of adenosine uptake, release, and inactivation. In: Jacobson KA, Jarvis MF (eds) Purinergic Approaches in Experimental Therapeutics. Wiley-Liss, New York, pp. 129–148

Ikeuchi Y, Nishizaki T (1996) P2 purinoceptor-operated potassium channel in rat cerebellar neurons. Biochem Biophys Res Commun 218:67–71

Ingall AH, Dixon J, Bailey A, Coombs ME, Cox D, McInally JI, Hunt SF, Kindon ND, Teobald BJ, Willis PA, Humphries RG, Leff P, Clegg JA, Smith JA, Tomlinson W (1999) Antagonists of the Platelet P_{2T} Receptor: A Novel Approach to Antithrombotic Therapy. J Med Chem. 42:213–220

Inscho EW, Cook AK, Mui VY, Miller J (1998) Direct assessment of renal microvascular responses to P2-purinoceptor agonists. Am J Physiol 274:F718–F727

Ishikawa J, Mitani H, Bandoh T, Kimura M, Totsuka T, Hayashi S (1998) Hypoglycemic and hypotensive effects of 6-cyclohexyl-2′-O-methyl-adenosine, an adenosine A_1 receptor agonist, in spontaneously hypertensive rat complicated with hyperglycemia. Diabetes Res Clin Pract 39:3–9

Jacobson KA, Barone S, Kammula U, Stiles GL (1989) Electrophilic derivatives of purines as irreversible inhibitors of A_1 adenosine receptors. J Med Chem 32:1043–1051

Jacobson KA, Gallo-Rodriguez C, Melman N, Fischer B, Maillard M, van Bergen A, van Galen PJM, Karton Y (1993a) Structure–activity relationships of 8-styrylxanthines as A_2-selective adenosine antagonists. J Med Chem 36:1333–1342

Jacobson KA, Guay-Broder C, van Galen PJM, Gallo-Rodriguez C, Melman N, Jacobson, MA, Eidelman O, Pollard HB (1995) Stimulation by alkylxanthines of chloride efflux in CFPAC-1 cells does not involve A_1 adenosine receptors. Biochemistry 34:9088–9094

Jacobson KA, Hoffmann C, Kim YC, Camaioni E, Nandanan E, Jang SY, Guo DP, Ji X-D, von Kügelgen I, Moro S, King BF, Brown SG, Wildman SS, Burnstock G, Boyer JL, Mohanram A, TK Harden (1999) Molecular recognition in P2 receptors: Ligand development aided by molecular modeling and mutagenesis. Prog Brain Res. 120:119–132

Jacobson KA, IJzerman AP, Linden, J (1999) 1,3-Dialkylxanthine derivatives having high potency as antagonists at human A_{2B} receptors. Drug Dev Res 47:45–53

Jacobson KA, Kim HA, Siddiqi SM, Olah ME, Stiles GL, von Lubitz DKJE (1995) A_3 adenosine receptors: design of selective ligands and therapeutic prospects. Drugs Fut 20:689–699

Jacobson KA, Kim Y-C, Camaioni E, van Rhee, AM (1997a) Structure activity relationships of P2 receptor agonists and antagonists. In: Turner JT, Weisman G, Fedan J (eds) The P2 Nucleotide Receptors. Humana Press, Clifton NJ, pp. 81–107

Jacobson KA, Kim Y-C, Wildman SS, Mohanram A, Harden TK, Boyer JL, King BF, Burnstock G (1998) A pyridoxine cyclic-phosphate and its 6-arylazo-derivative selectively potentiate and antagonize activation of $P2X_1$ receptors. J Med Chem 41:2201–2206

Jacobson KA, Kirk KL, Padgett WL, Daly JW (1985) Functionalized congeners of adenosine: preparation of analogs with high affinity for A_1-adenosine receptors. J Med Chem 28:1341–1346

Jacobson KA, Nikodijevic O, Padgett WL, Gallo-Rodriguez C, Maillard M, Daly JW (1993b) 8-(3-Chlorostyryl)caffeine (CSC) is a selective A_2-adenosine antagonist *in vitro* and *in vivo*. FEBS Lett 323:141–144

Jacobson KA, Pannell LK, Ji XD, Jarvis MF, Williams M, Hutchison AJ, Barrington WW, Stiles GL (1989) Agonist derived molecular probes for A_2 adenosine receptors. J Mol Recognit 2:170–178

Jacobson KA, Park K-S, Jiang J-L, Kim Y-C, Olah ME, Stiles GL, Ji X-D (1997b) Pharmacological characterization of novel A_3 adenosine receptor-selective antagonists. Neuropharmacology 36:1157–1165

Jacobson KA, van Rhee AM (1997) Development of Selective Purinoceptor Agonists and Antagonists. In: Jacobson KA, Jarvis MF (eds) Purinergic Approaches in Experimental Therapeutics. Wiley-Liss, New York, pp 101–128

Jacobson MA (1998) Novel selective non-xanthine A_3 adenosine receptor antagonists. Book of Abstracts, 215th ACS National Meeting, Dallas, March 29-April 2: MEDI-095

Janusz CA, Berman RF (1993) Adenosinergic modulation of the EEG and locomotor effects of the A_2 agonist CGS 21680. Pharm Biochem Behav 45:913–919

Jarvis MF, Schultz R, Hutchison AJ, Do UH, Sills MA, Williams M (1989) [^{3}H]-CGS 21680, a selective A_2 adenosine receptor agonist directly labels A_2 receptors in rat brain. J Pharmacol Exp Ther 251:888–893

Jarvis MF, Williams M, Do UH, Sills MA (1991) Characterization of the binding of a novel nonxanthine adenosine antagonist radioligand, [^{3}H]CGS 15943, to multiple affinity states of the adenosine A_1 receptor in the rat cortex. Mol Pharmacol 39:49–54

Ji X-D, Jacobson KA (1999) Use of the triazolotriazine [^{3}H]ZM 241385 as a radioligand at recombinant human A_{2B} adenosine receptors. Drug Des Discovery 16:217–226

Jiang J-l, van Rhee AM, Melman N, Ji X-D, Jacobson KA (1996) 6-Phenyl-1,4-dihydropyridine derivatives as potent and selective A_3 adenosine receptor antagonists. J Med Chem 39:4667–4675

Johansson B, Fredholm BB (1995) Further characterization of the binding of the adenosine receptor agonist [^{3}H]CGS 21680 to rat brain using autoradiography. Neuropharmacology 34:393–403

Jones PA, Smith RA, Stone TW (1998a) Protection against kainate-induced excitotoxicity by adenosine A_{2A} receptor agonists and antagonists. Neuroscience 85:229–237

Jones PA, Smith RA, Stone TW (1998b) Protection against hippocampal kainate excitotoxicity by intracerebral administration of an adenosine A_{2A} receptor antagonist. Brain Res 800:328–335

Kanda T, Tashiro T, Kuwana Y, Jenner P (1998) Adenosine A_{2A} receptors modify motor function in MPTP-treated common marmosets. Neuroreport 9:2857–2860

Kanda T, Jackson MJ, Smith LA, Pearce RKB, Nakamura J, Kase H, Kuwana Y, Jenner P (1998) Adenosine A_{2A} antagonist: a novel antiparkinsonian agent that does not provoke dyskinesia in Parkinsonian monkeys. Ann Neurol 43:507–513

Kanda T, Tashiro T, Kuwana Y, Jenner P (1998) Adenosine A_{2A} receptors modify motor function in MPTP-treated common marmosets. Neuroreport 9:2857–2860

Keddie JR, Poucher SM, Shaw GR, Brooks R, Collis MG (1996) In vivo characterization of ZM 241385, a selective adenosine A_{2A} receptor antagonist. Eur J Pharmacol 301:107–113

Kennedy C, Leff P (1995) How should P2X purinoceptors be classified pharmacologically. Trends Pharmacol Sci 16:168–174

Kennedy C, Qi A-I, Herold CL, Harden TK, Nicholas RA (2000) ATP, an agonist at the rat $P2Y_4$ receptor, is an antagonist at the human $P2Y_4$ receptor. Mol. Pharmacol 57:926–931

Khakh BS, Zhou X, Sydes J, Galligan JJ, Cester HA (2000) State-dependent cross-inhibition between transmitter-gated cation channels. Nature 406:405–410

Kikugawa K, Iizuka K, Ichino M (1973) Platelet aggregation inhibitors. 4. N^6-Substituted adenosines. J Med Chem 16:358–364

Kim Y-C, Ji, X-d, Melman N, Linden J, Jacobson KA (2000) Anilide derivatives of an 8-phenylxanthine carboxylic congener are highly potent and selective antagonists at human A_{2B} adenosine receptors. J Med Chem 43:1165–1172

Kim HO Ji, X-d, Siddiqi SM, Olah ME, Stiles GL, Jacobson KA (1994) 2-Substitution of N^6-Benzyladenosine-5′-uronamides Enhances Selectivity for A_3 Adenosine Receptors. J Med Chem 37:3614–3621

Kim YC, Camaioni E, Ziganshin AU, Ji XJ, King BF, Wildman SS, Rychkov A, Yoburn J, Kim H, Mohanram A, Harden TK, Boyer JL, Burnstock G, Jacobson KA (1998) Synthesis and structure activity relationships of pyridoxal-6-azoaryl-5'-phosphate and phosphonate derivatives as P2 receptor antagonists. Drug Dev Res 45:52–66

Kim Y-C, Ji X-D, Jacobson KA (1996) Derivatives of the triazoloquinazoline adenosine antagonist (CGS 15943) are selective for the human A_3 receptor subtype. J Med Chem 39:4142–4148

King, BF, Liu, M. Pintor, J. Gualix J, Miras-Portugal MT, Burnstock G. (1999) Diinosine pentaphosphate (Ip_5I) is a potent antagonist at recombinant rat $P2X_1$ receptors. Br J Pharmacol 128:981–988

Klitgaard H, Knutsen LJS, Thomsen C (1993) Contrasting effects of adenosine A_1 and A_2 receptor ligands in different chemoconvulsive models. Eur J Pharmacol 224:221–228

Klotz K-N, Hessling J, Hegler J, Owman C, Kull B, Fredholm BB, Lohse MJ (1998) Comparative pharmacology of human adenosine receptor subtypes – characterization of stably transfected receptors in CHO cells. Naunyn-Schmiedebergs Arch Pharmacol 357:1–9

Knutsen LJS, Lau J, Petersen H, Thomsen C, Weis JU, Shalmi M, Judge ME, Hansen A, Sheardown MJ (1999) N-Substituted adenosines as novel neuroprotective A_1 agonists with diminished hypotensive effects. J Med Chem 41:3463–3477

Knutsen LJS, Lau J, Eskesen K, Sheardown MJ, Thomsen C, Weis JU, Judge ME, Klitgaard H (1995) Anticonvulsant actions of novel and reference adenosine agonists. In Adenosine and Adenine Nucleotides: From Molecular Biology to Integrative Physiology, Belardinelli L and Pelleg A Eds Kluwer, Boston MA, pp. 479–487

Knutsen LJS, Murray TF (1997) Adenosine and ATP in Epilepsy. In: Jacobson KA, Jarvis MF (eds) Purinergic Approaches in Experimental Therapeutics. Wiley-Liss, New York, 1997, pp. 423–447

Knutsen LJS, Sheardown MJ, Roberts SM, Mogensen JP, Olsen UB, Thomsen C, Bowler AN (1998) Adenosine A_1 and A_3 Selective N-Alkoxypurines as novel Cytokine Modulators and Neuroprotectants. Drug Dev Res 45:214–221

Kohno Y, Sei Y, Koshiba M, Kim HO, Jacobson KA (1996) Induction of apoptosis in HL-60 human promyelocytic leukemia cells by adenosine A_3 receptor agonists. Biochem Biophys Res Commun 219:904–910

Koizumi S, Inoue K (1997) Inhibition by ATP of calcium oscillations in rat cultured hippocampal neurons. Br J Pharmacol 122:51–58

Krahl SE, Treas LM, Castle JD, Berman RF (1995) Attenuation of *in vivo* and *in vitro* seizure activity using the adenosine agonist metrifudil. Drug Dev Res 34:30–34

Kugler G, Westermann KW II (1974) Effects of adenosine on metabolic and electrocardiographic parameters during a trial pacing in patients with coronary heart disease. Z Kardiol 63:987–1000

Lazarowski ER, Watt WC, Stutts MJ, Brown HA, Boucher RC, Harden TK (1996) Enzymatic-synthesis of UTP-g-S, a potent hydrolysis resistant agonist of P2U-purinoceptors. Br J Pharmacol 117:203–209

Lindstroem K, Ongini E, Fredholm, BB (1996) The selective adenosine A_{2A} receptor antagonist SCH 58261 discriminates between two different binding sites for [^{3}H]-CGS 21680 in the rat brain. Naunyn-Schmiedebergs Arch Pharmacol 354:539–541

Louttit JB, Hunt AAE, Maxwell MP, Drew GM (1999) The time course of cardioprotection induced by GR79236, a selective adenosine A_1-receptor agonist, in myocardial ischemia-reperfusion injury in the pig. J Cardiovasc Pharmacol 33:285–291

Lozza G, Conti A, Ongini E, Monopoli A (1997) Cardioprotective effects of adenosine A_1 and A_{2A} receptor agonists in the isolated rat heart. Pharmacol Res 35:57–64

Macgregor DG, Graham DI, Jones PA, Stone TW (1998) Protection by an adenosine analog against kainate-induced extrahippocampal neuropathology. Gen Pharmacol 31:233–238

Malcolm KC, Trammell SE, Exton JH (1995) Purinergic agonist and G protein stimulation of phospholipase D in rat liver plasma membranes. Independence from phospholipase C activation. Biochim Biophys Acta 1268:152–158

Mally J, Stone TW (1998) Potential of adenosine A_{2A} receptor antagonists in the treatment of movement disorders. CNS Drugs 10:311–320

Martin PL, Wysocki RJ Jr, Barrett RJ, May JM, Linden J (1996) Characterization of 8-(N-methylisopropyl)amino-N^6-(5′-endohydroxy-endonorbornyl)-9-methyladenine (WRC-0571), a highly potent and selective, non-xanthine antagonist of A_1 adenosine receptors. J Pharmacol Exp Ther 276:490–499

Mathot RAA, van Den Aarsen FM, Soudijn W, Breimer DD, IJzerman AP, Danhof M (1995) Pharmacokinetic-pharmacodynamic modeling of the cardiovascular effects of R- and S-N^6-phenylisopropyladenosine in conscious normotensive rats. J Pharmacol Exp Ther 273:405–414

Merkel LA, Hawkins ED, Colussi DJ, Greenland BD, Smits GJ, Perrone MH, Cox BF (1995) Cardiovascular and antilipolytic effects of the adenosine agonist GR 79236. Pharmacology 51:224–236

Michel AD, Lundstroem K, Buell GN, Surprenant A, Valera S, Humphrey PPA (1996) A comparison of the binding characteristics of recombinant $P2X_1$ and $P2X_2$ purinoceptors. Br J Pharmacol 118:1806–1812

Mogensen JP, Roberts SM, Bowler AN, Thomsen C, Knutsen LJS (1998) The Synthesis Of New Adenosine A_3 Selective Ligands Containing Bioisosteric Isoxazoles. Bioorg Med Chem Lett 8:1767–1770

Monopoli A, Conti A, Zocchi C, Casati C, Volpini R, Cristalli G, Ongini E (1994) Pharmacology of the new selective A_{2a} adenosine receptor agonist 2-hexynyl-5′-N-ethylcarboxamidoadenosine. Arzneim -Forsch 44:1296–1304

Moos H, Szotek DS, Bruns RF (1985) N^6-Cycloalkyladenosines. Potent, A_1-selective adenosine agonists. J Med Chem 28:1383–1384

Nakazawa K, Inoue K, Ito K, Koizumi S (1995) Inhibition by suramin and reactive blue 2 of GABA and glutamate receptor channels in rat hippocampal neurons. Naunyn-Schmiedebergs Arch Pharmacol 351:202–208

Nandanan E, Camaioni E, Jang SY, Kim Y-C, Cristalli G, Herdewijn P, Secrist, JA, Tiwari KN, Mohanram A, Harden TK, Boyer JL, and Jacobson KA (1999) Structure activity relationships of bisphosphate nucleotide derivatives as $P2Y_1$ receptor antagonists and partial agonists. J Med Chem 42:1625–1638

Neely CF, DiPierro FV, Kong M, Greelish JP, Gardner TJ (1996) A_1 adenosine receptor antagonists block ischemia-reperfusion injury of the heart. Circulation Suppl 94:II376–II380

Nikodijevic O, Daly JW, Jacobson KA (1990) Characterization of the locomotor depression produced by an A_2-selective adenosine agonist. FEBS Lett: 261:67–70

North RA, Barnard EA (1997) Nucelotide receptors, Current Opinion in Neurobiology 7:346–357

Olah ME, Gallo-Rodriguez C, Jacobson KA, Stiles GL (1994) ^{125}I-4-aminobenzyl-5′-N-methylcarboxamido adenosine, a high affinity radioligand for the rat A_3 adenosine receptor. Mol Pharmacol 45:978–982

Olsson RA, Kusachi S, Thompson RD, Ukena D, Padgett W, Daly JW (1986) N^6-Substituted N-alkyladenosine-5′-uronamides: bifunctional ligands having recognition groups for A_1 and A_2 adenosine receptors. J Med Chem 29:1683–1689

Ongini E (1997) SCH 58261: a selective A_{2A} adenosine receptor antagonist. Drug Dev Res 42:63–70

Ongini E, Adami M, Ferri C, Bertorelli R (1998) Adenosine A_{2A} receptors and Neuroprotection. Ann NY Acad Sci 30–48

Palmer RK, Boyer JL, Schachter JB, Nicholas RA, Harden TK (1998) Agonist action of adenosine triphosphates at the human $P2Y_4$ receptor. Mol Pharmacol 54:1118–1123

Palmer TM, Poucher SM, Jacobson KA, Stiles GL (1995) ^{125}I-4(2-[7-amino-2-{2-furyl}{1,2,4}triazolo{2,3-a}{1,3,5}triazin-5-yl-amino]ethyl)phenol, a high affinity antagonist radioligand selective for the A_{2a} adenosine receptor. Mol Pharmacol 48:970–974

Petit P, Hillaire-Buys D, Manteghetti M, Debrus S, Chapal J, Loubatieres-Mariani MM (1998) Evidence for two different types of P2 receptors stimulating insulin secretion from pancreatic B cell. Br J Pharmacol 125:1368–1374

Pfister JR, Belardinelli L, Lee G, Lum RT, Milner P, Stanley WC, Linden J, Baker SP, Schreiner G (1997) Synthesis and Biological Evaluation of the Enantiomers of the Potent and Selective A_1-Adenosine Antagonist 1,3-Dipropyl-8-[2-(5,6-epoxy)norbonyl]- xanthine. J Med Chem 40:1773–1778

Pintor J, Puche JA, Gualix J, Hoyle CHV, Miras-Portugal MT (1997) Diadenosine polyphosphates evoke Ca^{2+} transients in guinea – pig brain via receptors distinct from those for ATP. J Physiol. (Cambridge, UK) 504:327–335

Popoli P, Reggio R, Pezzola A, Fuxe K, Ferre S (1998) Adenosine A_1 and A_{2A} receptor antagonists stimulate motor activity: evidence for an increased effectiveness in aged rats. Neurosci Lett 251:201–204

Poucher SM, Keddie JR, Brooks R, Shaw GR, McKillop D (1996) Pharmacodynamics of ZM 241385, a potent A_{2a} adenosine receptor antagonist, after enteric administration in rat, cat and dog. 4-(2-[7-Amino-2-(2-furyl)[1,2,4]triazolo[2,3-a][1,3,5]triazin-5-ylamino]ethyl)phenol (ZM 241385) is currently the most selective for the A_{2a} adenosine receptor antagonist. J Pharm Pharmacol 48:601–606

Poucher SM, Keddie JR, Singh P, Stoggall SM, Caulkett PWR, Jones G, Collis MG (1995) The *in vitro* pharmacology of ZM 241385, a potent, non-xanthine, A_{2a} selective adenosine receptor antagonist. Br J Pharmacol 115:1096–1102

Poulsen S-A, Quinn RJ (1998) Adenosine receptors: new opportunities for future drugs. Bioorg Med Chem 6:619–641

Raberger G, Schuetz W, Kraupp O (1977) Coronary dilatory action of adenosine analogs: a comparative study. Arch Int Pharmacodyn Ther 230:140–149

Rae MG, Rowan EG, Kennedy C (1998) Pharmacological properties of $P2X_3$-receptors present in neurons of the rat dorsal root ganglia. Br J Pharmacol 124:176–180

Ralevic V, Burnstock G (1998) Receptors for purines and pyrimidines. Pharmacol Rev. 50:413–492

Ribes G, Bertrand G, Petit P, Loubatieres-Mariani MM (1988) Effects of 2-methylthio ATP on insulin secretion in the dog in vivo. Eur J Pharmacol 155:171–174

Rivkees SA, Reppert SM (1992) RFL9 encodes an A_{2b}-adenosine receptor. Mol Endocrinol 6:1598–1604

Robertson SJ, Edwards FA (1998) ATP and glutamate are released from separate neurones in the rat medial habenula nucleus: frequency dependence and adenosine-mediated inhibition of release. J Physiol (Lond) 508:691–701

Ross FM, Brodie MJ, Stone TW (1998) Modulation by adenine nucleotides of epileptiform activity in the CA3 region of rat hippocampal slices. Br J Pharmacol 123:71–80

Rubino A, Burnstock G (1996) Evidence for a P2-purinoceptor mediating vasoconstriction by UTP, ATP and related nucleotides in the isolated pulmonary vascular bed of the rat. Br J Pharmacol 118:1415–1420

Rudolphi KA, Schubert P, Parkinson FE, Fredholm BB (1992) Neuroprotective role of adenosine in cerebral ischaemia. Trends Pharmacol Sci 13:439–445

Sajjadi FG, Takabayashi K, Foster AC, Domingo RC, Firestein GS (1996) Inhibition of TNF-a expression by adenosine. Role of A_3 adenosine receptors. J Immunol 156:3435–3442

Sansum AJ, Chessell IP, Hicks GA, Trezise DJ, Humphrey PPA (1998) Evidence that P2X purinoceptors mediate the excitatory effects of α,β-methylene-ADP in rat locus ceruleus neurons. Neuropharmacology 37:875–885

Satoh S, Matsumura H, Hayaishi O (1998) Involvement of adenosine A_{2A} receptor in sleep promotion. Eur J Pharmacol 351:155–162

Schaumann E, Kutscha W (1972) Clinical-pharmacological studies with a new orally active adenosine derivative. Drug Res 22:783–790

Schaumann E, Schlierf G, Pfleiderer T, Weber E (1972) Effect of repeated doses of phenylisopropyladenosine on lipid and carbohydrate metabolism in healthy fasting subjects. Arzneim-Forsch 22:593–596

Schingnitz G, Kuefner-Muehl U, Ensinger H, Lehr E, Kuhn FJ (1991) Selective A_1-antagonists for treatment of cognitive deficits. Nucleosides Nucleotides 10:1067–1076

Schwabe U, Trost T (1980) Characterization of adenosine receptors in rat brain by (−)-[^{3}H]N^6-phenylisopropyladenosine. Naunyn-Schmiedebergs Arch Pharmacol 313: 179–187

Sei Y, von Lubitz DKJE, Abracchio MP, Ji XD, Jacobsen KA (1997) Adenosine A_3 receptor agonist-induced neurotoxicity in rat cerebellar granule neurons. Drug Dev Res 40:267–273

Sheardown MJ, Hansen AJ, Thomsen C, Judge ME, Knutsen LJS (1985) Novel adenosine agonists: a strategy for stroke therapy. In: Grotta J, Miller L, Buchan AM, (eds) Ischemic Stroke: Recent Advances in Understanding and Therapy. International Business Communications pp. 187–214

Sheardown MJ, Knutsen LJS (1996). Unexpected neuroprotection observed with the adenosine A_{2A} receptor agonist CGS 21680. Drug Dev Res 39:108–114

Shearman LP, Weaver DR (1997) [^{125}I]4-Aminobenzyl-5′-N-methylcarboxamidoadenosine ([^{125}I]AB-MECA) labels multiple adenosine receptor subtypes in rat brain. Brain Res 745:10–20

Shibuya I, Tanaka K, Hattori Y, Uezono Y, Harayama N, Noguchi J, Ueta Y, Izumi F, Yamashita H (1999) Evidence that multiple P2X purinoceptors are functionally expressed in rat supraoptic neurones. J Physiol (Lond) 514 (Pt 2):351–367

Shimada J, Koike N, Nonaka H, Shiozaki S, Yanagawa K, Kanda T, Kobayashi H, Ichimura, M, Nakamura J, Kase H, Suzuki F (1997) Adenosine A_{2A} antagonists with potent anti-cataleptic activity. Bioorg Med Chem Lett 7:2349–2352

Shneyvays V, Nawrath H, Jacobson KA, Shainberg A (1998) Induction of apoptosis in cardiac myocytes by an A_3 adenosine receptor agonist. Exp Cell Res 243:383–397

Shryock JC, Ozeck MJ, Belardinelli L (1998) Inverse agonists and neutral antagonists of recombinant human A_1 adenosine receptors stably expressed in Chinese hamster ovary cells. Mol Pharmacol 53:886–893

Siddiqi SM, Jacobson KA, Esker JL, Olah ME, Li X-D, Melman N, Tiwari KN, Secrist III, JA, Schneller S, Cristalli G, Stiles GL, Johnson CR, IJzerman AP (1995) Search for new purine- and ribose-modified adenosine analogues as selective agonists and antagonists at adenosine receptors. J Med Chem 38:1174–1178

Soltoff SP, McMillian MK, Talamo BR (1989) Coomassie brilliant blue G is a more potent antagonist of P2 purinergic responses than reactive blue 2 (cibacron blue 3GA) in rat parotid acinar cells. Biochem Biophys Res Commun 165:1279–1285

Song SL, Chueh SH (1996) P2 purinoceptor-mediated inhibition of cyclic AMP accumulation in NG108-15 cells. Brain Res 734:243–251

Stoop R, Surprenant A, North RA (1997) Different sensitivities to pH of ATP-induced currents at four cloned P2X receptors. J Neurophysiol 78:1837–1840

Strong P, Anderson R, Coates J, Ellis F, Evans B, Gurden MF, Johnstone J, Kennedy I Martin DP (1993). Suppression of non-esterified fatty acids and triacylglycerol in experimental animals by the adenosine analog GR79236. Clin Sci 84:663–669

Thomsen C, Valsborg JS, Foged C, Knutsen LJS (1997) Characterization of [^{3}H]-N-[R-(2-Benzothiazolyl)thio-2-propyl]-2-chloroadenosine ([^{3}H]-NNC 21–0136) binding to rat brain: profile of a novel selective adenosine receptors for adenosine A_1 receptors. Drug Dev Res 42:86–97

Trivedi BK, Bruns RF (1989) C^2,N^6-Disubstituted adenosines: synthesis and structure–activity relationships. J Med Chem 32:1667–1673

Tuluc F, Bultmann R, Glanzel M, Frahm AW, Starke K (1998) P2-receptor antagonists: IV. Blockade of P2-receptor subtypes and ecto-nucleotidases by compounds related to reactive blue 2. Naunyn-Schmiedebergs Arch Pharmacol 357:111–120

Ukena D, Jacobson KA, Padgett WL, Ayala C, Shamim MT, Kirk KL, Olsson RA, Daly JW (1986) Species differences in structure–activity relationships of adenosine agonists and xanthine antagonists at brain A_1 adenosine receptors. FEBS Lett 209:122–128

Uneyama H, Uneyama C, Ebihara S, Akaike, N (1994) Suranim and reactive blue 2 are antagonists for a newly identified purinoceptor on rat megakaryocyte. Br J Pharmacol 111:245–249

Usune S, Katsuragi T, Furukawa T (1996) Effects of PPADS and suramin on contractions and cytoplasmic Ca^{2+} changes evoked by AP4A, ATP and α,β-methylene ATP in guinea pig urinary bladder. Br J Pharmacol 117:698–702

van Rhee AM, van der Heijden MPA, Beukers MW, IJzerman AP, Soudijn W, Nickel P, (1994) Novel competitive antagonists for P2 purinoceptors. Eur J Pharmacol 268:1–7

van Schaick EA, Kulkarni C, von Frijtag Drabbe Kunzel JK, Mathot RAA, Cristalli G, IJzerman AP, Danhof M (1997) Time course of action of three adenosine A_1 receptor agonists with differing lipophilicity in rats: comparison of pharmacokinetic, hemodynamic and EEG effects. Naunyn-Schmiedebergs Arch Pharmacol 356:827–837

van Schaick EA, Jacobson KA, Kim HO, IJzerman AP, Danhof M (1996) Hemodynamic effects and histamine release elicited by the selective adenosine A_3 receptor agonist 2-Cl-IB-MECA in conscious rats. Eur J Pharmacol 308:311–314

Varani K, Borea PA, Guerra L, Dionisotti S, Zocchi C, Ongini E (1995) Binding of the adenosine A_{2a} receptor ligand [^{3}H]CGS 21680 to human platelet membranes. Res Commun Mol Pathol Pharmacol 87:109–110

Varani K, Cacciari B, Baraldi PG, Dionisotti S, Ongini E, Borea PA (1998) Binding affinity of adenosine receptor agonists and antagonists at human cloned A_3 adenosine receptors. Life Sci 63:PL81–PL87

Varani K, Merighi S, Gessi S, Klotz KN, Leung E, Baraldi PG, Cacciara B, Romagnoli R, Spalluto G, Borea PA (2000) [^{3}H]MRE 3008F20: A Novel Antagonist Radioligand for the Pharmacological and Biochemical Characterization of Human A_3 Adenosine Receptors. Mol Pharmacol 57:968–975

Virginio C, Robertson G, Surprenant A, North RA (1998) Trinitrophenyl-substituted nucleotides are potent antagonists selective for $P2X_1$, $P2X_3$, and heteromeric $P2X_{2/3}$ receptors. Mol Pharmacol 53:969–973

Volonte C, Merlo D (1997) Biological effects of P2 purinoceptor modulators in cultured primary cerebellar granule neurons. Proc Eur Soc Neurochem Meeting 11:357–360

von Heijne M, Hao J-X, Yu W, Sollevi A, Xu X-J, Wiesenfeld-Hallin Z (1998) Reduced anti-allodynic effect of the adenosine A_1-receptor agonist R-phenylisopropyladenosine on repeated intrathecal administration and lack of cross-tolerance with morphine in a rat model of central pain. Anesth Analg 87:1367–1371

von Lubitz DKJE, Beenhakker M, Lin RC-S, Carter MF, Paul IA, Bischofberger N, Jacobson KA (1996) Reduction of postischemic brain damage and memory deficits following treatment with the selective adenosine A_1 receptor agonist. Eur J Pharmacol 302:43–48

von Lubitz DKJE, Lin RC-S, Boyd M, Bischofberger N, Jacobson KA (1999) Chronic administration of adenosine A_3 receptor agonist and cerebral ischemia: neuronal and glial effects. Eur J Pharmacol 367:157–163

von Lubitz DKJE, Lin RCS, Jacobson KA (1995b) Cerebral ischemia in gerbils: effects of acute and chronic treatment with adenosine A_{2A} receptor agonist and antagonist. Eur J Pharmacol 287:295–302

von Lubitz DKJE, Lin RC-S, Melman N Ji X-d, Carter MF, Jacobson KA (1994) Chronic administration of selective adenosine A_1 receptor agonist or antagonist in cerebral ischemia. Eur J Pharmacol 256:161–167

von Lubitz DKJE, Lin RC-S, Paul IA, Beenhakker M, Boyd M, Bischofberger N, Jacobson KA (1996b) Postischemic administration of adenosine amine congener (ADAC): analysis of recovery in gerbils. Eur J Pharmacol 316:171–179

Wainwright CL, Kang L, Ross S (1997) Studies on the mechanism underlying the antifibrillatory effect of the A_1-adenosine agonist, R-PIA, in rat isolated hearts. Cardiovasc Drugs Ther 11:669–678

Wagner H, Milavec-Krizman M, Gadient F, Menninger K, Schoeffter P, Tapparelli C, Pfannkuche H-J, Fozard JR (1995) General pharmacology of SDZ WAG 994, a potent selective and orally active adenosine A_1 receptor agonist. Drug Dev Res 34:276–288

Webb RL, Barclay BW, Graybill SC (1991) Cardiovascular effects of adenosine A_2 agonists in the conscious spontaneously hypertensive rat: a comparative study of three structurally distinct ligands. J Pharmacol Exp Ther 259:1203–1212

Wildman SS, Brown SG, King BF, Burnstock G (1999) Selectivity of diadenosine polyphosphates for rat P2X receptor subunits. Eur J Pharmacol 367:119–123

Wiley S, Gargett CE, Zhang W, Snook MB, Jamieson GP (1998) Partial agonists and antagonists reveal a second permeability state of human lymphocyte $P2Z/P2X_7$ channel. Am J Physiol 275:C1224–C1231

Williams M (1996) Challenges in developing P2 purinoceptor-based therapeutics. Ciba Found Symp 1996, 198:309–321

Williams M (2000) Purines: From Premise to promise. J. Autonom. Nervous System. In press

Windscheif U, Pfaff O, Ziganshin AU, Hoyle CHV, Bäumert HG, Mutschler E, Burnstock G, Lambrecht G (1995) Inhibitory action of PPADS on relaxant responses to adenine-nucleotides or electrical-field stimulation in guinea-pig taenia-coli and rat duodenum. Br J Pharmacol 115:1509–1517

Wolff AA, Skettino SL, Beckman E, Belardinelli L (1998) Renal effects of BG9719, a specific A_1 adenosine receptor antagonist, in congestive heart failure. Drug Dev Res 45:166–171

Xu H, Bianchi B, Metzger R, Lynch KJ, Kowaluk EA, Jarvis MF, van Biesen T (1999) Lack of specificity of [^{35}S]-ATPγS and [^{35}S]-ADPβS as radioligands for inotropic and metabotropic P2 receptor binding. Drug Dev Res 48:84–93

Yagil Y, Miyamoto M (1995) The hypotensive effect of an oral adenosine analog with selectivity for the A_2 receptor in the spontaneously hypertensive rat. Am J Hypertens 8:509–515

Yang CM, Tsai YJ, Pan SL, Tsai CT, Wu WB, Chiu CT, Luo SF, Ou JT (1997) Purinoceptor-stimulated phosphoinositide hydrolysis in Madin-Darby canine kidney (MDCK) cells. Naunyn-Schmiedebergs Arch Pharmacol 356:1–7

Yao Y, Sei Y, Abbracchio MP, Jiang J-L, Kim Y-C, Jacobson KA (1997) Adenosine A_3 receptor agonists protect HL-60 and U-937 cells from apoptosis induced by A_3 antagonists. Biochem Biophys Res Commun 232:317–322

Ziganshin AU, Hoyle C, Bo XN, Lambrecht G, Mutschler E, Bäumert HG, Burnstock G (1993) PPADS selectively antagonizes P2X purinoceptor-mediated responses in the rabbit urinary-bladder. Br J Pharmacol 110:1491–1495

Zimmet J, Järlebark L, van Galen PJM, Jacobson, KA, Heilbronn E (1993) Synthesis and biological activity of novel 2-thio derivatives of ATP. Nucleosides Nucleotides 12:1–20

Ziyal R, Ralevic V, Ziganshin AU, Nickel P, Adanuy U, Mutschler E, Lambrecht G, Burnstock G (1996) NF023, a selective P2X-purinoceptor antagonist in rat hamster and rabbit isolated blood vessels. Drug Dev Res 37:113

Section II
Neurotransmission

CHAPTER 7
Regulation of Purine Release

B. Sperlágh and E.S. Vizi

A. Introduction

With the unequivocal demonstration that purine and pyrimidine nucleosides and nucleotides function as key extracellular messengers in all mammalian tissue systems (Ralevic and Burnstock 1998), a major challenge has been to understand the factors that govern purine availability in the extracellular space and the dynamics of the process. The focus of the present chapter is to summarize recent knowledge on purine release under various physiological and pathophysiological conditions.

In the first section, neuronal synthesis and storage of purines, e.g., intracellular ATP and adenosine pools available for the release process, are described and in subsequent sections the nature of purine release elicited by various stimuli is reviewed with an emphasis on the possible origin of ATP released by physiological neuronal activity, receptor activation, and pathological stimuli.

B. Releasable Purine Stores in Neuronal and Non-Neuronal Cells

I. Releasable ATP Stores

ATP, utilized as an universal energy "currency" in biological systems, is present in all living cells including neurons, and is theoretically available for release. Although nerve terminals cannot synthesize the purine ring de novo, they can take up adenosine and other nucleosides via nucleoside carriers to use in the production of ATP. Adenine nucleotides and other purines can be formed at various steps in the intermediary metabolic pathways; however, the main regulator of ATP levels in the neurons is mitochondrial oxidative phosphorylation, which generates ATP from adenosine diphosphate (ADP) and ATP production is tightly controlled by the so-called respiratory control, i.e., by ADP availability (Fig. 1). Under normal metabolic conditions the cytoplasmic level of ATP in neurons is 10 mmol/l (McMahon and Nicholls 1991) which is essential for neuronal function and also provide a potential source of

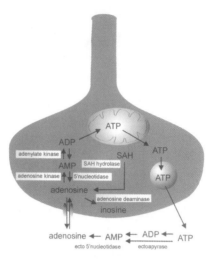

Fig. 1. Release and interconversion of ATP and adenosine. Adenosine is taken up to the nerve terminal by the bidirectional nucleoside carrier. Subsequently, it is rephosphorylated to AMP and ADP by adenosine kinase and adenylate kinase enzymes, respectively. The major pathway of ATP production is the mitochondrial oxidative phosphorylation which converts ADP to ATP according to the energy demand of the cell, i.e., to the actual amount of ADP. ATP is then transported out from the mitochondrion and is taken up by synaptic vesicles and released by vesicular exocytosis. In the extracellular space ATP is dephosphorylated to ADP and AMP by ectoapyrase enzyme, and AMP is hydrolysed to adenosine by ecto 5′-nucleotidase enzyme. If ADP and AMP is accumulated in the cytosol due to the shortage of oxidative phosphorylation 5′-nucleotidase enzyme may produce adenosine from AMP and adenosine could leave the cell via the nucleoside carrier. An additional source of cytoplasmic adenosine might be S-adenosyl-homocysteine (SAH) which is converted to adenosine by SAH hydrolase enzyme. Adenosine could be also deaminated by adenosine deaminase enzyme giving rise to formation of inosine

releasable ATP. However, the majority of cytoplasmic ATP is utilized for energy consuming processes such as the supply of Na^+/K^+ pump, and other transporters of the plasma membrane and intracellular organelles; for biosynthesis of proteins involved in exocytosis and signal transduction, and therefore probably is not directly available for release (McMahon and Nicholls 1991). Even if cytoplasmic ATP pools are involved in the release process, it is unclear whether they are compartmentalized for the signaling process, and what governs the size of any such pool. ATP is also a key constituent of all types of synaptic vesicles in nerve terminals, including cholinergic (Dowdall et al. 1974; Zimmermann 1978, 2000; Volknandt and Zimmermann 1986), large dense cored (Klein 1982; Lagercrantz and Stjarne 1974), and small dense cored (Smith 1979) catecholamine-containing vesicles, and chromaffin vesicles (Winkler and Westhead 1980; Aberer et al. 1978). Therefore, the major intracellular source of extracellular ATP is generally thought to be the synaptic vesicle, which takes up ATP from the cytoplasm by a Mg^{2+}-activated ATPase generating a proton gradient as the driving force for the uptake (ABERER et al. 1978). Vesicular storage of storage ratios of 3:1 to 50:1 with ACh and noradrenaline to ATP have been established, depending upon the type of vesicles, corresponding to a 1–200mmol/l concentration of ATP inside the vesicle (STJARNE 1989). Other nucleotides, e.g., ADP, AMP, UTP, Ap4A, Ap5A, and guanine nucleotides are also stored in synaptic vesicles and play a role in neurotransmission (RALEVIC and BURNSTOCK 1998; ZIMMERMANN, Chap. 8, this volume). Although their concentrations are lower than that of ATP, they are in the millimolar range and sufficient to serve as a releasable pool.

II. Releasable Adenosine Stores

Recent advances in the molecular biology of nucleoside transporters identified two protein families responsible for the transmembrane transport of nucleosides. These two families have been designated as the equilibrative and concentrative nucleoside transporter families (CASS et al. 1998). Among them, the concentrative nucleoside transporters are primarily expressed in epithelial cells, while the equilibrative transporters are found in most mammalian cell types (CASS et al. 1998). Equilibrative transporters have different isoforms and the members of this family correspond to nucleoside transporters identified biochemically in earlier papers in brain synaptosomes (LEE and JARVIS 1988; THORN and JARVIS 1996), dissociated cells (GEIGER et al. 1988). These carriers might be sensitive or insensitive to nanomolar concentrations of the transport inhibitor nitrobenzylthioinosine (NBI), the latter corresponding to [^{3}H]NBI binding sites (GEIGER and NAGY 1984), and they form the target site of different nucleoside uptake inhibitors used in the clinical practice, such as dypiridamole and dilazep (THORN and JARVIS 1996). Equilibrative transporters can function bidirectionally, i.e., they are able to transport adenosine in both inward and outward directions depending on the chemical gradient. Although different isoforms of this family exhibit different substrate specificities, and other nucleosides, like inosine and uridine could be also transported by this carrier, the preferred substrate appears to be adenosine (GEIGER et al. 1998; THORN and JARVIS 1996).

When adenosine is taken up into the nerve terminal via nucleoside carrier transport it is readily incorporated into ATP stores via phosphorylation by adenosine kinase (EC 2.7.1.20) and adenylate kinase (EC 2.7.4.3) enzymes (FISCHER and NEWSHOLME 1984; WU and PHILLIS 1984) such that intracellular adenosine concentrations are at the submicromolar level under physiological energetic conditions (NEWMAN and MCILWAIN 1977; FREDHOLM et al. 1984) (Fig. 1). The same pathway, however, may also work in the opposite direction and give rise to the formation of adenosine by the cytosolic 5′-nucleotidase enzyme when ADP and AMP levels are increased due to decreased ATP production or increased ATP utilization (Fig. 1; see also Sect. E.I.). Adenosine might as well be produced from S-adenosylhomocysteine by S-adenosylhomocysteine hydrolase enzyme as part of an essential transmethylation pathway (LLOYD et al. 1988), and a third important mechanism to balance the intracellular adenosine level is the adenosine deaminase enzyme, which metabolizes adenosine to inosine by deamination and prevents any increase in intracellular adenosine above the normal level (McCAMAN and McAFEE 1986) (Fig. 1).

C. Release of Purines During Physiological Neuronal Activity Simulated by Electrical or Chemical Depolarization

I. Release of Adenosine

The first demonstration of adenosine release during neuronal activity has been presented by PULL and McILWAIN (1972) which was followed by numerous other studies showing that different kinds of depolarization stimuli promote the efflux of adenosine and other purines from synaptosomes (FREDHOLM and VERNET 1979; DAVAL and BARBERIS 1981; BENDER et al. 1981) and brain slices (SCHUBERT et al. 1976; LEE et al. 1982; JONZON and FREDHOLM 1985; WOJCIK and NEFF 1983; PEDATA et al. 1990). In addition a stimulation-dependent release of adenosine has been reported in in vivo models (JHAMANDAS and DUMBRILLE 1980; BARBERIS et al. 1983; CHEN et al. 1992). Since adenosine release from brain slices (HOEHN and WHITE 1990b; PEDATA et al. 1991) and in in vivo studies (CHEN et al. 1992) is inhibited partly by glutamate receptor antagonists and NMDA and other glutamate receptor agonists are able to release adenosine (MANZONI et al. 1994; HOEHN and WHITE 1990a; CHEN et al. 1992), it seems likely that release of adenosine is coupled to glutamate receptor activation, i.e., to the activity of the excitatory pathways in the brain. Other mediators, such as NO (FISCHER et al. 1995; FALLAHI et al. 1996; JURÁNYI et al. 1999) have also been shown to be involved in adenosine release from the brain under particular conditions, such as ischemia, when the extracellular NO level is elevated (JURÁNYI et al. 1999). Although depolarization-induced adenosine release has been proved to be Ca^{2+}-dependent in a number of model systems (e.g., PEDATA et al. 1991; LATINI et al. 1997; CAHILL et al. 1993), the exact route whereby it enters the extracellular space is still controversial. Nucleoside carriers are one of the major candidates thought to drive adenosine release under various physiological and pathological conditions. This notion is supported by observations in which adenosine release was unaffected by the inhibitors of extracellular catabolism of nucleotides or reduced by nucleoside transport inhibitors (SCHOUSBOE et al. 1989; FREDHOLM 1980; SWEENEY et al. 1993; LLOYD et al. 1993; GU et al. 1995). However, in other studies inhibitors of nucleoside transporters such as dypiridamole or soluflazine did not reduce, but increased, the release of nucleosides (VAN BELLE et al. 1987; PHILLIS et al. 1989), indicating that the volume of inward transport by the nucleoside carrier might surpass the volume of outward transport and the effect uptake inhibitors mirror the net effect of bidirectional transport. The alternative hypothesis for the release mechanism of adenosine release is that the ATP is released primarily and its subsequent extracellular catabolism gives rise the accumulation of adenosine, reflecting a purinergic cascade (RICHARDSON et al. 1987; CUNHA et al. 1996; ABBRACCHIO and WILLIAMS, Chap. 29, second volume). The arguments on this side are that the free cytoplasmic concentration of adenosine is low (FREDHOLM et al. 1990), and studies where inhibitors of extracellular nucleotide breakdown inhibited adenosine release (MACDONALD and WHITE 1985; CRAIG and WHITE 1993; CUNHA et al. 1996). ATP is also released upon depolarization (e.g., RICHARDSON and BROWN 1987; SPERLÁGH et al. 1995) and the nucleotide catabolizing ectoenzyme chain shows widespread expression in the nervous system (ZIMMERMANN 2000). Presumably, both mechanisms play a role, but their relative contribution to the overall purine outflow depends on the conditions of stimulation (see also Sect. E.I).

II. Release of ATP

Since HOLTON (1959) first demonstrated the release of adenyl purines during antidromic nerve stimulation, a large body of evidence has accumulated, supporting the concept of neuronal ATP release following depolarizing stimuli and includes release studies in synaptosomal preparations from elasmobranchs (MOREL and MEUNIER 1981), brain (WHITE 1977, 1978; POTTER and WHITE 1980; RICHARDSON and BROWN 1987) and peripheral neurons (SILINSKY and HUBBARD 1973; WHITE and LESLIE 1982; WHITE and AL-HUMAYYD 1983), cultured neurons (ZHANG et al. 1988; VON KUGELGEN et al. 1994), brain slices (WIERASZKO et al. 1989; SPERLAGH et al. 1995), isolated organs innervated by sympathetic (LEW and WHITE 1987; WESTFALL et al. 1978, 1987; VIZI et al. 1992) or parasympathetic nerves (TAKAHASHI et al. 1971), and a limited number of in vivo studies (WU and PHILLIS 1978; SPERLAGH et al. 1992).

From a physiological point of view, the more critical studies are those obtained in preparations in which ATP meets the criteria necessary to function as neurotransmitter, i.e., it is synthesized and stored in nerves, released during nerve activity, and its interaction with specific receptors in the postsynaptic membrane leads to changes in postsynaptic activity (BURNSTOCK 1976). These tissues include the vas deferens (KIRKPATRICK et al. 1987; VIZI et al. 1992; VON KUGELGEN et al. 1991), medial habenula (SPERLÁGH et al. 1995), and hippocampus (WIERASZKO et al. 1989; CUNHA et al. 1996). The demonstration of stimulation-dependent ATP release from such preparations is an important building block to support the hypothesis that ATP acts as a signaling substance in the central and peripheral nervous systems.

While traditional, Ca^{2+}-dependent, tetrodotoxin sensitive exocytosis seems to be the most relevant mechanism for ATP release, Ca^{2+} independent (HAMANN and ATTWELL 1996) and tetrodotoxin-insensitive release (HAMANN and ATTWELL 1996; JURÁNYI et al. 1997, 1999) has also been reported although the physiological significance of this type of release has not yet been elucidated. The amount of ATP released during resting conditions is about 1–10 pmol/g/min, while stimulation-evoked release is in the 1–100 pmol/g range. The resting release of ATP is similar in most tissue preparations; however, the amount of stimulation-evoked release is greatest in pure and compact neuronal tissues, e.g., superior cervical ganglion (VIZI et al. 1997) and lower in more heterogeneous preparations, e.g., vas deferens or isolated blood vessels (VIZI et al. 1992; SEDAA et al. 1989).

Once released into the extracellular space ATP is degraded by various ectoenzymes that are localized to the tissue (ZIMMERMANN, Chap. 8, this volume). The breakdown rate of released ATP is dependent on the activity of the cell surface ectoATPases in a given tissue. The manipulations of tissue preparations to study ATP release may also affect these inactivating systems, requiring parallel measurements of the release and catabolism of extracellular ATP (e.g., SPERLAGH et al. 1995; NITAHARA et al. 1995) or an increase in the temporal resolution of outflow detection (TODOROV et al. 1996) to provide more accurate estimates of the dynamics of ATP release. Until selective and efficacious ectoATPase inhibitors that do not modulate receptor function become available, some portion of the released ATP will escape detection even under the most rigorous experimental conditions.

1. Source of ATP Release in the Central Nervous System (Presynaptic Origin)

Releasable ATP pools are present in both neurons and non-neuronal cells, further complicating the identification of the origin of ATP accumulating in the extracellular fluid during neuronal activity. In the neural tissue the possible sources of extracellular ATP could be the following: (i) catecholaminergic, cholinergic, glutamatergic, dopaminergic, serotonergic, GABAergic nerve terminals; (ii) purinergic nerve terminals; (iii) glial cells.

In order to address this question, preparations containing exclusively neuronal elements, e.g., cultured neurones or isolated nerve terminals, can be used. As an example, Ca^{2+}-dependent ATP release from brain-derived, affinity-purified cholinergic nerve terminals following K^+ and veratridine depolarization (RICHARDSON and BROWN 1987) and electrical stimulation induced ATP release from cultured chick sympathetic neurons (VON KUGELGEN et al. 1994) has been reported. While neuronal purine release can be isolated in these preparations, cultured cells and synaptosomes do not necessarily preserve all the properties of the tissue of origin, e.g., receptor distribution or synaptic organization. Thus, another approach is to use intact preparations (isolated organs, brain slices) and to destroy surgically or chemically the input pathways to study the release of ATP before and after the lesion. Early studies on synaptosomes derived from different parts of the rat brain showed that chemical denervation of central catecholaminergic pathways by i.c.v. 6-hydroxydopamine (6-OHDA) pretreatment had no affect on the amount of

ATP released by depolarization in brain synaptosomes (POTTER and WHITE 1982), suggesting that the origin of extraneuronal ATP in the brain is heterogeneous. In the rat medial habenula nucleus, the first central nucleus where clearcut electrophysiological evidence has been obtained that ATP acts as a fast transmitter (EDWARDS et al. 1992), ATP is released in a stimulation-dependent, tetrodotoxin-sensitive manner and undergoes metabolism by ectoATPases (SPERLÁGH et al. 1995). The main excitatory input to this nucleus originates from two septal nuclei – the nucleus triangularis septi (TS) and the nucleus septofimbrialis (SFi) via the stria medullaris, which together provide a massive, topographically organized projection ending on the dendrites of substance P- and choline acetyltransferase-containing neurons. In a combined morphological-neurochemical approach, the neuronal tracer *Phaseolus vulgaris* leucoagglutinin (PHAL) was used to trace the septohabenular pathway, which was injected into the TS and resulted in the labeling of a dense fiber network in the ipsilateral medial habenula (SPERLÁGH et al. 1998b). These PHAL-labeled terminals contained numerous, densely packed, round synaptic vesicles and formed asymmetric synapses with a pronounced postsynaptic density (Fig. 2). In subsequent experiments electrolytic lesion of the TS and SFi was performed and field stimulation-induced release of ATP was measured in superfused habenula slices of lesioned and sham operated animals. The lesion had no effect on the resting release of ATP but, in contrast, stimulation-evoked release was decreased by 81% in the lesioned rats as compared to controls (Fig. 3), suggesting that ATP is released stimulation-dependently from the nerve terminals of septohabenular projection neurons. This pathway predominantly uses glutamate as an excitatory neurotransmitter and was immunonegative for other neurotransmitters, indicating that glutamate was the most likely co-transmitter candidate for ATP, a conclusion also supported by the asymmetrical synapses formed by these terminals (Fig. 2). However, recent electrophysiological analysis showed that synaptic currents isolated from individual medial habenula neurons are either purinergic or glutamatergic, favoring the idea that ATP and glutamate are released from different subpopulations of septohabenular projection (ROBERTSON and EDWARDS 1998).

There are also other regions in the brain, e.g., the hippocampus, where ATP acts as a fast neurotransmitter eliciting a non-glutamatergic excitatory synaptic current (PANKRATOV et al. 1998). ATP has been shown to be released in a stimulation-dependent manner from rat hippocampal slices following electrical stimulation of the Schaffer collateral pathway with high frequency stimulation mimicking long term potentiation (LTP) phenomenon (WIERASZKO et al. 1989) and in a depolarization-induced Ca^{2+}-dependent manner from purified mossy fiber synaptosomes (TERRIAN et al. 1989), raising the possibility that ATP is derived from glutamatergic excitatory pathways in this region. However, low frequency stimulation, characteristic of conventional neurotransmission, failed to release ATP in a Ca^{2+}-dependent fashion in the study of WIERASZKO et al. (1989). In contrast, the use of a different high

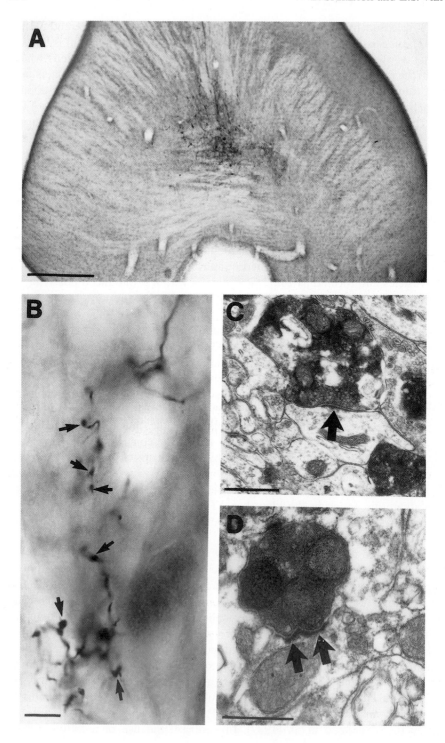

frequency burst and low frequency field stimulation paradigm to mimic LTP and long-term depression (LTD) elicited Ca^{2+}-dependent ATP and [^{3}H]adenosine release (CUNHA et al. 1996). Interestingly, the amount of ATP release evoked by stimulation was greater under high frequency conditions, while

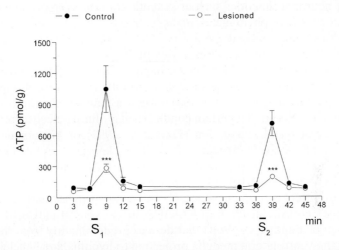

Fig. 3. Effect of TS lesion on the electrical field stimulation (2 Hz, 2.5 msec, 360 shocks) – induced release of endogenous ATP from the rat habenula preparation. The preparations were superfused for 60 min at a rate of 0.7 ml/min. Following the preperfusion period, 3-min samples were collected and assayed for ATP by the luciferin-luciferase assay. ATP was released in a significant quantity in response to electrical field stimulation (S_1, S_2). *Filled circles* represent ATP release, measured in sham-operated rats, *open circles* show ATP release measured in the lesioned rats. The release of ATP was expressed in pmol/g. The values show the mean ± SEM of 6–8 identical experiments. *Asterisks* indicate significant difference between sham-operated and lesioned animals, calculated by two way analysis of variance (ANOVA) (***$p < 0.001$) [Reprinted from Neuroscience, 86, SPERLÁGH B, MAGLÓCZKY Z, VIZI ES, FREUND T (1998a) The triangular septal nucleus as the major source of ATP release in the rat habenula: a combined neurochemical and morphological study, pp 1195–1207, with permission from Elsevier Science]

◄

Fig. 2A–D. Light and electronmicroscopic morphology of the septohabenular pathway that releases ATP upon field stimulation of habenula slices by anterograde tracing: **A** the injection site of the anterograde tracer, PHAL, iontophorised into the triangular septal nucleus. The sections are counterstained by cresyl violet. The PHAL injection involved only a small restricted part of the triangular septal nucleus; **B** the terminal cluster of a single PHAL-labeled axon. The fibers bear numerous large, complex boutons (*arrows*); **C,D** electron micrographs of triangulo-habenular terminals – PHAL-labeled triangular septal terminal in the medial habenula establishes asymmetric synaptic contact with a different dendritic profile (*arrow*) (**C**) – show a degenerating terminal after triangular septal lesion, making two asymmetric synaptic contacts (*arrows*) with a dendrite (**D**). The terminals and their postsynaptic profiles are similar to the PHAL-labeled boutons shown in C. Scales – **A**: 0.5 mm, **B**: 10 μm, **C**: 1.5 μm, **D**: 1 μm.

[³H]adenosine release was higher under conditions of low frequency stimulation. The proportion of extracellular nucleotides as the source of [³H]adenosine also varied: at high stimulation frequency extracellular ATP provided a greater contribution to extracellular adenosine than at low frequency (CUNHA et al. 1996). Therefore it seems likely that the ATP and adenosine involved in different neuronal functions such as synaptic transmission, presynaptic modulation or plasticity phenomena might have different sources depending on the pattern of neuronal activity.

In the caudal medulla oblongata of the brainstem, a catecholaminergic pathway originating from the A1 cell group provides a well-defined excitatory input to the neurosecretory cells in the supraoptic and paraventricular nuclei of the hypothalamus (PALKOVITS 1981), using a neurotransmitter other than noradrenaline (NA) under certain conditions, e.g., during electrical (GARTSIDE et al. 1995) or vagal stimulation (DAY et al. 1993) or haemorrhage stress (BULLER and KHANNA 1996), stimuli known to relay via the A1 cell group, and ATP has been suggested to be the signalling substance. ATP is co-released with [³H]NA from superfused hypothalamical slices, and the release of both neurotransmitters is frequency- and Ca^{2+}-dependent (SPERLAGH et al. 1998b). Bilateral stereotaxic injection of 6-OHDA to the ventral part of the ventral noradrenergic bundle (vVNAB) that derives predominantly from the A1 cell group in the ventrolateral medulla projecting to hypothalamic nuclei resulted in a 55% reduction of endogenous NA content of the hypothalamic slices, consistent with the contribution of the pathway to the total noradrenergic input of the whole hypothalamus (approximately 20%–60%), and uptake and the stimulation-evoked release of [³H]NA was also markedly reduced. While the basal release of ATP was unaffected, evoked release was diminished by 72% by this treatment, indicating a dominant contribution of this particular pathway to total ATP outflow from the tissue (Fig. 4). Since the evoked release of both ATP and [³H]NA was reduced after 6-OHDA pretreatment by more than 50%, these data indicate that ATP and NA are co-released from central noradrenergic nerves. Further analysis (SPERLÁGH et al. 1998b) of the potential subcellular source(s) of released ATP in rat hypothalamic slices used the $α_2$-adrenoceptor antagonists CH–38083 (VIZI et al. 1986) and BRL 44408 which increased the stimulation-evoked overflow of tritium, indicating that NA release from hypothalamic catecholaminergic axon terminals is subject to autoinhibition by endogenous NA acting via $α_2$-adrenoceptors. In contrast, stimulation-evoked ATP release was not affected by this treatment. As an explanation of this finding one can assume that nerve terminals that release ATP and NA in the hypothalamus, although both originate predominantly in the lower brain stem, are only partly overlapping and not evenly subject to inhibition via $α_2$-autoreceptors. Moreover, the existence of purely purinergic nerve terminals that are independent of the influence of NA cannot be excluded (SPERLÁGH et al. 1998b).

Although glial cells have different excitability properties compared to neurons, being devoid of action potential propagation and synapses, they

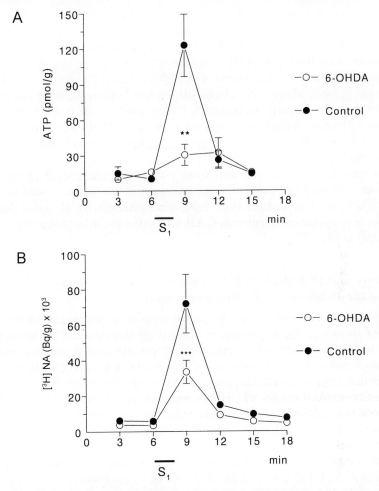

Fig. 4A,B. Effect of 6-OHDA lesion of vVNAB on the electrical stimulation induced release of ATP and [^{3}H]NA from hypothalamic slices: **A** ATP release, expressed in pmol/g; **B** [^{3}H]NA release, expressed in Bq/g. *Filled circles* represent results obtained in sham-operated animals, injected with the vehicle, *open circles* show results from lesioned animals injected with $2 \times 8\mu g$ 6-OHDA. Stereotaxic lesion was performed according to the method of GAILLET et al. (1993). Values show the mean ± SEM of six identical experiments. *Asterisks* represent significant differences between control and 6-OHDA treated animals calculated by the two way ANOVA (**$p < 0.01$, ***$p < 0.001$) [From Neuroscience 82, SPERLÁGH B, SERSHEN H, LAJTHA A, VIZI ES (1998b) Co-release of endogenous ATP and [^{3}H]noradrenaline from rat hypothalamic slices: origin and modulation by α_2-adrenoceptors, pp 511–520, with permission from Elsevier Science]

represent another possible source of ATP involved in neurotransmission responding to transmitters released from nerve terminals. Cultured astrocytes release ATP in response to NMDA and AMPA/kainate receptor activation (QUEIROZ et al. 1997) and ATP, acting via $P2X_7$ receptors, can evoke purine release from astrocytes (BALLERINI et al. 1997). Furthermore, in the neural lobe of the hypophysis, where a stimulation-dependent but tetrodotoxin-insensitive ATP release has recently been reported, pituicytes, modified glial cells of the posterior pituitary, have been proposed as the most likely source of extracellular ATP, controlling hormone secretion via P2 receptors (SPERLAGH et al. 1999; TROADEC et al. 1998).

In summary, ATP is released from distinct populations of cholinergic, glutamatergic, and catecholaminergic pathways in the brain as well as from glial cells upon neuronal activity. There is no evidence for the co-release of purines with dopamine, serotonin, GABA, or other putative transmitters (e.g., peptides) in the CNS.

2. Source of ATP Release in the Peripheral Nervous System: Cascade Transmission (Postsynaptic Origin)

The source of ATP released by nerve stimulation in organs containing nonneural tissues, e.g., the autonomic neuroeffector junction and the neuromuscular junction, is more heterogeneous, with smooth muscle, skeletal muscle, or endothelial cells being potential ATP sources. Although Ca^{2+}-dependent, neuronal release of ATP has been convincingly demonstrated in isolated organs innervated by the sympathetic nervous system (cf. BURNSTOCK 1990; cf. SPERLAGH and VIZI 1996), electrical stimulation-induced ATP outflow was decreased by α_1-adrenoceptor antagonists like prazosin to blocking the actions of the NA released from sympathetic terminals in response to axonal activity on postsynaptic α_1-adrenoceptors (VIZI and BURNSTOCK 1988; VON KÜGELGEN et al. 1991; VIZI et al. 1992). Since α_1-adrenoceptor agonists, like noradrenaline or methoxamine (see in details in Sect. D), can induce ATP release from various preparations following sympathetic denervation (VIZI et al. 1992) and α_1-adrenoceptor antagonists inhibit this α_1-agonist-induced ATP release, it is reasonable to assume that the α_1-adrenoceptor antagonist-sensitive part of field-stimulation induced ATP release originated from the activated target cell in response to action of NA on α_1-adrenoceptors (Fig. 4). Thus, in addition to neuronal and glial ATP stores, the source of ATP in these tissues could be the postsynaptic cell, e.g., the smooth muscle cell or the endothelial cell in the cardiovascular system. Moreover, not only NA but ATP or other nucleotides, acting on P2X receptors, can elicit further purine release, as seen in guinea-pig vas deferens (VON KÜGELGEN and STARKE 1991; VIZI and SPERLAGH 1999), longitudinal muscle strips of guinea-pig ileum (KATSURAGI et al. 1991, 1996; MATSUO et al. 1997), and in cardiac endothelial cells (YANG et al. 1994) providing a further contribution to extracellular purine levels. The action of neuronal NA and ATP on their postsynaptic receptors therefore results in a mixture of pre- and post-synaptically derived ATP in the junctional cleft. This phenomenon when the primary transmitter releases ATP from the target cell was named cascade transmission (VIZI et al. 1992). The relative contribution of these pools to local ATP concentration at the pre- and post-synaptic receptors depends on:

1. The amount of ATP and NA released from neurons (primary transmitters)
2. The amount of ATP released in response to α_1-adrenoceptor and P2X receptor activation (secondary transmitter)
3. The distribution of pre- and post-synaptic receptors
4. The width of the junctional cleft in individual tissues

In some tissues, where the junctional cleft is narrow, neuronal ATP release prevails over postsynaptic release (TODOROV et al. 1996), and in other tissues, e.g., in large diameter blood vessels, where the distance between nerve terminals and postsynaptic receptors is long, stimulation-induced ATP outflow might be almost entirely postsynaptic (SEDAA et al. 1989). The primary transmitter in these tissues should diffuse a long distance from its release site to reach its postsynaptic receptors, and its concentration decreases along the distance. The postsynaptic release of a secondary transmitter (ATP) triggered by the primary transmitter could result in an increased local concentration of ATP in the vicinity of postsynaptic receptors and thus provide an economic way of signal-amplification. Thus ATP released postsynaptically can take part in chemical transmission in three ways (Fig. 5):

1. Contributing to fast synaptic transmission
2. By potentiating the effect of primary transmitters on postsynaptic receptors
3. Modulating prejunctional aspects of neurotransmission, via P2 or A_1 receptors, after breakdown to adenosine

Postsynaptic ATP release has been demonstrated in tissues innervated by the sympathetic nervous system and in tissues where the principal neurotransmitter is ACh. Retrograde release of ATP was shown in Torpedo synaptosomes (ISRAEL and MEUNIER 1978) and postsynaptic ATP release contributes to stimulation-induced ATP release in guinea pig ileum (KATSURAGI et al. 1993), rat superior cervical ganglion (VIZI et al. 1997), and mouse neuromuscular junction (VIZI et al. 2000). As the primary neurotransmitter in these tissues is ACh, its action on postsynaptic nicotinic or muscarinic cholinergic receptors can elicit additional ATP release. In contrast, it appears that this trans-synaptic mechanism plays a less important role in central synapses as α_1-adrenoceptor agonists could not stimulate ATP release in rat hypothalamic slices (SPERLAGH et al. 1998b).

If ATP is co-released with a classical neurotransmitter within the same neuronal pathway, a major question is whether they are released from the same terminals/varicosities of those neurons and, if so, whether they originate

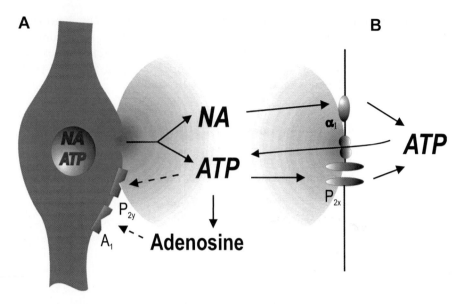

Fig. 5A,B. Cascade transmission (VIZI et al. 1992). Release of ATP from target cells in response to postsynaptic receptor activation by primary transmitters. ATP has been shown: **A** to be co- released with NA from sympathetic nerve varicosities upon neuronal activity; and **B** in response to their action on postsynaptic P2X and α_1-receptors in a carrier-mediated manner, which results in a non-uniform mixture of pre-and postsynaptically derived ATP in the junctional cleft. The relative contribution of this two pools to local concentration of ATP at the pre-and postsynaptic receptors will depend on (i) the amount of ATP and NA released from neurons, (ii) the amount of ATP released in response to α_1- and P2X receptor activation, (iii) the distribution of pre- and postsynaptic receptors, and (iv) the width of junctional cleft in individual tissues and junctions

from the same population of vesicles. Numerous studies have demonstrated ATP co-release with classical neurotransmitters, e.g., NA from vas deferens (KIRKPATRICK et al. 1987; KASAKOV et al. 1988; VIZI and BURNSTOCK 1988; VON KÜGELGEN and STARKE 1991; VIZI et al. 1992), from rat tail artery (WESTFALL et al. 1987; SHINOZUKA et al. 1991), and rabbit aorta (SEDAA et al. 1989). In addition, K^+-depolarization induced ATP release from pure cholinergic synaptosomes has been demonstrated in *Torpedo* (SCHWEITZER 1987), *Narcine brasiliensis* (UNSWORTH and JOHNSON 1990), and rat striatum (RICHARDSON and BROWN 1987), with constant ACh/ATP ratios similar to vesicular ACh/ATP ratios supporting the co-transmitter hypothesis. However, co-release may also occur from common vesicle populations or, partly or entirely, from independent vesicle populations. The lack of specific tools to identify ATP neurotransmitter pools inside the nerve terminal precludes a direct answer to this question. Instead, the common approach used to study this issue is to examine the parallel modulation of the release of ATP and its co-transmitter, or their electrophysiological response(s) using compounds known to act at presynaptic nerve terminals. Indeed, the differential or even opposite modulation of ATP and putative co-transmitter release have been described for a variety of receptors including α_2-adrenoceptors (SPERLAGH and VIZI 1992; DRIESSEN et al. 1994a), β-adrenoceptors (GONCALVES et al. 1996; DRIESSEN et al. 1996a), P1 purinergic (DRIESSEN et al. 1994b), angiotensin-II (ELLIS and BURNSTOCK 1989; DRIESSEN and STARKE 1994), and neuropeptide Y (ELLIS and BURNSTOCK 1990) receptors, even when postjunctional release was excluded using α_1 and P2 receptor antagonists (SPERLAGH and VIZI 1992; DRIESSEN et al. 1994a,b; DRIESSEN and STARKE 1994), showing that the releasable vesicular pool of ATP and its co-transmitter do not overlap completely with each other or are independent. In the guinea-pig vas deferens, a complete temporal dissociation of NA and ATP overflow was observed, corresponding to fast twitch and slow components of the biphasic contractile response (TODOROV et al. 1996). Furthermore, neurotoxins, like ω-conotoxin that interfere with voltage-dependent Ca^{2+}-channels (FARINAS et al. 1992) or directly with the exocytotic process, such as botulinum toxin A (MARSAL et al. 1987) and tetanus toxin (RABASSEDA et al. 1987), blocked ACh but not ATP release, from cholinergic synaptosomes of Torpedo indicating the existence of a co-transmitter-independent neuronal ATP pool. However findings, derived from electrophysiological and neurochemical techniques, are not always congruent. For instance, GONCALVES et al. (1996) and later DRIESSEN et al. (1996a) observed an opposite modulation of ATP and [^{3}H]NA overflow from the sympathetic neuroeffector junction, [^{3}H]NA overflow being augmented while ATP overflow was decreased by β-adrenoceptor activation supporting a separate vesicular origin of released NA and ATP. In contrast, using the ejp (excitatory junction potential) to monitor ATP release and measuring NA release using continuous amperometry and slow depolarization, BROCK et al. (1988) found that both the ejp and slow depolarization were increased by β-adrenoceptor activation consistent with the hypothesis that NA and ATP are released from the same population of nerve terminal and vesicles. The obvious advantage of electrophysiological methods is their real-time monitoring capacity and that they are sensitive to the junctionally active transmitter measuring indirectly the response of a given transmitter. On the other hand, they cannot detect the release process itself, and therefore any factor lying between the release and the functional response (e.g., uptake, metabolism) remains undetected. Furthermore, these methods do not identify the released transmitter. The question then arises whether junctionally active transmitters represent the same entity as the transmitter released from nerve terminals and captured by overflow experiment. The different results obtained by these two approaches would suggest that the answer is no. Nerve terminal depolarization, presumably, does not result in a uniformly distributed purine release, and local purine accumulation to postsynaptic receptors and presynaptic nerve terminals might have different functions: they might be involved in synaptic transmission, pre- or postjunctional modulation of transmitter release, or even non-synaptic transmission (Fig. 5).

D. Release of Purines by the Stimulation of Pre- or Post-Synaptic Receptors

Activation of α- and β-adrenoceptors (Vizi and Burnstock 1988; Westfall et al. 1987; Vizi et al. 1992; Tokunaga et al. 1993), nicotinic (White 1982; Von Kügelgen and Starke 1991; Sperlagh et al. 1992, 1998b) and muscarinic (Nitahara et al. 1995; Vizi et al. 1995), cholinoceptors, NMDA- and non-NMDA ionotropic glutamate (Hoehn and White 1990; Craig and White 1993; Manzoni et al. 1994; Chen et al. 1992), P2 purinergic (Katsuragi et al. 1991, 1996; Yang et al. 1994; Matsuo et al. 1997; Vizi and Sperlágh 1999), bradykinin, H_2 histamine, and substance P receptors (Tamesue et al. 1998), elicits purine release, most frequently adenine nucleotide release, in a variety of preparations. Receptor-evoked ATP release can occur both pre- and post-synaptically. For instance, nicotine acting on mecamylamine-sensitive nicotinic receptors release ATP from brain slices (Sperlágh et al. 1992, 1998b) and from the periphery (von Kügelgen and Starke 1991), and this release appears to be neuronal and exocytotic.

Activation of postsynaptic receptors in a given tissue can also elicit ATP release in a dose-dependent, receptor mediated manner. However, since postsynaptic target cells (endothelium, smooth muscle cell, dendrites, and somata of postsynaptic neurons) are devoid of neurotransmitter packaged vesicles, this release is most likely non-exocytotic. It has only recently recognized that, unlike exocytotic transmitter release, neurotransmitter uptake systems are inhibited by low temperature with 12°C being the cut-off point where vesicular and carrier-mediated release can be separated (Vizi 1998). Receptor evoked, postsynaptic ATP release is temperature-dependent, and thus it is reasonable to assume that a carrier-mediated mechanism is involved (Vizi and Sperlágh 1999). Carrier-mediated membrane transport systems for nucleotides have been reported in crayfish motor nerves and brain synaptosomes (Luqmani 1981; Chaudry et al. 1985; Sun and Lee 1985). In the guinea-pig ileum-myenteric plexus preparation ATP release evoked by carbamylcholine was strictly temperature-dependent, indicating the involvement of a carrier-mediated effect (Nitahara et al. 1995). In the guinea-pig ileum-myenteric plexus preparation it was found that ATP release by carbamylcholine was strictly temperature-dependent, indicating the involvement of a yet unidentified membrane carrier in this effect (Nitahara et al. 1995). A clearly defined need for the progress in this area is to isolate and identify membrane transporters able to carry purine nucleotides from other surface located ATP binding proteins. Expressing the carrier in cellular systems would help to understand better its functional characteristics and role in purinergic signaling. Nevertheless, as Zaidi and Matthews (1997) observed dendritic exocytosis in rat superior cervical ganglion upon stimulation of the ganglia by cholinergic drugs, the possibility that vesicular exocytosis also participate in the initiation of agonist-evoked postsynaptic ATP release cannot be entirely excluded.

Finally, an indirect mechanism, whereby a neuronally released principal transmitter initiates ATP release from a secondary pool by activating its receptors has been also reported. Tyramine, an indirectly acting sympathomimetic drug when taken up in nerve terminals, can release NA from the cytoplasm which in turn gives rise to ATP release via α_1-adrenoceptors (KIRKPATRICK and BURNSTOCK 1994; DRIESSEN et al. 1996b).

Although a physiological role for agonist-evoked purine release has yet to be established, this mechanism has to be considered as a potential target to modify extracellular purine levels under various pathophysiological conditions. In addition, a wide variety of drugs used in clinical practice act on receptors that have the potential to initiate purine release and may contribute to the therapeutic and/or side effects of these drugs.

E. Release of Purines by Pathological Stimuli

Extracellular purines, in addition to functioning as neurotransmitters/modulators, may also act as pathological signaling substances contributing to disease pathophysiology. Examples are ischemia-related diseases and inflammatory disease. The question arises how purine release is altered under these conditions.

I. Release of Purines by Hypoxia/Hypoglycemia/Energy Deprivation

Extracellular purine levels are increased when the supply/demand ratio of energy inside the cell is imbalanced (BRUNS 1990) as occurs following ischemia when the main substrates necessary for mitochondrial oxidative phosphorylation and energy production, e.g., oxygen and glucose, are cut off from the nervous tissue. Because of the shortage of these substrates, neurons cannot synthesize high-energy phosphates, and a massive reduction of tissue ATP content occurs (MILUSHEVA et al. 1996). The relative ADP and AMP levels are therefore increased and the energy charge used to define the energetic condition of the cell (defined as [ATP] + 0.5[ADP]/[ATP] + [ADP] + [AMP], (ATKINSON 1968)) falls (MILUSHEVA et al. 1996). Increased AMP levels in turn activate 5′-nucleotidase and give rise to increased formation of adenosine, which might leave the cell via the nucleoside transport system (WORKU and NEWBY 1983; MEGHJI and NEWBY 1990). Cytoplasmic adenosine levels therefore are coupled to small changes of ATP levels and energy charge and this mechanism provided the basis of the concept on the role of adenosine as a "retaliatory" (NEWBY et al. 1990) or "homeostatic" (WILLIAMS 1989; SCHRADER 1990; CUNHA et al. 2000) regulator.

Indeed, besides neuronal activity, the major stimulus which elevates intercellular adenosine levels is energy deprivation simulated by O_2 and glucose deprivation in in vitro experiments (LLOYD et al. 1993; JURANYI et al. 1999) and blood vessel occlusion in in vivo models (PHILLIS et al. 1994; HAGBERG et al.

1987). Under these conditions, cytoplasmic adenosine levels, which typically do not surpass micromolar levels, now increase up to 30 μmol/l (LATINI et al. 1999) and exert neuroprotective action on multiple target sites (SWEENEY 1997; PICANO and ABBRACCHIO 2000). The increased extracellular accumulation of adenosine during energy deprivation is well documented, and previous studies emphasized the intracellular formation of adenosine (HAGBERG et al. 1987; LLOYD et al. 1993). However, intracellular ATP stores are replenished within a relatively short period of time after the ischemic insult (DOOLETTE 1997), and the possibility that ATP release might occur during reperfusion also exists. ATP is released from heart cells in response to hypoxia (FORRESTER and WILLIAMS 1977; VIAL et al. 1987) and ischemia-induced release of purine nucleotides in the isolated perfused guinea pig Langendorff heart has been also reported (BORST and SCHRADER 1991). Combined hypoxia-hypoglycemia for 5 min resulted in a massive efflux of [^{3}H]purines from rat hippocampal slices with a net efflux of ATP (Fig. 6) (JURANYI et al. 1999). During ischemia, purines may be released in a receptor-mediated manner in response to the action of other mediators released either during or after ischemia. These include excitatory amino acids or nitric oxide. Interestingly, the P2 receptor antagonists, suramin and PPADS, reduced purine efflux induced by ischemia-like conditions in the hippocampal slice, indicating that purines are released in response to an initial purine trigger via P2 receptors in a self-amplifying manner (JURANYI et al. 1999). As the non-selective nitric oxide synthase inhibitor L-NAME inhibited ischemia evoked [^{3}H]purine release in the hippocampus while the selective neuronal NOS inhibitor, 7-nitroindazole was without effect (JURANYI et al. 1999), it seems reasonable to assume that the endothelial isoform of NOS was also responsible for this release, and endothelial cells appear to be a likely source of ATP release under these conditions. This assumption is supported by the observation that endothelial cells are able to release adenine nucleotides in response to ischemic-like conditions (BORST and SCHRADER 1991). ATP is released at a higher rate from platelets obtained from patients undergoing cerebral thrombosis, suggesting that if cerebral ischemia is caused by thrombosis, platelets may also contribute to higher purine levels inside the blood vessels (TOHGI et al. 1991).

II. Release of Purines by Inflammatory Stimuli

Purines are also involved in the activation of the immune system during inflammatory processes. P1 and P2 receptors are present on the surface of various immune cells, and they are expressed in an activity-dependent manner in response to their activation by bacterial lipopolysaccharide (LPS) or other antigenic stimuli (cf. DUBYAK and EL-MOATASSIM 1995; DUBYAK, Chap. 25, this volume). ATP can reach high local concentrations at the inflammation site by release from sympathetic nerve terminals together with NA or from immune cells as a part of immune activation. ATP is released from mast cells in response to antigenic stimulus (OSIPCHUK and CAHALAN 1992), from

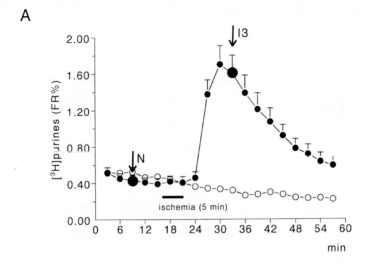

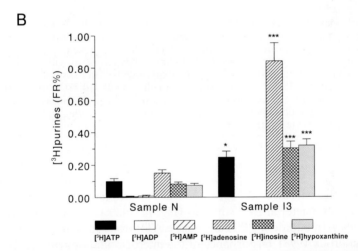

Fig. 6A,B. Effect of ischemic-like conditions on the outflow of [^{3}H]purines from superfused rat hippocampal slices: **A** *open circles* ($n = 8$) show spontaneous outflow of tritium without any treatment. Five-min treatment of the slices with combined hypoxia and hypoglycemia (ischemic-like conditions) increased the outflow of [^{3}H]purines significantly (*filled circles*, $n = 8$). [^{3}H] Purine outflow is expressed as a percentage of the actual tritium content (FR%); **B** the composition of [^{3}H] labeled purines during the normoxic period (sample N) and the peak of the ischemia induced response (Sample I3), as indicated by *arrows* on **A**. The amounts of [^{3}H]ATP, [^{3}H]adenosine, [^{3}H]inosine, and [^{3}H]hypoxanthine increased significantly in the samples collected during the peak response to the ischemic insult. [^{3}H]ADP and [^{3}H]AMP were hardly detectable. Samples were analyzed by HPLC ($n = 8$). *Asterisks* indicate significant differences between sample N and sample I3 (*$p < 0.05$, **$p < 0.01$, ***$p < 0.001$, calculated by one way ANOVA followed by Bonferroni test)

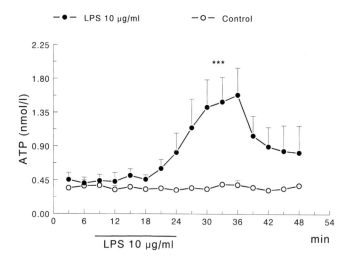

Fig. 7. Effect of LPS on ATP release measured from superfused RAW 264.7 cells. Cells were perfused with 0.2 ml/min and after a 60-min preperfusion 3-min samples were collected and assayed for ATP. ATP levels were measured by the luciferin luciferase assay and expressed in nmol/l, as a function of time. LPS was applied into the perfusion solution in 10 µg/ml for 15 min as indicated. *Open circles* represent the results in the absence of LPS, *filled circles* show values obtained in experiments with LPS treatment. Data show the mean ± SEM of 5–5 separate experiments. *Asterisks* indicate significant differences between LPS treated and control cells, calculated by two way ANOVA. [From Neurochemistry International, 33, SPERLÁGH B, HASKÓ G, NÉMETH ZH, VIZI ES (1998c) ATP released by lipopolysaccharide increases nitric oxide production in RAW 264.7 macrophages via $P2_Z/P2X_7$ receptors, pp 209–215, with permission from Elsevier Science]

lymphocytes (FILIPPINI et al. 1990), and from macrophages and microglial cells, in response to bacterial LPS stimulus (Fig. 7) (FERRARI et al. 1997; SPERLÁGH et al. 1998c). In addition, LPS can release ATP from endothelial cells (BODIN and BURNSTOCK 1998). Disturbance of the microcirculation may result in local ischemia and subsequent purine release as discussed in Sect. E.I. Cell necrosis at the inflammatory site may also cause ATP efflux into the extracellular space.

III. Release of Purines by Cellular Hypotonia

ATP release may also occur as the result of hypotonia. Cultured liver or ocular ciliary epithelial cells release ATP in response to hypotonic solution, a response blocked by chloride channel blockers (MITCHELL et al. 1998; WANG et al. 1996), indicating that an increase in cell volume leads to ATP efflux through the opening of a conductive pathway, which in turn stimulates P2 receptors and initiates chloride secretion. ATP may therefore function as an autocrine factor that couples increases in cell volume to opening of Cl⁻

channels (WEGNER, Chap. 23, this volume). Since hypotonic swelling occurs in pathophysiological states, e.g. ischemia-related neurodegeneration and cystic fibrosis, this mechanism may be relevant in these conditions.

IV. Purine Release by Cell Death (Apoptosis)

Since ATP is present in the cytoplasm of every living cell, disruption of the cell membrane results in ATP leakage to the extracellular space. ATP released following cytolysis is several orders of magnitude higher in concentration than that occurring as the result of either stimulus-evoked or receptor-mediated purine release. Additionally, it is irregular, variable, and independent of the functional activity of the tissue. Therefore it is unlikely that a significant cytolysis occurs under physiological conditions. Cellular integrity measurements, such as LDH assay, also indicate that the contribution of cytolytic release of ATP to the total release of ATP is minor upon neuronal activity (CUNHA et al. 1996; SPERLÁGH et al. 1996). On the other hand, cytolytic release of ATP may gain significance under any kind pathological condition which result in substantial cell damage. This might occur in neurodegenerative diseases, ischemia-related diseases, inflammation, disruption of tumors, etc. An indirect indication of permanent ATP accumulation in disease states was found in recent studies of BRAUN et al. (1998) who observed a strong upregulation of extracellular nucleotide metabolizing enzymes (ectoapyrase and 5′-nucleotidase) in the CA1 region, 24h after a transient ischemic insult exposed to rat hippocampus. Taking into account that ATP receptors are abundant in the nervous system (cf. RALEVIC and BURNSTOCK 1998), and cytoplasmic ATP concentration is in the millimolar range, ATP released by cell necrosis may reach local concentrations high enough to activate physiologically silent receptors and alter neurotransmission (FRANCESCHI et al. 1996).

F. Concluding Remarks

The ubiquitous cellular distribution of ATP provides a unique plasticity for purinergic signaling. Functional ATP release occurs from various type of neurons, e.g., catecholaminergic, cholinergic, glutamatergic neurons, and from glial cells, epithelial and endothelial, skeletal, cardiac, and smooth muscle, and immune cells. Purine release potentially occurs via quantal vesicular exocytosis, carrier mediated release, or by membrane leakage and has different characteristics under normal physiological conditions and in pathological states, such as hypoxia, hypoglycemia, and inflammation, providing an "inducible" response under disease conditions that can have both detrimental and beneficial effects depending on the disease state. A recognition of the dynamic nature of purine availability in the extracellular milieu and further studies to understand the role of adenosine and ATP and the dynamic nature of P1 and P2 receptor expression will be critical to focusing on the therapeutic potential of both selective receptor ligands and newer agents that modulate extracellular levels of the nucleoside and nucleotide. An additional challenge will be defining the roles of UTP and UDP as signaling molecules (CONNOLLY, Chap. 14, this volume) and the factors regulating their availability and reutilization.

List of Abbreviations

BRL44408	(±)2-((4,5-dihydro-1H-imidazol-2-yl)methyl)-2,3-dihydro-1-methyl-1H-isoindole
6-OHDA	6-hydroxydopamine
ACh	acetylcholine
ADP	adenosine diphosphate
ANOVA	analysis of variance
ATP	adenosine 5'-triphosphate
CH-38083	7,8-(methylenedioxi)-14? -alloberbanol
LPS	bacterial lipopolysaccharide
LTD	long-term depression
LTP	long term potentiation
NA	noradrenaline
PHAL	*Phaseolus vulgaris* leucoagglutinin
SFi	*nucleus septofimbrialis*
TS	*nucleus triangularis septi*
vVNAB	ventral part of the ventral noradrenergic bundle

References

Aberer W, Kostron H, Huber E, Winkler H (1978) A characterization of the nucleotide uptake by chromaffin granules of bovine adrenal medulla. Biochem J 172:353–360
Atkinson DE (1968) The energy charge of the adenylate pool as a regulatory parameter. Interaction with feedback modifiers. Biochemistry 7:4030–4034
Ballerini P, Rathbone MP, Di Iorio P, Renzetti A, Giuliani O, D'Alimonte I, Trubiani O, Caciagli F, Ciccarelli R (1996) Rat astroglial P2z (P2x7) receptors regulate intracellular calcium and purine release. Neuroreport 7:2533–2537
Barberis C, Daudet F, Charriere B, Guibert B, Leviel V (1983) Release of adenosine in vivo from cat caudate nucleus. Neurosci Lett 41:179–182
Bender AS, Wu PH, Phillis JW (1981) The rapid uptake and release of [^{3}H]adenosine by rat cerebral cortical synaptosomes. J Neurochem 36:651–660
Borst MM, Schrader J (1991) Adenine nucleotides release from isolated perfused guinea-pig hearts and extracellular formation of adenosine. Circ Res 68:797–806
Bodin P, Burnstock G (1998) Increased release of ATP from endothelial cells during acute inflammation. Inflamm Res 47:351–354
Braun N, Zhu Y, Krieglstein J, Culmsee C, Zimmermann H (1998) Upregulation of the enzyme chain hydrolyzing extracellular ATP after transient forebrain ischemia in the rat. J Neurosci 18:4891–4900
Brock JA, Bridgewater M, Cunnane TC (1997) Beta-adrenoceptor mediated facilitation of NA and adenosine 5'-triphosphate release from sympathetic nerves supplying the rat tail artery. Br J Pharmacol 120:769–776
Bruns RF (1990) Adenosine receptors. Roles and pharmacology. Ann N Y Acad Sci 603:211–225

Buller KM, Khanna S, Sibbald JR, Day TA (1996) Central noradrenergic neurons signal via ATP to elicit vasopressin responses to haemorrhage. Neuroscience 73:637–642

Burnstock G (1976) Do some cells release more than one transmitter? Neuroscience 1:239–248

Burnstock G (1990) Noradrenaline and ATP as cotransmitters in sympathetic nerves. Neurochem Int 17:357–368

Cahill CM, White TD Sawynok J (1993) Involvement of calcium channels in depolarization-evoked release of adenosine from spinal cord synaptosomes J Neurochem 60:886–893

Cass CE, Young, JD, Baldwin SA (1998) Recent advances in the molecular biology of nucleoside transporters of mammalian cells. Biochem Cell Biol 76:761–770

Chaudry IH, Clemens MG Baue AE (1985) Uptake of ATP by tissues. In: Stone TW (ed) Purines: pharmacology and physiological roles. Macmillan, London pp 115–124

Chen Y, Graham DI, Stone TW (1992) Release of endogenous adenosine and its metabolites by the activation of NMDA receptors in the rat hippocampus in vivo. Br J Pharmacol 106:632–638

Craig CG, White TD (1993) N-Methyl-D-aspartate- and non-N-methyl-D-aspartate-evoked adenosine release from rat cortical slices: distinct purinergic sources and mechanisms of release. J Neurochem 60:1073–1080

Cunha RA, Vizi ES, Ribeiro JA, Sebastiao AM (1996) Preferential release of ATP and its extracellular catabolism as a source of adenosine upon high- but not low-frequency stimulation of rat hippocampal slices. J Neurochem 67:2180–2187

Cunha RA (2000) Adenosine as a neuromodulator and as a homeostatic regulator in the nervous system: different roles, different sources and different receptors Neurochem Int (2000) in press

Daval J, Barberis C (1981) Release of radiolabelled adenosine derivatives from super-fused synaptosome beds. Biochem Pharmacol 30:2559–2567

Day TA, Sibbald JR, Khanna S (1993) ATP mediates an excitatory noradrenergic neuron input to supraoptic vasopressin cells. Brain Res 607:341–344

Doolette DJ (1997) Mechanism of adenosine accumulation in the hippocampal slice during energy deprivation. Neurochem Int 30:211–223

Dowdall MJ, Bony AF Whittier VP (1974) Adenosine triphosphate – a constituent of cholinergic synaptic vesicles. Biochem J 140:1–12

Driessen B, von Kugelgen I, Starke K (1994a) Neural ATP release and its α_2-adrenoceptor-mediated modulation in guinea-pig vas deferens. Naunyn-Schmiedebergs Arch Pharmacol 348:358–366

Driessen B, von Kugelgen I, Starke K (1994b) P1-purinoceptor-mediated modulation of neural noradrenaline and ATP release in guinea-pig vas deferens. Naunyn-Schmiedebergs Arch Pharmacol 350:42–48

Driessen B, Starke K (1994) Modulation of neural noradrenaline and ATP release by angiotensin II and prostaglandin E2 in guinea-pig vas deferens. Naunyn-Schmiedebergs Arch Pharmacol 350:618–625

Driessen B, Bultmann R, Goncalves J, Starke K (1996) Opposite modulation of noradrenaline and ATP release in guinea-pig vas deferens through prejunctional beta-adrenoceptors: evidence for the beta 2 subtype. Naunyn Schmiedebergs Arch Pharmacol 353:564–571

Driessen B, Goncalves J, Szabo B (1996) Failure of tyramine to release neuronal ATP as a cotransmitter of noradrenaline in the guinea-pig vas deferens. Naunyn Schmiedebergs Arch Pharmacol 353:175–183

Dubyak GR, El-Moatassim C (1995) Signal transduction via P2-purinergic receptors for extracellular ATP and other nucleotides. Am J Physiol 265: C577–C606

Edwards FA, Gibb AJ, Colquhoun D (1992) ATP receptor-mediated synaptic currents in the rat central nervous system. Nature 359:144–147

Ellis JL, Burnstock G (1989) Angiotensin neuromodulation of adrenergic and purinergic cotransmission in the guinea-pig vas deferens. Br J Pharmacol 97:1157–1164

Ellis JL, Burnstock G (1990) Neuropeptide Y neuromodulation of the co-transmitters ATP and noradrenaline in the guinea-pig vas deferens. Br J Pharmacol 100: 457–462

Fallahi N, Broad RM, Jin S, Fredholm BB (1996) Release of adenosine from rat hippocampal slices by nitric oxide donors. J Neurochem 67:186–193

Farinas I, Solsona C, Marsal J (1992) Omega-conotoxin differentially blocks acetylcholine and adenosine triphosphate releases from torpedo synaptosomes. Neurosci 47:641–648

Ferrari D, Chiozzi P, Falzoni S, et al. (1997b) Extracellular ATP triggers IL-1β release by activating the purinergic P2z receptor of human macrophages. J Immunol 159:1451–1458

Filippini A, Taffs RF, Sitkovsky MV (1990) Extracellular ATP in T-lymphocyte activation: possible role in effector functions. Proc Natl Acad Sci USA 87:8267–8271

Fischer H, Prast H, Philippu A (1995) Adenosine release in the ventral striatum of the rat is modulated by endogenous nitric oxide. Eur J Pharmacol 275:R5–R6

Fischer MN, Newsholme EA (1984) Properties of rat heart adenosine kinase. Biochem J 221:521–528

Forrester T, Williams CA (1977) Release of adenosine triphosphate from isolated adult heart cells in response to hypoxia. J Physiol (Lond) 268:371–390

Franceschi C, Abbracchio MP, Barbier D, Ceruti S, Ferrari D, Iliou JP, Rounds S, Schubert P, Schulze-Lohoff E, Rassendren FA, Staub M, Volonte C, Wakade AR, Burnstock G (1996) Purines and cell death, Drug Dev Res 39:442–449

Fredholm BB, Vernet L (1979) Release of [^{3}H]nucleotides from [^{3}H]adenine labelled hypothalamic synaptosomes. Acta Physiol Scand 106:97–107

Fredholm, BB, Sollevi A, Vernet L, Hedquist P (1980) Inhibition by dipyridamole of stimulated purine release. Naunyn Schmiedebergs Arch Pharmacol 313, R18

Fredholm BB, Dunwiddie TV, Bergman B, Lindström K (1984) Levels of adenosine and adenine nucleotides in slices of rat hippocampus, Brain Res 295:127–136

Gaillet S, Alonso G, Le Borgne R, Barbanel G, Malaval F, Assenmacher I, Szafarczyk A (1993) Effects of discrete lesions in the ventral noradrenergic ascending bundle on the corticotropic stress response depend on the site of the lesion and on the plasma levels of adrenal steroids. Neuroendocrinol 58:408–419

Gartside SE, Suaud-Chagny MF, Tappaz M (1995) Evidence that activation of the hypothalamo-pituitary-adrenal axis by electrical stimulation of the noradrenergic A1 group is not mediated by noradrenaline. Neuroendocrinology 62:2–12

Geiger JD, Nagy JI (1984) Heterogeneous distribution of adenosine transport sites labelled by ^{3}H-nitrobenzylthioinosine in rat brain: an autoradiographic and membrane binding study. Brain Res Bull 13:657–666

Geiger JD, Johnston ME, Yago V (1988) Pharmacological characterization of rapidly accumulated adenosine by dissociated brain cells from adult rat. J Neurochem 51:283–291

Goncalves J, Bultmann R, Driessen B (1996) Opposite modulation of cotransmitter release in guinea-pig vas deferens: increase of noradrenaline and decrease of ATP release by activation of prejunctional beta-adrenoceptors. Naunyn Schmiedebergs Arch Pharmacol 353:184–192

Gu JG, Foga IO, Parkinson FE, Geiger JD (1995) Involvement of bidirectional adenosine transporters in the release of L-[^{3}H]adenosine from rat brain synaptosomal preparations. J Neurochem 64:2105–2110

Hagberg H, Andersson P, Lacarewicz J, Jacobson I, Butcher S, Sandberg M, (1987) Extracellular adenosine inosine hypoxanthine and xanthine in relation to tissue nucleotides and purines in rat striatum during transient ischemia J Neurochem 49:227–231

Hamann M, Attwell D (1996) Non-synaptic release of ATP by electrical stimulation in slices of rat hippocampus, cerebellum and habenula. Eur J Neurosci 8:1510–1515

Hoehn K, White TD (1990a) N-Methyl-D-aspartate, kainate and quisqualate release endogenous adenosine from rat cortical slices. Neuroscience 39:441–450

Hoehn K, White TD (1990b) Role of excitatory amino acid receptors in K+ and glutamate-evoked release of endogenous adenosine from rat cortical slices. J Neurochem 54:256–226

Holton P (1959) The liberation of adenosine triphosphate on antidromic stimulation of sensory nerves. J Physiol (Lond) 145:494–504

Israel M, Meunier FM (1978) The release of ATP triggered by transmitter action and its possible physiological significance: retrograde transmission. J Physiol (Paris) 74:485–490

Jhamandas K, Dumbrille A (1980) Regional release of [^{3}H]adenosine derivatives from rat brain in vivo: effect of excitatory amino acids, opiate agonists, and benzodiazepines. Can J Physiol Pharmacol 58:1262–1278

Jonzon B, Fredholm BB (1985) Release of purines, noradrenaline and GABA from rat hippocampal slices by field stimulation J Neurochem 44:217–224

Juranyi Z, Sperlagh B, Vizi ES (1999) Involvement of P2 purinoceptors and the nitric oxide pathway in [^{3}H]purine outflow evoked by short-term hypoxia and hypoglycemia in rat hippocampal slices. Brain Res 823:183–190

Juranyi Z, Orso E, Janossy A, Szalay KS, Sperlagh B, Windisch K, Vinson GP, Vizi ES (1997) ATP and [^{3}H]noradrenaline release and the presence of ecto-Ca($^{2+}$)-ATPases in the capsule-glomerulosa fraction of the rat adrenal gland. J Endocrinol 153:105–114

Kasakov L, Ellis J, Kirkpatrick K, Milner P, Burnstock G (1988) Direct evidence for concomitant release of noradrenaline, adenosine 5'-triphosphate and neuropeptide Y from sympathetic nerve supplying the guinea-pig vas deferens. J Auton Nerv Syst 22:75–82

Katsuragi T, Tokunaga T, Ogawa S, Soejima O, Sato C, Furukawa T (1991) Existence of ATP-evoked ATP release system in smooth muscle. J Pharmacol Exp Ther 259:513–518

Katsuragi T, Matsuo K, Sato C, Honda K, Kamiya H, Furukawa T (1996) Non-neuronal release of ATP and inositol 1,4,5-trisphosphate accumulation evoked by P2- and M-receptor stimulation in guinea pig ileal segments. J Pharmacol Exp Ther 277:747–752

Kirkpatrick K, Burnstock G (1987) Sympathetic nerve-mediated release of ATP from the guinea-pig vas deferens is unaffected by reserpine. Eur J Pharmacol 138:207–214

Kirkpatrick KA, Burnstock G (1994) Release of endogenous ATP from the vasa deferentia of the rat and guinea pig by the indirect sympathomimetic tyramine. J Auton Pharmacol 14:325–335

Klein RL (1982) Chemical composition of the large noradrenergic vesicles. In: Klein RL, Lagercrantz H Zimmermann H (eds) Neurotransmitter vesicles. Academic Press, London pp 174–188

Lagercrantz H, Stjarne L (1974) Evidence that most noradrenaline is stored without ATP in sympathetic large dense cored vesicles. Nature 249:843–845

Latini S, Pedata F, Pepeu G (1997) The contribution of different types of calcium channels to electrically-evoked adenosine release from rat hippocampal slices. Naunyn Schmiedebergs Arch Pharmacol 355:250–255

Latini S, Bordoni F, Pedata F, Corradetti (1999) Extracellular adenosine concentrations during in vitro ischaemia in rat hippocampal slices. Br J Pharmacol 127:729–739

Lee CW, Jarvis SM (1988) Kinetic and inhibitor specificity of adenosine transport in guinea-pig cerebral cortical synaptosomes: evidence for two nucleoside transporters. Neurochem. Int. 12:483–492

Lee K, Schubert P, Gribkoff V, Sherman B, Lynch G (1982) A combined in vivo/in vitro study of the presynaptic release of adenosine derivatives in the hippocampus. J Neurochem 38:80–83

Lew MJ, White TD (1987) Release of endogenous ATP during sympathetic nerve stimulation. Br J Pharmacol 92:349–355

Lloyd HG, Deussen A, Wuppermann H, Schrader J (1988) The transmethylathion pathway as a source of adenosine in the isolated guinea-pig hearts, Biochem J 252:489–494

Lloyd HG, Lindstrom K, Fredholm BB (1993) Intracellular formation and release of adenosine from rat hippocampal slices evoked by electrical stimulation or energy depletion, Neurochem Int 23: 173–185

Luqmani YA (1981) Nucleotide uptake by isolated cholinergic synaptic vesicles: evidence for a carrier of adenosine 5'-triphosphate. Neurosci 6: 1011–1021

MacDonald WF, White TD (1985) Nature of extasynaptosomal accumulation of endogenous adenosine evoked by K+ and veratridine. J Neurochem 45:791–797

Manzoni OJ, Manabe T, Nicoll RA (1994) Release of adenosine by activation of NMDA receptors in the hippocampus. Science 265:2098–2101

Marsal J, Solsona C, Rabasseda X, Blasi J Casanova A (1987) Depolarization-induced release of ATP from cholinergic synaptosomes is not blocked by botulinum toxin type A. Neurochem Int 10:295–302

Matsuo K, Katsuragi T, Fujiki S, Sato C, Furukawa T (1997) ATP release and contraction mediated by different P2-receptor subtypes in guinea-pig ileal smooth muscle. Br J Pharmacol 121:1744–1748

McCaman MW, McAfee DA (1986) Effects of synaptic activity on the metabolism and release of purines in the rat superior cervical ganglion Cell Molec Neurobiol 6:349–362

McMahon HT, Nicholls DG (1991) The bioenergetics of neurotransmitter release. Biochim. Biophys. Acta 1059:243–264

Meghji P, Newby AC (1990) Sites of adenosine formation, action and inactivation in the brain. Neurochem Int 16:227–232

Meghji P, Tuttle JB, Rubio R (1989) Adenosine formation and release by embryonic chick neurons and glia in cell culture. J Neurochem 53:1852–1860

Milusheva EA, Doda M, Baranyi M, Vizi ES (1996) Effect of hypoxia and glucose deprivation on ATP level adenylate energy charge and $[Ca^{2+}]_o$-dependent and independent release of [^{3}H]dopamine in rat striatal slices. Neurochem Int 28: 501–507

Mitchell CH, Carre DA, McGlinn AM, Stone RA, Civan MM (1998) A release mechanism for stored ATP in ocular ciliary epithelial cells. Proc Natl Acad Sci USA 95:7174–7178

Morel N, Meunier FM (1981) Simultaneous release of acetylcholine and ATP from stimulated cholinergic synaptosomes. J Neurochem 36:1766–1773

Newby AC, Worku Y, Meghji P, Nakazawa M, Skladanowski AC (1990) Adenosine: a retaliatory metabolite or not? News in Phys Sci 5:67–70

Newman M, McIlwain H (1977) Adenosine as a constituent of the brain and of isolated cerebral tissues and its relationship to the generation of cyclic AMP Biochem J 164:131–137

Nitahara K, Kittel A, Liang SD Vizi ES (1995) A_1-receptor mediated effect of adenosine on the release of acetylcholine from the myenteric plexus: role and localization of ectoATPase and 5'-nucleotidase. Neurosci 67:159–168

Osipchuk Y, Cahalan M (1992) Cell-to-cell spread of calcium signals mediated by ATP receptors in mast cells. Nature 359: 241–244

Palkovits M Catecholamines in the hypothalamus: an anatomical review. (1981) Neuroendocrinol 33:123–128

Pankratov Y, Castro E, Miras-Portugal MT, Krishtal O (1998) A purinergic component of the excitatory postsynaptic current mediated by P2X receptors in the CA1 neurons of the rat hippocampus. Eur J Neurosci 10:3898–3902

Pedata F, Pazzagli M, Tilli S, Pepu G (1990) Regional differences in the electrically stimulated release of endogenous and radioactive adenosine and purine derivatives from rat brain slices. Naunyn Schmiedebergs Arch Pharmacol 342:447–453

Pedata F, Pazzagli M, Pepeu G (1991) Endogenous adenosine release from hippocampal slices excitatory amino acid agonists stimulate release, antagonists reduce

the electrically-evoked release. Naunyn-Schmiedebergs Arch Pharmacol 344:538–543
Picano E, Abbracchio MP (2000) Adenosine, the imperfect endogenous anti-ischemic cardio-neuroprotector. Brain Res Bull 52:75–82
Phillis JW, O'Regan MH, Walter GA (1989) Effects of two nucleoside transport inhibitors, dypiridamole and soluflazine, on purine release from the rat cerebral cortex
Phillis JW, Smith-Barbour M, O'Regan MH, Perkins LM (1994) Amino acid and purine release in rat brain following temporary middle cerebral artery occlusion. Neurochem Res 19:1125–1130
Potter P, White TD (1980) Release of adenosine 5'-triphosphate from synaptosomes from different regions of rat brain. Neurosci 5:1351–1356
Potter PE, White TD (1982) Lack of effect of 6-hydroxydopamine pretreatment on depolarization-induced release of ATP from rat brain synaptosomes. Eur J Pharmacol 80:143–147
Pull I, McIlwain H (1972) Adenine derivatives as neurohumoral agents in the brain. The quantities liberated on excitation of superfused cerebral tissues. Biochem J 130: 975–981
Richardson PJ, Brown SJ, Bailyes EM, Luzio JP (1987) Ectoenzymes control adenosine modulation of immunoisolated cholinergic synapses Nature 327:232–234
Queiroz G, Gebicke-Haerter PJ, Schobert A, Starke K, von Kugelgen I Release of ATP from cultured rat astrocytes elicited by glutamate receptor activation. Neuroscience 1997 78:1203–1208
Rabasseda X, Solsona C, Marsal J, Egea G Bizzini B (1987) ATP release from pure cholinergic synaptosomes is not blocked by tetanus toxin. FEBS Lett 213:337–340
Ralevic V, Burnstock G (1998) Receptors for purines and pyrimidines. Pharmac Rev 50:413–492
Richardson PJ, Brown SJ (1987) ATP release from affinity-purified rat cholinergic nerve terminals. J Neurochem 48:622–630
Robertson SJ, Edwards FA (1998) ATP and glutamate are released from separate neurones in the rat medial habenula nucleus: frequency dependence and adenosine-mediated inhibition of release. J Physiol (Lond) 508:691–701
Schousboe A, Frandsen A, Drejer J (1989) Evidence for evoked release of adenosine and glutamate from cultured cerebellar granule cells Neurochem Res 14:871–875
Schrader J (1990) Adenosine. A homeostatic metabolite in cardiac energy metabolism. Circulation 81: 389–391
Schweitzer E (1987) Coordinated release of ATP and ACh from cholinergic synaptosomes and its inhibition by calmodulin antagonists. J Neurosci 7:2948–2956
Schubert P, Lee K, West M, Deadwyler S, Lynch G (1976) Stimulation-dependent release of ^{3}H adenosine derivatives from central axon terminals to target neurones. Nature 260:541–542
Sedaa KO, Bjur RA, Shinozuka K, Westfall DP (1989) Nerve and drug-induced release of adenine nucleosides and nucleotides from rabbit aorta. J Pharmacol Exp Ther 252:1060–1067
Shinozuka K, Sedaa KO, Bjur RA, Westfall DP (1991) Participation by purines in the modulation of norepinephrine release by methoxamine. Eur J Pharmacol 192:431–434
Silinsky EM, Hubbard JI (1973) Release of ATP from motor nerve terminals. Nature 243:404–405
Smith AD (1979) Biochemical studies on the mechanism of release. In: Paton DM (ed) The release of catecholamines from adrenergic neurons. Pergamon Press, Oxford, pp 1–15
Sperlágh B, Tóth E, Lajtha A, Vizi ES (1992) Effect of (−)-nicotine on the endogenous ATP release from rat striatum and cerebral cortex. Int J Purine Pyrimidine Res 3:105

Sperlágh B, Vizi ES (1992) Is the neuronal ATP release from guinea-pig vas deferens subject to α_2-adrenoceptor-mediated modulation? Neurosci 51:203–209

Sperlágh B, Kittel A, Lajtha A, Vizi ES (1995) ATP acts as fast neurotransmitter in rat habenula: neurochemical and enzyme cytochemical evidence. Neurosci 66:915–920

Sperlágh B, Vizi ES (1996) Neuronal synthesis, storage and release of ATP. Semin Neuroscis 8:175–186

Sperlágh B, Maglóczky Z, Vizi ES, Freund T (1998a) The triangular septal nucleus as the major source of ATP release in the rat habenula: a combined neurochemical and morphological study. Neurosci 86:1195–1207

Sperlágh B, Sershen H, Lajtha A, Vizi ES (1998b) Co-release of endogenous ATP and [^{3}H]noradrenaline from rat hypothalamic slices: origin and modulation by α_2-adrenoceptors. Neurosci 82:511–520

Sperlágh B, Haskó G, Németh ZH, Vizi ES (1998c) ATP released by lipopolysaccharide increases nitric oxide production in RAW 264.7 macrophages via P2Z/P2X7 receptors. Neurochem Int 33:209–215

Sperlágh B, Mergl Z, Jurányi Z, Vizi ES, Makara GB (1999) Local regulation of vasopressin and oxytocin secretion by extracellular ATP in the isolated posterior lobe of the rat hypophysis. J Endocrinol 160:343–350

Stjaerne L (1989) Basic mechanisms and local modulation of nerve impulse-induced secretion of neurotransmitters from individual sympathetic nerve varicosities. Rev Physiol Biochem Pharmacol 112:4–122

Sun AY, Lee DZ (1985) Synaptosomal ADP uptake. J Neurochem 44:S90

Sweeney MI (1997) Neuroprotective effects of adenosine in cerebral ischemia: window of opportunity. Neurosci Biobehav Rev 21:207–217

Sweeney MI, White TD, Sawynok J (1993) Morphine-evoked release of adenosine from the spinal cord occurs via nucleoside carrier with differential sensitivity to dipyridamole and nitrobenzylthioadenosine. Brain Res 614:301–307

Takahashi T, Kusunoki M, Ishikawa Y, Kantoh M, Yamamura T, Utsunomiya J (1987) Adenosine 5'-triphosphate release evoked by electrical nerve stimulation from the guinea-pig gallbladder. Eur J Pharmacol 134:77–82

Tamesue S, Sato C, Katsuragi T (1998) ATP release caused by bradykinin, substance P and histamine from intact and cultured smooth muscles of guinea-pig vas deferens. Naunyn Schmiedebergs Arch Pharmacol 357:240–244

Terrian DM, Hernandez PG, Rea MA, Peters RI (1989) ATP release, adenosine formation and modulation of dynorphin and glutamic acid release by adenosine analogues in rat hippocampal mossy fiber synaptosomes. J Neurochem 53:1390–1399

Troadec JD, Thirion S, Nicaise G, Lemos JR, Dayanithi G 1998 ATP-evoked increases in [Ca^{2+}]i and peptide release from rat isolated neurohypophysial terminals via a $P2X_2$ purinoceptor. J Physiol (Lond) Aug 511:89–103

Thorn JA, Jarvis SM (1996) Adenosine transporters. Gen Pharmacol 27:613–620

Todorov L, Mihaylova-Todorova S, Craviso GL, Bjur RA, Westfall TD (1996) Evidence for the differential release of the cotransmitters ATP and noradrenaline from sympathetic nerves of the guinea-pig vas deferens. J Physiol (Lond.) 496:731–748

Tohgi H, Suzuki H, Tamura K, Kimura B (1991) Platelet volume, aggregation and adenosine triphosphate release in cerebral thrombosis. Stroke 22:17–21

Tokunaga T, Katsuragi T, Sato C, Furukawa T (1993) ATP release evoked by isoprenaline from adrenergic nerves of guinea-pig atrium. Neurosci Lett 186:95–98

Unsworth CD, Johnson RG (1990) Acetylcholine and ATP coreleased from the electromotor nerve terminals of Narcine brasiliensis by an exocytotic mechanism. Proc Natl Acad Sci USA 87:553–557

Van Belle H, Goosens F, Wynants J (19879 Formation and release of purine catabolites during hypoperfusion, anoxia and ischemia. Am J Physiol 252:H886–893

Vial C, Owen P, Opie LH, Posel D (1987) Significance of release of adenosine triphosphate and adenosine induced by hypoxia or adrenaline in perfused rat heart. J Mol Cell Cardiol 19:187–197

Vizi ES, Hársing LG Jr, Gaal J, Kapocsi J, Bernath S, Somogyi GT (1986) CH-38083, a selective, potent antagonist of alpha-2 adrenoceptors. J Pharmacol Exp Ther 238:701–706
Vizi ES, Burnstock G (1988) Origin of ATP release in the rat vas deferens: concomitant measurement of ^{3}H-noradrenaline and ^{14}C-ATP. Eur J Pharmacol 158:69–77
Vizi ES, Sperlágh B, Baranyi M (1992) Evidence that ATP, released from the postsynaptic site by noradrenaline, is involved in mechanical responses of guinea-pig vas deferens: cascade transmission. Neurosci 50:455–465
Vizi ES, Liang SD, Sperlágh B, Kittel A, Jurányi Z (1997) Studies on the release and extracellular metabolism of endogenous ATP in rat superior cervical ganglion: support for neurotransmitter role of ATP. Neurosci 79:893–903
Vizi ES, Sperlágh B (1999) Receptor- and carrier-mediated release of ATP of postsynaptic origin: cascade transmission. Progr Brain Res 20:159–169
Vizi ES, Nitahara K, Sato K, Sperlágh B (2000) Stimulation-dependent release, breakdown, and action of endogenous ATP in mouse hemidiaphragm preparation: the possible role of ATP in neuromuscular transmission. J Auton Nerv Syst 81:278–284
Vizi ES (1998) Different temperature dependence of carrier-mediated (cytoplasmic) and stimulus-evoked (exocytotic) release of transmitter: a simple method to separate the two types of release. Neurochem Int 33:359–366
Volknandt W, Zimmermann H (1986) Acetylcholine, ATP, and proteoglycan are common to synaptic vesicles isolated from the electric organs of electric eel and electric catfish as well as from rat diaphragm. J Neurochem 47:1449–1461
von Kügelgen I, Starke K (1991) Release of noradrenaline and ATP by electrical stimulation and nicotine in guinea-pig vas deferens. Naunyn-Schmiedebergs Arch Pharmacol 344:419–429
von Kügelgen I, Allgaier C, Schobert C Starke K (1994) Co-release of noradrenaline and ATP from cultured sympathetic neurons. Neurosci 61:199–202
Wang Y, Roman R, Lidofsky SD, Fitz JG (1996) Autocrine signaling through ATP release represents a novel mechanism for cell volume regulation. Proc Natl Acad Sci USA, 93:12020–12025
Westfall DP, Stitzel RE, Rowe JN (1978) The postjunctional effect and the neural release of purine compounds in the guinea-pig vas deferens. Eur J Pharmacol 50:27–38
Westfall DP, Sedaa K, Bjur RA (1987) Release of endogenous ATP from rat caudal artery. Blood Vessels 4:125–127
Wieraszko A, Goldsmith G, Seyfried TN (1989) Stimulation-dependent release of adenosine triphosphate from hippocampal slices. Brain Res 485:244–250
Worku Y, Newby AC (1983) The mechanism of adenosine production in rat polymorphonuclear leucocytes. Biochem J 214:325–330
White TD (1977) Direct detection of depolarization-induced release of ATP from synaptosomal preparation. Nature 267:67–79
White TD (1978) Release of ATP from a synaptosomal preparation by elevated extracellular K^+ and by veratridine. J Neurochem 30:329–336
White TD (1982) Release of ATP from isolated myenteric varicosities by nicotinic agonists. Eur J Pharmacol 79:333–334
White TD, Leslie RA (1982) Depolarization-induced release of adenosine 5'-triphosphate from isolated varicosities derived from the myenteric plexus of the guinea-pig small intestine. J Neurosci 2:206–215
White TD, Al-Hummayd M (1983) Acetylcholine releases ATP from varicosities isolated from guinea-pig myenteric plexus. J Neurochem 40:1069–1075
Williams M (1989) Adenosine: the prototypic neuromodulator. Neurochem Int 14:249–264
Winkler H, Westhead E (1980) The molecular organization of adrenal chromaffin granules. Neurosci 5:1803–1823
Wojcik WJ, Neff NH (1983) Location of adenosine release and adenosine A2 receptors to rat striatal neurons. Life Sci 33:755–763

Wu PH, Phillis JW (1978) Distribution and release of adenosine triphosphate in rat brain. Neurochem Res 3:563–571

Wu PH, Phillis JW (1984) Uptake by central nervous tissues as a mechanism for the regulation of extracellular adenosine concentrations. Neurochem Int 6:613–632

Yang SY, Cheek DJ, Westfall DP, Buxton ILO (1994) Purinergic axis in cardiac blood vessels: agonist-mediated release of ATP from cardiac endothelial cells. Circ Res 74:401–407

Zaidi ZF, Matthews, MR (1997) Exocytotic release from neuronal cell bodies, dendrites and nerve terminals in sympathetic ganglia of the rat, and its differential regulation. Neurosci 80:861–891

Zhang J, Kornecki E, Jackman J, Ehrlich YH (1988) ATP secretion and extracellular protein phosphorylation by CNS neurons in primary culture. Brain Res Bull 21:459–464

Zimmermann H (1978) Turnover of adenine nucleotides in cholinergic synaptic vesicles of the Torpedo electric organ. Neurosci 3:827–836

CHAPTER 8
Ecto-Nucleotidases

H. ZIMMERMANN

A. Introduction

Ecto-nucleotidases are cell surface-located enzymes which catalyze extracellular nucleotide hydrolysis. An extracellular hydrolysis pathway for nucleotides has been detected in essentially all tissues and also in a large variety of cell culture systems. Its general features include the following:

1. Nucleoside 5'-triphosphates are sequentially metabolized to the nucleoside with nucleoside 5'-diphosphate and nucleoside 5'-monophosphate appearing as intermediate products. The nucleoside may then be further deaminated to inosine by adenosine deaminase (FRANCO et al. 1997).
2. Not only ATP, ADP, and AMP but essentially all physiologically occurring purine and pyrimidine nucleotides are hydrolyzed.
3. Extracellular hydrolysis of nucleotides is not inhibited by known inhibitors of intracellular ATPases such as P-type, F-type, and V-type ATPases.
4. Nucleotide hydrolysis depends on divalent cations, generally millimolar concentrations of either Ca^{2+} or Mg^{2+}.
5. Nucleotide hydrolysis has an alkaline pH optimum.
6. A major function of the extracellular enzyme chain appears to be the termination of the physiological action of nucleotides released from cells.

To date no information is available as to whether the hydrolysis of ATP is used to drive energy-dependent processes (for reviews of the earlier work see ARCH and NEWSHOLME 1978; FOX 1978; PEARSON 1985; DHALLA and ZHAO 1988; ZIGANSHIN et al. 1994a; PLESNER 1995; SARKIS et al. 1995; BEAUDOIN et al. 1996; ZIMMERMANN 1996a,b; PLESNER et al. 1997; ZIMMERMANN and PEARSON 1998).

It was originally assumed that single and defined enzymes exist for the hydrolysis of either ATP (ecto-ATPase), ADP (ecto-ADPase), or, also, ATP and ADP (ecto-ATP diphosphohydrolase, ecto-apyrase). This type of nomenclature is therefore still prevailing in the current literature. More recent work has considerably complicated the issue and shown that ecto-nucleotidases comprise a heterogeneous group of enzymes with differing but partially overlapping catalytic activities. The currently known ecto-nucleotidases include members of the E-NTPDase family, E-NPP family, ecto-alkaline phosphatases,

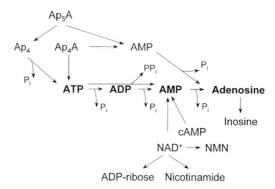

Fig. 1. Multiplicity of extracellular pathways in the metabolism of nucleotides. Note that not only adenine nucleotides but essential all physiologically occurring purine and pyrimidine nucleotides are extracellularly metabolized (modified from ZIMMERMANN 1996b)

and ecto-5′-nucleotidase which all have a broad tissue distribution. Ectoenzymes hydrolyzing diadenosine polyphosphates, NAD^+ as well as nucleotide converting enzymes such as ecto-nucleoside diphosphokinase or myokinase have also been described (ZIMMERMANN 1992, 1999a; PINTOR et al. 1997; GODING et al. 1998; RESTA et al. 1998; ZIMMERMANN and BRAUN 1999) (Fig. 1). In addition, ATPase activity has been found in association with the neural cell adhesion molecule (NCAM) (DZHANDZUGAZYAN and BOCK 1993, 1997) and with the muscle-associated alpha-sarcoglycan (BETTO et al. 1999). Early work identifying the rat hepatocyte cell adhesion molecule C-CAM (C-CAM 105) as being identical to the rat liver ecto-ATPase (LIN and GUIDOTTI 1989; AURIVILLIUS et al. 1990) has not been confirmed (SAWA et al. 1994). The fact that members of several enzyme families can hydrolyze nucleoside 5′-triphosphates such as ATP makes it difficult to interpret many of the previous data on the functional characterization of "ecto-ATPases" and also attempts to purify and characterize these enzymes. While these data are of considerable value in describing the general features of nucleotide hydrolysis pathways in individual tissues or cell types they reveal limited information about the biochemical identity of the enzymes involved. This also refers to the cellular localization by enzyme histochemical procedures (KITTEL 1997). Enzyme-specific antibodies as well as nucleic acid probes can now be applied to evaluate the tissue and cellular distribution of identified enzymes. In the following the individual groups of enzymes, their molecular and biochemical characteristics, and tissue distribution will briefly be discussed. In addition, examples will be given for defined functional roles of ecto-nucleotidases and the potential of enzyme inhibitors to elucidate the physiological relevance of extracellular nucleotide hydrolysis.

B. The E-NTPDase Family

I. From Yeast to Vertebrates

In the current literature this protein family is also referred to as ecto-apyrase, NTPase, E-NTPDase, or E-ATPase family. The unifying nomenclature recently suggested for the enzyme family and its members (ZIMMERMANN et al. 2000) will be applied here. Other names are given in parentheses. Members of this protein family can hydrolyze nucleoside 5'-triphosphates and nucleoside 5'-diphosphates. From a functional point of view they represent nucleoside 5'-triphosphate diphosphohydrolases (NTPDases) albeit with varying preference for nucleoside 5'-tri- or diphosphates. Once the first primary structures had been obtained it became clear that the gene family has members not only within vertebrates but also in invertebrates, plants, yeast, and protozoans (references in HANDA and GUIDOTTI 1996; VASCONCELOS et al. 1996; SMITH et al. 1997; ZIMMERMANN 1999a; ZIMMERMANN and BRAUN 1999). It has been recognized that all members of the family share five highly conserved sequence domains ("apyrase conserved regions," HANDA and GUIDOTTI 1996; SCHULTE and ESCH 1999). Not all of the proteins are surface-located ecto-enzymes and soluble forms exist in mammals, plants, and protozoans. All sequences have in common highly conserved domains that presumably are of major relevance for their catalytic activity.

II. Vertebrate Isoforms

Several related types of cell surface-located mammalian members of the E-NTPDase family have been characterized in both molecular and functional terms. The first sequence obtained was that of NTPDase 1, the human lymphocyte surface protein CD39 which was identified as an "ecto-ATP diphosphohydrolase" (CD39, apyrase) (MALISZEWSKI et al. 1994; CHRISTOFORIDIS et al. 1995; KACZMAREK et al. 1996; WANG and GUIDOTTI 1996; MARCUS et al. 1997). At present human, mouse, rat, and bovine sequences are available (ZIMMERMANN and BRAUN 1999). This was followed by the identification of the phylogenetically related NTPDase 2 (ecto-ATPase, CD39L1) that revealed a high preference for the hydrolysis of ATP over ADP (KEGEL et al. 1997; KIRLEY 1997; CHADWICK and FRISCHAUF 1997) (sequences from human, rat, mouse and chicken, ZIMMERMANN and BRAUN 1999). NTPDase 3 (CD39L3, HB6) was identified in man and chicken and revealed sequence identity to both previous members of the protein family (SMITH and KIRLEY 1998; NAGY et al. 1998).

Interestingly, members of the E-NTPDase family can be located not only at the cell surface but also in the Golgi apparatus (NTPDase 4α, UDPase) (WANG and GUIDOTTI 1988a) or in lysosomal/autophagic vacuoles (NTPDase 4β, LALP70) (BIEDERBICK et al. 1999). The Golgi enzyme (NTPDase 4α) reveals highest activity with UDP as a substrate. It also hydrolyzes a number of other nucleoside 5'-di- and triphosphates but not ATP and ADP. It functions presumably in the import of nucleotide sugars into Golgi cisternae.

Recently an additional member of the E-NTPDase family with differing membrane topography has been identified. NTPDase 5 (CD39 L4, CHADWICK and FRISCHAUF 1997, 1998) derived from a human cDNA library is secreted when transfected into COS-7 cells (MULERO et al. 1999). The enzyme has a high preference for nucleoside 5′-diphosphates, especially for UDP. The sequence-related put. NTPDase 6 has been cloned from a human cDNA library (CD39L2, CHADWICK and FRISCHAUF 1998) but has not yet been expressed. The existence of soluble members of the E-NTPDase family is of particular interest since the stimulation-induced release from nerve endings of soluble nucleotidases with undefined molecular structure has been reported (TODOROV et al. 1997).

The application of a unifying nomenclature is not devoid of complications. For example, the sequence of the functionally identified human NTPDase HB6 (SMITH and KIRLEY 1998) is identical to that of the human NTPDase CD39L3 (CHADWICK and FRISCHAUF 1998) except for two amino acid residues at the C-terminus. It is not known whether these differences reflect allelic structures and whether the functional properties of the proteins would be different. For simplicity, they are both referred to as NTPDase 3 in this context. Similarly the Golgi UDPase lacks nine amino acid residues that are present in the lysosomal LALP70. They are referred to here as (NTPDase 4α and NTPDase 4β, respectively). The human genes for NTPDase 1 (CD39), NTPDase 2 (CD39L1), and NTPDase 3 (CD39L3) are located at chromosomes 10, 9 and 3, respectively (MALISZEWSKI et al. 1994; CHADWICK and FRISCHAUF 1998). The mouse NTPDase 1 (CD39) gene consists of ten coding exons separated by nine introns (SCHOENBORN et al. 1998).

III. General Structural Properties

NTPDase 1 (CD39), NTPDase 2 (CD39L1), NTPDase 3 (CD39L3), and NTPDase 4 (UDPase) are all highly glycosylated ecto-enzymes with a single putative transmembrane domain at the N- and at the C-terminus (Fig. 2). Their predicted membrane topography thus resembles that of P2X receptors. The molecular masses predicted from their primary structures and the deglycosylated forms (CHRISTOFORIDIS et al. 1996) are in the order of 55–60 kDa; those of the glycosylated proteins after heterologous expression in COS-7 or CHO cells are 70–80 kDa (WANG et al. 1997; SMITH and KIRLEY 1998; NAGY et al. 1998; HEINE et al. 1999). The enzymes exist as homooligomers (dimers, trimers, or even tetramers) and the state of oligomerization may be essential for catalytic activity (STOUT and KIRLEY 1996; CARL et al. 1998; WANG et al. 1998; SMITH and KIRLEY 1999a). The primary structures of the NTPDase 1 (CD39) from mouse (MALISZEWSKI et al. 1994) and rat (WANG et al. 1997) share 75% sequence identity with the human sequence. The amino acid identity between NTPDase 1 (CD39) and NTPDase 2 (CD39L1) in the rat amounts to 40% (KEGEL et al.

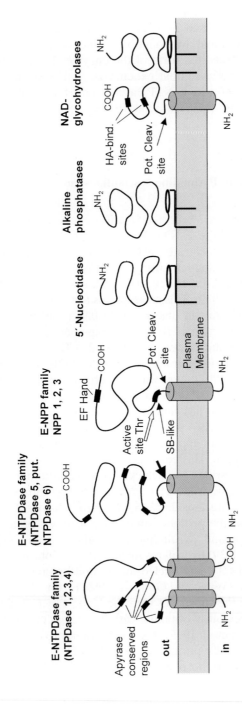

Fig. 2. Predicted membrane topography of key enzymes involved in the extracellular hydrolysis of nucleotides. 5′-Nucleotidase and alkaline phosphatases are anchored to the plasma membrane via a glycosylphosphatidyl inositol anchor. Some of the enzymes may occur as dimers or possibly also trimers and tetramers. NTPDase 5 occurs as a soluble protein (*arrow* indicates cleavage site). *HA-bind. sites*, hyaluronate binding sites; *Pot. Cleav. Site*, potential cleavage site; *SB like*, somatomedin B-like domain. (Modified from Zimmermann and Braun 1999)

1997; WANG et al. 1997). NTPDase 3 (CD39L3) is more closely related to the NTPDase 2 (CD39L1) than to the NTPDase 1 (CD39) (SMITH and KIRLEY 1998). The macrophage-expressed NTPDase 4 lacks the C-terminal hydrophobic domain. Its N-terminal hydrophobic leader sequence is cleaved, resulting in a soluble and secreted form of the enzyme (Fig. 2). The cleaved NTPDase 4 thus represents a soluble exoenzyme. It is less related to the surface-bound members of the E-NTPDase family but contains all five "apyrase-conserved regions" (MULERO et al. 1999).

IV. Catalytic Properties

1. Hydrolysis of Purine and Pyrimidine Nucleotides and Cation Dependence

The differences in sequence are reflected by differences in catalytic properties. Individual members of the E-NTPDase family differ in the hydrolysis rates for ATP and ADP as substrates. NTPDase 1 (CD39) has the highest ATP/ADP ratio. After heterologous expression it hydrolyzes ATP and ADP at a molecular ratio of about 1:0.5 to 1:0.9 (KACZMAREK et al. 1996; WANG and GUIDOTTI 1996; HEINE et al. 1999). This corresponds to the values obtained for the enzymes purified from human placenta (1 = 0.9) (CHRISTOFORIDIS et al. 1995), bovine aorta (1:1) (PICHER et al. 1996) or pig pancreas (1:0.9) (LEBEL et al. 1980). In contrast, rat and chicken NTPDase 2 (CD39L1) have a strong preference for ATP with molecular ratios of ATP:ADP of 1:0.03 or less (KEGEL et al. 1997; KIRLEY 1997). A recently described human splice variant of NTPDase 2 (CD39L1) that contains an additional 23 amino acid stretch in the putative extracellular loop has similar catalytic properties (MATEO et al. 1999). Human NTPDase 3 (HB6) is a functional intermediate and reveals a molecular ratio of ATP:ADP of approx. 1:0.3 (SMITH and KIRLEY 1998, 1999b) after heterologous expression. This corresponds to the ratio of 1:0.23 determined for the NTPDase 3 purified from chicken oviduct. Mutation of only two amino acid residues (D218E/W459A) of NTPDase 3 (HB6) considerably increases its preference for nucleoside 5'-triphosphates (SMITH et al. 1999).

All three members of the protein family have a rather broad substrate specificity towards purine and pyrimidine nucleotides. After heterologous expression NTPDase 1 (CD39) hydrolyzes nucleoside 5'-triphosphates such as ATP, GTP, UTP, ITP, and CTP and nucleoside 5'-diphosphates such as ADP, GDP, UDP, IDP, and CDP (WANG and GUIDOTTI 1996; HEINE et al. 1999). Generally, ATP and ADP are the substrates with the highest hydrolysis rate. This corresponds to results previously obtained with purified NTPDase 1 (LEBEL et al. 1980; CHRISTOFORIDIS et al. 1995). The same applies to nucleoside 5'-tri- and diphosphate hydrolysis by purified chicken NTPDase 3 (STROBEL et al. 1996) and for nucleoside 5'-triphosphate hydrolysis by NTPDase 2 (CD39L1) (HEINE et al. 1999; MATEO et al. 1999). The activity of all three types of ectonucleotidases depends on millimolar concentrations of divalent cations such as Ca^{2+} or Mg^{2+} (CHRISTOPHORIDIS et al. 1995; KACZMAREK et al. 1996; STROBEL et al. 1996; MARCUS et al. 1997; HEINE et al. 1999).

It is difficult to explain why in some cases catalysis proceeds to AMP and in others only to ADP. One possibility is that NTPDase 1 and NTPDase 3 that effectively hydrolyze nucleoside 5'-diphosphates would prevent ADP/UDP-receptor activation whereas NTPDase 2 with its high preference for nucleoside 5'-triphosphates would support this by generating ADP or UDP (VIGNE et al. 1998).

2. K_m-Values

Of major importance regarding the potency of ecto-nucleotidases in terminating the function of extracellular nucleotides are their K_m-values. K_m-values for ATP and ADP vary between investigations. They tend to be in the lower micromolar range (Table 1). NTPDase 1 (CD39) isolated from human placenta has a broad and alkaline pH optimum (pH 7–8) (CHRISTOFORIDIS et al. 1995). The K_m-values for ATP and ADP of $10\mu mol/l$ and $20\mu mol/l$, respectively, correspond to those obtained for an "ecto-apyrase" isolated from bovine aorta (PICHER et al. 1996). A K_m-value for ATP of $75\mu mol/l$ has been reported for human NTPDase-1 (CD39) expressed in COS-7 cells (WANG et al. 1998). Soluble recombinant human NTPDase 1 (CD39) lacking the N-and C-terminal hydrophobic domains revealed K_m-values of $220\mu mol/l$ (WANG et al. 1998) and $2.1\mu mol/l$ (GAYLE et al. 1998), respectively. The corresponding K_m-value for ADP was $5.9\mu mol/l$ (GAYLE et al. 1998). After heterologous expression the human splice variant of NTPDase 2 (CD39L1) (MATEO et al. 1999) revealed K_m-values for ATP and ADP of $394\mu mol/l$ and $102\mu mol/l$, respectively and those for NTPDase 3 (HB6) (SMITH and KIRLEY 1999b) were $128\mu mol/l$ and $96\mu mol/l$, respectively. The values obtained for the NTPDase 3 isolated from chicken oviduct were even higher with K_m-values for MgATP and CaATP of $480\mu mol/l$ and $120\mu mol/l$, respectively whereas

Table 1. Properties of ecto-nucleotidases

Enzyme	K_m-value (μ mol/l)	Cation dependence	pH Optimum	Tissue distribution	Apparent molecular mass (kDa)
E-NTPDases	10–200	Ca^{2+}, Mg^{2+}	7–8	Wide	80
E-NPPs	10–150	Ca^{2+}, Mg^{2+}	9	Wide	110–125
E-5'-nucleotidase	10–50	Zn^{2+}	8	Wide	70
Alkaline phosphatases	$1–4 \times 10^3$	Zn^{2+}, Mg^{2+}	8–9	Wide	70
Ap_nAase	0.4–4	Mg^{2+}	8.5–9	Wide	?
Ecto-nucleotide diphosphokinase	20–100	Mg^{2+}	–	Wide?	?

References are provided in the text.

that for MgADP amounted to 2020 µmol/l (STROBEL et al. 1996). K_m-values increase with increasing pH (KNOWLES and NAGY 1999). This suggests that the kinetic properties of the enzyme may be influenced by changes in extracellular pH. K_m-values for the expressed enzymes thus tend to be higher than those of the purified or solubilized proteins.

In this context it is of interest that many measurements on intact cells or tissues capable of hydrolyzing ATP and ADP (by unidentified ecto-nucleotidases) reported higher K_m-values. K_m-values for ATP range from 50 µmol/l to almost 1 mmol/l. Low K_m-values were obtained for isolated synaptosome preparations and human astrocytoma cells (approx. 50–120 µmol/l) (NAGY et al. 1986; BATTASTINI et al. 1991; SARKIS and SATO 1991; JAMES and RICHARDSON 1993; LAZAROWSKI et al. 1997) and high values (400–880 µmol/l) for, e.g., cultured rat coronary endothelial cells (MEGHJI et al. 1995), cultured rat astrocytes (STEFANOVIC et al. 1976), equine epididymal spermatozoa (MINELLI et al. 1997), or guinea pig urinary bladder and vas deferens (WELFORD et al. 1987; ZIGANSHIN et al. 1994b). Intermediate values (130–370 µmol/l) were reported for cultured chromaffin cells (TORRES et al. 1990), pig arterial smooth muscle cells (GORDON et al. 1989), or guinea-pig taenia coli (WELFORD et al. 1986). For comparison, rat Sertoli cells that hydrolyze only ATP but not ADP reveal a K_m-value for ATP of 114 µmol/l (BARBACCI et al. 1996). Presumably these differences are not simply due to variations in experimental protocol but reflect endogenous differences in molecular types of ecto-nucleotidases and their functional characteristics. Since synaptic vesicles and chromaffin granules store ATP at high millimolar concentrations (ZIMMERMANN 1994) the initial effective extracellular concentration of the released nucleotide may be in the same concentration range and would thus require a high K_m-ecto-nucleotidase for effective hydrolysis. The extracellular concentrations of ATP or UTP outside the synaptic cleft are difficult to estimate. High K_m-ecto-nucleotidases would be less suited for an effective degradation of nucleotides beyond low micromolar concentrations that are effective at nucleotide receptors (CRACK et al. 1995).

V. Apyrase Conserved Regions and the β- and γ-Phosphate Binding Motif

The catalytic domain structure of members of the E-NTPDase family is unknown. Furthermore it is uncertain whether nucleotide hydrolysis involves a phosphorylated intermediate. Members of the E-NTPDase family contain no consensus Walker ATP binding motifs. The high phylogenetic conservation, however, of the five "apyrase conserved regions" implies a major role of these sequence motifs in catalysis. The position of the "apyrase conserved region" suggests that the nucleotide binding site involves several distantly located sequence motifs and appropriate protein folding. The production of recombinant soluble forms of NTPDase 1 (CD39) (GAYLE et al. 1998; WANG et al. 1998) may result in the crystallization and complete structural analysis of the protein. It has been noted (ASAI et al. 1995; HANDA and GUIDOTTI 1996) that all nucleotidases belonging to the E-NTPDase family contain the actin-hsp 70-hexokinase β and γ-phosphate binding motif (A[IL]DLGG[TS]) (SCHWAB and WILSON 1989; FLAHERTY et al. 1990). Rat NTPDase 1 (CD39) contains the full sequence motif in the fourth "apyrase conserved region" (KEGEL et al. 1997). A marginal similarity to the motif is observed in the first "apyrase conserved region." Site-directed mutagenesis of conserved amino acids residues in the apyrase conserved regions I and IV severely disrupt nucleotidase activity of the human NTPDase 3 (CD39L3) (SMITH and KIRLEY 1999b). It is therefore likely that members of the E-NTPDase family and actin/HSP70/sugar kinases are derived from common ancestors. That the quaternary structure of members of the E-NTPDase family plays a role in controlling enzyme activity is concluded from crosslinking studies and investigations with soluble recombinant proteins (STOUT and KIRLEY 1996; CARL et al. 1998; WANG et al. 1998; SMITH and KIRLEY 1999a).

It is not known how substrate specificity between nucleoside diphosphates and nucleoside triphosphates is controlled. Do E-NTPDases have one common or two different binding sites for nucleoside diphosphates and nucleoside triphosphates? It is of interest that hydrolysis of ATP by rat NTPDase 1 (CD39) proceeds largely to AMP without releasing ADP as an intermediate (HEINE et al. 1999). This suggests that the ADP remains bound to the enzyme and becomes dephosphorylated at the identical or at a nearby catalytic site. An analysis of soluble E-NTPDase family members from the protozoan parasite *Toxoplasma gondii* suggests that the "apyrase conserved regions" need not be involved in the determination of substrate specificity.

VI. Inhibitors of Ecto-Nucleotidases

1. Search for Potency and Specificity

Inhibition of ecto-nucleotidase activity is mandatory for both studies of nucleotide release and analysis of the potency of externally applied ATP or its hydrolyzable analogues on subtypes of P2 receptors. In many tissues stable analogues of ATP are very potent in causing contractions up to a hundred times more effectively than ATP (ZIGANSHIN et al. 1994a; BAILEY and HOURANI 1995). Thus, it seems likely that the effects of exogenously applied ATP on P2-receptors is limited by its enzymatic degradation. Inhibitors of ecto-nucleotidases could be potential drugs that increase the lifetime of extracellular ATP in situ. They should preferably have no or only a small effect on P2-receptors (ZIGANSHIN et al. 1994a). In the past years diverse groups of compounds have been tested for their potency to inhibit the extracellular catabolism of ATP or ADP. Most of the compounds reveal only a mild inhibitory potency and in addition often affect receptor function. Of the compounds presently available the structural analogue of ATP, ARL 67156 (FPL 67156) (6-N,N-diethyl-D-β,γ-dibromomethylene ATP), is the only selective inhibitor of "ecto-ATPase" activity (albeit with moderate potency) and acts only weakly at purinergic receptors (CRACK et al. 1995; KENNEDY and LEFF 1995; KENNEDY et al. 1996). The compound potentiates purinergic synaptic transmission supporting the notion that endogenous ecto-nucleotidases reduce the effective concentrations of released ATP.

2. Nucleotide Analogues

An analysis of the structure-activity requirements of ecto-nucleotidase inhibitors suggest that substitutions on the purine ring have virtually no effect on the removal of the terminal phosphate from the ATP analogues. Thus, 2-methylthio-ATP, 2-chloro-ATP, 8-bromo-ATP, or N^6-phenyl-ATP are substrates of "ecto-ATPase". Methylene isoesters of ATP such as α,β-methylene ATP (α,βmeATP) or β,γ-methylene ATP (β,γmeATP) proved to resist degradation except for homo-ATP that is dephosphorylated at the same rate as ATP. Similarly, substitution on the terminal phosphate of nucleotides conferred resistance to dephosphorylation. ATP-γ-F, ADP-β-F or AMPS, ADP-β-S, and ATP-γ-S were not degraded. As purified ecto-nucleotidases were not available these investigations were performed by application of the compounds to various intact tissues such as guinea-pig urinary bladder, taenia coli, vas deferens, rat colon muscularis mucosae, or the innervated frog sartorius muscle (WELFORD et al. 1986, 1987; HOURANI et al. 1991; CASCALHEIRA and SEBASTIAO 1992; JOHNSON et al. 1996; BROWNHILL et al. 1997). The analyses thus do not refer to a defined type of ecto-nucleotidase or a defined cellular element but rather to the total activity in the tissue. Since similar results were obtained with an ATP-diphosphohydrolase purified from bovine aorta (PICHER et al. 1996) the inhibitory properties in the tissues previously investigated might reflect mainly those of an ecto-ATP-diphosphohydrolase (NTPDase). Differences encountered between tissues presumably result from the expression of different types of ecto-nucleotidases. In the innervated frog sartorius muscle ATP-γ-S and β,γ-methylene-ATP were metabolized (CASCALHEIRA and SEBASTIAO 1992). It should be noted in this context that ATP-γ-S can be used by kinases to thiophosphorylate proteins irreversibly (WIERASZKO and EHRLICH 1994). Thus, both the apparent hydrolysis of ATP-γ-S and other effects of the compound could be due to thiophosphorylation of proteins.

3. Inhibitors of P2-Purinergic Receptors as Ecto-Nucleotidase Inhibitors

Many inhibitors of P2-receptors also act as inhibitors of activity of "ecto-ATPase" or (where investigated) of "ecto-ATP diphosphohydrolase." These include the trypanocidal drug suramin and related compounds, a number of textile and protein dyes such as Evans blue, reactive blue 2, reactive red 2, Cibacron blue, Trypan blue, Condo red, Coomassie Brilliant blue, pyridoxylphosphate analogues such as pyridoxal phosphate-6-azophenly-2',4'-disulphonic acid (PPADS), isothiocyanato sulfonates of the aliphatic 2-isothiocyanatoethene-1-sulfonate (IES), or also the acetylcholinesterase inhibitor 9-amino-1,2,3,4-tetrahydroacridine (THA) (HOURANI and CHOWN 1989; BARRET et al. 1993; CRACK et al. 1994; BAILY and HOURANI 1995; BEUKERS et al. 1995; BÜLTMANN and STARKE 1995; BÜLTMANN et al. 1995, 1996a,b; MEGHJI and BURNSTOCK 1995; STOUT and KIRLEY 1995; STROBEL et al. 1996; ZIGANSHIN et al. 1995a,b, 1996; CHEN et al. 1996; MARTÍ et al. 1996a; JOHNSON et al. 1996; WITTENBURG et al. 1996; BONAN et al. 1997; DELGADO et al. 1997; DAMER et al. 1998; SUD'INA et al. 1998; TULUC et al. 1998). In most cases concentrations in the range of 100μmol/l or higher induce only partial enzyme inhibition. The use in particular of suramin and the textile dyes in intact tissues or cell cultures may be severely hampered by their non-specific interaction with various types of membrane proteins (PEOPLES and LI 1998).

Differences exist in inhibitor potency between individual members of the E-NTPDase family as shown by the comparative analysis of rat NTPDase 1 (CD39) and rat NTPDase 2 (CD39L1) expressed in CHO cells (HEINE et al. 1999). At a concentration of 100μmol/l Evans blue completely inhibited activity of NTPDase 1 (CD39) whereas activity of NTPDase 2 (CD39L1) was inhibited by only 55%. Suramin (100μmol/l) had little effect on NTPDase 2 (CD39) activity but inhibited NTPDase 2 (CD39L1) by 45%. PPADS (100μmol/l) was the least effective inhibitor and inhibited both enzymes to a similar extent. These results suggest that it might be possible to develop inhibitors with differential potency towards individual members of the E-NTPDase family.

4. Others

Other compounds that reveal moderate inhibitory effects on the hydrolysis of ATP and ADP include the structural analogue of ATP 5'-p-fluorosulfonylbenzoyladenosine (5'FSBA) that inhibits several ATPases (DOMBROWSKI et al. 1995; MARTÍ et al. 1996b; SÉVIGNY et al. 1997a), gramicidin, chlorpromazine, trifluoperazine, and thioridazine (MEGHJI and BURNSTOCK 1995), α,β-unsaturated sulphones and phosphonium salts (ZIGANSHIN et al. 1995b), cyclopiazonic acid, an inhibitor of sarcoplasmic ATPase (ZIGANSHIN et al. 1994b), imipramine, desimipramine, and amitryptiline (BARCELLOS et al. 1998), or heparin (SCHETINGER et al. 1998a). Azide concentrations of 10mmol/l and higher partially inhibit activity of NTPDase 1 (COTÉ et al. 1992; PICHER et al. 1994; CHRISTOFORIDIS et al. 1995; WANG and GUIDOTTI 1996; MARCUS et al. 1997) but not of "ecto-ATPase" activity (KNOWLES and NAGY 1999). A variety of metal cations such as Cu^{2+}, Ni^{2+}, Zn^{2+}, and La^{3+} inhibit extracellular hydrolysis of ATP (ZIGANSHIN et al. 1995c).

VII. Wide and Overlapping Tissue Distribution

1. Overall Distribution

The broad distribution of members of the E-NTPDase family corresponds to the variety of tissues and cellular systems in which extracellular hydrolysis of

ATP and/or ADP have previously been studied. Current evidence relies mainly on immunocytochemical and Northern hybridization analyses. Owing to the sequence identities between the closely related proteins, the immunocytochemical studies face the difficulty of possible cross-reactions of antibodies with other members of the same family. Cross-hybridization of nucleic acid probes between families is also a potential problem, and conditions need to be established that allow specific hybridization of individual probes with single family members. NTPDase 1 to NTPDase 3 appear to be coexpressed in many tissues (Table 2). These include for example brain, colon, lung, ovary, placenta, skeletal muscles, and small intestine. NTPDase 1 (CD39) and NTPDase 2 (CD39L1) may be colocalized not only within the same tissue (Lewis Carl and Kirley 1997) but also within identical cells (Kegel et al. 1997). The Golgi-located NTPDase 4 (UDPase) presumably represents a ubiquitous enzyme. According to Northern-blotting data put. NTPDase 6 (CD39L2) is present in all tissues investigated whereas NTPDase 5 (CD39L4) is more restricted, excluding, for example, heart, brain, placenta, and lung. The enzyme has been allocated to macrophages (Mulero et al. 1999).

The similarity in molecular mass and general biochemical properties (Table 1) and the overlapping distribution between tissues make it difficult to separate the various enzymes by biochemical means. An "ATP diphosphohydrolase" that revealed cross-immunoreactivity with antibodies against potato apyrase has been characterized in brain synaptic membranes (Battastini et al. 1995). Similarly, a mammalian "ecto-ATPase" cross-reacts with antibodies generated against chicken gizzard ecto-ATPase (Dombrowski et al. 1997). Several subtypes of "ATP diphosphohydrolase" (ATPDase) that vary slightly in biochemical properties have been isolated from mammalian tissues. They have been referred to as ATPDase Type I (pig pancreas), ATPDase Type II (bovine aorta), and ATPDase Type III (bovine lung) (for references see Picher et al. 1993). At present it is not clear whether these three types of enzymes are products of different E-NTPDase genes or whether they reflect splice variants or tissue differences in glycosylation. Using immunohistochemistry the distribution of "ecto-ATPase" has been investigated in bovine blood vessels (Sévigny et al. 1997b), heart (Beaudoin et al. 1997), lungs (Sévigny et al. 1997a), or rat brain (Wang and Guidotti 1998b). The immunolocalization will be more conclusive when the potential cross reactivity of antibodies with the new members of the E-NTPDase family has been excluded.

2. The Cardiovascular System

Both, ATP and ADP can be hydrolyzed by microvascular endothelial cells, aortic endothelial cells, smooth-muscle cells, and cardiac myocytes (Gordon et al. 1989; Meghji et al. 1995; James et al. 1996; de Oliveira et al. 1997). NTPDase 1 is associated with caveolae in cultured endothelial cells (Kittel et al. 1999). NTPDase 1 and NTPDase 2 are expressed in the heart (Table 2). The extracellular hydrolysis pathways for nucleotides in the vascular system have

Table 2. Tissue distribution of members of the E-NTPDase gene family as determined by Northern Blotting and immunoblotting

Tissue	NTPDase 1 (CD39, ecto-apyrase)	NTPDase 2 (CD39L1, ecto-ATPase)	NTPDase 3 (CD39L3, HB6)	NTPDase 4 (UDPase)	NTPDase 5 (CD39L4)	Put. NTPDase 6 (CD39L2)
Brain	+–	+(+–)	+	+	–	+
Bladder		(+)				
Colon	+	+(+)	+		+	+
Heart	+	+(+–)	–	+	–	+
Kidney	+–	+(+)	+	+	+	+
Liver	+–	+(+)	–	+	+	+
Lung	+–	+–(–)	+	+	–	+
Ovary	+	+(+)	+		+–	+
Pancreas	–			+	–	+
Periph. blood leukocyte	+	–	–		–	
Placenta	+	+	+	+	–	+
Prostate	+	+	+		+–	+
Skeletal muscle	+–	+(+)	+	+	+	+
Small intestine	+	+(+)	+		+	+
Spleen	+	+–(+)	+		–	+
Stomach		(+)				
Testis	+–	+(+)	+–		+	+
Thymus	+	+–(+–)	–		–	+

Data derived by Northern Blotting are from Kaczmarek et al. (1996) (human CD39); Kegel et al. (1997) (rat NTPDase 1, NTPDase 2); Wang et al. (1997) (rat NTPDase 1); Chadwick and Frischauf (1998) (human NTPDase 1–3, 5, 6); Wang and Guidotti (1998a) (human NTPDase 4). The Golgi-located NTPDase 4 presumably represents a ubiquitous enzyme. Data for immunoblotting (in parentheses) are from Smith et al. (1998) (rat NTPDase 2).

recently been reviewed (KACZMAREK et al. 1997; SLAKEY et al. 1997; ZIMMERMANN and PEARSON 1998). Extracellular hydrolysis of ATP and ADP in the vascular system is of interest since ADP plays a key role in platelet aggregation. NTPDase activity at the surface of the vascular endothelial cells controlling the lifetime of extracellular nucleotides may be important in the pathogenesis of intravascular thrombosis and arteriosclerosis. An antibody against an apyrase consensus sequence immunostained both endothelial and smooth muscle cells in the bovine aorta (SÉVIGNY et al. 1997b). In addition, soluble members of alkaline phosphatase and the PNDP family are present in serum that would be capable of hydrolyzing ATP or ADP (KAPLAN 1986; MEERSON et al. 1998). NTPDase 5 can be released from macrophages (MULERO et al. 1999). Ecto-nucleotidases can also be expressed on blood cells. Activity of "ecto-ATPase" has been identified on many cells of hematopoietic (both lymphoid and erythroid) and non-hematopoietic origin (for a review see DOMBROWSKI et al. 1998). NTPDase 1 (CD39) had previously been shown to be expressed on activated lymphocytes (KANSAS et al. 1991). A potential role of NTPDase 1 (CD39) in cell adhesion is implicated from the observation that treatment of NTPDase 1-positive B cell lines with subsets of anti-NTPDase 1 monoclonal antibodies induce homotypic adhesion. The phenomena involves tyrosine kinase activity but is unaffected by EDTA, suggesting that catalytic activity of NTPDase 1 is not involved. Studies with a variety of inhibitors suggest that the ecto-enzyme may play a role in controlling lymphocyte function including antigen recognition and/or the activation of effector activities of cytotoxic T cells (DOMBROWSKI et al. 1995). Activity of NTPDase has also been identified on rat and human blood platelets (FRASSETTO et al. 1993; PILLA et al. 1996).

3. The Nervous System

The functional and biochemical aspects of extracellular nucleotide hydrolysis in the nervous system have recently been reviewed (NAGY 1997; ZIMMERMANN and BRAUN 1999). Hydrolysis of extracellular nucleoside di- and triphosphates has been extensively studied in a variety of intact cellular systems including cultured neural cells, neuroblastoma cells, chromaffin cells, and the related PC12 cells, astrocytes, oligodendrocytes, isolated synaptosomes from a variety of vertebrate sources, cultured endothelial cells, and the neuromuscular junction and superfused peripheral organ preparations such as the vas deferens or the taenia coli (references in ZIMMERMANN 1996a,b). Histochemical staining identified NTPDase activity mainly on microglia and blood vessels (DALMAU et al. 1998). Generally, ATP is degraded and ADP, AMP, and adenosine sequentially appear in the assay medium. Where investigated, all physiologically occurring purine and pyrimidine nucleoside 5'-tri, di- and mono-phosphates were hydrolyzed. Presently available enzyme cytochemical analyses at the electron microscopical level do not identify individual species of ecto-nucleotidases but they further define the location of the enzyme cascade. For example, reaction product identifying hydrolysis of ATP is located at the outer surface of the plasma membrane surrounding neurosecretory nerve endings and pituicytes in the neurohypophysis (THIRION et al. 1996), at glial and neuronal plasma membranes including the synaptic cleft in the medial habenula (SPERLAGH et al. 1995; KITTEL et al. 1996), or at the neuronal plasma membrane/Schwann cell interface of the frog neuromuscular junction (PAPPAS and KRIHO 1988). The presence of an extracellular hydrolysis cascade for the degradation of ATP in the synaptic cleft is further supported by recent physiological studies. Monitoring the postsynaptic effects of adenosine apparently derived from released ATP at hippocampal synapses suggests that complete hydrolysis can occur in the synaptic cleft within less than 200 ms (DUNWIDDIE et al. 1997; CUNHA et al. 1988). This comes close to the extracellular hydrolysis rate of acetylcholine by acetylcholinesterase at the neuromuscular junction (MILEDI et al. 1984). The observation that soluble nucleotidases are released from stimulated but not from unstimulated sympathetic nerves could help to explain the relatively slow degradation of nucleotides by resting neural tissues (KENNEDY et al. 1997).

VIII. Regulation of Expression

Members of the E-NTPDase family are constitutively expressed in many tissues but evidence is now accumulating that the expression can be regulated. Little is known about the factors controlling the expression or the turnover of the enzyme-containing membrane compartments. No coherent picture can be derived at the moment but a few individual examples are cited here. NTPDase 2 (CD39L1) was found to be inducible at the transcription level by 2,3,7,8-tetrachlorodibenzo-p-dioxin in mouse hepatoma cells (GAO et al. 1998). The response is direct and requires both AhR and Arnt, members of a novel class of basic helix-loop-helix PAS proteins, which also mediate responses to other environmental stimuli, such as hypoxia. Expression of NTPDase 1 (CD39), the major ecto-ATP-hydrolyzing enzyme in human melanomas, is increased in differentiating melanomas followed by a gradual decrease with tumor progression (DZHANDZHUGAZYAN et al. 1998). Activity of "ecto-ATPDase" in human endothelial cells in-vitro is increased by aspirin (CHEUNG et al. 1994) and glomerular "ecto-ATPDase" immunoreactivity is modulated by estradiol (FAAS et al. 1997). Human epidermoid carcinoma cells increase the cascade for extracellular nucleotide hydrolysis when periodically treated with extracellular ATP, suggesting that the substrate itself may affect the expression of its hydrolysis chain (WIENDL et al. 1998).

IX. Physiological and Pathological Implications

1. Role in Thrombosis and Vascular Reperfusion

Since the physiological roles of extracellular nucleotides and their metabolic products display great variance in individual tissues this must also apply to the functional roles of the ecto-nucleotidases. Via its capacity to hydrolyze either ATP or ADP, the endothelial NTPDase 1 (CD39) could play an important role in the regulation of platelet aggregation. There is strong evidence that endothelial NTPDase 1 (CD39), by converting pro-aggregatory ADP to anti-aggregatory adenosine, limits the extent of intravascular platelet aggregation (KACZMAREK et al. 1996; MARCUS et al. 1997; IMAI et al. 1999). Recombinant soluble and catalytically active CD39 blocks ADP-induced platelet aggregation in vitro, and inhibits collagen-induced platelet reactivity. Soluble forms of NTPDase 1 (CD39) are therefore potential therapeutic agents for inhibition of platelet-mediated thrombotic diatheses (GAYLE et al. 1998). Intravenous administration of soluble apyrase prolongs discordant xenograft survival in rats undergoing heterotopic cardiac xenografting from guinea pigs. This is presumably due to the systemic antiaggregatory effects of the administered enzyme. Indeed, histological analyses revealed that administration of apyrase abrogates local platelet aggregation and activation (KOYAMADA et al. 1996). The generation of NTPDase 1-deficient mice (cd39–/–) corroborated the role of NTPDase 1 as a chief vascular ecto-nucleotidase and affirmed its importance in the control of purinergic signaling in both hemostasis and thromboregulation (ENJYOJI et al. 1999; ZIMMERMANN 1999b). Activation of endothelial cells as it accompanies inflammatory diseases results in a loss of "ecto-ATP diphosphohydrolase" and may thus be crucial for the progression of vascular injury (ROBSON et al. 1997). Activity of ecto-ATP diphosphohydrolase is highly sensitive to oxidative stress and greatly reduced in vivo with reperfusion injury (KACZMAREK et al. 1996). Similarly, glomeruli of transplanted kidneys reveal diminished immunostaining for ecto-ATP diphosphohydrolase, presumably reflecting the extent of ischemic tissue damage (VAN SON et al. 1997). Immunostaining of glomeruli is also reduced after perfusion with 100KF, a plasma serine protease that induces plasma protein leakage (CHEUNG et al. 1996).

2. Upregulation Following Global Forebrain Ischemia

Transient global cerebral ischemia results in an increase in extracellular nucleotide hydrolysis (BRAUN et al. 1998; SCHETINGER et al. 1998b). As revealed by Northern-blotting, NTPDase 1 (CD39) and ecto-5′-nucleotidase are upregulated during the days following transient forebrain ischemia in the rat (BRAUN et al. 1998). This suggests that the entire ecto-nucleotidase chain for the hydrolysis of ATP to adenosine becomes upregulated. An analysis of the tissue distribution of the nucleotidases (hydrolysis of ATP, UTP, ADP, AMP) by enzyme cytochemistry revealed a strong postischemic increase in staining intensity in the lesioned regions of the hippocampus, including the pyramidal cell layer of CA1, the stratum lacunosum-moleculare, and the dentate gyrus. A comparison with markers for activated astrocytes and microglia revealed a close similarity between the pattern of enzyme staining and that of markers for microglia. These data suggest that the increased expression of ecto-nucleotidases in the regions of damaged nerve cells is associated with activated glia, mainly microglia. The upregulation of the ecto-nucleotidase chain is suggestive of an ischemia-induced increased and sustained cellular release of nucleotides. The upregulation could have several functional implications. It would support the formation of extracellular adenosine as a neuroprotective agent and facilitate purine salvage. It could also relieve any potential cytotoxic, Ca^{2+}-mediated effects of ATP on neurons or glial cells. Since microglial cells express the cytolytic P2Z ($P2X_7$) subtype of ATP receptors (FERRARI et al. 1997) they may be particularly endangered by increased levels of extracellular ATP.

3. Alterations Following Plastic Changes in the Nervous System

Additional experiments suggest that changes in neural plasticity can be paralleled by changes in ecto-ATPase activity. Enzyme activity is reduced following avoidance learning (BONAN et al. 1998) and status epilepticus (NAGY et al. 1990, 1997).

C. The E-NPP Family

I. Three Types and their Structural Properties

Members of this gene family possess a surprisingly broad substrate specificity. They reveal alkaline phosphodiesterase as well as nucleotide pyrophosphatase activity, which are properties of the same enzyme molecule (GODING et al. 1998). In the current literature the enzyme family is also addressed as ecto-phosphodiesterase/pyrophosphatase, PC-1, or phosphodiesterase/nucleotide pyrophosphatase (PDNP) family (ZIMMERMANN 1999a; STEFAN et al. 1999). The nomenclature adopted here for the enzyme family and its members is that of ZIMMERMANN et al. (2000).

The first member of the enzyme family identified in molecular terms was the murine plasma cell differentiation antigen NPP 1 (PC-1) (VAN DRIEL et al. 1985; VAN DRIEL and GODING 1987). Three members of the family have now been cloned from human sources and related sequences are present in mouse and rat. In the current literature individual enzymes or cDNA sequences are referred to as PC-1 (NPP 1), PD-Iα, autotaxin (NPP 2) and PD-Iβ, B10, gp130^{RB13-6} (NPP 3) (VAN DRIEL et al. 1985; VAN DRIEL and GODING 1987; BUCKLEY et al. 1990; STRACKE et al. 1992; NARITA et al. 1994; DEISSLER et al. 1995; JIN-HUA et al. 1997; SCOTT et al. 1997). Whereas PD-Iβ, B10, and gp130^{RB13-6} denominate an identical protein, human PDNP-Iα and autotaxin presumably represent splice variants of the same gene (MURATA et al. 1994; KAWAGOE et al. 1995). They are not further differentiated here. Additional, functionally as yet uncharacterized alternatively spliced mRNAs have been described (FUSS et al. 1997). Related sequences have been identified in plants, yeast and *C. elegans* (ZIMMERMANN and BRAUN 1999).

The sequence of all members of the enzyme family predicts type II membrane proteins that carry a single transmembrane domain and an intracellular N-terminus (Fig. 2). The mammalian members of the E-NPP family have apparent molecular weights in the order of 110–125 kDa. Murine NPP 1 (PC-1) contains 905 amino acid residues, has an apparent molecular weight of 115 kDa, and occurs as a dimer of two identical 115 kDa subunits (VAN DRIEL et al. 1985; VAN DRIEL and GODING 1987). NPP 1 (PC-1) reveals 45% and 50% amino acid identity with NPP 2 (autotaxin) and NPP 3 (PD-Iβ), respectively (LEE et al. 1996a; JIN-HUA et al. 1997). The more closely related human NPP 1 and NPP 3 genes reside on the identical chromosome (number 6) (BUCKLEY et al. 1990; JIN-HUA et al. 1997). NPP 2 has been allocated to chromosome 8 (KAWAGOE et al. 1995; LEE et al. 1996a).

NPP 1 (PC-1) and NPP 2 (autotaxin) occur in a membrane-bound as well as in a soluble form (BELLI et al. 1993; CLAIR et al. 1997a). Soluble forms are presumably generated by proteolytic cleavage above a cysteine-rich stalk near the transmembrane domain. NPP 2 (autotaxin) has in fact first been identified as a proteolytically cleaved soluble protein (CLAIR et al. 1997a). So far a soluble form has not yet been described for NPP 3 but its existence may be inferred from the fact that all three forms are found in serum (MEERSON et al. 1998). Members of this protein family may thus occur both as ectoenzymes attached to the plasma membrane as well as proteolytically cleaved exoenzymes.

II. Catalytic Properties

1. Substrates and Inhibitors

The enzymes are apparently capable of hydrolyzing 3',5'-cAMP to 5'-AMP, ATP to AMP and PP_i, ADP to AMP and P_i, or NAD^+ to AMP and nicotinamide mononucleotide (Fig. 1). Both purine and pyrimidine nucleotides serve as substrates. Furthermore they can hydrolyze phosphodiester bonds of nucleic acids and the pyrophosphate linkages of nucleotide sugars (GODING et al. 1998). The simultaneous presence of phosphodiesterase and nucleotide pyrophosphatase activity on the same enzyme molecule has been confirmed by heterologous expression in COS-7 cells (REBBE et al. 1993). NPP 2 (autotaxin) has an even wider catalytic capacity than the other members of the family (CLAIR et al. 1997a). It can hydrolyze the phosphodiester bond on either side of the β-phosphate group of nucleoside 5'-triphosphates. This includes the hydrolysis of ATP and GTP to ADP and GDP, respectively. Furthermore, it hydrolyzes AMP to adenosine, a catalytic property shared with ecto-5'-nucleotidase, and PP_i to P_i.

Members of the E-NPP family are not affected by inhibitors of intracellular phosphodiesterases. NPP 1 (PC-1) is however inhibited by low concentrations (20 µmol/l) of PPADS, a frequently used inhibitor of P2 receptors as well as of ecto-nucleotidases (CHEN et al. 1996; WITTENBURG et al. 1996). The potential of glycosaminoglycans such as heparin and heparan sulfate to inhibit catalytic activity suggests that the enzymes may be under the control of proteins of the extracellular matrix (HOSODA et al. 1999). Phosphodiesterase activity of NPP 2 (autotaxin) is effectively inhibited by ATP, α,β-methylene ATP (α,βmeATP), and to a lower extent by β,γ-methylene ATP (β,γmeATP) and α,β-methylene ADP (α,βmeADP) (CLAIR et al. 1997).

2. Cation Dependence and K_m-Values

Catalytic activity of the E-NPP family members depends on divalent cations. It is increased in the presence of Ca^{2+} or Mg^{2+} and reduced in the presence of EDTA (REBBE et al. 1991; BELLI et al. 1994; DEISSLER et al. 1995; GROBBEN et al. 1999; HOSODA et al. 1999). Purified rat NPP 3 has a very alkaline pH optimum (pH 9.7), a K_m-value of 160 μmol/l for the hydrolysis of thymidine 5'-monophosphate nitrophenylester, and it can be inhibited by EDTA (DEISSLER et al. 1995). For comparison, rat C6 glioma cells expressing a NPP 1 (PC-1)-like ecto-nucleotide pyrophosphatase/phosphodiesterase revealed a K_m-value for ATP of 17 μmol/l (GROBBEN et al. 1999) and that of purified NPP 1 (PC-1) is in the range of 50 μmol/l (HOSODA et al. 1999). Apparent K_m-values are thus similar to those of members of the E-NTPDase family (Table 1).

3. The Catalytic Cycle

The catalytic site is assumed to be situated around Thr238 (mouse NPP 1 [PC-1]) that is conserved in all species of the E-NPP family. Mutation of this Thr to Tyr, Ser, or Ala abolished enzyme activity (BELLI et al. 1995). Presumably, NPP 1 (PC-1) is transiently covalently adenylated as part of the catalytic cycle (LANDT and BUTLER 1978). Hydrolysis of ATP is assumed to involve the attachment of AMP in a phosphodiester bond to the active site Thr238 with release of PPi followed by transfer of AMP to water. Correspondingly, ADP would be converted to AMP and P_i. In addition NPP 1 (PC-1) has been suggested to catalyze a threonine autophosphorylation reaction which involves the transfer of the γ-phosphate of ATP to the identical Thr238. Autophosphorylation of NPP 1 (PC-1) may serve as an autoregulatory mechanism for the phosphodiesterase-I/nucleotide pyrophosphatase activity (STEFAN et al. 1996). The question whether NPP 1 (PC-1) can also act as a protein kinase phosphorylating other proteins has been controversially discussed (ODA et al. 1991, 1993; BELLI et al. 1995; STEFAN et al. 1996; GODING et al. 1998). NPP 2 (autotaxin) becomes autophosphorylated but reveals no significant protein kinase activity towards a number of protein kinase substrates. GTP, NAD, AMP, and PP_i can all serve as phosphate donors in the phosphorylation (CLAIR et al. 1997a,b).

III. EF-Hand, Somatomedin B-Like Domain and RGD-Tripeptide

In addition to their complex nucleotide hydrolyzing activities, members of the E-NPP family share a number of additional interesting features. They contain the somatomedin B-like domain of vitronectin, the function of which is unknown. The domain may act as a stable "stalk" that holds the catalytic domain away from the membrane. Furthermore, the enzymes contain the consensus sequence of an EF-hand putative calcium-binding region (DXDXDGXXDXXE; X for an amino acid). This sequence is essential for enzymatic activity and is present in its complete form in NPP 1 (PC-1) and, with minor alterations, also in the other mammalian E-NPPs. In the case of NPP 1 (PC-1) this motif has been implicated in the stabilization of the proteins by divalent cations. A monoclonal antibody that is specific for the unfolded form of this region only binds to the protein in the absence of divalent cations or when the protein is denatured (BELLI et al. 1994; KAWAGOE et al. 1995; GODING et al. 1998). Another unique sequence is that of the RGD-tripeptide which is potentially recognizable by several integrins. The sequence is contained in the NPP 2 and NPP 3 but not in the NPP 1 sequence. The functional role of these motifs is not yet understood but they imply a functional Ca^{2+}-dependency and a potential interaction with cell surface receptors.

IV. Cell and Tissue Distribution

Members of the E-NPP family have a similarly wide tissue distribution as members of the E-NTPDase family (Table 3). Since there are high sequence identities between family members, the same caveat as for the E-NTPDases has to be applied concerning immunological and Northern-blotting data. Due to apparent differences between species the results obtained for human, rat, and mouse tissues are separately marked in Table 3. The data presently available imply that NPP 3 has the most restricted distribution and may be present in human liver, prostate, and uterus. NPP 1 is present in almost every tissue investigated, in accordance with early immunocytochemical investigations (MORLEY et al. 1987). Furthermore, their distribution can overlap with that of members of the E-NTPDase family.

Members of the protein family have been detected in various cellular systems. The murine plasma alloantigen NPP 1 (PC-1) is selectively expressed on B-lymphocytes in their terminal phase of differentiation into antibody-secreting cells (STEARNE et al. 1985). NPP 1 is also expressed by osteoblasts, osteoblast-like bone cells (OYAJOBI et al. 1994), osteoblast-like osteosarcoma cells (SOLAN et al. 1996), and rat C6 glioma cells (GROBBEN et al. 1999). NPP 2 has been identified in fibroblast-like synoviocytes (SANTOS et al. 1996), neuroblastoma cells (KAWAGOE et al. 1997), or the human tetracarcinoma cell line NTera2D1 (LEE et al. 1996a). The mRNA of NPP 2 (autotaxin) is located to oligodendrocytes and the choroid plexus in the brain (KAWAGOE et al. 1995). Unidentified ecto-phosphodiesterase/pyrophosphatase activity was observed in a variety of cellular systems including HeLa cells (KÜBLER et al. 1980) and human glomerular epithelial cells (STEFANOVIC et al. 1995). Hepatocytes reveal a surprising differential distribution of NPP 1 (PC-1) and NPP 3 (MEERSON et al. 1998; SCOTT et al. 1997). NPP 1 is exclusively located at the basolateral (sinu-

Table 3. Distribution of members of the E-NPP family as revealed by Northern blotting and immunological determination

Tissue	NPP 1 (PC-1)	NPP 2 (autotaxin, PD-Iα)	NPP 3 (GP130^{RB13-6}, PD-Iβ, B10)
Brain	$-_h, -_m (+_r)$	$+_h, +_r (+_r)$	$-_h, -_r$
Chondrocytes	$+_m$		
Colon	$-_h$		$-_h$
Epididymis	$+_m$		
Heart	$+_m, +-_r$	$+_r$	$-_h, +-_r$
Kidney	$-_h, +_m, +-_r (-_r)$	$+_h, +_r (-_r)$	$-_h, +_r$
Liver	$+_h, +_r$	$+_r$	$+_h, +_r$
Lung	$-_h, +_r$	$+_r$	$-_h, -_r$
Pancreas	$+_h, -_r$	$-_r$	$-_h, -_r$
Placenta	$+_h$	$+_h$	$-_h$
Ovary		$+_h$	
Prostate	$+-_h$	$+_r$	$+_h$
Salivary gland	$+_m$		
Skeletal muscle	$+-_h, +-_r$	$+-_r$	$-_h, -_r$
Small intestine	$-_h$	$+_h$	$-_h, +_r$
Spleen	$-_h, +_m (-_r)$	$+-_r (-_r)$	$-_h$
Testis	$+_h, +-_r$		$-_h$
Thymus	$+-_r$		$-_h$
Uterus	$+_h$		$+_h$
Vas deferens	$+_m$		

Data are derived from VAN DRIEL et al. (1985) (mouse NPP 1, Northern blotting); HARAHAP and GODING (1988) (mouse NPP 1, immunocytochemistry); KAWAGOE et al. (1995) (human NPP 2, Northern blotting); NARITA et al. (1994) (rat NPP 2, Northern blotting and immunoblotting); LEE et al. 1996a) (Northern blotting, human NPP 2); FUSS et al. (1997) (rat NPP 2, Northern Blotting); JIN-HUA et al. (1997) (human NPP 1 and 3, Northern Blotting); SCOTT et al. (1997) (rat NPP 3, Northern blotting and immunoblotting); STEFAN et al. (1998, 1999) (rat NPP 1, Northern blotting). Immunochemical data are given in parentheses.

soidal) surface whereas NPP 3 is exclusively associated with the bile canalicular surface. Presently neither the functional role of the differential subcellular distribution of the two enzymes nor the cellular sorting mechanisms are understood. The fact that the extracellular domains of NPP 1 and NPP 3 are highly conserved while their transmembrane and cytoplasmic domains are highly divergent (GODING et al. 1998; SCOTT et al. 1997) suggests that studies of the subcellular localization of chimeric proteins might help to reveal the sorting signals. NPP 1 is not expressed in the neonatal rat liver and its expression increases during development (STEFAN et al. 1998).

V. Functional Roles in Physiology and Pathology

The termination of signaling via extracellular nucleotides presumably represents one of the major functional roles of members of this enzyme family.

Nevertheless, the definition of their physiological substrates in the various tissues remains a major challenge. For example, phosphodiesterases may act as "guard dogs" to prevent subversion of the cell by destroying incoming DNA or RNA (GODING et al. 1998), hydrolyze free nucleic acids liberated on tissue injury and cell death, or simply serve purine salvage from extracellular fluids.

First examples identify a functional role of E-NPP-type ecto-nucleotidases. NPP 2 (autotaxin) has the ability to promote tumor cell motility (MURATA et al. 1994). Motility stimulation requires an intact 5′-nucleotide phosphodiesterase active site (STRACKE et al. 1995; LEE et al. 1996b). The enzyme is expressed by oligodendrocytes of the brain and may play an important role in oligodendrocyte function. The expression of NPP 2 (autotaxin) during development reveals an intermediate peak around the time of active myelination. On the other hand, the dysmyelinating mouse mutant *jimpy* has decreased levels of NPP 2 mRNA. Furthermore, mRNA levels are reduced in the CNS at onset of clinical symptoms in experimental autoimmune encephalomyelitis (FUSS et al. 1997).

NPP 1 (PC-1) may be an essential part of a mechanism that controls the balance between calcification and inhibition of calcification in skeletal tissues. It is highly expressed in chrondrocytes and appears to regulate soft-tissue calcification and bone mineralization by producing PPi, a major inhibitor of calcification (GODING et al. 1998; BÄCHNER et al. 1999). Interestingly, mice carrying the naturally occurring mutation Ttw (tip-toe walking), which is accompanied by an ossification of the spinal ligaments, as well as abnormal ossification at many other sites, have a nonsense mutation in the NPP 1 (PC-1) gene. The protein is truncated and inactive (OKAWA et al. 1998). Soluble E-NPPs are contained in the synovial fluids and the serum levels are elevated in patients with degenerative arthritis (CARDENAL et al. 1998). In addition, NPP 1 (PC-1) has been implicated in the pathogenesis of insulin resistance in type II diabetes. NPP 1 (PC-1) is thought to be an inhibitor of insulin signaling by inhibiting insulin receptor phosphorylation. Cells overexpressing NPP 1 produce insulin resistance (MADDUX et al. 1995; KUMAKURA et al. 1998). The issue is controversially discussed (GODING et al. 1998).

D. Extracellular Hydrolysis of Diadenosine Polyphosphates

Diadenosine polyphosphates are degraded extracellularly in an asymmetrical way, yielding AMP and Ap_{n-1} as products (Fig. 1). Three isoenzymes splitting Ap_4A and Ap_3A have been identified in human serum (LÜTHJE and OGILVIE 1987). The introduction of etheno-derivatives of Ap_nA and the combination of HPLC with a fluorimetric assay has greatly facilitated the determination of Ap_nA hydrolysis (RAMOS et al. 1995). Cultured bovine chromaffin cells degrade members of the diadenosine polyphosphate family from Ap_3A to Ap_6A (RODRIGUEZ-PASCUAL et al. 1992; RAMOS et al. 1995). Since guanosine polyphosphates are hydrolyzed equally well, the responsible enzyme may represent dinucleoside polyphosphatases (Np_nNases) rather than diadenosine polyphosphatases (Ap_nAases). The pH optimum is in the alkaline range (8.5–9.0) and K_m-values are in the order of 2–4 µmol/l (Table 1). The K_m values for the hydrolysis of extracellular ATP and ADP on chromaffin cells are thus about 50 times higher than those for the diadenosine polyphosphates. Similar results have been obtained for presynaptic plasma membranes of *Torpedo* electric organ (MATEO et al. 1997a). The kinetic data differ for the ecto-enzymes in cultured vascular endothelial cells (OGILVIE et al. 1989; MATEO et al. 1997b) implicating the existence of an additional form of the enzyme. K_m-values for Ap_3A to Ap_5A are in the order of 0.4 µmol/l. Mg^{2+}-ions are generally required for enzymatic activity and F^- is an inhibitor. The Ap_nAase in synaptic membranes from the *Torpedo* electric organ or on chromaffin cells are activated by Ca^{2+}-ions whereas the enzyme of cultured endothelial cells is inhibited. ATP or ATPγS can act as inhibitors of Ap_nA hydrolysis.

PPADS, and in particular suramin, are effective inhibitors of catalytic activity. The K_i-value of about 2 µmol/l observed for suramin on the ecto-Ap_nAase from *Torpedo* synaptic membranes is considerably lower than that of its antagonistic effect on P2 receptors (MATEO et al. 1996).

A major question is that of the molecular identity of the Np_nN-hydrolyzing enzymes. E-NPPs carry the potential to hydrolyze the phosphodiester bonds of diadenosine polyphosphates. Plasma membrane fractions derived from bovine adrenal medulla hydrolyze Ap_4A with a K_m-value of 2 µmol/l. Enzyme activity co-chromatographed with a protein immunoreactive for an anti-NPP 1 (PC-1) antibody (GASMI et al. 1998). Rat C6 glioma cells carrying NPP activity hydrolyze extracellular Ap_4A, Gp_4G, and NAD^+ (GROBBEN et al. 1999). It is possible that extracellular Np_nNs are generally hydrolyzed by members of the E-NPP family but this requires a more detailed biochemical analysis. The notion would not be compatible with the observation that ecto-Ap_nAase activity can be partially released from chromaffin cells by phosphatidyl inositol-specific phospholipase C suggesting membrane anchorage via a glycosylphosphatidyl inositol (GPI)-anchor (RODRIGUEZ-PASCUAL et al. 1992; RAMOS et al. 1995). In this context it is of interest that the release of undefined alkaline phosphodiesterase activity by phosphatidyl inositol-specific phospholipase C has also been reported for other cell types (NAKABAYASHI et al. 1993, 1994; ITAMI et al. 1997). At present it is not clear whether there exist additional variants of the enzyme or whether members of the E-NPP family may be proteolytically cleaved under these conditions.

E. Ecto-5′-Nucleotidase

The GPI-anchored ecto-5′-nucleotidase occurs in essentially all tissues. It is also known as the lymphocyte surface protein CD73 and represents a maturation marker for both T and B lymphocytes. A soluble form cleaved from the

GPI-anchor has been described. The enzyme catalyzes the final step of extracellular nucleotide degradation, the hydrolysis of nucleoside 5′-monophosphates to the respective nucleosides and Pi (Fig. 1). The K_m value for AMP is in the low micromolar range (Table 2). ATP and ADP act as very effective competitive inhibitors with K_i-values in the low micromolar range or even below. The enzyme is competitively inhibited by the nucleotide analogue α,βmeADP. Ecto-5′-nucleotidase is a zinc binding metalloenzyme and metal ion chelators such as EDTA inhibit enzyme activity. The X-ray structure of the related *Escherichia coli* periplasmic 5′-nucleotidase has recently been determined (KNÖFEL and STRÄTER 1999). The mammalian enzyme is upregulated following cerebral ischemia (BRAUN et al. 1997, 1998) and can be induced in neuroblastoma cells by phorbol ester and retinoic acid (KOHRING and ZIMMERMANN 1998). Catalytically active soluble ecto-5′-nucleotidase purified after heterologous expression in insect cells is now available for drug screening (SERVOS et al. 1999). In contrast to the other ecto-nucleotidases suramin, Evans blue and PPADS have little effect on enzyme activity. The molecular and biochemical properties of the enzyme (ZIMMERMANN 1992, 1996b) and its functional role in a variety of tissues including the vascular system (ZIMMERMANN and PEARSON 1998), nervous system (ZIMMERMANN 1996a; ZIMMERMANN and Braun 1999), and immune system (RESTA and THOMPSON 1997; CHRISTENSEN 1997; AIRAS 1998; RESTA et al. 1998) have recently been reviewed.

F. Alkaline Phosphatase

Alkaline phosphatases represent a protein family of non-specific ecto-phosphomonoesterases and thus reveal a broad substrate specificity (references in ZIMMERMANN 1996b). Like ecto-5′-nucleotidase, alkaline phosphatase is GPI-anchored and also occurs in serum (Fig. 2). The enzyme is capable of releasing inorganic phosphate from a variety of organic compounds including the degradation of nucleoside 5′-tri-, di-, and monophosphates and also the hydrolysis of PPi (FERNLEY 1971; COLEMAN 1992) (Fig. 1, Table 4). One single enzyme

Table 4. Ecto-nucleotidases and their potential substrates

Substrate	Enzyme capable of metabolizing substrate
ATP (NTP)	E-NTPDases, E-NPPs, alkaline phosphatases, nucleotide diphosphokinase, myokinase
ADP (NDP)	E-NTPDases, E-NPPs, alkaline phosphatases, nucleotide diphosphokinase, myokinase
AMP (NMP)	Ecto-5′-nucleotidase, NPP 2, alkaline phosphatases, myokinase
Ap_nA (Np_nN)	Ap_nAase?, NPP 1 (and other E-NPPs?)
PP_i	E-NPPs, alkaline phosphatases
NAD^+	NAD-glycohydrolase, NPP 1 (and other E-NPPs?), ecto-ADP-ribosyltransferase

could thus catalyze the entire hydrolysis chain from the nucleoside-5′-triphosphate to the respective nucleoside. K_m-values for a variety of substrates are in the low millimolar range. The various isoenzymes reveal a similarly broad tissue distribution as the other ecto-nucleotidases. There is evidence that alkaline phosphatase is coexpressed within the same tissues or on the same cells with other ecto-nucleotidases. It is presumably responsible for the "non-specific" phosphatase activity observed in many cellular systems. It has been implicated in a variety of tissue functions including skeletal mineralization (WHYTE 1996). Unfortunately, the role of the enzyme in purinergic signaling and extracellular nucleotide metabolism has received very little attention.

G. Ecto-Nucleoside Diphosphokinase and Ecto-Myokinase

Ecto-nucleoside diphosphokinase catalyses the transphosphorylation of nucleoside diphosphates utilizing nucleoside triphosphates as the γ-phosphate donor. While an intracellular form of the enzyme has previously been known the corresponding catalytic activity has now been also identified at the cell surface (LAZAROWSKI et al. 1997; HARDEN et al. 1997). The enzyme can interconvert nucleoside 5′-di- and triphosphates such as ATP and UDP to ADP and UTP. This is of interest since the parallel cellular release and presence in the extracellular space of several nucleotides can lead to an interconversion resulting in the mutual activation or inactivation of receptors for, e.g., ATP, ADP, UTP, and UDP. The enzyme is inhibited by EDTA and K_m-values for the nucleotides are in the order of 20–100 µmol/l. Activity of ecto-nucleoside diphosphokinase has been identified on human astrocytoma cells and human airway epithelia but its molecular identity and overall tissue distribution have not yet been determined.

In addition the presence of an ecto-ATP:AMP phosphotransferase (myokinase) reaction needs to be considered. The enzyme can lead to the extracellular formation of ATP and AMP from ADP and vice versa. In the nervous system the presence of this catalytic activity has been suggested for rat brain synaptosomes (NAGY et al. 1989) and hippocampal mossy fiber synaptosomes (TERRIAN et al. 1989).

H. Extracellular Hydrolysis of NAD^+

Functions of extracellular NAD^+ include depression of synaptic activity in brain tissues, an influence on the proliferative and cytolytic functions of mouse cytotoxic T cells, or a decrease in heart rate (for references see ZIMMERMANN 1996b). NAD^+ could be detected in microdialysis samples of rat cerebellum (DE FLORA et al. 1996). Extracellular NAD^+ can be degraded with a number of metabolites appearing. These include ADP-ribose, nicotinamide, nicotinamide mononucleotide, AMP, and adenosine (SNELL et al. 1984; PAIN and HENDRICK 1996). This implies the involvement of several types of ecto-nucleotidases. A major enzyme responsible for the extracellular hydrolysis of NAD^+ is CD38. It acts as an NAD^+-glycohydrolase (NADase) catalyzing the formation of nicotinamide and ADP-ribose. On the other hand, the cleavage of NAD^+ to AMP and nicotinamide mononucleotide may be catalyzed by a nucleotide pyrophosphatase. Hydrolysis of NAD^+ by an ecto-nucleotide pyrophosphatase associated with rat C6 glioma cells has been reported (GROBBEN et al. 1999). The ADP-ribose formed by CD38 may be further degraded by ecto-nucleotide pyrophosphatase (EVANS 1974; DETEREE et al. 1996) or also be taken up by the cells (ALEO et al. 1996).

CD38 is a member of a novel class eukaryotic gene family of NAD^+-converting enzymes (FERRERO and MALAVASI 1997; BRAREN et al. 1998). It represents a monomeric type II transmembrane glycoprotein of 45 kDa with a short intracellular N-terminus (Fig. 2). Soluble forms have been described (FUNARO et al. 1996; MALLONE et al. 1998). Related enzymes include BST-1, a GPI-anchored enzyme with identical catalytic activity (KAISHO et al. 1994; HIRATA et al. 1994) and the *Aplysia* ADP ribosyl cyclase whose crystal structure has been determined (PRASAD et al.1996). CD38 is widely expressed in humans both within the hematopoietic system (e.g., bone marrow progenitor cells, monocytes, platelets, and erythrocytes) and a large variety of tissues such as brain, prostate, kidney, gut, heart, and skeletal muscle (MEHTA et al. 1996; FERRERO and MALAVASI 1997).

CD38 is a multifunctional protein. First it acts as a bifunctional enzyme. It catalyzes an ADP-ribosyl cyclase reaction: NAD^+ is degraded to cyclic ADP-ribose (cADPR) and nicotinamide; cADPR is then rapidly converted to ADP-ribose by a cyclic ADP-ribose hydrolase reaction. Intracellular cADPR is a Ca^{2+}-mobilizing agent. However, it is still an open question whether extracellularly-formed cADPR can act as intracellular messenger (DA SILVA et al. 1998). CD38 enhances B-lymphocyte proliferation in the presence of other costimuli (FUNARO et al. 1995; LUND et al. 1995) and is thought to act as a cell surface receptor with signal transduction properties (PEOLA et al. 1996; SILVENNOINEN et al. 1996; DEAGLIO et al. 1998). CD38 has an extracellular hyaluronate binding motif and binds hyaluronate (NISHINA et al. 1994). Hyaluronate is involved in the receptor-mediated regulation of cell adhesion, motility, growth and differentiation. Deletion of CD38 results only in mild phenotypic alterations (COCKAYNE et al. 1998; LUND et al. 1998).The role of CD38 in the immune system has recently been reviewed (FUNARO et al. 1995; LUND et al. 1995, 1998; FLORA et al. 1997).

I. Ecto-Phosphorylation and Ecto-ADP Ribosylation

The metabolism of extracellular nucleotides can lead to posttranslational modifications of surface located or extracellular proteins such phosphorylation and

ADP-ribosylation (references in ZIMMERMANN 1996b). Ecto-protein kinase activity for the phosphorylation of both neighboring membrane proteins and extracellular proteins has been described in a wide variety of cellular systems. Both cAMP-dependent and cAMP-independent types have been described. ATP and GTP may serve as phosphoryl group donors. The presently identified surface-located protein kinases correspond to soluble intracellular forms of protein kinase CK1 and CK2 or protein kinase C. The functional properties of ecto-protein kinases have recently been reviewed (EHRLICH and KORNECKI 1999; REDEGELD et al. 1999). Ecto-phosphorylation implicates ecto-dephosphorylation. This issue has not yet been addressed but ecto-alkaline phosphatase would be a potential candidate to catalyze this reaction.

Another possibility for extracellular protein modification is that the ADP-ribose moiety of NAD^+ is transferred to specific acceptor proteins with the release of nicotinamide. This can lead to the ribosylation of surface proteins, a principal mechanism of protein modification well studied for bacterial exoenzymes such as cholera, pertussis, and diphtheria toxins. Within the immune system ADP-ribosylation has been implicated in the modulation of the cytotoxic and proliferative potentials of cytotoxic lymphocytes. Mono(ADP-ribosyl)transferases have been cloned from several vertebrates including rabbit and human skeletal muscle and rat T cells (HOLLMANN et al. 1996; HAAG et al.1997; BRAREN et al. 1998). In the absence of an appropriate acceptor protein the enzymes may also act as NAD^+-glycohydrolases resulting in transfer of ADP-ribose to water and net hydrolytic cleavage of NAD^+.

J. Outlook

The potential for the metabolism of extracellular nucleotides is impressive. Not only nucleoside 5′-tri-, di-, and monophosphates but also a variety of complex nucleotides can be degraded. Metabolism of extracellular nucleotides such as ATP and NAD^+ can also be involved in posttranslational protein modification. Recent work has resulted in the molecular identification of many of the surface-located ecto-nucleotidases and revealed novel gene families. There are several noteworthy features of the ecto-nucleotidase pathways:

1. Many of the enzymes have a very broad spectrum of catalytic activities and can metabolize several types of nucleotides. The same type of nucleotide can be metabolized by different types of enzymes (Table 4).
2. There appears to be an overlapping expression in many tissues for most of the enzymes. This implies multiple presence of the ecto-nucleotidase pathway and highlights the physiological importance of ubiquitous and highly effective extracellular nucleotide metabolism. It complicates the development of effective inhibitors. It also renders difficult any attempts to identify the enzyme function by the deletion of individual ecto-nucleotidase genes.

3. Several of the enzymes also occur in soluble forms suggesting that they can diffuse within the extracellular medium (or are transported in the blood) to their sites of action.
4. There is evidence that several of the enzymes may be multifunctional proteins. In addition to their catalytic activity they may act in cell adhesion or in transmembrane receptor functions.

Ecto-nucleotidases are thus related not only to purinergic signaling or extracellular purine salvage but appear also to be involved in a broad context in the complex of cell/cell interactions. Future research needs to identify the three-dimensional structure of individual ecto-nucleotidases and their catalytic site. Of equal importance is the exact cellular localization of the individual enzymes and a further characterization of their tissue-related physiological properties.

References

Airas L (1998) CD73 and adhesion of B-Cells to follicular dendritic cells. Leuk Lymphoma 29:37–47

Aleo MF, Sestini S, Pompucci G, Preti A (1996) Enzymatic activities affecting exogenous nicotinamide adenine dinucleotide in human skin fibroblasts. J Cell Physiol 167:173–176

Arch JRS, Newsholme EA (1978) The control of the metabolism and the hormonal role of adenosine. In: Campbell PN, Aldridge WN (eds) Essays in biochemistry 14. Academic Press, London, pp 82–123

Asai T, Miura S, Sibley LD, Okabayashi H, Takeuchi T (1995) Biochemical and molecular characterization of nucleoside triphosphate hydrolase isozymes from the parasitic protozoan *Toxoplasma gondii*. J Biol Chem 270:11391–11397

Aurivillius M, Hansen OC, Lazrek MBS, Bock E, Öbrink B (1990) The cell adhesion molecule Cell-CAM 105 is an ecto-ATPase and a member of the immunoglobulin superfamily. FEBS Lett 264:267–269

Bächner D, Ahrens M, Betat N, Schröder D, Gross G (1999) Developmental expression analysis of murine autotaxin (ATX). Mech Develop 84:121–125

Bailey SJ, Hourani SMO (1995) Effects of suramin on contractions of the guinea-pig vas deferens induced by analogues of adenosine 5'-triphosphate. Br J Pharmacol 114:1125–1132

Barbacci E, Filippini A, Decesaris P, Ziparo E (1996) Identification and characterization of an ecto-ATPase activity in rat Sertoli cells. Biochem Biophys Res Commun 222:273–279

Barcellos CK, Schetinger MRC, Dias RD, Sarkis JJF (1998) In vitro effect of central nervous system active drugs on the ATPase-ADPase activity and acetylcholinesterase activity from cerebral cortex of adult rats. Gen Pharmacol 31:563–567

Barret J-M, Ernould A-P, Rouillion M-H, Ferry G, Genton A, Boutin JA (1999) Studies of the potency of protein kinase inhibitors on ATPase activities. Chem Biol Interact 86:17–27

Battastini AMO, da Rocha JBT, Barcellos CK, Dias RD, Sarkis JJF (1991) Characterization of an ATP diphosphohydrolase (EC 3.6.1.5.) in synaptosomes from cerebral cortex of adult rats. Neurochem Res 16:1303–1310

Battastini AMO, Oliveira EM, Moreira CM, Bonan CD, Sarkis JJF, Dias RD (1995) Solubilization and characterization of an ATP diphosphohydrolase (EC 3.6.1.5) from rat brain synaptic plasma membranes. Biochem Mol Biol Int 37:209–219

Beaudoin AR, Sévigny J, Grondin G, Daoud S, Levesque FP (1997) Purification, characterization, and localization of two ATP diphosphohydrolase isoforms in bovine heart. Am J Physiol 273:H673–H681

Beaudoin AR, Sévigny J, Picher M (1996) ATP-diphosphohydrolases, apyrases, and nucleotide phosphohydrolases: Biochemical properties and functions. Biomembranes 5:369–401

Belli SI, Mercuri FA, Sali A, Goding JW (1995) Autophosphorylation of PC-1 (alkaline phosphodiesteraseI/nucleotide pyrophosphatasse) and analysis of the active site. Eur J Biochem 228:669–676

Belli SI, Sali A, Goding JW (1994) Divalent cations stabilize the conformation of plasma cell membrane glycoprotein PC-1 (alkaline phosphodiesterase I). Biochem J 304:75–80

Belli SI, van Driel IR, Goding JW (1993) Identification and characterization of soluble form of the plasma cell membrane glycoprotein PC-1 (5'-nucleotide phosphodiesterase). Eur J Biochem 217:421–428

Betto R, Senter L, Ceoldo S, Tarricone E, Biral D, Salviati G (1999) Ecto-ATPase activity of alpha-sarcoglycan (Adhalin). J Biol Chem 274:7907–7912

Beukers MW, Kerkhof CJM, Van Rhee MA, Ardanuy U, Gurgel C, Widjaja H, Nickel P, Ijzerman AP, Soudijn W (1995) Suramin analogs, divalent cations and ATP gamma S as inhibitors of ecto-ATPase. Naunyn-Schmiedebergs Arch Pharmacol 351:523–528

Biederbick A, Rose S, Elsässer HP (1999) A human intracellular apyrase-like protein, LALP70, localizes to lysosomal/autophagic vacuoles. J Cell Sci 112:2473–2484

Bonan CD, Battastini AMO, Schetinger MRC, Moreira CM, Frassetto SS, Dias RD, Sarkis JJF (1997) Effects of 9-amino-1,2,3,4-tetrahydroacridine (THA) on ATP diphosphohydrolase (EC 3.6.1.5) and 5'-nucleotidase (EC 3.1.3.5) from rat brain synaptosomes. Gen Pharmacol 28:761–766

Bonan CD, Dias MM, Battastini AMO, Dias RD, Sarkis JJF (1998) Inhibitory avoidance learning inhibits ectonucleotidase activities in hippocampal synaptosomes of adult rats. Neurochem Res 23:977–982

Braren R, Glowacki G, Nissen M, Haag F, Koch-Nolte F (1998) Molecular characterization and expression of the gene for mouse NAD^+: arginine ecto-mono(ADP-ribosyl)transferase. Biochem J 336:561–568

Braun N, Lenz C, Gillardon F, Zimmermann M, Zimmermann H (1997) Focal cerebral ischemia enhances glial expression of 5'-nucleotidase. Brain Res 766:213–226

Braun N, Zhu Y, Krieglstein J, Culmsee C, Zimmermann H (1998) Upregulation of the enzyme chain hydrolyzing extracellular ATP following transient forebrain ischemia in the rat. J Neurosci 18:4891–4900

Brownhill VR, Hourani SMO, Kitchen I (1997) Ontogeny of P2-purinoceptors in the longitudinal muscle and muscularis mucosae of the rat isolated duodenum. Br J Pharmacol 122:225–232

Buckley MF, Loveland KA, McKinstry WJ, Garson OM, Goding JW (1990) Plasma membrane glycoprotein PC-1: cDNA cloning of the human molecule, amino acid sequence and chromosomal location. J Biol Chem 265:17506–17511

Bültmann R, Starke K (1995) Reactive red 2: A P2y-selective purinoceptor antagonist and an inhibitor of ecto-nucleotidase. Naunyn-Schmiedebergs Arch Pharmacol 352:477–482

Bültmann R, Driessen B, Goncalves J, Starke K (1995) Functional consequences of inhibition of nucleotide breakdown in rat vas deferens: A study with Evans blue. Naunyn-Schmiedebergs Arch Pharmacol 351:555–560

Bültmann R, Pause B, Wittenburg H, Kurz G, Starke K (1996a) P2-purinoceptor antagonists .1. Blockade of P2-purinoceptor subtypes and ecto-nucleotidases by small aromatic isothiocyanato-sulphonates. Naunyn Schmiedebergs Arch Pharmacol 354:481–490

Bültmann R, Wittenburg H, Pause B, Kurz G, Nickel P, Starke K (1996b) P-purinoceptor antagonists. e3. Blockade of P2-purinoceptor subtypes and ecto-nucleotidases by compounds related to suramin. Naunyn Schmiedebergs Arch Pharmacol 354: 498–504

Cardenal A, Masuda I, Ono W, Haas AL, Ryan LM, Trotter D, Mccarty DJ (1998) Serum nucleotide pyrophosphohydrolase activity; elevated levels in osteoarthritis, calcium pyrophosphate crystal deposition disease, scleroderma, and fibromyalgia. J Rheumatol 25:2175–2180

Carl SAL, Smith TM, Kirley TL (1998) Cross-linking induces homodimer formation and inhibits enzymatic activity of chicken stomach ecto-apyrase. Biochem Mol Biol Int 44:463–470

Cascalheira JF, Sebastiao AM (1992) Adenine nucleotide analogues, including γ-phosphate-substituted analogues, are metabolized extracellularly in innervated frog sartorius muscle. Eur J Pharmacol 222:49–59

Chadwick BP, Frischauf AM (1997) Cloning and mapping of a human and mouse gene with homology to ecto-ATPase genes. Mamm Genome 8:668–672

Chadwick BP, Frischauf AM (1998) The CD39-like gene family: Identification of three new human members (CD39L2, CD39L3, and CD39L4), their murine homologues, and a member of the gene family from *Drosophila* melanogaster. Genomics 50:357–367

Chadwick BP, Williamson J, Sheer D, Frischauf AM (1998) cDNA cloning and chromosomal mapping of a mouse gene with homology to NTPases. Mamm Genome 9:162–164

Chen BC, Lee CM, Lin WW (1996) Inhibition of ecto-ATPase by PPADS, suramin and reactive blue in endothelial cells, C-6 glioma cells and RAW 264.7 macrophages. Br J Pharmacol 119:1628–1634

Cheung PK, Visser J, Bakker WW (1994) Upregulation of antithrombotic ectonucleotidases by aspirin in human endothelial cells in-vitro. J Pharm Pharmacol 46:1032–1034

Cheung PK, Klok PA, Bakker WW (1996) Minimal change-like glomerular alterations induced by a human plasma factor. Nephron 74:586–593

Christensen LD (1997) CD73 (ecto-5'-nucleotidase) on blood mononuclear cells. Regulation of ecto-5'-nucleotidase activity and antigenic heterogeneity of CD73 on blood mononuclear cells from healthy donors and from patients with immunodeficiency. APMIS 105:5–28:5–28

Christoforidis S, Papamarcaki T, Galaris D, Kellner R, Tsolas O (1995) Purification and properties of human placental ATP diphosphohydrolase. Eur J Biochem 234:66–74

Christoforidis S, Papamarcaki T, Tsolas O (1996) Human placental ATP diphosphohydrolase is a highly *N*-glycosylated plasma membrane enzyme. Biochim Biophys Acta (Biomemb) 1282:257–262

Clair T, Lee HY, Liotta LA, Stracke ML (1997a) Autotaxin is an exoenzyme possessing 5'-nucleotide phosphodiesterase/ATP pyrophosphatase and ATPase activities. J Biol Chem 272:996–1001

Clair T, Krutzsch HC, Liotta LA, Stracke ML (1997b) Nucleotide binding to autotaxin: Crosslinking of bound substrate followed by lysC digestion identifies two labeled peptides. Biochem Biophys Res Commun 236:449–454

Cockayne DA, Muchamuel T, Grimaldi JC, Muller-Steffner H, Randall TD, Lund FE, Murray R, Schuber F, Howard MC (1998) Mice deficient for the ecto-nicotinamide adenine dinucleotide glycohydrolase CD38 exhibit altered humoral immune responses. Blood 92:1324–1333

Coleman JE (1992) Structure and mechanism of alkaline phosphatase. Ann Rev Biophys Biomol Struct 21:441–483

Coté YP, Quellet S, Beaudoin A (1992) Kinetic properties of type-II ATP diphosphohydrolase from the tunica media of the bovine aorta. Biochim Biophys Acta 1160:246–250

Crack BE, Beukers MW, McKechnie KCW, Ijzerman AP, Leff P (1994) Pharmacological analysis of ecto-ATPase inhibition: evidence for combined enzyme inhibition

and receptor antagonism in P2X purinoceptor ligands. Br J Pharmacol 113: 1432–1438
Crack BE, Pollard CE, Beukers MW, Roberts SM, Hunt SF, Ingall AH, McKechnie KCW, Ijzerman TP, Leff P (1995) Pharmacological and biochemical analysis of FPL 67156, a novel, selective inhibitor of ecto-ATPase. Br J Pharmacol 114: 475–481
Cunha RA, Sebastiao AM, Ribeiro JA (1998) Inhibition by ATP of hippocampal synaptic transmission requires localized extracellular catabolism by ecto-nucleotidases into adenosine and channeling to adenosine A(1) receptors. J Neurosci 18: 1987–1995
da Silva CP, Schweitzer K, Heyer P, Malavasi F, Mayr GW, Guse AH (1998) Ectocellular CD38-catalyzed synthesis and intracellular Ca^{2+}- signalling activity of cyclic ADP-ribose in T- lymphocytes are not functionally related. FEBS Lett 439: 291–296
Dalmau I, Vela JM, Gonzalez B, Castellano B (1998) Expression of purine metabolism-related enzymes by microglial cells in the developing rat brain. J Comp Neurol 398:333–346
Damer S, Niebel B, Czeche S, Nickel P, Ardanuy U, Schmalzing G, Rettinger J, Mutschler E, Lambrecht G (1998) NF279: a novel potent and selective antagonist of P2X receptor-mediated responses. Eur J Pharmacol 350:R5–R6
De Flora A, Guida L, Franco L, Zocchi E, Pestarino M, Usai C, Marchetti C, Fedele E, Fontana G, Raiteri M (1996) Ectocellular in vitro and in vivo metabolism of cADP-ribose in cerebellum. Biochem J 320:665–672
De Flora A, Guida L, Franco L, Zocchi E (1997) The CD38/cyclic ADP-ribose system: A topological paradox. Int J Biochem Cell Biol 29:1149–1166
de Oliveira EM, Battastini AMO, Meirelles MNL, Moreira CM, Dias RD, Sarkis JJF (1997) Characterization and localization of an ATP diphosphohydrolase activity (EC 3.6.1.5) in sarcolemmal membrane from rat heart. Mol Cell Biochem 170:115–123
Deaglio S, Morra M, Mallone R, Ausielleo CM, Prager E, Garbarino G, Dianzani U, Stockinger H, Malavasi F (1998) Human CD38 (ADP-ribosyl cyclase) is a counter receptor of CD31, an Ig superfamily member. J Immunol 160:395–402
Deissler H, Lottspeich F, Rajewsky MF (1995) Affinity purification and cDNA cloning of rat neural differentiation and tumor cell surface antigen gp130[RB13-6] reveals relationship to human and murine PC-1. J Biol Chem 270:9849–9855
Delgado J, Moro G, Saborido A, Megias A (1997) T-tubule membranes from chicken skeletal muscle possess an enzymic cascade for degradation of extracellular ATP. Biochem J 327:899–907
Deterre P, Gelman L, Gary-Gouy H, Arrieumerlou C, Berthelier V, Tixier EM, Ktorza S, Goding L, Schmitt C, Bismuth G (1996) Coordinated regulation in human T cells of nucleotide-hydrolyzing ecto-enzymatic activities, including CD38 and PC-1. Possible role in the recycling of nicotinamide adenine dinucleotide metabolites. J Immunol 157:1381–1388
Dhalla NS, Zhao D (1988) Cell membrane Ca^{2+}/Mg^{2+} ATPase. Prog Biophys Mol Biol 52:1–37
Dombrowski KE, Ke Y, Thompson LF, Kapp JA (1995) Antigen recognition by CTL is dependent upon ectoATPase activity. J Immunol 154:6227–6237
Dombrowski KE, Brewer KA, Maleckar JR, Kirley TL, Thomas JW, Kapp JA (1997) Identification and partial characterization of ectoATPase expressed by immortalized B lymphocytes. Arch Biochem Biophys 340:10–18
Dombrowski KE, Ke Y, Brewer KA, Kapp JA (1998) Ecto-ATPase: an activation marker necessary for effector cell function. Immunol Rev 161:111–118
Dunwiddie TV, Diao LH, Proctor WR (1997) Adenine nucleotides undergo rapid, quantitative conversion to adenosine in the extracellular space in rat hippocampus. J Neurosci 17:7673–7682

Dzhandzhugazyan K, Bock E (1993) Demonstration of (Ca^{2+}-Mg^{2+})-ATPase activity of the neural cell adhesion molecule. FEBS Lett 336:279–283

Dzhandzhugazyan K, Bock E (1997) Demonstration of an extracellular ATP-binding site in NCAM: Functional implications of nucleotide binding. Biochemistry 36:15381–15395

Dzhandzhugazyan KN, Kirkin AF, Straten PT, Zeuthen J (1998) Ecto-ATP diphosphohydrolase/CD39 is overexpressed in differentiated human melanomas. FEBS Lett 430:227–230

Ehrlich YH, Kornecki E (1999) Ecto-protein kinases as mediators for the action of secreted ATP in the brain. Prog Brain Res 120:411–426

Enjyoji K, Sévigny J, Lin Y, Frenette P, Christie PD, Schulte am Esch J, Imai M, Edelberger JM, Rayburn H, Lech M, Beeler DM, Csizmadia E, Wagner DD, Robson SC, Rosenberg RD (1999) Targeted disruption of *cd39*/ATP diphosphohydrolase results in disordered hemostasis and thromboregulation. Nature Med 5: 1010–1017

Evans WH (1974) Nucleotide pyrophosphatase, a sialoglycoprotein located on the hepatocyte surface. Nature 250:391–394

Faas MM, Bakker WW, Klok PA, Baller JFW, Schuiling GA (1997) Modulation of glomerular ECTO-ADPase expression by oestradiol. A histochemical study. Thromb Haemost 77:767–771

Fernley HN (1971) Mammalian alkaline phosphatases. In: Boyer PD (ed) The enzymes. Academic Press, New York, pp 417–447

Ferrari D, Chiozzi P, Falzoni S, Dal Susino M, Collo G, Buell G, Di Virgilio F (1997) ATP-mediated cytotoxicity in microglial cells. Neuropharmacology 36:1295–1301

Ferrero E, Malavasi F (1997) Human CD38, a leukocyte receptor and ectoenzyme, is a member of a novel eukaryotic gene family of nicotinamide adenine dinucleotide(+)-converting enzymes: extensive structural homology with the genes for murine bone marrow stromal cell antigen 1 and aplysian ADP-ribosyl cyclase. J Immunol 159:3858–3865

Flaherty KM, DeLuca-Flaherty C, Mckay DB (1990) Three-dimensional structure of the ATPase fragment of a 70K heat-shock cognate protein. Nature 346:623–628

Fox IH (1978) Degradation of purine nucleotides. In: Kelley WN, Weiner IM (eds) Uric acid. Springer, Berlin Heidelberg New York, pp 93–124

Franco R, Casado V, Ciruela F, Saura C, Mallol J, Canela EI, Lluis C (1997) Cell surface adenosine deaminase: much more than an ectoenzyme. Prog Neurobiol 52:283–294

Frassetto SS, Dias RD, Sarkis JJF (1993) Characterization of an ATP disphosphohydrolase activity (APYRASE, EC 3.6.1.5) in rat blood platelets. Mol Cell Biochem 129:47–55

Funaro A, Horenstein AL, Malavasi F (1995) Human CD38: a versatile leukocyte molecule with emerging clinical prospects. Fund Clin Pharmacol 3:101–113

Funaro A, Horenstein AL, Calosso L, Morra M, Tarocco RP, Franco L, De Flora A, Malavasi F (1996) Identification and characterization of an active soluble form of human CD38 in normal and pathological fluids. Int Immunol 8:1643–1650

Fuss B, Baba H, Phan T, Tuohy VK, Macklin WB (1997) Phosphodiesterase I, a novel adhesion molecule and/or cytokine involved in oligodendrocyte function. J Neurosci 17:9095–9103

Gao L, Dong LQ, Whitlock JP (1998) A novel response to dioxin – Induction of ecto-ATPase gene expression. J Biol Chem 273:15358–15365

Gasmi L, Cartwright JL, McLennan AG (1998) The hydrolytic activity of bovine adrenal medullary plasma membranes towards diadenosine polyphosphates is due to alkaline phosphodiesterase-I. Biochim Biophys Acta 1405:121–127

Gayle RB, Maliszewski CR, Gimpel SD, Schoenborn MA, Caspary RG, Richards C, Brasel K, Price V, Drosopoulos JHF, Islam N, Alyonycheva TN, Broekman MJ, Marcus AJ (1998) Inhibition of platelet function by recombinant soluble ecto-ADPase/CD39. J Clin Invest 101:1851–1859

Goding JW, Terkeltaub R, Maurice M, Deterre P, Sali A, Belli SI (1998) Ecto-phosphodiesterase/pyrophosphatase of lymphocytes and non-lymphoid cells: structure and function of the PC-1 family. Immunol Rev 161:11–26

Gordon EL, Pearson JD, Dickinson ES, Moreau D, Slakey LL (1989) The hydrolysis of extracellular adenine nucleotides by arterial smooth muscle cells; Regulation of adenosine production at the cell surface. J Biol Chem 264:18986–18992

Grobben B, Anciaux K, Roymans D, Stefan C, Bollen M, Esmans EL, Slegers H (1999) An ecto-nucleotide pyrophosphatase is one of the main enzymes involved in the extracellular metabolism of ATP in rat C6 glioma. J Neurochem 72:826–834

Haag F, Koch-Nolte F, Gerber A, Schroder J, Thiele HG (1997) Rat T cell differentiation alloantigens RT6.1 and RT6.2 are NAD^+-metabolizing ecto-enzymes that differ in their enzymatic activities. Transplant Proc 29:1699–1700

Handa M, Guidotti G (1996) Purification and cloning of a soluble ATP-diphosphohydrolase (apyrase) from potato tubers (*Solanum tuberosum*). Biochem Biophys Res Commun 218:916–923

Harahap AR, Goding JW (1988) Distribution of the murine plasma cell antigen PC-1 in non-lymphoid cells. J Immunol 141:2317–2320

Harden TK, Lazarowski ER, Boucher RC (1997) Release, metabolism and interconversion of adenine and uridine nucleotides: Implications for G protein-coupled P2 receptor agonist selectivity. Trends Pharmacol Sci 18:43–46

Heine P, Braun N, Zimmermann H (1999) Functional characterization of rat ecto-ATPase and ecto-ATP diphosphohydrolase after heterlogous expression in CHO cells. Eur J Biochem 262:102–107

Hirata O, Kimura N, Sato K, Ohsugi Y, Takasawa S, Okamoto H, Ishikawa J, Kaisho T, Ishihara K, Hirano T (1994) ADP ribosyl cyclase activity of a novel bone marrow stromal cell surface molecule, BST-1. FEBS Lett 356:244–248

Hollmann C, Haag F, Schlott M, Damaske A, Bertuleit H, Matthes M, Kuhl M, Thiele HG, Koch-Nolte F (1996) Molecular characterization of mouse T-cell ecto-ADP-ribosyltransferase Rt6: Cloning of a second functional gene and identification of the Rt6 gene products. Mol Immunol 33:807–817

Hosoda N, Hoshino S, Kanda Y, Katada T (1999) Inhibition of phosphodiesterase/pyrophosphatase activity of PC-1 by its association with glycosaminoglycans. Eur J Biochem 265:763–770

Hourani SMO, Chown JA (1989) The effects of some possible inhibitors of ectonucleotidases on the breakdown and pharmacological effects of ATP in the guinea-pig urinary bladder. Gen Pharmacol 4:413–416

Hourani SMO, Bailey SJ, Nicholls J, Kitchen I (1991) Direct effects of adenylyl 5'-(β,-γ-methylene)diphosphonate, a stable ATP analogue, on relaxant P_1-purinoceptors in smooth muscle. Br J Pharmacol 104:685–690

Imai M, Kaczmarek E, Koziak K, Sévigny J, Goepfert C, Guckelberger O, Csizmadia E, Esch JSA, Robson SC (1999) Suppression of ATP diphosphohydrolase/CD39 in human vascular endothelial cells. Biochemistry 38:13473–13479

Itami C, Taguchi R, Ikezawa H, Nakabayashi T (1997) Release of ectoenzymes from small intestine brush border membranes of mice by phospholipases. Biosci Biotechnol Biochem 61:336–340

James S, Richardson PJ (1993) Production of adenosine from extracellular ATP at the striatal cholinergic synapse. J Neurochem 60:219–227

James SG, Appleby GJ, Miller KA, Steen JT, Colquhoun EQ, Clark MG (1996) Purine and pyrimidine nucleotide metabolism of vascular smooth muscle cells in culture. Gen Pharmacol 27:837–844

Jin-Hua P, Goding JW, Nakamura H, Sano K (1997) Molecular cloning and chromosomal localization of PD-I? (PDNP3), a new member of the human phosphodiesterase I genes. Genomics 45:412–415

Johnson CR, Charlton SJ, Hourani SMO (1996) Responses of the longitudinal muscle and the muscularis mucosae of the rat duodenum to adenine and uracil nucleotides. Br J Pharmacol 117:823–830

Kaczmarek E, Koziak K, Sévigny J, Siegel JB, Anrather J, Beaudoin AR, Bach FH, Robson SC (1996) Identification and characterization of CD39 vascular ATP diphosphohydrolase. J Biol Chem 271:33116–33122

Kaczmarek E, Siegel JB, Sévigny J, Koziak K, Hancock WW, Beaudoin A, Bach FH, Robson SC (1997) Vascular ATP diphosphohydrolase (CD39/ATPDase). In: Plesner L, Kirley TL, Knowles AF (eds) Ecto-ATPases: recent progress on structure and function. Plenum Press, New York, pp 171–185

Kaisho T, Ishikawa J, Oritani K, Inazawa J, Tomizawa H, Muraoka O, Ochi T, Hirano T (1994) BST-1, a surface molecule of bone marrow stromal cell lines that facilitates pre-B-cell growth. Proc Natl Acad Sci USA 91:5325–5329

Kansas GS, Wood GS, Tedder TF (1991) Expression, distribution, and biochemistry of human CD39: Role in activation-associated homotypic adhesion of lymphocytes. J Immunol 146:2235–2244

Kaplan MM (1986) Serum alkaline phosphatase–another piece is added to the puzzle. Hepatol 6:526–528

Kawagoe H, Soma O, Goji J, Nishimura N, Narita M, Inazawa J, Nakamura H, Sano K (1995) Molecular cloning and chromosomal assignment of the humans brain-type phosphodiesterase I/nucleotide pyrophosphatase gene (*PDNP2*). Genomics 30: 380–384

Kawagoe H, Stracke ML, Nakamura H, Sano K (1997) Expression and transcriptional regulation of the PD-I alpha/autotaxin gene in neuroblastoma. Cancer Res. 57: 2516–2521

Kegel B, Braun N, Heine P, Maliszewski CR, Zimmermann H (1997) An ecto-ATPase and an ecto-ATP diphosphohydrolase are expressed in rat brain. Neuropharmacology 36:1189–1200

Kennedy C, Leff P (1995) How should P2X purinoceptors be classified pharmacologically? Trends Pharmacol Sci 16:168–174

Kennedy C, Westfall TD, Sneddon P (1996) Modulation of purinergic neurotransmission by ecto-ATPase. Semin Neurosci 8:195–199

Kennedy C, Todorov LD, Mihaylova-Todorova S, Sneddon P (1997) Release of soluble nucleotidases: A novel mechanism for neurotransmitter inactivation? Trends Pharmacol Sci 18:263–266

Kirley TL (1997) Complementary DNA cloning and sequencing of the chicken muscle Ecto-ATPase – Homology with the lymphoid cell activation antigen CD39. J Biol Chem 272:1076–1081

Kittel A (1997) Role of ecto-ATPases, based on histochemical investigations: Evidences and doubts. In: Plesner L, Kirley TL, Knowles AF (eds) Ecto-ATPases: Recent progress on structure and function. Plenum Press, New York, pp 65–72

Kittel A, Siklós L, Thuróczy G, Somosy Z (1996) Qualitative enzyme histochemistry and microanalysis reveals changes in ultrastructural distribution of calcium and calcium-activated ATPases after microwave irradiation of the medial habenula. Acta Neuropath 92:362–368

Kittel A, Kaczmarek E, Sevigny J, Lengyel K, Csizmadia E, Robson SC (1999) CD39 as a caveolar-associated ectonucleotidase. Biochem Biophys Res Commun 262: 596–599

Knowles AE, Nagy AK (1999) Inhibition of an ecto-ATP-diphosphohydrolase by azide. Eur J Biochem 262:349–357

Knöfel T, Sträter N (1999) X-ray structure of the *Escherichia coli* periplasmic 5'-nucleotidase containing a dimetal catalytic site. Nat Struct Biol 6:448–453

Kohring K, Zimmermann H (1998) Upregulation of ecto-5'-nucleotidase in human neuroblastoma SH-SY5Y cells on differentiation by retinoic acid or phorbolester. Neurosci Lett 258:127–130

Koyamada N, Miyatake T, Candinas D, Hechenleitner P, Siegel J, Hancock WW, Bach FH, Robson SC (1996) Apyrase administration prolongs discordant xenograft survival. Transplantation 62:1739–1743

Kumakura S, Maddux BA, Sung CK (1998) Overexpression of membrane gylcoprotein PC-1 can influence insulin action at a post-receptor site. J Cell Biochem 68:366–377

Kübler D, Pyerin W, Kinzel V (1980) Generation of pyrophosphate from extracellular ATP at the surface of HeLa cells. Eur J Cell Biol 21:231–233

Landt M, Butler LG (1978) 5'-nucleotide phosphodiesterase: Isolation of covalently bound 5'-adenosine monophosphate, an intermediate in the catalytic cycle. Biochemistry 17:4130–4135

Lazarowski ER, Homolya L, Boucher RC, Harden TK (1997) Identification of an ecto-nucleoside diphosphokinase and its contribution to interconversion of P2 receptor agonists. J Biol Chem 272:20402–20407

LeBel D, Poirier GG, Phaneuf S, St.-Jean P, Laliberté JF, Beaudoin AR (1980) Characterization and purification of a calcium-sensitive ATP diphosphohydrolase from pig pancreas. J Biol Chem 255:1227–1233

Lee HY, Murata J, Clair T, Polymeropoulos MH, Torres R, Manrow RE, Liotta LA, Stracke ML (1996a) Cloning chromosomal localization, and tissue expression of autotaxin from human tetracarcinoma cells. Biochem Biophys Res Commun 218:714–719

Lee HY, Clair T, Mulvaney PT, Woodhouse EC, Aznavoorian S, Liotta LA, Stracke ML (1996b) Stimulation of tumor cell motility linked to phosphodiesterase catalytic site of autotaxin. J Biol Chem 271:24408–24412

Lewis Carl S, Kirley TL (1997) Immunolocalization of the ecto-ATPase and ecto-apyrase in chicken gizzard and stomach – Purification and N-terminal sequence of the stomach ecto-apyrase. J Biol Chem 272:23645–23652

Lin S-H, Guidotti G (1989) Cloning and expression of a cDNA coding for a rat liver plasma membrane ecto-ATPase. The primary structure of the ecto-ATPase is similar to that of the human biliary glycoprotein I. J Biol Chem 264:14408–14414

Lund F, Solvason N, Grimaldi JC, Parkhouse RME, Howard M (1995) Murine CD38: An immunoregulatory ectoenzyme. Immunol Today 16:469–473

Lund FE, Cockayne DA, Randall TD, Solvason N, Schuber F, Howard MC (1998) CD38: A new paradigm in lymphocyte activation and signal transduction. Immunol Rev 161:79–93:79–93

Lüthje J, Ogilvie A (1987) Catabolism of AP$_4$A and Ap$_3$A in human serum. Identification of isoenzymes and their partial characterization. Eur J Biochem 169:385–388

Maddux BA, Sbraccia P, Kumakura S, Sasson S, Youngren J, Fisher A, Soencer S, Grupe A, Henzel W, Stewart TA (1995) Membrane glycoprotein PC-1 and insulin resistance in non-insulin-dependent diabetes mellitus. Nature 373:448–451

Maliszewski CR, DeLepesse GJT, Schoenborn MA, Armitage RJ, Fanslow WC, Nakajima T, Baker E, Sutherland GR, Poindexter K, Birks C, Alpert A, Friend D, Gimpel SD, Gayle RB (1994) The CD39 lymphoid cell activation antigen: Molecular cloning and structural characterization. J Immunol 153:3574–3583

Mallone R, Ferrua S, Morra M, Zocchi E, Mehta K, Notarangelo LD, Malavasi F (1998) Characterization of a CD38-like 78-kilodalton soluble protein released from B cell lines derived from patients with X-linked agammaglobulinemia. J Clin Invest 101:2821–2830

Marcus AJ, Broekman MJ, Drosopoulos JHF, Islam N, Alyoncheva TN, Safier LB, Hajjar KA, Posnett DN, Schoenborn MA, Schooley KA, Gayle RB, Maliszewski CR (1997) The endothelial cell ecto-ADPase responsible for inhibition of platelet function is CD39. J Clin Invest 99:1351–1360

Martí E, Canti C, Gomez de Aranda I, Miralles F, Solsona C (1996a) Action of suramin upon ecto-apyrase activity and synaptic depression of *Torpedo* electric organ. Br J Pharmacol 118:1232–1236

Martí E, Gomez de Aranda I, Solsona C (1996b) Inhibition of ATP-diphosphohydrolase (apyrase) of *Torpedo* electric organ by 5'-*p*-fluorosulfonylbenzoyladenosine. Biochim Biophys Acta (Biomembranes) 1282:17–24

Mateo J, Rotllán P, Miras-Portugal MT (1996) Suramin – a powerful inhibitor of neural ecto-adenosine polyphosphate hydrolase. Br J Pharmacol 119:1–2

Mateo J, Rotllan P, Martí E, Gomez de Aranda I, Salsona C, Miras-Portugal MT (1997a) Diadenosine polyphosphate hydrolase from presynaptic plasma membranes of *Torpedo* electric organ. Biochem J 323:677–684

Mateo J, Miras-Portugal MT, Rotllan P (1997b) Ecto-enzymatic hydrolysis of diadenosine polyphosphates by cultured adenomedullary vascular endothelial cells. Am J Physiol Cell Physiol 42:C918–C927

Mateo J, Harden TK, Boyer JL (1999) Functional expression of a cDNA encoding a human ecto-ATPase. Br J Pharmacol 128:396–402

Meerson NR, Delautier D, Durand-Schneider A-M, Moreau A, Schilsky ML, Sternlieb I, Feldmann G, Maurice M (1998) Identification of B10, an alkaline phosphodiesterase of the apical plasma membrane of hepatocytes and biliary cells, in rat serum: Increased levels following bile duct ligation and during the development of cholagiocarcinoma. Hepatol 27:563–568

Meghji P, Burnstock G (1995) Inhibition of extracellular ATP degradation in endothelial cells. Life Sci 57:763–771

Meghji P, Pearson JD, Slakey LL (1995) Kinetics of extracellular ATP hydrolysis by microvascular endothelial cells from rat heart. Biochem J 308:725–731

Mehta K, Shahid U, Malavasi F (1996) Human CD38, a cell-surface protein with multiple functions. FASEB J 10:1408–1417

Miledi R, Molenaar PC, Polak RL (1984) Acetylcholinesterase activity in intact and homogenized skeletal muscle of the frog. J Physiol London 349:663–686

Minelli A, Moroni M, Trinari D, Mezzasoma I (1997) Hydrolysis of extracellular adenine nucleotides by equine epidydimal spermatozoa. Comp Biochem Physiol [B] 117:531–534

Morley DJ, Hawley DM, Ulbright DM, Butler LG, Culp JS, Hodes ME (1987) Distribution of phosphodiesterase I in normal human tissues. J Histochem Cytochem 35:75–82

Mulero JJ, Yeung G, Nelken ST, Ford JE (1999) CD39-L4 is a secreted human apyrase, specific for the hydrolysis of nucleoside diphosphates. J Biol Chem 29:20064–20067

Murata J, Lee HJ, Clair T, Krutzsch HC, Arestad AA, Sobel ME, Liotta LA, Stracke ML (1994) cDNA cloning of the human motility-stimulating protein, autotaxin, reveals a homology with phosphodiesterases. J Biol Chem 269:30479–30484

Nagy AK, Shuster TA, Delgado-Escueta AV (1986) Ecto-ATPase of mammalian synaptosomes: Identification and enzymic characterization. J Neurochem 47:976–986

Nagy AK, Shuster TA, Delgado-Escueta V (1989) Rat brain synaptosomal ATP:AMP-phosphotransferase activity. J Neurochem 53:1166–1172

Nagy AK, Houser CR, Delgado-Escueta AV (1990) Synaptosomal ATPase activities in temporal cortex and hippocampal formation of humans with focal epilepsy. Brain Res 529:192–201

Nagy AK (1997) Ecto-ATPases of the nervous system. In: Plesner L, Kirley TL, Knowles AF (eds) Ecto-ATPases: recent progress in structure and function. Plenum Press, New York, pp 1–13

Nagy AK, Walton NY, Treiman DM (1997) Reduced cortical ecto-ATPase activity in rat brains during prolonged status epilepticus induced by sequential administration of lithium and pilocarpine. Mol Chem Neuropathol 31:135–147

Nagy AK, Knowles AF, Nagami GT (1998) Molecular cloning of the chicken oviduct ecto-ATP-diphosphohydrolase. J Biol Chem 273:16043–16049

Nakabayashi T, Matsuoka Y, Taguchi R, Ikezawa H, Kimura Y (1993) Proof of alkaline phosphodiesterase-I as a phosphatidylinositol-anchor enzyme. Int J Biochem 25:689–696

Nakabayashi T, Matsuoka Y, Ikezawa H, Kimura Y (1994) Alkaline phosphodiesterase I release from eucaryotic plasma membranes by phosphatidylinositol-specific phospholipase C. 4. The release from *Cacia porcellus* organs. Int J Biochem 26:171–179

Narita M, Goji J, Nakamura H, Sano K (1994) Molecular cloning, expression, and localization of a brain-specific phosphodiesterase I/nucleotide pyrophosphatase (PD-I alpha) from rat brain. J Biol Chem 269:28235–28242

Nishina H, Inageda K, Takahashi K, Hoshino S, Ikeda K, Katada T (1994) Cell surface antigen CD38 identified as ecto-enzyme of NAD glycohydrolase has hyaluronate-binding activity. Biochem Biophys Res Commun 203:1318–1323

Oda Y, Kuo MD, Huang SS, Huang JS (1991) The plasma cell membrane glycoprotein, PC-1, is a threonine-specific protein kinase stimulated by acidic fibroblast growth factor. J Biol Chem 266:16791–16795

Oda Y, Kuo MD, Huang SS, Huang JS (1993) The major acidic fibroblast growth factor (αFGF)-stimulated phosphoprotein from bovine liver plasma membranes has αFGF-stimulated kinase, autoadenylation, and alkaline phosphodiesterase activities. J Biol Chem 268:27318–27326

Ogilvie A, Lüthje J, Pohl U, Busse R (1989) Identification and partial characterization of an adenosine(5')tetraphospho(5')adenosine hydrolase on intact bovine aortic endothelial cells. Biochem J 259:97–103

Okawa A, Nakamura I, Goto S, Moriya H, Nakamura Y, Ikegawa S (1998) Mutation in *Npps* in a mouse model of ossification of the posterior longitudinal ligament of the spine. Nature Genet 19:271–273

Oyajobi BO, Russell RGG, Caswell AM (1994) Modulation of ecto-nucleoside triphosphate pyrophosphatase activity of human osteoblast-like bone cells by 1α,25-dihydroxyvitamin D_3, 24R,25-dihydroxyvitamin D_3, parathyroid hormone, and dexamethasone. J Bone Miner Res 9:1259–1266

Pain T, Headrick JP (1996) Effects and metabolites of NAD in the perfused rat heart. Drug Develop Res 37:150

Pappas GD, Kriho V (1988) Fine structural localization of Ca^{2+}-ATPase activity at the forg neuromuscular junction. J Neurocytol 17:417–423

Pearson JD (1985) Ectonucleotidases. Measuremem of activities and use of inhibitors. In: Paton DM (ed) Methods in pharmacology 6. Plenum Press, New York, pp 83–107

Peola S, Borrione P, Matera L, Malavasi F, Pileri A, Massaia M (1996) Selective induction of CD73 expression in human lymphocytes by CD38 ligation: a novel pathway linking signal transducers with ecto-enzyme activities. J Immunol 157:4354–4362

Peoples RW, Li CY (1998) Inhibition of NMDA-gated ion channels by the P2 purinoceptor antagonists suramin and reactive blue 2 in mouse hippocampal neurones. Br J Pharmacol 124:400–408

Picher M, Coté YP, Béliveau R, Potier M, Beaudoin AR (1993) Demonstration of a novel type of ATP-diphosphohydrolase (EC 3.6.1.5) in the bovine lung. J Biol Chem 268:4699–4703

Picher M, Beliveau R, Potier M, Savaria D, Rousseau E, Beaudoin AR (1994) Demonstration of an ecto-ATP-diphosphohydrolase (EC 3.6.1.5.) in non-vascular smooth muscles of the bovine trachea. Biochim Biophys Acta (Gen Subj) 1200:167–174

Picher M, Sévigny J, D'Orléans-Juste P, Beaudoin AR (1996) Hydrolysis of P2-purinoceptor agonists by a purified ectonucleotidase from the bovine aorta, the ATP- diphosphohydrolase. Biochem Pharmacol 51:1453–1460

Pilla C, Emanuelli T, Frassetto SS, Battastini AMO, Dias RD, Sarkis JJF (1996) ATP diphosphohydrolase activity (apyrase, EC 3.6.1.5) in human blood platelets. Platelets 7:225–230

Pintor J, Hoyle CHV, Gualix J, Miras-Portugal MT (1997) Mini-Review: Diadenosine polyphosphates in the central nervous system. Neurosci Res Commun 20:69–78

Plesner L (1995) Ecto-ATPases: identities and functions. Int Rev Cytol 158:141–214

Plesner L, Kirley TL, Knowles AF (eds) (1997) Ecto-ATPases: Recent progress on structure and function. Plenum Press, New York

Prasad GS, McRee DE, Stura EA, Levitt DG, Lee HC, Stout CD (1996) Crystal structure of aplysia ADP ribosyl cyclase, a homologue of the bifunctional ectozyme CD38. Nat Struct Biol 3:957–964

Ramos A, Pintor J, Miras-Portugal MT, Rotllan P (1995) Use of fluorogenic substrates for detection and investigation of ectoenzymatic hydrolysis of diadenosine polyphosphates: a fluorometric study on chromaffin cells. Anal Biochem 228:74–82

Rebbe NF, Tong BD, Finley EM, Hickman S (1991) Identification of nucleotide pyrophosphatase/alkaline phosphodiesterase I activity associated with the mouse plasma cell differentiation antigen PC-1. Proc Natl Acad Sci USA 88:5192–5196

Rebbe NF, Tong BD, Hickman S (1993) Expression of nucleotide pyrophosphatase and alkaline phosphodiesterase I activities of PC-1, the murine plasma cell antigen. Mol Immunol 30:87–93

Redegeld FA, Caldwell CC, Sitkovsky MV (1999) Ecto-protein kinases: ectodomain phosphorylation as a novel target for pharmacological manipulation? Trends Pharmacol Sci 20:453–459

Resta R, Thompson LF (1997) T cell signalling through CD73. Cell Signal 9:131–139

Resta R, Yamashita Y, Thompson LF (1998) Ecto-enzyme and signaling functions of lymphocyte CD73. Immunol Rev 161:95–109:95–109

Robson SC, Kaczmarek E, Siegel JB, Candinas D, Koziak K, Millan M, Hancock WW, Bach FH (1997) Loss of ATP diphosphohydrolase activity with endothelial cell activation. J Exp Med 185:153–163

Rodriguez-Pascual F, Torres M, Rotllan P, Miras-Portugal MT (1992) Extracellular hydrolysis of diadenosine polyphosphates, Ap_nA, by bovine chromaffin cells in culture. Arch Biochem Biophys 297:176–183

Santos AN, Riemann D, Kehlen A, Thiele K, Langner J (1996) Treatment of fibroblast-like synoviocytes with IFN-g results in the downregulation of autotaxin mRNA. Biochem Biophys Res Commun 229:419–424

Sarkis JJF, Battastini AMO, Oliveira EM, Frassetto SS, Dias RD (1995) ATP diphosphohydrolases: an overview. Ciencia e Cultura 47:131–136

Sarkis JJF, Salto C (1991) Characterization of a synaptosomal ATP diphosphohydrolase from the electric organ of *Torpedo marmorata*. Brain Res Bull 26:871–876

Sawa H, Kamada K, Sato H, Sendo S, Kondo A, Saito I, Edlund M, Öbrink B (1994) C-CAM expression in the developing rat central nervous system. Brain Res Dev Brain Res 78:35–43

Schetinger MRC, Falquembach F, Michelot F, Mezzomo A, Rocha JBT (1998a) Heparin modulates adenine nucleotide hydrolysis by synaptosomes from cerebral cortex. Neurochem Int 33:243–249

Schetinger MRC, Bonan CD, Schierholt R, Webber A, Arteni N, Emanuelli T, Dias RD, Sarkis JJF, Netto CA (1998b) Nucleotide hydrolysis in rats submitted to global cerebral ischemia: A possible link between preconditioning and adenosine production. J Stroke Cerebrovasc Dis 7:281–286

Schoenborn MA, Jenkins NA, Copeland NG, Gilbert DJ, Gayle RB, Maliszewski CR (1998) Gene structure and chromosome location of mouse Cd39 coding for an ecto-apyrase. Cytogenet Cell Genet 81:287–289

Schulte am Esch JSA, Sévigny J, Kaczmarek E, Siegel JB, Imai M, Koziak K, Beaudoin AR, Robson SC (1999) Structural elements and limited proteolysis of CD39 influence ATP diphosphohydrolase activity. Biochemistry 38:2248–2258

Schwab DA, Wilson JE (1989) Complete amino acid sequence or rat brain hexokinase, deduced from the cloned cDNA, and a proposed structure of a mammalian hexokinase. Proc Natl Acad Sci USA 86:2563–2567

Scott LJ, Delautier D, Meerson NR, Trugnan G, Goding JW, Maurice M (1997) Biochemical and molecular identification of distinct forms of alkaline phosphodiesterase I expressed on the apical and basolateral plasma membrane surfaces of rat hepatocytes. Hepatol 25:995–1002

Servos J, Reiländer H, Zimmermann H (1998) Catalytically active soluble ecto-5'-nucleotidase purified after heterologous expression as a tool for drug screening. Drug Develop Res 45:269–276

Sévigny J, Picher M, Grondin G, Beaudoin AR (1997a) Purification and immunohistochemical localization of the ATP diphosphohydrolase in bovine lungs. Am J Physiol 272:L939–L950
Sévigny J, Levesque FP, Grondin G, Beaudoin AR (1997b) Purification of the blood vessel ATP diphosphohydrolase, identification and localisation by immunological techniques. Biochim Biophys Acta (Gen Subj) 1334:73–88
Silvennoinen O, Nishigaki H, Kitanaka A, Kumagai M-A, Ito C, Malavasi F, Lin Q, Conley ME, Campana D (1996) CD38 signal transduction in human B cell precursors. Rapid induction of tyrosine phosphorylation, activation of *syk* tyrosine kinase, and phosphorylation of phospholipase Cγ and phosphytidylinotisdol 3 kinase. J Immunol 156:100–107
Slakey LL, Dickinson ES, Goldman SJ, Gordon EL, Meghji P, Pearson JD (1997) The hydrolysis of extracellular adenine nucleotides by cultured vascular cells and cardiac myocytes. In: Plesner L, Kirley TL, Knowles AF (eds) Ecto-ATPases: recent progress on structure and function. Plenum Press, New York, pp 27–32
Smith TM, Kirley TL, Hennessey TM (1997) A soluble ecto-ATPase from *Tetrahymena thermophila*: Purification and similarity to the membrane-bound ecto-ATPase of smooth muscle. Arch Biochem Biophys 337:351–359
Smith TM, Kirley TL (1998) Cloning, sequencing, and expression of a human brain ecto-apyrase related to both the ecto-ATPases and CD39 ecto- apyrases. Biochim Biophys Acta 1386:65–78
Smith T, Carl SAL, Kirley TL (1998) Immunological detection of ecto-ATPase in chicken and rat tissues: Characterization, distribution, and a cautionary note. Biochem Mol Biol Int 45:1057–1066
Smith TM, Kirley TL (1999a) Glycosylation is essential for functional expression of a human brain ecto-apyrase. Biochemistry 38:1509–1516
Smith TM, Kirley TL (1999b) Site-directed mutagenesis of a human brain ecto-apyrase: evidence that the E-type ATPases are related to the actin/heat shock 70/sugar kinase superfamily. Biochemistry 38:321–328
Smith TM, Carl SAL, Kirley TL (1999) Mutagenesis of two conserved tryptophan residues of the E- type ATPases: inactivation and conversion of an ecto-apyrase to an ecto-NTPase. Biochemistry 38:5849–5857
Snell CR, Snell PH, Richards CD (1984) Degradation of NAD by synaptosomes and its inhibition by nicotinamide mononucleotide: implications for the role of NAD as a synaptic modulator. J Neurochem 43:1610–1615
Solan JL, Deftos LJ, Goding JW, Terkeltaub RA (1996) Expression of the nucleoside triphosphate pyrophosphohydrolase PC-1 is induced by basic fibroblast growth factor (BFGF) and modulated by activation of the protein kinase A and C pathways in osteoblast-like osteosarcoma cells. J Bone Miner Res 11:183–192
Sperlagh B, Kittel A, Lajtha A, Vizi ES (1995) ATP acts as fast neurotransmitter in rat habenula: neurochemical and enzymecytochemical evidence. Neuroscience 66:915–920
Stearne PA, van Driel IR, Grego B, Simpson RJ, Goding JW (1985) The murine plasma cell antigen PC-1: purification and partial amino acid sequence. J Immunol 134:443–448
Stefan C, Stalmans W, Bollen M (1996) Threonine autophosphorylation and nucleotidylation of the hepatic membrane protein PC-1. Eur J Biochem 241:338–342
Stefan C, Stalmans W, Bollen M (1998) Growth-related expression of the ectonucleotide pyrophosphatase PC-1 in rat liver. Hepatol 28:1497–1503
Stefan C, Gijsbers R, Stalmans W, Bollen M (1999) Differential regulation of the expression of nucleotide pyrophosphatases phosphodiesterases in rat liver. Biochim Biophys Acta 1450:45–52
Stefanovic V, Ledig M, Mandel P (1976) Divalent cation-activated ecto-nucleoside triphosphatase activity of nervous system cells in tissue culture. J Neurochem 27:799–805

Stefanovic V, Vlahovic P, Ardaillou R (1995) Characterization and control of expression of cell surface alkaline phosphodiesterase I activity in rat mesangial glomerular cells. Renal Physiol Biochem 18:12–20

Stout JG, Kirley TL (1995) Inhibition of purified chicken gizzard smooth muscle ecto-ATPase by P2 purinoceptor antagonists. Biochem Mol Biol Int 36:927–934

Stout JG, Kirley TL (1996) Control of cell membrane ecto-ATPase by oligomerization state: intermolecular cross-linking modulates ATPase activity. Biochemistry 35:8289–8298

Stracke ML, Krutzsch HC, Unsworth EJ, Arestad AA, Cioce V, Schiffmann E, Liotta LA (1992) Identification, purification, and partial sequence analysis of autotaxin, a novel motility-stimulating protein. J Biol Chem 267:2524–2529

Stracke ML, Arestad A, Levine M, Krutzsch HC, Liotta LA (1995) Autotaxin is an N-linked glycoprotein but the sugar moieties are not needed for its stimulation of cellular motility. Melanoma Res 5:203–209

Strobel RS, Nagy AK, Knowles AF, Buegel J, Rosenberg MD (1996) Chicken oviductal ecto-ATP-diphosphohydrolase. Purification and characterization. J Biol Chem 271:16323–16331

Sud'ina F, Mirzoeva OK, Galkina SI, Pushkareva MA, Ullrich V (1998) Involvement of ecto-ATPase and extracellular ATP in polymorphonuclear granulocyte-endothelial interactions. FEBS Lett 423:243–248

Terrian DM, Hernandez PG, Rea MA, Peters RI (1989) ATP release, adenosine formation, and modulation of dynorphin and glutamic acid release by adenosine analogues in rat hippocampal mossy fiber synaptosomes. J Neurochem 53:1390–1399

Thirion S, Troadec JD, Nicaise G (1996) Cytochemical localization of ecto-ATPses in rat neurohypophysis. J Histochem Cytochem 44:103–111

Todorov LD, Mihaylova Todorova S, Westfall TD, Sneddon P, Kennedy C, Bjur RA, Westfall DP (1997) Neuronal release of soluble nucleotidases and their role in neurotransmitter inactivation. Nature 387:76–79

Torres M, Pintor J, Miras-Portugal MT (1990) Presence of ectonucleotidases in cultured chromaffin cells: hydrolysis of extracellular adenine nucleotides. Arch Biochem Biophys 279:37–44

Tuluc F, Bültmann R, Glanzel M, Frahm AW, Starke K (1998) P2-receptor antagonists. 4. Blockade of P2-receptor subtypes and ecto-nucleotidases by compounds related to reactive blue 2. Naunyn Schmiedebergs Arch Pharmacol 357:111–120

van Driel IR, Goding JW (1987) Plasma cell membrane glycoprotein PC-1: primary structure deduced from cDNA clones. J Biol Chem 262:4882–4887

van Son WJ, Wit F, van Balen OLB, Tegzess AM, Ploeg RJ, Bakker WW (1997) Decreased expression of glomerular ecto-ATPase in kidney grafts with delayed graft function. Transplant Proc 29:352–354

van Driel IR, Wilks AF, Pietersz GA, Goding JW (1985) Murine plasma cell membrane antigen PC-1: Molecular cloning of cDNA and analysis of expression. Proc Natl Acad Sci USA 82:8619–8623

Vasconcelos EG, Ferreira ST, Carvalho TMU, de Souza W, Kettlun AM, Mancilla M, Valenzuela MA, Verjovski-Almeida S (1996) Partial purification and immunohistochemical localization of ATP diphosphohydrolase from *Schistosoma mansoni*. Immunological cross-reactivities with potato apyrase and *Toxoplasma gondii* nucleoside triphosphate hydrolase. J Biol Chem 271:22139–22145

Vigne P, Breittmayer JP, Frelin C (1998) Analysis of the influence of nucleotidases on the apparent activity of exogenous ATP and ADP at $P2Y_1$ receptors. Br J Pharmacol 125:675–680

Wang TF, Guidotti G (1996) CD39 is an ecto-(Ca^{2+},Mg^{2+})-apyrase. J Biol Chem 271:9898–9901

Wang TF, Rosenberg PA, Guidotti G (1997) Characterization of brain ecto-apyrase: evidence for only one ecto-apyrase (CD39) gene. Mol Brain Res 47:295–302

Wang TF, Guidotti G (1998a) Golgi localization and functional expression of human uridine diphosphatase. J Biol Chem 273:11392–11399

Wang TF, Guidotti G (1998b) Widespread expression of ecto-apyrase (CD39) in the central nervous system. Brain Res 790:318–322

Wang TF, Ou Y, Guidotti G (1998) The transmembrane domains of ectoapyrase (CD39) affect its enzymatic activity and quaternary structure. J Biol Chem 273: 24814–24821

Welford LA, Cusack NJ, Hourani MO (1986) ATP analogues and the guinea-pig taenia coli: A comparison of the structure-activity relationships of ectonucleotidases with those of the P2-purinoceptor. Eur J Pharmacol 129:217–224

Welford LA, Cusack NJ, Hourani SMO (1987) The structure-activity relationships of ectonucleotidases and the excitatory P2-purinoceptors: Evidence that dephosphorylation of ATP analogues reduces pharmacological potency. Eur J Pharmacol 141:123–130

Whyte PP (1996) Hypophosphatasia: natures window on alkaline phosphatase function in man. In: Bilezkian J, Raisz L, Rodan G (eds) Principles of bone biology. Academic Press, San Diego, pp 951–968

Wiendl HS, Schneider C, Ogilvie A (1998) Nucleotide metabolizing ectoenzymes are upregulated in A431 cells periodically treated with cytostatic ATP leading to partial resistance without preventing apoptosis. Biochim Biophys Acta (Mol Cell Res) 1404:282–298

Wieraszko A, Ehrlich YH (1994) On the role of extracellular ATP in the induction of long- term potentiation in the hippocampus. J Neurochem 63:1731–1738

Wittenburg H, Bültmann R, Pause B, Ganter C, Kurz G, Starke K (1996) P2-purinoceptor antagonists: II. Blockade of P2-purinoceptor subtypes and ecto-nucleotidases by compounds related to Evans blue and trypan blue. Naunyn Schmiedebergs Arch Pharmacol 354:491–497

Ziganshin AU, Hoyle CHV, Burnstock G (1994a) Ecto-enzymes and metabolism of extracellular ATP. Drug Develop Res 32:134–146

Ziganshin AU, Hoyle CHV, Ziganshina LE, Burnstock G (1994b) Effects of cyclopiazonic acid on contractility and ecto- ATPase activity in guinea-pig urinary bladder and vas deferens. Br J Pharmacol 113:669–674

Ziganshin AU, Ziganshina LE, Bodin P, Bailey D, Burnstock G (1995a) Effects of P2-purinoceptor antagonists on ecto-nucleotidase activity of guinea-pig vas deferens cultured smooth muscle cells. Biochem Mol Biol Int 36:863–869

Ziganshin AU, Berdnikov EA, Ziganshina LE, Tantasheva FR, Hoyle CHV, Burnstock G (1995b) Effects of α,β-unsaturated sulphones and phosphonium salts on ecto-ATPase activity and contractile responses mediated via P2x-purinoceptors. Gen Pharmacol 26:527–532

Ziganshin AU, Ziganshina LE, Hoyle CHV, Burnstock G (1995c) Effects of divalent cations and La^{3+} on contractility and ecto- ATPase activity in the guinea-pig urinary bladder. Br J Pharmacol 114:632–639

Ziganshin AU, Ziganshina LE, King BF, Pintor J, Burnstock G (1996) Effects of P2-purinoceptor antagonists on degradation of adenine nucleotides by ectonucleotidases in folliculated oocytes of *Xenopus laevis*. Biochem Pharmacol 51:897–901

Zimmermann H (1992) 5'-Nucleotidase: molecular structure and functional aspects. Biochem J 285:345–365

Zimmermann H (1994) Signalling via ATP in the nervous system. Trends Neurosci 17:420–426

Zimmermann H (1996a) Biochemistry, localization and functional roles of ectonucleotidases in the nervous system. Prog Neurobiol 49:589–618

Zimmermann H (1996b) Extracellular purine metabolism. Drug Develop Res 39: 337–352

Zimmermann H, Pearson J (1998) Extracellular metabolism of nucleotides and adenosine in the cardiovascular system. In: Burnstock G, Dobson JG, Liang BT, Linden J (eds) Cardiovascular biology of purines. Kluwer, London pp 342–358

Zimmermann H (1999a) Two novel families of ecto-nucleotidases: Molecular structures, catalytic properties, and a search for function. Trends Pharmacol Sci 20: 231–236

Zimmermann H (1999b) Nucleotides and cd39: principal modulatory players in hemostasis and thrombosis. Nature Med 5:987–988

Zimmermann H, Braun N (1999) Ecto-nucleotidases: molecular structures, catalytic properties, and functional roles in the nervous system. Prog Brain Res 120:371–385

Zimmermann H, Beaudoin AR, Bollen M, Goding JW, Guidotti G, Kirley TL, Robson SC, Sano K (2000) Proposed nomenclature for two novel nucleotide hydrolyzing enzyme families expressed on the cell surface. In: Vanduffel L, Lemmens R (eds) Ecto-ATPases and related ectonucleotidases, Shaker Publishing BV, Maastricht, pp 1–8

CHAPTER 9
Role of Purines and Pyrimidines in the Central Nervous System

S.A. MASINO and T.V. DUNWIDDIE

A. Introduction

Purines and pyrimidines present an interesting paradox in the nervous system. Molecules such as adenosine, ATP, and UTP play an essential role in cellular metabolism, and are found in every cell in the brain. Nevertheless, it is now well-established that both of these molecules play an important secondary role as extracellular signaling molecules, and that there are specific receptors for both adenosine and ATP (and in some cases for other nucleotides such as ADP and UTP) on neurons as well as glial cells. The widespread distribution of ligands and receptors, the relatively recent cloning of multiple subtypes of nucleoside and nucleotide receptors, and their diverse effects throughout the nervous system, have provided the stimuli to accelerate research in this area. There are many lines of circumstantial evidence to suggest that the two major roles of purines may be linked, i.e., that activation of at least some purinergic receptors may be a way of communicating to other cells information about metabolic state. However, while there is a great deal known about purinergic signaling, some of the more important functional questions remain unresolved. For example, there is abundant evidence that some adenosine receptors in the brain are tonically activated by the basal concentrations of adenosine in the extracellular space. However, the significance of this inhibition is unclear and, in many cases, blockade of this tonic inhibitory influence (e.g., with caffeine) seems to have beneficial rather than deleterious effects on the overall function of the nervous system.

The primary focus of this chapter will be on rapidly developing areas of research, with emphasis on functional studies that provide insights into the potential role played by these signaling molecules in the central nervous system.

B. P1 Adenosine Receptors in Nervous Tissue

In very general terms, A_1 receptors in the CNS have a largely inhibitory action on neural activity, A_{2A} receptors generally facilitate neurotransmission, while the A_{2B} and A_3 receptors have more subtle effects on electrophysiological activity that are not usually observed as a direct modulation of ion channel activity or synaptic transmission. A_{2A} receptors reside predominantly in dopamine-rich areas, such as the striatum, and modulate the effects of several neurotransmitters, including dopamine. A_{2B} and A_3 receptors are both widely distributed at a relatively low level, but the physiological role of these relatively low-affinity adenosine receptors is neither as well understood nor as extensively investigated as that of A_1 and A_{2A} receptors.

I. Physiological Responses

1. Electrophysiological Actions

Adenosine (P1) receptors are located both pre- and postsynaptically in the CNS, and their activation causes a number of direct effects on ion channels and second messenger systems. In addition, adenosine receptor activation may modulate or interact with the effects of other neurotransmitters. One mechanism coupled to adenosine A_1 receptors is activation of a G-protein-coupled inwardly-rectifying K^+ channel (GIRK), the opening of which results in an outward K^+ current (DASCAL 1997). Because outward K^+ currents cause membrane hyperpolarization, activation of GIRKs is a general way of regulating cell excitability (EHRENGRUBER et al. 1997). While opening of this K^+ channel is a prominent neuronal consequence of A_1 receptor activation, GIRKs are not exclusively coupled to adenosine receptors but can be activated by a wide range of neurotransmitter receptors, including the $5HT_{1A}$ receptor, opioid receptors, muscarinic receptors, and the $GABA_B$ receptor. As discussed below with respect to receptor interactions(see Sect. B.I.1.d), a novel and synergistic effect of low concentrations of adenosine and GABA on GIRKs has been described (SODICKSON and BEAN 1998).

Another effect linked to adenosine receptor activation is modulation of Ca^{2+} channel activity. All subtypes of adenosine receptors have been reported to influence the activity of voltage-dependent calcium channels (VDCCs). A_1 receptor activation has been reported to inhibit Ca^{2+} influx in several neuronal systems. This reduction in Ca^{2+} channel activity has been observed in hippocampal (AMBROSIO et al. 1997) and cerebrocortical (VAZQUEZ and SANCHEZ-PRIETO 1997) synaptosomes. Fluorescent measurement of Ca^{2+} concentration in nerve terminals has been used to demonstrate a presynaptic effect of adenosine A_1 receptors on calcium channels (QIAN et al. 1997). In the rat the N-type Ca^{2+} channel and unknown "non-LNPQ" Ca^{2+} channels were both inhibited by adenosine. Inhibition of Ca^{2+} channels by adenosine has also been observed in the soma and proximal dendrites of hippocampal neurons, and even more dramatically in the distal dendrites (CHEN and LAMBERT 1997). Thus, available evidence demonstrates both pre- and postsynaptic inhibition of several types of VDCCs by A_1 receptor activation.

In keeping with their largely excitatory role, both A_{2A} and A_{2B} receptors increase Ca^{2+} channel activity, and the balance between the inhibitory and excitatory effects of A_1 vs A_{2A} receptor activation is likely due to regional distributions of these receptor subtypes. In support of this, A_{2A} receptor activation in the hippocampus facilitated calcium influx through class A (P/Q type) calcium channels in CA3 synaptosomes, but not in synaptosomes from the CA1 region, an area where A_1 receptors predominate (GONCALVES et al. 1997). A_{2B} receptor activation has been shown to increase Ca^{2+} currents and/or intracellular calcium levels both in neurons (MOGUL et al. 1993) and in astrocytes (PILITSIS and KIMELBERG 1998). In acutely prepared cortical astrocytes from very young rats (P4–P12) A_{2B} receptor activation stimulated a rise in intracellular calcium which did not depend on extracellular calcium, suggesting release from intracellular stores (PILITSIS and KIMELBERG 1998). Like the A_{2A} and A_{2B} adenosine receptor subtypes, A_3 receptors have been also been reported to enhance Ca^{2+} currents, specifically those mediated by high threshold Ca^{2+} channels (FLEMING and MOGUL 1997). Thus, modulation of Ca^{2+} channel activity by adenosine, both in a facilitatory and in an inhibitory fashion, can result in alterations in cellular activity that can be tied to changes in this important intracellular messenger.

a) Modulation of Transmission

The primary consequence of adenosine receptor activation in many systems is an inhibitory modulation of synaptic transmission (for reviews, see FREDHOLM and DUNWIDDIE 1988; SEBASTINO and RIBERIO 2000), an effect that is primarily mediated via the inhibition of Ca^{2+} influx into the presynaptic nerve terminal, and a consequent inhibition of transmitter release. There are also likely to be inhibitory mechanisms that do not involve Ca^{2+} influx, because neurotransmitter release that occurs in the complete absence of extracellular Ca^{2+} can also be inhibited (SILINSKY 1984; SCANZIANI et al. 1992; SCHOLZ and MILLER 1992). In addition to a tonic inhibitory influence on neurotransmission, there is evidence that adenosine can modulate excitatory synaptic transmission in a more specific fashion at the level of individual neurons. Loading adenosine via a whole-cell patch pipette into an individual hippocampal pyramidal cell results in an increase in extracellular adenosine, and an adenosine-mediated reduction in excitatory transmission at synapses onto that neuron, but not onto other neighboring cells (BRUNDEGE and DUNWIDDIE 1997). This result suggests that the potential exists for local, cell-specific modulation of transmission by adenosine. Thus, adenosine may in some instances be a retrograde messenger that provides a means for a postsynaptic cell to regulate the strength of its excitatory inputs.

Although the inhibitory modulation of excitatory transmission mediated by A_1 receptors is quite common, inhibition of GABA release is relatively rare; nevertheless there are a few instances where this does occur. SHEN and JOHNSON (1997) reported that A_1 receptor activation inhibits GABA release in the substantia nigra, and CHEN and VAN DEN POL (1997) found A_1- as well as A_2-like receptor mediated inhibition of GABA release in the hypothalamus. There may also be postsynaptic mechanisms for regulation of receptor sensitivity to consider as well, as is suggested by a report that in isolated dorsal root ganglion neurons, adenosine inhibited GABA-activated currents in the majority of cells (75%), but potentiated these currents in 18% of the neurons recorded (Hu and Li 1997).

As previously discussed, A_2 receptors often facilitate transmission. Li and Henry (1998) found both pre- and post-synaptic excitatory effects of A_2-like receptor activation in CA1 pyramidal cells. Although synaptic transmission in many areas is tonically inhibited by endogenous adenosine (acting via A_1 receptors), in vivo recordings in the superior colliculus revealed that an excitatory influence of adenosine predominates (Hirai and Okada 1995; Ishikawa et al. 1997). This excitatory effect of endogenous adenosine is mediated by A_{2A} receptors, despite the higher expression of A_1 receptors in this region (Ishikawa et al. 1997).

b) Modulation of Plasticity

In accord with its role in modulating intracellular second messengers as well as synaptic transmission, adenosine has also been shown to influence long-term modifications of transmission, which include processes such as long-term potentiation (LTP) and long-term depression (LTD). The effects of adenosine are quite complex, and include both enhancement and antagonism of LTP, a reduction in LTD (Kemp and Bashir 1997), mediation of short-term depression (Lovinger and Choi 1995), and depotentiation (Abraham and Huggett 1997). The reported interactions between adenosine and neuronal plasticity are extensive, and have been reviewed by de Mendonça and Ribeiro (1996). As with other physiological effects, the role of adenosine in synaptic plasticity is likely determined by the regional expression of adenosine receptor subtypes, so that a "unitary" role for adenosine in synaptic plasticity may not be forthcoming.

c) Atypical A_{2A} Receptors

The primary expression of the A_{2A} receptor is in dopaminergic areas of the brain, and early reports suggested that these were the only regions that had detectable levels of receptor binding. However, although the highest concentrations of A_{2A} receptors (as defined by ^{3}H-CGS 21680 binding) are found in the striatum, nucleus accumbens, and olfactory tubercle, more recent studies have demonstrated that specific binding is also detectable in the cerebral cortex and hippocampus (Fredholm et al. 1993), and that the binding site displays somewhat different characteristics from the A_{2A} site in dopaminergic regions. An A_{2A} receptor with "typical" A_{2A} pharmacology is found in the hippocampus and cortex with low abundance, but the predominant binding site for ^{3}H-CGS 21680 in the cortex and hippocampus displays different pharmacological characteristics. These A_{2A} receptors, which have an unusually low affinity for several of the classical antagonists that are effective at striatal A_{2A} receptors [such as KF17837 and chlorostyryl-caffeine (CSC)], are referred to as "atypical" A_{2A} receptors. The molecular explanation for the "atypical" site is not clear, but might reflect alternative splicing, post-translational processing of the receptor, or some other unknown process. The two different binding sites for the A_{2A} receptor have been characterized by the binding of both A_{2A}-specific agonists and antagonists (CUNHA et al. 1996, 1997) and physiological responses attributable to the atypical A_{2A} receptor have been described in the hippocampus (JIN and FREDHOLM 1997). While A_{2A} receptor activation did not affect neurotransmitter release in the striatum, it facilitated acetylcholine release in the hippocampus (JIN and FREDHOLM 1997). Consistent with these results, A_{2A} receptor activation in the hippocampus caused activation of class A calcium channels (GONCALVES et al. 1997), which provides a cellular mechanism for this facilitation of neurotransmitter release. This was found in synaptosomes prepared from the CA3 region of the hippocampus, but not in synaptosomes prepared from CA1. Because A_{2A} receptors bind adenosine with somewhat lower affinity than A_1 receptors, the activation of A_{2A} receptors, and possible interactions between A_1 and A_{2A} activation (O'KANE and STONE 1998), may not occur at normal "basal" concentrations of adenosine, but would be apparent when adenosine concentrations are elevated, and would be brain region specific (e.g., the CA3 but not the CA1 region of the hippocampus). Further characterization of this atypical site may reveal novel effects of atypical A_{2A} receptor activation in other areas of the hippocampus, or in areas such as the cerebral cortex.

d) Interaction between Receptors

In addition to their direct actions, adenosine receptors have been shown to interact with each other, as well as with receptors for other neurotransmitters, and in some areas (e.g., in the striatum) there are extensive biochemical and physiological data to support such interactions. While many studies have tried to identify direct actions linked to activation of adenosine receptors, sometimes without success, their primary role in some brain regions may be largely related to influences on other neurotransmitter receptors. These kinds of interactions have been documented at the cellular, physiological, and even behavioral levels.

A_{2A}/A_1 interactions constitute one such locus of action, and physiological interactions between these receptors have been reported in both striatum and hippocampus. In striatal synaptosomes A_{2A} receptor activation has been shown to desensitize A_1 receptors by reducing the binding affinity of A_1 receptors in a PKC-dependent manner (DIXON et al. 1997). Two studies have suggested that there may be an antagonistic A_{2A}/A_1 interaction in the hippocampus (CUNHA et al. 1994; O'KANE and STONE 1998), although other studies have reported no interactions (DUNWIDDIE et al. 1997). An important difference between these studies is that the latter focused on presynaptic, release-modulating A_1 receptors, whereas the two previous studies recorded population spikes, which are significantly affected by the hyperpolarizing effects of activation of postsynaptic A_1 receptors. These results would suggest that the locus of interaction between A_{2A} receptors and A_1 receptors in hippocampus may be exclusively postsynaptic. A_3 receptor activation has also been shown to cause a heterologous desensitization of presynaptic A_1 receptor-mediated responses in hippocampal slices (DUNWIDDIE et al. 1997), as well as of presynaptic metabotropic receptors (MACEK et al. 1998). The desensitization was specific to these two presynaptic receptors, as presynaptic modulation mediated via $GABA_B$ and muscarinic receptors was unaffected by pretreatment with the A_3 agonist (DUNWIDDIE et al. 1997). The receptor interactions between A_3 and A_1 receptors may be relevant under particularly high concentrations of extracellular adenosine, such as during hypoxia or ischemia, because the concentrations of adenosine required to activate A_3 receptors are quite high (DUNWIDDIE et al. 1997).

In addition to cross-talk between adenosine receptors, interactions occur between adenosine receptors and other neurotransmitter receptors including dopamine, glutamate, and GABA receptors. Adenosine A_{2A} and dopamine D_2 receptors are co-localized on striatopallidal neurons, and exert opposing effects on locomotor behavior and synaptic transmission (MAYFIELD et al. 1996). There is a similarly antagonistic interaction between A_1 and D_1 receptors. The relationship between these receptors at the cellular, physiological, and behavioral level have been reviewed by FERRE et al. (1997), FUXE et al. (1998), and FERRE (1997), who note that there may be some interesting clinical implications in terms of therapeutic opportunities for diseases related to altered dopaminergic function. The adenosine antagonist caffeine, consumed as coffee, has been shown to reduce the incidence of Parkinson's disease in humans in a 30-year longitudinal study of Japanese-American males (WEBSTER et al. 2000). There is also evidence to suggest that adenosine receptors interact with glutamate receptors. During hypoxia, transmitter release is inhibited by adenosine acting at A_1 receptors (discussed in more detail in Sect. B.I.3.a), and this inhibition is not affected by glutamate receptor antagonists. However, when A_1 receptors in the hippocampus are blocked during hypoxia, an inhibitory effect linked to activation of metabotropic glutamate receptors is revealed (DE MENDONÇA and RIBEIRO 1997), suggesting that the A_1 receptor may normally suppress metabotropic receptor activation. In the striatum, A_{2A} receptors may inhibit the conductance of the NMDA-receptor channel via a G-protein-dependent mechanism (NORENBERG et al. 1997). Thus, there may be multiple interactions in different regions of the nervous system between the different adenosine and glutamate receptor subtypes.

A novel synergistic interaction has recently been described between A_1 and $GABA_B$ receptors in acutely dissociated CA3 neurons. Both receptor subtypes activate a G-protein-activated inward-rectifier potassium channel (GIRK), and at saturating concentrations of agonist the effects of simultaneous application of agonists for both receptor subtypes is completely occlusive, and no larger than the response to a saturating concentration of the most efficacious agonist (usually GABA). However at relatively low (e.g., physiologically relevant) concentrations of agonists, the effect of co-application of

2-chloroadenosine (adenosine receptors) and baclofen ($GABA_B$ receptors) is supra-additive (SODICKSON and BEAN 1998). Such cooperative interactions between neurotransmitters may be common, and could affect other cellular responses in addition to the GIRK conductance examined here. These findings suggest that although there are no known adenosine synapses in brain, adenosine may be an endogenous regulator of the sensitivity of other G protein coupled receptors, such as the $GABA_B$ receptor, which do mediate clearly defined postsynaptic responses.

e) Adenosine–Nitric Oxide Interactions

Recent reports have suggested that there are interactions between adenosine receptor activation and nitric oxide (NO) release. While the effects of adenosine, particularly during cell stress, are viewed as largely neuroprotective (see Sect. B.I.3.a), there are at least some situations in which this is not the case. BARTH et al. (1997) reported that application of rather high concentrations of the stable adenosine analogue, 2-chloroadenosine (2-CADO), caused widespread neuronal damage when paired with anoxia/hypoglycemia in organotypic hippocampal slices, and showed this damage was blocked by both adenosine receptor antagonists and inhibitors of NO production. While this might appear to be at odds with reports that increased concentrations of extracellular adenosine during oxygen or glucose deprivation reduced, rather than enhanced, neuronal damage, it should be noted that BARTH et al. (1997) were unable to initiate cellular damage with adenosine itself, even in very high (500 μmol/l) concentrations. A previous report also demonstrated adenosine receptor-mediated NO release from cultures of cortical astrocytes, and suggested that glial cells may be the site of the interaction between adenosine receptor activation and NO release in the nervous system (JANIGRO et al. 1996). However, another recent report (BRODIE et al. 1998) found seemingly opposite results – activation of A_{2A} receptors in cultured astrocytes *inhibited* the production of NO as well as the inducible form of NO synthase. Because the preceding studies did not use receptor selective agonists, a possible explanation is that different receptor subtypes have antagonistic effects on NO production. Finally, there has also been a report that NO donors can stimulate adenosine release (FALLAHI et al. 1996). Although the mechanism underlying this effect is not understood, it does imply that there may be reciprocal interactions between these modulatory agents. More work is clearly needed to understand the interaction between these two common and important molecules in different regions and between different cell types in the nervous system.

2. Behavioral Actions

A complex variety of behavioral actions have been attributed to the effects of adenosine acting at specific receptor subtypes. These effects, which include analgesia, sleep, anxiolysis, drug addiction, and reduced locomotor behavior, are discussed below. It should also be noted that adenosine receptor antagonists, such as caffeine and theophylline, are widely used psychoactive drugs that act primarily if not exclusively by inhibiting the actions of endogenous adenosine. The actions of these non-selective adenosine receptor antagonists are diverse, and include effects such as behavioral activation and respiratory stimulation. Under normal conditions, it is likely that the only adenosine receptors substantially activated by endogenous adenosine are the A_1 receptors; thus, the effects of these antagonists are probably most informative concerning the role of A_1 receptors, and probably give little insight into the actions of other lower affinity receptors such as the A_{2B} and the A_3. The behavioral and physiological consequences of adenosine receptor antagonist administration in nonhuman primates has been recently reviewed by HOWELL et al. (1997).

An alternative approach to identifying the role of adenosine in the brain has been the development of knockout mice which lack the adenosine A_{2A} receptor ($A_{2a}R^{-/-}$) (LEDENT et al. 1997). The behavior of these mice, which show reduced exploratory behavior, increased anxiety, increased aggression in the males, high blood pressure, and reduced response to pain, is generally consistent with the conclusions of more conventional pharmacological experiments using A_{2A} antagonists that have previously been conducted with mice. It seems likely that this approach will be used increasingly to probe not only the functional role played by adenosine receptors, but also of other molecules (e.g., adenosine transporters) that can indirectly regulate the level of adenosine receptor activation in the brain.

a) Pain

Adenosine receptors have been implicated in analgesia, and this is an area where the clinical potential for manipulating adenosinergic mechanisms is being actively explored. In general, central administration of adenosine seems to have an antinociceptive effect via A_1 receptors, whereas specific A_3 receptor activation appears to increase nociception. SOLLEVI (1997), SAWYNOK (1998), and SALTER and SOLLEVI (Chap. 13, this volume) discuss the relationship between adenosine and pain in detail.

b) Sleep/Arousal

Considerable circumstantial evidence, not the least of which is the alerting effects of caffeinated beverages, has long suggested that adenosine may be important in reducing arousal and promoting sleep. Adenosine receptor antagonists, such as caffeine and theophylline, are well known to reduce or disrupt sleep, and increase arousal, and there have been numerous studies that have suggested that adenosine agonists enhance sleep (RADULOVACKI 1985; TICHO and RADULOVACKI 1991). Nevertheless, until recently there has been little compelling evidence that *endogenous* adenosine may play a causal role in sleep. Although systemic injections of stable adenosine analogs do modulate sleep, the profound peripheral side effects of most of these agents makes it difficult to draw conclusions regarding their effects. However, electroencephalographic (EEG) arousal has been shown to be correlated with discharge activity of mesopontine cholinergic neurons, and their tonic level of activity is regulated by endogenous adenosine levels. Both the firing rate and the excitability of these neurons are reduced by adenosine (RAINNIE et al. 1994), and these actions thus provide a direct mechanism by which reductions in adenosine concentrations in the brain could translate into EEG arousal. Subsequent work in which adenosine was injected into two cholinergic areas (basal forebrain or the laterodorsal tegmental nucleus) showed that an increase in extracellular adenosine in either area produced a 50% decrease in waking, and injection of adenosine into the basal forebrain resulted in a significant increase in rapid eye movement (REM) sleep (PORTAS et al. 1997). This finding is corroborated by a recent finding that microinjection of either a cholinergic or an adenosine agonist into the pontine reticular formation caused a dramatic increase in REM sleep (MARKS and BIRABIL 1998). As the effects of the two compounds were not additive, it suggests that they recruit similar cellular mechanisms.

Microdialysis studies have directly measured extracellular adenosine levels in the basal forebrain with respect to sleep onset and sleep deprivation. When cats are deprived of sleep, adenosine levels continue to rise, and drop slowly only during recovery sleep. The link between sleep/wake cycles and endogenous adenosine seems to hold specifically for the basal forebrain, as changes in adenosine levels in the thalamus were not significantly associated with behavioral state (PORKKA-HEISKANEN et al. 1997). However, a link between extracellular adenosine levels and the sleep/wake cycle has also been found in the rat hippocampus (HUSTON et al. 1996). During the active period, extracellular adenosine levels increase and are associated with an increase in sleep-like behaviors, such as yawning. The adenosine levels fall within an hour of sleep onset, and these researchers suggest that adenosine promotes sleep as a daily restoration (HUSTON et al. 1996); increases in endogenous adenosine would reduce the level of neural activity, and might mitigate neuronal damage to which the hippocampus is particularly vulnerable.

Related to the sleep-promoting role of adenosine is its ability to reduce anxiety or decrease arousal. Higher doses of a stable A_1 adenosine agonists can cause sedation (DUNWIDDIE and WORTH 1982), and administration of an A_1 receptor agonist has an anxiolytic effect in mice (FLORIO et al. 1998; JAIN et al. 1995). Systemic administration of an A_1 receptor agonist in rats tends to increase low frequency EEG activity (FULGA and STONE 1998), an indication of reduced arousal. A_{2A} receptor agonists, known to produce hypolocomotion, have also been shown to increase both slow wave and REM sleep when injected into the ventral portion of the rostral basal forebrain, an area that includes the ventral striatum (SATOH et al. 1996). In addition to increases in sleep behavior by direct administration of A_{2A} receptor agonists, activation of this receptor subtype seems to be specifically involved in the sleep-promoting pathways activated by prostaglandin D_2 (SATOH et al. 1996). At this point there is strong evidence for the sleep-promoting and arousal-reducing effects of adenosine. However future studies may establish the involvement of specific brain regions and cellular mechanisms underlying these actions. The role of adenosine in wakefulness has been reviewed (STRECKER et al. 2000).

c) Thermoregulation

Sleep behavior and arousal levels change in parallel with a change in brain temperature, and there is evidence that endogenous adenosinergic activity may also change in accordance with these circadian temperature changes. Adenosine levels (PORKKA-HEISKANEN et al. 1997) and brain temperature (REFINETTI and MENAKER 1992) both rise prior to sleep onset, and decrease during sleep. Caffeine administration suppresses the normal nighttime decrease in temperature (WRIGHT et al. 1997), suggesting that adenosine may mediate this temperature decrease.

Increasing the recording temperature of hippocampal slices by several degrees from an initial recording temperature of 32°C causes a decrease in synaptic transmission mediated by increased A_1 receptor activation (MASINO and DUNWIDDIE 1999). This study suggests a very sensitive and direct relationship between extracellular adenosine and brain temperature that occurs within a physiologically relevant temperature range. GABRIEL et al. (1998) found a similar adenosine-mediated depression of synaptic transmission upon warming hippocampal slices starting from a recording temperature of 22°C in rats. Interestingly, this adenosine-mediated decrease did not occur upon warming hippocampal slices from the golden hamster, a hibernating animal. Large increases in extracellular brain adenosine concentrations in response to increases in brain temperature would be maladaptive in an animal emerging from hibernation, but might have protective effects in an animal where such temperature changes do not normally occur. In non-hibernators such as the rat, however, small ongoing temperature changes may regulate to some extent to the level of adenosinergic inhibition. The increase in extracellular adenosine could become more extreme during larger temperature changes (e.g., fever or heatstroke), and prevent excessive excitatory and potentially neurotoxic activity during the elevated temperature, via the mechanisms discussed in the section on neuroprotection(see Sect. B.I.3.a). In addition, increased adenosine levels may be a mechanism for reducing temperature, because injections of adenosine agonists result in profound hypothermia (DUNWIDDIE and WORTH 1982; PROCTOR and DUNWIDDIE 1984), and A_1 agonists prevent pyrogen-induced fever (MATUSZEK and GAGALO 1997).

d) Addiction

While most studies of drug addiction have focused on the central role played by dopamine and/or opiate receptors, there is evidence for modification of adenosinergic mechanisms related to drug addiction or withdrawal from chronic drug use. Systemic administration of either adenosine A_1 (cyclopentyladenosine; CPA) or A_{2A} (CGS 21680) selective agonists decreased, and adenosine antagonists increased, the behavioral symptoms of naloxone-precipitated opiate withdrawal (SALEM and HOPE 1997). After withdrawal from chronic administration of either cocaine or morphine, slices of the ventral tegmental area exhibit increased adenosinergic tone (i.e., there is an increased

A_1-mediated presynaptic inhibition of $GABA_B$ responses), which appears to be due to a change in the extracellular adenosine concentration. Furthermore, there is a reversal of the modulatory effect of dopamine D_1 receptors on this synapse; in control slices, D_1 activation increases the GABA IPSC, whereas in withdrawn slices, D_1 activation *inhibits* transmission, and this inhibitory effect is mediated via adenosine (BONCI and WILLIAMS 1996). In contrast, in the nucleus accumbens there was a *decreased* presynaptic effect of adenosine on glutamate transmission one week after chronic cocaine use, due to increased adenosine uptake (MANZONI et al. 1998). A previous study provided evidence that chronic opioid administration increased adenosine uptake, in that increased binding for the adenosine transporter was found in the striatum (12% increase) and particularly in the hypothalamus (37% increase) after withdrawal from chronic morphine treatment (KAPLAN and LEITE-MORRIS 1997). Taken together, these results suggest an important, regionally-specific role for adenosine in drug addiction, and one that may be exploited in reducing the long-term effects of drug addiction, such as continued drug craving and drug-seeking behavior during withdrawal.

e) Ethanol/Adenosine

One area of continuing interest insofar as the behavioral actions of adenosine are concerned relates to a role for adenosine as a mediator of some of the intoxicating effects of ethanol. There are obvious similarities in the actions of adenosine and ethanol, in that they both have hypothermic, sedative, and anticonvulsant actions, and both induce vasodilation. Moreover, some (but certainly not all) of the physiological effects of ethanol can be completely antagonized by adenosine receptor antagonists (CARMICHAEL et al. 1988), lending further support to this hypothesis. There are three basic mechanisms that have been proposed to account for the interactions between ethanol and adenosine. First, when ethanol is metabolized, acetate is formed, and when acetate is incorporated into acetyl-CoA, significant amounts of 5'-AMP are formed. Subsequent hydrolysis by 5'-nucleotidases could result in the formation of adenosine (CARMICHAEL et al. 1991). A second potential mechanism involves adenosine transporters; transport mediated by the *es* subtype of nucleoside transporter, which is inhibited by nitrobenzylthioinosine, is also inhibited by ethanol (KRAUSS et al. 1993). A third possibility involves G-protein coupled effector mechanisms; ethanol can facilitate the receptor-mediated activation of adenylate cyclase by various hormones and neurotransmitters (RABIN and MOLINOFF 1981; LUTHIN and TABAKOFF 1984; RABIN 1990; HOFFMAN and TABAKOFF 1990), and thus ethanol might enhance actions mediated via A_2 receptors. The evidence for a "purinergic" link in the actions of ethanol has been reviewed elsewhere (DUNWIDDIE 1995, 1999), and will not be covered in detail here. However, a few recent developments will be summarized.

One approach that has been used previously to examine the role of adenosine in the actions of ethanol has been to use adenosine receptor antagonists to try to reverse the effects of ethanol, but the interpretation of these experiments is often problematic because of the effects of antagonists on baseline measures of activity. An alternative approach that has been tried has been to inject antisense oligonucleotides for A_1 receptors into the striatum, and then determine whether ethanol sensitivity is altered. One recent study reported no effect of such treatment (BIGGS and MYERS 1997), whereas another very similar study reported that antisense injection antagonized the motor incoordination induced by ethanol (PHAN et al. 1997). Given the vagaries of the antisense approach, and the fact that the former study did not demonstrate directly (via ligand binding) or indirectly (via assessment of the effects of A_1 agonists on motor function) that A_1 receptor number had been successfully reduced by the antisense treatment, the positive result obtained in the study by PHAN et al. (1997) seems somewhat more convincing. However, further studies, perhaps with knockout animals, will provide additional evidence to resolve these differences.

There is additional evidence to support a role for A_1 receptors in the effects of ethanol in a recent study by CAMPISI et al. (1997). They reported that the abilities of ethanol, and the ethanol metabolite acetate, to potentiate the effects of general anesthetics were centrally mediated effects that involved adenosine receptors, because both actions could be antagonized by administration of an adenosine receptor antagonist into the brain. The effects of acetate were fully reversed (suggesting that they were completely mediated via adenosine receptors), whereas the effects of ethanol were only partially antagonized by antagonists. These results provide additional support for the hypothesis that a component of the centrally mediated effects of ethanol can be attributed to increased activation of A_1 receptors, and that conversion of ethanol to acetate, and the subsequent increase in adenosine formation, may be the mechanism that underlies the increased level of A_1 receptor activation. Attempts to evaluate this hypothesis at the cellular level in brain have been less supportive; although there has been one positive report (CULLEN and CARLEN 1992), other studies suggested that treatment of brain slices with acetate did not lead to any increases in adenosine receptor occupation (BRUNDEGE and DUNWIDDIE 1995).

Further behavioral evidence to support a link between ethanol and adenosine has come from the group of Dar, who have characterized in great detail the interactions between ethanol and other receptor systems, including adenosine, on motor incoordination. This work has provided suggestive evidence that adenosine may mediate some of the effects of ethanol. The fact that pertussis toxin injected into the cerebellum can antagonize the effects of ethanol (DAR 1998) is consistent with the hypothesis that A_1 receptors are involved, because the physiological effects of A_1 receptor activation are often mediated via pertussis toxin sensitive G proteins. The effects of ethanol can also be antagonized by intracerebellar injections of a metabolically stable cAMP analog (DAR 1997), or by intrastriatal injections of forskolin (MENG et al. 1998), which has been taken as further evidence for A_1 receptor involvement. If this is the case, however, it would suggest that neither the A_1 receptor-mediated activation of K^+ channels, nor inhibition of Ca^{2+} channels, is involved in this response, since neither effect is mediated via changes in cAMP levels. The way in which an ethanol-induced decrease in cAMP levels (MENG et al. 1998) is translated into alterations in physiological activity that result in motor incoordination remains to be determined.

As has been concluded in previous reviews, it appears likely that adenosine contributes significantly to the behavioral effects of ethanol, and some of the effects of ethanol may be mediated entirely via purinergic mechanisms. However, the cellular mechanisms by which ethanol affects the extracellular concentrations of adenosine in brain and/or adenosine receptors remain somewhat obscure.

f) Interactions with Dopamine Systems

The interaction between adenosine and dopamine is an extensive and very active area of research. Several recent reviews (FERRE et al. 1997; FUXE et al. 1998; FERRE 1997) have summarized these results. As presented in Sect. B.I.1.d, an antagonistic relationship has been proposed between D_1 and A_1 receptors, and D_2 and A_{2A} receptors, and a variety of behavioral data including human data (WEBSTER et al. 2000) is consistent with this hypothesis (HAUBER and MUNKLE 1997; FENU et al. 1997; RIMONDINI et al. 1998). These interactions between adenosine and dopamine may provide at least in part the cellular basis for the experimental observations that adenosine receptor antagonists, such as caffeine and theophylline, influence dopaminergic transmission. The role of dopamine in the behavioral effects of caffeine has been recently reviewed (GARRETT and GRIFFITHS 1997).

One novel and interesting finding regarding the interaction between adenosine and dopamine is that these receptors may interact with each other at the level of the cell membrane. Results obtained in cell culture stably co-transfected with the different receptor subtypes suggest that receptor heterodimers may form containing an A_1/D_1 or an A_{2A}/D_2 complex (YANG et al. 1995). This is unusual in that most receptor interactions occur at the level of second messenger systems within the cell. However, the formation of functionally interacting A_1/D_1 receptor heteromeric complexes has been recently confirmed (GINES et al. 2000) Receptor-receptor interactions at the membrane level may alter directly the affinity of agonist binding, or subsequent signal transduction pathways upon binding of one or both agonists.

α) Role of Adenosine in Pathological Conditions

The interactions between adenosine and dopamine in the basal ganglia have suggested that adenosinergic drugs may have therapeutic potential in diseases that are often considered to be "dopaminergic" disorders, such as Parkinson's disease and schizophrenia. Because adenosine and dopamine receptors induce opposing physiological effects, adenosine antagonists have promise in promoting dopaminergic transmission, and may lack the side effects and drug tolerance that results from chronic treatment with dopamine agonists. Parkinson's disease is characterized by a loss of dopaminergic transmission, and a consequent loss of motor function. Several studies have shown that A_{2A}

antagonists like KW 6002 and KF 17837 promote motor behavior in models of Parkinson's, either in rats (HAUBER et al. 1998; MANDHANE et al. 1997) or monkeys (KANDA et al. 1998). The potential clinical relevance of A_{2A} receptor antagonists in the treatment of Parkinson's disease has been discussed (RICHARDSON et al. 1997; WILLIAMS, Chap. 29, this volume).

In contrast to the loss of dopamine in Parkinson's disease, schizophrenia is characterized by an overabundance of dopamine, and the opposing effects of adenosine and dopamine may also be clinically useful in treating this disorder. In this manner, A_{2A} agonists have been proposed to have antipsychotic potential because of their ability to reduce dopamine transmission (RIMONDINI et al. 1997; FERRE 1997). One such agent, CI-936, was advanced to clinical trials in the early 1980s but was discontinued for unknown reasons.

β) Locomotor Effects

Administration of non-selective adenosine agonists results in decreased spontaneous motor activity. Several reports have indicated that this is most likely due to central activation of A_2 receptors (JAIN et al. 1995; FLORIO et al. 1997), and it has recently been attributed more specifically to the A_{2A} receptor because hypolocomotion occurred after administration of A_{2A}-specific agonists (HAUBER and MUNKLE 1997; MARSTON et al. 1998). The motor depressant effects induced by an A_{2A} agonist (CGS 21680) alone were similar to those induced by a D_2 antagonist, raclopride (HAUBER and MUNKLE 1997), and are a clear example at the behavioral level of the opposing effects of these two receptors. In agreement with these results, blockade of endogenous A_{2A} receptor activation with A_{2A}-specific antagonists caused motor activation (SVENNINGSSON et al. 1997), and also blocked catalepsy induced by dopamine receptor antagonists (HAUBER et al. 1998). Some evidence also supports a role for central A_1 receptors in behavioral activation; in rats, hypolocomotion induced by the A_1 receptor agonist, N^6-cyclopentyladenosine (CPA) was not antagonized by the A_{2A}-specific antagonist, KF17837 (MARSTON et al. 1998), suggesting a specific involvement of the A_1 receptor. Likewise, the hypolocomotion induced by the A_{2A} receptor agonist APEC was not antagonized by an A_1-receptor specific antagonist (DPCPX), and these two receptor types may be able to influence locomotor activity independently (MARSTON et al. 1998). A_3 receptors have also been suggested to reduce locomotion (JACOBSON et al. 1993). Nevertheless, although A_1 or A_3 receptors may play a role in behavioral activation under some circumstances, most of the current evidence suggests that the hypolocomotor effect of adenosine occurs predominantly via A_{2A} receptors.

3. Functional Role(s) of Adenosine

There are a wide range of diverse physiological events influenced by adenosine receptor activation, including (but not exclusive to) regulation of the cerebral vasculature, control of respiration, protection of cells during potentially pathological events like hypoxia and ischemia, termination of epileptic seizures, and initiation of apoptosis. Several of these topics have been the focus of recent reviews, and will not be summarized here: the role of adenosine in the control of respiration (MONIN 1997; HERLENIUS et al. 1997), dilation of cerebral vasculature (SAWYNOK 1995; CONEY and MARSHALL 1998), and apoptosis (ABBRACCHIO et al. 1997; CHOW et al. 1997) have all been recently reviewed.

a) Neuroprotection

Many studies have reported an interaction between adenosine receptor activation and the extent of neuronal damage associated with oxygen and/or glucose deprivation. In general, activating A_1 receptors ameliorate, and antagonizing A_1 receptors exacerbate neuronal damage, as determined using morphological, physiological, and behavioral measures of damage. This is in accord with the generally inhibitory role of A_1 receptors on neuronal activity and, more specifically, appears to be linked to the ability of A_1 receptors to inhibit the release of glutamate, which is a key mediator of excitotoxic damage. In this context, enhancing the binding of endogenous adenosine to A_1 receptors with the allosteric modulator PD 81,273 significantly reduces the damaging effects of a hypoxic episode (HALLE et al. 1997). In contrast to the protection offered by A_1 receptor activation, reducing the excitatory action of A_{2A} receptors with A_{2A}-specific antagonists may afford neuroprotective benefits. The role of adenosine and its clinical potential in neuroprotection has been reviewed recently by BISCHOFBERGER et al. (1997) (A_1 receptors), ONGINI et al. (1997) (A_{2A} receptors), and FREDHOLM (1997). The neuroprotective effects of adenosine are also dependent on the timing of administration; chronic administration of adenosine agonists can have the opposite effects of acute administration (JACOBSON et al. 1996), most probably because of the up- or downregulation of receptors in response to chronic drug administration.

There is extensive evidence that during deprivation of oxygen or glucose, there are large increases in extracellular adenosine, which influences both pre- and postsynaptic activity. Presynaptically A_1 receptor activation inhibits neurotransmitter release by inhibiting Ca^{2+} influx and/or opening K^+ channels and, at least in the hippocampus, this action is exerted at glutamatergic (but not GABAergic) synapses. The ability of endogenous adenosine to act via A_1 receptors to inhibit neurotransmitter release has been demonstrated during aglycemia (CALABRESI et al. 1997; JIN and FREDHOLM 1997) or hypoxia (JIN and FREDHOLM 1997) in the striatum, and during a variety of situations causing energy deprivation in the hippocampus (GRIBKOFF and BAUMAN 1992; FOWLER 1993; LLOYD et al. 1993; ZHU and KRNJEVIC 1997a,b). Postsynaptically, adenosine A_1 receptors hyperpolarize neurons by opening K^+ channels, and may also have protective effects via their ability to inhibit Ca^{2+} influx via voltage-dependent Ca^{2+} channels. Although under normoxic conditions endogenous adenosine exerts relatively minor effects on the membrane properties of CA1 neurons (ZHU and KRNJEVIC 1997b), additional adenosine released during cyanide superfusion or during hypoxia has been shown to induce both a presynaptic suppression of transmitter release and a postsynaptic hyperpolarization (ZHU and KRNJEVIC 1997a,b). However, there may be some interesting regional specificity in this effect; DOHERTY and DINGLEDINE (1997) found that although adenosine can inhibit transmission at excitatory synapses onto dentate interneurons, it does not mediate the depression of excitatory input that occurs during hypoxia.

Although acute hypoxia or ischemia induces cellular damage, several studies have shown that a sublethal episode can protect against a subsequent period of hypoxia or ischemia. This effect, well documented in the heart and termed "ischemic preconditioning" (reviewed by LINDEN 1995; SCHWARZ et al. 1997) occurs in the brain as well, and in the hippocampus appears to involve adenosine and A_1 receptors in vivo (HEURTEAUX et al. 1995) and in vitro (PÉREZ-PINZÓN et al. 1996). A possible additional role of A_3 receptors in this phenomenon has been suggested in heart (JACOBSEN et al. 2000), although recent evidence appears to emphasize the involvement of A_1 receptors (HILL et al. 1998); nevertheless, a role for A_3 receptors in brain must be considered a possibility as well.

While experimental evidence suggests that adenosine formed during energy deprivation is generated intracellularly and released into the extracellular space as adenosine, it is interesting to note a recent study showing that the mRNA for ecto-nucleotidases are upregulated after transient ischemia. These effects are also greatly delayed relative to the ischemia, such that increased staining for apyrase and ecto-5'-nucleotidase was detectable at 2 days and persisted at 28 days after an ischemic episode, particularly in the CA1 region of the hippocampus (BRAUN et al. 1998). The significance of these observations is unclear; it is possible that the enhanced breakdown of nucleotides could provide an additional source of extracellular adenosine following the initial insult, and would promote the removal of ATP released as a result of the ischemia from the extracellular space. On the other hand, it is possible that these increases reflect the migration of cell types that have higher levels of these enzymes (e.g., microglia) to the region of tissue damage, and do not necessarily reflect a protective or compensatory type of response per se. ATP receptors (see Section C.II.) mediate fast excitatory events that might exacerbate excitotoxicity, so an increase in ecto-nucleotidase activity might have additional protective effects both by reducing P_2 receptor activation, and by increasing P_1 activation.

The mechanisms by which hypoxia and/or ischemia cause brain damage in neonates may be somewhat different from those in the adult. Although enhancing binding to adenosine A_1 receptors reduced neonatal brain damage (HALLE et al. 1997), it has been reported that antagonizing adenosine receptors with the non-selective antagonist theophylline in neonatal rats (postnatal day 7) actually *reduced* damage in a dose-dependent manner, with the lowest doses being the most effective (20mg/kg) (BONA et al. 1997). Because the A_1-

specific antagonist DPCPX was not protective, A_1 receptors appear uninvolved in this effect; it is more likely that the effect of theophylline in this study is related to its antagonistic effects at A_2 receptors. Interestingly, specifically antagonizing A_{2A} receptors *after* the hypoxia-ischemia modestly reduced damage as well. The absence of a relationship between A_1 receptor activation and hypoxia-induced damage in this study is surprising, as most studies of adult neuronal tissue support an important role for the A_1 receptor in neuroprotection. However, the protective effects of an A_{2A} receptor-selective antagonist is consistent with studies in the adult, and with the generally excitatory effects mediated by these receptors. At this point there is not a definitive resolution of the roles of adenosine receptor subtypes during or after hypoxia in neonatal animals. In addition, region-specific neuroprotective or neurotoxic effects of adenosine receptors may still be revealed.

b) Epilepsy

Due to their powerful inhibitory influence over neuronal activity, adenosine and adenosine receptors have been identified and experimentally evaluated with respect to an endogenous and/or therapeutic role as anticonvulsants (reviewed by DRAGUNOW 1988). Indeed, extracellular adenosine levels are increased during seizure activity and may be involved in terminating the seizure and prolonging the period of postictal depression (DURING and SPENCER 1993). Commonly ingested adenosine receptor antagonists such as caffeine and theophylline have long been noted for their pro-convulsant actions (AULT et al. 1987; DUNWIDDIE 1980).

In general, most studies have suggested that the anticonvulsive effect of adenosine can be attributed to A_1 receptor activation (ZHANG et al. 1994; MURRAY et al. 1992; MALHOTRA and GUPTA 1997). A_1 receptor antagonists enhance epileptic discharges and inhibit the depressant effect of adenosine on burst frequency in hippocampal slices (DUNWIDDIE 1980; AULT and WANG 1986; Ross et al. 1998a,b), initiate seizures in genetically epilepsy-prone rats (DE SARRO et al. 1997), shorten the post-ictal depression, prolong EEG afterdischarge, and increase the number of spikes when administered to normal rats (KULKARNI et al. 1997). In contrast, adenosinergic agonists reduce the incidence of both pentylenetetrazol (DUNWIDDIE and WORTH 1982; GEORGIEV and TCHEKALAROVA 1998) and amygdala-kindled seizures (ABDUL-GHANI et al. 1997; POURGHOLAMI et al. 1997). In other recent developments, activation of A_3 adenosine receptors has been shown to potentiate the anticonvulsant activity of a variety of antiepileptic agents (BOROWICZ et al. 1997; VON LUBITZ et al. 1995). A recent study has also addressed the issue of the relative importance of pre- vs postsynaptic actions of A_1 receptors in terms of the antiepileptic actions of adenosine (TANCREDI et al. 1998). The results of this study supported the conclusion that the inhibition of glutamate release is likely to be of primary importance in suppressing seizure discharges, although the activation of K^+ channels postsynaptically may play a role as well.

II. Regulation of Extracellular Adenosine Concentrations in Brain

Extracellular adenosine appears to be ubiquitous in the nervous system, at concentrations sufficient to activate tonically at least A_1 receptors. Several mechanisms are known to influence the level of endogenous extracellular adenosine. These mechanisms fall into two general categories:

1. Direct flux of adenosine into/out of the extracellular space
2. Extracellular formation of adenosine via the metabolism of adenine nucleotides (see BRUNDEGE and DUNWIDDIE 1997; SPERLÁGH and VIZI, Chap. 7, this volume; ZIMMERMANN, Chap. 8, this volume)

Nevertheless, a number of key issues in this area remain unresolved, such as the relative importance of each pathway in regulating extracellular adenosine under different conditions, and the extent to which adenosine levels are actively regulated. It is also unclear whether adenosine may also be a local messenger (i.e., there can be significant local and temporal changes in its concentration), or whether it is a more general signal that in a sense "integrates" the overall activity of the nervous system, and only changes significantly in response to very widespread changes in activity (e.g., during prolonged wakefulness). Given the fact that there are peripheral adenosine receptors that can profoundly affect heart rate, blood pressure, etc., most pharmacological studies aimed at developing therapeutic drugs seem to be moving away from the study of direct agonists/antagonists, which would have profound side effects unrelated to activation of central nervous system receptors, and towards the development of agents that may enhance the processes that lead to the local generation of adenosine (e.g., adenosine kinase inhibitors). Thus, the characterization of the regulation of extracellular adenosine concentrations continues to be an area of very active investigation.

1. Adenosine Transporters

Adenosine can be generated intracellularly either by metabolism of adenine nucleotides (ATP → ADP → 5'-AMP → adenosine, or cAMP → 5'-AMP → adenosine) or cleavage of S-adenosylhomocysteine (SAH) into adenosine and homocysteine. Bi-directional facilitated diffusion transporters appear to play the largest role in regulating extracellular adenosine because they essentially equilibrate extracellular and intracellular adenosine levels and provide a mechanism for direct flux of adenosine into/out of the extracellular space. Under normal conditions these transporters function primarily to remove extracellular adenosine; extracellular levels are usually greater than intracellular levels, which are kept low by the action of adenosine kinase. However, under metabolically stressful conditions, the concentration of intracellular adenosine rises above the extracellular concentration, and the net flow of adenosine through the equilibrative transporter is out of the cell and into the extracellular space. A dramatic increase in extracellular adenosine has been shown in in vitro hippocampal slices following electrical stimulation or energy depletion (oxygen/

glucose removal) (LLOYD et al. 1993) or energy deprivation (inhibition of mitochondrial function) (DOOLETTE 1997). Although adenosine levels might rise in these situations due to an inability to rephosphorylate adenosine into adenine nucleotides, recent studies have shown that it may not be this straightforward. Rather, it seems that manipulations that inhibit mitochondrial function (DOOLETTE 1997; ZHU and KRNJEVIC 1997a) result in an efflux of adenosine even if ATP levels are preserved (DOOLETTE 1997). Inhibiting oxidative phosphorylation alone did not produce synaptic depression, which would be expected if there were an increase in adenosine levels and efflux through the transporter into the extracellular space (DOOLETTE 1997).

2. Activity-Dependent Adenosine Release

Another mechanism that alters extracellular adenosine concentrations is glutamate receptor activation. Thus, increases in excitatory synaptic transmission can themselves cause an increase in extracellular adenosine concentrations. At high levels of activity an increasing contribution of adenosinergic inhibition via A_1 receptors has been hypothesized to afford a protective inhibitory threshold against excitotoxic levels of activity. In cortical slices, activation of both NMDA and non-NMDA receptors results in an increase in extracellular adenosine (HOEHN and WHITE 1990), and low levels of NMDA receptor activation are particularly effective at increasing extracellular adenosine levels (HOEHN et al. 1990). Studies in cortical slices have shown that NMDA receptor activation increases extracellular adenosine levels indirectly via the release and subsequent extracellular conversion of adenine nucleotides to adenosine, whereas non-NMDA receptor activation increases adenosine levels directly through the release of adenosine (CRAIG and WHITE 1993). The cellular mechanism by which NMDA releases nucleotides, however, is not well understood. NMDA receptor-mediated increases in extracellular adenosine have been proposed to provide an inhibitory threshold which must be overcome to allow additional NMDA receptor activation to proceed (HOEHN et al. 1990; CRAIG and WHITE 1992). Adenosine can interfere with NMDA responses either by inhibiting glutamate release or by inhibiting NMDA receptor-mediated currents postsynaptically. For example, application of 2-chloroadenosine (CADO) has been shown to inhibit NMDA receptor-mediated currents in acutely dissociated pyramidal cells (DE MENDONÇA et al. 1995), and a similar type of antagonistic effect involving A_{2A} receptors has been described in striatum as well (NORENBERG et al. 1997). Electrophysiological studies in hippocampal slices have also shown NMDA-receptor dependent purine release, possibly of adenosine per se, as it was not reduced by an inhibitor of 5'-nucleotidase (MANZONI et al. 1994). Because stimulation-induced adenosine release was antagonized by enkephalin, an opioid agonist that specifically inhibits interneurons, this study suggests that interneuron activity may play a role in modulating extracellular adenosine levels (MANZONI et al. 1994).

3. cAMP Efflux

Efflux of cAMP, and subsequent extracellular conversion to adenosine by ecto-phosphodiesterases and 5'-nucleotidase, has been identified as a significant source of extracellular adenosine using both biochemical and physiological techniques. Activation of adenylyl cyclase, and the consequent conversion of ATP into cAMP, has been shown to elevate extracellular adenosine levels in several in vitro systems. Direct activation of adenylate cyclase by forskolin caused an adenosine-mediated decrease in synaptic transmission in hippocampal slices, similar to that seen with bath superfusion with cAMP (GEREAU and CONN 1994; BRUNDEGE et al. 1997). Forskolin application also results in extracellular adenosine accumulation in primary cultures of cortical neurons (ROSENBERG and LI 1996). β-Adrenergic receptor-mediated activation of adenylate cyclase, either in cortical cell cultures (ROSENBERG and LI 1995) or in hippocampal brain slices (GEREAU and CONN 1994) results in an efflux of cAMP and subsequent extracellular degradation to adenosine. Thus, extracellular adenosine formation and activation of adenosine receptors may be a significant secondary mediator of responses to activation of adenylate cyclase-coupled receptors.

An example of the physiological relevance of the pathway from cAMP to adenosine receptor activation has been provided by BONCI and WILLIAMS (1996). Chronic treatment with either cocaine or morphine results in a reversal of the dopamine D_1 receptor modulation of GABA IPSPs in the ventral tegmental area. After drug treatment, dopamine inhibited, rather than augmented, GABA IPSPs. The cellular mechanisms underlying this change were shown to be altered metabolism of cAMP, and a consequent increased activation of adenosine receptors. This long-term consequence of drug treatment on adenosine levels underscores the importance of metabolic pathways which result in altered extracellular adenosine levels.

4. Extracellular Nucleotide Metabolism

In addition to extracellular degradation of cAMP to adenosine, extracellular metabolism of nucleotides by ecto-nucleotidases can provide a source of extracellular adenosine (ZIMMERMANN 1998, 1999, Chap. 8, this volume). Recent studies have shown that these nucleotidases will rapidly convert nucleotides to adenosine, and in quantities that are sufficient to exert an inhibitory effect on synaptic transmission. Extracellularly applied ATP, ADP, or AMP, as well as many "metabolically stable" nucleotides, are all converted to adenosine and can then activate A_1 receptors (DUNWIDDIE et al. 1997; CUNHA et al. 1998). The extracellular conversion is extremely rapid and efficient (<1s for nearly complete conversion of small local applications of these nucleotides) (DUNWIDDIE et al. 1997). Differences between biochemical measurements of nucleotide conversion to adenosine (which show very low levels of conversion of "metabolically stable" nucleotides to adenosine), and electrophysiological studies of the effects of nucleotides (which seem to suggest the opposite conclusion), has led to the hypothesis that the nucleotidases may specifically "channel" the adenosine formed from the action of ecto-nucleotidases to the receptors that mediate its physiological effects (CUNHA et al. 1998).

5. Adenosine Phosphorylation and Degradation

Extracellular adenosine levels ultimately reflect the rate of 5′-AMP conversion to adenosine by 5′-nucleotidase (both extracellularly and intracellularly), as well as the rates of adenosine rephosphorylation and degradation, mediated by adenosine kinase and adenosine deaminase, respectively. In some instances, interfering with these pathways may not have large effects per se, but can modulate responses initiated by other agents. For example, inhibition of adenosine kinase contemporaneously with activation of excitatory amino acid receptors greatly potentiates the formation of extracellular adenosine, and this has been proposed as a potentially useful way to augment the neuroprotective effects of adenosine specifically in locations where there is excessive excitation (WHITE 1996). Co-administration of inhibitors of adenosine kinase and adenosine deaminase also produces a supra-additive effect on NMDA receptor-mediated adenosine formation (HEBB and WHITE 1998).

The pharmacological manipulation of these pathways has helped to underscore their importance in terms of regulating extracellular adenosine concentrations, but leave unanswered some of the most important questions concerning the way in which physiological events may regulate key metabolic enzymes, and lead to increases in endogenous adenosine. The fact that the activity of adenosine kinase is dependent on both pH and the levels of adenosine present (YAMADA et al. 1980), and the fact that intracellular acidification occurs during hypoxia (FUJIWARA et al. 1992), suggests at least one means by which adenosine levels could be increased. A better understanding of the regulation of metabolic pathways involving these enzymes may help to understand not only the way in which extracellular adenosine is regulated, but may also suggest potential targets for clinical interventions.

III. General Conclusions and New Directions

Physiological responses mediated by adenosine were first observed over 70 years ago, and more recently its role in the brain as an endogenous modulator of synaptic transmission has become well-established. At the systemic level, the critical role played by adenosine in behavioral state changes, such as from waking to sleep, has come to be recognized, and other roles, such as alterations in purinergic systems following chronic drug use, seem to be developing as well. There is accumulating evidence that adenosine levels may be determined by, and can determine, brain temperature, and the interactions of adenosine with other neurotransmitters in a regional or cell-type specific manner is another area of developing interest.

The general picture of adenosine as an inhibitory modulator of the activity of the nervous system has arisen from many studies that have focused on its effects on excitatory glutamatergic transmission. At these synapses, the most prominent effects of adenosine are linked to presynaptic A_1 receptors inhibiting excitatory neurotransmitter release. However, this general picture has become more complex, and recent evidence has demonstrated that adenosine can presynaptically inhibit GABAergic transmission, or interfere with some kinds of excitatory responses at the postsynaptic level. Furthermore, in brain areas such as the superior colliculus, the effect of endogenous adenosine on synaptic transmission is excitatory, not inhibitory. Thus, the concept of adenosine as a purely inhibitory modulator is in need of revision. Many of the still controversial issues in adenosine pharmacology may also reflect differences between brain regions, cell types, methodologies, or the pharmacological specificities of different adenosine receptors. Because the subtypes of adenosine receptors differ significantly in their affinities for adenosine, they will be activated under widely differing conditions, and the effects of antagonists in particular will be very dependent upon the conditions under which they are tested. The proposed role for adenosine as an endogenous neuroprotectant, however, still appears to have strong support. Adenosine concentrations rise dramatically under conditions of cell stress, and in many cases adenosine appears to protect cells from excessive excitation, both through its inhibitory action on synaptic transmission, but also by virtue of its ability to hyperpolarize neurons and reduce postsynaptic excitability.

The powerful modulation of synaptic transmission by adenosine, as well as the ubiquitous nature of adenosine receptors, has made it an appealing candidate for a therapeutic agent in many different clinical conditions. The neurodegeneration that occurs after episodes of hypoxia or ischemia, the pathological overactivity of epilepsy, the persistent activity underlying chronic pain, and the specific changes in dopaminergic activity in Parkinson's disease or schizophrenia, all present possible therapeutic opportunities for purinergic drugs. The challenge, however, will be to develop strategies that have some degree of selectivity for the neuronal systems that underlie each of these conditions.

C. P2 Nucleotide Receptors

I. P2 Receptor Distribution

P2X receptors are expressed to varying degrees in either the central or peripheral nervous system. $P2X_4$ and $P2X_6$ receptors show particularly high expression in the central nervous system (COLLO et al. 1996; CHESSELL et al., Chap. 3, this volume), as inferred using in situ hybridization techniques, but $P2X_1$ (VALERA et al. 1994; KIDD et al. 1995), $P2X_2$ (BRAKE et al. 1994; COLLO et al. 1996; KIDD et al. 1995), and $P2X_5$ receptors (COLLO et al. 1996; KHAKH et al. 1997) are also found with more limited distributions. $P2Y_1$ (BOARDER and WEBB, Chap. 4, this volume) receptors have been localized using radioligand binding techniques, and also show wide distribution in the nervous system (SIMON et al. 1997), where they are expressed in both neuronal and non-neuronal cells (WEBB et al. 1998). The presence of P2Y$_2$ receptors has also been reported in brain (LUSTIG et al. 1993), as has the P2Y$_4$ receptor, albeit at relatively low levels (WEBB et al. 1998). There is good evidence that some P2X responses are mediated via P2X$_2$/P2X$_3$ heteromeric receptors (LEWIS et al. 1995; see below), and that other combinations of P2X receptors form functioning heteromers in heterologous expression systems (LE et al. 1998; TORRES et al. 1998), but to date there has been little biochemical characterization of these heteromers, and where they are expressed in the nervous system.

II. Physiological Responses to Application of ATP and Other Nucleotides

One approach to identifying sites where ATP may serve as a transmitter has been to characterize ATP (or other nucleotide) mediated responses that might reflect activation of P2 receptors. While there are numerous positive reports of such effects in the literature, there are at least two potential pitfalls that need to be kept in mind in interpreting these experiments. First, ATP itself is very rapidly converted to adenosine in the extracellular space of brain; experiments in hippocampal slices suggest that the half-time for such conversion is approximately 200 msec (DUNWIDDIE et al. 1997), and these estimates are consistent with the known extracellular localization of multiple enzymes for dephosphorylating adenine nucleotides (ZIMMERMANN, Chap. 8, this volume). Thus, it is probably rather difficult for any ATP added to the bath to ever reach P2 receptors in a brain slice; this conclusion is supported by the observation that in slices of the habenula, which is the only brain region where P2X synapses have been clearly demonstrated to date (EDWARDS et al. 1992), there is no response to bath superfusion of ATP. A second potential problem is that ATP is a substrate for ecto-kinases, and it has been suggested at least that a number of responses to ATP may reflect phosphorylation of extracellular proteins, rather than activation of P2 receptors. In particular, it has been suggested that ecto-phosphorylation may underlie some aspects of LTP, since hydrolyzable ATP analogs induce an LTP-like effect, non-hydrolyzable analogs block stimulation-induced LTP (WIERASZKO 1996; WIERASZKO and EHRLICH 1994), and ecto-kinase inhibitors block LTP (FUJII et al. 1995).

Despite these caveats, there are numerous reports where the application of ATP, ADP, UTP, or other nucleotide analogs elicit physiological responses, and where the evidence is relatively strong that P2 receptors mediate these responses. One system where there appears to be clearly defined P2 receptors is on locus coeruleus neurons, which are depolarized by ATP and stable ATP analogs (HARMS et al. 1992), and these responses are blocked by suramin (TSCHOPL et al. 1992), which blocks several subtypes of P2X receptors. These responses are somewhat complex in that they involve both an activation of a cationic current, and a decrease in a K$^+$ current (SHEN and NORTH 1993; HARMS et al. 1992), and although the pharmacological evidence is not definitive, it seems likely that these reflect activation of both P2X and P2Y receptors (NIEBER et al. 1997; FROHLICH et al. 1996). Unlike locus coeruleus neurons, cells in the ventral tegmental area (VTA) appear to lack P2 receptors (POELCHEN et al. 1998), despite the fact that P2 agonists appear to stimulate the release of dopamine from the terminals of VTA neurons.

Some of the other systems where ATP and its stable analogs have been reported to have effects are in cells of the nucleus of the trigeminal nerve (KHAKH et al. 1997; P2X receptors), cultured hippocampal neurons (INOUE et al. 1992), neurons of the tuberomammillary nucleus (FURUKAWA et al. 1994; possibly P2Y), and cerebellar Purkinje neurons (MATEO et al. 1998; possibly $P2X_2$). In addition, although it has not been demonstrated with native receptors, it has been shown that when the $P2Y_2$ receptor is expressed in transfected neurons, it can inhibit the activity of both Ca^{2+} channels and K^+ channels (FILIPPOV et al. 1998).

A rather unusual response that has been nominally attributed to activation of a P2Y receptor is the activation of an outwardly rectifying K^+ channel in a number of different brain regions (IKEUCHI et al. 1995, 1996; IKEUCHI and NISHIZAKI 1995). It has been suggested that this response is mediated via P2Y receptors because it can be elicited by a variety of nucleotides, but is completely blocked by GDPβS, which implies mediation via a G protein coupled receptor (IKEUCHI et al. 1996a,b). Although adenosine also activates a K^+ channel in many of these same cells via a P1 receptor, the response described by Ikeuchi and colleagues differs in that:

1. The associated K^+ channel is *outwardly* rectifying (as opposed to inwardly rectifying for the A_1/GIRK mediated response; TRUSSELL and JACKSON 1987).
2. The response is not elicited by A_1- or A_2-like receptor selective agonists (IKEUCHI et al. 1996a,b).
3. The response is not blocked by either A_1- or A_2-like receptor antagonists (IKEUCHI et al. 1996a,b). It has been proposed that responses such as these that are activated by both adenine nucleotides and nucleosides may represent activation of a novel "P3" receptor (RALEVIC and BURNSTOCK 1998), but this remains conjectural at this point.

In addition to direct effects upon neuronal excitability, another significant role of P2 receptors in the nervous system may be to modulate transmitter release. For example, norepinephrine release in the hippocampus appears to be modulated by both A_1 and P2 receptors (KOCH et al. 1997), and the two responses can be clearly differentiated by the use of selective antagonists. In the trigeminal mesencephalic motor nucleus, ATP appears to stimulate the release of glutamate via actions on glutamatergic nerve terminals (KHAKH and HENDERSON 1998). Likewise, the release of 5HT (VON KÜGELGEN et al. 1997), and dopamine (ZHANG et al. 1995) may be modulated by P2 receptors. It is not entirely clear that all of these effects are direct, because in some instances

ATP appears to release another transmitter (e.g., glutamate), which then acts on nerve terminals to modulate the release of a second transmitter; this possibility has been suggested to underlie at least some of the reported effects of ATP (POELCHEN et al. 1998).

III. Synaptic Responses Mediated via ATP

The phenomenon of transmission mediated by P2 receptors is well established in the peripheral nervous system, and there is good evidence that ATP may play a role in transmission in sensory systems. For example, in the dorsal horn, ATP can stimulate the release of glutamate from presynaptic nerve terminals (GU and MACDERMOTT 1997). However, in the brain there are only two reports of synapses where transmission is mediated by nucleotide receptors. EDWARDS et al. (1992) were the first to report on synapses in the central nervous system that utilized ATP as the primary neurotransmitter. In the medial habenula, local stimulation evokes small, fast, inward currents that are not blocked by antagonists of other ligand-gated ion channels, but can be inhibited by superfusion with suramin, and are reduced by the desensitizing agonist, α,β-methylene ATP. Spontaneous events having the same kinetic and pharmacological properties as the evoked responses were also observed. Given the relatively limited pharmacological experiments involving these receptors, it is not possible to determine the subtype of receptor involved. However, their rapid kinetics are consistent with a P2X type of receptor, and their localization and antagonism by suramin would suggest a $P2X_1$, $P2X_2$, or $P2X_5$ receptor, or heteromeric receptors involving these subunits. The channels mediating these synaptic responses showed either low, or quite high Ca^{2+} permeability (EDWARDS et al. 1997), which suggests that multiple subtypes of receptors are involved. Finally, a variety of approaches have demonstrated that ATP and glutamate are released from different neurons in the habenula, so ATP does not appear to be a co-transmitter at glutamatergic synapses in this region (ROBERTSON and EDWARDS 1998).

NIEBER et al. (1997) have reported that electrical stimulation of the locus coeruleus leads to the activation of a fast inward current that is mediated via P2 receptors. Although it was not possible to isolate pharmacologically separate components of these responses corresponding to activation of the P2X and P2Y receptors that are thought to be present on these cells (see above), the kinetics of the response would suggest that the rapid onset current must be mediated via P2X receptors, and the sensitivity to α,β-methylene ATP, suramin, and PPADS, a P2 receptor antagonist, are most consistent with involvement of a $P2X_1$ receptor or a heteromer containing this subunit. It was not possible to determine whether P2Y receptors are also synaptically activated by this type of stimulation.

There is also a recent report suggesting that ATP-induced currents may mediate as much as 25% of the fast excitatory current evoked by electrical stimulation in the hippocampus (PANKRATOV et al. 1998); this is perhaps not unexpected, given the relatively high level of expression of P2X receptors in hippocampus, but it is not clear why previous studies have not reported a component of fast excitatory responses that is not sensitive to the glutamatergic antagonists such as DNQX and APV.

IV. Physiological Effects of Diadenosine Polyphosphates

Considerable evidence exists to suggest that diadenosine polyphosphates, such as diadenosine tetraphosphate (Ap4A), may function as neurotransmitters. Synthesis, storage in vesicles, Ca^{2+} dependent release from synaptosomes, and specific high-affinity binding sites for diadenosine polyphosphates have all been reported, and have been the subject of recent reviews (PINTOR et al. 1997a; PINTOR and MIRAS-PORTUGAL 1995). The receptors that mediate responses to dinucleotides have not proved amenable to cloning, so it is difficult at this point to assess their relationship to other purine and pyrimidine receptors. However, in guinea pig synaptosomes, where diadenosine polyphosphate receptors are linked to increases in intracellular Ca^{2+}, these receptors appear to be pharmacologically distinct from the receptors with which "mononucleotides" such as ATP or UTP interact (PINTOR et al. 1997b), and binding studies suggest a similar specificity in the ability of dinucleotides vs mononucleotides to interact with these binding sites (PINTOR et al. 1993). A variety of actions have been ascribed to activation of these receptors, including increases in intracellular Ca^{2+}, activation of PKC, and enhancement of the affinity of ATP at P2X receptors. In some instances, the increases in intracellular Ca^{2+} do not appear be the result of a direct action on dinucleotide-activated channels, but rather reflect the facilitation of the activation of N-type Ca^{2+} channels (PANCHENKO et al. 1996). At this point the existence of multiple adenine dinucleotide-activated receptors appears likely, but whether they are metabotropic, ionotropic, or both remains to be definitively established.

V. Responses Mediated by Pyrimidine Receptors

The cloning of the P2Y receptors, the demonstration that a number of these receptors are readily activated by UTP (P2Y$_2$ and P2Y$_4$) or UDP (P2Y$_6$), and that they are expressed in the nervous system (LUSTIG et al. 1993), raise the interesting possibility that pyrimidine nucleotides may serve as neurotransmitters in much the same fashion as does ATP. It has also been proposed that by analogy with adenosine, uridine might serve a modulatory role in the nervous system as well (CONNOLLY and DULEY 1999), but currently there is no direct evidence for uridine-sensitive receptors. At the present time, the evidence to support a neurotransmitter/neuromodulator role for uridine and its nucleotides is weak; in particular, candidate synapses have yet to be identified where uridine, UTP, or UDP might mediate or modulate postsynaptic responses. Physiological responses to exogenous UTP can be demonstrated in various systems; for example, in the rat superior cervical ganglion, it has been shown that UTP has a depolarizing effect, which appears to be mediated via a site that is distinct from the one at which ATP has its actions (CONNOLLY and HARRISON 1995; CONNOLLY 1994), and UTP can stimulate the release of norepinephrine as well (BOEHM et al. 1995), again through a site not activated by ATP. Uridine itself has been demonstrated to have a number of actions in intact animals and in humans that suggest that it has its action on the nervous system (reviewed by CONNOLLY and DULEY 1999), but in the absence of identified receptors, and metabolically stable agents with which to activate them, it is difficult to determine the mechanism of action underlying these effects.

D. Conclusions

The existence of physiological responses in the central nervous system to both purine nucleotides and nucleosides is unequivocally established. While the existence of all four P1 receptor subtypes has been established in the CNS, it cannot be said that there are any examples of "adenosinergic" synapses in brain, in that adenosine does not appear to be located in synaptic vesicles and released in a conventional, Ca^{2+}-dependent fashion as are other neurotransmitters. Thus, the focus of much recent work in this area is concerned with understanding more about the role played by adenosine in the brain, and characterizing the way in which adenosine is released in response to a variety of stimuli.

As far as P2 receptors are concerned, they are abundant in the central nervous system and, unlike adenosine, there are at least a few examples of purinergic synapses where ATP appears to be the transmitter. Nevertheless, they are clearly expressed in many brain regions in which synaptic responses mediated by ATP have not been demonstrated; thus, the immediate challenge is to identify more such synapses, and to understand the way in which the complex properties of these synapses relate to the multiple subtypes of homomeric and heteromeric P2 receptors.

References

Abbracchio MP, Ceruti S, Brambilla R, Franceschi C, Malorni W, Jacobson, KA, Von Lubitz DK, Cattabeni F (1997) Modulation of apoptosis by adenosine in the central nervous system: a possible role for the A_3 receptor. Pathophysiological significance and therapeutic implications for neurodegenerative disorders. Ann NY Acad Sci 825:11–22

Abdul-Ghani AS, Attwell PJE, Bradford HF (1997) The protective effect of 2-chloroadenosine against the development of amygdala kindling and on amygdala-kindled seizures. Eur J Pharmacol 326:7–14

Abraham WC, Huggett A (1997) Induction and reversal of long-term potentiation by repeated high-frequency stimulation in rat hippocampal slices. Hippocampus 7:137–145

Ambrosio AF, Malva JO, Carvalho AP, Carvalho CM (1997) Inhibition of N-,P/Q- and other types of Ca^{2+} channels in rat hippocampal nerve terminals by the adenosine A_1 receptor. Eur J Pharmacol 340:301–310

Ault B, Olney MA, Joyner JL, Boyer CE, Notrica MA, Soroko FE, Wang CM (1987) Pro-convulsant actions of theophylline and caffeine in the hippocampus: implications for the management of temporal lobe epilepsy. Brain Res 426:93–102

Ault B, Wang CM (1986) Adenosine inhibits epileptiform activity arising in hippocampal area CA3. Br J Pharmacol 87:695–703

Barth A, Newell DW, Nguyen LB, Winn HR, Wender R, Meno JR, Janigro D (1997) Neurotoxicity in organotypic hippocampal slices mediated by adenosine analogues and nitric oxide. Brain Res 762:79–88

Biggs TAG, Myers RD (1997) Adenosine A_1 receptor antisense infused in striatum of rats: actions on alcohol-induced locomotor impairment, blood alcohol, and body temperature. Alcohol 14:617–621

Bischofberger N, Jacobson KA, Von Lubitz DK (1997) Adenosine A_1 receptor agonists as clinically viable agents for treatment of ischemic brain disorders. Ann NY Acad Sci 825:23–29

Boehm S, Huck S, Illes P (1995) UTP- and ATP-triggered transmitter release from rat sympathetic neurones via separate receptors. Br J Pharmacol 116:2341–2343

Bona E, Ådén U, Gilland E, Fredholm BB, Hagberg H (1997) Neonatal cerebral hypoxia-ischemia: The effect of adenosine receptor antagonists. Neuropharmacol 36:1327–1338

Bonci A, Williams JT (1996) A common mechanism mediates long-term changes in synaptic transmission after chronic cocaine and morphine. Neuron 16:631–639

Borowicz KK, Kleinrok Z, Czuczwar SJ (1997) N^6-2-(4-aminophenyl)ethyl-adenosine enhances the anticonvulsive activity of antiepileptic drugs. Eur J Pharmacol 327:125–133

Brake AJ, Wagenbach MJ, Julius D (1994) New structural motif for ligand-gated ion channels defined by an ionotropic ATP receptor. Nature 371:519–523

Braun N, Zhu Y, Krieglstein J, Culmsee C, Zimmermann H (1998) Upregulation of the enzyme chain hydrolyzing extracellular ATP after transient forebrain ischemia in the rat. J Neurosci 18:4891–4900

Brodie C, Blumberg PM, Jacobson KA (1998) Activation of the A_{2A} adenosine receptor inhibits nitric oxide production in glial cells. FEBS Letters 429:139–142

Brundege JM, Dunwiddie TV (1995) The role of acetate as a potential mediator of the effects of ethanol in the brain. Neurosci Lett 186:214–218

Brundege JM, Diao LH, Proctor WR, Dunwiddie TV (199) The role of cyclic AMP as a precursor of extracellular adenosine in the rat hippocampus. Neuropharmacol 36:1201–1210

Brundege JM, Dunwiddie TV (1997) Role of adenosine as a modulator of synaptic activity in the central nervous system. Adv Pharmacol 39:353–391

Burnstock G (1997) The past, present and future of purine nucleotides as signalling molecules. Neuropharmacology 36:1127–1139

Calabresi P, Centonze D, Pisani A, Bernardi G (1997) Endogenous adenosine mediates the presynaptic inhibition induced by aglycemia at corticostriatal synapses. J Neurosci 17:4509–4516

Campisi P, Carmichael FJL, Crawford M, Orrego H, Khanna JM (1997) Role of adenosine in the ethanol-induced potentiation of the effects of general anesthetics in rats. Eur J Pharmacol 325:165–172

Carmichael FJ, Israel Y, Crawford M, Minhas K, Saldivia V, Sandrin S, Campisi P, Orrego H (1991) Central nervous system effects of acetate: contribution to the central effects of ethanol. J Pharmacol Exp Ther 259:403–408

Carmichael FJ, Saldivia V, Varghese GA, Israel Y, Orrego H (1988) Ethanol-induced increase in portal blood flow: role of acetate and A_1- and A_2-adenosine receptors. Am J Physiol 255:G417–G423

Chen G, van den Pol AN (1997) Adenosine modulation of calcium currents and presynaptic inhibition of GABA release in suprachiasmatic and arcuate nucleus neurons. J Neurophysiol 77:3035–3047

Chen H, Lambert NA (1997) Inhibition of dendritic calcium influx by activation of G-protein-coupled receptors in the hippocampus. J Neurophysiol 78:3484–3488

Chow SC, Kass GE, Orrenius S (1997) Purines and their roles in apoptosis. Neuropharmacology 36:1149–1156

Collo G, North RA, Kawashima E, Merlo-Pich E, Neidhart, Surprenant A, Buell G (1996) Cloning of $P2X_5$ and $P2X_6$ receptors and the distribution and properties of an extended family of ATP-gated ion channels. J Neurosci 16:2495–2507

Coney AM, Marshall JM (1998) Role of adenosine and its receptors in the vasodilatation induced in the cerebral cortex of the rat by systemic hypoxia. J Physiol 509:507–518

Connolly GP (1994) Evidence from desensitization studies for distinct receptors for ATP and UTP on the rat superior cervical ganglion. Br J Pharmacol 112:357–359

Connolly GP, Duley JA (1999) Uridine and its nucleotides: biological actions, therapeutic potentials. Trends Pharmacol 20:218–225

Connolly GP, Harrison PJ (1995) Structure-activity relationship of a pyrimidine receptor in the rat isolated superior cervical ganglion. Br J Pharmacol 116:2764–2770

Craig CG, White TD (1992) Low-level N-methyl-D-aspartate receptor activation provides a purinergic inhibitory threshold against further N-methyl-D-aspartate-mediated neurotransmission in the cortex. J Pharmacol Exp Ther 260:1278–1284

Craig CG, White TD (1993) N-methyl-D-aspartate- and non-N-methyl-D-aspartate-evoked adenosine release from rat cortical slices: distinct purinergic sources and mechanisms of release. J Neurochem 60:1073–1080

Cullen N, Carlen PL (1992) Electrophysiological action of acetate, a metabolite of ethanol, on hippocampal dentate granule neurons: interaction with adenosine. Brain Res 588:49–57.

Cunha RA, Constantino MD, Ribeiro JA (1997) ZM241385 is an antagonist of the facilitatory responses produced by the A2 A adenosine receptor agonists CGS21680 and HENECA in the rat hippocampus. Br J Pharmacol 122:1279–1284

Cunha RA, Johansson B, Constantino MD, Sebastiao AM, Fredholm BB (1996) Evidence for high affinity binding sites for the adenosine A_{2A} receptor agonist [^{3}H] CGS 21680 in the rat hippocampus and cerebral cortex that are different from striatal A_{2A} receptors. Naunyn-Schmiedebergs Arch Pharmacol 353:261–271

Cunha RA, Johansson B, Van der Ploeg I, Sebastiao AM, Ribeiro JA, Fredholm BB (1994) Evidence for functionally important adenosine A_{2A} receptors in the rat hippocampus. Brain Res 649:208–216

Cunha RA, Sebastiao AM, Ribeiro JA (1998) Inhibition by ATP of hippocampal synaptic transmission requires localized extracellular catabolism by ecto-nucleotidases into adenosine and channeling to adenosine A_1 receptors. J Neurosci 18:1987–1995

Dar MS (1997) Mouse cerebellar adenosinergic modulation of ethanol-induced motor incoordination: possible involvement of cAMP. Brain Res 749:263–274

Dar MS (1998) Involvement of κ-opioids in the mouse cerebellar adenosinergic modulation of ethanol-induced motor incoordination. Alcohol Clin Exp Res 22:444–454

Dascal N (1997) Signalling via the G protein-activated K+ channels. Cell Signal 9:551–573

de Mendonça A, Ribeiro JA (1997) Contribution of metabotropic glutamate receptors to the depression of excitatory postsynaptic potentials during hypoxia. Neuroreport 8:3667–3671

de Mendonça A, Ribeiro JA (1996) Adenosine and neuronal plasticity. Life Sci 60:245–251

de Mendonça A, Sebastiao AM, Ribeiro JA (1995) Inhibition of NMDA receptor-mediated currents in isolated rat hippocampal neurones by adenosine A_1 receptor activation. Neuroreport 6:1097–1100

de Sarro A, Grasso S, Zappala M, Nava F, De, Sarro G (1997) Convulsant effects of some xanthine derivatives in genetically epilepsy-prone rats. Naunyn-Schmiedebergs Arch Pharmacol 356:48–55

Di Iorio P, Ballerini P, Caciagli F, Ciccarelli R (1998) Purinoceptor-mediated modulation of purine and neurotransmitter release from nervous tissue. Pharmacol Res 37:169–178

Dixon AK, Widdowson L, Richardson PJ (1997) Desensitisation of the adenosine A_1 receptor by the A_{2A} receptor in the rat striatum. J Neurochem 69:315–321

Doherty J, Dingledine R (1997) Regulation of excitatory input to inhibitory interneurons of the dentate gyrus during hypoxia. J Neurophys 77:393–404

Doolette DJ (1997) Mechanism of adenosine accumulation in the hippocampal slice during energy deprivation. Neurochem Int 30:211–223

Dragunow M (1988) Purinergic mechanisms in epilepsy. Prog Neurobiol 31:85–108

Dunwiddie TV (1980) Endogenously released adenosine regulates excitability in the in vitro hippocampus. Epilepsia 21:541–548

Dunwiddie TV (1995) Acute and chronic effects of ethanol on the brain: interactions of ethanol with adenosine, adenosine transporters, and adenosine receptors. In: Deitrich RA, Erwin VG (eds) Pharmacological effects of ethanol on the nervous system. CRC Press, Boca Raton, pp 147–161

Dunwiddie TV (1999) Adenosine and ethanol: is there a caffeine connection in the actions of ethanol? In: Liu Y, Hunt WA (eds) The "drunken" synapse: studies of alcohol-related disorders. Plenum Publishing, New York

Dunwiddie TV, Diao L, Kim HO, Jiang JL, Jacobson KA (1997) Activation of hippocampal adenosine A_3 receptors produces a desensitization of A1 receptor-mediated responses in rat hippocampus. J Neurosci 17:607–614

Dunwiddie TV, Diao LH, Proctor WR (1997) Adenine nucleotides undergo rapid, quantitative conversion to adenosine in the extracellular space in rat hippocampus. J Neurosci 17:7673–7682

Dunwiddie TV, Worth TS (1982) Sedative and anticonvulsant effects of adenosine analogs in mouse and rat. J Pharmacol Exp Ther 220:70–76

During MJ, Spencer DD (1993) Adenosine: a potential mediator of seizure arrest and postictal refractoriness. Ann Neurol 32:618–624

Edwards FA, Gibb AJ, Colquhoun D (1992) ATP receptor-mediated synaptic currents in the central nervous system. Nature 359:144–147

Edwards FA, Robertson SJ, Gibb AJ (1997) Properties of ATP receptor-mediated synaptic transmission in the rat medial habenula. Neuropharmacol 36:1253–1268

Ehrengruber MU, Doupnik CA, Xu Y, Garvey J, Jasek MC, Lester HA, Davidson N (1997) Activation of heteromeric G protein-gated inward rectifier K+ channels overexpressed by adenovirus gene transfer inhibits the excitability of hippocampal neurons. Proc Nat Acad Sci USA 94:7070–7075

Fallahi N, Broad RM, Jin S, Fredholm BB (1996) Release of adenosine from rat hippocampal slices by nitric oxide donors. J Neurochem 67:186–193

Fenu S, Pinna A, Ongini E, Morelli M (1997) Adenosine A2A receptor antagonism potentiates L-DOPA-induced turning behaviour and c-fos expression in 6-hydroxydopamine-lesioned rats. Eur J Pharmacol 321:143–147

Feoktistov I, Biaggioni I (1997) Adenosine A_{2B} receptors. Pharmacol Rev 49:381–402

Ferre S (1997) Adenosine-dopamine interactions in the ventral striatum. Implications for the treatment of schizophrenia. Psychopharmacology 133:107–120

Ferre S, Fredholm BB, Morelli M, Popoli P, Fuxe K (1997) Adenosine-dopamine receptor-receptor interactions as an integrative mechanism in the basal ganglia. Trends Neurosci 20:482–487

Filippov AK, Webb TE, Barnard EA, Brown DA (1998) P2Y$_2$ nucleotide receptors expressed heterologously in sympathetic neurons inhibit both N-type Ca^{2+} and M-type K^+ currents. J Neurosci 18:5170–5179

Fleming KM, Mogul DJ (1997) Adenosine A_3 receptors potentiate hippocampal calcium current by a PKA-dependent/PKC-independent pathway. Neuropharmacol 36:353–362

Florio C, Prezioso A, Papaioannou A, Vertua R (1998) Adenosine A_1 receptors modulate anxiety in CD1 mice. Psychopharmacol 136:311–319

Florio C, Rosati AM, Traversa U, Vertua R (1997) Inhibitory and excitatory effects of adenosine antagonists on spontaneous locomotor activity in mice. Life Sci 60:1477–1486

Fowler JC (1993) Purine release and inhibition of synaptic transmission during hypoxia and hypoglycemia in rat hippocampal slices. Neurosci Lett 157:83–86

Fredholm BB (1997) Adenosine and neuroprotection. International Rev Neurobiol 40:259–280

Fredholm BB, Dunwiddie TV (1988) How does adenosine inhibit transmitter release? Trends Pharmacol 9:130–134

Fredholm BB, Gerwins P, Johansson B, Parkinson FE, Van der Ploeg I (1993) Signalling via G-protein coupled receptors–using adenosine receptors as an example. Drug Des Discov 9:189–197

Frohlich R, Boehm S, Illes P (1996) Pharmacological characterization of P2 purinoceptor types in rat locus coeruleus neurons. Eur J Pharmacol 315:255–261

Fujii S, Ito K, Osada H, Hamaguchi T, Kuroda Y, Kato H (1995) Extracellular phosphorylation of membrane protein modifies theta burst-induced long-term potentiation in CA1 neurons of guinea-pig hippocampal slices. Neurosci Lett 187:133–136

Fujiwara N, Abe T, Endoh H, Warashina A, Shimoji K (1992) Changes in intracellular pH of mouse hippocampal slices responding to hypoxia and/or glucose depletion. Brain Res 572:335–339

Fulga I, Stone TW (1998) Comparison of an adenosine A_1 receptor agonist and antagonist on the rat EEG. Neurosci Lett 244:55–59

Furukawa K, Ishibashi H, Akaike N (1994) ATP-induced inward current in neurons freshly dissociated from the tuberomammillary nucleus. J Neurophys 71:868–873

Fuxe K, Ferre S, Zoli M, Agnati LF (1998) Integrated events in central dopamine transmission as analyzed at multiple levels. Evidence for intramembrane adenosine A_{2A}/dopamine D_2 and adenosine A_1/dopamine D_1 receptor interactions in the basal ganglia. Brain Res Brain Res Rev 26:258–273

Gabriel A, Klussmann FW, Igelmund P (1998) Rapid temperature changes induce adenosine-mediated depression of synaptic transmission in hippocampal slices from rats (non-hibernators) but not in slices from golden hamsters (hibernators). Neuroscience 86:67–77

Garrett BE, Griffiths RR (1997) The role of dopamine in the behavioral effects of caffeine in animals and humans. Pharmacol Biochem Behav 57:533–541

Georgiev VP, Tchekalarova JD (1998) Interaction of angiotensin II and adenosine receptors in pentylenetetrazol-induced kindling in mice. Brain Res 779:259–261

Gereau RW, Conn PJ (1994) Potentiation of cAMP responses by metabotropic glutamate receptors depresses excitatory synaptic transmission by a kinase-independent mechanism. Neuron 12:1121–1129

Gines S, Hillion J, Torvinen M, Le Crom S, Casado V, Canela EI, Rondin S, Lew JY, Watson S, Zoli M, Agnati LF, Verniera P, Lluis C, Ferre S, Fuxe K, Franco R (2000) Dopamine D1 and adenosine A1 receptors form functionally interacting heteromeric complexes. Proceedings of the National Academy of Sciences of the United States of America. 97:8606–8611

Goncalves ML, Cunha RA, Ribeiro JA (1997) Adenosine A_{2A} receptors facilitate 45Ca^{2+} uptake through class A calcium channels in rat hippocampal CA3 but not CA1 synaptosomes. Neurosci Lett 238:73–77

Gribkoff VK, Bauman LA (1992) Endogenous adenosine contributes to hypoxic synaptic depression in hippocampus from young and aged rats. J Neurophys 68:620–628

Gu JG, MacDermott AB (1997) Activation of ATP P2X receptors elicits glutamate release from sensory neuron synapses. Nature 389:749–753

Halle JN, Kasper CE, Gidday JM, Koos BJ (1997) Enhancing adenosine A1 receptor binding reduces hypoxic-ischemic brain injury in newborn rats. Brain Res 759: 309–312

Harms L, Finta EP, Tschopl M, Illes P (1992) Depolarization of rat locus coeruleus neurons by adenosine 5′-triphosphate. Neuroscience 48:941–952

Hauber W, Munkle M (1997) Motor depressant effects mediated by dopamine D_2 and adenosine A_{2A} receptors in the nucleus accumbens and the caudate-putamen. Eur J Pharmacol 323:127–131

Hauber W, Nagel J, Sauer R, Muller CE (1998) Motor effects induced by a blockade of adenosine A_{2A} receptors in the caudate-putamen. Neuroreport 9:1803–1806

Hebb MO, White TD (1998) Co-administration of adenosine kinase and deaminase inhibitors produces supra-additive potentiation of N-methyl-D-aspartate-evoked adenosine formation in cortex. Eur J Pharmacol 344:121–125

Herlenius E, Lagercrantz H, Yamamoto Y (1997) Adenosine modulates inspiratory neurons and the respiratory pattern in the brainstem of neonatal rats. Pediatr Res 42:46–53

Heurteaux C, Lauritzen I, Widmann C, Lazdunski M (1995) Essential role of adenosine, adenosine A_1 receptors, and ATP-sensitive K^+ channels in cerebral ischemic preconditioning. Proc Natl Acad Sci USA 92:4666–4670

Hill RJ, Oleynek JJ, Magee W, Knight DR, Tracey WR (1998) Relative importance of adenosine A_1 and A_3 receptors in mediating physiological or pharmacological protection from ischemic myocardial injury in the rabbit heart. J Mol Cell Cardiol 30:579–585

Hirai H, Okada Y (1995) Adenosine facilitates in vivo neurotransmission in the superior colliculus of the rat. J Neurophys 74:950–960

Hoehn K, Craig CG, White TD (1990) A comparison of N-methyl-D-aspartate-evoked release of adenosine and [^{3}H]-norepinephrine from rat cortical slices. J Pharmacol Exp Ther 255:174–171

Hoehn K, White TD (1990) N-methyl-D-aspartate, kainate and quisqualate release endogenous adenosine from rat cortical slices. Neuroscience 39:441–450

Hoffman PL, Tabakoff B (1990) Ethanol and guanine nucleotide binding proteins: a selective interaction. FASEB J 4:2612–2622

Howell LL, Coffin VL, Spealman RD (1997) Behavioral and physiological effects of xanthines in nonhuman primates. Psychopharmacology (Berl) 129:1–14

Hu H-Z, Li Z-W (1997) Modulation by adenosine of GABA-activated current in rat dorsal root ganglion neurons. J Physiol 501:67–75

Huston JP, Haas HL, Boix F, Pfister M, Decking U, Schrader J, Schwarting RK (1996) Extracellular adenosine levels in neostriatum and hippocampus during rest and activity periods of rats. Neuroscience 73:99–107

Ikeuchi Y, Nishizaki T (1995) ATP-evoked potassium currents in rat striatal neurons are mediated by a P_2 purinergic receptor. Neurosci Lett 190:89–92

Ikeuchi Y, Nishizaki T, Mori M, Okada Y (1995) Adenosine activates the potassium channel via a P_2 purinoceptor but not via an adenosine receptor in cultured rat superior colliculus neurons. Neurosci Lett 198:205–208

Ikeuchi Y, Nishizaki T, Mori M, Okada Y (1996a) Adenosine activates the K+ channel and enhances cytosolic Ca^{2+} release via a P2Y purinoceptor in hippocampal neurons. Eur J Pharmacol 304:191–199

Ikeuchi Y, Nishizaki T, Mori M, Okada Y (1996b) Regulation of the potassium current and cytosolic Ca^{2+} release induced by 2-methylthio ATP in hippocampal neurons. Biochem Biophys Res Commun 218:428–433

Illes P, Nieber K, Norenberg W (1996) Electrophysiological effects of ATP on brain neurones. J Auton Pharmacol 16:407–411

Inoue K, Koizumi S, Ueno S (1996) Implication of ATP receptors in brain functions. Prog Neurobiol 50:483–492

Inoue K, Nakazawa K, Fujimori K, Watano T, Takanaka A (1992) Extracellular adenosine 5'-triphosphate-evoked glutamate release in cultured hippocampal neurons. Neurosci Lett 134:215–218

Ishikawa S, Saijoh K, Okada Y (1997) Endogenous adenosine facilitates neurotransmission via A_{2A} adenosine receptors in the rat superior colliculus in vivo. Brain Res 757:268–275

Jacobson KA, Nikodijevic O, Shi D, Gallo-Rodriguez C, Olah ME, Stiles GL, Daly JW (1993) A role for central A_3 adenosine receptors: mediation of behavioral depressant effects. FEBS Lett 336:57–60

Jacobson KA, von Lubitz DKJE, Daly JW, Fredholm BB (1996) Adenosine receptor ligands: differences with acute and chronic treatment. Trends Pharmacol 17: 108–113

Jacobsen KA, Xie R, Young L, Chang L, Liang BT (2000) A Novel Pharmacological Approach to Treating Cardiac Ischemia. Binary conjugates of A_1 and A_3 adenosine receptor agonists. J Biol Chem 275:30272–30279

Jain N, Kemp N, Adeyemo O, Buchanan P, Stone TW (1995) Anxiolytic activity of adenosine receptor activation in mice. Br J Pharmacol 116:2127–2133

Janigro D, Wender R, Ransom G, Tinklepaugh DL, Winn HR (1996) Adenosine-induced release of nitric oxide from cortical astrocytes. Neuroreport 7:1640–1644

Jin S, Fredholm BB (1997) Adenosine A_1 receptors mediate hypoxia-induced inhibition of electrically evoked transmitter release from rat striatal slices. Eur J Pharmacol 329:107–113

Jin S, Fredholm BB (1997) Adenosine A_{2A} receptor stimulation increases release of acetylcholine from rat hippocampus but not striatum, and does not affect catecholamine release. Naunyn-Schmiedebergs Arch Pharmacol 355:48–56

Jin S, Fredholm BB (1997) Glucose deprivation increases basal and electrically evoked transmitter release from rat striatal slices. Role of NMDA and adenosine A_1 receptors. Eur J Pharmacol 340:169–175

Kanda T, Jackson MJ, Smith LA, Pearce RK, Nakamura J, Kase H, Kuwana Y, Jenner P (1998) Adenosine A_{2A} antagonist: a novel antiparkinsonian agent that does not provoke dyskinesia in parkinsonian monkeys. Ann Neurol 43:507–513

Kaplan GB, Leite-Morris KA (1997) Up-regulation of adenosine transporter-binding sites in striatum and hypothalamus of opiate tolerant mice. Brain Res 763:215–220

Kemp N, Bashir ZI (1997) A role for adenosine in the regulation of long-term depression in the adult rat hippocampus in vitro. Neurosci Lett 225:189–192

Khakh BS, Henderson G (1998) ATP receptor-mediated enhancement of fast excitatory neurotransmitter release in the brain. Mol Pharmacol 54:372–378

Khakh BS, Humphrey PP, Henderson G (1997) ATP-gated cation channels (P2X purinoceptors) in trigeminal mesencephalic nucleus neurons of the rat. J Physiol (Lond) 498:709–715

Khakh BS, Kennedy C (1998) Adenosine and ATP: progress in their receptors' structures and functions. Trends Pharmacol Sci 19:39–41

Kidd EJ, Grahames CB, Simon J, Michel AD, Barnard EA, Humphrey PP (1995) Localization of P2X purinoceptor transcripts in the rat nervous system. Mol Pharmacol 48:569–573

Koch H, Kugelgen I, Starke K (1997) P2-receptor-mediated inhibition of noradrenaline release in the rat hippocampus. Naunyn-Schmiedebergs Arch Pharmacol 355:707–715

Krauss SW, Ghirnikar RB, Diamond I, Gordon AS (1993) Inhibition of adenosine uptake by ethanol is specific for one class of nucleoside transporters. Mol Pharmacol 44:1021–1026

Kulkarni C, David J, Joseph T (1997) Influence of adenosine, dipyridamole, adenosine antagonists and antiepileptic drugs on EEG after discharge following cortical stimulation. Indian J Exp Biol 35:342–347

Le KT, Babinski K, Seguela P (1998) Central P2X4 and P2X6 channel subunits coassemble into a novel heteromeric ATP receptor. J Neurosci 18:7152–7159

Ledent C, Vaugeois JM, Schiffmann SN, Pedrazzini T, El Yacoubi M, Vanderhaeghen JJ, Costentin J, Heath JK, Vassart G, Parmentier M (1997) Aggressiveness, hypoalgesia and high blood pressure in mice lacking the adenosine A_{2A} receptor. Nature 388:674–678

Lewis C, Neidhart S, Holy C, North RA, Buell G, Surprenant A (1995) Coexpression of $P2X_2$ and $P2X_3$ receptor subunits can account for ATP-gated currents in sensory neurons. Nature 377:432–435

Li H, Henry JL (1998) Adenosine A_2 receptor mediation of pre- and postsynaptic excitatory effects of adenosine in rat hippocampus in vitro. Eur J Pharmacol 347:173–182

Lloyd HGE, Lindström K, Fredholm BB (1993) Intracellular formation and release of adenosine from rat hippocampal slices evoked by electrical stimulation or energy depletion. Neurochem Int 23:173–185

Lovinger DM, Choi S (1995) Activation of adenosine A_1 receptors initiates short-term synaptic depression in rat striatum. Neurosci Lett 199:9–12

Lustig KD, Shiau AK, Brake AJ, Julius D (1993) Expression cloning of an ATP receptor from mouse neuroblastoma cells. Proc Nat Acad Sci USA 90:5113–5117

Luthin GR, Tabakoff B (1984) Activation of adenylate cyclase by alcohol requires the nucleotide-binding protein. J Pharmacol Exp Ther 228:579–587

Macek TA, Schaffhauser H, Conn PJ (1998) Protein kinase C and A_3 adenosine receptor activation inhibit presynaptic metabotropic glutamate receptor (mGluR) function and uncouple mGluRs from GTP-binding proteins. J Neurosci 18:6138–6146

Malhotra J, Gupta YK (1997) Effect of adenosine receptor modulation on pentylenetetrazole-induced seizures in rats. Br J Pharmacol 120:282–288

Mandhane SN, Chopde CT, Ghosh AK (1997) Adenosine A_2 receptors modulate haloperidol-induced catalepsy in rats. Eur J Pharmacol 328:135–141

Manzoni O, Pujalte D, Williams J, Bockaert J (1998) Decreased presynaptic sensitivity to adenosine after cocaine withdrawal. J Neurosci 18:7996–8002

Manzoni OJ, Manabe T, Nicoll RA (1994) Release of adenosine by activation of NMDA receptors in hippocampus. Science 265:2098–2101

Marks GA, Birabil CG (1998) Enhancement of rapid eye movement sleep in the rat by cholinergic and adenosinergic agonists infused into the pontine reticular formation. Neuroscience 86:29–37

Marston HM, Finlayson K, Maemoto T, Olverman HJ, Akahane A, Sharkey J, Butcher SP (1998) Pharmacological characterization of a simple behavioral response mediated selectively by central adenosine A_1 receptors, using in vivo and in vitro techniques. J Pharmacol Exp Ther 285:1023–1030

Masino SA, Dunwiddie TV (1999) Temperature-dependent modulation of excitatory transmission in hippocampal slices is mediated by extracellular adenosine. J Neurosci 19:1932–1939

Mateo J, Garcia-Lecea M, Miras-Portugal MT, Castro E (1998) Ca^{2+} signals mediated by P2X-type purinoceptors in cultured cerebellar Purkinje cells. J Neurosci 18:1704–1712

Matuszek MT, Gagalo IT (1997) The antipyretic activity of adenosine agonists. Gen Pharmacol 29:371–374

Mayfield RD, Larson G, Orona RA, Zahniser NR (1996) Opposing actions of adenosine A_{2A} and dopamine D_2 receptor activation on GABA release in the basal ganglia: Evidence for an A_{2A}/D_2 receptor interaction in globus pallidus. Synapse 22:132–138

Meng ZH, Pennington SN, Dar MS (1998) Rat striatal adenosinergic modulation of ethanol-induced motor impairment: Possible role of striatal cyclic AMP. Neuroscience 85:919–930

Mogul DJ, Adams ME, Fox AP (1993) Differential activation of adenosine receptors decreases N-type but potentiates P-type Ca^{2+} current in hippocampal CA3 neurons. Neuron 10:327–334

Monin P (1997) Pharmacology of respiratory control in neonates and children. Pediatr Pulmonol Suppl 16:222–224

Murray TF, Franklin PH, Zhang G, Tripp E (1992) A_1 adenosine receptors express seizure-suppressant activity in the rat prepiriform cortex. Epilepsy Res Suppl 8:255–261

Nieber K, Poelchen W, Illes P (1997) Role of ATP in fast excitatory synaptic potentials in locus coeruleus neurones of the rat. Br J Pharmacol 122:423–430

Norenberg W, Wirkner K, Illes P (1997) Effect of adenosine and some of its structural analogues on the conductance of NMDA receptor channels in a subset of rat neostriatal neurones. Br J Pharmacol 122:71–80

Norenberg W, Wirkner K, Illes P (1997) Effect of adenosine and some of its structural analogues on the conductance of NMDA receptor channels in a subset of rat neostriatal neurones. Br J Pharmacol 122:71–80

North RA, Barnard EA (1997) Nucleotide receptors. Current Opinion in Neurobiology 7:346–357

O'Kane EM, Stone TW (1998) Interaction between adenosine A_1 and A_2 receptor-mediated responses in the rat hippocampus in vitro. Eur J Pharmacol 362:17–25

Ongini E, Adami M, Ferri C, Bertorelli R (1997) Adenosine A_{2A} receptors and neuroprotection. Ann NY Acad Sci 825:30–48

Panchenko VA, Pintor J, Tsyndrenko AY, Miras-Portugal MT, Krishtal OA (1996) Diadenosine polyphosphates selectively potentiate N-type Ca^{2+} channels in rat central neurons. Neuroscience 70:353–360

Pankratov Y, Castro E, Miras-Portugal MT, Krishtal O (1998) A purinergic component of the excitatory postsynaptic current mediated by P2X receptors in the CA1 neurones of the rat hippocampus. Eur J Neurosci 10:3898–3902

Pérez-Pinzón MA, Mumford PL, Rosenthal M, Sick TJ (1996) Anoxic preconditioning in hippocampal slices: role of adenosine. Neuroscience 75:687–694

Phan TA, Gray AM, Nyce JW (1997) Intrastriatal adenosine A_1 receptor antisense oligodeoxynucleotide blocks ethanol-induced motor incoordination. Eur J Pharmacol 323:R5–R7

Pilitsis JG, Kimelberg HK (1998) Adenosine receptor mediated stimulation of intracellular calcium in acutely isolated astrocytes. Brain Res 798:294–303

Pintor J, Diaz-Rey MA, Miras-Portugal MT (1993) Ap_4A and ADP-β-S binding to P2 purinoceptors present on rat brain synaptic terminals. Br J Pharmacol 108:1094–1099

Pintor J, Hoyle CHV, Gualix J, Miras-Portugal MT (1997a) Diadenosine polyphosphates in the central nervous system. Neurosci Res Commun 20:69–78

Pintor J, Miras-Portugal MT (1995) P_2 purinergic receptors for diadenosine polyphosphates in the nervous system. Gen Pharmacol 26:229–235

Pintor J, Puche JA, Gualix J, Hoyle CHV, Miras-Portugal MT (1997b) Diadenosine polyphosphates evoke Ca^{2+} transients in guinea-pig brain via receptors distinct from those for ATP. J Physiol (Lond) 504:327–335

Poelchen W, Sieler D, Inoue K, Illes P (1998) Effect of extracellular adenosine 5'-triphosphate on principal neurons of the rat ventral tegmental area. Brain Res 800:170–173

Porkka-Heiskanen T, Strecker RE, Thakkar M, Bjorkum AA, Greene RW, McCarley RW (1997) Adenosine: a mediator of the sleep-inducing effects of prolonged wakefulness. Science 276:1265–1268

Portas CM, Thakkar M, Rainnie DG, Greene RW, McCarley RW (1997) Role of adenosine in behavioral state modulation: a microdialysis study in the freely moving cat. Neuroscience 79:225–235

Pourgholami MH, Rostampour M, Mirnajafi-Zadeh J, Palizvan MR (1997) Intra-amygdala infusion of 2-chloroadenosine suppresses amygdala-kindled seizures. Brain Res 775:37–42

Proctor WR, Dunwiddie TV (1984) Behavioral sensitivity to purinergic drugs parallels ethanol sensitivity in selectively bred mice. Science 224:519–521

Qian J, Colmers WF, Saggau P (1997) Inhibition of synaptic transmission by neuropeptide Y in rat hippocampal area CA1: modulation of presynaptic Ca^{2+} entry. J Neurosci 17:8169–8177

Rabin RA (1990) Direct effects of chronic ethanol exposure on beta-adrenergic and adenosine-sensitive adenylate cyclase activities and cyclic AMP content in primary cerebellar cultures. J Neurochem 55:122–128

Rabin RA, Molinoff PB (1981) Activation of adenylate cyclase by ethanol in mouse striatal tissue. J Pharmacol Exp Ther 216:129–134

Radulovacki M (1985) Role of adenosine in sleep in rats. Rev Clin Basic Pharmacol 5:327–339

Rainnie DG, Grunze HC, McCarley RW, Greene RW (1994) Adenosine inhibition of mesopontine cholinergic neurons: implications for EEG arousal. Science 263:689–692

Ralevic V, Burnstock G (1998) Receptors for purines and pyrimidines. Pharmacol Rev 50:413–492

Refinetti R, Menaker M (1992) The circadian rhythm of body temperature. Physiol Behav 51:613–637

Richardson PJ, Kase H, Jenner PG (1997) Adenosine A_{2A} receptor antagonists as new agents for the treatment of Parkinson's disease. Trends Pharmacol Sci 18:338–344

Rimondini R, Ferre S, Gimenez-Llort L, Ogren SO, Fuxe K (1998) Differential effects of selective adenosine A_1 and A_{2A} receptor agonists on dopamine receptor agonist-induced behavioural responses in rats. Eur J Pharmacol 347:153–158

Rimondini R, Ferré S, Ögren SO, Fuxe K (1997) Adenosine A_{2A} agonists: A potential new type of atypical antipsychotic. Neuropsychopharmacology 17:82–91

Robertson SJ, Edwards FA (1998) ATP and glutamate are released from separate neurones in the rat medial habenula nucleus: frequency dependence and adenosine-mediated inhibition of release. J Physiol 508:691–701

Rosenberg PA, Li Y (1995) Adenylyl cyclase activation underlies intracellular cyclic AMP accumulation, cyclic AMP transport, and extracellular adenosine accumulation evoked by β-adrenergic receptor stimulation in mixed cultures of neurons and astrocytes derived from rat cerebral cortex. Brain Res 692:227–232

Rosenberg PA, Li Y (1996) Forskolin evokes extracellular adenosine accumulation in rat cortical cultures. Neurosci Lett 211:49–52

Ross FM, Brodie MJ, Stone TW (1998a) Modulation by adenine nucleotides of epileptiform activity in the CA3 region of rat hippocampal slices. Br J Pharmacol 123:71–80

Ross FM, Brodie MJ, Stone TW (1998b) The effects of adenine dinucleotides on epileptiform activity in the CA3 region of rat hippocampal slices. Neuroscience 85:217–228

Salem A, Hope W (1997) Effect of adenosine receptor agonists and antagonists on the expression of opiate withdrawal in rats. Pharmacol Biochem Behav 57:671–679

Satoh S, Matsumura H, Suzuki F, Hayaishi O (1996) Promotion of sleep mediated by the A2a-adenosine receptor and possible involvement of this receptor in the sleep induced by prostaglandin D2 in rats. Proc Nat Acad Sci 93:5980–5984

Sawynok J (1995) Pharmacological rationale for the clinical use of caffeine. Drugs 49:37–50

Sawynok J (1998) Adenosine receptor activation and nociception. Eur J Pharmacol 347:1–11

Scanziani M, Capogna M, Gahwiler BH, Thompson SM (1992) Presynaptic inhibition of miniature excitatory synaptic currents by baclofen and adenosine in the hippocampus. Neuron 9:919–927

Scholz KP, Miller RJ (1992) Inhibition of quantal transmitter release in the absence of calcium influx by a G protein-linked adenosine receptor at hippocampal synapses. Neuron 8:1139–1150

Schwarz ER, Whyte WS, Kloner RA (1997) Ischemic preconditioning. Curr Opin Cardiol 12:475–481

Sebastião AM, Ribeiro JA (2000) Fine-tuning neuromodulation by adenosine. Trends Pharmacol Sci 21:341–346

Shen KZ, Johnson SW (1997) Presynaptic $GABA_B$ and adenosine A_1 receptors regulate synaptic transmission to rat substantia nigra reticulata neurones. J Physiol 505:153–163

Shen KZ, North RA (1993) Excitation of rat locus coeruleus neurons by adenosine 5′-triphosphate: ionic mechanism and receptor characterization. J Neurosci 13:894–899

Silinsky EM (1984) On the mechanism by which adenosine receptor activation inhibits the release of acetylcholine from motor nerve endings. J Physiol (Lond) 346:243–256

Simon J, Webb TE, Barnard EA (1997) Distribution of [^{35}S]dATP alpha S binding sites in the adult rat neuraxis. Neuropharmacology 36:1243–1251

Sodickson DL, Bean BP (1998) Neurotransmitter activation of inwardly rectifying potassium current in dissociated hippocampal CA3 neurons: interactions among multiple receptors. J Neurosci 18:8153–8162

Sollevi A (1997) Adenosine for pain control. Acta Anaesthesiol Scand Suppl 110:135–136

Strecker RE, Morairty S, Thakkar MM, Porkka-Heiskanen T, Basheer R, Dauphin LJ, Rainnie DG, Portas CM, Greene RW, McCarley RW (2000) Adenosinergic modulation of basal forebrain and preoptic/anterior hypothalamic neuronal activity in the control of behavioral state. Behav Brain Res 113:183–204

Svenningsson P, Nomikos GG, Ongini E, Fredholm BB (1997) Antagonism of adenosine A_{2A} receptors underlies the behavioural activating effect of caffeine and is associated with reduced expression of messenger RNA for NGFI-A and NGFI-B in caudate-putamen and nucleus accumbens. Neuroscience 79:753–764

Tancredi V, D'Antuono M, Nehlig A, Avoli M (1998) Modulation of epileptiform activity by adenosine A1 receptor-mediated mechanisms in the juvenile rat hippocampus. J Pharmacol Exp Ther 286:1412–1419

Ticho SR, Radulovacki M (1991) Role of adenosine in sleep and temperature regulation in the preoptic area of rats. Pharmacol Biochem Behav 40:33–40

Torres GE, Haines WR, Egan TM, Voigt MM (1998) Co-expression of P2X1 and P2X5 receptor subunits reveals a novel ATP-gated ion channel. Mol Pharmacol 54:989–993

Trussell LO, Jackson MB (1987) Dependence of an adenosine-activated potassium current on a GTP-binding protein in mammalian central neurons. J Neurosci 7:3306–3316

Tschopl M, Harms L, Norenberg W, Illes P (1992) Excitatory effects of adenosine 5′-triphosphate on rat locus coeruleus neurones. Eur J Pharmacol 213:71–77

Valera S, Hussy N, Evans RJ, Adami N, North RA, Surprenant A, Buell G (1994) A new class of ligand-gated ion channel defined by P2X receptor for extracellular ATP. Nature 371:516–519

Vazquez E, Sanchez-Prieto J (1997) Presynaptic modulation of glutamate release targets different calcium channels in rat cerebrocortical nerve terminals. Eur J Neurosci 9:2009–2018

von Kügelgen I, Koch H, Starke K (1997) P2-receptor-mediated inhibition of serotonin release in the rat brain cortex. Neuropharmacology 36:1221–1227

von Lubitz DKJE, Carter MF, Deutsch SI, Lin RCS, Mastropaolo J, Meshulam Y, Jacobson KA (1995) The effects of adenosine A_3 receptor stimulation on seizures in mice. Eur J Pharmacol 275:23–29

Webb TE, Henderson DJ, Roberts JA, Barnard EA (1998) Molecular cloning and characterization of the rat $P2Y_4$ receptor. J Neurochem 71:1348–1357

Webb TE, Simon J, Barnard EA (1998) Regional distribution of [^{35}S]2'-deoxy 5'-O-(1-thio) ATP binding sites and the P2Y1 messenger RNA within the chick brain. Neuroscience 84:825–837

Webster RG, Abbott RD, Petrovitch H, Morens DM, Grandinetti A, Tung K-H, Tanner CM, Maskai KH, Blanchette PL, Curb DJ, Popper JS, White LR (2000) Association of coffee and caffeine with the risk of Parkinson Disease J Amer Med Assoc 283:2674–2679

White TD (1996) Potentiation of excitatory amino acid-evoked adenosine release from rat cortex by inhibitors of adenosine kinase and adenosine deaminase and by acadesine. Eur J Pharmacol 303:27–38

Wieraszko A (1996) Extracellular ATP as a neurotransmitter: its role in synaptic plasticity in the hippocampus. Acta Neurobiol Exp (Warsz) 56:637–648

Wieraszko A, Ehrlich YH (1994) On the role of extracellular ATP in the induction of long-term potentiation in the hippocampus. J Neurochem 63:1731–1738

Wright KPJ, Badia P, Myers BL, Plenzler SC, Hakel M (1997) Caffeine and light effects on nighttime melatonin and temperature levels in sleep-deprived humans. Brain Res 747:78–84

Yamada Y, Goto H, Ogasawara N (1980) Purification and properties of adenosine kinase from rat brain. Biochim Biophys Acta 616:199–207

Yang SN, Dasgupta S, Lledo PM, Vincent JD, Fuxe K (1995) Reduction of dopamine D_2 receptor transduction by activation of adenosine A_{2A} receptors in stably A_{2a}/D_2 (long- form) receptor co-transfected mouse fibroblast cell lines: Studies on intracellular calcium levels. Neuroscience 68:729–736

Zhang G, Franklin PH, Murray TF (1994) Activation of adenosine A_1 receptors underlies anticonvulsant effect of CGS21680. Eur J Pharmacol 255:239–243

Zhang YX, Yamashita H, Ohshita T, Sawamoto N, Nakamura S (1995) ATP increases extracellular dopamine level through stimulation of P2Y purinoceptors in the rat striatum. Brain Res 691:205–212

Zhu PJ, Krnjevic K (1997a) Adenosine release mediates cyanide-induced suppression of CA1 neuronal activity. J Neurosci 17:2355–2364

Zhu PJ, Krnjevic K (1997b) Endogenous adenosine on membrane properties of CA1 neurons in rat hippocampal slices during normoxia and hypoxia. Neuropharmacology 36:169–176

CHAPTER 10
The Role of Purines in the Peripheral Nervous System

C. KENNEDY

A. Introduction

The pioneering work on neurotransmission in the peripheral nervous system carried out earlier in the century led to the widely accepted idea that noradrenaline (NA) was the sole neurotransmitter released by postganglionic sympathetic nerves and that acetylcholine (ACh) was the sole neurotransmitter released by postganglionic parasympathetic nerves (DALE 1935). However, by the early 1970s evidence was accumulating to support the idea that adenosine 5'-triphosphate (ATP) was also a neurotransmitter from "purinergic" nerves in some parts of the autonomic nervous system (BURNSTOCK 1972). Several years later BURNSTOCK (1976) developed a further understanding of peripheral neurotransmission by suggesting that these nerves could in fact release more than one substance as a neurotransmitter. Since then, cotransmission has become accepted as the rule rather than the exception, with NA, ACh, ATP, nitric oxide (NO) and numerous peptides all having been shown to act as cotransmitters in numerous types of peripheral tissue.

A variety of experimental techniques, such as organ bath pharmacology, electrophysiology and biochemical methods for measuring neurotransmitter release have been employed to characterise extensively the actions of ATP as a cootransmitter with NA in sympathetic nerves innervating smooth muscle preparations such as the vas deferens and most arteries (Fig. 1). Neuropeptide Y is also present in sympathetic nerves and its release can also be measured. The main effects of neuropeptide Y appear to be to modulate the effects of ATP and NA and are discussed elsewhere (see SNEDDON et al. 1996; KENNEDY et al. 1997a). On the other hand, the cotransmitter actions of ATP with ACh from parasympathetic nerves have been most extensively characterised in the urinary bladder (Fig. 2). In this chapter, evidence that supports these cotransmitter roles is described, concentrating mainly on more recent developments. Previous reviews describe the historical development of this field in greater detail (BURNSTOCK and KENNEDY 1986; VON KÜGELGEN and STARKE 1991; KENNEDY 1998; KENNEDY et al. 1997a,b).

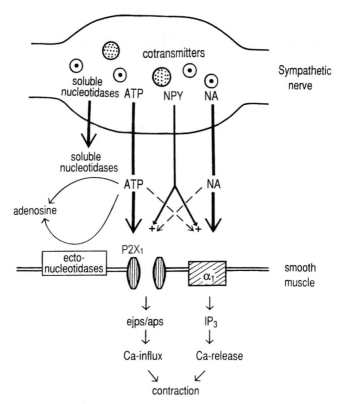

Fig. 1. Schematic representation of cotransmission by ATP, noradrenaline and neuropeptide Y in sympathetic nerves and breakdown of ATP by ecto- and soluble nucleotidases. *NA*, noradrenaline; *NPY*, neuropeptide Y; *P2X$_1$*, P2X$_1$ receptor; α_1, α_1-adrenoceptor; *ejps*, excitatory junction potentials; *IP$_3$*, inositol 1,4,5-trisphosphate

B. ATP and NA as Sympathetic Neurotransmitters

I. Storage in and Release of ATP from Sympathetic Nerves

The earliest evidence that ATP could perhaps function as a neurotransmitter from sympathetic nerves came from biochemical studies that showed that ATP was co-stored with NA in sympathetic synaptic vesicles (see BURNSTOCK and KENNEDY 1986). In these studies NA was always present in much higher amounts than ATP.

Recent evidence suggests that there may well also be further populations of vesicles that store ATP and NA separately. In the most detailed study to date, TODOROV et al. (1996) monitored simultaneously by HPLC the release of ATP and NA during stimulation of sympathetic nerves innervating the guinea-pig, isolated vas deferens, and found that NA release peaked 40s after the start

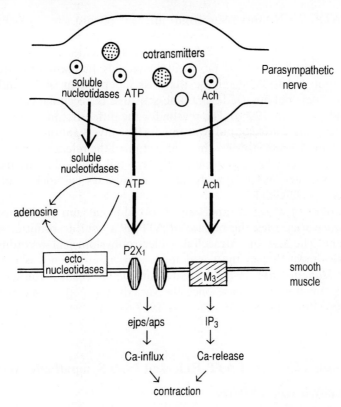

Fig. 2. Schematic representation of cotransmission by ATP and acetylcholine in parasympathetic nerves and breakdown of ATP by ecto- and soluble nucleotidases. *ACh*, acetylcholine; *P2X₁*, P2X$_1$ receptor; *M3*, M3-muscarinic receptor; *ejps*, excitatory junction potentials; *IP$_3$*, inositol 1,4,5-trisphosphate

of stimulation and then remained constant. However, ATP release peaked more quickly, 20s after the start of stimulation and then decreased, even though the nerves were still being stimulated. This is consistent with the phasic nature of the purinergic component of neurogenic contractions in this tissue.

The calcium-dependent release of ATP and NA during sympathetic nerve stimulation into the perfusate has been measured in a variety of tissues (see BURNSTOCK and KENNEDY 1986). Several groups have studied the role of individual subtypes of voltage-dependent calcium channel in this release. Using a sensitive HPLC system, WESTFALL et al. (1996b) found that, during sympathetic nerve stimulation at 8 Hz, ATP release was inhibited more than NA release by blockade of the P-type calcium channels, whereas blockade of the N-type calcium channels inhibited NA release more than ATP release. These results support this group's previous claim that in the guinea-pig vas deferens at least, ATP and NA originate from different release sites in sympathetic nerves. However, SMITH and CUNNANE (1996a) found that ATP released electrophysiologically during sympathetic nerve stimulation at 1 Hz could be abolished by blockade of N-type calcium channels. Increasing the frequency of stimulation led to ATP release that was independent of N- and P-type calcium channels, but involved ryanodine-sensitive, intracellular calcium stores (SMITH and CUNNANE 1996b). The reason for the difference in results between the two groups is not clear and warrants further investigation.

BROCK and CUNNANE (1999) have also studied the release of ATP and NA from sympathetic nerves innervating the rat tail artery. In this tissue, the release of ATP and NA was modified in similar ways by blockers of N- and P-types calcium channels.

Regardless of which channels are involved in calcium entry into sympathetic nerve varicosities, the release of ATP as a neurotransmitter is highly intermittent. The use of extracellular electrophysiological recording techniques has shown that in the vas deferens this intermittence is not due to failure of the neuronal action potential to invade varicosities. Rather it is due to a low probability of release once the varicosity is depolarised (CUNNANE and STJÄRNE 1984; BROCK and CUNNANE 1987, 1988).

II. Functional Effects of ATP Released from Sympathetic Nerves

1. Electrophysiological Effects

a) Excitatory Junction Potentials

The first functional response that can be measured during sympathetic nerve stimulation is rapid, transient excitatory junction potentials (ejps) in the postjunctional smooth muscle cells of the vas deferens and most blood vessels. When the stimulation intensity or frequency is high enough, the ejps summate and the membrane depolarises enough to trigger the opening of voltage-dependent calcium channels. This in turn initiates a calcium action potential and contraction.

Ejps are resistant to α-adrenoceptor blockade, but abolished by tetrodotoxin and guanethidine, indicating that they are due to stimulation of sympathetic nerves. In most blood vessels NA released from sympathetic nerves evokes contraction without changing the membrane potential of smooth muscle cells. Several explanations were proposed to explain how NA could nonetheless mediate the ejps. However, in the 1980s it became clear that ATP in fact mediates them. In the guinea-pig vas deferens, the P2 receptor antagonist, ANAPP3, inhibited the ejps. Subsequently, they were also shown to be inhibited by other P2 receptor antagonists, such as suramin and PPADS and by desensitisation of the P2 receptor by the agonist α,β-meATP (see SNEDDON et al. 1996). Similar studies using mainly desensitisation of the P2 receptor by α,β-meATP or suramin clearly demonstrate that ejps in blood vessels are also due to ATP (see VON KÜGELGEN et al. 1991; MCLAREN et al. 1995).

b) P2X Receptors

Ejps are mediated by ATP acting at $P2X_1$ receptors. P2X receptor subtypes are discussed elsewhere in this volume (CHESSELL et al., Chap. 3); therefore only a brief description of their properties will be given here. P2X receptors are permeable to sodium, potassium and calcium ions and so when activated cause depolarisation and excite cells. To date, seven subtypes of P2X receptor have been cloned ($P2X_{1-7}$). In situ hybridisation or RT-PCR of mRNA (VALERA et al. 1994; COLLO et al. 1996; ERLINGE et al. 1998) and binding of antibodies to expressed receptors (Bo et al. 1998a,b; CHAN et al. 1998) show that the predominant subtype present in the vas deferens and blood vessels is the $P2X_1$ receptor. Several recent studies using RT-PCR have also identified mRNA for other subtypes, particularly the $P2X_4$, in several blood vessels (NORI et al. 1998; PHILLIPS et al. 1998), but the functional relevance of these findings is still unclear.

When functionally expressed in mammalian cell lines or *Xenopus* oocytes, the cloned $P2X_1$ receptor carries an inward, depolarising cationic current (VALERA et al. 1994; NORTH and BARNARD 1997). ATP activates similar currents via native $P2X_1$ receptors in isolated smooth muscle cells from the vas deferens and numerous blood vessels, including the rabbit ear artery and portal vein, rat tail artery, aorta, portal vein and pulmonary artery and human saphenous vein (see KENNEDY 1997a).

The current carried by $P2X_1$ receptors activates with a latency of a few milliseconds, (consistent with the action of a ligand-gated cation channel) and rapidly reaches a peak, then decays. Suramin and PPADS are both antagonists at cloned (VALERA et al.,1994; NORTH and BARNARD 1997) and native (EVANS and KENNEDY 1994; MCLAREN et al. 1998) $P2X_1$ receptors.

2. Smooth Muscle Contraction

a) Calcium Influx

In general, contraction of smooth muscle during sympathetic nerve stimulation is due to $P2X_1$ receptor-dependent calcium influx and $\alpha 1$-adrenoceptor-dependent release of calcium stores, via inositol trisphosphate, in the postjunctional smooth muscle cells. $P2X_1$ receptor mediated influx occurs via two routes. Initial influx occurs through the $P2X_1$ receptor ion channel itself, as it is permeable to calcium ions. Thus, in the absence of extracellular calcium, $P2X_1$ receptor stimulation does not cause intracellular calcium levels to rise (see KENNEDY 1997a for references). In some cases calcium-induced calcium release may also occur. Secondly, the depolarising current carried by $P2X_1$ receptors may also be sufficient to increase the opening of the L-type of voltage-dependent calcium channel.

b) Contraction Profile

In the guinea-pig vas deferens neurogenic contractions are biphasic. ATP largely mediates the initial rapid, transient component, with a small contribution by NA. In contrast, NA predominantly mediates the secondary, slow tonic component. The vas deferens from other species contract with different temporal profiles, but two clearly defined phases can always be seen. The initial response is always predominantly purinergic and the second phase predominantly noradrenergic (see SNEDDON et al. 1996).

In most blood vessels, neurogenic contractions tend to be monophasic rather than biphasic and so the involvement of a non-adrenergic neurotransmitter was not seriously considered for many years. However, in the 1980s, an increasing number of studies reported that a component of neurogenic vasoconstriction was in fact resistant to α-adrenoceptor blockade and mediated by ATP (for review, see VON KÜGELGEN and STARKE 1991). Since then, ATP has been shown to be a neurotransmitter in most blood vessels studied.

The contribution of ATP to perivascular sympathetic neurotransmission is vessel-dependent. In the rat tail artery neurotransmission is predominantly noradrenergic, with ATP mediating only about 10% of the peak response (BAO et al. 1993). This increases to 20%–60% in the rabbit central ear artery (KENNEDY et al. 1986) and up to 100% in the rabbit mesenteric artery (RAMME et al. 1987). The contribution of ATP is also frequency-dependent, being greater at low than high frequencies (KENNEDY et al. 1986; BAO et al. 1993).

III. Termination of the Neurotransmitter Actions of ATP and NA

The neurotransmitter actions of NA are terminated by uptake back into sympathetic nerves via a well-characterised transporter. In contrast, ATP is broken down in the extracellular space. Breakdown is very rapid and the majority of ATP released is detected in the superfusate as adenosine (KENNEDY and LEFF 1995; TODOROV et al. 1996). Extracellular membrane-bound nucleotidases (ecto-nucleotidases) which sequentially dephosphorylate ATP to adenosine 5'-diphosphate, adenosine 5'-monophosphate and adenosine are ubiquitous in smooth muscle tissues and have long been considered to terminate the neurotransmitter actions of ATP (ZIMMERMANN, Chap. 8, this volume). However, we recently found that when the sympathetic nerves innervating the guinea-pig vas deferens are stimulated, they release not only neuronal ATP, but also soluble nucleotidases which breakdown ATP to adenosine (TODOROV et al. 1997; KENNEDY et al. 1997b). This release of specific nucleotidases may represent a novel mechanism for terminating the actions of a neurotransmitter.

IV. Issues to Be Resolved

It is now widely accepted that ATP and NA function as cotransmitters from sympathetic nerves. However, there are still some areas of debate.

1. Source of ATP Released During Sympathetic Nerve Stimulation

As discussed above, ATP is present in synaptic vesicle present in sympathetic nerves and its release can be evoked by nerve stimulation in a calcium-dependent manner. However, several groups have reported that a substantial component of the ATP released during sympathetic nerve stimulation in the guinea-pig isolated vas deferens came from a non-neuronal source (KATSURAGI et al. 1990; VIZI et al. 1992; VON KÜGELGEN and STARKE 1994). Thus, the neurotransmitter role of ATP has been questioned. In contrast, TODOROV et al. (1996) found no evidence for postjunctional release of ATP in the same preparation.

The reasons for this discrepancy have not been resolved, but may involve methodological factors. In the studies where a postsynaptic component of ATP release was seen, the luciferin-luciferase enzyme system was used. Although very sensitive, this system can only measure intact ATP, not its breakdown products. This greatly underestimates the release of ATP since, as discussed above, ATP is rapidly broken down upon release.

In contrast, TODOROV et al. (1996) employed an HPLC system that could monitor ATP and all its metabolites. This system, although less sensitive than luciferin-luciferase for ATP per se, will give a more accurate measure of the total amount of ATP released. Consequently, the amount of ATP measured by TODOROV et al. (1996) was much greater than measured by VIZI et al. (1992) under the same stimulation conditions. It may be that the luciferin-luciferase method requires greater stimulation intensities in order to produce high enough levels of intact ATP in the superfusate and these greater intensities cause postjunctional release of ATP. Clearly this topic requires further study.

Whatever the reason for these differences, it is unlikely that the postjunctional release of ATP contributes to its neurotransmitter role, as NA-induced release of ATP peaks after 2–6 min (VIZI et al. 1992), whereas contractions evoked by endogenous or exogenous ATP reach a peak within 5 s (SNEDDON et al. 1996).

2. Transient or Maintained Release of ATP

The purinergic component of sympathetic neurotransmission is transient and several mechanisms have been proposed to account for this transience. KENNEDY et al. (1986) proposed that a postjunctional mechanism served to limit the effects of ATP and this was supported when the cloned $P2X_1$ receptor was found to display profound desensitisation (NORTH and BARNARD 1997). However, DRIESSEN et al. (1994) have claimed that inactivation of L-type calcium channels may be the underlying mechanism. Finally, as discussed above, TODOROV et al. (1996) have reported that neuronal release of ATP is transient, so implicating a prejunctional mechanism for transience. It should be noted that none of these proposed mechanisms are mutually exclusive and so one or more may be involved.

C. ATP and ACh as Parasympathetic Neurotransmitters

Activation of the parasympathetic nerves that innervate the urinary bladder leads to contraction of the smooth muscle and so voiding of urine. Dysfunction of the urinary bladder is associated with incontinence and so control of bladder motor function by endogenous factors and exogenous drugs has been studied extensively (see BRADING 1987; ANDERSON 1993; FERGUSON and CHRISTOPHER 1996; HOYLE 1996 for reviews).

In most species the neurogenic contractions are partially resistant to blockade of muscarinic receptors by atropine and substantial evidence supports a neurotransmitter role for ATP (Fig. 2). This evidence is largely functional and much less is known about vesicular storage of ATP and its calcium-dependent release from parasympathetic nerves than we know about sympathetic nerves (see HOYLE 1996).

I. Functional Effects of ATP Released from Parasympathetic Nerves

1. Excitatory Junctional Potentials

The initial response seen on stimulation of parasympathetic nerves is ejps in the postjunctional smooth muscle cells of the urinary bladder. In the guinea-pig these are resistant to atropine, but blocked by desensitisation of the $P2X_1$ receptor by α,β-meATP (FUJII 1988; BRADING and MOSTWIN 1989; CREED et al. 1994; BRAMICH and BRADING 1996) and by the $P2X_1$ antagonist suramin (HASHITANI and SUZUKI 1995). Similar effects are seen in the urinary bladder of the rabbit (HOYLE and BURNSTOCK 1985; FUJII 1988; CREED et al. 1994), pig (FUJII 1988) and sheep (CREED et al. 1994).

2. Smooth Muscle Contraction

Contractions of urinary bladder smooth muscle evoked by parasympathetic nerve stimulation are biphasic, although the two phases are not as well defined temporarily as in the vas deferens or as well maintained as in the vas deferens or blood vessels. A rapid transient contraction is followed by an irregular contraction, which tends to decay rather than reach a steady level.

In most species the $P2X_1$ receptor-mediated ejps summate to evoke depolarisation, calcium influx and contraction, in a manner similar to that described above for the vas deferens. Thus, in the guinea-pig the atropine-resistant component of neurogenic contractions is inhibited by desensitisation of the $P2X_1$ receptor with α,β-meATP or by the $P2X_1$ antagonists suramin and PPADS (KASAKOV and BURNSTOCK 1983; MOSS and BURNSTOCK 1985; BRADING and MOSTWIN 1989; BRADING and WILLIAMS 1990; HOYLE et al. 1990; CREED et al. 1994). Similar effects are seen in the urinary bladder of the rat (BRADING and WILLIAMS 1990; LUHESHI and ZAR 1990; PARIJA et al. 1991), rabbit (LONGHURST et al. 1984; SNEDDON and MCLEES 1992; ZIGANSHIN et al. 1993;

Creed et al. 1994), mouse (Waterman 1996) and ferret (Moss and Burnstock 1985).

ATP may also act at a receptor other than the $P2X_1$ receptor to evoke contraction. In the rat urinary bladder a receptor has been identified that is activated by ADP-β-S and α,β-meADP, but not by other adenine nucleotides and which evokes contraction via activation of phospholipase C, production of inositol trisphosphate and release of calcium stores (Suzuki and Kokubun 1994; Naramatsu et al. 1997). This receptor is very likely a subtype of the P2Y receptor family, which are heptahelical receptors coupled to phospholipase C (North and Barnard 1997). Hashimoto and Kokubun (1995) have further suggested that about 20% of the neurogenic contraction of rat urinary bladder is mediated by this receptor. The presence of such a P2 receptor may be species-dependent as release of calcium stores is also seen in the urinary bladder of the rabbit (Oike et al. 1998), but not the dog (Suzuki and Kokubun 1994).

In all species studied to date, ACh acts at M3-muscarinic receptors (Noronha-Blob et al. 1991; Longhurst et al. 1995) to stimulate production of 1,4,5-inositol trisphosphate and release of internal calcium stores (Iacovou et al. 1990). Many M2-muscarinic receptors are also present in urinary bladder smooth muscle (Longhurst et al. 1995) and these inhibit adenylate cyclase activity. This appears to increase contraction indirectly by counteracting the increase in cAMP levels evoked by NA acting at β-adrenoceptors following release from sympathetic nerves (Hegde et al. 1997).

The studies discussed above were all carried out on bladder muscle strips in vitro. Several studies have confirmed that ATP and ACh also function as excitatory cotransmitters in vivo, in pithed rats (Hegde et al. 1998), in anaesthetised guinea-pigs (Peterson and Noronha-Blob 1989) and rabbits (Levin et al. 1986) and even in unanaesthetised rats (Igawa et al. 1993).

3. Calcium Channels Involved in Neurotransmitter Release

Multiple subtypes of voltage-dependent calcium channels are involved in the release of ATP and ACh from parasympathetic nerves and in a species- and frequency-dependent manner. In the mouse urinary bladder, ATP release involves predominantly P- and Q-type channels, with the P-type component being more prominent at low frequencies. In contrast, ACh release depends primarily on N-type channels and to a lesser extent on P- and Q-type channels (Waterman 1996). Similarly, in the rat (Maggi 1991) and rabbit (Zygmunt et al. 1993), ACh release appears to depend more on N-type channels than does ATP release.

The role of individual types of calcium channel may be species dependent. Whereas P-channels play a large role in ATP release in the mouse urinary bladder, they appear to have a negligible role in either ATP or ACh release in the rat (Frew and Lundy 1995). Also, in the guinea-pig urinary bladder, both purinergic ejps and contractions are substantially inhibited by blockade of N-type calcium channels (MAGGI et al. 1988; BRAMICH and BRADING 1996). Which type of calcium channel mediates the remainder of the responses remains to be determined, but presynaptic L-type channels are not involved (HASHITANI and SUZUKI 1995).

4. Termination of the Neurotransmitter Actions of ATP and ACh

Both ACh and ATP are inactivated by breakdown in the extracellular space. ACh is degraded by acetylcholinesterase. Similar to the vas deferens, ATP can be sequentially dephosphorylated to adenosine by ecto-nucleotidases and also soluble nucleotidases (WESTFALL et al. 1997; KENNEDY et al. 1997b). The degradation of ATP by both types of enzyme can be inhibited by the ATPase inhibitor, ARL 67156 in the guinea-pig vas deferens (WESTFALL et al. 1996a) and urinary bladder (WESTFALL et al. 1997). In both tissues ARL 67156 potentiates the purinergic component of neurogenic contractions and increases the amplitude of contractions to exogenous ATP.

II. Parasympathetic Neurotransmission in the Human Urinary Bladder

1. The Healthy Bladder

As discussed above, in most species tested, neurogenic contractions of the urinary bladder comprise of cholinergic and non-cholinergic components. However, in humans, baboons and rheusus monkeys (i.e. new world primates), these neurogenic contractions are abolished or virtually abolished by blockade of muscarinic receptors (SJÖGREN et al. 1982; SIBLEY 1984; CRAGGS et al. 1986; CORSI et al. 1991; PALEA et al. 1993; CHEN et al. 1994). Similar to other species, the M3-muscarinic receptor mediates the neurotransmitter action of ACh (NEWGREEN and NAYLOR 1996).

The lack of function for ATP in human urinary bladder is not due to the absence of P2X receptors. The human $P2X_1$ receptor was first isolated from this tissue (VALERA et al. 1995) and exogenous α,β-meATP and ATP cause contraction of isolated bladder muscle strips (HUSTED et al. 1983; HOYLE et al. 1989; PALEA et al. 1994).

2. Pathological Conditions

Although ATP does not appear to be an excitatory neurotransmitter from parasympathetic nerves innervating the healthy human urinary bladder, this is not the case in pathological conditions. Interstitial cystitis is a chronic bladder disorder which involves inflammation, pain in the bladder and pelvic musculature and bladder motor dysfunction. Its aetiology is unknown, but is probably multi-factorial. In muscle strips prepared from patients suffering from interstitial cystitis, more than 50% of the electrically-evoked contraction was atropine-resistant (PALEA et al. 1993). Desensitisation of the $P2X_1$

receptor by α,β-meATP abolished this non-cholinergic component of neurotransmission.

Both exogenous ACh and histamine had a lower contractile potency in the affected tissue compared with controls. However, α,β-meATP was more potent in the tissues from interstitial cystitis patients, suggesting that the $P2X_1$ receptor may be upregulated in this condition.

Data consistent with a change in $P2X_1$ receptor function in pathological conditions have been obtained in other species. Histamine is a major inflammatory mediator released from mast cells. In the guinea-pig urinary bladder, histamine, acting at H_1-receptors, increased the amplitude of spontaneous contractions, contractions evoked by exogenous ATP and the purinergic component of neurogenic contractions. The cholinergic component of neurogenic contractions was unaffected by histamine (PATRA and WESTFALL 1994).

The change in bladder function in ageing has been studied in the rat urinary bladder (SAITO et al. 1991). Contractions to ACh and several other contractile agents were found to be similar in young and old rats. However, the potency of ATP, NA and 5-HT was significantly greater in the older rats. Thus, it was suggested that this might contribute to the development of an unstable bladder in elderly people.

D. Concluding Remarks

The concept of cotransmission was controversial when first proposed in the 1970s, but it is now accepted to be the norm rather than the exception for nerves in the peripheral nervous system. As described above, ATP clearly has a major role as an excitatory cotransmitter in both sympathetic and parasympathetic nerves, but the clinical implications of these roles have not yet been resolved. In part this is due to a lack of selective antagonists for the P2 receptors which mediate the actions of ATP in visceral and vascular smooth muscle. Currently, the most promising possibility is an involvement of ATP and P2 receptors in pathological dysfunctions of the urinary bladder. Clearly, more work is necessary to confirm such an involvement, but the studies discussed above do suggest that this may be a rewarding area of research.

Acknowledgements. This work was supported by grants from Astra plc and the Wellcome Trust.

List of Abbreviations

ACh	Acetylcholine
ARL 67156	6-N,N-Diethyl-D-β,γ-dibromomethyleneATP
ATP	Adenosine 5'-triphosphate
ejps	Excitatory junction potentials
NA	Noradrenaline

PPADS   Pyridoxalphoshphate-6-azophenyl-2´,4´-disulphonic acid
RT-PCR   Reverse transcriptase polymerase chain reaction

References

Anderson KE (1993) Pharmacology of lower urinary tract smooth muscles and penile erectile tissues. Pharmacol Rev 45:253–308
Bao JX (1993) Sympathetic neuromuscular transmission in rat tail artery: A study based on electrochemical, electrophysiological and mechanical recording. Acta Physiol Scand Suppl 610:1–58
Bo X, Sexton A, Xiang Z, Nori SL, Burnstock G (1998a) Pharmacological and histochemical evidence for P2X receptors in human umbilical vessels. Eur J Pharmacol 353:59–65
Bo X, Karoon P, Nori SL, Bardini M, Burnstock G (1998b) P2X purinoceptors in post-mortem cerebral arteries. J Card Pharmacol 31:749–799
Brading A (1987) Physiology of bladder smooth muscle. In: Torrens M, Morrison J (eds) The physiology of the lower urinary tract. Springer, Berlin Heidelberg New York, pp 161–191
Brading AF, Mostwin JL (1989) Electrical and mechanical responses of guinea-pig bladder muscle to nerve stimulation. Br J Pharmacol 98:1083–1090
Brading AF, Williams JH (1990) Contractile responses of smooth muscle strips from rat and guinea-pig urinary bladder to transmural stimulation: effects of atropine and α,β-methylene ATP. Br J Pharmacol 99:493–498
Bramich NJ, Brading AF (1996) Electrical properties of smooth muscle in the guinea-pig urinary bladder. J Physiol 492:185–198
Brock JA, Cunnane TC (1987) Relationship between the nerve action potential and transmitter release from sympathetic postganglionic nerve terminals. Nature 326:605–607
Brock JA, Cunnane TC (1988) Electrical activity at the sympathetic neuroeffector junction in the guinea-pig vas deferens. J Physiol 399:607–632
Brock JA, Cunnane TC (1999) Effects of Ca^{2+} concentration and Ca^{2+} channel blockers on noradrenaline release and purinergic neuroeffector transmission in rat tail artery. Br J Pharmacol 126:11–18
Buell G, Collo G, Rassendren F (1996) P2X Receptors, an emerging channel family. Eur J Neurosci 8:2221–2228
Burnstock G (1972) Purinergic nerves. Pharmacol Rev 24:509–581
Burnstock G (1976) Do some nerve cells release more than transmitter? Neurosci 1:239–248
Burnstock G, Kennedy C (1986) A dual function for adenosine 5′-triphosphate in the regulation of vascular tone. Excitatory cotransmitter with noradrenaline from perivascular nerves and locally released inhibitory intravascular agent. Circ Res 1986 58:319–330
Chan CM, Unwin RJ, Bardini M, Oglesby IB, Ford APD, Townsend-Nicholson A, Burnstock G (1998) Localisation of $P2X_1$ purinoceptors by autoradiography and immunohistochemistry in rat kidneys. Am J Physiol 274:F799–F804
Chen TF, Doyle PT, Ferguson DR (1994) Inhibition in the human urinary bladder by gamma-amino-butyric acid. Br J Urol 73:250–255
Collo G, North AN, Kawashima E, Merlo-Pich E, Neidhart S, Surprenant A, Buell G (1996) Cloning of $P2X_5$ and $P2X_6$ receptors and the distribution and properties of an extended family of ATP-gated ion channels. J Neurosci 16:2495–2507
Corsi M, Pietra C, Toson G, Trist D, Tuccitto G, Artibani W (1991) Pharmacological analysis of 5-hydroxytryptamine effects on electrically stimulated human isolated urinary bladder. Br J Pharmacol 104:719–725

Craggs MD, Rushton DN, Stephenson JD (1986) A putative non-cholinergic mechanism in urinary bladders of new but not old world primates. J Urol 136:1348–1350

Creed KE, Callahan SM, Ito Y (1994) Excitatory neurotransmission in the mammalian bladder and the effects of suramin. Br J Urol 74:736–743

Cunnane TC, Stjärne L (1984) Transmitter secretion from individual varicosities of guinea-pig and mouse vas deferens: highly intermittent and monoquantal. Neurosci 13:1–20

Dale HH (1935) Pharmacology and nerve endings. Proc Roy Soc Med 28:319–322

Dreissen B, Von Kügelgen, I, Bultmann R, Elrick DB, Cunnane TC, Starke K (1994) The fade of the purinergic neurogenic contraction of the guinea-pig vas deferens: analysis of possible mechanisms. Naunyn Schmiedebergs Arch Pharmacol 350: 482–490

Erlinge D, Hou MH, Webb TE, Barnard EA, Moller S (1998) Phenotype changes of the vascular smooth muscle cell regulate P2 receptor expression as measured by quantitative RT-PCR. Biochem Biophys Res Comm 248:864–870

Evans RJ, Kennedy C (1994) Characterisation of P2-purinoceptors in the smooth muscle of the rat tail artery; a comparison between contractile and electrophysiological responses. Br J Pharmacol 113:853–860

Ferguson D, Christopher N (1996) Urinary bladder function and drug development. Trends Phramacol Sci 17:161–165

Frew R, Lundy PM (1995) A role for Q type Ca^{2+} channels in neurotransmission in the rat urinary bladder. Br J Pharmacol 116:1595–1598

Fujii K (1988) Evidence for adenosine triphosphate as an excitatory transmitter in guinea-pig, rabbit and pig urinary bladder. J Physiol 404:39–52

Hashimoto M, Kokubun S (1995) Contribution of P2-purinoceptors to neurogenic contraction of rat urinary bladder smooth muscle. Br J Pharmacol 115:636–640

Hashitani H, Suzuki H (1995) Electrical and mechanical responses produced by nerve stimulation in detrusor smooth muscle of the guinea-pig. Eur J Pharmacol 284: 177–183

Hegde SS, Choppin A, Bonhaus D, Briaud S, Loeb M, Moy TM, Loury D, Eglen RM (1997) Functional role of M2 and M3 muscarinic receptors in the urinary bladder of rats in vitro and in vivo. Br J Pharmacol 120:1409–1418

Hegde SS, Mandel DA, Wilford MR, Briaud S, Ford APDW, Eglen RM (1998) Evidence for purinergic neurotransmission in the urinary bladder of pithed rats. Eur J Pharmacol 349:75–82

Hoyle CHV (1996) Purinergic cotransmission: parasympathetic and enteric nerves. Sem Neurosci, Purinerg Neurotrans 8:207–215

Hoyle CHV, Burnstock G (1985) Atropine-resistant excitatory junction potentials in rabbit bladder are blocked by α,β-methyleneATP, Eur J Pharmacol 114:239–240

Hoyle CHV, Chapple C, Burnstock G (1989) Isolated human bladder: evidence for an adenine dinucleotide acting on P2X purinoceptors and for purinergic transmission. Eur J Pharmacol 174:115–118

Hoyle CHV, Knight GE, Burnstock G (1990) Suramin antagonizes responses to P_2-purinoceptor agonists and purinergic nerve stimulation in the guinea-pig urinary bladder and taenia coli. Br J Pharmacol 99:617–621

Husted S, Sjögren C, Andersson KE (1983) Direct effects of adenosine and adenine nucleotides on isolated human urinary bladder and their influence on electrically induced contractions. J Urol 130:392–398

Iacovou JW, Hill SJ, Birmingham AT (1990) Agonist-induced contraction and accumulation of inositol phosphates in the guinea-pig detrusor: evidence that muscarinic and purinergic receptors raise intracellular calcium by different mechanisms. J Urol 144:775–779

Igawa Y, Mattiasson A, Karl EA (1993) Functional importance of cholinergic and purinergic neurotransmission for micturition contraction in the normal, unanaesthetized rat. Br J Pharmacol 109:473–479

Kasakov L, Burnstock G (1983) The use of the slowly degradable analog α,β-methyleneATP, to produce desensitisation of the P_2-purinoceptor: effect on non-adrenergic, non-cholinergic responses of the guinea-pig urinary bladder. Eur J Pharmacol 86:291–294

Katsuragi T, Tokunaga T, Ogawa S, Osamu S, Chiemi S, Furukawa T (1991) Existence of ATP-evoked ATP release system in smooth muscles. J Pharmacol Exp Ther 259:513–518

Kennedy C (1998) Vascular P_2-receptors and their possible roles in hypertension. In: Burnstock G, Dobson JG, Liang BT, Linden J (eds) Cardiovascular biology of purines. Kluwer, London, pp 243–256

Kennedy C, Leff P (1995) How should P2X-purinoceptors be characterised pharmacologically? Trends Pharmacol Sci 16:168–174

Kennedy C, McLaren GJ, Sneddon P (1997a) ATP as a cotransmitter with noradrenaline in sympathetic nerves: a target for hypertensive therapy? In: Jacobson K, Jarvis M (eds) Purinergic approaches in experimental therapeutics. John Wiley, New York, pp 173–184

Kennedy C, Saville V, Burnstock G (1986) The contributions of noradrenaline and ATP to the responses of the rabbit central ear artery to sympathetic nerve stimulation depend on the parameters of stimulation. Eur J Pharmacol 122:291–300

Kennedy C, Todorov LD, Mihaylova-Todorova S, Sneddon P (1997b) Release of soluble nucleotidases; a novel mechanism for neurotransmitter inactivation? Trends Pharmacol Sci 18:263–266

Levin RM, Ruggieri MR, Wein AJ (1986) Functional effects of the purinergic innervation of the rabbit urinary bladder. J Pharmacol Exp Ther 236:452–457

Longhurst PA, Belis JA, O'Donnell JP, Galie JR, Westfall DP (1995) A study of the atropine-resistant component of the neurogenic response of the rabbit urinary bladder. Eur J Pharmacol 99:295–302

Longhurst PA, Leggett RE, Briscoe JAK (1995) Characterization of the functional muscarinic receptors in the rat urinary bladder. Br J Pharmacol 116:2279–2285

Luheshi G, Zar A (1990) Purinoceptor desensitisation impairs, but does not abolish the non-cholinergic motor transmission in rat isolated urinary bladder. Eur J Pharmacol, 185:203–208

Maggi CA (1991) Omega conotoxin and prejunctional modulation of the biphasic response of the rat isolated urinary bladder to single pulse electrical field stimulation. J Auton Pharmacol 11:295–304

Maggi A, Patacchini R, Santicoli P, Irmgard TL, Gepetti P, Bianco D, Selleri S, Meli A (1988) The effect of omega conotoxin GVIA, a peptide modulator of the N-type voltage sensitive calcium channels, on motor responses produced by activation of efferent and sensory nerves mammalian smooth muscle. Naunyn Schmiedebergs Arch Pharmacol 338:107–113

McLaren GJ, Kennedy C, Sneddon P (1995) The effects of suramin on purinergic and noradrenergic neurotransmission in the rat isolated tail artery. Eur J Pharmacol 277:57–61

McLaren GJ, Sneddon P, Kennedy C (1998) Comparison of the actions of ATP and UTP at the $P2X_1$-receptor in smooth muscle cells of the rat tail artery. Eur J Pharmacol 351:139–144

Moss HE, Burnstock G (1985) A comparative study of electrical field stimulation of the guinea-pig, ferret and marmoset urinary bladder. Eur J Pharmacol 114:311–316

Naramatsu M, Yamashita T, Kokubun S (1997) The signalling pathway which causes contraction via P2-purinoceptors in rat urinary bladder smooth muscle. Br J Pharmacol 122:558–562

Newgreen DT, Naylor AM (1996) Characterisation of functional muscarinic receptors in human bladders. Br J Pharmacol 119:45P

Nori S, Fumagalli L, Bo X, Bogdanov Y, Burnstock G (1998) Coexpression of mRNAs for $P2X_1$, $P2X_2$ and $P2X_4$ receptors in rat vascular smooth muscle: an in situ hybridisation and RT-PCR study. J Vasc Res 35:179–185

Noronha-Blob L, Prosser JC, Sturm BL, Lowe VC, Enna S (1991) (±)-Terodiline: an M_1-selective muscarinic receptor antagonist. In vivo effects at muscarinic receptors mediating urinary bladder contraction, mydriasis and salivary secretion. Eur J Pharmacol 201:135–142

North RA, Barnard EA (1998) Nucleotide receptors. Curr Op Neurobiol 7:346–357

Oike M, Creed KE, Onoue H, Hiroyuki T, Ito Y (1998) Increase in calcium in smooth muscle cells of the rabbit bladder induced by acetylcholine and ATP. J Auton Nerv Sys 69:141–147

Palea S, Artibani W, Ostardo E, Trist DG, Pietra C (1993) Evidence for purinergic neuro-transmission in human urinary bladder affected by interstitial cystitis. J Urol 150:2007–2012

Palea S, Corsi M, Pietra W, Artibani A, Calpista A, Gaviraghi G, Trist DG (1994) ADPβS induces contraction of the human isolated urinary bladder through a purinoceptor subtype different from P2X and P2Y. J Pharmacol Exp Ther 269:193–197

Parija SC, Raviprakash V, Mishra SK (1991) Adenosine and α,β-methyleneATP-induced differential inhibition of cholinergic and non-cholinergic responses in rat urinary bladder. Br J Pharmacol 102:396–400

Patra PB, Westfall DP (1994) Potentiation of purinergic neurotransmission in guinea-pig urinary bladder by histamine. J Urol 151:787–790

Peterson JS, Nononha-Blob L (1989) Effects of selective cholinergic antagonists and α,β-methyleneATP on guinea-pig urinary bladder contractions in vivo following pelvic nerve stimulation. J Auton Pharmacol 9:303–313

Phillips JK, McLean AJ, Hill CE (1998) Receptors involved in nerve-mediated vasoconstriction in small arteries of the rat hepatic mesentery. Br J Pharmacol 124:1403–1412

Ramme D, Regenold JT, Starke K, Busse R, Illes P (1987) Identification of the neuroeffector transmitter in jejunal branches of the rabbit mesenteric artery. Naunyn Schmiedebergs Arch Pharmacol 336:267–273

Saito M, Gotoh K, Kondu A (1991) Influence of ageing on the rat urinary bladder function. Urol Int 47 [Suppl 1]:39–42

Sibley GNA (1984) A comparison of spontaneous and nerve-mediated activity in bladder muscle from man, pig and rabbit, J Physiol 354:431–443

Sjögren C, Andersson KE, Husted S, Mattiasson A, Moller-Madsen B (1982) Atropine resistance of transmurally stimulated isolated human bladder muscle, J Urol 128:1368–1371

Smith AB, Cunnane TC (1996a) ω-Conotoxin GVIA-resistant neurotransmitter release in postganglionic sympathetic nerve terminals. Neurosci 70:817–824

Smith AB, Cunnane TC (1996b) Ryanodine-sensitive calcium stores involved in neurotransmitter release from sympathetic nerve terminals of the guinea-pig, J Physiol 497:657–664

Sneddon P, McLaren GJ, Kennedy C (1996) Purinergic cotransmission: sympathetic nerves. Sem Neurosci, Purinergic Neurotransmission 8:201–206

Sneddon P, McLees A (1992) Purinergic and cholinergic contractions in adult and neonatal rabbit bladder. Eur J Pharmacol 214:7–12

Suzuki I, Kokubun S (1994) Subtypes of purinoceptors in rat and dog urinary bladder smooth muscles. Br J Pharmacol, 112:117–123

Todorov LD, Mihaylova-Todorova SM, Craviso GL, Bjur RA, Westfall DP (1996) Evidence for the differential release of the cotransmitters ATP and noradrenaline from sympathetic nerves of the guinea-pig vas deferens. J Physiol 496:731–748

Todorov LD, Mihaylova-Todorova S, Westfall TD, Sneddon P, Kennedy C, Bjur RA, Westfall DP (1997) Neuronal release of soluble nucleotidases and their role in neurotransmitter inactivation. Nature 387:76–79

Valera S, Hussy N, Evans RJ, Adami N, North RA, Surprenant A, Buell G (1994) A new class of ligand-gated ion channel defined by P2X receptor for extracellular ATP. Nature 371:519–523

Valera S, Talbot F, Evans RJ, Gos A, Antonarakis E, Morris MA, Buell GN (1995) Characterization and chromosomal localization of a human P2X receptor from the urinary bladder. Receptors Channels 3:283–289

Vizi ES, Sperlagh B, Baranyi M (1992) Evidence that ATP, released from the postsynaptic site by noradrenaline, is involved in mechanical responses of guinea-pig vas deferens: cascade transmission. Neurosci 50:455–465

Von Kügelgen I, Starke KS (1991) Noradrenaline and ATP cotransmission in the sympathetic nervous system. Trends Pharmacol Sci 12:319–324

Von Kügelgen I, Starke K (1994) Corelease of noradrenaline and ATP by brief pulse trains in guinea-pig vas deferens. Naunyn-Schmiedebergs Arch Pharmacol 350: 123–129

Waterman S (1996) Multiple subtypes of voltage-gated calcium channel mediate transmitter release from parasympathetic neurons in the mouse bladder. J Neurosci 16:4155–4161

Westfall TD, Kennedy C, Sneddon P (1996a) Enhancement of sympathetic purinergic neurotransmission in the guinea-pig isolated vas deferens by the novel ecto-ATPase inhibitor ARL 67156. Br J Pharmacol 117:867–872

Westfall TD, Kennedy C, Sneddon P (1997) The ecto-ATPase inhibitor ARL 67156 enhances parasympathetic neurotransmission in the guinea-pig urinary bladder. Eur J Pharmacol 329:169–173

Westfall DP, Todorov LD, Mihaylova-Todorova SM, Bjur RA (1996b) Differences between the regulation of noradrenaline and ATP release. J Auton Pharmacol 16:393–395

Ziganshin AU, Hoyle CHV, Bo X, Lambrecht G, Mutschler E, Bäumert HG, Burnstock G (1994) PPADS selectively antagonizes P2X-purinoceptor-mediated responses in the rabbit urinary bladder. Br J Pharmacol 110:1491–1495

Zygmunt PM, Zygmunt PKE, Hogestatt E, Anderson E (1993) Effects of ω-conotoxin on adrenergic, cholinergic and NANC neurotransmission in the rabbit urethra and detrusor. Br J Pharmacol 110:1285–1290

CHAPTER 11
Trophic Roles of Purines and Pyrimidines

J.T. NEARY and M.P. ABBRACCHIO

A. Introduction

In the central, peripheral and enteric nervous systems, extracellular purines and, to a lesser extent, pyrimidines, are key modulators of neurotransmission (RALEVIC and BURNSTOCK 1998) acting via cell surface receptors. Receptors for extracellular nucleosides (the P1 receptors, subdivided into the A_1, A_{2A}, A_{2B}, and A_3 subtypes) or nucleotides (the P2-receptors, subdivided into the ion-gated P2X and G-protein-coupled P2Y families) (ABBRACCHIO and BURNSTOCK 1994) are found on the surface of virtually all tissues in higher animals (ABBRACCHIO and BURNSTOCK 1998; LORENZEN and SCHWABE, Chap. 2, this volume; CHESSELL et al., Chap. 3, this volume; BOARDER and WEBB, Chap. 4, this volume).

More recent work (NEARY et al. 1996a) has identified another fundamental role for extracellular nucleotides and nucleosides functioning as trophic factors to exert long-term effects on their target cells. They thus have the potential to regulate diverse cellular functions including cell growth, differentiation, and cell death, and the release of hormones, neurotransmitters and cytokines (ABBRACCHIO 1996, 1997; ABBRACCHIO and BURNSTOCK 1998). Changes in the physiological signaling role of these molecules may lead to altered regulation of cell proliferation and differentiation, that may in turn result in the development of various pathologies including tumor growth, immune system disorders, inflammation and pain, neurodegenerative diseases, and osteoporosis.

The characterization of the specific P1 and/or P2 receptors involved in these effects and an understanding of their signal transduction pathways will be critical in defining novel pharmacological approaches to diseases characterized by pathological alterations in cell proliferation and differentiation.

Nucleosides and nucleotides can also influence cell viability and differentiation via mechanisms that are independent of the activation of extracellular receptors, but instead are mediated via the modulation of specific intracellular targets and/or events. In this chapter, an overview of the positive and negative trophic roles of purines and pyrimidines in various organs and systems is given in the context of the involvement of specific receptors and/or intracellular pathways, that may be amenable to therapeutic targeting by specific ligands in order to develop novel therapeutic entities.

B. Effects on Muscle Cells

I. Effects on Vascular Smooth Muscle and Endothelial Cells

In both the cardiovascular and central nervous systems, large amounts of ATP and adenosine are released into the circulation from both lysed cells and degranulating platelets following vascular injury (e.g., after a thrombotic event resulting in stroke). This release may have implications in endothelial repair and in the induction of angiogenesis, a long-term trophic autoregulatory response that reestablishes the metabolic equilibrium between blood flow and oxygen demand (ABBRACCHIO 1996; see also below). Consistent with this hypothesis are results from animal studies showing that, in both coronary and brain vessels, nucleosides and nucleotides are involved in hypoxia-induced angiogenesis. Chronic administration of adenosine, at doses that increase coronary and muscle blood flow, result in the enhancement of myocardial and skeletal muscle capillary density (ZIADA et al. 1984). Both adenosine and the adenosine reuptake inhibitor, nitrobenzylthioinosine (NBTI) increase vessel density in the chick chorioallantoic membrane (DUSSEAU and HUTCHINS 1988), suggesting a role for purines in the vasal compensatory mechanisms consequent to blood flow reduction.

In vascular smooth muscle cells (VSMCs) maintained in culture, both adenosine and ATP have mitogenic activity (ZIADA et al. 1984; WANG D-J et al. 1992; MALAM-SOULEY et al. 1993; ERLINGE et al. 1995), acting via A_{2A}-like and P2Y receptors, respectively. Adenosine can also stimulate endothelial cell proliferation (DUSSEAU and HUTCHINS 1988; MEININGER and GRANGER 1990; FISHER et al. 1995). In cerebral microvascular endothelial cells, adenosine induces vascular endothelial growth factor (VEGF) mRNA and increases its stability, so that new peptide can be synthesized without the need for new transcription (FISCHER et al. 1995), a process that may significantly contribute to re-establishment of the brain microvasculature following ischemic or traumatic insults (see also below).

The transductional and intracellular mechanisms specifically involved in ATP-induced proliferation of VSMCs involve a specific modulation of the cell cycle (MALAM-SOULEY et al. 1993). Exposure of VSMCs to ATP induced the chronological activation of cell cycle-dependent genes that are usually expressed 1–23h after mitogenic stimulation. ADP and 2meSATP, but not AMP or α,β-meATP, increased c-*fos* mRNA levels. Since ATP was unable to increase late G_1 gene mRNA levels, it was concluded that this nucleotide induces a limited cell cycle progression through the G_1 phase, but is unable by itself to induce the crossing over of the G_1-S boundary, and consequently DNA synthesis (MALAM-SOULEY et al. 1993). ATP-induced cell-cycle progression was hypothesized to sequentially involve:

1. Increased intracellular Ca^{2+} concentrations as a result of phospholipase C-linked P2Y receptor activation (GONZALES et al. 1990).
2. Ca^{2+}-dependent activation of protein kinase C and of the arachidonic acid pathway occurring within minutes after ATP exposure (GONZALES et al. 1990; WANG D-J et al. 1992).

The observation that ATP itself is insufficient in triggering DNA synthesis suggests that this nucleotide may act more as a "competence" rather than as a "progression" factor in the cell cycle, and may therefore need to act in combination with other known growth factors to trigger proliferation. However, in primary cultures of VSMCs using monoclonal antibodies directed against cell-cycle specific nuclear antigens (PCNA and Ki-67) (KOBAYASHI et al. 1994) to identify the cell cycle phase of each single cell, activation of the P_{2U} (= $P2Y_2/P2Y_4$) receptor by both UTP and ATP resulted in the transition from the G_1 to the S and M phases but not from the G_0 to G_1 phase (MIYAGI et al. 1996), suggesting that nucleotides may specifically promote progression of cells along the cell cycle rather than acting as competence factors. These data contrasted with those of MALAM-SOULEY et al. (1993), suggesting that different P2 receptor subtypes regulate the cell cycle at different levels and may be differentially expressed during the cell cycle depending upon the culture conditions and the presence of co-mitogens and growth factors.

Modulation of VSMC and endothelial cell proliferation and growth by nucleotides and nucleosides may play a crucial role(s) in the response to various insults (e.g., trauma, stress, and hypoxia), when their concentrations are significantly increased at the site of damage. They may thus contribute to the initiation of healing mechanisms involved in tissue repair and regeneration. The relevance of the angiogenic effects induced by nucleosides and nucleotides to the human situation is supported by recent data from anginal patients treated chronically with the adenosine uptake blocker, dypiridamole ("The adenosine collateral hypothesis", PICANO and MICHELASSI 1997), where coronary angiogenesis was demonstrated, and also in the European Stroke Prevention Study 2 (DIENER et al. 1996) where 6602 patients with prior stroke or transient ischemic attack were randomized to treatment with low dose acetylsalicylic acid (ASA) alone, modified release dipyridamole alone (400mg daily), the two agents in a combined formulation, or placebo. The primary endpoints in this study were stroke recurrence, death, or stroke and death together. Highly significant effects for both ASA and dipyridamole in reducing the risk of stroke ($P < 0.001$) and stroke and death combined ($P < 0.01$) were observed. The risk of stroke and death in comparison to placebo was reduced by 13% with ASA alone, 15% by dipyridamole alone, and by 24% with the combination therapy. The differential actions of these two agents on platelet function were claimed to be the basis of the additive efficacy of the two treatments, e.g., ASA inhibition of platelet thromboxane formation and dipyridamole reduction of platelet aggregation via increases in intracellular cAMP levels (DIENER et al. 1996). However, dipyridamole can also induce a sustained and long-lasting increase of plasma adenosine levels (DI PERRI and LAGHI-PASINI 1991) and its efficacy may also be explained by an angiogenic effect associated with increases in endogenous adenosine levels. This mechanism may explain why dipyridamole was so efficacious despite its rather weak antiplatelet activity (FITZGERALD 1987), and would also be consistent with the time delay necessary to detectable beneficial effects since a simple effect on platelet function would be immediately beneficial (see PICANO and ABBRACCHIO 1998).

The pivotal role exerted by nucleotides and nucleosides in the control of vascular cell growth and neointima formation may also have implications in the pathogenesis of hypertension, renal vascular injury, and atherosclerosis. In the latter instance, abnormal VSMC proliferation is a key event in the development of atherosclerotic plaques. It may be hypothesized that alterations of the normal control of growth exerted by nucleosides and nucleotides on these cells may have a role in the development of these diseases and, consequently, that modulation of these events via selective P1 or P2 receptor ligands may lead to the development of novel therapeutic agents of potential utility in cardiovascular disorders.

II. Effects on Myocardial Cells

In cardiomyocytes, activation of adenosine A_3 receptors results in both cell protection and cell death depending upon the degree of activation and specific pathophysiological conditions. Exposure of ventricular cardiomyocytes to nanomolar concentrations of selective A_3 receptor agonists results in protection against subsequent noxious stimuli (STAMBAUGH et al. 1997; LIANG and JACOBSON 1998; BROAD and LINDEN, Chap. 16, second volume), whereas micromolar concentrations of these same compounds resulted in apoptotic cell death (SHNEYVAIS et al. 1998). Whereas the effects mediated by nanomolar A_3 agonist concentrations seems to depend really on the activation of the A_3 receptor subtype (see also below), the possibility that, at micromolar agonist concentrations, other adenosine receptor subtypes may also contribute to the detected effects cannot be completely ruled out. The cardioprotective effects induced by low concentrations of A_3 agonists play a key role in "ischemic preconditioning," a phenomenon according to which a brief ischemia period protects the heart from a subsequent longer ischemic period. The specificity of these effects is supported by the finding that the selective A_3 receptor antagonist, MRS1191 inhibited A_3 agonist-induced protection of ventricular myocytes (LIANG and JACOBSON 1998). Moreover, transfection of atrial cells lacking native A_3 receptors and that exhibit a shorter duration of cardioprotection with cDNA encoding the human adenosine A_3 receptor resulted in a sustained A_3-agonist-mediated cardioprotection.

It has been hypothesized that the adenosine-induced protective effects associated with ischemic preconditioning are mediated by protein kinase C (PKC) activation and phosphorylation of an unidentified cytoskeletal protein

(DOWNEY and COHEN 1995). While this remains to be established, similar cytoprotective effects involving a stabilization of the cytoskeleton have been demonstrated in astroglial cells following exposure to nanomolar concentrations of the same A_3 receptor agonists (ABBRACCHIO et al. 1997, 1998; see below). In support of the concept that the cytoskeleton may represent a primary target for both positive and negative regulation of cell survival by adenosine, apoptosis of cardiomyocytes induced by micromolar A_3 agonist concentrations (SHNEYVAIS et al. 1998) appeared to occur via a loss of highly organized myofibril ultrastructure in differentiated cardiac cells and the accumulation of intermediate filaments.

III. Effects on Skeletal Muscle Cells

Both P1 and P2 receptors are expressed in skeletal muscle cells, although their functional roles have yet to be determined. Adenosine can regulate muscle cell viability and may play a role in the pathogenesis of some degenerative muscular diseases, e.g., Duchenne Muscular Dystrophy (DMD), congenital muscular dystrophy, and acute infantile spinal muscular atrophy. Cell death by apoptosis is a normal developmental event in both proliferating myoblasts and in postmitotic muscle fibers (GARCIA-MARTINEZ et al. 1993; SCHWARTZ 1992; TIDBALL et al; 1995). In vivo, apoptotic nuclei are also observed in differentiated fibers of several muscular degenerative diseases (TIDBALL et al. 1995; FIDZIANSKA et al. 1990) as well as in loaded muscle after prolonged exercise (SANDRI et al. 1995). A potential role for adenosine in the onset of muscle degeneration may occur:

1. As a result of abnormal catabolism of purines (an accelerated degradation of ATP coupled to a greatly decreased adenylate kinase activity) and adenosine levels are markedly increased in both the blood and muscles of DMD patients (CASTRO-GAGO et al. 1987; CAMINA et al. 1995).
2. Both adenosine and the stable adenosine analog, 2-chloroadenosine (2-CADO) induce apoptosis of myogenic cells (both mononucleated resting myoblasts and polynucleated differentiated myotubes) via a mechanism targeting the microfilament system (RUFINI et al. 1997; CERUTI et al. 2000).

In ADA immunodeficiency syndrome (COHEN et al. 1978), a genetic disorder characterized by lack of adenosine deaminase, the toxic accumulation of adenosine results in thymocyte deletion and dramatic immunodeficiency. Similarly, excessive accumulation of adenosine in the muscles of DMD patients may represent a unique pathogenetic pathway that contributes to a progressive deterioration of muscle cells and, eventually, to cell death.

Adenosine-induced cytotoxic effects in myoblasts and myotubules does not appear to involve extracellular P1 receptors, but rather blockade of intracellular methylation reactions, that in turn leads to inhibition of DNA and RNA synthesis and cell death (CERUTI et al. 2000), in a similar manner to adenosine-induced apoptosis in pulmonary artery endothelial cells (ROUNDS et al. 1998).

C. Effects on Brain Cells

Nucleotides and nucleosides have potent trophic effects on brain cells (NEARY et al. 1996a) including astrocyte stellation and proliferation and neurite outgrowth. Because of the fundamental importance of these processes during development and in brain injury and repair, considerable interest has focused on characterizing the types of purinergic receptors and the signal transduction mechanisms that mediate these trophic actions. Nucleotides and nucleosides have also been implicated in apoptosis, aging, and amelioration of cognitive dysfunction.

I. Effects on Astrocytes

Reactive gliosis is a hypertrophic and hyperplastic response of astrocytes to brain injury, e.g., trauma, stroke, seizure, and degenerative and demyelinative disorders. The gliotic response is characterized by the generation and elongation of cellular processes (stellation), increased expression of the astrocyte intermediate filament protein, glial fibrillary acidic protein (GFAP), and cellular proliferation. These hallmark events can be induced by application of ATP or its analogues to astrocytes in culture. In rat cortical and striatal astrocytes, stellation and proliferation as well as increased GFAP and GFAP mRNA have been observed (NEARY and NORENBERG 1992; NEARY et al. 1994a,c; ABBRACCHIO et al. 1994; BOLEGO et al. 1997). In chick astrocytes, guanosine and adenosine, as well as their nucleotides, induced proliferation (RATHBONE et al. 1992; KIM et al. 1991). In human fetal astrocytes, ATP and adenosine analogues stimulated mitogenesis (NEARY et al. 1998). The trophic activity of the ATP analog, α,β-meATP (ABBRACCHIO et al. 1994; BOLEGO et al. 1997) and the relatively slow formation of AMP and adenosine from ATP by astrocytic ectonucleotidases (LAI and WONG 1991; NEARY et al. 1994a) indicates that these gliotic effects are stimulated by P2 receptors. Adenosine can also stimulate the stellation of rat cortical astrocytes (ABE and SAITO 1998).

These in vitro studies suggest a role for nucleotides and nucleosides in reactive gliosis, but only a small percentage of acutely isolated astrocytes responded to ATP stimulation with a calcium signal (KIMELBERG et al. 1997), leading to the suggestion that expression of P2Y receptors in culture may not reflect the in vivo situation. While astroglial P2Y receptors mediating differentiation and proliferation may couple to transduction mechanisms that do not involve calcium signaling (BOLEGO et al. 1997; CENTEMERI et al. 1997; NEARY et al. 1999a), the low calcium responsiveness in the acutely isolated astrocytes could be caused by P2Y receptor desensitization due to the ATP released during their preparation; mild perturbations such as fluid flow can evoke a tenfold increase in ATP release from astrocytes (SHITTA-BEY and NEARY 1998) and other cells (OSTROM et al. 2000). Moreover, a number of studies indicate the presence of P2Y receptors in vivo: functional P2Y responses occur in glial cells in slice preparations from Bergmann glia (KIRISCHUK et al. 1995), corpus callosum (BERNSTEIN et al. 1996), and retina (SUGIOKA et al. 1996; NEWMAN and ZAHS 1997); UTP stimulated calcium waves in astrocytes in hippocampal slices (NEARY et al. 1999b); and P2Y receptors were localized in neurons and glia by in situ hybridization (WEBB et al. 1998). Microinfusion of the P2Y receptor agonist, 2MeSATP, into rat nucleus accumbens increased GFAP immunoreactivity, astrocyte hypertrophy, and proliferation. The effects were reduced by the antagonists, reactive blue 2 and PPADS, indicating a role for P2 receptors in mediating astrogliosis in vivo (FRANKE et al. 1997, 1999). GFAP immunostaining was also increased after infusion of an adenosine analogue into rat brain, supporting a role for P1 receptors in vivo (HINDLEY et al. 1994). Thus, from both in vitro and in vivo studies, it is clear that the release of nucleotides and nucleosides from damaged or dying cells could play an important role in brain response to injury.

Polypeptide growth factors and cytokines are also increased after brain injury, and several of these have been implicated in reactive gliosis because of their ability to induce astrocyte stellation and proliferation (EDDLESTON and MUCKE 1993; RIDET et al. 1997; ASENSIO and CAMPBELL 1999). Of particular interest is basic fibroblast growth factor (b-FGF; FGF-2) which has profound effects on astrocyte growth. In rat astrocytes, ATP synergistically enhanced FGF-2-induced proliferation (NEARY et al. 1994b).

The specificity of this synergism was demonstrated by the failure of ATP to enhance mitogenesis induced by epidermal growth factor (EGF), platelet-derived growth factor (PDGF), or nerve growth factor (NGF). The synergistic interactions between ATP and FGF-2 were mediated via P2 receptor activation since ATP was more effective than adenosine and because the synergism was also found with ATP hydrolysis-resistant analogues. Extracellular ATP also potentiated FGF-2-induced mitogenesis in human fetal astrocytes (NEARY et al. 1999b). These findings are consistent with a role for extracellular ATP, acting alone or together with FGF-2, in the development of reactive gliosis.

Purines can also induce the synthesis of polypeptide growth factors in astrocytes. Addition of GTP or guanosine to mouse astrocytes resulted in release of NGF and an increase in NGF mRNA (MIDDLEMISS et al. 1995b). The ability of nucleotides and nucleosides to stimulate neurotrophin synthesis may be beneficial in enhancing neuronal regeneration after brain injury. Conversely, some astroglial purinergic receptors may activate targets that directly contribute to ischemia- and trauma-associated brain damage. For example, in astrocytic primary cultures from rat brain, the ATP analogs, α,β-meATP and β,γ-meATP, activate a novel P2Y receptor coupled to induction of cyclooxygenase-2 (COX-2) (BRAMBILLA et al. 1999), an enzyme that plays a pathogenetic role in various acute and chronic neurodegenerative diseases including ischemia and Alzheimer's disease (OHTSUKI et al. 1996; Tocco et al. 1997). A key role for COX-2 in neurological disorders characterized by a marked inflammatory component and excessive activation of astroglial cells is also supported by epidemiological studies demonstrating that steroidal as well as non-steroidal anti-inflammatory drugs (NSAIDs) lower the risk of developing Alzheimer's disease (BREITNER 1996). ATP-evoked reactive gliosis (evaluated as increases of the length of GFAP-positive astrocytic processes) was reversed by the COX-2 inhibitor, NS-398 (BRAMBILLA et al. 1999). Reactive astrogliosis may have both beneficial and detrimental effects; similarly activation of astroglial cells by purines and pyrimidines may be associated with both beneficial effects (e.g., synthesis of neurotrophins and pleiotrophins, see above), and potentially damaging events (e.g., induction of COX-2). The reason for these biphasic responses is unclear and it may be hypothesized that the beneficial effects represent a homeostatic response by the brain to restore normal condition after ischemic and traumatic insults, whereas the detrimental effects reflect a disregulation of these mechanisms at the level of reactive astrogliosis as a consequence of excessive and chronic inflammatory events. Future studies aimed at correlating the above in vitro data with the time-course of post-ischemic events in vivo are likely to disclose the exact functional significance of purine-mediated reactive astrogliosis.

In some instances purines, especially adenosine, induce biphasic effects in astroglial cells, depending upon specific pathophysiological conditions and ligand concentrations. Adenosine A_3 receptor agonists can mediate both trophic effects and cell death (see Sect. B.II for clarification). Exposure of rat primary astrocytes and human astrocytoma cells to nanomolar A_3 agonist concentrations elicited trophic actions (elongation of GFAP-positive processes, formation of actin-positive stress fibers, protection against spontaneous apoptosis), while at micromolar concentrations, the same compounds induced cell death (ABBRACCHIO et al. 1997, 1998). These opposing effects may be relevant to the in vivo exposure to different concentrations of endogenous purines, suggesting that a moderate release of purines may homeostatically sustain astroglial cell survival, whereas exposure to high purine concentrations resulting from an acute trauma may trigger the death of irreversibly damaged cells in order to spare energy and space for recovering ones (NEARY et al. 1996a). These data support the concept that purines may be signals for both cell survival and cell death, depending on the degree of brain/nervous tissue damage and/or the specific time after damage induction.

II. Effects on Microglia

Microglia are rapidly activated after brain injury (DAVIS et al. 1994; ALDSKOGIUS and KOZLOVA 1998). In microglia, adenosine agonists and propentofylline, a P1 receptor modulator, inhibited the production of reactive oxygen species and interfered with the generation of phagocytic macrophagic cells (SCHUBERT et al. 1997). Microglia also express pore forming $P2X_7$ receptors that render these cells susceptible to ATP-mediated apoptosis (FERRARI et al. 1997; DI VIRGILIO et al., Chap. 26, second volume) blunting the inflammatory response to brain injury.

III. Effects on Other Glial Cells

The effects of nucleotides and nucleosides on oligodendrocytes and Schwann cells have been less well studied, although ATP can trigger a rapid increase in intracellular calcium in these glial cells (KASTRITSIS and MCCARTHY 1993; LYONS et al. 1994). In immortalized Schwann cells, extracellular ATP stimulated phosphatidylinositol hydrolysis and inhibited adenylate cyclase activity (BERTI-MATTERA et al. 1996). Activation of P2Y receptors in oligodendrocytes and Schwann cells resulted in mobilization of calcium from intracellular stores, as has been observed in astrocytes (NEARY et al. 1988; PEARCE et al. 1989; NEARY et al. 1991; KASTRITSIS et al. 1992; SALTER and HICKS 1994, 1995; KING et al. 1996; CENTEMERI et al. 1997) playing a potentially important role in neuron-glial and glial-glial communication (CORNELL-BELL et al. 1990; NEDERGAARD 1994; GUTHRIE et al. 1999).

IV. Effects on Neuronal Cells

Purines also elicit trophic effects on neurons, neuronal-like pheochromocytoma (PC12) cells, and neuroblastoma cells. For example, activation of adenylate cyclase by the action of adenosine on A_2-like receptors stimulated differentiation in neuroblastoma cells (ABBRACCHIO et al. 1989). In contrast, A_1 receptors appeared to inhibit neurite outgrowth in adult mouse dorsal root ganglion cells (SHABAN et al. 1998). This suggests that A_1- and A_2-like receptors may exert opposing effects on neuritogenesis. A factor stimulating the secretion of catecholamines from hypothalamic dopaminergic neurons was purified from pituitary extracts and identified as adenosine (PORTER et al. 1995), thereby indicating secretory roles for purines in the brain (MASINO and DUNWIDDIE, Chap. 9, this volume). Guanosine and GTP caused an increase in the number of neurites in primary cultures of mouse hippocampal neurons, suggesting that nucleotides and nucleosides released by damaged or dying cells may play a role in stimulating outgrowth and branching of axons and dendrites (RATHBONE and JUURLINK 1993). Evidence has been recently generated for a guanosine receptor (TRAVERSÀ et al. 2000). The effects of the laminrelated protein netrin-1 on axon outgrowth may be mediated via allosteric interaction with the A_{2B} receptor (CORSET et al. 2000).

Recent evidence also suggests a possible role for ATP in the regulation of proliferation and differentiation during neurogenesis (SUGIOKA et al. 1996, 1999). In chick epithelium, ATP elicits calcium mobilization through the activation of $P2Y_2$- or $P2Y_4$-like receptors. Interestingly, this response declines in parallel with the decrease in mitotic activity of retinal precursor cells: response to ATP is maximal at embryonic day 3 and drastically declines towards embryonic day 8 when proliferation ceases. It is known that neural precursor cells (stem cells) proliferate until their birth-date when they divide for the last time and start to differentiate and migrate towards their final destination in the developing nervous system. It is tempting to speculate that ATP may crucially regulate the shift from proliferation to differentiation and hence determine the developmental age at which neuronal cells are born. If confirmed in further experimental models, this hypothesis may carry invaluable implications in brain formation as well as in congenital pathologies characterized by abnormal nervous system development.

Synergistic interactions between purines and polypeptide growth factors also occur in neurons. In cultured myenteric neurons, 2-CADO synergistically potentiated the neurite outgrowth induced by FGF-2 (SCHAFER et al. 1995). In addition, the P1 receptor agonist, 5'-N-ethylcarboxamideadenosine (NECA) (GUROFF et al. 1981) as well as guanosine and GTP enhanced NGF-induced neurite outgrowth in PC12 cells (GYSBERS and RATHBONE 1996). Guanosine and GTP also synergistically enhanced neuritogenesis induced by NECA, suggesting that guanine- and adenine-based purines work through different receptors and signaling mechanisms to potentiate neurite outgrowth. Importantly, the synthetic purine, AIT-082, which reduces memory deficits in aging mice (GLASKY et al. 1994) can also enhance NGF-induced neurite outgrowth and stimulate increases in neurotrophin mRNAs in the cerebral cortex and hippocampus (MIDDLEMISS et al. 1995a; GLASKY et al. 1996). By providing a link between the trophic actions of extracellular purines and memory disorders, these studies may lead to pharmacological applications in aging and dementia. AIT-082 (Neotrofin) is currently in Phase II clinical trials for the treatment of Alzheimer's disease (WILLIAMS, Chap. 29, second volume).

D. Effects on Immune Cells

Immune cells express both P1 and P2 receptors, the latter including both P2Y and P2X receptor families (see HOURANI et al. 1997; DUBYAK, Chap. 25, second volume) and activation of these receptors has been linked to a variety of effects, ranging from priming of cells to subsequent stimulation with other mediators, release of lysosomal enzymes, and upregulation of lymphocyte adherence. The specific effects induced by nucleosides and nucleotides on these cells are covered elsewhere in the monograph (MONTESINOS and CRONSTEIN, Chap. 24, second volume; DUBYAK, Chap. 25, second volume) such that the present discussion will focus on the positive and negative trophic effects exerted by these compounds on cells of the immune lineage.

Both adenosine and ATP potently induce immune cell apoptosis. For adenosine, apoptosis appears to occur via both activation of extracellular P1 receptors (mainly the A_{2A} and A_3 subtypes) (SZONDY 1994, 1995; BARBIERI et al. 1998) and also via direct cellular entry of nucleosides. In several instances, an adenosine-activated apoptotic program could be inhibited by suppression of nucleoside uptake, suggesting a direct intracellular action. This was shown for the transport blocker, dipyridamole, in HL-60 cells with adenosine (TANAKA et al. 1994), for 2-CADO in thymocytes (SZONDY 1995), and in the rat myelocytic leukemia cell line, IPC-81, with tubercidin (RUCHAUD et al. 1995). In the human promyelocytic leukemia cell lines (NB4) 2-chloro, 6-chloro- and 8-chloro-adenosine induce apoptosis as shown via tunuel analysis and flow-cytometry (HOFFMAN et al. 1996), an effect that was blocked by dipyridamole or nitrobenzylthioinosine but not by P1 receptor antagonists

(HOFFMANN 1996). In some instances, both receptor-dependent and receptor-independent apoptosis can be activated by adenosine even in the same cells (BARBIERI et al. 1998), suggesting that the choice of the pathway to cell death is dependent on specific pathophysiological conditions (see also below). One possible intracellular site of adenosine-induced apoptosis demonstrated using 3-deazaadenosine is poly-(ADP-ribose)-polymerase (TANAKA et al. 1994), an enzyme that is cleaved by caspases during apoptosis (PORTER et al. 1997). This pathway has been excluded in thymocytes (SZONDY 1995), suggesting that the modalities utilized by adenosine in inducing cell death may depend upon the degree of cell differentiation and specific development stages. The apoptotic action of 3-deazaadenosine in HL-60 cells is reduced by application of homocysteine (ENDRESEN et al. 1993) indicating that altered methylation reactions may play a role in apoptosis.

Adenosine-dependent apoptosis may also be involved in the elimination of autoreactive clones during intrathymic T cell selection (ABBRACCHIO 1996). It may therefore be speculated that an early disregulation of adenosine-dependent apoptosis may have serious consequences for immune cell function in adulthood. A dramatic confirmation of this hypothesis occurs in the ADA-immunodeficiency syndrome previously discussed, where the excessive accumulation of adenosine and 2'-deoxyadenosine leads to the early and massive deletion of thymocytes and the consequent dramatic immunodeficiency.

The apoptotic effects evoked by ATP involve the activation of the $P2X_7$ receptor (DI VIRGILIO et al., Chap. 26, second volume). In J774 mouse macrophages that spontaneously undergo cell death in culture upon reaching confluence, the incidence of "spontaneous" apoptosis has been correlated with $P2X_7$ receptor expression, suggesting that this receptor functions as a cytotoxic entity following autocrine/paracrine stimulation by ATP (CHIOZZI et al. 1996). Activation of the $P2X_7$ receptor on both macrophages and microglial cells results in release of IL-1β in its mature form (DI VIRGILIO et al. 1996). Efflux of cytoplasmic K^+ through the $P2X_7$ receptor pore is crucial for IL-1β release, which has intriguing implications in terms of regulation of cell survival. In fact, the intracellular maturation of IL-1β involves a cysteine protease (interleukin-IL-1β converting enzyme, ICE) (THORNBERRY et al. 1992), a member of a family of intracellular cysteine proteases that are directly involved in the activation of the nuclear enzymes that initiate DNA fragmentation and also trigger early phases of apoptosis. It is not currently known whether ATP also stimulates caspase 3 activity, thus implying a role for this enzyme in ATP-dependent cytotoxic mechanisms, but this is clearly an appealing hypothesis. Purine-induced apoptosis is being explored therapeutically for the treatment of advanced cancer states (BRYSON and SORKIN 1993; BEUTLER 1992; WOOD 1994; WILLIAMS, Chap. 29, second volume).

In addition to affecting immune cell viability, nucleotides and nucleosides also influence differentiation and cell specification. P2Y receptor subtypes are differentially expressed during leukocyte differentiation when studied in ex vivo bone marrow myeloblasts and promyelocytes/promonocytes or in HL60 and THP1 cell lines (WEISMAN et al. 1996). The $P2Y_2$ receptor is transiently up-regulated in activated thymocytes and acts as an immediate early gene in the T-cell differentiation process (KOSHIBA et al. 1997). This suggests that, besides affecting the function of mature immune cells, ATP may also play a role in their maturation by addressing their differentiation via transiently expressed P2Y receptors. These data suggest that both nucleosides and nucleotides profoundly affect immune cell function, by (a) regulating their development, maturation, and physiological functions, which further supports a role for these compounds in the inflammatory reaction, and (b) in the immune response in cancer, autoimmune, and neurodegenerative diseases.

E. Effects on Secretory Cells

I. Effects on Exocrine Cells

1. Salivary Cells

Differentiating and de-differentiating effects of nucleotides occur in salivary cells. $P2Y_1$-like receptors are expressed in submandibular salivary glands from immature rats, but are absent in adults (TURNER et al. 1998). UTP responses were also lacking in cells isolated from adult rats but were observed after these cells had grown in culture for several days and were paralleled by an increase in steady-state $P2Y_2$ receptor mRNA levels, thereby suggesting that de-differentiation promotes the expression of $P2Y_2$ receptors.

In wound healing, differentiating and de-differentiating phenomena are important cellular responses to tissue injury and may involve expression of nucleotide receptors. Ligation of the main excretory duct of rat submandibular glands resulted in an increase in $P2Y_2$ receptor mRNA levels (TURNER et al. 1998). Additionally, disruption of autonomic innervation in rat parotid acinar cells resulted in three- to seven-fold increase in ATP-evoked calcium responses that were characteristic of $P2X_4$ receptor activation (TENNETI et al. 1998). Expression of $P2X_4$ receptors was increased after this parasympathetic denervation.

The secretion and composition of parotid saliva is regulated by sympathetic and parasympathetic nerves; thus the increase in sensitivity to ATP and the increase in $P2X_4$ receptor expression after denervation implies that changes in synaptic activity related to physiologically relevant secretory responses can be mediated by extracellular ATP acting on specific P2 receptors. These interesting findings also suggest that upregulation of $P2X_4$ and perhaps other nucleotide receptors may take place in other cellular locations after tissue injury.

2. Kidney Cells

Proliferation of glomerular mesangial cells is a hallmark of glomerulonephritis. Nucleotides are released after damage of glomerular cells and aggregation of thrombocytes. The mitogenic effects of extracellular ATP on mesangial cells have been extensively studied (SCHULZE-LOHOFF et al. 1996 and references therein). As in the case for other cells, ATP also synergizes with polypeptide growth factors to potentiate mitogenesis of these cells (SCHULZE-LOHOFF et al. 1995). Whereas ATP and ATPγS were mitogenic, UTP was not, thereby implicating ATP-preferring P2 receptors. ATP also stimulated mitogenesis in cells cultured from rat renal inner medullary collecting ducts (ISHIKAWA et al. 1997) and in renal tubular epithelial cells (CHENG and LI 1997). In vivo infusion of ATP during reperfusion after renal artery occlusion lead to an increase in renal DNA synthesis, suggesting a role for purinergic receptors in renal ischemia (PALLER et al. 1998). In experimental mesangioproliferative disease, administration of hydrolysis-resistant ATP analogs aggravated, and 2-CADO alleviated, the course of the disease, thereby suggesting that nucleotides are proinflammatory whereas nucleosides are anti-inflammatory. Because glomerular diseases often cause chronic kidney failure leading to the need for dialysis and transplantation, the development of selective P2 receptor antagonists could be of considerable benefit.

II. Effects on Endocrine Cells

Receptors for nucleotides and nucleosides are present on a variety of endocrine cells including pancreatic β (LOUBATIERES-MARIANI et al. 1997), pituitary (DAVIDSON et al. 1990; CHEN et al. 1996), adrenal medullary chromaffin (DIVERSE-PIERLUISSI et al. 1991; CASTRO et al. 1992; GANDIA et al. 1993; HOUCHI et al. 1995; LIN et al. 1995; OTSUGURO et al. 1995), PC12 phaeochromocytoma (DE SOUZA et al. 1995; YAKUSHI et al. 1996) and thyroid cells (NAKAMURA and OHTAKI 1990; SATO et al. 1992; SMALLRIDGE and GIST 1994; TORNQUIST et al. 1996), luteal and granulosa cells of the ovary (KAMADA et al. 1994; SOODAK et al. 1988) and Sertoli cells (FILIPPINI et al. 1994). Although the exact functional role of these receptors is largely unknown, their widespread expression clearly supports a role in the regulation of endocrine function at various levels. ADPβS acting via pancreatic P2Y receptors stimulated insulin secretion, improved glucose tolerance, and was effective after oral administration. Moreover, this ATP analog also retained its insulin-stimulatory effects in streptozotocin-diabetic rats, clearly supporting the hypothesis that P2Y receptors of the pancreatic β cells may be a novel potential target for the development of new orally-active antidiabetic drugs (PETIT et al., Chap. 27, second volume).

Nucleotides and nucleosides may also regulate the proliferation and differentiation of endocrine cells. ATP acts as a mitogenic agent on ovarian tumor cells (POPPER and BATRA 1993) and as a co-mitogen on thyroid FRTL-5 cells (TORNQUIST et al. 1996). In the latter study, extracellular ATP (but not thyrotrophin) in the presence of insulin enhanced DNA synthesis and increased cell number (TORNQUIST et al. 1996), suggesting that, as also reported for SSMCs, astrocytes, fibroblasts and other cells (see other sections), ATP may act in concert with classical growth-promoting agents to regulate mitosis.

Adenosine can inhibit follicle-stimulating hormone-induced functional differentiation of granulosa cells (KNECHT et al. 1984), likely via an A_2-like receptor subtype. The elucidation of the exact role of nucleotides and nucleosides in regulating the growth and differentiation of endocrine cells is likely to reveal additional targets for the pharmacological modulation of diseases involving endocrine dysfunction.

F. Effects on Fibroblasts

Purines can enhance the trophic actions of growth factors, especially in fibroblasts. Adenosine potentiates the proliferative response of quiescent fibroblasts to serum (e.g., SCHOR and ROZENGURT 1973) an effect that was thought to be related to an increase in intracellular purine nucleotide pools rather than a receptor-mediated phenomenon. However, low concentrations (0.1–1 µmol/l) of the P1 agonist, NECA, in the presence of insulin, stimulated cell division in quiescent cultures of Swiss 3T3 cells (ROZENGURT 1982). Furthermore, the mitogenic effect occurred when a large excess of hypoxanthine was present in the culture medium to ensure sufficient intracellular nucleotide pools. Since NECA is not metabolized through the same metabolic pathways of adenosine, its mitogenic activity appeared to be related to alterations in cAMP since NECA stimulated adenylate cyclase activity, an effect that was potentiated by phosphodiesterase inhibitors. The methylxanthine aminophylline inhibited the effects of NECA on adenylate cyclase activity and DNA synthesis. Interestingly, NECA-stimulated cAMP elevation was not dependent on insulin, but only a slight stimulation of DNA synthesis occurred in the absence of insulin indicating that the increase in cAMP was insufficient to activate mitogenesis, further suggesting a synergy between NECA and insulin.

Synergistic mitogenic effects of P2 agonists and polypeptide growth factors on fibroblasts have also been noted. On its own, in Swiss 3T3 and 3T6 mouse fibroblasts, ATP had very little mitogenic activity; however, in the presence of EGF, PDGF, transforming growth factor α (TGFα), or insulin, the nucleotide caused synergistic stimulation of DNA synthesis and cell number (HUANG et al. 1989; WANG et al. 1990). ADP was equivalent in activity to ATP, while UTP and ITP were much less active. These effects of ATP appeared to be mediated by P2 rather than P1 receptors since the metabolically stable ATP analog, adenosine 5′-[β,γ-imido]triphosphate stimulated mitogenesis and the effects of ATP were not blocked by the P1 receptor antagonist, aminophylline. Moreover, 60 min after addition of ATP, only 10% was converted to ADP by fibroblast ectonucleotidases and no adenosine was detected. ATP was described as a competence factor as its presence for 1 h was sufficient to elicit synergy.

An early event in wound repair involves the release of PDGF and TGFα from platelets which also release ATP and ADP after injury. Thus, the finding of synergism between polypeptide growth factors and adenine nucleotides led

Heppel's group to speculate that the combined actions of these agents may lead to fibroblast proliferation and stimulate synthesis of new extracellular matrix at the wound site, thereby suggesting an important role for nucleotides in wound repair (HUANG et al. 1989; WANG et al. 1990).

The mitogenic activity of ATP in fibroblasts is not a universal phenomenon. Growth inhibition in the presence of extracellular ATP occurs in Swiss 3T6 mouse fibroblasts, an effect attributed to the slow generation of adenosine from nucleotides by enzymes on the cell surface and in serum (WEISMAN et al. 1988). ATP does not stimulate DNA synthesis in cardiac fibroblasts or myocytes and, in fact, inhibited norepinephrine-induced hypertrophy (ZHENG et al. 1998) suggesting that different fibroblast responses may be attributed to cell-type specific differences in P2 receptor expression and related signaling mechanisms.

An interesting in vitro aging phenomenon was reported by Heppel and coworkers (HUANG et al. 1993). Basal levels of cAMP as well as arachidonic acid and prostaglandin E2 release rose during in vitro aging in IMR-90 human lung fibroblasts, and the mitogenic responsiveness to growth factors was reduced. Addition of ATP, in combination with EGF or insulin, restored the greatly reduced mitogenic responsiveness of aged cells nearly to the level of young cells. Extracellular ATP suppressed the aging effects on cAMP, arachidonic acid, and prostaglandin E2 levels, suggesting that the mitogenic actions of ATP on aged cells may result from a reduction in arachidonate metabolism.

G. Effects on Bone Cells

ATP and UTP participate in the physiological processes involved in bone formation and resorption, and bone repair following injuries and fractures. The potential involvement of specific P2 receptors raises the possibility of pharmacologically modulating these processes in pathologies characterized by excessive bone demineralization (e.g., osteoporosis), Bone resorption is the resultant of three successive steps (VAES 1988; ZAIDI et al. 1993). The first involves formation of osteoclast progenitors in hematopoietic tissues, followed by their vascular dissemination and the generation of resting osteoclasts in bone. The second consists of the activation of osteoclasts at the contact of mineralized bone. In the third step, osteoclasts resorb both the mineral and the organic of mineralized bone by secreting ions and agents in the segregated zone underlying their ruffled border. In particular, a membrane H^+/ATPase extrudes protons into the extracellular bone resorption compartment between the osteoclast and the mineralized surface, while a Cl^-/ HCO_3^- exchanger is responsible for an efflux of HCO_3^- across the opposite cell membrane. ATP can modulate these complex processes at various levels. Osteoclasts are particularly sensitive to the cytotoxic actions of extracellular ATP^{4-} (MODDERMANN et al. 1994; NIJWEIDE et al. 1995), which is without effect on the viability of other bone cells of non-hematopoietic origin, e.g., osteoblasts and chondrocytes. A $P2X_7$ receptor linked to membrane permeabilization may therefore be involved in the pathophysiological regulation of the number of osteoclast progenitors in hematopoietic tissues. Osteoclasts also express a P2 receptor coupled to enhancement of Cl^-/HCO_3^- exchange (YU and FERRIER 1993), suggesting that ATP may participate in the bone demineralization process by regulating HCO_3^- efflux. Several P2Y receptors including the elusive P_{2T} and $P2Y_2$ subtypes have been identified on both chondrocytes (LEONG et al. 1993) and osteoblasts (YU and FERRIER 1993). In chondrocytes, cytokines, which play a major role in the loss of cartilage extracellular matrix in rheumatoid arthritis and osteoarthritis, appear to act through enhancement of responsiveness to ATP (LEONG et al. 1993). In osteoblasts, P2 receptor activation activates phospholipase C leading to the formation of inositol-phosphates and the release of Ca^{2+} from intracellular stores (FERRIER, Chap. 28, second volume). Since indirect stimulation of osteoclastic resorption is thought to be triggered by an initial elevation of Ca^{2+} in osteoblasts, it is intriguing to speculate that this ATP-driven effect is a physiological signal for osteoclastic activation, and, hence, that blockade of the osteoblast receptor via selective P2Y receptor antagonists may counteract excessive osteoporosis-associated bone loss.

Globally, these findings support the idea that nucleotides play an important role in regulating the equilibrium between osteoclasts and osteoblasts in bone formation and turnover by both directly affecting the mineralizing and demineralization processes via specific extracellular receptors and by indirectly influencing the actions evoked on these cells by cytokines and specific hormones.

H. Effects on Reproduction

The characterization of the role of nucleotides and nucleosides in reproduction is at an early stage. Adenosine influences the differentiation of luteal and granulosa cells of the ovary (see Sect. E.II). Moreover, both P1 (MINELLI et al. 1998) and P2 receptors (YEUNG 1986) are expressed in sperm and Sertoli cells (see also Sect. E.II). ATP is obligatory for sperm movement (YEUNG 1986) and is a trigger for capacitation, the acrosome reaction necessary to fertilize the egg (FORESTA et al. 1992). Moreover, extracellular ATP promotes a rapid increase in the Na^+ permeability of the fertilized egg membrane via activation of a specific P2 receptor (KUPITZ and ATLAS 1993). These data, together with the observation that ATP-activated spermatozoa show very high success rates in fertilization tests (FORESTA et al. 1992), support a role for ATP as the key sperm-to-egg signal in the process of fertilization. Knock-out mice lacking the $P2Y_1$ receptor show decreased male fertlitily due to effects on smooth muscle tone in the vas deferens (MULRYAN et al. 2000).

I. Mechanisms

Compounds that influence the trophic actions of purines and pyrimidines could have a major impact on many disorders, including neurodegeneration of the central and peripheral nervous systems, cardiovascular disease, and imbalances in exocrine and endocrine functions. The various components of signaling pathways also provide novel targets for therapeutic intervention. Accordingly, considerable effort has focused on defining the signal transduction pathways coupled to P1 and P2 receptors that regulate changes in the expression of genes underlying cell growth and cell death.

P1 and P2Y receptors in many cells are coupled to second messenger pathways involving phosphatidylinositol-specific phospholipase C (PI-PLC) and adenylate cyclase (for reviews see BURNSTOCK 1990; LINDEN et al. 1991; HARDEN et al. 1995). P2Y receptors are linked to phospholipase D (PLD) in endothelial cells (MARTIN and MICHAELIS 1989; PIROTTON et al. 1990; PURKISS and BOARDER 1992), granulocytes (XIE et al. 1991), kidney cells (PFEILSCHIFTER and MERRIWEATHER 1993; BALBOA et al. 1994), and astrocytes (GUSTAVSSON et al. 1993; NEARY et al. 1999a) and to phospholipase A_2 (PLA_2) in endothelial cells (VAN COEVORDEN and BOEYNAEMS 1984), astrocytes (BRUNER and MURPHY 1990, 1993; BOLEGO et al. 1997), and fibroblasts (HUANG et al. 1991). Recent studies on trophic signaling pathways activated by purines and pyrimidines have focused on protein kinase (PK) cascades that are down-stream or independent of second messenger-activated systems and are known to mediate cellular proliferation, differentiation, growth arrest, and apoptosis. These signal transduction pathways, termed mitogen-activated protein kinase (MAPK) cascades, are composed of at least three cytoplasmic PKs that are activated sequentially – MAPK kinase kinase, MAPK kinase, and MAPK (SEGER and KREBS 1995; NEARY 1997). Three major types of MAPKs have been identified – extracellular signal regulated protein kinase (ERK), stress activated protein kinase (SAPK, also known as c-Jun NH_2-terminal kinase, JNK), and p38/MAPK. Trophic factors can stimulate these cascades via receptor tyrosine kinases (e.g., EGF, FGF, and PDGF receptors) and G-protein coupled receptors (GPCRs) such as some types of adrenergic, muscarinic, serotonergic, and purinergic receptors; stress and inflammation signals also activate MAPK cascades. The activated MAPKs translocate to the nucleus where they activate or induce transcription factors and complexes, thereby leading to the up- or down-regulation of genes involved in cell growth or death (Fig. 1).

Stimulation of P2Y receptors activates ERK in many cell types, including rat and human astrocytes (NEARY and ZHU 1994; NEARY 1996; KING et al. 1996; NEARY et al. 1998, 1999a), endothelial cells (GRAHAM et al. 1996; ALBERT et al. 1997), smooth muscle cells (HARPER et al. 1998; WILDEN et al. 1998), kidney cells (HUWILER and PFEILSCHIFTER 1994; GAO et al. 1999; ISHIKAWA et al. 1997; CHENG and LI 1997), thyroid cells (TORNQUIST et al. 1996), and PC12 cells (SOLTOFF et al. 1998). Importantly, inhibition of the ERK cascade blocks ATP-induced mitogenesis of human (NEARY et al. 1998) and rat astrocytes (NEARY et al. 1999a) and UTP-induced proliferation of aortic smooth muscle cells derived from spontaneously hypertensive rats (HARPER et al. 1998). These studies indicate the crucial role of ERK in purine and pyrimidine receptor-mediated mitogenic signaling.

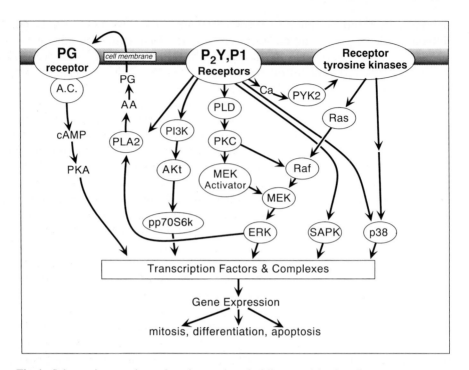

Fig. 1. Schematic overview of purine and pyrimidine trophic signaling pathways. Not all pathways operate in all cell types or with all P2Y or P1 receptors. Emerging cell-specific and receptor subtype-specific differences are discussed in the text

The signaling pathways coupling P1 and P2Y receptors to the ERK cascade are now beginning to be elucidated. For some GPCRs, e.g., α_{1A}-adrenergic receptors (VAN BIESEN et al. 1996), protein kinase C (PKC) is upstream of ERK, and in rat and human astrocytes inhibition or downregulation of PKC greatly reduced signaling from P2Y receptors to ERK (NEARY 1996; NEARY et al. 1998 1999a). Since PKC can be activated by diacylglycerol and calcium, second messengers that are generated by the PI-PLC pathway, and since PI-PLC can be activated by P2Y receptors, it appeared that stimulation of PI-PLC, calcium, and PKC by P2Y receptors would lead to ERK activation. However, several lines of evidence involving both astrocyte stellation and mitogensis suggested that the PI-PLC/calcium pathway may not be upstream of ERK. For example, α,βmeATP effectively stimulated astrocyte process elongation without increasing intracellular calcium, and neomycin inhibition of PI-PLC did not block the differentiating effect (BOLEGO et al. 1997; CENTEMERI et al. 1997). In addition, mitogenic signaling to ERK was unaffected by chelation of intracellular calcium or inhibition of PI-PLC by U-73122 (NEARY et al. 1999a). Additional studies indicated that P2Y receptors in rat cortical astrocytes recruit a calcium-independent PKC isoform, PKCδ, and that the diacylglycerol required to activate PKC is generated by hydrolysis of phosphatidylcholine, as catalyzed by PLD (NEARY et al. 1999a). Thus, P2Y receptors in astrocytes can be coupled independently to PI-PLC/calcium and ERK pathways.

As in astrocytes, trophic signaling by P2Y receptors in rat glomerular mesangial cells is independent of the PI-PLC/calcium pathway. The calcium antagonists TMB-8 and verapamil did not block ATP-induced mitogenesis; moreover, ATP was mitogenic in a mesangial cell line in which ATP does not stimulate PI-PLC (SCHULZE-LOHOFF et al. 1995). However, in HEK-293 cells, inhibition of PI-PLC partially reduced signaling from A_{2B} and $P2Y_2$ receptors to ERK, indicating that in these cells there is at least partial dependence on PI-PLC (GAO et al. 1999). As discussed below, cell-specific or receptor subtype-specific differences are likely to account for variations in trophic signaling pathways.

ERK has additional targets besides transcription factors, indicating that ERK may affect other cellular functions or may be involved in trophic signaling by other pathways. For example, PLA_2 is activated by ERK (LIN et al. 1993; NEMENOFF et al. 1993), and PLA_2-stimulated arachidonic acid release followed by production of prostaglandin E2 and elevated cAMP levels has been implicated in P2Y receptor activation of fibroblast proliferation (HUANG et al. 1991) suggesting that ERK-mediated mitogenic signaling may also involve PLA_2 in some cells. In rat striatal astrocytes, inhibition of PLA_2 blocked elongation of astrocytic processes induced by α,βmeATP (BOLEGO et al. 1997) implicating the ERK cascade in astrocyte differentiation.

The PI 3-kinase pathway has also been implicated in mitogenic signaling by P2Y receptors since inhibition of PI 3-kinase blocked ATP-induced increases in DNA synthesis and cell number in porcine coronary artery smooth muscle cells (WILDEN et al. 1998). Inhibition of the ERK cascade also blocked ATP-induced mitogenesis but this pathway is independent of the PI 3-kinase pathway because inhibition of PI 3-kinase did not block ERK activation by ATP. Thus, it appears that both ERK and PI 3-kinase pathways mediate mitogenic signaling in these smooth muscle cells. The PI 3-kinase pathway is also involved in nucleotide-enhanced mitogenesis in NIH 3T3 fibroblasts (CHUNG et al. 1997). ATP releases choline phosphate from these cells and can act as a co-mitogen, with choline phosphate and insulin, to synergistically enhance proliferation. This mitogenic pathway involves PI 3-kinase as well as pp70 S6 kinase. The co-mitogenic actions of choline phosphate and ATP may be independent of ERK whereas the co-mitogenic effects of sphingosine 1-phosphate and ATP appeared to be ERK-dependent. As a co-mitogen, UTP was equally as effective as ATP, while 2MeSATP was less effective, suggesting the involvement of P_{2U}-like receptors.

Cell-specific and receptor subtype-specific differences are emerging as components of the trophic signaling pathways that couple purine and pyrimidine receptors to the ERK cascade, particularly regarding the role of PKC and its isoforms. In HEK-293 cells, signaling from P_{2U}-like pyrimidine receptors to ERK depends on PKCζ, an atypical PKC isoform that is not activated by calcium or diacylglycerol (Gao et al. 1999). In rat renal mesangial cells (Huwiler and Pfeilschifter 1994) and cortical astrocytes (Neary et al. 1999b), P_{2U}-like receptors are coupled to ERK by a calcium-independent, rather than atypical, PKC isoform. In addition, NECA signaling to ERK was independent of PKC in HEK-293 cells (Gao et al. 1999) but dependent on PKC in human fetal astrocytes (Neary et al. 1998). Both PKC-dependent and -independent pathways are present in smooth muscle cells (Wang et al. 1992) and PC12 cells (Soltoff et al. 1998). In contrast to the situation in cortical astrocytes in which signaling to ERK did not involve calcium, signaling from P_{2U}-like receptors in PC12 cells was calcium-dependent, perhaps due to cell-specific or receptor-subtype specific differences in ERK signaling. In PC12 cells signaling from P_{2U}-like receptors to ERK involved the transactivation of EGF receptors (Soltoff et al. 1998), thereby adding another level of complexity to purine and pyrimidine signaling. In the endothelial cell line EAhy 926, signaling from P_{2U}-like receptors to ERK involved PKC and was partially dependent on calcium (Graham et al. 1996), whereas in 3T3 and 3T6 fibroblasts (Huang et al. 1991), and aortic smooth muscle cells derived from spontaneously hypertensive rats (Harper et al. 1998), PKC was not involved in ATP-induced mitogenesis. Collectively, these studies demonstrate that purines and pyrimidines can utilize a complex variety of trophic signaling pathways and pathway components, depending on differences in cell types and in receptor expression.

Purinergic receptors are also coupled to other MAPK cascades. Adenosine coupled with a brief period of ischemia, e.g., preconditioning stimuli, activated p38/MAPK in a rat myoblast cell line (Nagarkatti and Sha'afi 1998), suggesting a role for P1 receptors and p38/MAPK in the protective effect of preconditioning stimuli against lethal ischemic insult. ATP activated p38/MAPK, and inhibition of this MAPK blocked mitogenic signaling in human astrocytes (Neary et al. 1999b). In rat vascular smooth muscle cells, ATP and adenosine stimulated SAPK (Hamada et al. 1998), and in rat renal mesangial cells, purines and pyrimidines activated SAPK by a PKC-independent pathway (Huwiler et al. 1997). These findings suggest that both P1 and P2Y receptors can initiate coordinated signaling among MAPK cascades related to the trophic actions of purines and pyrimidines.

Stimulation of MAPK cascades leads to activation or induction of a variety of transcription factors and complexes involved in gene expression. Induction of the immediate early response gene *c-fos* after stimulation of purinergic receptors occurs in many cells including endothelial (Boutherin-Falson et al. 1990), smooth muscle (Malam-Souley et al. 1996; Erlinge et al. 1996), thyroid (Tornquist et al. 1996), and kidney cells (Paller et al. 1998), fibroblasts (Zheng et al. 1998), and myocytes (Zheng et al. 1994). Increased expression of *c-jun*, *jun-b*, and/or *c-myc* have been reported in myocytes

(Zheng et al. 1994), kidney (Paller et al. 1998), and endothelial cells (Boutherin-Falson et al. 1990). ATP also elicited changes in mRNA levels of skeletal α-actin and sarcoplasmic reticulum Ca^{2+}ATPase, genes associated with norepinephrine-induced hypertrophy in cardiac myocytes (Zheng et al. 1996). In astrocytes, α,βmeATP increased nuclear Fos and Jun immunostaining (Bolego et al. 1997), and extracellular ATP and UTP stimulated formation of functional AP-1 transcriptional complexes composed of Fos and Jun families of transcription factors (Neary et al. 1996b). AP-1 DNA binding activity was partially blocked by inhibition of PKC and of the ERK cascade, thereby indicating that PKC and ERK are upstream of the AP-1 complex formation in astrocytes (Neary et al. 1996b, 1999b). The P2Y receptor agonist, ADPβS stimulated rapid and transient increases in mRNA levels of the immediate early genes *c-fos, c-jun, junB*, and *Tis11* in astrocytes (Priller et al. 1998) and microglia (Priller et al. 1995). Interestingly, ADPβS, in combination with calcitonin gene-related peptide (CGRP), synergistically increased *c-fos* mRNA in astrocytes (Priller et al. 1998). Since CGRP has been implicated in glial cell activation after motoneuron injury (Reddington et al. 1995), the release of ATP from damaged cells may enhance the gliotic response to CRGP in brain injury. These studies demonstrate that purinergic receptors can elicit nuclear events leading to changes in gene expression, but additional work is needed to determine the array of transcription factors, complexes, and genes that mediate the trophic actions of purines and pyrimidines.

Efforts to determine the subtype(s) of P2Y receptor that activates trophic responses, as well as other cellular responses regulated by nucleotides, are often complicated by the difficulties in correlating properties of cloned receptors to those of native receptors, particularly in cells where multiple P2Y receptor subtypes are expressed (King et al. 1998). Examples of this problem are evident in recent astrocyte studies. The ability of pertussis toxin to block α,βmeATP induction of process elongation in rat striatal astrocytes (Bolego et al. 1997) supports the involvement of P2Y receptors, yet α,βmeATP is not a typical agonist of any of the currently cloned P2Y receptors. Another example involves the P2Y subtypes coupled to the ERK cascade in rat cortical astrocytes. These cells express $P2Y_1$, $P2Y_2$, and $P2Y_4$ receptors (Webb et al. 1996; Lenz et al. 2000), and the P2Y agonists ATP, UTP, and 2MeSATP can activate ERK (King et al. 1996). However, whereas UTP and 2MeSATP recruit cRaf-1, the first kinase in the ERK cascade, ATP does not activate this kinase or other common MEK activators (Lenz et al. 2000). ATP was reported to activate cRaf-1 in vascular smooth muscle cells (Yu et al. 1996), but this paper has been retracted (Wu et al. 1998). Thus, the effect of these P2Y agonists on cRaf-1 activation does not correspond to agonist profiles of the known P2Y subtypes expressed in rat cortical astrocytes. These findings in striatal and cortical astrocytes, as well as those in other tissues (King et al. 1998), may suggest the existence of novel as yet uncloned P2Y receptors involved in trophic signaling.

J. Conclusions and Future Perspectives

The data reviewed in the present chapter indicate that purine and pyrimidine nucleotides and nucleosides in addition to functioning as neurotransmitters also exert profound long-term trophic actions on target cells. In many organs and systems they act in concert with conventional growth factors, cytokines, and other growth-regulatory molecules to regulate cell proliferation, apoptosis, and differentiation. Evidence has been accumulating during the years to support a role for nucleotides and nucleosides in development, ischemia, and reperfusion injury, inflammation, aging-associated neurodegenerative diseases, wound healing, bone resorption, and cancer. The studies summarized above have also demonstrated that these trophic actions involve the activation of specific extracellular receptors or intracellular mechanisms that may represent targets for selective ligands to develop novel drugs.

It has also become evident that nucleotides and nucleosides have biphasic effects on cell viability, can be both protective and detrimental, and can mediate both life-saving and life-threatening effects, depending upon the experimental setting. In many instances, both positive and negative trophic actions have been reported even in the same target cells, suggesting that modulation of cell viability and differentiation by nucleotides and nucleosides is a highly-regulated phenomenon, which may involve different receptor subtypes and/or intracellular pathways depending upon specific pathophysiological conditions. Several hypotheses have attempted to reconcile these conflicting and paradoxical actions.

First, in several instances, cell death mediated by nucleotides and nucleosides may indeed represent a protective rather than a detrimental action. This may apply to adenosine-induced apoptosis of thymocytes (which may play a key role in the elimination of auto-reactive clones during immune system development), and to purine-induced neuronal and astroglial cell death, which may instead participate in the elimination of irreversibly-damaged cells in post-ischemic and post-traumatic brain (NEARY et al. 1996a); such an event may be crucial to re-routing energy and resources to damaged brain areas that still retain a chance for recovery. Second, the detrimental effects mediated by nucleotides and nucleosides under specific pathological conditions may simply represent a dysregulation of the physiological control normally exerted by these compounds on specific functions. This may apply to P2 receptor-mediated induction of COX-2 in reactive astrogliosis (that may represent the consequence of excessive and chronic brain inflammatory events) or to abnormal purine-mediated proliferation of VSMCs, that may instead play a role in the development of vasal atherosclerotic lesions. Third, the trophic activity of nucleotides on astrocytes and neurons may be beneficial in promoting neuronal regeneration. While an excess of reactive astrocytes may inhibit regeneration and remyelination, astrocytes also possess neurotrophic properties needed for axonal growth and guidance. In the latter case, the potentiation of conventional growth factor activity by purines and pyrimidines may stimulate astrocyte and neuronal functions to aid in repair and regeneration.

Several new areas are just beginning to emerge that include defining the role(s) of nucleotides and nucleosides in regulating the viability of skeletal muscle cells, oligodendroglia, endocrine cells, and reproductive cells. The more extensive use of antibodies to the various P1 and P2 receptor subtypes, antisense oligonucleotides, and transgenic mice for P1 and P2 receptor subtypes will aid in better defining the key role of these receptors in specific situations as well as providing information on possible system redundancies.

Finally, the development of potent, selective, and bioavailable agonists and antagonists for the various P1 and P2 receptor subtypes will be key, not only as novel pharmacological tools to validate the role of these receptors in the various in vitro and in vivo experimental settings but also as potential leads for drug discovery.

Acknowledgements. MPA has been a partner in the shared-cost project "Nucleotides, a Novel Class of Extracellular Signalling Substances in the Nervous System" (BMHT CT96–0676) supported by the European Union within the BIOMED 2 programme and is currently Principal Investigator and Coordinator of the Project *"Abnormal purine metabolism in Duchenne-type muscular dystrophy: does adenosine play a role in promoting muscular degeneration and loss of function?"* supported by Telethon (Grant No. 1050), whose financial support is gratefully acknowledged. MPA is also supported by the Ministero dell'Universita' e della Ricerca Scientifica e Tecnologica (60%, MURST No. 1201008) and by the Consiglio Nazionale delle Ricerche (CNR No. 98.01047.CT04). JTN is supported by the Department of Veterans Affairs, USA.

References

Abbracchio MP, Cattabeni F, Clementi F, Sher E (1989) Adenosine receptors linked to adenylate cyclase activity in human neuroblastoma cells: modulation during cell differentiation. Neurosci 30:819–825

Abbracchio MP, Burnstock G (1994) Purinoceptors: are there families of P2X and P2Y purinoceptors? Pharm Ther 64:445–475

Abbracchio MP, Saffrey MJ, Hopker V, Burnstock G (1994) Modulation of astroglial cell proliferation by analogues of adenosine and ATP in primary cultures of rat striatum. Neurosci. 59:67–76

Abbracchio MP (1996) P1 and P2 receptors in cell growth and differentiation. Drug Dev Res 39:393–406

Abbracchio MP (1997) ATP in brain function. In: Jacobson KA, Jarvis MF (eds) Purinergic Approaches in Experimental Therapeutics, Wiley-Liss, New York, pp 383–404

Abbracchio MP, Rainaldi G, Giammarioli AM, Ceruti S, Brambilla R, Cattabeni F, Barbieri D, Franceschi C, Jacobson KA, Malorni W (1997) The A_3 adenosine receptor mediates cell spreading, reorganization of actin cytoskeleton, and distribution of Bcl-x_L. Studies in human astroglioma cells. Biochem Biophys Res Commun 241: 297–304

Abbracchio MP, Burnstock G (1998) Purinergic signalling: Pathophysiological roles. Jpn J Pharmacol 78:113–145

Abbracchio MP, Ceruti S, Brambilla R, Barbieri D, Camurri A, Franceschi C, Giammarioli AM, Jacobson KA, Cattabeni F, Malorni W (1998) Adenosine A_3 receptors and viability of astrocytes. Drug Devel Res 45:379–386

Abe K, Saito H (1998) Adenosine stimulates stellation of cultured rat cortical astrocytes. Brain Res 804:63–71

Albert JL, Boyle JP, Roberts JA, Challiss J, Gubby SE, Boarder MR (1997) Regulation of brain capillary endothelial cells by P2Y receptors coupled to Ca^{2+}, phospholipase C and mitogen-activated protein kinase. Brit J Pharmacol 122:935–941

Aldskogius H, Kozlova EN (1998) Central neuron-glial and glial-glial interactions following axon injury. Prog. Neurobiol. 55:1–26

Asensio VC, Campbell IL (1999) Chemokines in the CNS: plurifunctional mediators in diverse states. Trends Neurosci. 22:504–512

Barbieri D, Abbracchio MP, Salvioli S, Monti D, Cossarizza A, Ceruti S, Brambilla R, Cattabeni F, Jacobson KA, Franceschi C (1998) Apoptosis by 2-chloro-2'-deoxyadenosine and 2-chloro-adenosine in human peripheral blood mononuclear cells. Neurochem Int. 32:493–504

Balboa MA, Firestein BL, Godson C, Bell KS, Insel PA (1994) Protein kinase $C\alpha\alpha$ mediates phospholipase D activation by nucleotides and phorbol ester in Madin-Darby canine kidney cells. J Biol Chem 269:10511–10516

Bernstein M, Lyons SA, Moller T, Kettenmann H (1996) Receptor-mediated calcium signalling in glial cells from mouse corpus callosum slices. J Neurosci Res 46:152–163

Berti-Mattera LN, Wilkins PL, Madhun Z, Suchovsky D (1996) P2-purinergic receptors regulate phospholipase C and adenylate cyclase activities in immortalized Schwann cells. Biochem J 314:555–561

Beutler E (1992) Cladribine (2-chlorodeoxyadenosine). Lancet 340:952–956

Bolego C, Ceruti S, Brambilla R, Puglisi L, Cattabeni F, Burnstock G, Abbracchio MP (1997) Characterization of the signalling pathways involved in ATP and basic fibroblast growth factor-induced astrogliosis. Brit J Pharmacol 121:1692–1699

Boutherin-Falson O, Reuse S, Dumont JE, Boeynaems J-M (1990) Increased levels of c-fos and c-myc mRNA in ATP-stimulated endothelial cells. Biochem Biophys Res Commun 172:306–312

Brambilla R, Burnstock G, Bonazzi A, Ceruti S, Cattabeni F, Abbracchio MP (1999). Cyclo-oxygenase-2 mediates P2Y receptor-induced reactive astrogliosis. Br J Pharmacol 126:563–567

Brietner HS (1996) The role of anti-inflammatory drugs in the prevention and treatment of Alzheimer's disease. Ann Rev Med 47:401–411

Bruner G, Murphy S (1990) ATP-evoked arachidonic acid mobilization in astrocytes is via a P2y-purinergic receptor. J Neurochem 55:1569–1575

Bruner G, Murphy S (1993) Purinergic P2Y receptors on astrocytes are directly coupled to phospholipase A2. Glia 7:219–224

Bryson HM, Sorkin EM (1993) Cladribine. A review of its pharmacodynamic and pharmacokinetic properties and therapeutic potential in haematological malignancies. Drugs 46:872–894

Burnstock G (1990) Overview: Purinergic mechanisms. Ann NY Acad Sci 603:1–18

Camina F, Novo-Rodriguez MI, Rodriguez-Segade S, Castro-Gago M. (1995) Purine and carnitine metabolism in muscle of patients with Duchenne muscular dystrophy. Chim Clin Acta 243:151–164

Castro E., Pintor J, Miras-Portugal MT (1992) Ca^{2+}-stores mobilization by diadenosine tetraphosphate, Ap_4A, through a putative P_{2Y} purinoceptor in adrenal chromaffin cells. Br J Pharmacol 106:833–837

Castro-Gago M, Lojo S, Novo I, del Rio R, Pena J, Rodriguez-Segade S (1987) Effects of chronic allopurinol therapy on purine metabolism in Duchenne muscular dystrophy. Biochem Biophys Res Commun 147:152–157

Centemeri C, Bolego C, Abbracchio MP, Cattabeni F, Puglisi L, Burnstock G, Nicosia S (1997) Characterization of the Ca^{2+} responses evoked by ATP and other nucleotides in mammalian brain astrocytes. Brit J Pharmacol 121:1700–1706

Ceruti S, Giammarioli AM, Camurri A, Falzano L, Rufini S, Frank C, Fiorentini C, Malorni W, Abbracchio MP (2000) Adenosine- and 2-chloro-adenosine-induced cytopathic effects on myoblastic cells and myotubes: involvement of different intracellular mechanisms. Neuromusc Dis 10:436–446

Chen Z-P, Kratzmeier M, Poch A, Xu S, McArdle CA, Levy A, Mukhopadhyay AK, Lightman SL (1996) Effects of extracellular nucleotides in the pituitary: adenosine triphosphate receptor-mediated intracellular responses in gonadotrope-derived acT3–1 cells. Endocrinology 137:248–256

Cheng X, Li X (1997) Exogenous adenosine triphosphate promotes proliferation of renal tubular epithelial cells [Chinese]. Chinese Med J 77:351–354

Chiozzi P, Murgia M, Falzoni S, Ferrari D, Di Virgilio F. (1996) Role of the purinergic P2Z receptor in spontaneous cell death in J774 macrophage cultures. Biochem Biophys Res Commun 218:176–181

Chung T, Crilly KS, Anderson WH, Mukherjee JJ, Kiss Z (1997) ATP-dependent choline phosphate-induced mitogenesis in fibroblasts involves activation of pp70 S6 kinase and phosphatidlyinositol 3'-kinase through an extracellular site. Synergistic mitogenic effects of choline phosphate and sphingosine 1-phosphate. J Biol Chem 272:3064–3072

Cohen A, Hirschhorn R, Horowitz SD, Rubinstein A, Palmar SH, Hong R, Martin DW Jr (1978) Deoxyadenosine triphosphate as a potentially toxic metabolite in adenosine deaminase deficiency. Proc Natl Acad Sci USA 75:471–476

Cornell-Bell AH, Finkbeiner SM, Cooper MS, Smith SJ (1990) Glutamate induces calcium waves in cultured astrocytes: long-range glial signaling. Science 247:470–473

Corset V, Nguyen-Ba-Charvet KT, Forcet C, Moyse E, Chedotal A, Mehlen P (2000) Netrin-1-mediated axon outgrowth and cAMP production requires interaction with adenosine A2b receptor Nature, 290:131–134

Davidson JS, Wakefield IK, Sohnius U, Van Der Merwe PA, Millar RP (1990) A novel extracellular nucleotide receptor coupled phosphoinositidase-C in pituitary cells. Endocrinology 126:80–87

Davis EJ, Foster TD, Thomas WE (1994) Cellular forms and functions of brain microglia. Brain Res.Bull. 34:73–78

De Souza, LR, Moore H, Raha S, Reed JK (1995) Purine and pyrimidine nucleotides activate distinct signalling pathways in PC12 cells. J Neurosci Res 41:753–763

Di Perri T, Laghi-Pasini F eds (1991) In Fisiologia e Farmacologia del Sistema Adenosinico, Luigi Pozzi, Italy, pp 179–189

Di Virgilio F, Ferrari D, Chiozzi P, Falzoni S, Sanz JM, Dal Susino M, Mutini C, Hanau S, Baricordi OR (1996) Purinoceptor function in the immune system. Drug Dev Res 39:319–329

Diener HC, Cunha L, Forbes C, Sivenius J, Smets P, Lowenthal A (1996) European stroke prevention study 2. Dipyridamole and acetylsalicylic acid in the secondary prevention of stroke. J Neurol Sci 143:1–13

Diverse-Pierluissi, M, Dunlap K, Westhead EW (1991) Multiple actions of extracellular ATP on calcium currents in cultured bovine chromaffin cells. Proc Natl Acad Sci USA 88:1261–1265

Downey JM, Cohen MV (1995) Role of adenosine in the phenomenon of ischemic preconditioning. In: Belardinelli L, Pelleg A (eds) Adenosine and adenine nucleotides: from molecular biology to integrative physiology. Kluwer Academic Press, Norwell, pp 461–475

Dusseau JW, Hutchins PM (1988) Hypoxia-induced angiogenesis in chick chorioallantoic membranes: a role for adenosine. Resp Physiol 71:33–44

Eddleston M, Mucke L (1993) Molecular profile of reactive astrocytes: implications for their role in neurologic disease. Neurosci 54:15–36

Endresen PC, Eide TJ, Aarbakke J (1993) Cell death initiated by 3-deazaadenosine in HL-60 cells is apoptosis and is partially inhibited by homocysteine. Biochem Pharmacol 46:1893–1901

Erlinge D, Yoo H, Edvinsson L, Reis DJ, Wanlestedt C (1993) Mitogenic effects of ATP

on vascular smooth muscle cells vs. other growth factors and sympathetic cotransmitters. Am J Physiol 265:H1089–H1097
Erlinge D, Helig M, Edvinsson L (1996) Tyrophostin inhibition of ATP-stimulated DNA synthesis, cell proliferation and Fos-protein expression in vascular smooth muscle cells. Brit J Pharmacol 118:1028–1034
Ferrari D, Chiozzi P, Falzoni S, Dal Susino M, Collo G, Buell G, Di Virgilio F (1997) ATP-mediated cytotoxicity in microglial cells. Neuropharmacol. 36:1295–1301
Fidzianska A, Goebel HH, Warlo I (1990) Acute infantile spinal muscular atrophy. Muscle apoptosis as a proposed pathogenetic mechanism. Brain 113:433–445
Fischer S, Sharma HS, Karliczek GF, Schaper W (1995) Expression of vascular permeability factor/vascular endothelial growth factor in pig cerebral microvascular endothelial cells and its upregulation by adenosine. Mol Brain Res 28:141–148
Filippini A, Riccioli A, De Cesaris P, Paniccia R, Teti A, Stefanini M, Conti M, Ziparo E (1994) Activation of inositol phospholipid turnover and calcium signaling in rat Sertoli cells by P_2-purinergic receptors: modulation of follicle-stimulating hormone responses. Endocrinology 134:1537–1545
Fitzgerald GA (1987) Dipyridamole. New Engl J Med 316:1247–1257
Foresta C, Rossato M, Di Virgilio F (1992) Extracellular ATP is a trigger for the acrosome reaction in human spermatozoa. J Biol Chem 267:19443–19447
Franke H, Krügel U, Illes P (1997) Immunohistochemical studies on P2 purinoceptor-mediated proliferative effects on astrocytes in vivo. Naunyn-Schmiedebergs Arch Pharmacol 356:R33
Franke H, Krügel U, Illes P (1999) P2 receptor-mediated proliferative effects on astrocytes in vivo. Glia, 28:190–200
Gandia L, Garcia AG, Morad M (1993) ATP modulation of calcium channels in chromaffin cells. J. Physiol. 470:55–72
Gao Z, Chen T, Weber MJ, Linden J (1999) A_{2B} adenosine and $P2Y_2$ receptors stimulate mitogen-activated protein kinase in human embryonic kidney-293 cells. Cross-talk between cyclic AMP and protein kinase C pathways. J Biol Chem 274: 5972–5980
Garcia-Martinez V, Macias D, Ganan Y, Garcia-Lobo JM, Francia MV, Fernandez-Teran MA, Hurle JM (1993) Internucleosomal DNA fragmentation and programmed cell death (apoptosis) in the interdigital tissue of the embryonic chick leg bud. J Cell Sci 106:201–208
Glasky AJ, Melchoir CL, Pirzadeh B, Heydari N, Ritzmann RF (1994) Effect of AIT-082, a purine analog, on working memory in normal and aged mice. Pharmacol Biochem Behav 47:325–329
Glasky AJ, Ritzmann RF, Prisecaru I, Santos S, Kafi K, Rathbone MP, Middlemiss PJ, Crocker C (1996) Elevation of brain mRNAs for neurotrophins by oral AIT-082 in mice. Soc Neurosci Abstr 22:751
Gonzalez FA, Wang D-J, Huang N-N, Heppel LA (1990) Activation of early events of the mitogenic response by a P_{2Y} purinoceptor with covalently bound 3′-O-(4-benzoyl)-benzoyladenosine 5′-triphosphate. Proc Natl Acad Sci USA 87:9717–9721
Graham A, McLees A, Kennedy C, Gould GW, Plevin R (1996) Stimulation by the nucleotides ATP and UTP of mitogen-activated protein kinase in EAhy 926 endothelial cells. Brit J Pharmacol 117:1341–1347
Guroff G, Dickens G, End D, Londos C (1981) The action of adenosine analogs on PC12 cells. J Neurochem 37:1431–1439
Gustavsson L, Lundqvist C, Hansson E (1993) Receptor-mediated phospholipase D activity in primary astroglial cultures. Glia 8:249–255
Guthrie PB, Knappenberger J, Segal M, Bennett MVL, Charles AC (1999) ATP released from astrocytes mediates glial calcium waves. J Neurosci 19:520–528
Gysbers JW, Rathbone MP (1996) GTP and guanosine synergistically enhance NGF-induced neurite outgrowth from PC12 cells. Int J Devl Neurosci 14:19–34

Hamada K, Takuwa N, Yokoyama K, Takuwa Y (1998) Stretch activates Jun N-terminal kinase/stress-activated protein kinase in vascular smooth muscle cells through mechanisms involving autocrine ATP stimulation of purinoceptors. J Biol Chem 273:6334–6340

Harden TK, Boyer JL, Nicholas RA (1995) P_2-Purinergic receptors: subtype-associated signaling responses and structure. Ann Rev Pharmacol Toxicol 35:541–579

Harper S, Webb TE, Charlton SJ, Ng LL, Boarder MR (1998) Evidence that $P2Y_4$ nucleotide receptors are involved in the regulation of rat aortic smooth muscle cells by UTP and ATP. Brit J Pharmacol. 124:703–710

Hindley S, Herman MAR, Rathbone MP (1994) Stimulation of reactive gliosis in vivo by extracellular adenosine diphosphate or an adenosine A_2 receptor agonist. J Neurosci Res 38:399–406

Hoffmann C (1996) Nucleotid-Analoge als molekulare Sonden zur Aufklärung von Mechanismen der Zellteilung, Differenzierung und der Apoptosis Dissertation, Universität Bremen

Hoffmann C, Raffel S, Ruchaud S, et al. (1996) Chloro-substituted cAMP analogues and their adenosine metabolites induce apoptosis of the human promelocytic leukemia cell line NB4: molecular basis for cell type selectivity. Cell Pharmacol 3:417–427

Houchi H, Okuno M, Yoshizumi M, Oka M (1995) Regulatory mechanism of calcium efflux from cultured bovine adrenal chromaffin cells induced by extracellular ATP. Neurosci Lett 198:177–180

Hourani S, Di Virgilio F, Loubatieres-Mariani M-M (1997) Physiological roles for P2 receptors in platelets, visceral smooth muscle, and the immune and endocrine systems. In: Turner, JT, Weisman GA, Fedan JS (eds) The P2 nucleotide receptors, Humana Press, Totowa, NJ, pp 361–411

Huang N, Wang D, Heppel LA (1989) Extracellular ATP is a mitogen for 3T3, 3T6, and A431 cells and acts synergistically with other growth factors. Proc Natl Acad Sci USA 86:7904–7908

Huang N-N, Wang D-J, Gonzalez F, Heppel LA (1991) Multiple signal transduction pathways lead to extracellular ATP-stimulated mitogenesis in mammalian cells: II. A pathway involving arachidonic acid release, prostaglandin synthesis, and cyclic AMP accumulation. J Cell Physiol 146:483–494

Huang N, Wang D, Heppel LA (1993) Stimulation of aged human lung fibroblasts by extracellular ATP via suppression of arachidonate metabolism. J Biol Chem 268: 10789–10795

Huwiler A, Pfeilschifter J (1994) Stimulation by extracellular ATP and UTP of the mitogen-activated protein kinase cascade and proliferation of rat renal mesangial cells. Brit J Pharmacol 113:1455–1463

Huwiler A, van Rossum G, Wartmann M, Pfeilschifter J (1997) Stimulation by extracellular ATP and UTP of the stress-activated protein kinase cascade in rat renal mesangial cells. Brit J Pharmacol 120:807–812

Ishikawa S, Higashiyama M, Kusaka I, Saito T, Nagasaka S, Fukuda S (1997) Extracellular ATP promotes cellular growth of renal inner medullary collecting duct cells mediated via P_{2U} receptors. Nephron 76:208–214

Kamada S, Blackmore PF, Oehninger S, Gordon K, Hodgen GD (1994) Existence of P_2-purinoceptors on human and porcine granulosa cells. J Clin Endocrinol Metabol 78:650–656

Kastritsis CH, Salm AK, McCarthy K (1992) Stimulation of the P2y purinergic receptor on type 1 astroglia results in inositol phosphate formation and calcium mobilization. J Neurochem 58:1277–1284

Kastritsis CH, McCarthy KD (1993) Oligodendroglial lineage cells express neuroligand receptors. Glia 8:106–113

Kim J-K, Rathbone MP, Middlemiss PJ, Hughes DW, Smith RW (1991) Purinergic stimulation of astroblast proliferation: guanosine and its nucleotides stimulate cell division in chick astroblasts. J Neurosci Res 28:442–455

Kimelberg HK, Cai Z, Rastogi P, Charniga CJ, Goderie S, Dave V, Jalonen TO (1997) Transmitter-induced calcium responses differ in astrocytes acutely isolated from rat brain and in culture. J Neurochem 68:1088–1098

King BF, Neary JT, Zhu Q, Wang S, Norenberg MD, Burnstock G (1996) P2 purinoceptors in rat cortical astrocytes: expression, calcium-imaging and signalling studies. Neurosci 74:1187–1196

King BF, Townsend-Nicholson A, Burnstock G (1998) Metabotropic receptors for ATP and UTP: exploring the correspondence between native and recombinant nucleotide receptors. Trends Pharmacol Sci 19:506–514

Kirischuk S, Moller T, Voitenko N, Kettenmann H, Verkhratsky A (1995) ATP-induced cytoplasmic calcium mobilization in Bergmann glial cells. Neurosci 15:7861–7871

Knecht M, Darbon J-M, Ranta T, Baukal A, Catt KJ (1984) Inhibitory actions of adenosine on follicle-stimulating hormone-induced differentiation of cultured granulosa cells. Biol Reprod 30:1082–1090

Kobayashi S, Nishimura J, Kanaide H (1994) Cytosolic Ca^{2+} transients are not required for platelet-derived growth factor to induce cell cycle progression of vascular smooth muscle cells in primary culture. Actions of tyrosine kinase. J Biol Chem 269:9011–9018

Koshiba M, Apasov S, Sverdlov V, Chen P, Erb L, Turner JT, Weisman GA, Sitkovski MV (1997). Transient up-regulation of P2Y2 nucleotide receptor mRNA expression is an immediate early gene response in activated thymocytes. Proc Natl Acad Sci USA 94:831–6

Kupitz Y, Atlas D (1993) A putative ATP-activated Na^+ channel involved in sperm-induced fertilization. Science 261:484–486

Lai K-M, Wong PCL (1991) Metabolism of extracellular adenine nucleotides by cultured rat brain astrocytes. J Neurochem 57:1510–1515

Lenz G, Gottfried C, Luo Z, Avruch J, Rodnight R, Nie W-J, Kang Y, Neary JT (2000) P2Y purinoceptor subtypes recruit different MEK activators in astrocytes. Brit J Pharmacol 129:927–936

Leong WS, Russell GG, Caswell AM (1993) Induction of enhanced responsiveness of human articular chondrocytes to extracellular ATP by tumour necrosis factor α. Clin Sci 85:569–575

Liang BT, Jacobson KA (1998) A physiological role of the adenosine A_3 receptor: sustained cardioprotection. Proc Natl Acad Sci USA 95:6995–6999

Lin LF, Bott MC, Kao L-S, Westhead EW (1995) ATP stimulated catecholamine secretion: response in perfused adrenal glands and a subpopulation of cultured chromaffin cells. Neurosci Lett 183:147–150

Lin L-L, Wartmann M, Lin AY, Knopf JL, Seth A, Davis RJ (1993) cPLA2 is phosphorylated and activated by MAP kinase. Cell 72:269–278

Linden J, Tucker AL, Lynch KR (1991) Molecular cloning of adenosine A_1 and A_2 receptors. Trends Pharmacol Sci 12:326–328

Loubatieres-Mariani M-M, Hillaire-Buys D, Chapal J, Bertrand G, Petit P (1997) P_2 purinoceptor agonists: new insulin secretagogues potentially useful in the treatment of non-insulin-dependent diabetes mellitus. In: Jacobson KA, Jarvis MF (eds) Purinergic Approaches in Experimental Therapeutics, Wiley-Liss, New York, pp 253–260

Lyons SA, Morell P, McCarthy KD (1994) Schwann cells exhibit P2Y purinergic receptors that regulate intracellular calcium and are up-regulated by cyclic AMP analogues. J Neurochem 63:552–560

Malam-Souley R, Campan M, Gadeau A-P, Desgranges C (1993). Exogenous ATP induces limited cell cycle progression of arterial smooth muscle cells. Am J Physiol 264:C783–C788

Malam-Souley R, Seye C, Gadeau A-P, Loirand G, Pillois X, Campan M, Pacaud P, Desgranges C (1996) Nucleotide receptor P_{2U} partially mediates ATP-induced cell cycle progression of aortic smooth muscle cells. J Cell Physiol 166:57–65

Martin TW, Michaelis K (1989) P2-purinergic agonists stimulate phosphodiesteratic cleavage of phosphatidylcholine in endothelial cells. Evidence for activation of phospholipase D. J Biol Chem 264:8847–8856

Meininger CJ, Granger JH (1990) Mechanisms leading to adenosine-stimulated proliferation of microvascular endothelial cells. Am J Physiol 258 (Health Circ Physiol 27): H198–H206

Meloche S, Seuwen K, Pages G, Pouyssegur J (1992) Biphasic and synergistic activation of p44^{mapk} (ERK1) by growth factors: correlation between late phase activation and mitogenicity. Mol Endocrinol 5:845–854

Middlemiss PJ, Glasky AJ, Rathbone MP, Werstuik E, Hindley S, Gysbers J (1995a) AIT-082, a unique purine derivative, enhances nerve growth factor mediated neurite outgrowth from PC12 cells. Neurosci Lett 199:131–134

Middlemiss PJ, Gysbers JW, Rathbone MP (1995b) Extracellular guansoine and guanosine-5′-triphosphate increase NGF synthesis and release from cultured mouse neopallial astrocytes. Brain Res 677:152–156

Minelli A, Allegruzzi C, Rosati R, Mezzasoma I (1998) Regulation of agonist-receptor binding by G proteins and divalent cations in spermatozoa solubilized A_1 adenosine receptors. Mol Gen Metab 63:183–190

Miyagi Y., Kobayashi S, Ahmed A, Nishimura J, Fukui M, Kanaide H (1996). P_{2U} purinergic activation leads to the cell cycle progression from the G_1 to the S and M phases but not from the G_0 to G_1 phase in vascular smooth muscle cells in primary culture. Biochem Biophys Res Comm 222:652–658

Modderman WE, Weidema AF, Vrijheid-Lammers T, Wassenaar AM, Nijweide PJ (1994) Permeabilization of cells of hemopoietic origin by extracellular ATP^{4-}: elimination of osteoclasts, macrophages, and their precursors from isolated bone cell populations and fetal bone rudiments. Calcif Tissue Int 55:141–150

Mulryan K, Gitterman DP, Lewis CJ, Vial C, Leckle BJ, Cobb AL, Broen JE, Conley EC, Buell G, Pritchard CA, Evans RJ (2000) Reduced vas deferens contraction and male infertility in mice lacking $P2X_1$ receptors. Nature 403:86–89

Nagarkatti DS, Sha'afi RI (1998) Role of p38 MAP kinase in myocardial stress. J Mol Cell Cardiol 30:1651–1664

Nakamura Y, Ohtaki S (1990) Extracellular ATP-induced production of hydrogen peroxide in porcine thyroid cells. J Endocrinol 126:283–287

Neary JT, van Breemen C, Forster E, Norenberg LOB, Norenberg MD (1988) ATP stimulates calcium influx in primary astrocyte cultures. Biochem Biophys Res Commun 157:1410–1416

Neary JT, Laskey R, van Breemen C, Blicharska J, Norenberg LOB, Norenberg MD (1991) ATP-evoked calcium signal stimulates protein phosphorylation/dephosphorylation in astrocytes. Brain Res 566:89–94

Neary JT, Norenberg MD (1992) Signalling by extracellular ATP: Physiological and pathological considerations in neuronal-astrocytic interactions. Prog Brain Res 94:145–151

Neary JT, Baker L, Jorgensen SL, Norenberg MD (1994a) Extracellular ATP induces stellation and increases GFAP content and DNA synthesis in primary astrocyte cultures. Acta Neuropathol 87:8–13

Neary JT, Whittemore SR, Zhu Q, Norenberg MD (1994b) Synergistic activation of DNA synthesis in astrocytes by fibroblast growth factor and extracellular ATP. J Neurochem 63:490–494

Neary JT, Whittemore SR, Zhu Q, Norenberg MD (1994c) Destabilization of glial fibrillary acidic protein mRNA in astrocytes by ammonia and protection by extracellular ATP. J Neurochem 63:2021–2027

Neary JT, Zhu Q (1994) Signaling by ATP receptors in astrocytes. NeuroReport 5:1617–1620

Neary JT (1996) Trophic actions of extracellular ATP on astrocytes, synergistic interactions with fibroblast growth factors and underlying signal transduction mechanisms. In: P2 purinoceptors: localization, function and transduction mecha-

nisms. Ciba Foundation Symposium 198. J Wiley & Sons, Chichester, pp 130–141
Neary JT, Rathbone MP, Cattabeni F, Abbracchio MP, Burnstock G (1996a) Trophic actions of extracellular nucleotides and nucleosides on glial and neuronal cells. Trends Neurosci 19:13–18
Neary JT, Zhu Q, Kang Y, Dash PK (1996b) Extracellular ATP induces formation of AP-1 complexes in astrocytes via P2 purinoceptors. NeuroReport 7:2893–2896
Neary JT (1997) MAPK cascades in cell growth and death. News in Physiol Sci 12:286–293
Neary JT, McCarthy M, Kang Y, Zuniga S (1998) Mitogenic signaling from P1 and P2 purinergic receptors to mitogen-activated protein kinase in human fetal astrocytes. Neurosci Lett 242:159–162
Neary JT, Kang Y, Bu Y, Yu E, Akong K, Peters CM (1999a) Mitogenic signaling by ATP/P2Y purinergic receptors in astrocytes: involvement of a calcium-independent protein kinase C, extracellular signal regulated protein kinase pathway distinct from the phosphatidylinositol-specific phospholipase C, calcium pathway. J Neurosci 19:4211–4220
Neary JT, McCarthy M, Cornell-Bell AH, Kang Y (1999b) Trophic signaling pathways activated by purinergic receptors in rat and human astroglia. Prog Brain Res 120:323–332
Nedergaard M (1994) Direct signaling from astrocytes to neurons in cultures of mammalian brain cells. Science 263:1768–1771
Nemenoff RA, Winitz S, Qian N-X, Van Putten V, Johnson GL, Heasley LE (1993) Phosphorylation and activation of a high molecular weight form of phospholipase A2 by p42 microtubule-associated protein 2 kinase and protein kinase C. J Biol Chem 268:1960–1964
Newman EA, Zahs KR (1997) Calcium waves in retinal glial cells. Science 275:844–847
Nijweide PJ, Modderman WE, Hagenaars CE (1995) Extracellular adenosine triphosphate. A shock to hemopoietic cells. Clin Orthop 313:92–102
Ohtsuki T, Kitagawa K, Yamagata K, Mandai K, Mabuchi T, Matsushita K, Yanagihara T, Matsumoto M (1996) Induction of cyclo-oxygenase-2 mRNA in gerbil hippocampal neurons after transient forebrain ischemia. Brain Res 736:353–356
Ostrom RS, Gregorian C, Insel PA (2000) Cellular release of and response to ATP as key determinants of the set-point of signal transduction pathways. J Biol Chem 275:11735–11739
Otsuguro K-I, Asano T, Ohta T, Ito S, Nakazato Y (1995) ATP-evoked membrane current in guinea pig adrenal chromaffin cells. Neurosci Lett 187:145–148
Paller MS, Schnaith EJ, Rosenberg ME (1998) Purinergic receptors mediate cell proliferation and enhanced recovery from renal ischemia by adenosine triphosphate. J Lab Clin Med 131:174–183
Pearce B, Murphy S, Jeremy J, Morrow C, Dandona P (1989) ATP-evoked Ca2+ mobilization and prostanoid release from astrocytes: P2-purinergic receptors linked to phosphoinositide hydrolysis. J Neurochem 52:971–977
Pfeilschifter J, Merriweather C (1993) Extracellular ATP and UTP activation of phospholipase D is mediated by protein kinase C-ε in rat renal mesangial cells. Brit J Pharmacol 110:847–853
Picano E, Abbracchio MP (1998) European stroke prevention study-2 results: serendipitous demonstration of neuroprotection induced by endogenous adenosine accumulation? Trends Pharmacol Sci 19:14–16
Picano E., Michelassi C (1997) Chronic oral dipyridamole as a novel antianginal drug: the collateral hypothesis. Cardiovasc Res 33:666–670
Pirotton S, Robaye B, Lagneau C, Boeynaems J-M (1990) Adenine nucleotides modulate phosphatidylcholine metabolism in aortic endothelial cells. J Cell Physiol 142:449–457

Popper LD, Batra S (1993) Calcium mobilization and cell proliferation activated by extracellular ATP in human ovarian tumor cells. Cell Calcium 14:P209

Porter JC, Ijames CF, Wang TC, Markey SP. (1995) Purification and identification of pituitary cytotropic factor. Proc Natl Acad Sci USA 92:5351–5355

Porter AG, Ng P, Jänicke RU (1997) Death substrates come alive. BioAssay 19:501–507

Priller J, Haas CA, Reddington M, Kreutzberg GW (1995) Calcitonin gene-regulated peptide and ATP induce immediate early gene expression in cultured rat microglial cells. Glia 15:447–457

Priller J, Reddington M, Haas CA, Kreutzberg GW (1998) Stimulation of P_{2Y}-purinoceptors on astrocytes results in immediated early gene expression and potentiation of neuropeptide action. Neurosci 85:521–525

Purkiss JR, Boarder MR (1992) Stimulation of phosphatidate synthesis in endothelial cells in response to P2-receptor activation. Biochem J 287:31–36

Ralevic V, Burnstock G (1998) Receptors for purines and pyrimidines. Pharm Rev 50:413–492

Rathbone MP, Middlemiss PJ, Kim J-L, Gysbers JW, DeForge SP, Smith RW, Hughes DW (1992) Adenosine and its nucleotides stimulate proliferation of chick astrocytes and human astrocytoma cells. Neurosci Res 13:1–17

Rathbone MP, Juurlink B (1993) Hippocampal neurons in vitro respond to extracellular guanosine and GTP with neurite outgrowth and branching. Soc Neurosci Abstr 19:38

Reddington M, Priller J, Treichel J, Haas CA, Kreutzberg GW (1995) Astrocytes and microglia as potential targets for calcitonin gene related peptide in the central nervous system. Can J Physiol Pharmacol 73:1047–1049

Ridet JL, Malhotra SK, Privat A, Gage FH (1997) Reactive astrocytes: cellular and molecular cues to biological function. Trends Neurosci 20:570–577

Rounds S, Yee WL, Dawicki DD, Harrington E, Parks N, Cutaia MV (1998) Mechanisms of extracellular ATP- and adenosine-induced apoptosis of cultures pulmonary artery endothelial cells. Am J Physiol 275:L379–L388.

Rozengurt E (1982) Adenosine receptor activation in quiescent Swiss 3T3 cells: Enhancement of cAMP levels, DNA synthesis and cell division. Exp Cell Res 139:71–78

Ruchaud S, Zorn M, Davilar-Villar E, et al. (1995) Evidence for several pathways of biological response to hydrolysable cAMP analogues using a model system of apoptosis in IPC-81 leukaemia cells. Cell Pharmacol 2:127–140

Rufini S, Rainaldi G, Abbracchio MP, Fiorentini C, Capri M, Franceschi C, Malorni W (1997) Actin cytoskeleton as a target for 2-chloro-adenosine: evidence for induction of apoptosis in C2C12 myoblastic cells. Biochem Biophys Res Comm 238:361–366

Salter MW, Hicks JL (1994) ATP-evoked increases in intracellular calcium in neurons and glia from the dorsal spinal cord. J Neurosci 14:1563–1575

Salter MW, Hicks JL (1995) ATP causes release of intracellular Ca^{2+} via the phospholipase Cbeta/IP3 pathway in astrocytes from the dorsal spinal cord. J Neurosci 15:2961–2967

Sandri M, Carraro U, Podhorska-Okolow M, Rizzi C, Arslan P, Monti D, Franceschi C (1995) Apoptosis, DNA damage and ubiquitin expression in normal and mdx muscle fibers after exercise. FEBS Letters 373:291–295

Sato K, Okajima F, Kondo Y (1992) Extracellular ATP stimulates three different receptor-signal transduction systems in FRTL-5 thyroid cells. Biochem. J 283:281–287

Schafer K-H, Saffrey MJ, Burnstock G (1995) Trophic actions of 2-chloroadenosine and bFGF on cultured myenteric neurones. Neuro Report 6:937–941

Schor S, Rozengurt E (1973) Enhancement by purine nucleosides and nucleotides of serum-induced DNA synthesis in quiescent 3T3 cells. J Cell Physiol 81:339–346

Schubert P, Ogata T, Rudolphi K, Marchini C, McRae A, Ferroni S (1997) Support of homeostatic glial cell signaling: a novel therapeutic approach by propentofylline. Ann NY Acad Sci 826:337–347

Schulze-Lohoff E, Schagerl S, Oglivie A, Sterzel RB (1995) Extracellular ATP augments mesangial cell growth induced by multiple growth factors. Nephrology, Dialysis, Transplantation 10:2027–2034

Schulze-Lohoff E, Ogilvie A, Sterzel RB (1996) Extracellular nucleotides as signalling molecules for renal mesangial cells. J Auton Pharmacol 16:381–384

Schwartz LM (1992) Insect muscle as a model for programmed cell death. J Neurobiol 23:1312–1326

Seger R, Krebs EG (1995) The MAPK signaling cascade. FASEB J 9:726–735

Shaban M, Smith PA, Stone TW (1998) Adenosine receptor-mediated inhibition of neurite outgrowth from cultured sensory neurons is via an A_1 receptor and is reduced by nerve growth factor. Dev Brain Res 105:167–173

Shitta-Bey A, Neary JT (1998) Fluid shear stress stimulates release of ATP from astrocyte cultures. FASEB J 12:A1469

Shneyvais V, Nawrath H, Jacobson KA, Shainberg A (1998) Induction of apoptosis in cardiac myocytes by an A_3 adenosine receptor agonist. Exp Cell Res 243:383–397

Smallridge RC, Gist ID (1994) P_2-purinergic stimulation of iodide efflux in FRTL-5 rat thyroid cells involves parallel activation of PLC and PLA_2. Am J Physiol 267 (Endocrinol Metab 30): E323–E330

Soodak LK, Macdonald GJ, H.R. Behrman HR (1988) Luteolysis is linked to luteinizing hormone-induced depletion of adenosine triphosphate in vivo. Endocrinology 122:187–193

Soltoff SP, Avraham H, Avraham S, Cantley LC (1998) Activation of P2Y2 receptors by UTP and ATP stimulates mitogen-activated kinase activity through a pathway that involves related adhesion focal tyrosine kinase and protein kinase C. J Biol Chem 273:2653–2660

Stambaugh K, Jacobson KA, Jiang J-l, Liang, BT (1997) A novel cardioprotective function of adenosine A_1 and A_3 receptors during prolonged stimulated ischemia. Am J Physiol 273:H501–H505

Sugioka M, Fukuda Y, Yamashita M (1996) Ca2+ responses to ATP via purinoceptors in the early embryonic chick retina. J Physiol 493:855–863

Sugioka M, Zhou W-L, Hofmann H-D, Yamashita M (1999) Involvement of P2 purinoceptors in the regulation of DNA synthesis in the neural retina of chick embryo. Int J Dev Neurosci 17:135–144

Szondy Z (1994) Adenosine stimulates DNA fragmentation in human thymocytes by Ca^{2+}-mediated mechanisms. Biochem J 304:877–885

Szondy Z (1995) The 2-chlorodeoxyadenosine-induced cell death signalling pathway in human thymocytes is different from that induced by 2-chloroadenosine. Biochem J 311:585–588

Tanaka Y, Yoshihara K, Tsuyuki M, Kamiya T (1994) Apoptosis induced by adenosine in human leukemia HL-60 cells. Exp Cell Res 213:242–252

Tenneti L, Gibbons SJ, Talamo BR (1998) Expression and trans-synaptic regulation of P2X4 and P2Z receptors for extracellular ATP in parotid acinar cells: effects of parasympathetic denervation. J Biol Chem 273:26799–26808

Thornberry NA, Bull HG, Calaycay JR, Chapman KT, Howard AD, Kostura MJ, Miller DK, Molineau SM, Weidner JR, Aunins J, et al. (1992) A novel heterodimeric cystein protease is required for interleukin-1 bp processing in monocytes. Nature 356:768–774

Tidball JG, Albrecht DE, Lokensgard BE, Spencer MJ (1995) Apoptosis precedes necrosis of dystrophin-deficient muscle. J Cell Sci 108:2197–2204

Tocco G, Freire-Moar J, Schreiber SS, Sakhi SH, Aisen PS, Pasinetti GM (1997) Maturational regulation and regional induction of cyclooxygenase-2 in rat brain: implications for Alzheimer's disease. Exp Neurol 144:339–349

Tornquist K, Ekokoski E, Dugue B (1996) Purinergic agonist ATP is a comitogen in thyroid FRTL-5 cells. J Cell Physiol 166:241–248

Traversa U, Florio T, Virgilio A, Caciagli F, Rathbone MP (2000) Are neuroprotective

effects of guanosine mediated by guanosine receptors? Soc Neurosci Abstr 26: 148.15
Turner JT, Park M, Camden JM, Weisman GA (1998) Salivary gland nucleotide receptors. Changes in expression and activity related to development and tissue damage. Ann NY Acad Sci 842:70–75
Van Biesen T, Hawes BE, Raymond JR, Luttrell LM, Koch WJ, Lefkowitz RJ (1996) Go-protein alpha-subunits activate mitogen-activated protein kinase via a novel protein kinase C-dependent mechanism. J Biol Chem 271:1266–1269
Van Coevorden A, Boeynaems J-M (1984) Physiological concentrations of ADP stimulate the release of prostacyclin from bovine aortic endothelial cells. Prostaglandins 27:615–626
Vaes G (1998) Cellular biology and biochemical mechanism of bone resorption. Clin Orthop 231:239–271
Wang D-J, Huang N-N, Heppel LA (1990) Extracellular ATP shows synergistic enhancement of DNA synthesis when combined with agents that are active in wound healing or as neurotransmitters. Biochem Biophys Res Commun 166:251–258
Wang D-J, Huang N-N, Heppel LA (1992) Extracellular ATP and ADP stimulate proliferation of porcine aortic smooth muscle cells. J Cell Physiol 153:221–233
Wang Y, Simonson MS, Pouyssegur J, Dunn MJ (1992) Endothelin rapidly stimulates mitogen-activated protein kinase activity in rat mesangial cells. Biochem J 287:589–594
Webb TE, Feolde E, Vigne P, Neary JT, Runbert A, Frelin C, Barnard EA (1996) The P2Y purinoceptor in rat brain microvascular endothelial cells couples to inhibition of adenylate cyclase. Br J Pharmacol 119:1385–1392
Webb TE, Simon J, Barnard EA (1998) Regional distribution of [^{35}S]2′-deoxy 5′-O-(1-thio)ATP binding sites and the $P2Y_1$ messenger RNA within the chick brain. Neurosci 84:825–837
Weisman GA, Gonzalez FA, Eeb L, Garrad RC, Turner JT (1997) The cloning and expression of G protein-coupled P2Y nucleotide receptors. In: Turner, JT, Weisman GA, Fedan JS (eds) The P2 nucleotide receptors, Humana Press, Totowa, NJ, pp 63–79
Weisman GA, Lustig KD, Lane E, Huang N-N, Belzer H, Friedberg I (1988) Growth inhibition of transformed mouse fibroblasts by adenine nucleotides occurs via generation of extracellular adenosine. J Biol Chem 263:12367–12372
Wilden PA, Agazie YM, Kaufman R, Halenda SP (1998) ATP-stimulated smooth muscle cell proliferation requires independent ERK and PI3 K signaling pathways. Am J Physiol 275:H1209–H1215
Wood AJJ (1994) New purines analogues for the treatment of hairy-cell leukemia. N Engl J Med 330:691–697
Wu D, Yang C-M, Lau Y-T, Chen J-C (1998) Retraction of publication. Mol Pharmacol 53:346
Xie M, Jacobs LS, Dubyak GR (1991) Regulation of phospholipase D and primary granule secretion by P2-purinergic- and chemotactic peptide-receptor agonists is induced during granulocytic differentiation of HL-60 cells. J Clin Invest 88:45–54
Yakushi Y, Watanabe A, Murayama T, Nomura Y (1996) P2 purinoceptor-mediated stimulation of adenylyl cyclase in PC12 cells. Eur J Pharmacol 314:243–248
Yeung CH (1986) Temporary inhibition of the initiation of motility of demembranated hamster sperm by high concentrations of ATP. Int J Androl 9:359–370
Yu H, Ferrier J (1993) Osteoblast-like cells have a variable mixed population of purino/nucleotide receptors. FEBS Lett 328:209–214
Yu S-, Chen S-F, Lau Y-T, Yang C-M, Chen J-C (1996) Mechanism of extracellular ATP-induced proliferation of vascular smooth muscle cells. Mol Pharmacol 50:1000–1009
Zaidi M, Alam AS, Shankar CS, Bax BE, Bax CM, Moonga BS, Bevis PJ, Stevens C,

Blake DR, Pazianas M (1993) Cellular biology of bone resorption. Biol Rev Cambridge Philosophic Soc 68:197–264

Zheng J-S, Boluyt MO, Long X, O'Neill L, Lakatta EG, Crow MT (1996) Extracellular ATP inhibits adrenergic agonist-induced hypertrophy of neonatal cardiac myocytes. Circ Res 78:525–535

Zheng J-S, Boluyt MO, O'Neill L, Crow MT, Lakatta EG (1994) Extracellular ATP induces immediate early gene expression but not cellular hypertrophy in neonatal cardiac myocytes. Circ Res 74:1034–1041

Zheng J-S, O'Neill L, Long X, Webb TE, Barnard EA, Lakatta EO, Boluyt MO (1998) Stimulation of P2Y receptors activates *c-fos* gene expression and inhibits DNA synthesis in cultured cardiac fibroblasts. Cardio Res 37:718–728

Ziada AMAR, Hudlicka O, Tyler KR, Wright AJA (1984). The effect of long-term vasodilatation on capillary growth and performance in rabbit heart and skeletal muscle. Cardiovasc Res 18:724–732

CHAPTER 12
Nucleoside and Nucleotide Transmission in Sensory Systems

G.D. HOUSLEY

A. Introduction

Nucleotide and nucleoside signaling occurs via autocrine, paracrine, and neurotransmission modes, mediated by P2 (nucleotide) and P1 (adenosine A_1–A_3) receptor transduction paths, dynamically linked to purine and pyrimidine release (ATP, associated hydrolysis products, adenosine, diadenosine polyphosphates (Ap_nA), and pyrimidines (particularly UTP), and extracellular hydrolysis, phosphorylation and reuptake pathways.

This chapter seeks to integrate data from a number of studies investigating nucleotide and nucleoside signaling in sensory systems. Analysis of nucleotide signaling associated with pain pathways is described in detail in SALTER and SOLLEVI (Chap. 13, this volume). Nucleoside actions (adenosine) are diverse and well established in the visual system, whereas elements of nucleotide signaling are only becoming apparent. Studies in the inner ear have demonstrated diverse roles for extracellular purine and pyrimidine nucleotide-mediated transmission in sensory systems. A comparison of purinergic and pyrimidinergic signaling in these organs, and data from other sensory systems, show interesting parallels in physiologically significant features of sensory function. In many cases the experimental data provide new avenues for treatment of pathophysiology.

I. Sensory Transmission Mediated by Nucleotides and Nucleosides

Early reports of purinergic involvement in afferent sensory pathways included release of ATP following antidromic stimulation of the sensory innervation to the rabbit ear vascular bed associated with vasodilatation (HOLTON 1959). This likely involves afferent axon colaterals having purinergic vasomotor feedback action (BURNSTOCK 1977, 1993; MAGGI and MELI 1988). Adenosine may have a similar function in the mesenteric arterial bed (RELEVIC et al. 1996). This compliments the purinergic autonomic vasomotor control process (BURNSTOCK 1997).

Fast nucleotide-mediated afferent neurotransmission occurs via ATP-gated non-selective cation channels (BEAN 1990; KHAKH et al. 1995; KRISHTAL et al. 1988, 1983). These channels assemble as multimers of different P2X receptor subtypes, $P2X_2$ and $P2X_3$ heteromultimeric assembly providing the functional characteristics of purinergic responses in dorsal root ganglion nociceptive fibers (BURNSTOCK and WOOD 1996; CHEN et al. 1995; LEWIS et al. 1995). In the dorsal root ganglia $P2X_3$ receptor mRNA expression may be biased to nociceptive afferent transmission as it is not among the P2X receptor mRNA subtypes expressed by muscle stretch receptor primary afferent neurones (COOK et al. 1997). Sensory afferents expressing the $P2X_3$ receptor have synaptic input localized to lamina II of the dorsal horn (BRADBURY et al. 1998; VULCHANOVA et al. 1998), while $P2X_2$ receptor immunostaining is more distributed in this region (KANJHAN et al. 1999). Data from frog mechanosensory fibers indicates that P2Y1 receptor mediated neuromodulation of mechanotransduction also occurs (NAKAMURA and STRITTMATTER 1996).

In the rabbit urinary bladder, altered hydrostatic pressure leads to release of ATP from the serosal surface of the urothelial layer, providing a novel physiological sensory pathway for signaling bladder distension (FERGUSON et al. 1997). It appears likely that amiloride sensitive mechanoreceptor channels on the mucosal surface of the epithelial cells respond to pressure changes, resulting in a release of ATP to the sensory nerve fibers which form a closely apposed meshwork close to the serosal surface of the urothelia (GABELLA and DAVIS 1998). A non-neural element of this sensory system may involve released ATP acting directly on non-neural P2Y receptors on smooth muscle fibers to increase bladder compliance (McMURRAY et al. 1998), independent of the purinergic motor innervation which acts via P2X receptors to stimulate contraction.

Nociception by DRG neurones, as well as chemosensory transduction by vagal epicardial afferent neurones, occurs by ATP acting directly as a stimulus signal (ARMOUR et al. 1994; COOK et al. 1997). P2 receptor mediated signaling is also implicated in vagal afferent input to central cardiorespiratory reflexes (McQUEEN et al. 1998). Proprioceptive primary afferent neurones located within the mesencephalic trigeminal nucleus of the CNS express a number of P2X receptor mRNA subtypes (COLLO et al. 1996; KHAKH et al. 1997), including developmentally linked expression of $P2X_3$ (KIDD et al. 1998). Nucleotide afferent transmission into the CNS clearly involves pre-synaptic receptor actions which modulate glutamate-mediated synaptic transmission (GU and MACDERMOTT 1997; LI and PERL 1995; LI et al. 1998). Post-synaptic actions of nucleotides–as well as inhibitory actions via adenosine arising from nucleotidase activity, and nucleotide receptor based glia–neurone signaling–are also functionally significant within the CNS regions receiving purinergic sensory afferent input (JAHR and JESSELL 1983; LI and PERL 1995; SALT and HILL 1983; SALTER et al. 1993).

B. Olfactory and Gustatory Systems

Sensory involvement of nucleotide transmission is also apparent in the olfactory and gustatory systems. ATP, and particularly the breakdown products of ATP (5'-AMP) act as stimuli for olfactory transduction in the model spiny lobster antennule (Derby et al. 1996; Gleeson et al. 1992) and catfish olfactory neurones (Kang and Caprio 1995). Centrally, $P2X_2$ and $P2X_4$ receptor expression is significant in the mammalian olfactory bulb, the termination site of the olfactory neurones (Bo et al. 1995; Kanjhan et al. 1999; Kidd et al. 1995; Le et al. 1998). Evidence for nucleoside-involvement in taste arises from the association of the adenosine receptor antagonist caffeine with taste sensation (Chou 1992), although the psychophysical data are equivocal (Brosvic and Rowe 1992). Both adenosine and ATP have been proposed as peripheral gustatory transmitters based on adenosine storage in taste buds (Borisy et al. 1993), and $P2X_2$ and $P2X_3$ receptor immunolabeling associated with the afferent fibers innervating the taste buds of rat circumvillate and fungiform papillae (Bo et al. 1999). ATP has also been implicated as a central gustatory neurotransmitter on the basis of extensive expression of the $P2X_2$ receptor in the rostral region of the rat nucleus of the tractus solitarius (NTS) (Kanjhan et al. 1999). Taste stimulation in mouse taste receptor cells is triggered by P2Y receptor activation by ATP, suggesting that the ATP content of food may be a taste-related signal (Kim et al. 2000).

C. Visual System

Extracellular nucleoside and nucleotide signaling has a role in a number of key physiological processes associated with vision. These include neural roles within the retina, as well as humoral affects on retinal blood flow and intraocular pressure via action at the retinal choroid and ciliary body. In the case of adenosine, many of the physiological effects appear "protective" in nature. Until recently little attention was paid to the potential for nucleotide signaling, the focus being the roles adenosine played as a modulator of GABAergic and cholinergic neurotransmission, and actions on chloride secretion within the retinal pigmented epithelium (Maruiwa et al. 1995). Experiments have now shown that adenosine has roles in a number of additional processes, including retinal pigmented epithelium (RPE)-mediated rod outer segment phagocytosis (Gregory et al. 1994), retinal choroidal bloodflow, regulation of intraocular pressure via the ciliary body and regulation of corneal hydration via decreased endothelium permeability (Riley et al. 1996, 1998), angiogenesis mediated by upregulation of vascular endothelial growth factor (VEGF) expression, and upregulation of glucose transport.

Release of adenosine from retinal tissues exposed to hypoxia has been demonstrated in many studies (Blazynski and Perez 1991, 1992). In addition to its protective role on retinal neurotransmission, adenosine also acts at sites maintaining the fluid and electrolyte balances in the eye. However hypoxia-induced release of purine nucleotides and adenosine formation is associated with metabolism (for example xanthine production via adenosine deaminase

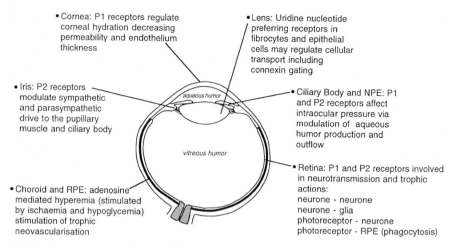

Fig. 1. Sites of nucleotide and nucleoside signaling in the eye. (Drawn by Nick Raybould)

activity) which may be significant as a precursor for harmful oxygen free radical formation (ROTH et al. 1997a,b). Thus nucleotide and nucleoside release, degradation, and re-uptake processes are balanced by additional metabolism of adenosine or AMP (BLAZYNSKI and PEREZ 1992). This diversity of signaling targets is outlined in Fig. 1.

I. Adenosine Actions on Retinal Neurotransmission

There is considerable evidence that purines have a role in retinal neurotransmission. Earlier reviews (BLAZYNSKI and PEREZ 1991, 1992; THORNE and HOUSLEY 1996) summarized evidence for key indicators of adenosine neurotransmitter action in this tissue, namely storage and release of adenosine and adenine nucleotides, signal termination via metabolism (adenosine deaminase) and adenosine transporter-mediated re-uptake, and action via A_1 and A_2 receptors. More recently, in vivo measurements have confirmed ischemia-induced increases of adenosine and its metabolites in the rat retina (ROTH et al. 1996). Purine release is partially $[Ca^{2+}]$-dependent and can be elicited by K^+-induced depolarization, excitatory neurotransmission, or cessation of flickering light (BLAZYNSKI and PEREZ 1992). Physiological data supports adenosine actions based on electrophysiological recording of adenosine agonist and antagonist responses in RPE, inner retina and ganglion cells (BLAZYNSKI et al. 1989; DAWIS and NIEMEYER 1987; NIEMEYER and FRÜH 1989). In keeping with the inhibition of excitatory transmitter release by adenosine in the CNS, chick and rabbit retina amacrine cells exhibit inhibition of K^+- and light-evoked release of ACh (A_1 receptor mediated) (BLAZYNSKI and PEREZ 1992; BLAZYNSKI et al. 1992; PEREZ and EHINGER 1989a; SANTOS et al. 1998). A P2 receptor-mediated mechanism was also apparent (see below). Both adenosine and purine nucleotides were also released by chick amacrine-like neurone cultures, with ATP degradation by ectonucleotidases contributing to the P1 receptor-mediated inhibition of ACh release (SANTOS et al. 1998). This inhibition is reciprocal (PEREZ and EHINGER 1989b). Similarly, in recognition of the inhibitory "neuroprotective" actions of adenosine on CNS synaptic regions, interactions with other neurotransmitter systems arising from A_1 receptors present in rabbit inner plexiform layer (IPL) are evident (BLAZYNSKI and PEREZ 1992; CROSSON et al. 1994) and likely contribute to the maintenance of β-wave activity after rat eye ischemic-reperfusion injury when adenosine deaminase was inhibited with erythro-9-(2-hydroxyl-3-nonyl)adenine (EHNA) or with addition of the A_1 receptor agonist R-N6-(2-phenylisopropyl)adenosine (R-PIA) (LARSEN and OSBORNE 1996).

Identification of the sites of P1 receptor expression in the retina have been crucial for determining likely actions. A_1 receptor responses are mediated by inhibition of adenylate cyclase via G_i, while A_2 receptor effects are G_s coupled. Inhibition of adenylate cyclase in rabbit retinas by 1–10nmol/l cyclohexyladenosine (CHA) and PIA has been reported (BLAZYNSKI 1987) as has activation of adenylate cyclase in rabbit retinas by much higher concentrations of adenosine, compatible with a low affinity A_2-like receptor (e.g., A_{2B}) acting independently of dopamine or norepinephrine pathways (BLAZYNSKI 1990; BLAZYNSKI et al. 1986; BLAZYNSKI and PEREZ 1992).

Autoradiographic analysis of adenosine receptor binding in rabbit and mouse retina shows distinct demarcation between A_1 receptors (detected using the agonists [^{3}H]PIA and [^{3}H]CHA) and A_2-like receptors detected using the mixed A_1-A_2 receptor agonist [^{3}H] ethycarboxamidoadenosine ([^{3}H]NECA). These data indicate A_1 receptor expression within the inner retina, especially the inner nuclear layer (INL) and IPL, whereas the A_2-like receptor ([^{3}H]NECA) binding was confined to the RPE and outer and inner photoreceptor segments (BLAZYNSKI 1990). Binding studies in isolated bovine rod outer segment using 2-p-(2-[^{3}H]-carboxyethyl)phenylamino-5′-N-ethylcarboxamido adenosine (CGS 21680) identified high-affinity A_{2a} receptors (K_D = 9.4nmol/l), in addition to low affinity A_{2B} receptors labeled with [^{3}H]NECA (MCINTOSH and BLAZYNSKI 1994). The A_{2B} receptors were the only adenosine receptor type identified in bovine retina NPE (BLAZYNSKI 1993).

Cloning studies have now identified four classes of adenosine receptor: A_1, A_{2A}, A_{2B}, A_3; mRNAs for all but the last subtype have been detected and localized in rat retina and ciliary body (KVANTA et al. 1997). These molecular data are in reasonable agreement with the ligand binding studies in so far as A_1 and A_{2A} receptor mRNA expression was biased towards the inner and outer retina respectively. A_1 receptor mRNA expression was highest in the INL and ganglion cell layer (GCL; approximately 50% of the neurones) while A_{2A} receptor expression was present in most INL cells, but limited in the GCL and ONL, NPE, and choriocapillaries. A_{2B} receptor mRNA expression was only detected in the epithelial cells of the ciliary body processes along with A_{2A} and

A_1 receptor mRNA. The localization of the A_{2A} receptors in the inner retina raised the possibility of interactions with the A_1 receptors, and also for an A_{2A} receptor mediated enhancement of transmitter release (SEBASTIAO and RIBEIRO 1996).

II. Nucleosides and Nucleotides in Retinal Ontogeny

Adenosine likely contributes to signaling associated with neural ontogeny in the retina (PAES de CARVALHO 1990; PAES DE CARVALHO and DE MELLO 1982, 1985). In the chick retina, A_1 receptor expression (detected by autoradiographic analysis of R-[^{3}H]PIA binding) and adenosine immunoreactivity were absent at E8 when retinal cell migration is becoming established, but were detected by E12. A_1 receptor expression was focused within the plexiform regions, while the adenosine-immunoreactivity was present in the GCL, INL, and photoreceptor cell layer, with increasing intensity by E18. The latter supports predetermination of adenosine-storing neural elements (PAES DE CARVALHO et al. 1992). Adenosine storage and uptake processes were identified in approximately a quarter of the embryonic retinal ganglion cells, whereas most photoreceptors were positively labeled (PAES DE CARVALHO et al. 1990). Adenosine uptake sites, labeled using [^{3}H]nitrobenzylthioinosine (NBI) were diffusely distributed at E8, with more focal expression in the plexiform layers from E12 (PAES DE CARVALHO et al. 1992).

In an analogous manner to the recognized cholinergic trophic signaling during ontogeny (WESSLER et al. 1998), extracellular nucleotide signaling also seems likely to contribute to retinal development. In the early developing chick retina both mucarinic receptor and P2Y receptor activation of thapsigargin-sensitive Ca^{2+} stores have been detected. Both systems utilize the phospholipase C (PLCβ)–inositol trisphosphate (IP$_3$) pathway, although the P2Y receptor pathway is pertussis toxin (PTX)-sensitive, whereas the muscarinic acetylcholine receptor (mAChR) pathway is not (SAKAKI et al. 1996). ATP-induced [Ca^{2+}]-mobilization was most pronounced at E3 and declined towards E13 (SUGIOKA et al. 1996), indicative of function during early differentiation and mobilization of the neural plexus. The response was independent of extracellular Ca^{2+} (precluding P2X receptor involvement) and the pharmacological profile UTP > ATP >> 2methylthioATP (2MeSATP) was compatible with a P2 receptor classification.

III. Nucleotide Signaling in the Retina

Physiological data strongly support roles for extracellular ATP as a transmission signal used by retinal neurons and glial cells. In the rabbit retina, endogenous release of ATP with ACh during flickering light stimulus was blocked by the P2 receptor antagonist pyridoxalphosphate-6-azophenyl-2',4'-disulphonic acid (PPADS). ATP acted to inhibit further transmitter release via a strychnine-sensitive glycinergic negative feedback (NEAL and CUNNINGHAM 1994).

Inhibition of GABA uptake, putatively attributed to the rat retinal glial (Muller) cells, was elicited using ATP analogs, including α,β-methyleneATP (α,βMeATP) and 2MeSATP (NEAL et al. 1998) and blocked by PPADS. Adenosine and the P2Y receptor selective agonist, 2-chloroATP had no effect. These data were supported by fura-2 Ca^{2+} imaging studies of cultured rabbit retinal Muller cells (LIU and WAKAKURA 1998) which also provide evidence for P2Y receptor and P1 (A_2-like receptor) receptor expression. Putative P2Y receptor-mediated Ca^{2+} release has also been reported in tiger salamander retinal Muller cells (KEIRSTEAD and MILLER 1997), while preparations of astrocytes and Muller cells from rat retina confirm this signaling pathway, showing Ca^{2+} signal coupling between the two types of glial cells which was independent of extracellular Ca^{2+}, and blocked by PPADS, intracellular heparin (IP_3-gated Ca^{2+} channel blocker) and thapsigargin [sarco/endoplasmic reticulum Ca^{2+}-ATPase (SERCA) inhibitor]. This observation has been proposed as a model of extraneuronal signaling in the CNS (NEWMAN and ZAHS 1997). Rabbit retina oligodendrocytes exhibit $P2Y_2$-like responses (UTP = ATP > ADP, adenosine ineffective) which were largely confined to the processes of the cells rather than their soma (KIRISCHUK et al. 1995).

The physiological studies on nucleotide signaling in the retina are supported by molecular analysis showing diverse expression of P2X receptor subunits in rat retinal tissues. Of the seven known P2X receptor subtypes, only $P2X_1$ and $P2X_6$ receptor mRNAs were not detected by RT-PCR (BRÄNDLE et al. 1998a,b; GREENWOOD et al. 1997). $P2X_1$ receptor mRNA was reported in retina by Northern Blot analysis (Valera et al. 1994) but this may not be as reliable as RT-PCR analysis. The heterogeneity of P2X receptor subunit expression, including evidence for splice variants of the $P2X_2$ subtype (BRÄNDLE et al. 1998a) suggest that purinergic neurotransmission is likely to have complexity arising from the different kinetics and desensitization rates of the ion channels assembled from these subunits. The detection of $P2X_7$ receptor expression (BRÄNDLE et al. 1998b) may be particularly significant given the likely role of the channels formed from this subunit in development, via apoptotic processes (ABBRACCHIO and BURNSTOCK 1998). The expression of the $P2X_2$ receptor subunit mRNA has been localized using in situ RT-PCR – in situ hybridization and immunocytochemistry (GREENWOOD et al. 1997). These data show extensive expression in the photoreceptors, inner and outer nuclear layers, and the retinal ganglion neurones. This seems to parallel the data from the cochlea, where sensory cells, supporting cells, and primary auditory neurones all express the $P2X_2$ receptor subunit, and where both humoral actions and neurotransmission processes have been linked to nucleotide signaling. Localization of $P2X_2$ receptor immunoreactivity to the outer segments of the photoreceptors supports an additional non-neural signaling role for extracellular ATP in the retina (GREENWOOD et al. 1997), possibly associated with recycling of photopigment at the RPE though phagocytosis of distal photoreceptor outer segment, and of significance with respect to retinitis pigmentosa. Recent whole-cell recordings from cultured neonatal rat retinal ganglion neurones has confirmed functional expression of ATP-gated channels in approximately 65% of the cell population (TASCHENBERGER et al. 1999). Interestingly the ATP-sensitive multipolar neurones were glutamatergic and corresponded to the sub-population delineated by anti-Thy1.1 immmunolabeling, used as a marker of retinal ganglion neurones rather than amacrine neurones. Amacrine cells were unresponsive to P2X receptor agonists. In the retinal ganglion neurones, variability in P2X receptor agonist sensitivity (α,βMeATP and ADP), and differing desensitization rates, support the likely heterogeneous nature of P2X receptor subunit expression. These data are similar to other analyses of P2X receptor expression in cochlear primary afferent neurones (SALIH et al., 1999; see below).

IV. Nucleoside and Nucleotide Signaling in Ciliary Ganglion, Ciliary Body, and Non-Pigmented Epithelium

The role of the ciliary body is to generate the aqueous humor and maintain intraocular pressure which underpins the dynamic optics of the eye. A number of studies have provided unequivocal evidence that adenosine and nucleotide signaling are key elements in these processes.

1. Regulation of Intraocular Pressure (IOP): Actions on the Ciliary Body

The aqueous humor is secreted from the stroma into the posterior chamber via the ciliary processes (COLE 1977), a bilayer of pigmented and non-pigmented epithelial cells. IOP arises from a balance between aqueous humor production from the NPE, and outflow resistance within the canals of Schlemm in the anterior chamber of the eye (CIVAN 1998). Glaucoma, a major cause of blindness through progressive optic nerve damage, is associated with increased ocular pressure. Given that glaucoma treatments are directed towards reducing ocular hypertension, the role of adenosine and nucleotide signaling in this process is clearly of considerable clinical importance.

Based on quinacrine fluorescence labeling, ATP is stored in both the pigmented (PE) and NPE layers of the rabbit ciliary body (MITCHELL et al. 1998). Hypo-osmotic-induced release of ATP has been detected in transformed bovine PE and NPE ciliary epithelial cells using the luciferin-luciferase assay. The ATP release occurred via two processes; a 5-nitro-2-(3-phenylpropylamino)benzoic acid (NPPB)-sensitive Cl$^-$ channel pathway, and putative Ca^{2+}-dependent exocitosis sensitive to KN-62, a Ca^{2+}/calmodulin kinase II blocker (MITCHELL et al. 1998). Interestingly, KN-62 is a selective antagonist of the murine and human $P2X_7$ receptor (GARGETT and WILEY 1997; HUMPHREYS et al. 1998). The $P2X_7$ ($P2_Z$) receptor has been associated with ATP-induced cell lysis (MACKENZIE et al. 1999) and may therefore have been involved in inducing a positive-feedback release of ATP in this tissue. In both NPE and PE, adenosine production arose at least in part from ATP hydrolysis via ectonucleotidase-activity. Adenosine levels at 321 nmol/l in the aqueous humor and 210nmol/l in the vitreous humor of porcine eyes (HOWARD et al. 1998) likely represent physiologically significant stimulus levels given the IC_{50} values of A_1 and A_2 receptors. Thus adenosine and ATP are available for autocrine and paracrine stimulation of P1 and P2 receptors respectively in ciliary body epithelium.

Early evidence for P1 and P2 receptor expression includes the identification of CPA inhibition (IC_{50} = 1nmol/l) of forskolin-simulated adenylate cyclase activity in cultures of human ciliary NPE and bovine ciliary epithelial cells (WAX et al. 1993), indicating A_1 receptor action. Low affinity (A_2-like receptor) stimulation of adenylate cyclase was also detected using CGS 21680, as well as P2Y receptor activity based on the ability of UTP and ATP to induce IP_3 production (2MeSATP and α,βMeATP were ineffective). The latter is compatible with a $P2Y_2$ receptor classification and is supported by demonstration of rapid nucleotide-activated intracellular Ca^{2+} mobilization (UTP = ATP > ADP > AMP) in cultured non-transformed bovine ciliary epithelial cells under Ca^{2+}-free conditions (SHAHIDULLAH and WILSON 1997).

As described previously, A_1, A_{2A}, and A_{2B} receptor mRNA expression was detected in the rat ciliary body, and particularly in epithelial cells of the associated ciliary processes (KVANTA et al. 1997). In normotensive monkeys, application of PIA or CHA (A_1 receptor selective agonists) elicited an initial hypertensive pressure response which was 3,7-dimethyl-1-propargylxanthine-sensitive (DMPX; A_2-like receptor antagonist), followed by a prolonged (DMPX-insensitive) A_1 receptor-mediated hypotensive response arising from increased outflow, but not aqueous humor production (TIAN et al. 1997). Both hypertensive (A_2 receptor) and hypotensive (A_1 receptor) adenosine actions have been established in rabbit and cat iris-ciliary body, with the former arising to some extent from reduced aqueous humor flow in these species (CROSSON 1992, 1995; CROSSON and GRAY 1994, 1996). The effects are independent of sympathetic neural drive. An adenosine-induced increase in aqueous flow arises from enhanced ciliary body NPE transport, measured as increased short-circuit current of isolated rabbit iris-ciliary body; isosmotic cell shrinkage and increased whole cell currents of cultured human NPE cells; activated via P1 receptors (ATP and UTP insensitive) (CARRE et al. 1997). The effect is A_2-like receptor mediated, and occurs via increased Cl^- conductance. The coupling of an A_2-like receptor to increased Cl^- channel activation may occur via an intracellular PLC-coupled Ca^{2+} signaling pathway (CARRE et al. 1997; MITCHELL et al. 1997). In support of this, adenosine produces a synergistic release of Ca^{2+} stores in association with mAChR activation (FARAHBAKHSH and CILLUFFO 1997). Despite the insensitivity of the Cl^- transport mechanism in human NPE cultures to nucleotides, P2Y receptor signaling is apparent. Cultured bovine ciliary epithelium demonstrated mobilization of stored Ca^{2+} with order of potency UTP = ATP > ADP > AMP; adenosine, α,βMeATP, 2MeSATP, adrenaline, noradrenaline (NE), acetylcholine, and carbachol were ineffective (SHAHIDULLAH and WILSON 1997). These data suggest that Ca^{2+} signaling in ciliary epithelium is highly compartmentalized, with limited cross-talk between receptors, even those utilizing common activation processes. β,γ-MeATP has greater efficacy in reducing IOP (40%) than the muscarinic agonist, pilocarpine (25%) or the β-adrenoceptor blocker, timolol (30%; PINTOR et al. 2000).

2. Neural Signaling in the Ciliary Body and Iris

Nucleosides and nucleotides are associated with neural input to the ciliary body and iris meridional muscle controlling pupil dilation. Adenosine acts (via A1 receptors) to inhibit NE release from the sympathetic innervation in electrically stimulated rat and rabbit iris and ciliary body preparations (CROSSON and GRAY 1997; FUDER et al. 1992). Evidence for an associated ATP-mediated inhibition of NE release by P2Y receptors is also reported, although the contribution of ectonucleotidase activity in generating adenosine cannot be ruled out (FUDER et al. 1992; FUDER and MUTH 1993).

At the parasympathetic synapses within the ciliary ganglion, a similar mechanism occurs via adenosine autoreceptors that inhibit pre-synaptic Ω conotoxin GVIA-sensitive (N-type) Ca^{2+} channel activation, and hence reduce ACh release (SATO et al. 1993; YAWO and CHUHMA 1993). The process is a PTX-sensitive G_i/G_o-coupled A_1 receptor mechanism, and is most apparent in the rabbit ciliary body preparation at high stimulus frequencies. P2X receptor mediated pre- and post-synaptic modulation of cholinergic parasympathetic drive to the ciliary body and iris is also evident. The ciliary ganglion neurones are innervated via giant calyx endings of pre-ganglionic parasympathetic fibers derived from the accessory oculomotor nucleus. Whole cell voltage-clamp of the calyx endings of E15 chick embryos demonstrated desensitizing ATP-gated inward currents which are inwardly rectifying and have a pharmacological profile compatible with a complex P2X receptor classification (ATP = α,βMeATP > 2MeSATP; no response was seen with β,γMeATP or adenosine; 100μmol/l), suggesting complex P2X receptor subunit expression. The responses were blocked by suramin. Single channel data indicated a 17pS conductance, close to the reported unitary conductance of the recombinant $P2X_1$ receptor (EVANS 1996). Post-synaptic P2X receptor expression is also present. A desensitizing ATP-gated non-selective cation conductance which was inwardly rectifying and exhibited a different purine pharmacology (ATP > 2MeSATP; α,βMeATP insensitive) was detected using perforated patch whole-cell recordings of E14 chick ciliary ganglion cells (ABE et al. 1995); the described properties fit P2X receptors rather than the P2Y receptor designation assigned by the authors. An associated imaging study demonstrated significant Na^+ and Ca^{2+} entry into the ciliary ganglion neurones through ATP-gated ion channels (SORIMACHI et al. 1995). This post-synaptic action is compatible with the co-transmission of ACh (acting via nAChR) and ATP at the ciliary ganglion synapse.

V. Adenosine Regulation of Ocular Blood Flow: Protection During Ischemia and Hypoglycemia

Adenosine is a potent vasodilator of arterioles and venules in the vasculature of the eye including retina, iris, ciliary body and choroid (BRAUNAGEL et al. 1988). Indeed 200mg of caffeine, taken orally, produced a 13% decrease in human macular blood flow despite a 9% increase in mean diastolic blood pressure (LOTFI and GRUNWALD 1991). Both acute and prolonged retinal ischemia produce elevations in adenosine and the metabolite hypoxanthine in the retina and choroid (ROTH et al. 1997a). The adenosine-mediated hyperemia appears significant during ischemia and post-ischemic reperfusion subsequent to transient optic hypertension (GIDDAY and PARK 1993; OSTWALD et al. 1997; PORTELLOS et al. 1995; ROTH 1995). In the pig model, A_2-like adenosine receptor action elicits vasodilatation (CROSSON et al. 1994; GIDDAY and PARK 1993) and retinal hypofusion after transient asphyxia (GIDDAY and ZHU 1998) is reversed by the adenosine transporter inhibitor 4-nitrobenzyl-6-thioinosine (NBTI), supporting endogenous adenosine action. Similarly block of adenosine uptake by NBTI increased arteriolar vasodilatation induced by hemorrhagic hypotension (GIDDAY and PARK 1993). This hyperemic response appears to be independent of the adenylate cyclase-cAMP pathway; instead it involves K_{ATP} channels (GIDDAY et al. 1996), and is independent of nitric oxide-induced vasodilatation, or activation or suppression of adenylate cyclase (determined using forskolin and dideoxyadenosine, respectively).

Acute hypoglycemia in piglets causes hyperemia in the retina. This increased blood flow is reduced by 80% by 8-phenyltheophylline (8PT), a nonspecific adenosine receptor antagonist (ZHU and GIDDAY 1996).

VI. Nucleotide Signaling in the Lens

P2Y receptor signaling is evident in cultured lens epithelial cells, where ATP and UTP activate a PLC-coupled mobilization of intracellular Ca^{2+} (CHURCHILL and LOUIS 1997; RIACH et al. 1995). Analysis of P2Y receptor mRNA expression in the rat lens identified both UTP preferring ($P2Y_2$) and UTP insensitive ($P2Y_1$) transcripts (MERRIMAN-SMITH et al. 1998). Interestingly both $P2Y_1$ and $P2Y_2$ mRNA transcripts were localized to the fibrocytes rather than the epithelial cells by in situ RT-PCR. This leaves open the possibility that additional UTP-preferring P2Y receptor isoforms are expressed by the epithelial cells. $P2Y_2$ receptor mRNA expression preceded $P2Y_1$ receptor expression developmentally insofar as the transcript occurs in the bow region of the lens, with fibrocyte migration towards the nucleus occurring over time. The physiological significance of P2Y receptor signaling in the lens fibrocytes has not been determined; however the associated PLC-coupled Ca^{2+} signaling pathway may play an important role in regulation of connexin gating. Interestingly, connexins have recently been identified as a regulatory element in the control of ATP release (COTRINA et al. 1998). Given that disruption of gap-junctions has been proposed as a trigger for onset of cataract formation (DONALDSON et al. 1997), future research may well look at the potential interaction between nucleotide signaling, connexon function in the lens, and indeed other ocular tissues, such as retinal astrocytes, where IP_3 and Ca^{2+} signals are likely to spread between cells.

VII. Trophic Actions of Adenosine

1. Vascular Endothelial Growth Factor

Age-related macular degeneration and diabetic retinopathy are associated with subretinal neovascularization. The angiogenesis is linked to proliferation of RPE cells, choroidal fibroblasts, and vascular endothelial cells. Adenosine has been shown to upregulate expression of vascular endothelial growth factor (VEGF) which is a potent stimulator of this angiogenesis, acting via tyrosine kinase receptors (XIA et al. 1996). Cultured bovine retinal capillary endothelial cells and vascular pericytes exhibit increased VEGF mRNA levels when exposed to hypoxia; this process, which was associated with increased cAMP levels, was blocked by the A_{2a} receptor antagonist 8-(3-chlorostyryl)caffeine (CSC) and adenosine degradation using adenosine deaminase (TAKAGI et al. 1996). A_2 receptor agonists mimicked the effect; A_1 receptor agonists and antagonists were ineffective. A study using post-mortem human eye tissue demonstrated that VEGF expression was inducible in choroidal fibroblasts, but not RPE cells, via protein kinase C (PKC) activation using phorbol esters (KVANTA 1995). Activation of the adenylate cyclase pathway using forskolin had no effect on VEGF mRNA expression levels. The effects of VEGF on angiogenesis appear to be part of a negative regulatory feedback process involving the extracellular matrix protein thrombospondin-1 (SUZUMA et al. 1999). Thrombospondin-1 inhibits angiogenesis, and endothelial cell proliferation. Its expression is promoted by VEGF and has the effect of inhibiting VEGF-induced endothelial cell proliferation. VEGF protein and mRNA levels, and the mRNA for the receptors VEGFR-1 and VEGFR-2, are widely distributed, with higher expression apparent in streptozotocin-induced diabetic rat retina (GILBERT et al. 1998). This increased expression of VEGF is also apparent in patients with diabetic retinopathy (BOULTON et al. 1998). Given the trophic influence of adenosine on VEGF expression, this may well contribute to the changes in retinal expression induced by diabetes, and provides a promising investigative path for clinical treatment.

2. Glucose Transport

Release of adenosine into the retinal vasculature has recently been associated with onset of diabetic retinopathy (TAKAGI et al. 1998). This condition is linked to cytosolic glucose loading in retinal vascular cells and appears to be preceded by retinal hypoxia. Hypoxia-induced upregulation of glucose transporter (GLUT1) expression in cultured bovine retinal endothelial cells is activated by an A_{2a} receptor coupled adenylate cyclase–cAMP pathway (TAKAGI et al. 1998), although adenosine-independent activation via alternative intracellular signaling pathways are also apparent.

D. Cochlear and Vestibular Systems

Subsequent to the report of significant inhibition of auditory nerve activity when ATP was introduced into the perilymphatic compartment of the cochlea (BOBBIN and THOMPSON 1978), ATP was found to activate non-selective cation channels and mobilize intracellular Ca^{2+} in isolated guinea-pig outer hair cells (OHC) (ASHMORE and OHMORI 1990; NAKAGAWA et al. 1990). This work raised the possibility that ATP was acting as a co-transmitter with ACh at the efferent innervation to the OHC, a hypothesis that became less likely with the electrophysiological localization of the ATP-gated ion channels to the apical (endolymphatic surface) of these cells (HOUSLEY et al. 1992). This work and further in vivo and in vitro analysis (e.g., AUBERT et al. 1994; DULON et al. 1991; HOUSLEY et al. 1993; IKEDA et al. 1995; KUJAWA et al. 1994a,b; LIU et al. 1995; MUÑOZ et al. 1995; NIEDZIELSKI and SCHACHT 1992; RENNIE and ASHMORE 1993) raised the possibility that extracellular nucleotides were extensively involved in sound transduction in both types of hair cells, cochlear fluid homeostasis, and auditory and vestibular neurotransmission. Extracellular nucleotide signaling now features as a major element of these transmission and regulatory signaling systems in the inner ear (HOUSLEY 1998). Key elements of the roles for nucleotide signaling in the inner ear are summarized in this section (see Fig. 2), along with data on adenosine actions.

I. Modulation of Hair Cell Sound Transduction and Cochlear Micromechanics

1. Hair Cell Membrane Conductance

Both inner hair cells (IHC) and OHC express ATP-activated membrane conductances that show varying degrees of desensitization and inward rectification (HOUSLEY et al. 1992, 1998c; SUGASAWA et al. 1996c). The equipotency on these channels of 2MeSATP and ATP, insensitivity to α,βMeATP, and block by PPADS (HOUSLEY et al. 1998c, 1999; NAKAGAWA et al. 1990), indicate that the ATP-gated ion channels are assembled from $P2X_2$ receptor subunits, although molecular evidence suggests that more than one splice variant may be involved (HOUSLEY 1998). In addition, the IHC co-express a $K_{(Ca)}$ conductance which is activated via Ca^{2+} entry through the ATP-gated ion channels, leading to an apparent inactivation of the inward ATP-gated current (HOUSLEY et al. 1993). Thus OHC are depolarized by ATP while IHC membrane potential is transiently affected, but the conductance is increased. In both cases the magnitude of the ATP-activated conductance considerably outstrips the transducer current arising from stretch-activation of ion channels localized to the tip-link regions of the stereocilia, referred to as met, or mechano electrical

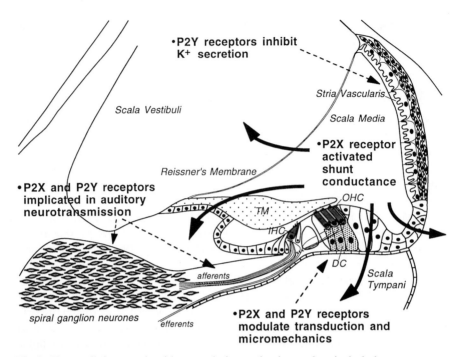

Fig. 2. Extracellular nucleotide regulation of electrochemical balance, sensory transduction, micromechanics and auditory neurotransmission. *DC*, Deiters' cells; *IHC*, inner hair cell; *OHC*, outer hair cells; *TM*, tectorial membrane. (Drawn by Nick Raybould)

transduction, channels (DENK et al. 1995; TORRE et al. 1995). Electrophysiological recordings, fluorescence imaging, and immunocytochemistry support a co-localization of ATP-gated ion channels in the same region of the hair cells (HOUSLEY et al. 1992, 1997, 1998a,c, 1999; MAMMANO et al. 1999; MOCKETT et al. 1994). Thus ATP in the endolymph likely activates an electrical shunt across the transducer region of the hair cells which would diminish the sound-generated receptor potential (HOUSLEY et al. 1992, 1998c). The significance of this effect is further supported by the observation that the OHC P2X receptor conductance increases in proportion to the tonotopic increase in background K^+ conductance towards the basal (high frequency encoding) region of the guinea-pig cochlea (RAYBOULD and HOUSLEY 1997). It seems likely that elevation in endolymphatic ATP in scala media, arising from noise or ischemia (THORNE et al. 1999), would therefore diminish the transduction potential and act to inhibit sound transduction.

2. Nucleotide Regulation of Cellular Mechanics

The effect of ATP-induced OHC depolarization on hearing function remains equivocal. OHC electromotility confers particular frequency selectivity and sensitivity to the mammalian ear (ASHMORE and KOLSTON 1994; DALLOS and EVANS 1995). Non-linearity in the gain of the OHC voltage-length converter means that minor ATP-activated depolarizations or ATP-induced volume loading may enhance the OHC cochlear amplification (HOUSLEY et al. 1995a). However, maximal P2X receptor activation, in association with global actions on P2X receptor conductances in other cells lining the endolymphatic fluid compartment (see next section), suggests that ATP released into scala media and activating the OHC ATP-gated ion channels may be protective in function due to a reduction in the electrochemical driving force and dampening of the OHC-derived mechanical stimulus to the IHC. Interestingly, iontophoresis of ATP into the scala media of the guinea-pig cochlea in vivo produced an enhanced electrically evoked otoacoustic emission derived from OHC electromotility (KIRK and YATES 1998). While this may support an enhanced OHC electromotility in vivo in the presence of ATP, the effect could possibly be an artifact of the increased current entering through ATP-gated ion channels which co-localize with the met channels, particularly since the current path for electrically evoked emissions occurs via the met channels in the absence of activation of P2X receptor channels (YATES and KIRK 1998).

Recently, an additional pathway for ATP-mediated alteration of OHC micromechanics has been identified, involving putative P2Y receptor-mediated Ca^{2+} signaling also localized to the stereocilia (MAMMANO et al. 1999). In this process, extracellular ATP elicits, via G-protein mediated IP production, release of Ca^{2+} stored in Hensen's body, a specialized endoplasmic reticulum localized at the base of the stereocilia bundle. This ATP-triggered elevation in Ca^{2+} is well placed to affect the stiffness of the OHC hair bundle and hence modulate sound transduction.

Cochlear mechanics are also likely altered by extracellular ATP-mediated changes in Deiters' cell stiffness (CHEN and BOBBIN 1998). Ultimately, altered mechanics equates to altered sound transduction, because the IHC stereocilia act as the principal forward transduction element integrating basilar membrane vibration derived from the active and passive elements of the organ of Corti. Deiters' cells, which support the OHC at the base and apex, and provide an additional site for nucleotide-mediated regulation of cochlear mechanics. P2X and P2Y receptors predominantly exposed to the perilymphatic milieu provide paths for increasing intracellular Ca^{2+}, a process which leads to shape and stiffness changes in these cells (CHEN and BOBBIN 1998; DULON 1995; DULON et al. 1993; HOUSLEY et al. 1999). Interestingly, the P2Y receptor mediated Ca^{2+} signaling in Deiters' cells is inhibited by a nitric oxide-cGMP-mediated process (MATSUNOBU and SCHACHT 1999). Thus the static tensioning of the basilar membrane, which the active OHC electromotility works upon, is likely to be a dynamic process.

Data supporting endogenous nucleotide regulation of cochlear mechanics in vivo comes from the PPADS-induced reduction in the decline in quadratic distortion product otoacoustic emission (DPOAE) evident during continuous sound stimulation at moderate intensity levels (CHEN et al. 1998).

Ectonucleotidase activity would clearly influence both hair cell and supporting cell functions by regulating ATP signal levels in both the enodolymphatic and perilymphatic compartments via expression of two isoforms of ecto-ATPase, and an ecto-ATPDase (VLAJKOVIC et al. 1998a,b, 1999).

3. Regulation of the Electrochemical Gradient across the Cochlear Partition

Introduction of ATP in scala media of the guinea-pig cochlea provides convincing evidence for a distributed ATP-gated shunt conductance in this compartment. Nanoliter quantities of ATP injected into the endolymph produced reversible reductions in endocochlear potential (EP) and cochlear microphonic (a measure of the transduction current passing through the met channels) (MUÑOZ et al. 1995, 1999). This observation has been repeated using iontophoretic application of ATP (KIRK and YATES 1998). Most recently, analysis of the cochlear partition resistance, obtained using constant current injections into scala media during ATP applications, showed a linear correlation between the magnitude of the ATP-induced falls in EP and the increase in conductance across the cochlear partition (HOUSLEY et al. 1997; THORNE et al. 1999). The endocochlear potential is a positive biopotential (up to approximately +90mV) in the rat cochlea endolymph, which, along with a high K^+ concentration (approximately 150mmol/l), provides the extracellular part of the electrochemical driving force for sound transduction (BOSHER and WARREN 1968). The intracellular component of the driving force for transduction comes from the negative resting potential of the hair cells (approximately –50mV to –85mV for IHC and OHC, respectively). Thus an ATP-induced drop in EP, along with an ATP-induced depolarization of the hair cell generates considerable compression of the driving force for the sound-evoked receptor current.

The substrate for a global fall in resistance in scala media in the presence of ATP is the activation of the P2X receptor conductance in all the cells lining this cochlear compartment which express ATP-gated ion channels. Analysis of P2X receptor expression outside of the hair cells revealed the complexity of non-neural signaling by ATP in the cochlea and suggests that the ATP-gated shunt conductance involves a significant fraction of the cells lining the endolymphatic compartment (HOUSLEY et al. 1998b). Selective removal of the hair cells using kanamycin had no significant effect on the endolympathic ATP-induced reduction in EP in the guinea-pig cochlea in vivo (THORNE et al. 1999). Riboprobe in situ hybridization analysis of $P2X_2$ receptor mRNA expression in the rat cochlea (HOUSLEY et al. 1998b; HOUSLEY and RYAN 1997) and immunocytochemistry in guinea-pig cochlea (HOUSLEY et al. 1999) showed high levels of expression in the sensory and supporting cells of the organ of Corti (cochlear sensory epithelium), and epithelial cells lining scala media, with the exception of the stria vascularis. $P2X_2$ receptor expression included Reissner's membrane, a membrane bilayer of significant surface area separating scala media from scala vestibuli, responsible for maintaining EP and the

K⁺ gradient (BOSHER and WARREN 1968; SCHMIDT and FERNANDEZ 1963) across these compartments. RT-PCR, immunocytochemistry and whole-cell patch clamp analysis all confirmed the expression of ATP-gated ion channels in the epithelial cells of this tissue (KING et al. 1998), demonstrating the activation of large non-selective cation conductances across the endolymphatic surface of these cells in situ.

II. Regulation of the K⁺ Flux into Endolymph

Nucleotide-based autocrine and paracrine regulation in scala media extends to the regulation of K⁺ secretion into the cochlear endolymph. A similar process is also apparent in the endolymphatic compartment of the vestibular system, mediated by vestibular dark cells. Studies using a micro-Üssing chamber and macropatches in gerbil strial tissue and vestibular dark cell epithelium have demonstrated tissue P_{2u} receptor expression, where UTP (and ATP) decrease the short-circuit current (LIU et al. 1995; MARCUS et al. 1998). The pharmacological profile of the nucleotide-mediated inhibition of the short circuit K⁺ current is compatible with the $P2Y_4$ type of uridine-sensitive receptor in that UTP = ATP >> Ap_4A; UDP was ineffective (MARCUS and LIU 1999). While $P2Y_4$ receptor mRNA expression has not been confirmed in cochlear or vestibular tissue, $P2Y_2$ mRNA (also a uridine-preferring subtype) was localized to this site in the guinea-pig cochlea (JÄRLEBARK et al. 1996). The pharmacology to date does not preclude $P2Y_2$ receptor involvement, and future studies may reveal a complexity of P2Y receptor expression associated with this regulation of K⁺-induced secretion from the marginal cells of the stria vascularis, and vestibular dark cells. Interestingly both the marginal cells and the vestibular dark cells provide an exception to the extensive $P2X_2$ receptor expression in the rat cochlear and vestibular systems (HOUSLEY et al. 1998b), thus demarcating the secretory tissue as a region independent of ATP-activated shunt conductance. This is logical if the role of the P2Y receptors is to close the K⁺ channels which are the conduit for K⁺ secretion into the endolymphatic compartment. I_{sK} knock-out mice are unable to generate transepithelial K⁺ secretion (VETTER et al. 1996). Presence of ATP-gated non-selective channels (P2X receptors) here would increase K⁺ secretion. The proposed mechanism for the $P2Y_4$ receptor mediated reduction in K⁺ secretion is via a PLC-protein kinase C (PKC) pathway, which acts to inhibit I_{sK}/K_vLQT1 channel activity in the gerbil and rat cochlea (MARCUS et al. 1996, 1997, 1998). Species differences have been noted, as the PKC phosphorylation site is non-functional in the guinea-pig I_{sK} regulatory subunit channel and hence PKC-mediated inhibition of K⁺ conductance does not occur; human I_{sK} channels conform to the rat and gerbil PKC consensus site and therefore probably exhibit this P2Y receptor inhibition of K⁺ secretion (MARCUS et al. 1998). Given the localization of ATP-like vesicular stores in the marginal cells of the guinea-pig stria vascularis (WHITE et al. 1995; THORNE et al. 1999), it is likely that the regulation of K⁺ secretion via the I_{sK} channels is under local autocrine or paracrine negative feedback (LIU et al. 1995). The stimulus for ATP release at this site remains to be determined.

III. Putative Role in Auditory Neurotransmission

ATP release from the guinea-pig organ of Corti is Ca^{2+}-dependent (WANGEMANN 1996). This is compatible with the possible role of ATP as a neurotransmitter in the peripheral auditory system. Further evidence to support this hypothesis includes the demonstration that a sub-population of rat spiral ganglion neurones (primary auditory neurones) express $P2X_2$ receptor mRNA (SALIH et al. 1998a). The expression of a number of P2X receptor subunit types has been detected in the soma of these neurones using immunocytochemistry (XIANG et al. 1999). Complex ATP-activated membrane conductances have been recorded in cultured neonatal rat spiral ganglion neurones, consistent with heteromultimeric P2X receptor subunit assembly (SALIH et al. 1999). Using fura-2 Ca^{2+} imaging, P2Y receptor mediated release of stored Ca^{2+} has been detected both on the soma and on processes of guinea-pig spiral ganglion neurones (CHO et al. 1997). This varies from the glutamate-mediated Ca^{2+} mobilization which is specific to the spiral ganglion neurone processes. The localization of $P2X_2$ receptor immunoreactivity to the inner spiral plexus, associated with the primary afferent innervation of neonatal rat inner hair cells (SALIH et al. 1999) further implicates ATP in hair cell neurotransmission. The data is most convincing in the guinea-pig cochlea where immunogold labeling has demonstrated $P2X_2$ receptor subunit expression localized to the postsynaptic thickenings of synapses between primary auditory neurones and both inner and outer hair cells (HOUSLEY et al. 1999). In the case of the inner hair cells, the data suggests that ATP acts at a subset of the inner hair cell synapses in conjunction with glutamate. The localization of $P2X_2$ receptor-like immunogold labeling at outer hair cell-type II spiral ganglion neuron synapses represents the first identification of a candidate fast neurotransmitter receptor for this class of auditory neurone. Interestingly these type II neurones represent less than 10% of the primary auditory neurone population and their physiological role in hearing remains enigmatic.

IV. Localization of P2X and P2Y Receptor Expression: A Role for Nucleotides in Cochlear Ontogeny

Extracellular nucleotides likely mediate trophic actions associated with development. Riboprobe-based in situ hybridization indicates that $P2X_2$ receptor expression reaches significant levels in the developing rat cochlea well before the functional onset of hearing (P8–P10) (HOUSLEY et al. 1998b). The presence of $P2X_2$ receptor mRNA expression in early embryonic stages of cochlear development suggests that extracellular ATP may have trophic actions. As early as P12, $P2X_2$ mRNA was detectable in the region of the developing spiral ganglion and the basal region of the cochlear duct. Expression increases in the lateral region around birth and migrates medially to encompass the sensory epithelium and the epithelial cells of the spiral limbus region. Peak expression occurs co-incidentally with the onset of hearing and then declines to a moderate level. $P2X_2$ receptor expression is also significant in the sensory and supporting epithelium of the vestibular end-organs, and likely acts in similar fashion there as in the cochlea to support the electrochemical gradient for transduction of motion and positional cues. Data for the expression of other P2X receptor subtypes in the inner ear is limited. Message RNA for the $P2X_1$ receptor (VALERA et al. 1994), known to play a role leucocyte apoptosis (OWENS et al. 1991), was detected at low levels by RT-PCR in the neonatal rat cochlea and vestibular sensory epithelium, but not in the adult inner ear tissues (HOUSLEY and RYAN 1997). Expression of other P2X receptor subunits in the cochlea have also been detected by this technique, but to date no localization data is available. P2Y receptor expression is also well established in cochlear tissues on the basis of electrophysiological and Ca^{2+} imaging data (IKEDA et al. 1995; SUGASAWA et al. 1996a,b); however, little has been published on the expression patterns of the different isoforms. Analysis of $P2Y_2$ receptor expression in the guinea-pig stria vascularis (JÄRLEBARK et al. 1996) is consistent with the previously described UTP-sensitive suppression of I_{sK}. No data on P2Y receptor expression during development has been published, although both P2X and P2Y receptor mediated pathways are likely to be significant elements regulating cochlear ontogeny. This is supported by in vivo experiments demonstrating a role for ATP in cytotoxicity and mitogenesis (BOBBIN et al. 1997).

V. Significance of Alternative Splicing of the $P2X_2$ Receptor mRNA in the Inner Ear

A feature of $P2X_2$ receptor expression in the inner ear is the extensive representation of splice variants in the tissues. By comparison with functional properties of spliced transcripts in vitro, this heterogeneity of $P2X_{2-n}$ splice variants likely confers a diversity of function to the ATP-gated ion channels assembled from these subunits. Thus in the rat spiral ganglion neurones, three different isoforms are expressed ($P2X_{2-1}$, $P2X_{2-2}$, and $P2X_{2-3}$) as a result of variations in intron-exon splicing (SALIH et al. 1998a). All three transcripts encode subunits which assemble as functional homomers characterized by differences in desensitization rate (BRAKE et al. 1994; BRÄNDLE et al. 1997; GREENWOOD et al. 1998). The expression of these isoforms, along with a fourth splice variant whose expression was noted in cochlear lateral wall tissue, but not sensory epithelium (HOUSLEY et al. 1995b), support an important functional role for mRNA editing of the $P2X_2$ receptor gene in the cochlea. This functional diversity of $P2X_2$ receptor transcripts also extends to the guinea-pig cochlea where three different splice variants have also been reported (PARKER et al. 1998). Two are homologues corresponding to rat $P2X_{2-1}$ and $P2X_{2-2}$, while the third, $gpP2X_{2-3}$, arises from splicing in of an intron in the region coding for the extracellular domain. The latter forms functional homo-oligomeric ATP-gated ion channels in HEK293 cells which exhibit desensitization (CHEN et al. 1999).

VI. Adenosine Actions on Afferent Neurotransmission, Free Radical Scavenging, Altered Cochlear Blood Flow, and Hair Cell Cation Channel Downregulation

Adenosine inhibits neurotransmission in the inner ear. Using a number of criteria including release and block of uptake, adenosine has been shown to suppress afferent activity in the frog semicircular canal (BRYANT et al. 1987). This mechanism is also apparent in the guinea-pig cochlea, where perilymphatic perfusion of adenosine and the A_1 receptor agonist 2-CADO (but not A_2 agonists) significantly reduced cochlear microphonics (indicative of suppressed OHC transduction), and auditory nerve compound action potential, but produced no change in EP (NARIO et al. 1994). Inhibition of A_1 receptor expression has been detected in rat cochlear tissue by RT-PCR, and biochemically characterized as a G_i coupled inhibition of adenylate cyclase activity (RAMKUMAR et al. 1994). Expression of both A_1 and A_3 receptors are also evident in the chinchilla cochlea where they have been implicated in the upregulation of antioxidant enzymes which may provide tissue protection from free radical damage during ischemic stress (FORD et al. 1997; RAMKUMAR et al. 1995). The report that R-PIA (A_1 receptor agonist) can protect noise-induced hearing losses in chinchillas (HU et al. 1997), with significant reductions in temporary and permanent threshold shifts, and hair cell damage, following exposure to broad-band noise at 105dB SPL for 4h provides further support for the involvement of adenosine in oxidative tissue damage responses in the cochlea.

A putative A_3 receptor mediated action has also been proposed as a mechanism by which perilymphatic adenosine increases cochlear blood flow (COBF) (MUÑOZ et al. 1999). In these experiments, adenosine-induced increase in COBF was blocked by the non-fastidious antagonist theophilline (1,3-dimethylxanthine) but not DMPX (A_2-like receptor antagonist) or 8-cyclopentyl-1,3-dipropylxanthine (DPCPX; A_1 receptor antagonist). Interestingly, extracellular ATP and ATPγS (an analog resistant to hydrolysis) facilitated COBF via a putative P2Y receptor pathway (blocked by PPADS and reactive blue 2). Infusion of ATP into the guinea-pig cochlear blood supply via the anterior inferior cerebellar artery produced a biphasic suppression, then enhancement, of cochlear blood flow (COBF), consistent with P2X and P2Y receptor mediated actions (REN et al. 1997). The latter may be linked to NO signaling as are other vasodilatatory actions in vascular smooth muscle, as N^Ω-nitro-L-arginine methyl ester (L-NAME) reduced the ATP-induced increase in COBF.

Experiments involving inside-out patches from guinea pig cochlear OHC have demonstrated inhibition of calcium activated non-selective cation channels by the A_2-like receptor agonist NECA, and by the catalytic subunit of protein kinase A (VAN DEN ABBEELE et al. 1996). These data are compatible with A_2-like receptor mediated downregulation of the OHC cation channels via cAMP-dependent phosphorylation, and may indicate adenosine modulation of OHC sound transduction.

E. Conclusion

Studies on peripheral sensory afferents, including nociceptor fibers, cardiac chemosensory fibers, and olfactory neurones, indicate that nucleosides and nucleotides act as physiological stimuli. Sensory-motor collaterals mediate vasoactive dilatation in vascular arterioles, while pre-and post-synaptic actions are key elements of central sensory neurotransmission. The visual system and the inner ear demonstrate considerable diversity of signaling by purines and pyrimidines, with actions extending beyond neurotransmission to involve the majority of sensory and supporting tissues in autocrine and paracrine actions which provide key regulatory functions at the core of homeostasis directed at maintaining transducer sensitivity and integrity. These tissues provide clear insight into the complexity of purinergic and pyrimidinergic signaling in sensory systems, with an interlinking of ionotropic and metabotropic receptors of both P_1 and P_2 classes, as well as the signal flow-down effects of nucleotide hydrolysis by ecto-enzymes, and interactions between intracellular signaling processes, particularly intracellular Ca^{2+}-linked systems.

Significant pathophysiological corollaries to purine and pyrimidinergic signaling are now emerging and new clinical treatments appear imminent for sensory organ dysfunctions. Examples in the eye include glaucoma, arising from ocular hypertension and under A_2-like receptor modulation, and diabetic retinopathy and macular degeneration arising from subretinal neovascularization. The latter are associated with vascular endothelial growth factor stimulation which is upregulated by adenosine. Adenosine (via A_1 receptor action) also upregulates antioxidant activity in the cochlea, protecting against noise-induced hearing loss. ATP action on P2 receptors in the inner ear underpins electrochemical homeostasis, with likely additional impact on micromechanics and neurotransmission, implicating nucleoside and nucleotide signaling with vertigo, Meniere's disease, tinnitus, and temporary and permanent hearing loss.

Acknowledgements. Dr. Paul Donaldson is thanked for discussions relating to the role of nucleotide signaling in the lens. Associate Professor Peter Thorne is thanked for his collaborative support. This work was funded by the Health Research Council (NZ), Marsden Fund, New Zealand Lottery Grants Board, and Deafness Research Foundation (N.Z).

References

Abbracchio MP, Burnstock G (1998) Purinergic signaling: pathophysiological roles Jpn J Pharmacol 78:113–145

Abe Y, Sorimachi M, Itoyama Y, Furukawa K, Akaike N (1995) ATP responses in the embryo chick ciliary ganglion cells. Neuroscience 64:547–551

Armour JA, Huang MH, Pelleg A, Sylven C (1994) Responsiveness of in situ canine nodose ganglion afferent neurones to epicardial mechanical or chemical stimuli. Cardiovasc Res 28:1218–1225

Ashmore JF, Kolston PJ (1994) Hair cell based amplification in the cochlea. Curr Opin Neurobiol 4:503–508

Ashmore JF, Ohmori H (1990) Control of intracellular calcium by ATP in isolated outer hair cells of the guinea-pig cochlea. J Physiol 428:109–131

Aubert A, Norris CH, Guth PS (1994) Influence of ATP and ATP agonists on the physiology of the isolated semicircular canal of the frog (*Rana pipiens*). Neuroscience 62:963–974

Bean BP (1990) ATP-activated channels in rat and bullfrog sensory neurons: concentration dependence and kinetics. J Neurosci 10:1–10

Blazynski C (1987) Adenosine A_1 receptor-mediated inhibition of adenylate cyclase in rabbit retina. J Neurosci 7:2522–2528

Blazynski C (1990) Discrete distributions of adenosine receptors in mammalian retina. J Neurochem 54:648–655

Blazynski C (1993) Characterization of adenosine A2 receptors in bovine retinal pigment epithelial membranes. Exp Eye Res 56:595–599

Blazynski C, Cohen AI, Fruh B, Niemeyer G (1989) Adenosine: autoradiographic localization and electrophysiological effects in the cat retina. Invest Ophthalamol Vis Sci 30:2533–2536

Blazynski C, Kinscherf DA, Geary KM, Ferrendelli JA (1986) Adenosine-mediated regulation of cyclic AMP levels in isolated incubated retinas. Brain Res 366:224–229

Blazynski C, Perez M-TR (1992) Neuroregulatory functions of adenosine in the retina. In: Osborne NN, Chader GJ (eds) Progress in retinal research, vol 11. Pergamon Press, Oxford, pp 293–332

Blazynski C, Perez MT (1991) Adenosine in vertebrate retina: localization, receptor characterization, and function. Cell Mol Neurobiol 11:463–484

Blazynski C, Woods C, Mathews GC (1992) Evidence for the action of endogenous adenosine in the rabbit retina: modulation of the light-evoked release of acetylcholine. J Neurochem 58:761–767

Bo X, Zhang Y, Nassar M, Burnstock G, Schoepfer R (1995) A P2x purinoceptor cDNA conferring a novel pharmacological profile. FEBS Letters 375:129–133

Bo X, Alavi A, Xiang Z, Oglesby I, Ford A, Burnstock G (1999) Localization of ATP-gated $P2X_2$ and $P2X_3$ receptor immunoreactive nerves in rat taste buds. Neuroreport 10, 1107–1111

Bobbin RP, Chu SH, Skellett RA, Campbell J, Fallon M (1997) Cytotoxicity and mitogenicity of adenosine triphosphate in the cochlea. Hear Res 113:155–164

Bobbin RP, Thompson MH (1978) Effects of putative transmitters on afferent cochlear transmission. Ann Otol Rhinol Laryngol 87:185–190

Borisy FF, Hwang PN, Ronnett GV, Snyder SH (1993) High-affinity cAMP phosphodiesterase and adenosine localized in sensory organs. Brain Res 610:199–207

Bosher SK, Warren RS (1968) Observations on the electro-chemistry of the cochlear endolymph of the rat. Proc Royal Soc Lond B 171:227–247

Boulton M, Foreman D, Williams G, McLeod D (1998) VEGF localisation in diabetic retinopathy. Br J Ophthalmol 82:561–568

Bradbury EJ, Burnstock G, McMahon SB (1998) The expression of P2X3 purinoreceptors in sensory neurons: effects of axotomy and glial-derived neurotrophic factor. Mol Cell Neurosci 12:256–268

Brake AJ, Wagenbach MJ, Julius D (1994) New structural motif for ligand-gated ion channels defined by an ionotropic ATP receptor. Nature 371:519–523

Brändle U, Guenther E, Irrle C, Wheeler-Schilling TH (1998a) Gene expression of the P2X receptors in the rat retina. Brain Res Mol Brain Res 59:269–272

Brändle U, Kohler K, Wheeler-Schilling TH (1998b) Expression of the P2X7-receptor subunit in neurons of the rat retina. Brain Res Mol Brain Res 62:106–109

Brändle U, Spielmanns P, Osteroth R, Sim J, Surprenant A, Buell G, Ruppersberg JP, Plinkert PK, Zenner HP, Glowatzki E (1997) Desensitization of the $P2X_2$ receptor controlled by alternative splicing. FEBS Lett 404:294–298

Braunagel SC, Xiao JG, Chiou GC (1988) The potential role of adenosine in regulating blood flow in the eye. J Ocul Pharmacol 4:61–73

Brosvic GM, Rowe MM (1992) Methyl xanthine, adenosine, and human taste responsivity. Physiol Behav 52:559–563

Bryant GM, Barron SE, Norris CH, Guth PS (1987) Adenosine is a modulator of hair cell-afferent neurotransmission. Hear Res 30:231–237

Burnstock G (1977) Autonomic neuroeffector junctions – reflex vasodilation of the skin. J Invest Dermat 69:47–57

Burnstock G (1993) Physiological and pathological roles of purines: an update. Drug Devel Res 28:195–206

Burnstock G (1997) The past, present and future of purine nucleotides as signaling molecules. Neuropharmacology 36:1127–1139

Burnstock G, Wood JN (1996) Purinergic receptors: their role in nociception and primary afferent neurotransmission. Curr Opin Neurobiol 6:526–532

Carre DA, Mitchell CH, Peterson-Yantorno K, Coca-Prados M, Civan MM (1997) Adenosine stimulates Cl⁻ channels of nonpigmented ciliary epithelial cells. Am J Physiol 273:C1354–1361

Chen C, Barnes AP, Parker MS, Deininger PL, Bobbin RP (1999) Functional expression of guinea pig cochlear $P2X_2$ ATP receptor variants in HEK 293 cells. Proc Assoc Res Otolaryngol, St. Petersburg Beach, Florida, vol 22, p 309

Chen C, Bobbin RP (1998) P2X receptors in cochlear Deiters' cells. Br J Pharmacol 124:337–344

Chen C, Skellett RA, Fallon M, Bobbin RP (1998) Additional pharmacological evidence that endogenous ATP modulates cochlear mechanics. Hear Res 118:47–61

Chen C-C, Akoplan AN, Sivilotti L, Colquhoun D, Burnstock G, Wood JN (1995) A P2x purinoceptor expressed by a subset of sensory neurons. Nature 377:428–431

Cho H, Harada N, Yamashita T (1997) Extracellular ATP-induced Ca^{2+} mobilization of type I spiral ganglion cells from the guinea pig cochlea. Acta Oto-Laryngologica 117:545–552

Chou T (1992) Wake up and smell the coffee. Caffeine, coffee, and the medical consequences. West J Med 157:544–553

Churchill GC, Louis CF (1997) Stimulation of P_{2u} purinergic or a1A adrenergic receptors mobilizes Ca^{2+} in lens cells. Invest Ophthalmol Vis Sci 38:855–865

Civan MM (1998) In: Civan MM (ed) The eye's aqueous humor – from secretion to glaucoma. Current topics in Membranes, vol 45. Academic Press, San Diego. pp 1–24

Cole DF (1977) Secretion of the aqueous humour. Exp Eye Res 25:161–176

Collo G, North RA, Kawashima E, Merlo-Pich E, Neidhart S, Surprenant A, Buell G (1996) Cloning of $P2X_5$ and $P2X_6$ receptors and the distribution and properties of an extended family of ATP-gated ion channels. J Neurosci 16:2495–2507

Cook SP, Vulchanova L, Hargreaves KM, Elde R, McCleskey EW (1997) Distinct ATP receptors on pain-sensing and stretch-sensing neurons. Nature 387:505–508

Cotrina ML, Lin JH, Alves-Rodrigues A, Liu S, Li J, Azmi-Ghadimi H, Kang J, Naus CC, Nedergaard M (1998) Connexins regulate calcium signaling by controlling ATP release. Proc Natl Acad Sci USA 95:15735–15740

Crosson CE (1992) Ocular hypotensive activity of the adenosine agonist (R)-phenylisopropyladenosine in rabbits. Curr Eye Res 11:453–458

Crosson CE (1995) Adenosine receptor activation modulates intraocular pressure in rabbits. J Pharmacol Exp Ther 273:320–326

Crosson CE, DeBenedetto R, Gidday JM (1994) Functional evidence for retinal adenosine receptors. J Ocul Pharmacol 10:499–507

Crosson CE, Gray T (1994) Modulation of intraocular pressure by adenosine agonists. J Ocul Pharmacol 10:379–383

Crosson CE, Gray T (1996) Characterization of ocular hypertension induced by adenosine agonists. Invest Ophthalmol Vis Sci 37:1833–1839

Crosson CE, Gray T (1997) Response to prejunctional adenosine receptors is dependent on stimulus frequency. Curr Eye Res 16:359–364

Dallos P, Evans BN (1995) High-frequency motility of outer hair cells and the cochlear amplifier. Science 267:2006–2009

Dawis S, Niemeyer G (1987) Theophylline abolishes the light peak in perfused cat eyes. Invest Ophthalmol Vis Sci 28:700–706

Denk W, Holt JR, Shepherd GMG, Corey DP (1995) Calcium imaging of single stereocilia in hair cells – localization of transduction channels at both ends of tip links. Neuron 15:1311–1321

Derby CD, Hutson M, Livermore BA, Lynn WH (1996) Generalization among related complex odorant mixtures and their components: analysis of olfactory perception in the spiny lobster. Physiol Behav 60:87–95

Donaldson P, Eckert R, Green C, Kistler J (1997) Gap junction channels: new roles in disease. Histol Histopathol 12:219–231

Dulon D (1995) Ca^{2+} signaling in Deiters cells of the guinea-pig cochlea: active process in supporting cells? In: Flock Å, Ottoson D, Ulfendahl M (eds) Active hearing. Pergamon, Oxford, pp 195–207

Dulon D, Moataz R, Mollard P (1993) Characterization of Ca^{2+} signals generated by extracellular nucleotides in supporting cells of the organ of Corti. Cell Calcium 14:245–254

Dulon D, Mollard P, Aran JM (1991) Extracellular ATP elevates cytosolic Ca^{2+} in cochlear inner hair cells. Neuroreport 2:69–72

Evans RJ (1996) Single channel properties of ATP-gated cation channels (P2X receptors) heterologously expressed in Chinese hamster ovary cells. Neurosci Lett 212:212–214

Farahbakhsh NA, Cilluffo MC (1997) Synergistic increase in Ca^{2+} produced by A_1 adenosine and muscarinic receptor activation via a pertussis-toxin-sensitive pathway in epithelial cells of the rabbit ciliary body. Exp Eye Res 64:173–179

Ferguson DR, Kennedy I, Burton TJ (1997) ATP is released from rabbit urinary bladder epithelial cells by hydrostatic pressure changes – a possible sensory mechanism? J. Physiol 505:503–511

Ford MS, Nie Z, Whitworth C, Rybak LP, Ramkumar V (1997) Up-regulation of adenosine receptors in the cochlea by cisplatin. Hear Res 111:143–152

Fuder H, Brink A, Meincke M, Tauber U (1992) Purinoceptor-mediated modulation by endogenous and exogenous agonists of stimulation-evoked [^{3}H]noradrenaline release on rat iris. Naunyn Schmiedebergs Arch Pharmacol 345:417–423

Fuder H, Muth U (1993) ATP and endogenous agonists inhibit evoked [^{3}H]-noradrenaline release in rat iris via A_1 and $P2_y$-like purinoceptors. Naunyn Schmiedebergs Arch Pharmacol 348:352–357

Gabella G, Davis C (1998) Distribution of afferent axons in the bladder of rats. J Neurocytol 27:141–155

Gargett CE, Wiley JS (1997) The isoquinoline derivative KN-62 a potent antagonist of the P2Z-receptor of human lymphocytes. Br J Pharmacol 120:1483–1490

Gidday JM, Maceren RG, Shah AR, Meier JA, Zhu Y (1996) KATP channels mediate adenosine-induced hyperemia in retina. Invest Ophthalmol Vis Sci 37:2624–2633

Gidday JM, Park TS (1993) Adenosine-mediated autoregulation of retinal arteriolar tone in the piglet. Invest Ophthalmol Vis Sci 34:2713–2719

Gidday JM, Zhu Y (1998) Endothelium-dependent changes in retinal blood flow following ischemia. Curr Eye Res 17:798–807

Gilbert RE, Vranes D, Berka JL, Kelly DJ, Cox A, Wu LL, Stacker SA, Cooper ME (1998) Vascular endothelial growth factor and its receptors in control and diabetic rat eyes. Lab Invest 78:1017–2107

Gleeson RA, Trapido-Rosenthal HG, McDowell LM, Aldrich HC, Carr WE (1992) Ecto-ATPase/phosphatase activity in the olfactory sensilla of the spiny lobster, *Panulirus argus*: localization and characterization. Cell Tissue Res 269:439–445

Greenwood D, Housley GD, Raybould NP, Birch NP, Greenwood DR (1998) Characterisation of a novel isoform of the P2X$_2$ receptor subunit of the ATP-gated ion channel. Proc. N.Z Mol. Biol. Biochem Soc., Masterton, New Zealand, p 11

Greenwood D, Yao WP, Housley GD (1997) Expression of the P2X$_2$ receptor subunit of the ATP-gated ion channel in the retina. Neuroreport 8:1083–1088

Gregory CY, Abrams TA, Hall MO (1994) Stimulation of A$_2$ adenosine receptors inhibits the ingestion of photoreceptor outer segments by retinal pigment epithelium. Invest Ophthalmol Vis Sci 35:819–825

Gu JG, MacDermott AB (1997) Activation of ATP P2X receptors elicits glutamate release from sensory neuron synapses. Nature 389:749–753

Holton P (1959) The liberation of adenosine triphosphate on antidromic stimulation of sensory nerves. J Physiol 145:494–504

Housley GD (1998) Extracellular nucleotide signaling in the inner ear. Mol Neurobiol 16:21–48

Housley GD, Connor BJ, Raybould NP (1995a) Purinergic modulation of outer hair cell electromotility. In: Flock Å, Ottoson D, Ulfendahl M (eds) Active hearing. Elsevier, Oxford, pp 221–238

Housley GD, Greenwood D, Ashmore JF (1992) Localization of cholinergic and purinergic receptors on outer hair cells isolated from the guinea-pig cochlea. Proc R Soc Lond B Biol Sci 249:265–73

Housley GD, Greenwood D, Bennett T, Ryan AF (1995b) Identification of a short form of the P2xR1-purinoceptor subunit produced by alternative splicing in the pituitary and cochlea. Biochem Biophys Res Commun 212:501–508

Housley GD, Greenwood D, Mockett BG, Muñoz DJB, Thorne PR (1993) Differential actions of ATP-activated conductances in outer and inner hair cells isolated from the guinea-pig organ of Corti: A humoral purinergic influence on cochlear function. In: Duifhuis H, Horst JW, van Dijk P, van Netten SM (eds) Biophysics of hair cell sensory systems. World Scientific, Singapore, pp 116–123

Housley GD, Greenwood D, Raybould NP, Kanjhan R, Salih SG, King M, Burton LD, Järlebark L, Thorne PR, Usami S, Matsubara, A, Yoshie H, Luo L, Ryan AF (1998a) Expression of the P2X$_2$ receptor subunit of the ATP-gated ion channel in the cochlea: Presence at stereociliary tips and upregulation following noise exposure. Proc Molecular Biology of Hearing & Deafness, Bethesda, Md., USA, p 75P.

Housley GD, Kanjhan R, Raybould NP, Greenwood D, Salih SG, Järlebark L, Burton LD, Setz VCM, Cannell MB, Soeller C, Christie DL, Usami S, Matsubara A, Yoshie H, Ryan AF, Thorne PR (1999) Expression of the P2X$_2$ receptor subunit of the ATP-gated ion channel in the cochlea: implications for sound transduction and auditory neurotransmission. J Neurosci 19:8377–8388

Housley GD, Luo L, Ryan AF (1998b) Localization of mRNA encoding the P2X$_2$ receptor subunit of the adenosine 5'-triphosphate-gated ion channel in the adult and developing rat inner ear by in situ hybridization. J Comp Neurol 393:403–414

Housley GD, Raybould NP, Thorne PR (1998c) Fluorescence imaging of Na$^+$ influx via P2X receptors in cochlear hair cells. Hear Res 119:1–13

Housley GD, Ryan AF (1997) Cholinergic and purinergic neurohumoral signaling in the inner ear: A molecular physiological analysis. Audiol Neuro-Otol 2:92–110

Housley GD, Thorne PR, Kanjhan R, Raybould NP, Muñoz DJB, Luo L, Ryan AF (1997) Regulation of the electrochemical gradient for sound transduction by ATP-gated ion channels on cochlear hair cell stereocilia. Soc Neurosci Abstr 23:731

Howard M, Sen HA, Capoor S, Herfel R, Crooks PA, Jacobson MK (1998) Measurement of adenosine concentration in aqueous and vitreous. Invest Ophthalmol Vis Sci 39:1942–1946

Hu BH, Zheng XY, McFadden SL, Kopke RD, Henderson D (1997) *R*-Phenylisopropyladenosine attenuates noise-induced hearing loss in the chinchilla. Hear Res 113:198–206

Humphreys BD, Virginio C, Surprenant A, Rice J, Dubyak GR (1998) Isoquinolines as antagonists of the P2X$_7$ nucleotide receptor: high selectivity for the human versus rat receptor homologues. Mol Pharmacol 54:22–32

Ikeda K, Suzuki M, Furukawa M, Takasaka T (1995) Calcium mobilization and entry induced by extracellular ATP in the non-sensory epithelial cell of the cochlear lateral wall. Cell Calcium 18:89–99

Jahr CE, Jessell TM (1983) ATP excites a subpopulation of rat dorsal horn neurones. Nature 304:730–733

Järlebark L, Bjorklund J, Heilbronn E (1996) P2Y$_2$ receptor expressed in rat and guinea pig cochlea – partial cloning and localization. Drug Devel Res 37:184

Kang J, Caprio J (1995) In vivo responses of single olfactory receptor neurons in the channel catfish, *Ictalurus punctatus*. J Neurophysiol 73:172–177

Kanjhan R, Housley GD, Burton LD, Christie DL, Kippenberger A, Thorne PR, Luo L, Ryan AF (1999) Distribution of the P2X$_2$ receptor subunit of the ATP-gated ion channels in the rat central nervous system. J Comp Neurol 407:11–32

Keirstead SA, Miller RF (1997) Metabotropic glutamate receptor agonists evoke calcium waves in isolated Muller cells. Glia 21:194–203

Khakh BS, Humphrey PP, Surprenant A (1995) Electrophysiological properties of P2X-purinoceptors in rat superior cervical, nodose and guinea-pig coeliac neurones. J Physiol Lond 484:385–395

Khakh BS, Humphrey PPA, Henderson G (1997) ATP-gated cation channels (P2X purinoceptors) in trigeminal mesencephalic nucleus of the rat. J Physiol 498:709–715

Kidd EJ, Grahames CBA, Simon J, Michel AD, Barnard EA, Humphrey PPA (1995) Localization of P2x purinoceptor transcripts in the rat nervous system. Mol Pharmacol 48:569–573

Kidd EJ, Miller KJ, Sansum AJ, Humphrey PP (1998) Evidence for P2X$_3$ receptors in the developing rat brain. Neuroscience 87:533–539

Kin YV, Bobkov YV, Kolesnikov SS (2000) Adenosine triphosphate mobilizes cytosolic calcium and modulates ionic currents in mouse taste receptor cells. Neurosci Letts 290:165–168

King M, Housley GD, Raybould NP, Greenwood D, Salih SG (1998) Expression of ATP-gated ion channels by Reissner's membrane epithelial cells. Neuroreport 9:2467–7244

Kirischuk S, Scherer J, Kettenmann H, Verkhratsky A (1995) Activation of P2-purinoreceptors triggered Ca^{2+} release from InsP$_3$-sensitive internal stores in mammalian oligodendrocytes. J Physiol 483:41–57

Kirk DL, Yates GK (1998) ATP in endolymph enhances electrically-evoked otoacoustic emissions from the guinea pig cochlea. Neurosci Lett 250:149–152

Krishtal OA, Marchenko SM, Obukhov AG (1988) Cationic channels activated by extracellular ATP in rat sensory neurons. Neuroscience 27:995–1000

Krishtal OA, Marchenko SM, Pidoplichko VI (1983) Receptor for ATP in the membrane of mammalian sensory neurones. Neurosci Lett 35:41–45

Kujawa SG, Erostegui C, Fallon M, Crist J, Bobbin RP (1994a) Effects of adenosine 5′-triphosphate and related agonists on cochlear function. Hear Res 76:87–100

Kujawa SG, Fallon M, Bobbin RP (1994b) ATP antagonists cibacron blue, basilen blue and suramin alter sound-evoked responses of the cochlea and auditory nerve. Hear Res 78:181–188

Kvanta A (1995) Expression and regulation of vascular endothelial growth factor in choroidal fibroblasts. Curr Eye Res 14:1015–1020

Kvanta A, Seregard S, Sejersen S, Kull B, Fredholm BB (1997) Localization of adenosine receptor messenger RNAs in the rat eye. Exp Eye Res 65:595–602

Larsen AK, Osborne NN (1996) Involvement of adenosine in retinal ischemia. Studies on the rat. Invest Ophthalmol Vis Sci 37:260326–11

Le KT, Villeneuve P, Ramjaun AR, McPherson PS, Beaudet A, Seguela P (1998) Sensory presynaptic and widespread somatodendritic immunolocalization of central ionotropic P2X ATP receptors. Neuroscience 83:177–190

Lewis C, Neidhart S, Holy C, North RA, Buell G, Surprenant A (1995) Coexpression of P2X2 and P2X3 receptor subunits can account for ATP- gated currents in sensory neurons. Nature 377:432–435

Li J, Perl ER (1995) ATP modulation of synaptic transmission in the spinal substantia gelatinosa. J Neurosci 15:3357–3365

Li P, Calejesan AA, Zhuo M (1998) ATP P2x receptors and sensory synaptic transmission between primary afferent fibers and spinal dorsal horn neurons in rats. J Neurophysiol 80:3356–3360

Liu J, Kozakura K, Marcus DC (1995) Evidence for purinergic receptors in vestibular dark cell and strial marginal cell epithelia of the gerbil. Audit Neurosci 1:331–340

Liu Y, Wakakura M (1998) P1-/P2-purinergic receptors on cultured rabbit retinal Muller cells. Jpn J Ophthalmol 42:33–40

Lotfi K, Grunwald JE (1991) The effect of caffeine on the human macular circulation. Invest Ophthalmol Vis Sci 32:3028–3032

MacKenzie AB, Surprenant A, North RA (1999) Functional and molecular diversity of purinergic ion channel receptors. Ann N Y Acad Sci 868:716–729

Maggi CA, Meli A (1988) The sensory-efferent function of capsaicin-sensitive sensory nerves. Gen Pharmacol 19:1–43

Mammano F, Frolenkov GI, Lagostena L, Belyantseva IA, Kurc M, Dodane V, Colavita A, Kachar B (1999) ATP-induced Ca^{2+} release in cochlear outer hair cells: localization of an inositol triphosphate-gated Ca^{2+} store to the base of the sensory hair bundle. J Neurosci 19:6918–6929

Marcus DC, Liu J, Sunose H, Shen Z (1996) Protein Kinase C Mediates I_{sK} (min K) channel down-regulation by apical P_{2u} purinoceptor in K^+-secretory epithelial cells of the inner ear. J Gen Physiol 108:26a

Marcus DC, Liu JZ (1999) Purinergic receptors of P2Y4 subtype regulate K secretion in the inner ear. Proc Assoc Res Otolaryngol, St. Petersberg, Florida, vol 22, p 308

Marcus DC, Sunose H, Liu J, Bennett T, Shen Z, Scofield MA, Ryan AF (1998) Protein kinase C mediates P2U purinergic receptor inhibition of K^+ channel in apical membrane of strial marginal cells. Hear Res 115:82–92

Marcus DC, Sunose H, Liu J, Shen Z, Scofield MA (1997) P2U purinergic receptor inhibits apical IsK/KvLQT1 channel via protein kinase C in vestibular dark cells. Am J Physiol 273:C2022–2029

Maruiwa F, Nao-i N, Nakazaki S, Sawada A (1995) Effects of adenosine on chick retinal pigment epithelium: membrane potentials and light-evoked responses. Curr Eye Res 14:685–691

Matsunobu T, Schacht J (1999) The nitric oxide/cGMP pathway attenuates ATP-evoked intracellular calcium increases in cochlear supporting cells. Proc Assoc Res Otolaryngol, vol 22, p 317

McMurray G, Dass N, Brading AF (1998) Purinoceptor subtypes mediating contraction and relaxation of marmoset urinary bladder smooth muscle. Br J Pharmacol 123:1579–1586

McIntosh HH, Blazynski C (1994) Characterization and localization of adenosine A2 receptors in bovine rod outer segments. J Neurochem 62:992–997

McQueen DS, Bond SM, Moores C, Chessell I, Humphrey PP, Dowd E (1998) Activation of P2X receptors for adenosine triphosphate evokes cardiorespiratory reflexes in anaesthetized rats. J Physiol 507:843–855

Merriman-Smith R, Tunstall M, Kistler J, Donaldson P, Housley G, Eckert R (1998) Expression profiles of P2-receptor isoforms $P2Y_1$ and $P2Y_2$ in the rat lens. Invest Ophthalmol Vis Sci 39:2791–2796

Mitchell CH, Carre DA, McGlinn AM, Stone RA, Civan MM (1998) A release mechanism for stored ATP in ocular ciliary epithelial cells. Proc Natl Acad Sci USA 95:7174–7178

Mitchell CH, Zhang JJ, Wang L, Jacob TJ (1997) Volume-sensitive chloride current in pigmented ciliary epithelial cells: role of phospholipases. Am J Physiol 272:C212–222

Mockett BG, Housley GD, Thorne PR (1994) Fluorescence imaging of extracellular purinergic receptor sites and putative ecto-ATPase sites on isolated cochlear hair cells. J Neurosci 14:6992–7007

Muñoz D, Thorne P, Housley G, Billett T, Battersby J (1995) Extracellular adenosine 5′-triphosphate (ATP) in the endolymphatic compartment influences cochlear function. Hear Res 90:106–111

Muñoz DJB, McFie C, Thorne PR (1999) Modulation of cochlear blood flow by extracellular purines. Hear Res 127:55–61

Muñoz DJB, Thorne PR, Housley GD (1999) P2X receptor-mediated changes in cochlear potentials arising from exogenous adenosine 5′-triphosphate in endolymph. Hear Res 138:56–64

Nakagawa T, Akaike N, Kimitsuki T, Komune S, Arima T (1990) ATP-induced current in isolated outer hair cells of guinea pig cochlea. J. Neurophysiology 63:1068–1074

Nakamura F, Strittmatter SM (1996) P2Y1 purinergic receptors in sensory neurons: contribution to touch- induced impulse generation. Proc Natl Acad Sci USA 93:10465–10470

Nario K, Kitano I, Mori N, Matsunaga T (1994) The effect of adenosine on cochlear potentials in the guinea pig. Eur Arch Otorhinolaryngol 251:428–433

Neal M, Cunningham J (1994) Modulation by endogenous ATP of the light-evoked release of ACh from retinal cholinergic neurones. Br J Pharmacol 113:1085–1087

Neal MJ, Cunningham JR, Dent Z (1998) Modulation of extracellular GABA levels in the retina by activation of glial P2X-purinoceptors. Br J Pharmacol 124:317–322

Newman EA, Zahs KR (1997) Calcium waves in retinal glial cells. Science 275:844–847

Niedzielski AS, Schacht J (1992) P2 purinoceptors stimulate inositol phosphate release in the organ of Corti. Neuroreport 3:273–275

Niemeyer G, Früh B (1989) Adenosine and cyclohexyladenosine inhibit the cat's optic nerve action potential. Experientia 45:A18

Ostwald P, Park SS, Toledano AY, Roth S (1997) Adenosine receptor blockade and nitric oxide synthase inhibition in the retina: impact upon post-ischemic hyperemia and the electroretinogram. Vision Res 37:3453–3461

Owens GP, Hahn WE, Cohen JJ (1991) Identification of mRNAs associated with programmed cell death in immature thymocytes. Mol Cell Biol 11:4177–4188

Paes de Carvalho R (1990) Development of A_1 adenosine receptors in the chick embryo retina. J Neurosci Res 25:236–242

Paes de Carvalho R, Braas KM, Adler R, Snyder SH (1992) Developmental regulation of adenosine A_1 receptors, uptake sites and endogenous adenosine in the chick retina. Brain Res Dev Brain Res 70:87–95

Paes de Carvalho R, Braas KM, Snyder SH, Adler R (1990) Analysis of adenosine immunoreactivity, uptake, and release in purified cultures of developing chick embryo retinal neurons and photoreceptors. J Neurochem 55:1603–1611

Paes de Carvalho R, de Mello FG (1982) Adenosine-elicited accumulation of adenosine 3′,5′-cyclic monophosphate in the chick embryo retina. J Neurochem 38:493–500

Paes de Carvalho R, de Mello FG (1985) Expression of A_1 adenosine receptors modulating dopamine-dependent cyclic AMP accumulation in the chick embyronic retina. J Neurochem. 44:845–851

Parker MS, Larroque ML, Campbell JM, Bobbin RP, Deininger PL (1998) Novel variant of the $P2X_2$ ATP receptor from the guinea pig organ of Corti. Hear Res 121:62–70

Perez MTR, Ehinger B (1989a) Adenosine inhibits evoked acetylcholine release from the rabbit retina. J. Neurochem 52:S157

Perez MTR, Ehinger B (1989b) Multiple neurotransmitter systems influence the release of adenosine derivatives from the rabbit retina. Neurochem Int 15:411–420

Pintor J, Peral A, Navas B, Gallar J, Pelaez T (2000) Therapeutic potential of nucleotides in the eye. Drug Dev Res 50:S14-05

Portellos M, Riva CE, Cranstoun SD, Petrig BL, Brucker AJ (1995) Effects of adenosine on ocular blood flow. Invest Ophthalmol Vis Sci 36:1904–1909

Ramkumar V, Nie Z, Rybak LP, Maggirwar SB (1995) Adenosine, antioxidant enzymes and cytoprotection. TiPS 16:283–285

Ramkumar V, Ravi R, Wilson MC, Gettys TW, Whitworth C, Rybak LP (1994) Identification of A_1 adenosine receptors in rat cochlea coupled to inhibition of adenylyl cyclase. Am J Physiol 267:C731–737

Raybould NP, Housley GD (1997) Variation in expression of the outer hair cell P2X receptor conductance along the guinea-pig cochlea. J Physiol 498.3:717–727

Relevic V, Rubino A, Burnstock G (1996) Augmented sensory-motor vasodilatation of the rat mesenteric arterial bed after chronic infusion of the P1-purinoceptor antagonist, DPSPX. Br J Pharmacol 118:1675–1680

Ren T, Nuttall AL, Miller JM (1997) ATP-induced cochlear blood flow changes involve the nitric oxide pathway. Hear Res 112:87–94

Rennie KJ, Ashmore JF (1993) Effects of extracellular ATP on hair cells isolated from the guinea-pig semicircular canals. Neurosci Lett 160:185–189

Riach RA, Duncan G, Williams MR, Webb SF (1995) Histamine and ATP mobilize calcium by activation of H1 and P2u receptors in human lens epithelial cells. J Physiol 486:273–282

Riley MV, Winkler BS, Starnes CA, Peters MI (1996) Adenosine promotes regulation of corneal hydration through cyclic adenosine monophosphate. Invest Ophthalmol Vis Sci 37:1–10

Riley MV, Winkler BS, Starnes CA, Peters MI, Dang L (1998) Regulation of corneal endothelial barrier function by adenosine, cyclic AMP, and protein kinases. Invest Ophthalmol Vis Sci 39:2076–2084

Roth S (1995) Post-ischemic hyperemia in the cat retina: the effects of adenosine receptor blockade. Curr Eye Res 14:323–328

Roth S, Osinski JV, Park SS, ostwald P, Moshfeghi AA (1996) Measurement of purine nucleoside concentration in the intact rat retina. J Neurosci Methods 68:87–90

Roth S, Park SS, Sikorski CW, Osinski J, Chan R, Loomis K (1997a) Concentrations of adenosine and its metabolites in the rat retina/choroid during reperfusion after ischemia. Curr Eye Res 16:875–885

Roth S, Rosenbaum PS, Osinski J, Park SS, Toledano AY, Li B, Moshfeghi AA (1997b) Ischemia induces significant changes in purine nucleoside concentration in the retina-choroid in rats. Exp Eye Res 65:771–779

Sakaki Y, Fukuda Y, Yamashita M (1996) Muscarinic and purinergic Ca^{2+} mobilizations in the neural retina of early embryonic chick. Int J Dev Neurosci 14:691–699

Salih SG, Housley GD, Burton LD, Greenwood D (1998) $P2X_2$ receptor subunit expression in a subpopulation of cochlear type I spiral ganglion neurones. Neuroreport 9:279–282

Salih SG, Housley GD, Raybould NP, Thorne PR (1999) ATP-gated ion channel expression in primary auditory neurones. Neuroreport 10:2579–2586

Salt TE, Hill RG (1983) Excitation of single sensory neurones in the rat caudal trigeminal nucleus by iontophoretically applied adenosine 5′-triphosphate. Neurosci Lett 35:53–57

Salter MW, De Koninck Y, Henry JL (1993) Physiological roles for adenosine and ATP in synaptic transmission in the spinal dorsal horn. Prog Neurobiol 41:125–156

Santos PF, Santos MS, Carvalho AP, Duarte CB (1998) Modulation of [3H]acetylcholine release from cultured amacrine-like neurons by adenosine A_1 receptors. J Neurochem 71:1086–1094

Sato K, Park NG, Kohno T, Maeda T, Kim JI, Kato R, Takahashi M (1993) Role of basic residues for the binding of omega-conotoxin GVIA to N- type calcium channels. Biochem Biophys Res Commun 194:1292–1296

Schmidt RS, Fernandez C (1963) Development of mammalian endocochlear potential. J Exp Zool 153:227–235

Sebastiao AM, Ribeiro JA (1996) Adenosine A_2 receptor-mediated excitatory actions on the nervous system. Prog Neurobiol 48:167–189

Shahidullah M, Wilson WS (1997) Mobilisation of intracellular calcium by P2Y2 receptors in cultured, non-transformed bovine ciliary epithelial cells. Curr Eye Res 16:1006–1016

Sorimachi M, Abe Y, Furukawa K, Akaike N (1995) Mechanism underlying the ATP-induced increase in the cytosolic Ca^{2+} concentration in chick ciliary ganglion neurons. J Neurochem 64:1169–1174

Sugasawa M, Erostegui C, Blanchet C, Dulon D (1996a) ATP activates a cation conductance and Ca^{2+}-dependent Cl^- conductance in Hensen cells of guinea pig cochlea. Am J Physiol 271:C1817–1827

Sugasawa M, Erostegui C, Blanchet C, Dulon D (1996b) ATP activates non-selective cation channels and calcium release in inner hair cells of the guinea-pig cochlea. J Physiol 491:707–718

Sugasawa M, Erostegui C, Blanchet C, Dulon D (1996c) Purinergic responses in supporting cells of the guinea-pig organ of corti. Proc Assoc Res Otolaryngol vol 19, p 132

Sugioka M, Fukuda Y, Yamashita M (1996) Ca^{2+} responses to ATP via purinoceptors in the early embryonic chick retina. J Physiol 493:855–863

Suzuma K, Takagi H, Otani A, Oh H, Honda Y (1999) Expression of thrombospondin-1 in ischemia-induced retinal neovascularization. Am J Pathol 154:343–354

Takagi H, King GL, Aiello LP (1998) Hypoxia upregulates glucose transport activity through an adenosine- mediated increase of GLUT1 expression in retinal capillary endothelial cells. Diabetes 47:1480–1488

Takagi H, King GL, Robinson GS, Ferrara N, Aiello LP (1996) Adenosine mediates hypoxic induction of vascular endothelial growth factor in retinal pericytes and endothelial cells. Invest Ophthalmol Vis Sci 37:2165–2176

Taschenberger H, Juttner R, Grantyn R, (1999) Ca^{2+}-permeable P2X receptor channels in cultured rat retinal ganglion cells. J Neurosci 19:3353–3366

Thorne PR, Housley GD (1996) Purinergic signaling in sensory systems. Sem Neurosci 8:233–246

Thorne PR, Muñoz DJB, Housley GD, Vlajkovic S, Kendrick IS, Rasam M (1999) Regulation of cochlear sensitivity by extracellular ATP. Proc Assoc Res Otolaryngol, vol 22, p 482

Tian B, Gabelt BT, Crosson CE, Kaufman PL (1997) Effects of adenosine agonists on intraocular pressure and aqueous humor dynamics in cynomolgus monkeys. Exp Eye Res 64:979–989

Torre V, Ashmore JF, Lamb TD, Menini A (1995) Transduction and adaptation in sensory receptor cells. J Neurosci 15:7757–7768

Valera S, Hussy N, Evans RJ, Adami N, North RA, Surprenant A, Buell G (1994) A new class of ligand-gated ion channel defined by P2x receptor for extracellular ATP. Nature 371:516–519

Van den Abbeele T, Tran Ba Huy P, Teulon J (1996) Modulation by purines of calcium-activated non-selective cation channels in the outer hair cells of the guinea-pig cochlea. J Physiol 494:77–89

Vetter DE, Mann JR, Wangemann P, Liu JZ, McLaughlin KJ, Lesage F, Marcus DC, Lazdunski M, Heinemann SF, Barhanin J (1996) Inner ear defects induced by null mutation of the Isk gene. Neuron. 17:1251–126

Vlajkovic SM, Housley GD, Greenwood D, Thorne PR (1999) Evidence for alternative splicing of ecto-ATPase associated with termination of purinergic transmission. Brain Res Mol Brain Res 73:85–92

Vlajkovic S, Thorne PR, Housley GD, Muñoz DJB, Kendrick IS (1998a) The pharmacology and kinetics of ecto-nucleotidases in the perilymphatic compartment of the guinea-pig cochlea. Hear Res 117:71–80

Vlajkovic SM, Thorne PR, Housley GD, Muñoz DJB, Kendrick IS (1998b) Ecto-nucleotidases terminate purinergic signaling in the cochlear endolymphatic compartment. Neuroreport 9:1559–1565

Vulchanova L, Riedl MS, Shuster SJ, Stone LS, Hargreaves KM, Buell G, Surprenant A, North RA, Elde R (1998) $P2X_3$ is expressed by DRG neurons that terminate in inner lamina II. Eur J Neurosci 10:3470–3478

Wangemann P (1996) Ca^{2+}-dependent release of ATP from the organ of Corti measured with a luciferin-luciferase bioluminescence assay. Audit Neurosci 2:187–192

Wax M, Sanghavi DM, Lee CH, Kapadia M (1993) Purinergic receptors in ocular ciliary epithelial cells. Exp Eye Res 57:89–95

Wessler I, Kirkpatrick CJ, Racke K (1998) Non-neuronal acetylcholine, a locally acting molecule, widely distributed in biological systems: expression and function in humans. Pharmacol Ther 77:59–79

White P, Thorne P, Housley G, Mockett B, Billett T, Burnstock G (1995) Quinacrine staining of marginal cells in the stria vascularis of the guinea-pig cochlea: a possible source of extracellular ATP? Hear Res 90:97–105

Xia P, Aiello LP, Ishii H, Jiang ZY, Park DJ, Robinson GS, Takagi H, Newsome WP, Jirousek MR, King GL (1996) Characterization of vascular endothelial growth factor's effect on the activation of protein kinase C, its isoforms, and endothelial cell growth. J Clin Invest 98:2018–2026

Xiang Z, Bo X, Burnstock G (1999) P2X receptor immunoreactivity in the rat cochlea, vestibular ganglion and cochlear nucleus. Hear Res 128:190–196

Yates GK, Kirk DL (1998) Cochlear electrically evoked emissions modulated by mechanical transduction channels J Neurosci 18:1996–2003

Yawo H, Chuhma N (1993) Preferential inhibition of omega-conotoxin-sensitive presynaptic Ca^{2+} channels by adenosine autoreceptors. Nature 365:256–258

Zhu Y, Gidday JM (1996) Hypoglycemic hyperemia in retina of newborn pigs. Involvement of adenosine. Invest Ophthalmol Vis Sci 37:86–92

CHAPTER 13
Roles of Purines in Nociception and Pain

M.W. SALTER and A. SOLLEVI

A. Introduction

Transmission of somatosensory information is normally initiated in the periphery through stimulation of the endings of primary afferent neurons which encode information as an action potential discharge that is propagated into the central nervous system (CNS). Whether this information is ultimately perceived as noxious or innocuous depends, in part, on the activation of specific peripheral sensory afferents and, as importantly, on the actions and interactions of numerous neurotransmitter/neuromodulator systems at peripheral and central sites. The first level of central processing for most somatosensory information is in the dorsal horn of the spinal cord or in the homologous region of the trigeminal nucleus. Multiple lines of evidence support roles for adenosine and adenosine 5′-triphosphate (ATP) in the transmission of sensory information in the periphery and in the dorsal horn (BURNSTOCK and WOOD 1996; SALTER et al. 1993; SAWYNOK 1998). In this chapter we review evidence that adenosine and ATP participate in sensory transmission, highlighting recent developments in our understanding of purine nucleotides and nucleosides in nociceptive, e.g., pain-related, and non-nociceptive sensory transmission at peripheral and spinal sites. Finally, we discuss results of humans studies which to date have focussed mainly on the use of adenosine as a novel therapeutic agent for the treatment of clinical pain.

B. ATP and Adenosine in Somatosensory Neurotransmission

I. Release of ATP

1. Release of ATP from Primary Afferent Neurons

Evidence first implicating ATP as a neurotransmitter in the somatosensory system came from studies in the mid- to late-1950s by Holton and Holton who demonstrated that ATP is released from peripheral endings of primary sensory neurons (HOLTON and HOLTON 1954; HOLTON 1959). Based upon Dale's principle (DALE 1935) they hypothesized that ATP may also be released from central terminals of primary afferents. Evidence for central release of ATP was provided by WHITE et al. (1985) who showed that ATP was released in a calcium-dependent fashion from dorsal horn synaptosomes by depolarizing concentrations of potassium. This release is greatly depressed by dorsal rhizotomy, suggesting that ATP may be released from central terminals of primary afferent neurons (SAWYNOK et al. 1993). Release of ATP from dorsal horn synaptosomes is not affected by neonatal treatment with capsaicin, which destroys many nociceptive primary afferents. Thus, ATP released from primary afferent terminals in the dorsal horn derives from a subpopulation of afferents which is capsaicin-insensitive.

2. Co-Release of ATP with GABA from Dorsal Horn Neurons

That ATP release from dorsal horn synaptosomes was not completely eliminated by dorsal rhizotomy implies that terminals of non-primary afferents within the dorsal horn may also release this nucleotide. Recently P2X-receptor mediated synaptic responses have been observed in about 50% cases of synapses between neurons in laminae I-III of the dorsal horn (Jo and SCHLICHTER 1999). These observations are electrophysiological evidence that ATP is released from the terminals of a subpopulation of dorsal horn neurons in these laminae. Unexpectedly, all neurons releasing ATP also released the inhibitory amino acid, GABA, whereas co-release of the excitatory amino acid, glutamate, was not observed. Cells from which ATP release was detected represented a subset of at least 70% of the GABAergic neurons. Thus, in laminae I-III ATP and GABA appear to be co-transmitters at many synapses. These transmitters have opposing postsynaptic actions which has significant implications for sensory processing in the dorsal horn (see Sect. B.III.1 and SALTER and DE KONINCK 1999). Whether such co-transmission occurs for GABAergic neurons in other regions of the CNS remains to be determined.

3. Release of ATP from Other Cell Types in the Periphery

There are a number of potential peripheral sources of ATP in addition to the terminals of primary afferent neurons. ATP may exit from cells damaged as a consequence of physical injury. ATP may also be released from peripheral endings of postganglionic sympathetic nerve terminals and this release has been proposed to contribute to peripheral excitation of primary afferent nociceptors in "sympathetically maintained pain" (BURNSTOCK 1996). Other potential sources of ATP include Merkel cells, vascular endothelial cells, and tumor cells (see BURNSTOCK and WOOD 1996).

II. P1 and P2 Purinergic Receptors

1. Purinergic Receptor Expression by Primary Afferent Neurons

The possibility that primary afferent neurons express ionotropic receptors for ATP, e.g., P2X purinergic receptors, was postulated first on the basis of responsiveness of subpopulations of these neurons to ATP and other nucleotides as described in Sect. B.III.1.a. A family of P2X receptor subunits has been cloned from different tissues (CHESSELL et al., Chap. 3, this volume). Sensory ganglia, which contain the cell bodies of primary afferent neurons, are known to express mRNA for $P2X_{1-6}$ subunits (COLLO et al. 1996). Of these subunits, $P2X_3$ has emerged as one with a potentially important role in nociception (CHEN et al. 1995; LEWIS et al. 1995). mRNA for the $P2X_3$ subunit is found only in sensory ganglia (CHEN et al. 1995; LEWIS et al. 1995), and labeling for $P2X_3$ in primary afferent cell bodies, in situ, occurs in cells which also express immunostaining for peripherin (CHEN et al. 1995) and other markers of C-fiber afferents. mRNA for $P2X_3$ is abolished following neonatal capsaicin treatment (CHEN et al. 1995), indicating that $P2X_3$ is expressed exclusively by the capsaicin-sensitive subpopulation of C-fiber afferents. $P2X_3$ receptors also show co-localization with vanilloid VR-1 receptors (GUO et al. 1999). Such neurons may be presumed to be nociceptive and this has been confirmed in a direct test of expression in nociceptive neurons. $P2X_3$ immunoreactivity is expressed by tooth-pulp afferents, a pure population of nociceptors, but not by muscle-stretch afferents, which are non-nociceptive (COOK et al. 1997). Interestingly, muscle-stretch afferents express other P2X subtypes. Moreover, both types of afferent respond to ATP, but the biophysical and pharmacological characteristics of the responses of the tooth pulp afferents matched only those of recombinant P2X receptors which contained $P2X_3$: in some cells ATP-evoked currents were rapidly inactivating like $P2X_3$ homomeric channels, in other cells the currents were sustained like $P2X_{2/3}$ heteromeric channels (LEWIS et al. 1995). This provides compelling evidence that $P2X_3$ is functionally expressed by nociceptive primary sensory neurons. More recently it has been found that immunoreactivity for $P2X_3$ extends into lamina II of the spinal cord dorsal horn (VULCHANOVA et al. 1998). Thus, $P2X_3$ appears to be targeted to both peripheral and central terminals of nociceptive afferents.

The expression of adenosine receptor by primary sensory neurons has largely been inferred from functional studies (see below and SAWYNOK 1998). Results of these studies suggest strongly that primary afferents express A_1 adenosine receptors. There are also data indicating that there are pronociceptive responses in the periphery mediated by A_{2A} receptors but direct evidence that expression of such receptors by primary afferent neurons is limited.

2. Purinergic Receptor Expression by Dorsal Horn Neurons

As was the case for primary afferents, the notion that dorsal horn neurons express receptors for ATP was first proposed on the basis of studies showing that ATP excites dorsal horn neurons (see Sect. B.III.2.a). Subsequently it has been found that of the cloned P2X receptor subunits, $P2X_2$, $P2X_4$, and $P2X_6$ receptor subunit mRNAs are expressed in the dorsal horn (BRAKE et al. 1994; BUELL et al. 1996; COLLO et al. 1996). Immunoreactivity for $P2X_2$ is limited to the very superficial portion of the dorsal horn in a pattern consistent with expression in primary afferent terminals (VULCHANOVA et al. 1996). On the other hand, $P2X_4$ immunoreactivity (LE et al. 1998) has been shown to be expressed in the spinal cord dorsal horn laminae I–III. These patterns of expression support the idea that postsynaptic P2X receptors, particularly $P2X_4$ and $P2X_6$, may participate in synaptic function in the superficial laminae of the dorsal horn, an important site for processing nociceptive information.

In terms of adenosine receptors, binding sites for A_1- and A_2-like receptors have been demonstrated in the spinal cord (GOODMAN and SNYDER 1982; MURRAY and CHENEY 1982; GEIGER et al. 1984; CHOCA et al. 1987, 1988) with binding occurring most heavily in the superficial laminae of the dorsal horn (GOODMAN and SNYDER 1982; CHOCA et al. 1988). Selective lesion studies show that adenosine receptor binding is only partially decreased following the destruction of spinal interneurons (GEIGER et al. 1984; CHOCA et al. 1988). Moreover, mRNA for A_1- and A_2-like receptors is expressed in the spinal cord (REPPERT et al. 1991; RIVKEES and REPPERT 1992; STEHLE et al. 1992; RIVKEES 1995) supporting the concept that adenosine receptors are expressed by intrinsic dorsal horn neurons. These findings are consistent with post-synaptic effects of adenosine on dorsal horn neurons, and that A_1- or A_2-like receptors may participate in the physiological actions of adenosine at spinal levels (see Sect. B.III.2.b).

III. Actions of ATP and Adenosine

1. Actions in Periphery

a) ATP

In addition to releasing ATP, primary afferent neurons are sensitive to this nucleotide. It is now well-known that ATP produces depolarization when applied to the cell bodies of primary afferent neurons located within the dorsal root ganglia (DRG) (JAHR and JESSELL 1983; KRISHTAL et al. 1983). The depolarizing effect of ATP results from activating a non-selective cation channel (BEAN 1990) and is blocked by P2-purinoceptor antagonists, indicating the excitation is mediated via P2X-purinoceptors. The effects of ATP on DRG cell bodies raised the possibility that terminals of primary afferents may also be excited by ATP and in this way ATP could be a sensory mediator in peripheral tissues. The work of McCleskey and colleagues showed that applying ATP to directly sensory nerve endings does indeed cause excitation (COOK et al. 1997). Moreover, it has been found by GU and MACDERMOTT (1997) that the depolarizing effects of ATP on sensory nerve terminals may be sufficient to evoke action potentials and to cause release of the terminals to release the excitatory amino acid, glutamate.

In behavioral testing, the actions of peripherally administered ATP are pro-nociceptive. ATP has long been known to induce nociceptive or algogenic responses following peripheral administration (COLLIER et al. 1966; BLEEHEN and KEELE 1977). Nociceptive responses induced by ATP have been suggested to be due to direct activation of peripheral nerve terminals (reviewed by ILLES and NÖRENBERG 1993) or to indirect effects through actions on inflammatory cells (DUBYAK et al. 1988; HANDWERKER and REEH 1991; RANG et al. 1991). Inflammatory cells release many mediators that may excite nociceptive primary afferents (DRAY 1995).

b) Adenosine

In electrophysiological recordings from DRG cell bodies adenosine has been shown to inhibit high voltage-gated Ca^{2+} currents (DOLPHIN et al. 1986; MACDONALD et al. 1986; GROSS et al. 1989), an effect mediated mainly through activating A_1 receptors. These findings indicate that A_1 receptors are expressed by primary sensory neurons. That A_1 receptors are trafficked to the peripheral endings of primary afferents has been argued from the effects of peripherally administered A_1 agonists and antagonists (see below). If A_1 receptors are also transported to the central terminals they would be in a position to affect transmitter release. Presynaptic inhibitory effects of adenosine are well-known in the CNS (MASINO and DUNWIDDIE, Chap. 9, this volume) and may involve suppression of Ca^{2+} currents and also inhibition at steps in the transmitter release process that do not require Ca^{2+} influx.

The peripheral actions of adenosine have been found to be pro- or antinociceptive in behavioral tests. In human subjects, pain is evoked when adenosine is administered locally, for example, to blister bases (BLEEHEN and KEELE 1977) or into the coronary artery (LAGERQVIST et al. 1990). Algogenic or pronociceptive effects have also been seen in animal models (TAIWO and LEVINE 1990; KARLSTEN et al. 1992; DOAK and SAWYNOK 1995). Pharmacological studies in animal models, but not in human studies (see Sect. C.I), indicate that the pro-nociceptive effects of peripherally-administered adenosine are mediated by actions at A_2-like adenosine receptors (TAIWO and LEVINE 1990, 1991; DOAK and SAWYNOK 1995). The A_{2A}-selective agonist, CGS 21680, enhances formalin-induced nociceptive behavior only during the latter phases of the test while the A_{2A}-selective antagonist, DMPX, has the opposite effect (DOAK and SAWYNOK 1995). Moreover, in this study the intradermal injection of adenosine did not induce nociceptive behavior in the absence of coadministered formalin. These results raise the possibility that adenosine may not activate sensory neurons directly but may enhance the stimulatory effects of other inflammatory mediators (DOAK and SAWYNOK 1995).

In contrast to the local algogenic effects of adenosine, the systemic administration of adenosine at low dosage has been shown to induce analgesia in humans as discussed in detail in Sect. C. In animal models the activation of peripheral A_1 adenosine receptors induces antinociceptive effects (TAIWO and LEVINE 1990; KARLSTEN et al. 1992; ALEY et al. 1995; DOAK and SAWYNOK 1995; SAWYNOK et al. 1998). Tissue adenosine levels are elevated in conditions of ischemia and inflammation (CRONSTEIN 1995). It has been suggested that this adenosine may activate peripheral A_1 receptors which may participate in reducing post-inflammatory pain since the peripheral administration of A_1 antagonists increase, and conversely A_1 agonists decrease, pain behaviors in nociceptive testing (ALEY and LEVINE 1997; DOAK and SAWYNOK 1995).

Additional antinociceptive effects of adenosine may come through its anti-inflammatory effects. The anti-inflammatory actions of certain pharmacological agents have also been attributed to the release of adenosine at inflamed sites (CRONSTEIN 1995; GADANGI et al. 1996). The anti-inflammatory effects of aspirin but not that of the glucocorticoid, dexamethasone, are mediated via adenosine acting as anti-inflammatory autacoid independent of inhibition of either COX-1 or COX-2 or of NFkB p105 (CRONSTEIN et al. 1999). The peripheral inhibition of adenosine kinase (AK) also reduce inflammation (FIRESTEIN et al. 1994; CRONSTEIN et al. 1995; ROSENGREN et al. 1995). The anti-inflammatory effects of adenosine have been attributed to peripheral activation of A_2-like receptors which, amongst other actions, inhibits neutrophil adhesion and prevents the secretion of inflammatory cytokines (GADANGI et al. 1996). Thus, while A_2-like receptor activation might be involved in pronociception, anti-inflammatory effects are also seen which may lead to diminished post-inflammatory pain.

It is possible that adenosine-mediated actions at A_3 adenosine receptors on immune cells may also participate in modulating peripheral somatosensory function. A predominant response to activating A_3 receptors is degranulation of mast cells (LINDEN 1994), causing release of multiple inflammatory mediators (HANDWERKER and REEH 1991). Indeed, it has been found that the A_3 agonist N^6-benzyl-NECA potentiates formalin-induced inflammation (SAWYNOK et al. 1997). Further involvement of A_3 receptors in inflammation and pain may be a function of adenosine-mediated inhibition of release of tumor necrosis factor α (TNF-α), a proinflammatory cytokine, from macrophages and monocytes (SAJJADI et al. 1996).

2. Actions in the Dorsal Horn

a) ATP

Central effects of ATP in somatosensory pathways were first investigated in the cuneate nucleus where iontophoretic application of ATP in vivo was found to excite cuneate neurons (GALINDO et al. 1967). Subsequently ATP application was shown also to excite neurons in the dorsal horn in vivo (SALT and HILL 1983; FYFFE and PERL 1984; SALTER and HENRY 1985). Each of these groups of neurons receive direct input from primary afferents and it was observed that the excitatory effects of ATP are correlated with input from non-nociceptive primary afferents (SALTER and HENRY 1985). Given the more recent findings that ATP evokes release of glutamate from primary afferent terminals, it is possible that some of the excitatory effects of in vivo administration of ATP may have occurred indirectly through glutamate release. However, ATP also directly affects dorsal horn neurons: ATP causes an inward current in cultured dorsal horn neurons (JAHR and JESSELL 1983) and in acutely isolated dorsal horn neurons (BARDONI et al. 1997), such currents not being affected by blocking glutamate receptors.

The effects of exogenously administered ATP would suggest that if it is released synaptically ATP should produce an excitatory post-synaptic response. Evidence for an excitatory postsynaptic current (EPSC) mediated by ATP in dorsal horn neurons has been obtained in a subpopulation of dorsal horn neurons in an in vitro slice preparation (BARDONI et al. 1997) and in primary cultures of laminae I–III (Jo and SCHLICHTER 1999). These EPSCs are resistant to blockade by glutamate receptor antagonists but are blocked by the P2-purinoceptor antagonists, suramin or PPADS. In a spinal cord slice preparation, P2X-mediated EPSCs are evoked by dorsal root stimulation which may indicate that there are ATP-mediated synaptic responses at a subgroup of primary afferent-dorsal horn synapses, possibly at non-nociceptive synapses. However, because ATP can be co-released from intrinsic GABAergic dorsal horn neurons (Jo and SCHLICHTER 1999) it is possible that P2X-mediated responses following dorsal root stimulation may also be produced disynaptically.

As $P2X_4$ and $P2X_6$ receptor subunits are the predominant receptor subtypes expressed by dorsal horn neurons, the effects of suramin and PPADS might seem paradoxical because homomeric assemblies of these receptor subunits are insensitive to either of these compounds. However, Seguela and colleagues have reported that heteromeric assemblies of $P2X_{4/6}$ are inhibited by suramin and by PPADS (LE et al. 1998). These heteromultimers are activated by α,β-meATP, whereas this ATP analogue excites less that 10% of the ATP-sensitive dorsal horn neurons (BARDONI et al. 1997). Thus, the subunit composition of native P2X receptors in dorsal horn neurons cannot be assigned unequivocally on the basis of the pharmacological profiles.

b) Adenosine

Considerable evidence from physiological, biochemical, and pharmacological studies indicates that adenosine is an inhibitory mediator in many CNS areas (see MASINO and DUNWIDDIE, Chap 9, this volume). In the spinal dorsal horn, adenosine has been shown to inhibit the firing discharge of nociceptive, as well as non-nociceptive, neurons in vivo (SALTER and HENRY 1985). This inhibition is mediated at least in part postsynaptically as adenosine causes hyperpolarization of dorsal horn neurons in vivo (DE KONINCK and HENRY 1992) and in vitro (LI and PERL 1994) via activation of a potassium conductance.

Adenosine-mediated actions in the spinal cord may also involve presynaptic actions. Evidence that adenosine might have presynaptic effects on primary afferent terminals in the spinal cord was first shown by PHILLIS and KIRKPATRICK (1978). Subsequent studies have indicated that adenosine may inhibit release of substance P and CGRP from dorsal horn slices (SANTICIOLI et al. 1992, 1993), but whether this was due to pre- or post-synaptic mechanisms was not determined. Adenosine decreases the rate of spontaneously occurring miniature EPSCs in dorsal horn neurons (LI and PERL 1994), suggesting that adenosine may act pre-synaptically to inhibit glutamate-mediated synaptic transmission in the dorsal horn. Moreover, it was found that bath-applied adenosine suppressed all polysynaptic EPSCs, and inhibited approximately 40% of the monosynaptic EPSCs, in dorsal horn neurons. REEVE and DICKENSON (1995a) reported that selective A_1 adenosine receptor agonists inhibit the C-fiber evoked responses in dorsal horn neurons, but that an A_{2A} selective agonist was without significant effects. In addition, the excitatory responses of dorsal horn neurons following subcutaneous application of formalin were suppressed by centrally-administered A_1 agonists. The inhibitory responses were blocked by prior treatment with adenosine receptor antagonists, suggesting that A_1 receptor activation might inhibit acute nociceptive transmission in the dorsal horn and central responses to peripheral inflammation.

In animal models of acute and chronic pain, intrathecal administration of adenosine or adenosine analogs induce anti-nociception (SAWYNOK and SWEENEY 1989; SAWYNOK 1998). Significant decreases in pain behavior are observed in the formalin test (MALMBERG and YAKSH 1993; DOAK and SAWYNOK 1995; POON and SAWYNOK 1995), in thermal hyperalgesic states following nerve injury (YAMAMOTO and YAKSH 1993), and in strychnine- (SOSNOWSKI and YAKSH 1989; SOSNOWSKI et al. 1989) or prostaglandin $F_2\alpha$- (MINAMI et al. 1992a,b) induced hyperesthetic/allodynic behavior. Activation of spinal adenosine receptors, therefore, induces antinociception in a wide number of animal models of acute, chronic and pathological pain states.

Studies using receptor-selective agonists indicate that activating A_1 adenosine receptors produces antinociception in behavioral models of pain (KARLSTEN et al. 1990, 1991; DELANDER et al. 1992; MALMBERG and YAKSH 1993; DELANDER and KEIL 1994; POON and SAWYNOK 1995). As discussed above, binding sites for both A_1- and A_2-like receptors are found in the spinal cord and mRNA for both receptor subtypes are found in intrinsic dorsal horn neurons, and spinally-administered non-selective A2 receptor agonists induce antinociception (SAWYNOK and SWEENEY 1989). Thus, it is possible that A_2-like receptors might also participate in antinociceptive effects of adenosine.

The actions of extracellular adenosine are terminated principally by uptake and intracellular metabolism (ZIMMERMANN, Chap. 8, this volume). Hence, pharmacological manipulations which decrease the clearance of extracellular adenosine in the dorsal horn would be expected either to induce antinociception or to enhance adenosine-mediated antinociceptive effects.

Inhibition of adenosine uptake by inhibitors of nucleoside transporters has been shown to potentiate and prolong adenosine-mediated antinociception (KEIL and DELANDER 1995). The possibility that other inhibitors of adenosine clearance induce antinociception was examined by KEIL and DELANDER (1992) who found antinociceptive effects of an inhibitor of adenosine kinase. Other studies have also shown that inhibition of adenosine kinase results in anti-nociception (KEIL and DELANDER 1994; POON and SAWYNOK 1995), and in vivo and in vitro studies indicate that adenosine kinase inhibitors lead to increases in extracellular adenosine levels (GOLEMBIOWSKA et al. 1995, 1996). Inhibition of adenosine deaminase did not induce demonstrable antinociception in the absence of exogenous adenosine (KEIL and DELANDER 1994; POON and SAWYNOK 1995), but antinociception induced by exogenous adenosine is potentiated following the inhibition of adenosine deaminase (KEIL and DELANDER 1994). Together, these studies indicate that pharmacological manipulations which increase endogenous adenosine levels may induce appreciable levels of antinociception, and potentiate physiological manipulations which increase extracellular adenosine levels.

In contrast to the antinociceptive effects which occur following spinal adenosine receptor agonist administration, spinally administered adenosine receptor antagonists produce behaviors indicating hyperalgesia – exaggerated pain to a normally painful stimuli – (JURNA 1984; SAWYNOK et al. 1986), allodynia – pain evoked by a non-painful stimuli – (KEIL and DELANDER 1996), or nociception without overt peripheral stimulation (NAGAOKA et al. 1993; KEIL and DELANDER 1996). The ability of spinal adenosine receptor antagonists to induce these behaviors suggests adenosine might tonically inhibit nociceptive pathways and hence processes which diminish spinal adenosine activity could potentially participate in or underlie certain pain states.

c) *Extracellular Conversion of ATP to Adenosine Mediates an Inhibitory Postsynaptic Response to Low Threshold Primary Afferent Input*

Application of ATP not only excites dorsal horn neurons but also produces inhibition through extracellular degradation of ATP to adenosine (SALTER and HENRY 1985; LI and PERL 1995). The effects of ATP in vivo are correlated strongly with the functional subtype of dorsal horn neuron, and are similar to the responses to physiologically stimulating non-nociceptive inputs to the neurons. This led to the suggestion that ATP might be released from non-nociceptive primary afferents and that conversion of ATP to adenosine may mediate some types of inhibitory responses of nociceptive dorsal horn neurons to stimulation of non-nociceptive inputs (SALTER and HENRY 1985). Subsequently, it was found that stimulating vibration-sensitive primary afferents, likely Pacinian corpuscle afferents, produced an inhibitory postsynaptic response in nociceptive dorsal horn neurons that could be accounted for by ATP-derived adenosine (SALTER and HENRY 1987); this inhibition was blocked by adenosine receptor antagonists and potentiated by adenosine uptake inhibitors. Moreover, the inhibition was shown to be produced postsynaptically by an increase in potassium conductance (DE KONINCK and HENRY 1992) due to activating K_{ATP} channels (SALTER et al. 1992). Nociceptive, but not non-nociceptive, dorsal horn neurons show this adenosine-mediated inhibitory postsynaptic potential (IPSP) and, therefore, adenosine acts physiologically, as well as pharmacologically, to decrease nociceptive transmission in the dorsal horn.

IV. Potential Therapeutic Implications from Basic Studies

Some potential roles for ATP and adenosine in modulating nociceptive signaling at peripheral and central sites are summarized in Fig. 1. As discussed above, these compounds have multiple functions in somatosensory processing which are dependent upon the site of action and the receptor subtype which is activated. Pharmacological agents or physiological manipulations which specifically target the antinociceptive actions of adenosine and ATP have the potential to provide analgesia. However, algogenic or pronociceptive results might occur if agents are used which are non-selective, either in their mechanism of action or in their primary site of action. Therefore, site- or receptor-targeting strategies are needed in order to maximize analgesic effects while minimizing any algogenic effects.

1. Targeting P2 Purinoceptors

As ATP may be involved in activating nociceptive sensory neurons through $P2X_3$-containing receptors, peripherally-acting antagonists selective for these

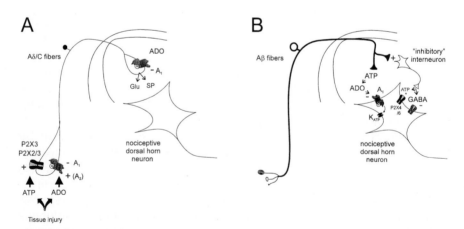

Fig. 1A,B. Potential peripheral and central roles of adenosine- and ATP-mediated modulation of nociceptive transmission. **A** Roles with respect to Aδ/C fibers. **B** Roles in functioning of Aβ fibers. + indicates effects that enhance nociceptive transmission; − indicates effects that inhibit nociceptive transmission. See text for details

receptors might be analgesics. Potential actions of such compounds administered peripherally would not likely be associated with significant side effects because $P2X_3$ subunits are only expressed by capsaicin-sensitive primary afferents.

The effect of $P2X_3$-selective compounds administered spinally might also lead to effects on transmission of nociceptive information. These receptor subunits are not expressed by dorsal horn neurons and thus selective $P2X_3$ agonists or antagonists would not be expected to have direct effects on these neurons. However, $P2X_3$ agonists might depolarize central terminals of nociceptive primary afferents and cause release of such excitatory transmitters as glutamate (Gu and MacDermott 1997). The resultant depolarization could potentially activate these afferents antidromically, leading to release of pronociceptive mediators in the periphery. Such antidromic activation has been implicated as being important in the pathogenesis of peripheral inflammation (Rees et al. 1995). Conversely, stimulating P2X receptors on primary afferent terminals will induce primary afferent depolarization which can inhibit central release of transmitters (Shapiro et al. 1980). Thus, the net effect of central agonists or antagonists at $P2X_3$ receptors might be difficult to predict.

The co-release of ATP and GABA from superficial dorsal horn neurons (Jo and Schlichter 1999) raises an interesting and novel possibility for the involvement of P2X receptors in nociception. This possibility relates to the clinical findings that Aβ fiber input, which normally inhibits pain, may be excitatory (i.e., allodynic) in a number of chronic pain conditions (Woolf and Doubell 1994). As described above Aβ-fiber mediated inhibition of nociceptive dorsal horn neurons may, in the case of stimulating Pacinian corpuscle afferents, come about through ATP-adenosine conversion. There is also strong evidence for a disynaptic GABAergic inhibition upon stimulating Aβ fibers (e.g., De Koninck and Henry 1994). The results of Jo and Schlichter imply that synaptic transmission by "GABAergic" dorsal horn neurons is ambiguous because of the co-release of ATP, and there are a variety of potential mechanisms where net inhibition could be converted to net excitation (Salter and De Koninck 1999). Were such conversion of inhibition to excitation to occur, it can be readily seen how Aβ fiber stimulation would now lead to enhancement, rather than inhibition, of nociceptive transmission by means of stimulating P2X receptors.

On balance, inhibiting P2X receptors, especially $P2X_3$ in the periphery, appears to have the potential to be analgesic or antinociceptive. In behavioral studies in animals, suramin and PPADS are reported to be antinociceptive (Ho et al. 1992; Driesen et al. 1994). Also, there is evidence that suramin may lower pain ratings when used as an antineoplastic agent (e.g., Eisenberger et al. 1993), but the mechanism for this analgesia is not known and could be related to decrease in tumor size. One important caveat relating to the use of suramin is that this compound has been recently found to block the NMDA subtype of glutamate receptor (Gu et al. 1998; Wang and Salter, unpublished), a receptor subtype known to be important in pain and nociception (Dickenson et al.

1997). Thus if suramin crosses the blood-brain barrier this effect on NMDA receptors, on the one hand, limits the interpretation of studies in terms of elucidating the mechanism underlying the antinociceptive/analgesic effects of suramin. On the other hand, the combined effects of suramin on P2X and NMDA receptors may be very desirable therapeutically. In contrast to suramin, PPADS has been found not to inhibit ionotropic glutamate receptors (Gu et al. 1998) and therefore its use would not be subject to the same issues.

2. Targeting Adenosine Receptors

Systemic administration of adenosine may induce analgesic effects through actions at spinal and peripheral A_1 receptors, and anti-inflammatory effects at peripheral A_2 or A_3 receptors. A potential limitation to the use of adenosine in the treatment of pain and inflammation, however, may be the peripheral algogenic effects of adenosine receptor stimulation. The overall effect of adenosine, therefore, may depend on the relative magnitude of opposing pro- and anti-nociceptive systems. The use of agents selective for a particular subtype of receptor could be considered a pharmacological targeting strategy which would avoid possible pro-nociceptive effects seen with non-selective agents. As indicated from animal studies, systemically administered A_1-selective agonists would be expected to induce analgesia at both peripheral and, if able to cross the blood brain barrier, central A_1 receptors. However, results from the clinical use of a selective A_1 antagonist indicate that stimulating peripheral A_1 receptors is pro-nociceptive in humans (PAPPAGALLO et al. 1993; GASPARDONE et al. 1995; see Sect. C.I). Consequently, species differences with respect to these receptor-mediated responses may complicate the prediction of clinical effects of selective adenosine agonist.

A_{2A}-selective agonists induce anti-inflammatory actions in the periphery, but concurrent algogenic effects might limit any clinical usage. In the spinal cord, A_{2A} agonists might induce analgesia; in animal models, however, these agonists often produce profound motor-impairing effects which may be expected to occur in human patients. Peripheral A_{2A}-selective antagonists may inhibit adenosine-mediated pro-nociception, but these agents might facilitate inflammatory processes. The effects of compounds with agonist activity at A_3 adenosine receptors may be expected to induce analgesia via A_3 receptor-mediated anti-inflammatory effects but additional studies are needed to clarify the activity of this receptor subtype in somatosensory processes.

3. Other Potential Therapeutic Implications

As described above, the antinociceptive effects of spinal adenosine may be mediated in part through the activation of K_{ATP} channels. Opioid-induced antinociceptive effects have also been attributed in part to activity of these channels (WELCH and DUNLOW 1993). Moreover, release of adenosine in the spinal cord has been shown to be involved in the antinociceptive actions of opioid compounds (see SAWYNOK 1998). Adenosine and opioid-mediated antinociception show overlap and this overlap might result in cross-tolerance in the spinal cord, similar to the cross-tolerance between opioids and adenosine analogs in the periphery (ALEY et al. 1995). However, in a rat model of central pain there was no cross-tolerance for the antinociceptive effect of adenosine receptor stimulation and morphine (VON HEIJNE et al. 1998). Also, supra-additive interactions between adenosine and opioid analgesics might occur in clinical settings, such that the overall efficacy in the treatment of certain pain states might be enhanced at lower total analgesic doses (DELANDER and KEIL 1994).

In the preceding sections the focus has been on the effects of ATP and adenosine on neurons. However, the effects of extracellular ATP are not restricted to neurons and ATP has been found to cause Ca^{2+} responses in nearly all astrocytes from the dorsal horn (SALTER and HICKS 1994). These cells express two pharmacologically distinct types of P2Y-purinergic receptor (Ho et al. 1995), both of which are linked to activation of the phospholipase $C\beta/IP_3$ pathway (SALTER and HICKS 1995; IDESTRUP and SALTER 1998). In astrocytes, as in most non-excitable cells, Ca^{2+} responses are a principal form of cell signaling. Astrocytes and other glial cells may have distinct roles in pathological processes which occur following peripheral nerve injury (HAJOS et al. 1990; GILMORE et al. 1990; GARRISON et al. 1991; SVENSSON et al. 1993) and these cells may contribute to abnormal nociceptive processing and pathological pain states (MELLER et al. 1994). Although much work is needed in this area, it is conceivable that pharmacological agents with actions at P2Y receptors might represent a novel class of analgesic agents.

C. Clinical Aspects of Nociception/Antinociception

I. General Considerations

As discussed above, adenosine and ATP have complex influences on pain transmission in the periphery as well as at central sites. ATP receptor (P2) activation is associated with nociception (algogenic effect), while adenosine receptor activation may both be algogenic as well as analgesic/antinociceptive. In humans there is limited information regarding ATP and its involvement in pain modulation (see Sect. B.III.1.a). This limited information is partly due to its rapid degradation as well as to lack of clinically available receptor antagonists. On the other hand, there is a considerable amount of information related to adenosine and pain modulation in humans. Therefore, this section will focus on human studies on adenosine and pain.

An early clinical observation, when adenosine was exogenously administered, was the occurrence of pain symptoms (SYLVÉN et al. 1986). These were localized to the chest, head, abdomen and/or extremities, were dose-dependent and observed both by intravenous (IV) as well as intraarterial injection and by IV infusion (SYLVÉN 1993). From those studies using relatively high doses of exogenous adenosine that produce pain, as well as results from the pain inhibitory effects of adenosine receptor antagonists (methylxanthines, i.e., caffeine and theophylline) in patients with ischemic heart disease, it was put forward that locally released endogenous adenosine sensitizes pain afferent nerves and thereby serves as an important messenger link for tissue ischemic pain (cf. SYLVÉN 1993). In this context, the A_{2A} adenosine receptor subtype has been considered to be involved in the peripheral pain-inducing effect in animal models (see Sect. B.III.1.b and SAWYNOK 1998). However, clinically applied bamiphylline, a relatively A_1-selective adenosine receptor antagonist, both inhibits pain induced by exogenous adenosine as well as pain induced by ischemia/hypoxia (PAPPAGALLO et al. 1993; GASPARDONE et al. 1995). Therefore, it is likely that peripheral algogenic effects of adenosine in humans involve A_1 adenosine receptor activation. Adding further to the complexity, virtually all animal experimental evidence related to the analgesic/antinociceptive action of adenosine suggests that the adenosine A_1 receptor is involved in this effect (see Sect. B.III.1.b and SAWYNOK 1998). To what extent human peripheral adenosine A_1 receptors contribute to the potential clinical analgesic/antinociceptive effect of adenosine still remains to be elucidated. However, recent clinical studies have generated novel information about prominent analgesic/antinociceptive actions of exogenous adenosine. These studies have been performed using dosages of adenosine that are below those previously reported to be painful (i.e., below $80\,\mu g\,kg^{-1}\,min^{-1}$ infusion). The following overview summarizes the analgesic/antinociceptive effects of adenosine in different clinical conditions, administered both by IV and intrathecal (IT) route (Tables 1 and 2).

II. Administration of Adenosine and Adenosine Analogues

1. Safety and Adverse Effects

a) IV Administration

When considering new compounds for pain treatment, knowledge about and minimization of adverse effects are crucial. In an IV dose range of 50–$70\,\mu g\,kg^{-1}\,min^{-1}$, subjects and non-anesthetized patients may feel a slight chest pressure and cutaneous flushing. In only one of all the healthy subjects exam-

Table 1. Clinical conditions, modes of administration and doses of adenosine in studies related to antinociception

	Intravenous (IV) dose ($\mu g\,kg^{-1}\,min^{-1}$)	Intrathecal (IT) dose (μg)
Healthy volunteers	50–70	500–2000
Surgical patients	80	500
Coronary artery disease patients	35	n.d.
Chronic neuropathic pain patients	50	500–1000

Table 2. Analgesic/antinociceptive effects of IV and IT adenosine in humans

	Healthy volunteers	Surgical patients	CAD[a] patients	Chronic neuropathic pain patients
Elevated skin heat-pain threshold	Yes IV (Ekblom et al. 1995; Sylvén et al. 1996)			
Reduced intensity of cold immersion pain	No IT (Rane et al. 1998)			
Elevated provoked tactile pain threshold	No IV (Segerdahl et al. 1995a); Yes IT (Rane et al. 1998)			Yes IV (Sollevi et al. 1995; Belfrage et al. 1995)
Reduced area of tactile allodynia	Yes IV (Segerdahl et al. 1995a; Sjölund et al. 1999); Yes IT (Rane et al. 1998)			Yes IV (Sollevi et al. 1995; Belfrage et al. 1995; KF Sjölund et al., unpublished)
Reduced spontaneous pain intensity		No IV (Segerdahl et al. 1995b, 1996, 1997)		Yes IV (Sollevi et al. 1995; Belfrage et al. 1995; KF Sjölund et al., unpublished)
Reduced experimental forearm ischemic pain	Yes IV (Segerdahl et al. 1994); Yes IT (Rane et al. 1998)			
Reduced ischemic pain (angina pectoris)			Yes IV Sylvén et al. 1996)	
Reduced intraoperative anesthetic requirement		Yes IV (Segerdahl et al. 1995b, 1996, 1997); No IT (Rane 2000)		
Reduced postoperative analgesic requirement		Yes IV (Segerdahl et al. 1995b, 1996, 1997); No IT (Rane 2000)		

[a] CAD, coronary artery disease.

ined ($n = 50$), who had an prior history of gastric ulcer, was epigastric pain experienced during infusion of $50\mu g\ kg^{-1}\ min^{-1}$ (FORSBERG et al. 1998). This reaction is previously described at higher doses (WATT et al. 1987). Reported adverse effects during IV adenosine infusion can be expected from the dose-dependent cardiovascular effects as well as the sensitizing action on peripheral nerves related to nociception (see section above). Consequently, there is a clinical IV dose range from approximately $35\mu g\ kg^{-1}\ min^{-1}$ to $70\mu g\ kg^{-1}\ min^{-1}$ where analgesic/antinociceptive effects of adenosine safely can be evaluated (Table 1). Further, constipation is a common adverse effect of many drugs used for pain relief. There are experimental data suggesting endogenous adenosine to be involved in regulation of lower intestine propulsion and exogenously administered adenosine analogs were also shown to inhibit upper gastrointestinal motility (FARGEAS et al. 1990). The effect on gastric emptying of IV adenosine was therefore investigated with a placebo-controlled cross-over design, applying the acetaminophen absorption test. An infusion dose of $50\mu g\ kg^{-1}\ min^{-1}$ (over 2h) did not affect the rate of gastric emptying in man (FORSBERG et al. 1998).

b) Intrathecal Administration

In animal models, a dose-dependent and reversible motor impairment by various adenosine analogs has been demonstrated, suggesting effects in the ventral horn of the spinal cord (HOLMGREN et al. 1986). However, motor impairment has not been reported after administration of the endogenous compound, adenosine to mice (DELANDER and HOPKINS 1986). Toxicity by chronic intrathecal (i.t.) administration of the analogue, R-PIA has also been investigated, and no change was seen in spinal cord vascularization, morphology, or quantitative morphometry (Karlsten et al. 1992, 1994). With respect to adenosine, potential toxic effects of chronic (two weeks) twice-daily i.t. administration ($100\mu g$, maximally deliverable dose) in rats has been evaluated with regard to behavior, spinal cord morphology, and quantitative morphometry. Long-term administration was not associated with any detectable neurotoxicity or adverse behavioral effects (RANE et al. 1999).

A safety and dose escalation study has been performed in healthy volunteers, injecting adenosine ($500\mu g$, $1000\mu g$, or $2000\mu g$) at the lower lumbar level. Dose escalation above $2000\mu g$ was interrupted due to the appearance of a transient (30min) dull pain in the lumbar region after administration of a dose of $2000\mu g$ in one subject (RANE et al. 1998). In a clinical study in chronic neurogenic pain patients, 5 out of 14 i.t. adenosine injections caused transient lumbar pain (BELFRAGE et al. 1999). These mild local pain reactions were *not* dose-dependent and were noted at both $500\mu g$ and $1000\mu g$. Local pain symptoms of i.t. adenosine in neurogenic pain patients were abolished upon repeated injection when combined with a low, and non-motor blocking, dose of a local anesthetic agent (A. Sollevi et al., unpublished). This lumbar pain adverse effect is possibly mediated via direct stimulation of the primary afferents of the dorsal root or by direct influence at superficial layers of the cord. Adenosine A_1- or A_2-like receptors, both receptor subtypes suggested to be involved in the peripheral algogenic effect of adenosine, may be involved in this transient pain sensation. Another possibility may be that adenosine induces meningeal vasodilatation via A_2-like receptor activation, leading to a migraine-like vascular pain. It has been reported that i.t. injection of an A_1 receptor agonist causes vasodilatation in the spinal cord (KARLSTEN et al. 1992).

Adenosine (up to $2000\mu g$) administered to healthy volunteers did not induce sedation or motor deficiencies and neurological examination revealed no disturbances in extremity reflexes or balance. Voiding reflexes were unaffected. No cardiovascular or other systemic side effects have been observed after i.t. administration to volunteers or patients (RANE et al. 1998; BELFRAGE et al. 1999).

c) Pharmacokinetics in CSF

Adenosine levels in CSF has also been determined in volunteers (RANE et al. 1998). In 12 subjects, concentrations of adenosine were analyzed at preinjection and 10 min after injection. Preinjection values were detectable at normal CSF levels of approximately $0.05\mu mol\ l^{-1}$ (SOLLEVI 1986). Peak concentrations showed marked increase, up to $200\mu mol\ l^{-1}$, corresponding to elevation of more than three orders of magnitude after $1000\mu g$ and $2000\mu g$. In two of the subjects, elimination half-life was determined by serial CSF samples. There was good semilogarithmic correction, with half-lives of approximately 10 min and 20 min. Thus, pharmacologically elevated adenosine levels in human CSF can remain above basal concentrations for a few hours, which is in marked contrast to the elimination within seconds in circulating blood.

2. Effects of Adenosine in Healthy Volunteers

An IV infusion of adenosine of $50-70\mu g\ kg^{-1}\ min^{-1}$ increases the heat pain threshold (C-fiber mediated) in normal hairy skin of healthy volunteers, with no effect on a suprathreshold heat pain provocation (EKBLOM et al. 1995). Similar results on elevated heat pain threshold were also reproduced by SYLVÉN et al. (1996). These findings are interesting since other analgesic agents like morphine and ketamine do not affect the skin heat pain threshold (EKBLOM et al. 1995). Adenosine does not seem to affect normal (non-noxious) cold (Ad-fiber mediated) or warmth (C-fiber mediated) perception. The studies lend support to a selective influence of IV adenosine on thermal C-fiber mediated nociceptive transmission in the normal non-sensitized state, without apparent influence on thermal non-noxious threshold perception.

A forearm ischemic pain model has also been used in humans. This well-documented tourniquet method was applied to induce a continuous nociceptive type of pain, mainly C-fiber mediated (CREWS et al. 1994). The experimental model may resemble clinical postoperative deep somatic pain.

The subjects rated pain every minute, while the tourniquet was inflated on their upper arm, for up to 30 min. Pain (visual analog scale 0–100) was assessed every minute during the test and the scores were then added to a sum of pain scores (SPS). Adenosine infusion ($70\,\mu g\,kg^{-1}\,min^{-1}$) reduced this SPS by 30% compared to placebo, i.e., to the same extent as a clinically effective IV dose of morphine (SEGERDAHL et al. 1994). In patients with coronary artery disease (CAD) and angina pectoris, a low dose of adenosine infusion (Table 2) has been shown to reduce exercise-induced pain rating in a double-blind placebo-controlled study (SYLVÉN et al. 1996). This latter study illustrates that IV adenosine can also reduce visceral pain.

In relation to i.t. adenosine administration, different models of pain were tested; e.g., tactile and thermal thresholds for perception and pain, acute cold immersion pain (foot in ice water for 1 min), the forearm ischemic (30 min) test, and sensory changes induced by experimental skin inflammation (RANE et al. 1998). In analogy with previous studies with IV adenosine administration, i.t. adenosine also reduced the ischemic forearm pain rating. However, the cold immersion test, representing acute but brief primary C-fiber mediated pain, was unaffected.

Two different experimental models of inducing inflammatory pain of the skin has been applied in relation to the action of adenosine in volunteers. In one model, inflammatory pain was induced by mustard oil (MuO), topically applied to the skin of the forearm. This will result in a chemical burn, strong enough to activate nociceptive afferent C-fibers and induce a surrounding area of secondary allodynia or hyperalgesia, as an expression of central sensitization (KOLTZENBURG et al. 1992). Another validated method of inducing skin burn with secondary allodynia/hyperalgesia is by applying a Peltier thermode firmly to the skin, keeping it at a temperature of 47°C for 7 min. In these models, double-blind placebo controlled crossover studies have been performed. Adenosine infused at 50–$60\,\mu g\,kg^{-1}\,min^{-1}$ during the 60-min test period attenuated development of the areas of allodynia by 30%–50% (SEGERDAHL et al. 1995a, SJÖLUND et al. 1999). Although adenosine infusion clearly reduces areas of hypersensitive skin, the reduction in tactile pain threshold within the allodynic area occurring after inflammation was not attenuated. When other volunteers were subjected to i.t. adenosine in an open study, the area of secondary allodynia after MuO application was reduced in a similar fashion to the IV infusion (RANE et al. 1998). This effect also occurred without influencing tactile pain thresholds in the remaining secondary allodynic area. Thus, experimental studies in healthy volunteers suggest that adenosine treatment primarily counteracts pain mechanisms that are involved in central sensitization.

3. Effects of Adenosine in Acute Surgical Pain

a) Intraoperative Pain

In three clinical studies investigating shoulder surgery (30 patients), breast surgery (72 patients) and hysterectomies (41 patients) (SEGERDAHL et al. 1995b, 1996, 1997), representing deep somatic, cutaneous/subcutaneous, and visceral pain, respectively, anesthetic requirements and postoperative analgesics requirements were compared in relation to adenosine/placebo infusion. The studies were randomized and double-blind. During surgery, anesthetic requirements were significantly reduced by adenosine. An antinociceptive effect of adenosine infusion is a likely explanation for the difference in anesthetic requirement. Intraoperatively, systolic blood pressure was increased in the placebo treatment groups, compared with presurgical values, while the adenosine treated groups, in all studies, showed a stable level of intraoperative systolic blood pressure. The significantly smaller increase in systolic blood pressure at commencement of surgery indicates a significantly weaker response to painful surgical stimulation in the adenosine groups. Further support for the theory of a pain reducing effect of intraoperative adenosine infusion is that fewer patients, 8/31 vs 19/32, perceived pain when regaining consciousness after surgery in the adenosine group compared to placebo (SEGERDAHL et al. 1995b). However, there was a notable difference between type of surgery and influence of adenosine. Anesthetic requirements were similar during placebo treatment in all studies, but the effect of adenosine was less pronounced during superficial and most pronounced during visceral surgery. Such a differential susceptibility to the action of agents with presumed analgesic activity is known clinically (ARNÉR and ARNÉR 1985) and this may reflect nociceptive activity from different tissues activating different mechanisms.

In order to test if i.t. adenosine administration in conjunction with surgical trauma could reduce intraoperative anesthetic requirement, a randomized double-blind placebo controlled study in 40 patients undergoing elective hysterectomies (visceral surgery) was performed (RANE 2000). Patients received an i.t. injection of 500 μg adenosine/placebo immediately prior to induction of anesthesia and surgery. Anesthetic requirement and postoperative analgesic demand was not reduced by the i.t. adenosine treatment.

b) Postoperative Pain – Analgesic Requirements

After breast surgery and hysterectomies, at a similar degree of pain relief, the 24-h opioid requirement after adenosine infusion was reduced by 27% and 18%, respectively (SEGERDAHL et al. 1995b, 1997). This indicates an extended antinociceptive effect of IV adenosine treatment. It has therefore been suggested that adenosine may affect neuronal mechanisms involved in central hyperexcitability (sensitization), and that such effects persist for longer than the period of direct exposure to the compound. However, as adenosine was only given in conjunction with surgery, the possibility of a peripheral antinociceptive effect of adenosine cannot be excluded, since adenosine has well-known anti-inflammatory properties (CRONSTEIN 1994). Such an effect may reduce the secondary inflammation and, consequently, reduce ongoing nociceptive stimulation. The latter is further supported by the fact that i.t. adenosine injection in conjunction with visceral surgery did not affect the postoperative analgesic requirement.

The rationale for administering adenosine intraoperatively is to reduce perioperative opioid related adverse effects, e.g., postoperative nausea and vomiting, respiratory disturbances, and sedation, and to improve recovery. Analysis of postoperative adverse effects in the 145 randomized surgical patients (SEGERDAHL et al. 1995b, 1996, 1997) demonstrated that the incidence of opioid-related adverse effects was unaffected (M. Segerdahl et al., unpublished data). It is therefore unlikely that the addition of adenosine infusion during general anesthesia would improve perioperative quality and recovery in unselected patients. Further, the potential risk of cardiovascular complications should be kept in mind in relation to patients with coronary heart disease, since adenosine induces intramyocardial flow redistribution and ischemia in a dose range above $80\mu g\ kg^{-1}\ min^{-1}$. On the other hand, intraoperative adenosine may be of use for pain relief in selected cases with a known history of chronic neuropathic pain where all efforts to counteract central sensitization should be used.

4. Effects of Adenosine in Chronic Neuropathic Pain Patients

A considerable proportion of the pain in cases of neuropathic pain is related to factors involved in central sensitization, expressed as hypersensitivity to stimulation of the skin or deep tissue (JENSEN 1996). In a first report of cases, two patients suffering neuropathic pain were treated with a low dose of adenosine infusion resulting in alleviation of pain (SOLLEVI et al. 1995). In one of these patients, 45 min of infusion of adenosine $50\mu g\ kg^{-1}\ min^{-1}$, but not placebo, abolished the preexisting allodynia to touch and warmth, dysesthesia to cold, increased the tactile pain threshold, and normalized the tactile perception and heat pain threshold. A follow-up randomized double-blind placebo controlled crossover study in seven patients suffering from chronic neuropathic pain (BELFRAGE et al. 1995) also indicated pain reduction. All patients had, as part of their pain syndrome, allodynia and hyperalgesia. Patients received adenosine $50\mu g\ kg^{-1}\ min^{-1}$ for 45–60 min. In six patients with spontaneous pain, pain ratings were reduced on average by 50%. The onset time for pain reduction range from 20 min to 60 min. Also tactile pain thresholds in the neuropathic areas were elevated, an indication of reduced hypersensitivity. Furthermore, the duration of the perceived pain relief extended from hours to 4 days, by far outlasting any direct action of the infused compound. Recently, a multi-center, placebo controlled randomized crossover study, involving 26 patients with intractable neuropathic pain of postsurgical or posttraumatic origin, confirmed these earlier results (K.F. Sjölund et al., unpublished data). Several patients have, on a clinical basis, received repeated IV adenosine infusion if the duration of pain relief was longer than 1 week. In these cases, repeated IV administration provides a method of pain treatment. Around 5%–10% of adenosine responsive patients with peripheral neuropathic pain seem to be permanently (>6 months) relieved by a single 60 min adenosine infusion. These effects also involve improvement in pathological tactile hyperphenomena, as assessed by quantitative methods of sensory testing (A. Sollevi et al., unpublished. data).

The first clinical case report on i.t. adenosine agonist administration to a patient with chronic pain relates to the adenosine analogue, R-PIA (KARLSTEN and GORDH 1995). The pain inhibitory and anti-allodynic effects of this A_1-receptor selective agonist (single spinal injection, $50\,\mu g$) lasted for several months. In a recent open tolerability study, adenosine ($500\,\mu g$ or $1000\,\mu g$) was injected i.t. at the lumbar level in 14 patients suffering intractable chronic neurogenic pain with tactile hyperphenomena (pain duration from 8 months to 27 years; BELFRAGE et al. 1999). A majority of patients demonstrated a reduction in spontaneous and evoked pain, including an increased tactile pain thresh-old. Areas of tactile hyperphenomena (allodynia/hyperalgesia) were also markedly reduced. The median duration of pain reduction was 24 h. Thus, in this patient population, i.t. adenosine administration reduces various aspects of pain, primarily via adenosine receptor activation at the spinal level. Randomized placebo-controlled studies are currently being performed for further elucidation of the potential use of i.t. adenosine in chronic pain patients.

5. Possible Sites of Action

An important question is whether or not systemically administered adenosine can reach the CNS and structures modulating nociception. The short elimination time in blood (half-life in seconds) may raise doubts as to whether IV adenosine reaches the CNS at adequate concentrations. The finding of an antinociceptive effect by IV infusion, in accordance with data from experimental as well as clinical studies on i.t. administration, nevertheless speaks in favor of a central site of action. The clear-cut onset time during IV infusion in chronic pain patients also suggests that the compound needs to be distributed for some time in order to reach its site(s) of action. The blood brain barrier is not present in all parts of the CNS and there may be supraspinal regions of penetration, especially at the brain stem level.

Consequently, central spinal and/or supraspinal sites of action seem probable for a major portion of the pain-reducing effects of adenosine in situations involving central sensitization, even though some peripheral anti-inflammatory effects may be involved after IV infusion. It could also be speculated that IV adenosine, by as yet unknown afferent/efferent nervous modulatory influence, could exert some reflex modulation of pain. Since the analgesic effect seen in patients with chronic neuropathic pain was essentially similar by the IV and i.t. routes, it is likely that a central site of action is involved when adenosine is infused for this type of pain relief. On the other hand, the antinociceptive effect of intraoperative adenosine administration is likely not mediated primarily by a central mechanism of action, since intraoperative i.t. adenosine had no effect either on intraoperative anesthetic requirement or on postoperative analgesic consumption (RANE et al. 2000). That study clearly indicates that lumbar spinal mechanisms are not involved. It is therefore more likely that the perioperative pain reducing effect of IV adenosine infusion is mediated via peripheral anti-inflammatory actions, or some supraspinal site of action not accessible via the IT adenosine injection.

III. Adenosine as a Mediator of Analgesia Produced by TENS

As described in Sect. B.III.2.c, physiological stimulation of non-nociceptive primary afferents evokes an adenosine-mediated IPSP in nociceptive dorsal horn. The inhibition of nociceptive dorsal horn neurons by activating non-nociceptive inputs may be the cellular basis for the analgesia produced by stimulation of these inputs by transcutaneous electrical nerve stimulation (TENS) or by innocuous mechanical stimuli such as vibration (WOOLF and THOMPSON 1994). Therefore, it has been hypothesized that analgesia produced by these types of stimulation may be mediated by adenosine which causes the postsynaptic inhibition of nociceptive dorsal horn neurons (SALTER et al. 1993). This possibility has been explored in a double blind test on thermal pain rating in 17 normal human subjects (MARCHAND et al. 1995). TENS reduced heat-pain intensity and unpleasantness but this TENS-induced analgesia was prevented by caffeine (200mg, p.o.). At this dose, caffeine is expected to antagonize central adenosine receptors (FREDHOLM 1980). Thus, the effect of caffeine provides strong support for the hypothesis that analgesia produced by TENS is mediated by adenosine.

These observations indicate that patients receiving TENS therapy for pain relief should refrain from consumption of caffeine or other methylxanthines. Moreover, the analgesic effects of dorsal column stimulation, which is mediated through activation of low-threshold primary afferent neurons, may also be mediated by adenosine (SALTER et al. 1993). The involvement of adenosine and GABAergic mediated mechanisms in the antinociceptive effect of dorsal column stimulation has recently been demonstrated in sciatic nerve lesioned mononeuropathic rats (CUI et al. 1998). Thus, similar considerations concerning the consumption of methylxanthines would also apply to these treatments. As the inhibition of nociceptive dorsal horn neurons by low-threshold primary afferent stimulation may be potentiated by inhibitors of adenosine uptake (SALTER and HENRY 1987), the extracellular adenosine concentration produced by the stimulation may not be sufficient to fully activate the receptors or downstream signaling elements. If this is the case in humans, then several novel pharmacological means for increasing the efficacy of pain treatment with

TENS, dorsal column stimulation or vibration may be suggested. Thus, it may be possible to use as adjuvants in these treatments inhibitors of adenosine uptake or degradation, allosteric enhancers of adenosine receptors, or agents that enhance the activity of K_{ATP} channels.

D. Summary and Conclusions

In this chapter the varied roles played by adenosine and ATP in the peripheral and spinal processing of nociceptive information have been reviewed. Clinical investigations to date have focussed on effects of adenosine in experimentally-induced pain and clinical pain. These studies have shown that adenosine administration reduces pain, primarily in situations that involve enhanced excitability and nociceptive transmission in the central nervous system. Since centrally-mediated enhanced excitability is considered an important factor in chronic pain conditions, the adenosine-induced relief of pain in patients with neuropathic pain suggests that adenosine and adenosine analogues are especially relevant for future development and research. P2X receptors, particularly $P2X_3$ receptors, may represent a potentially important novel target for the development of future analgesics. There is clearly, however, a requirement for better understanding of the mechanism(s) of action for adenosine and ATP in pain modulation. In conclusion, the multiple roles provide a rich number of potential therapeutic targets and, thus, there is a great potential for purinergic compounds as analgesics.

Acknowledgements. Work of the authors is supported by the Medical Research Council of Canada (MWS), Swedish Medical Research Council, and Karolinska Institutet (AS). Thanks to Conor Gallagher for comments on the manuscript and David Wong for assistance in preparation.

References

Aley KO, Green PG, Levine JD (1995). Opioid and adenosine peripheral antinociception are subject to tolerance and withdrawal. J Neurosci 15:8031–8038

Aley KO, Levine JD (1997) Multiple receptors involved in peripheral alpha 2, mu, and A1 antinociception, tolerance, and withdrawal. J Neurosci 17:735–744

Arnér S, Arnér B (1985) Differential effects of epidural morphine in the treatment of cancer-related pain. Acta Anaesthesiol Scand 29:32–36

Bardoni R, Goldstein PA, Lee CJ, Gu JG, MacDermott AB (1997) ATP P2X receptors mediate fast synaptic transmission in the dorsal horn of the rat spinal cord. J Neurosci 17:5297–5304.

Bean BP (1990) ATP-activated channels in rat and bullfrog sensory neurones. Concentration dependence and kinetics. J Neurosci 10:1–10

Belfrage M, Sollevi A, Segerdahl M, Sjölund K-F, Hansson P (1995) Systemic adenosine infusion alleviates spontaneous and stimulus evoked pain in patients with peripheral neuropathic pain. Anesth Analg 81:713–717

Belfrage M, Segerdahl M, Arnér S, Sollevi A (1999) Safety and efficacy of intrathecal adenosine in patients with chronic neuropathic pain. Anesth Analg 89:136–142

Biaggioni I, Olafsson B, Robertson RM, Hollister AS, Robertson D (1987) Cardiovascular and respiratory effects of adenosine in conscious man. Evidence for carotid body chemoreceptor activation. Circ Res 61:779–786

Bleehen T, Keele CA (1977) Observations on the algogenic actions of adenosine compounds on the human blister base preparation. Pain 3:367–377

Burnstock G (1978) A basis for distinguishing two types of purinergic receptor. In: Straub RW and Bolis L eds) Cell membrane receptors for drugs and hormones: A multidisciplinary approach. New York, Rave Press, pp107–118

Burnstock G (1996) A unifying purinergic hypothesis for the initiation of pain. Lancet 347:1604–1105

Burnstock G, Wood JN (1996) Purinergic receptors: their role in nociception and primary afferent neurotransmission. Curr Opin Neurobiol 6:526–532

Cahill CM, White TD, Sawynok J (1993) Morphine activates T-conotoxin-sensitive Ca^{2+} channels to release adenosine from spinal cord synaptosomes. J Neurochem 60:894–901

Cahill CM, White TD, Sawynok J (1995) Spinal opioid receptors and adenosine release: Neurochemical and behavioural characterization of opioid subtypes. J Pharmacol Exp Therap 275:84–93

Chen C-C, Akoplan AN, Sivilotti L, Colquhoun D, Burnstock G, Wood JN (1995) A P2X purinoceptor expressed by a subset of sensory neurons. Nature 377:428–431

Choca JI, Green RD, Proudfit HK (1988) Adenosine A1 and A2 receptors of the substantia gelatinosa are located predominantly on intrinsic neurones: An audoradiographic study. J Pharmacol Exp Therap 247:757–764

Choca JI, Proudfit HK, Green RD (1987) Identification of A1 and A2 adenosine receptors in the rat spinal cord. J Pharmacol Exp Therap 242:905–910

Collier HOJ, James GWL, Schneider C (1966) Antagonism by aspirin and femanates of bronchoconstriction and nociception by adenosine-5'-triphosphate. Nature 212:411–412

Collo G, North RA, Kawashima E, Merlo-Pich E, Neidhart S, Surprenant A, Buell G (1996) Cloning of $P2X_5$ and $P2X_6$ receptors and the distribution and properties of an extended family of ATP-gated ion channels. J Neurosci 16:2495–2507

Cook SP, Vulchanova L, Hargreaves KM, Elde R, McCleskey EW (1997) Distinct ATP receptors on pain–sensing and stretch–sensing neurons. Nature 387:505–508

Crews JC, Cahall M, Bebhani MM (1994) The neurophysiologic mechanisms of tourniquet pain. Anesthesiology 81:730–736

Cronstein BN (1994) Adenosine, an endogenous anti-inflammatory agent. J Appl Physiol 76(1):5–13

Cronstein BN (1995) A novel approach to the development of anti-inflammatory agents: Adenosine release at inflamed sites. J Investigative Med 43:50–57

Cronstein BN, Naime D, Firestein GS (1995) The antiinflammatory effects of an adenosine kinase inhibitor are mediated by adenosine. Arthritis Rheum 38:1040–1045

Cronstein BN, Montesinos, MC, Weissman G (1999) Salicylates and sulfasalazine, but not glucocorticoids, inhibit leukocyte accumulation by an adenosine-dependent mechanism that is independent of prostaglandin synthesis and p105 of NFkB. Proc. Natl Acad Sci USA 96:6377 –6381

Cui J-G, Meyersson B, Sollevi A, Linderoth B (1998) Effect of spinal cord stimulation on tactile hypersensitivity in mononeuropathic rats is potentiated by simultaneous $GABA_B$ and adenosine receptor activation. Neurosci Lett 247:183–186.

Dale HH (1935) Pharmacology and nerve-endings. Proc Roy Soc Med 28:319–322

DeKoninck Y, Henry JL (1992) Peripheral vibration causes an adenosine-mediated postsynaptic inhibitory potential in dorsal horn neurones in the cat spinal cord. Neurosci 50:435–443

De Koninck Y, Henry JL (1994) Prolonged GABAA–mediated inhibition following single hair afferent input to single spinal dorsal horn neurones in cats. J Physiol (Lond) 476:89–100

DeLander GE, Keil II GJ (1994) Antinociception induced by intrathecal coadministration of selective adenosine receptor and selective opioid receptor agonists in mice. J Pharmacol Exp Therap 268:943–951

DeLander GE, Mossberg HI, Porreca F (1992) Involvement of adenosine in antinociception produced by spinal or supraspinal receptor-selective opioid agonists: Dissociation from gastrointestinal effects in mice. J Pharmacol Exp Therap 263:1097–1104

DeLander GE, Hopkins CJ (1986) Spinal adenosine modulates descending antinociceptive pathways stimulated by morphine. J Pharmacol Exp Ther 239:88–93

Dickenson AH, Chapman V, Green GM (1997) The pharmacology of excitatory and inhibitory amino acid–mediated events in the transmission and modulation of pain in the spinal cord. Gen Pharmacol 28:633–638

Doak GJ, Sawynok J (1995) Complex role of peripheral adenosine in the genesis of the response to subcutaneous formalin in the rat. Eur J Pharmacol 281:311–318

Dolphin AC, Forda SR, Scott RH (1986) Calcium-dependent currents in cultured rat dorsal root ganglion neurones are inhibited by an adenosine analog. J Physiol (London) 373:47–61

Dray A (1995) Inflammatory mediators of pain. Br J Anaesth;75:125–131

Driessen B, Reimann W, Selve N, Friderichs E, Bultmann R (1994) Antinociceptive effect of intrathecally administered P2-purinoceptor antagonists in rats. Brain Res 666:182–188

Dubyak, GR, Cowen DS, Lazarus HM (1988) Activation of the inositol phospholipid signaling system by receptors for extracellular ATP in human neutrophils, monocytes and neutrophil/monocyte progenitor cells. Ann NY Acad Sci 551:218–237

Dubyak GR, El-Moatassim C (1993) Signal transduction via P_2-purinergic receptors for extracellular ATP and other nucleotides. Am J Physiol 265 (Cell Physiol 43).C577–606

Edwards, FA, Gibb AJ, Colquhoun D (1992) ATP receptor-mediated synaptic current in the central nervous system. Nature 359:144–146

Eisenberger MA, Reyno LM, Jodrell DI, Sinibaldi VJ, Tkaczuk KH, Sridhara R, Zuhowski EG, Lowitt MH, Jacobs SC, Egorin MJ J (1993) Suramin, an active drug for prostate cancer: interim observations in a phase I trial. Natl Cancer Inst Apr 85:611–621

Ekblom A, Segerdahl M, Sollevi A (1995) Adenosine but not ketamine or morphine increases the cutaneous heat pain threshold in healthy volunteers. Acta Anaesthesiol Scand 39:717–722

Evans RJ, Derkach V, Surprenant A (1992) ATP mediates fast synaptic transmission in mammalian neurons. Nature 357:503–505

Fargeas MJ, Fioramonti J, Bueno L (1990) Central and peripheral actions of adenosine and its analogues on intestinal myoelectric activity and propulsion in rats. J Gastroint Motility 2:121–127

Firestein GS, Boyle D, Bullough DA, Gruber HE, Sajjadi FG, Montag A, Sambol B, Mullare KM (1994) Protective effect of an adenosine kinase inhibitor in septic shock. J Immunol 152:5853–5859

Forsberg C, Sollevi A, Thörn SE, Segerdahl M (1998) Effects of adenosine infusion on gastric emptying in healthy volunteers. Acta Anaesthesiol Scand 43:87–90

Fredholm BB (1980) Are methylxanthine effects due to antagonism of endogenous adenosine? Trends Pharmacol Sci 1:129–132

Fredholm BB, Dunwiddie TV (1988) How does adenosine inhibit transmitter release? Trends Pharmacol Sci 9:130–134

Fyffe RE, Perl ER (1984) Is ATP a central synaptic mediator for certain primary afferent fibers from mammalian skin? Proc Natl Acad Sci 81:6890–6893

Gadangi P, Longaker M, Naime D, Levin RI, Recht PA, Montesinos MC, Buckley MT, Carlin G, Cronstein BN (1996) The anti-inflammatory mechanism of sulfasalazine is related to adenosine release at inflamed sites. J Immunol 156:1937–1941

Galindo, A, Krnjevic K, Schwartz S (1967) Microiontophoretic studies on neurones in the cuneate nucleus. J Physiol (London) 192:359–377

Garrison CJ, Dougherty PM, Kajander KC, Carlton SM (1991) Staining of GFAP in lumbar spinal cord increases following a sciatic nerve constriction injury. Brain Res 565:1–7

Gaspardone A, Crea P, Tomai F, Versaci F, Iamele M, Gioffré G, Chiariello, L, Gioffré P (1995) Muscular and cardiac adenosine-induced pain is mediated by A_1 receptors. JACC 25:251–257

Geiger JD, LaBella FS, Nagy JI (1984) Characterization and localization of adenosine receptors in rat spinal cord. J Neurosci 4:2303–2310

Gilmore SA, Sims TJ, Leiting JE (1990) Astrocytic reactions in spinal gray matter following sciatic axotomy. Glia 3:342–349

Golembiowska K, White TD, Sawynok J (1995) Modulation of adenosine release from rat spinal cord by adenosine deaminase and adenosine kinase inhibitors. Brain Res 699:315–320

Golembiowska K, White TD, Sawynok J (1996) Adenosine kinase inhibitors augment release of adenosine from spinal cord slices. Eur J Pharmacol 307:157–162

Goodman RR, Snyder SH (1982) Autoradiographic localization of adenosine receptors in rat brain using [^{3}H]cyclohexyladenosine. J Neurosci 2:1230–1241

Gross RA, Macdonald RL, Ryan-Jastrow T (1989) 2-Chloroadenosine reduces the N calcium current of cultured mouse sensory neurones in a pertussis toxin-sensitive manner. J Physiol 411:585–595

Gu JG, MacDermott AB (1997) Activation of ATP P2X receptors elicits glutamate release from sensory neuron synapses. Nature 389:749–753

Gu JG, Bardoni R, Magherini PC, MacDermott AB (1998) Effects of the P2–purinoceptor antagonists suramin and pyridoxal–phosphate–6–azophenyl–2',4'–disulfonic acid on glutamatergic synaptic transmission in rat dorsal horn neurons of the spinal cord. Neurosci Lett 253:167–170

Guo A Vulchanova L Wang J Li X, Elde R (1999) Immunocytochemical localization of the vanilloid receptor 1 (VR1): relationship to neuropeptides, the $P2X_3$ purinoceptor and IB4 binding sites, Eur J Neurosci 11: 946–958

Hajos F, Csillik B, Knyihar-Csillik E (1990) Alterations in flial fibrillary acidic protein immunoreactivity in the upper dorsal horn of the rat spinal cord in the course of transganglionic degenerative atrophy and regenerative proliferation. Neurosci Lett 117:8–13

Handwerker HO, Reeh PW (1991) Pain and inflammation. In: Bond MR, Charlton JE, Woolf CJ (eds) Pain research and clinical management. Elsevier Science, New York, pp59–70

Heijne von M, Hao J-X, Yu W, Sollevi A, Xu X-J, Wiesenfeld-Hallin Z (1998) Reduced anti-allodynic effect of the adenosine A_1-receptor agonist R-phenylisopropyladenosine on repeated intrathecal administration and lack of cross-tolerance with morphine in a rat model of central pain. Anesth Analg 87:1367–1371

Ho C, Hicks JL, Salter MW (1995) A novel P_2-purinoceptor expressed by a subpopulation of astrocytes from the dorsal spinal cord of the rat Br J Pharmacol 116: 2909–2918

Ho BT, Huo YY, Lu JG, Newman RA, Levin VA (1992) Analgesic activity of anticancer agent suramin Anticancer Drugs 3:91–4

Holmgren M, Hedner J, Mellstrand T, Nordberg G, Hedner T (1986) Characterization of the antinociceptive effects of some adenosine analogues in the rat. Naunyn-Schmiedeberg's Arch Pharmacol 334:290–293

Holton P (1959) The liberation of adenosine triphosphate on antidromic stimulation of sensory nerves. J Physiol (London) 145:494–504

Holton FA, Holton P (1954) The capillary dilator substances in dry powders of spinal roots; a possible role of adenosine triphosphate in chemical transmission from nerve endings. J Physiol (London) 126:124–140

Idestrup C, Salter MW (1998) P2Y- and P2U-receptors differentially release intracellular Ca^{2+} via the PLC/IP_3 pathway in astrocytes from the dorsal spinal cord. Neuroscience 86: 913–923

Illes P, Nörenberg W (1993) Neuronal ATP receptors and their mechanism of action. Trends Pharmacol Sci 14:50–54

Jahr, CE, Jessell TM (1983) ATP excites a subpopulation of rat dorsal horn neurones. Nature 304:730–733

Jensen TS (1996) Mechanisms of neuropathic pain. In Campbell JN (ed) Pain 1996 – an updated review. Seattle: IASP Press, pp. 77–86

Jessell TM, Jahr CE (1985) Fast and slow excitatory transmitters at primary afferent synapses in the dorsal horn of the spinal-cord. In: Advances in pain research and therapy. New York: Raven Press, pp. 31–39

Jo YH, Schlichter R (1999) Synaptic corelease of ATP and GABA in cultured spinal. Nature Neuroscience 2: 241–245

Jurna I (1984) Cyclic nucleotides and aminophylline produce different effects on nociceptive motor and sensory responses in the rat spinal cord. Archives Pharmacol 327:23–30

Karlsten R, Gordh T, Hartvig P, Post C (1991) Effects of intrathecal injection of the adenosine receptor agonist R-phenylisopropyl-adenosine and N-ethylcarboxamide-adenosine on nociception and motor function in the rat. Anesth Analg 71:60–64

Karlsten R, Gordh T, Post C (1992) Local antinociceptive and hyperalgesic effects in the formalin test after peripheral administration of adenosine analogues in mice. Pharmacol Toxicol 70:434–438

Karlsten R, Gordh T Jr, Svensson BA (1994) A neurotoxicological evaluation of the spinal cord after chronic intrathecal injection of R-phenylisopropyl adenosine (R-PIA) in the rat. Anesth Analg 77:731–736

Karlsten R, Kristensen J, Gordh T Jr (1992) R-Phenylisopropyl-adenosine increases spinal cord blood flow after intrathecal injection in the rat. Anesth Analg 75:972–976

Karlsten R, Post C, Hide I, Daly JW (1991) The antinociceptive effect of intrathecally administered adenosine analogs in mice correlates with the affinity for the A_1-adenosine receptor. Neurosci Lett 121:267–270

Karlsten R, Gordh T Jr (1995) An A_1-selective adenosine agonist abolishes allodynia elicited by vibration and touch after intrathecal injection. Anesth Analg 80:844–847

Keil II GJ, DeLander GE (1992) Spinally-mediated antinociception is induced in mice by an adenosine kinase-, but not by an adenosine deaminase-, inhibitor. Life Sci 51:PL171–176

Keil II GJ, DeLander GE (1994) Adenosine kinase and adenosine deaminase inhibition modulate spinal adenosine- and opioid agonist-induced antinociception in mice. Eur J Pharmacol 271:37–46

Keil II GJ, DeLander GE (1995) Time-dependent antinociceptive interactions between opioids and nucleoside transport inhibitors. J Pharmacol Exp Therap 274:1387–1392

Keil GJ 2nd, DeLander GE (1996) Altered sensory behaviors in mice following manipulation of endogenous spinal adenosine neurotransmission. Eur J Pharmacol 312:7–14

Koltzenburg M, Lundberg LER, Torebjörk HE (1992) Dynamic and static components of mechanical hyperalgesia in human hairy skin. Pain 51:207–219

Krishtal OA, Marchenko SM, Obukhov AG (1988) Cationic channels activated by extracellular ATP in rat sensory neurons. Neurosci 27:995–1000

Krishtal OA, Marchenko SM, Pidoplichko VI (1983) Receptor for ATP in the membrane of mammalian sensory neurones. Neurosci Lett 35:41–45

Lagerqvist B, Sylvén C, Beermann B, Helmius G, Waldenström A (1990) Intracoronary adenosine causes angina pectoris like pain – an inquiry into the nature of visceral pain. Cardiovasc Res 24:609–613

Le KT, Villeneuve P, Ramjaun AR, McPherson PS, Beaudet A, Seguela P (1998) Sensory presynaptic and widespread somatodendritic immunolocalization of central ionotropic P2X ATP receptors. Neuroscience 83:177–190

Lewis C, Neldhart S, Holy C, North RA, Buell G, Surprenant A (1995) Coexpression of $P2X_2$ and $P2X_3$ receptor subunits can account for ATP-gated currents in sensory neurons. Nature 377:432–435

Li J, Perl ER (1995) ATP modulation of synaptic transmission in the spinal substantia gelatinosa. J Neurosci 15:3357–3365

Li J, Perl ER (1994) Adenosine inhibition of synaptic transmission in the substantia gelatinosa. J Neurophysiol 72:1611–1621

Linden J (1994) Cloned adenosine A_3 receptors: Pharmacological properties, species differences and receptor functions. Trends Pharmacol Sci 15:298–306

Macdonald RL, Skerritt JH, Werz MA (1986) Adenosine agonists reduce voltage-dependent calcium conductance of mouse sensory neurones in cell culture. J Physiol (London) 370:75–90

Malmberg AB, Yaksh TL (1993) Pharmacology of the spinal action of ketorolac, morphine, ST-91, U50,488H, and L-PIA on the formalin test and an isobolographic analysis of the NSAID interaction. Anesthesiology 79:270–281

Marchand S, Li J, Charest J (1995) Effects of caffeine on analgesia from transcutaneous electrical nerve stimulation. NEJM 333:325–326

Meller, ST, Dykstra C, Grzybycki D, Murphy S, Gebhart GF (1994) The possible role of glia in nociceptive processing and hyperalgesia in the spinal cord of the rat. Neuropharmacol 33:1471–1478

Minami T, Uda R, Horiguchi S, Ito S, Hyodo M, Hayaishi O (1992a) Effects of clonidine and baclofen on prostaglandin $F_2\beta$-induced allodynia in conscious mice. Pain Res 7:129–134

Minami T, Uda R, Horiguchi S, Ito S, Hyodo M, Hayaishi O (1992b) Allodynia evoked by intrathecal administration of prostaglandin $F_2\beta$ to conscious mice. Pain 50:223–229

Murray TF, Cheney DL (1982) Neuronal location of N^6-cyclohexyl[^{3}H]adenosine binding sites in rat and guinea pig brain. Neuropharmacol 21:575–580

Nagaoka H, Sakurada S, Sakurada T, Takeda S, Nakagawa Y, Kisara K, Arai Y (1993) Theophylline-induced nociceptive behavioral response in mice: Possible indirect interaction with spinal N-methyl-D-aspartate receptors. Neurochem Int 22:69–74

Pappagallo M, Gaspardone A, Tomai F, Iamele M, Crea P, Gioffré P (1993) Analgesic effect of bambiphylline on pain induced by intradermal injection of adenosine. Pain 53:199–204

Phillis JW, Kirkpatrick JR (1978) The actions of adenosine and various nucleosides and nucleotides on the isolated toad spinal cord. Gen Pharmacol 9:239–247

Poon A, Sawynok J (1995) Antinociception by adenosine analogs and ana adenosine kinase inhibitor: Dependence on formalin concentration. Eur J Pharmacol 286:177–184

Ralevic V, Burnstock G (1998) Receptors for purines and pyrimidines. Pharmacol Rev 50:413–492

Rane K, Karlsten R, Sollevi A, Gordh T Jr, Svensson B (1999) Spinal cord morphology after chronic intrathecal administration of adenosine in the rat. Acta Anesth Scand 43:1035–1040

Rane K, Segerdahl M, Goiny M, Sollevi A (1998) Intrathecal adenosine administration – A phase 1 clinical safety study in healthy volunteers, with additional evaluation of its influence on sensory thresholds and experimental pain. Anesthesiology 89:1108–1115

Rane K, Sollevi A, Segerdahl M (2000) Intrathecal adenosine administration in abdominal hystepectomy lacks analgesic effect. Acta Anaesth Scand 44:868–872

Rang HP, Bevan S, Dray A (1991) Chemical activation of nociceptive peripheral neurones. British Med Bull 47:534–548

Rees H, Sluka KA, Westlund KN, Willis WD (1995) The role of glutamate and GABA receptors in the generation of dorsal root reflexes by acute arthritis in the anaesthetized rat. J Physiol (Lond) 484:437–445

Reeve AJ, Dickenson AH (1995a) The roles of spinal adenosine receptors in the control of acute and more persistent nociceptive responses of dorsal horn neurones in the anaesthetized rat. Br J Pharmacol 116:2221–2228

Reeve AJ, Dickenson AH (1995b) Electrophysiological study on spinal antinociceptive interactions between adenosine and morphine in the dorsal horn of the rat. Neurosci Lett 194:81–84

Reppert SM, Weaver DR, Stehle JH, Rivkees SA (1991) Molecular cloning and characterization of a rat A_1-adenosine receptor that is widely expressed in brain and spinal cord. Molec Endocrinol 5:1037–1048

Rivkees SA (1995) The ontogeny of cardiac and neural A_1 adenosine receptor expression in rats. Develop Brain Res 89:202–213

Rivkees SA, Reppert SM (1992) RFL9 encodes an A_{2b} adenosine receptor. Mol Endocrinol 6:1598–1604

Rosengren S, Bong GW, Firestein GS (1995) Anti-inflammatory effects of an adenosine kinase inhibitor: Decreased neutrophil accumulation and vascular leakage. J Immunol 154:5444–5451

Sajjadi FG, Takabayashi K, Foster AC, Domingo RC, Firestein GS (1996) Inhibition of TNF-α by adenosine. Role of A_3 adenosine receptors. J Immunol 156:3435–3442

Salt TE, Hill RG (1983) Excitation of single sensory neurones in the rat caudal trigeminal nucleus by iontophoretically applied adenosine 5'-triphosphate. Neurosci Lett 35:53–57

Salter MW, De Koninck Y (1999) An ambiguous fast synapse: a new twist in the tale of two transmitters. Nature Neuroscience 2:199–200

Salter MW, DeKoninck Y, Henry JL (1992) ATP-sensitive K^+ channels mediate an IPSP in dorsal horn neurones elicited by sensory stimulation. Synapse 11:214–220

Salter MW, DeKoninck Y, Henry JL (1993) Physiological roles for adenosine and ATP in synaptic transmission in the spinal dorsal horn. Prog Neurobiol 41:125–156

Salter MW, Henry JL (1985) Effects of adenosine 5'-monophosphate and adenosine 5'-triphosphate on functionally identified units in the cat spinal dorsal horn. Evidence for a differential effect of adenosine 5'-triphosphate on nociceptive vs non-nociceptive units. Neurosci 15:815–825

Salter MW, Henry JL (1987) Evidence that adenosine mediates the depression of spinal dorsal horn neurones induced by peripheral vibration in the cat. Neurosci 22:631–650

Salter MW, Hicks JL (1994) ATP-evoked increases in intracellular calcium in neurons and glia from the dorsal spinal cord. J Neurosci 14:1563–1575

Salter MW, Hicks JL (1995) ATP causes release of intracellular Ca^{2+} via the phospholipase $C\beta/IP_3$ pathway in astrocytes from the dorsal spinal cord. J Neurosci 15:2961–2971

Santicioli P, Del Bianco E, Maggi CA (1993) Adenosine A_1 receptors mediate the presynaptic inhibition of calcitonin gene-related peptide release in the rat spinal cord. Eur J Pharmacol 231:139–142

Santicioli P, Del Bianco E, Tramontana M, Maggi CA (1992) Adenosine inhibits action potential-dependent release of calcitonin gene-related peptide- and substance P-like immunoreactivities from primary afferents in rat spinal cord. Neurosci Lett 144:211–214

Sawynok J (1998) Adenosine receptor activation and nociception. Eur J Pharmacol 317:1–11

Sawynok J, Downie JW, Reid AR, Cahill CM, White TD (1993) ATP release from dorsal spinal cord synaptosomes: Characterization and neural origin. Brain Res 610:32–38

Sawynok J, Sweeney MI (1989) The role of purines in nociception. Neurosci 32:557–569

Sawynok J, Sweeney MI, White TD (1986) Classification of adenosine receptors mediating antinociception in the rat spinal cord. Br J Pharmacol 88:923–930

Sawynok J, Sweeney MI, White TD (1989) Adensine release may mediate spinal analgesia by morphine. Trends Pharmacol Sci 10:186–189

Sawynok J, Reid A, Poon A (1998) Pain Peripheral antinociceptive effect of an adenosine kinase inhibitor, with augmentation by an adenosine deaminase inhibitor, in the rat formalin test 74:75–81

Sawynok J, Zarrindast MR, Reid AR, Doak GJ (1997) Adenosine A_3 receptor activation produces nociceptive behaviour and edema by release of histamine and 5–hydroxytryptamine. Eur J Pharmacol 333:1–7

Segerdahl M, Ekblom A, Sjölund K-F, Belfrage M, Forsberg C, Sollevi A (1995) Systemic adenosine attenuates touch evoked allosynia induced by mustard oil in humans. NeuroReport 6:753–756

Segerdahl M, Ekblom A, Sandelin K, Wickman M, Sollevi A (1995b) Perioperative adenosine infusion reduces the requirements for isoflurane and postoperative analgesics. Anesth & Analg 80:1145–1149

Segerdahl M, Ekblom A, Sollevi A (1994) The influence of adenosine, ketamine and morphine on experimentally induced ischemic pain in healthy volunteers. Anesth Analg 79:787–791

Segerdahl M, Irestedt L, Sollevi A (1997) Antinociceptive effect of perioperative adenosine infusion in hysterectomy. Acta Anaesthiol Scand 41:473–479

Segerdahl M, Persson E, Ekblom A, Sollevi A (1996) Perioperative adenosine infusion reduces isoflurane requirements during general anesthesia for shoulder surgery. Acta Anaesthiol Scand 40:792–797

Shapiro E, Castellucci VF, Kandel ER (1980) Presynaptic inhibition in aplysia involves a decrease in the Ca^{2+} current of the presynaptic neuron. Proc Natl Acad Sci 77:1185–1189

Sjölund K-F, Segerdahl M, Sollevi A (1999) Adenosine reduces secondary hyperalgesia in two human models of cutaneous inflammatory pain. Anesth Analg 88:605–610

Sollevi A, Belfrage M, Lundeberg T, Segerdahl M, Hansson P (1995) Systemic adenosine infusion: A new treatment modality to alleviate neuropathic pain. Pain 61:155–158

Sollevi A (1986) Cardiovascular effects of adenosine in man; Possible clinical implications. Progr Neurobiol 27:319–349

Sosnowski M, Yaksh TL (1989) Role of spinal adenosine receptors in modulating the hyperesthesia produced by spinal glycine receptor antagonism. Anesth Analg 69:587–892

Sosnowski M, Stevens CW, Yaksh TL (1989) Assessment of the role of A_1/A_2 adenosine receptors mediating the purine antinociception, motor, and autonomic function in the rat spinal cord. J Pharmacol Exp Therap 250:915–922

Stehle JH, Rivkees SA, Lee JJ, Weaver DR, Deeds JD, Reppert SM (1992) Molecular cloning and expression of the cDNA for a novel A_2 adenosine receptor subtype. Molecular Endocrinol 6:384–393

Suprenant A, Buell G, North RA (1995) P_{2X} receptors bring new structure to ligand-gated ion channels. Trends Neurosci 18:224–229

Svensson M, Eriksson P, Pesson JKE, Molander C, Arvidsson Aldskogius H (1993) The response of central glia to peripheral nerve injury. Brain Res Bull 30:499–506

Sylvén C (1993) Mechanisms of pain in angina pectoris– a critical review of the adenosine hypothesis. Cardiovasc Drug Ther 7:745–759.

Sylvén C, Beermann B, Jonzon B, Brandt R (1986) Angina pectoris-like pain provoked by intravenous adenosine in healthy volunteers. Br Med J 293:227–230.

Sylvén C, Eriksson B, Jensen J, Geigant E, Hallin RG (1996) Analgesic effects of adenosine during exercise-provoked myocardial ischemia. NeuroReport 7:1521–1525

Sylvén C, Beermann B, Edlund A, Lewander R, Jonzon B, Mogensen L (1988a) Provocation of chest pain in patients with coronary insufficiency using the vasodilator adenosine. Eur Heart J 9:6–10

Sylvén C, Jonzon B, Fredholm BB, Kaijer L (1988b) Adenosine injected into the brachial artery produces ischaemia-like pain or discomfort in the forearm. Cardiovasc Res 22:674–678

Taiwo YO, Levine JD (1990) Direct cutaneous hyperalgesia induced by adenosine. Neurosci 38:757–762

Taiwo, YO, Levine JD (1991) Further confirmation of the role of adenyl cyclase and of cAMP-dependent protein kinase in primary afferent hyperalgesia. Neurosci 44:131–135

Vulchanova L, Riedl MS, Shuster SJ, Stone LS, Hargreaves KM, Buell G, Surprenant A, North RA, Elde R (1998) P2X3 is expressed by DRG neurons that terminate in inner lamina II. Eur J Neurosci 10:3470–3478

Watt AH, Lewis DJM, Horne JJ, Smith PM (1987) Reproduction of epigastric pain of duodenal ulceration by adenosine. Br Med J 294:10–12

Welch SP, Dunlow LD (1993) Antinociceptive activity of intrathecally administered potassium channel openers and opioid agonists: A common mechanism of action? J Pharmacol Exp Therap 267:390–399

White TD, Downie JW, Leslie RA (1985) Characteristics of K^+- and veratridine-induced release of ATP from synaptosomes prepared from dorsal and ventral spinal cord. Brain Res 334:372–374

Woolf CJ, Doubell TP (1994) The pathophysiology of chronic pain-increased sensitivity to low threshold A beta–fibre inputs. Curr Opin Neurobiol 4:525–534

Woolf CJ, Thompson JW (1994) Stimulation-induced analgesia: Transcutaneous electrical nerve stimulation (TENS) and vibration. In: PD Wall and R Melzack (eds) Textbook of Pain. Edinburgh: Churchill Livingstone, pp1191–1208

Yamamoto T, Yaksh TL (1993) Stereospecific effects of a nonpeptidic NK1 selective antagonist, CP-96,345: Antinociception in the absence of motor dysfunction. Life Sci 49:1955–1963

CHAPTER 14
Uridine and Pyrimidine Nucleotides in Cell Function

G.P. CONNOLLY

A. Introduction

The pyrimidine nucleosides, e.g. uridine and cytidine and the related nucleotides UDP and UTP, are endogenous cellular constituents that can modulate cell function at both the intra- and extra-cellular levels. Many disorders of pyrimidine metabolism are linked to altered cytidine and uracil metabolism, leading to systemic and neurological dysfunction. Pyrimidines can also alter cell activity, under physiological and pathological circumstances, via the activation of cell surface P2Y receptors (COMMUNI and BOEYNAMS 1997; BOARDER and WEBB, Chap. 4, this volume).

The present review focuses on the role(s) of uridine and its derivatives in the peripheral and central nervous system since these are the best studied. However, other likely therapeutic targets where uridine may be involved include the respiratory, circulatory, hepatic and reproductive and nervous system dysfunction. The use of UTP, acting via $P2Y_2$ receptors in lung tissue, in the treatment of cystic fibrosis, chronic bronchitis and chronic obstructive pulmonary disease (COPD) is well established (WEGNER, Chap. 23, second volume).

B. Synthesis and Salvage of Pyrimidines

In mammals, pyrimidines, like the related purines, are synthesised de novo, with the bioavailability of uridine being particularly critical to the synthesis of RNA and membranes, via the formation of pyrimidine nucleotide-lipid conjugates, and thus for normal cell function and growth. During cell turnover, DNA, RNA and their component nucleotides are broken down and the majority of the pyrimidines resulting from this process are salvaged rather than excreted. The relative contribution of salvage vs de novo synthesis appears dependent on the cell type, the age of the animal or tissue and its physiological state. For example, neuroblastoma cells in their stationary phase of growth rely predominately upon exogenous uridine, whereas in their exponential phase they switch to de novo synthesis.

De novo pyrimidine synthesis is high in fetal rat brain and declines rapidly with development (TREMBLAY et al. 1976). Similarly morphological differentiation of M1 neuroblastoma is accompanied by an increase in adenine nucleotides but a decrease of cytidine and uridine nucleotides (DIERICH et al. 1980).

In the isolated choroid plexus of conscious adult rabbits, [^{3}H]-uridine is transported from the blood into CSF, the extracellular space of brain and then into brain cells, where it is phosphorylated to UMP, UDP and UTP or catabolised to uracil (SPECTOR 1985). Thus the brain requires and maintains a constant supply of uridine, regulating the production of uridine nucleotides and eliminating any excess uridine via catabolism to uracil.

Uridine metabolites are critical for normal brain maturation and function. Young rats incorporate pyrimidine precursors into cerebral cortical RNA at rates many times higher than those observed in adult cells (GUROFF and BRODSKY 1971). After birth, de novo synthesis is decreased as is uridine salvage with a corresponding increase in uridine catabolism.

Pyrimidine metabolism in brain is different to that in other tissues. Low amounts of free cytidine and uridine prolong the survival time of the isolated brain (GEIGER and YAMASAKI 1956) and the brain maintains low levels of free cytidine nucleotides with most in polymeric form (MANDEL et al. 1966; MANDEL and EDEL-HARTH 1966) or in conjugates necessary for normal metabolism. The cytidine nucleotides, cytidine monophosphate (CMP) and cytidine diphosphate (CDP) are formed in rat brain by conversion of uridine nucleotides (DAWSON 1968).CMP is involved in the sialylation of cell membranes via conjugation to N-acetylneuraminic acid, while CDP is a constituent of CDP-choline (CDPC) and CDP-ethanolamine (CDPE) that are used in phosphoplipid synthesis, e.g. inositol trisphosphate. Sugar conjugates of uridine, e.g. UDP-glucose and UDP-galactose (UGPGal), are substrates for glucose and glycogen metabolism, and thus play a key role in regulating cellular energy balance. The involvement of uridine and cytidine nucleotides in regulating brain phospholipid metabolism (MANDEL and EDEL-HART 1966; MANDEL et al.,1966) and the fact that these pyrimidines are elevated in cells from patients with the neurological disorder, Lesch-Nyhan syndrome, and in hypoxanthine phosphoribosyltransferase (HGPRT)-deficient cells, may explain their role in abnormal phosphoinositol signalling (HUSSAIN et al. 1999) and the increased adhesiveness of such cells (STACEY et al. 1999).

Mammalian purine and pyrimidine metabolism differ from one another in that pyrimidines are salvaged and recycled as nucleosides, e.g. uridine and cytidine while purines are recycled from their bases, e.g. adenine and hypoxanthine. Mammalian cells, rather than salvaging uracil and thymine, catabolise them (CAPPIELLO et al. 1999). Patients with genetically blocked pyrimidine synthesis, e.g. oritidine phosphoribosyl transferase (OPRT) deficiency, respond to oral uridine but not uracil even though both compounds are transported across the gut. (BECROFT et al. 1969; SCRIVER et al. 1995). However, the neurological symptoms of OPRT-deficiency are not relived by exogenous uridine. Another difference between purine and pyrimidine utilisation is that, while humans do not utilise dietary purines, which become converted by the gut into the end-product uric acid, pyrimidine nucleosides can be absorbed into the bloodstream for utilisation by other tissues including the liver.

C. Plasma and CSF Uridine

Plasma, CSF and seminal fluid maintain uridine at metabolically active concentrations, which are then available for salvage by cells. Uridine concentrations in mammalian tissues or blood plasma, bone marrow and CSF remain within a fixed range of 1–10 µmol/l depending on the species (TRAUT 1994). High concentrations of uridine given i.v. rapidly equilibrate to normal levels and in humans and rats diet is not normally a significant source of uridine. Because the liver synthesises and degrades uridine, it plays a central role in maintaining plasma uridine. Consequently, liver failure may lead to altered pyrimidine nucleotide pools in other tissues, e.g. brain, and consequently their dysfunction (see below).

Subcellular storage organelles of rabbit platelets and bovine adrenal medulla contain ATP, GTP and UTP (GOERTZ et al. 1991) and may provide releasable pools of pyrimidines, e.g. uridine following catabolism. More likely, uridine may be made available to the body by erythrocytes which, unlike nucleated cells, do not store UTP or synthesise pyrimidines de novo, but rapidly take up orotic acid, converting it to UDP-glucose (VALENTINE et al. 1974). Thus, in peripheral tissues, erythrocyte stores of UDP-glucose may be catabolised to provide uridine and glucose to rapidly metabolising tissues such as brain and skeletal muscle.

D. Liver and Kidney Pathology

Uridine and cytidine are beneficial in treating hepatic coma (DRAGO et al. 1967) and impaired liver function due to cirrhosis was improved by UDP-glucose (UDRG; COLTORI 1975). Similar beneficial effects of UDPG in chronic hepatopathies have been reported (OKOLICSANYI et al. 1980b) while animal studies support a role for nucleotides in hepatic regeneration after hepatectomy (USAMI et al. 1996). Thus uridine and its nucleotides may prove helpful in promoting recovery from hepatic disorders. Increased membrane stabilisation may explain some of the hepatoprotective effects of uridine and UDPG. A UDP-glucose sensitive G-protein coupled receptor has been identified (CHAMBERS et al. 2000).

Elevated urinary levels of modified nucleosides, especially pseudouridine, occur in patients with a variety of malignant conditions, but not brain tumours. Pseudouridine is the major pyrimidine catabolite in urine and is normally present in low amounts in the CSF, being formed by isomerisation of uridine. In two cases where cancer was ruled out, e.g. pseudouridinuria (KIHARA 1967) and chronic renal failure (GERRITS et al. 1991), excessive urinary pseudouridine levels were found to be associated with mental retardation. Abnormally high levels of pseudouridine may interfere with normal biological functions, e.g. neuronal development and transmission within the CNS, leading to mental retardation (KIHARA 1967; GERRITS et al. 1990).

E. Uridine Nucleotides and Cystic Fibrosis

Cystic fibrosis (CF) is characterised by abnormal fluid transport, e.g. decreased chloride and increased sodium transport, across many epithelia including airways, pancreas, sweat glands and small intestine. Improving mucociliary clearance has the potential to prevent progressive lung damage due to infection (WEGNER, Chap. 23, second volume). CF is caused by an absence or dysfunction of the cystic fibrosis transmembrane conductance regulator (CFTR), a chloride channel expressed by epithelial cells, and a consequent increase in active sodium absorption.

Inhaled UTP, acting via $P2Y_2$ purinergic receptors, stimulates chloride secretion in epithelial cells and decreases sodium absorption (KNOWLES et al. 1991), bypassing the defective chloride secretion to activate an alternative calcium-dependent chloride secretory pathway. Extracellular UTP is susceptible to hydrolysis by ecto-nucleotidases and phosphatases present in human airways (ZIMMERMAN and BRAUN 1999). Stable UTP analogues like INS 365 in combination with other drugs, e.g. amiloride (Table 1), may be used in the treatment of CF and other airway diseases, including asthma and chronic bronchitis, both of which involve dysfunctional mucociliary clearance and dehydration (OLIVIER et al. 1996; WEGNER, Chap. 23, second volume).

The effects of uridine and its nucleotides on isolated blood vessels are complex, acting directly on smooth muscle cells or stimulating surrounding endothelial cells at physiological levels, e.g. micromolar concentrations (SEIFERT and SCHULTZ 1989; KUNAPULI and DANIEL 1998). The effects of exogenous uridine and pyrimidine nucleotides are not restricted to the vasculature per se with effects on perfusion of organs, e.g. the liver and brain, being apparent (SEIFERT and SCHULTZ 1989). The exact nature of the P2 receptors mediating these effects remains unknown.

Three pyrimidine pathologies of erythrocytes have been described. Pyrimidine 5'-nucleotidase deficiency usually presents as a mild haemolytic anaemia with basophilic stippling (SCRIVER 1995). This defect also occurs in lymphocytes and granulocytes and thus may affect other organs including the brain. The disease may be inherited or induced, e.g. by lead poisoning (which inhibits pyrimidine 5'-nucleotidase), and results in gross uridine and cytidine nucleotide accumulation in erythrocytes together with their conjugates. The latter also accumulate as a result of inherited putative deficiency of CDPC phosphotransferase (PAGLIA et al. 1983) which is also associated with anaemia. Conversely, classic orotic aciduria, an inherited deficiency of UMPS, results in markedly reduced uridine nucleotide levels, presenting with a severe megaloblastic anaemia that can be reversed by oral uridine (BECROFT et al. 1969;

Table 1. Clinical trials of uridine and UDPGlucose

Disease/Ailment	Symptoms	Treatment regimen	Effect and proposed mechanism of action
Hereditary orotic aciduria	Megaloblastic anaemia, retarded physical and mental development, orotic acid crystalluria in children	Life long administration of oral uridine	Rapid adsorption from gut, salvage of uridine to make nucleotides bypassing the metabolic block. ↑ UMP production
Pervasive developmental delay	Developmental delay, seizures, ataxia, recurrent infections, severe language deficit, hyperactivity, short attention span, poor social interaction and hyperuricosuria in children	Double-blind placebo trial of oral uridine	Six- to tenfold ↑- in cytosolic purine 5'-nucleotidase activity and reduced uridine uptake. Uridine ↑-synthesis of nucleotides
Schizophrenia	Paranoid schizophrenia in adults (DSM III-R)	Open trial of oral uridine combined with haloperidol on 20 patients	↓- time required before haloperidol is effective and its maintenance dose. Antagonises haloperidol-induced extra-pyrimidal evoked side effects, e.g., hypertonia, rigidity, dyskinesia
Epilepsy	Convulsions, paroxysmal discharge of cortical neurones in adults	Open trial of oral uridine	↓- seizures. Interaction with GABAergic systems
Cancer	Solid tumours in adults	Open trial with uridine and UDPGlucose	Decrease toxicity of 5-fluoruracil, possibly by preventing toxic metabolite production
Porphyria cutanea	Disturbance in porphyrin metabolism	UDPGlucose	↓- Porphyria, possibly due an increased RNA and DNA production by cells
Galactosaemia	Galactosaemic children with cognitive delay and decreased intellectual function	Longitudinal study of 35 patients for 2–5 years, given uridine (150 mg/kg, ibd)	Ineffective – uridine showed no beneficial effects
Diabetes-induced peripheral nerve neuropathy	Decreased neurotransmission in peripheral nerves of limbs	Double blind trial placebo trial (20 patients on oral uridine and 20 with placebo	↑- lipid metabolism, via conversion of uridine cytidine nucleotides conjugates of ethanolamine and phospholipids

See text for references.
↑, increase, ↓, decrease.

Table 1.).The aetiology of anaemia in each case remains unexplained, but the unifying feature is the deregulation of erythrocyte pyrimidine nucleotide levels, and a dysfunctional assembly of cell membranes (reliant upon pyrimidine nucleotide lipid and sugar conjugate production) is likely to be a factor in poor erythrocyte stability and survival.

F. Pyrimidines and Reproductive Function

Uridine and its nucleotides are involved in the initiation and success of the reproductive process (PERSSON et al. 1991; RONQUIST and NIKLASSON 1984; RONQUIST et al. 1985). As compared to other body fluids, seminal fluid contains 1000-fold greater concentrations (e.g. millimolar) of uridine, and these concentrations are increased significantly following vasectomy (RONQUIST and NIKLASSON 1984). The source of uridine in human seminal plasma probably results from the catabolism of uridine nucleotides by plasma soluble 5'-nucleotidases (TOMLINS et al. 1998), one of which has highly distinctive regulatory properties, e.g. lack of product regulation or inhibition by xanthines and is specific to human seminal plasma (MINELLI et al. 1997). In prostatitis, low concentrations of uridine are found in prostatic secretions with very high concentrations in seminal fluid (RONQUIST et al. 1985).

High levels of uridine may be involved in the promotion of sperm motility, as a correlation exists between seminal plasma uridine concentrations and percent sperm motility (RONQUIST et al. 1985). It is perhaps relevant, therefore, that regulation of uridine diphosphatase during spermatogenesis in rat is under hormonal control (XUMA and TURKINGTON 1972). The predominance of uridine in seminal fluids suggests important functions in control of fertilisation and implantation, although the precise role of UTP and its catabolites remains unresolved. Interestingly, although activation of human sperm by ATP involves a P2X receptor coupled to sodium channels (FORESTA et al. 1996), a P_{2U} receptor from rat spermatogenic cells which is activated by UTP > ATP > UDP has been expressed in *Xenopus oocytes* (WU et al. 1998). The role of these P_{2U} receptors which couple to calcium-activated potassium channels remains unknown, but given the key role of calcium in sperm activation and in sperm-egg interaction, a P_{2U} receptor and extracellular pyrimidines are likely to be involved.

G. Cancer and Antiviral Therapy

Marked changes in purine and pyrimidine metabolism occur in cancerous cells and it is noteworthy that many anticancer drugs are based on purine and pyrimidine pharmacophores (e.g. 5-flurouridine, 5-FU) being designed to interact with these pathways, thus inhibiting cell replication. Uridine and UDP-glucose have shown considerable promise as novel agents to counteract the unwanted toxicity of certain pyrimidine-based anticancer drugs. For instance, uridine is used as a "rescue" therapy for the myelotoxicity and gastrointestinal toxicity produced by 5-FU (VAN-GROENINGEN et al. 1986), while trials of UDP-glucose also show promise in this context (LEYVA et al. 1984; CODACCI-PISANELLI et al. 1997) (Table 1). Uridine and benzylacyclouridine protected mice against neurotoxic side effects of pyrimidine-based drugs, e.g. azidothymidine, used to treat HIV infections (FALCONE et al. 1990; CALABRESI et al. 1990) and in a neuroblastoma cell line used to model 2',3'-dideoxycytidine-induced neurotoxicity (KEILBAUGH et al. 1993).

The toxicity of fluropyrimidines like 5-FU varies with time of administration and mice exhibit a circadian rhythm of liver uridine phosphorylase activity and plasma concentration of uridine (EL KOUNI et al. 1990), demonstrating that the liver has an important role in the regulation of uridine levels in blood. These circadian events may be involved in the humoral control of sleep by uridine (see below) and may also be of clinical significance in enhancing the anti-tumour and -viral efficacy of fluoropyrimidines by modulating their therapeutic efficacy (PETERS et al. 1987c). Thus there is considerable interest in developing compounds to manipulate plasma uridine, e.g. uridine prodrugs like PN401 (KELSEN et al. 1997), or agents that modulate circadian rhythms by selective inhibition of uridine phosphorylase. Several efficacious pyrimidine-based antiviral and anticancer drugs, e.g. 6-azauracil and sorivudin, have been withdrawn from the clinic because of severe neurotoxicity or lethality which, retrospectively, involved interactions with pyrimidine metabolism (CONNOLLY 1994; GONZALEZ and FERNANDEZ-SALGUERO 1995; CONNOLLY et al. 1996). For example, 5-FU therapy can be lethal to patients deficient in dihydropyrimidine dehydrogenase (DPD) (WATABE 1996). This deficiency can be either inherited or iatrogenically induced by administration of bromovinyluridines such as sorivudine, and there is increasing evidence showing that uridine may relieve this effect (CONNOLLY et al. 1996).

A major lesson from studies of pyrimidine in use as antiviral and anticancer agents is that, while pyrimidine-based analogues are powerful and effective agents, thorough metabolic and pharmacological studies are often necessary to discover their full physiological and pharmacological actions. It is clear that uridine and its derivatives could have an important role in modulating the toxicity of new antiviral and anticancer drugs.

H. Peripheral Nervous System Modulation

Pyrimidine nucleotides can act as neuromodulators or neurotransmitters via activation of P2Y receptors (BOEYNAEMS et al. 1997; BOARDER and WEBB, Chap. 4, this volume). Both astrocytes and neurons have ecto-enzymes that catabolise (CONNOLLY et al. 1998; CONNOLLY and DULEY 1999; ZIMMERMANN, Chap. 8, this volume) or anabolise nucleoside triphosphates, e.g. ecto-nucleoside diphosphokinase on rat astrocytoma cells (HARDEN et al. 1997).

Uridine, UDP, UTP and UDPGlucose can depolarise or hyperpolarise amphibian ganglia at submicromolar concentrations (SIGGINS et al. 1978) and have similar actions on isolated rat superior cervical ganglia (SCG) (CONNOLLY et al. 1993; CONNOLLY and HARRISON 1995).

UDP and UTP increase intracellular calcium levels (G. Connolly, unpublished data) and thus modulate neurotransmitter release, e.g. noradrenaline from cultured rat SCG (BOEHM et al. 1995). Ecto-enzymes capable of degrading UTP to uridine are present on SCG neurones (CONNOLLY et al. 1998; CONNOLLY and DULEY 1999) supporting a functional role for these pyrimidines in the PNS. While UTP had an excitatory action, uridine hyperpolarised rat SCG (CONNOLLY et al. 1993) thus having an inhibitory action. Intracellular uridine nucleotide levels are in the millimolar range and tissue disruption and/or cell death may release sufficient UTP to attain micromolar concentrations even in the presence of ecto-enzymes, and thus are within a pharmacological range capable of activating P2 receptors. Hence uridine or its nucleotides are likely to be released under both physiological and pathological situations, e.g. from endothelial and tumour cells, but evidence for their release as neurotransmitters remains elusive.

A neuromodulatory role appears to be a more likely physiologic function of pyrimidines in synaptic transmission (CONNOLLY et al. 1993, 1998; CONNOLLY 1994; CONNOLLY and HARRISON 1995) where UTP is an excitatory agent at P2Y receptor sub-types (BOARDER and WEBB, Chap. 4, this volume) whereas uridine may potentially activate putative inhibitory "U" nucleoside receptors similar to the P1 receptor family (LORENZEN and SCHWABE, Chap. 2, this volume). Thus, as in the case of the ATP/adenosine purinergic cascade (WILLIAMS and JARVIS 2000), the initial excitatory effects of pyrimidine nucleotides would be counteracted by their metabolism to inhibitory pyrimidine nucleosides, e.g. uridine. However, no evidence currently exists for discrete uridine receptors However, uridine could activate a fast transmembrane Ca^{2+} ion flux into resealed plasma lemma fragments and nerve endings of rat cerebral cortex homogenates (KARDOS et al. 1999). In addition [^{3}H]-uridine labelled both high and low affinity specific binding sites in purified synaptosomal membranes, was taken up by synaptosomes in a dipyridamole-sensitive process and was released by depolarisation.

I. CNS Modulation

The seminal observations of GEIGER and YAMASAKI (1956) demonstrated that the mammalian brain relies on a supply of circulating pyrimidines such as uridine and cytidine for electrophysiological activity and for carbohydrate and phospholipid content, thus emphasising the important role of pyrimidines in the maintenance of normal CNS activity (CONNOLLY 1994; CONNOLLY et al. 1996).

Studies on the origin of pyrimidines for biosynthesis of RNA in the rat suggest that uridine was far superior to orotic acid in labelling RNA in brain slices, whereas both compounds were equally effective in hepatic or renal slices (HOGANS et al. 1971). This and other studies indicate that the mature brain, unlike other tissues, relies heavily on performed precursors from the 'salvage pathway' and to a lesser extent utilises de novo pyrimidine synthesis. There has however been little research in the area for over a quarter of century and it is likely that much remains to be discovered.

I. Clinical Studies

Abnormal levels of uridine and its catabolites in CSF and blood are indicators of inherited diseases of pyrimidine metabolism and are likely culprits in the aetiology of various neurological disorders (SCRIVER 1995; GERRITS et al. 1991; CONNOLLY 1994; GONZALEZ and FERNANDEZ-SALGUERO 1995; CONNOLLY et al. 1996).

The uridine-containing sugar nucleotides, UDPG and UDP-galactose (UDPGal), are important intermediates in galactose metabolism, and tissue UDPGal may be a key factor in the aetiology of the long-term clinical manifestations of patients with galactose-1-phosphate uridyltransferase deficient galactosaemia. In classical galactosaemia the brain and the ovaries are target organs of galactose toxicity. POPOV et al. (1984) reported a protective effect of uridine on D-galactosamine-induced deficiency in brain uridine phosphates and in D-galactosamine-induced impairment of retention performance of rats and (NG et al. 1989) suggested uridine may raise intracellular UDPGal in galactosaemia. However, ROGERS et al. (1991) suggested caution in administering uridine to patients with genetically limited transferase activity because it may inhibit residual enzyme in the tissues of affected subjects. KIRKMAN (1992) was unable to demonstrate a deficiency of UDPGal in galactosaemia and questioned the need for treating galactosaemic children with uridine. A recent clinical trial has failed to shown any clinical improvement in children with galactosaemia upon uridine treatment (MANIS et al. 1997).

The neurological consequences of disregulation of uridine salvage have been clearly demonstrated by PAGE et al. (1997) who found that superactivity of purine and pyrimidine 5'-nucleotidase is linked to a form of autism with seizures (Table 1). Skin fibroblasts from these children exhibited decreased salvage of uridine into pyrimidine nucleotides. Significantly, these children were successfully treated with uridine (Table 1), showing remarkable improvements in speech and behaviour as well as decreased seizure activity and frequency of infections. Replacement of uridine with placebo produced rapid regression of all patients to their pre-treatment states. Dysfunctional intra- and extra-cellular pyrimidine levels occur in a number of disorders such as hyperammonaemia, drug treatment, (e.g. with allopurinol, and anti-cancer drugs (see above)) and in patients lacking inborn errors of pyrimidine metabolism. However, the most prevalent incident of altered pyrimidine levels occurs because of genetic metabolic disorders of pyrimidine metabolism. Inherited disorders of pyrimidine metabolism presenting with CNS dysfunction are

UMPS deficiency (BECROFT et al. 1969; SCRIVER et al. 1995) and deficiencies of steps in the catabolic pathway, including DPD, dihydropyrimidinase (DP), γ-ureidopropionase, and putative γ-alanine aminotransferase activities (CONNOLLY 1994; SCRIVER 1995; GONZALEZ and FERNANDEZ-SALGUERO 1995; CONNOLLY et al. 1996).

These encompass many symptoms but all of these pyrimidinergic defects centre around abnormal CNS activity, including convulsions, mental retardation, autism and poor motor co-ordination. Such 'experiments of nature' suggest fundamental roles for pyrimidine catabolites in the regulation of CNS function. Why disorders of pyrimidine metabolism cause such profound pathologies is unknown, but they are likely to involve at least two distinct but not exclusive mechanisms.

First, these effects may be mediated by changes in altered levels of extracellular pyrimidines, e.g. uridine and UTP, which can interact with receptors for pyrimidines or neurotransmitters. Second, these effects may result in altered intracellular metabolism leading to abnormal physiological activity. An uneven regional distribution of nucleotide metabolism in human brain (KOVACS et al. 1998) and differences in regional uptake support a neurotransmitter/modulator role for uridine and its nucleotides in the CNS.

II. Animal Studies

1. Neuropeptide Interactions

Severe hypoglycaemia decreased somatostatin but not cholecystokinin (CCK)-like immunoreactivity in the dorsal hippocampal formation and frontoparietal cortex of rats treated with insulin. However, treatment with uridine significantly counteracted the insulin effects (AGNATI et al. 1986). The ability of uridine to prevent release and/or increase synthesis of cortical somatostatin in severe hypoglycaemia may prove beneficial for such disorders.

Rats given uridine in drinking water (0.5mg/ml) showed a selective decrease in neuronal somatostatin, CCK and galanin-like immunoreactivity in various brain areas of 12- but not 3-month-old rats. Uridine decreased peptide immunoreactivity in nerve terminals of some diencephalic areas, but increased CCK-like immunoreactivity of most telencephalic brain areas (AGNATI et al. 1986; Table 2). The ability of uridine to influence activity of the hypothalamic-hypophyseal axis, probably by interference with endogenous neurotransmitters, e.g. dopamine, is supported by observations by BERNARDINI et al. (1983) who found uridine itself was ineffective, but potentiated the inhibitory effect of dopamine on prolactin release in vitro, and the hypoprolactinaemic effect of glutamine (Table 2).

2. Dopamine Interactions

An interaction between uridine and dopamine receptors is suggested by the ability of chronic uridine treatment to reduce spiperone binding site numbers

Uridine and Pyrimidine Nucleotides in Cell Function

Treatment regimen	Effect of uridine	Proposed mechanism of action
Acute (15 mg/kg i.v.)	Ablation of hypoglycemia-induced ↓ in somatostatin in rat dorsal hippocampal formation and frontopaietal cortex of rats treated with insulin	Antagonism of disappearance of somatostatin-like immunoreactivity. Metabolic effect proposed
Acute (100 mg/kg i.v.)	Hypothermia induced in rats and mice	Unknown (possibly effects of uridine catabolites)
Acute (intra-carotid perfusion at $0.1\ ml\ min^{-1}\ kg^{-1}$ at 12.5 mmol/l)	Maintenance of dog brain metabolism under severe hypoglycemia	Changes in amino acid transmitter release and glycolytic metabolites
Acute (350 mg RNA hydrolysate)	Orally given pyrimidines increased alpha wave activity and increased EEG activity of human brain	Unknown (possible interaction with barbiturates)
Chronic (mean daily dose 12.5 mg/rat for 6 months)	↓ in rat brain somatostatin, CCK and galanin in 12 month but not 3 month old animals ↓ in galanin, neuropeptide Y and CCK in some diencephalic areas but ↑ in CCK in most telencephalic areas of rats brain	Selective action of uridine on activity of some neuropeptides, e.g. CCK
Chronic (15 mg/kg per day i.p. for 14 days)	↓ spiperone binding in striatum of young rats ↑ rate of recovery of striatal spiperone-labelled dopamine receptors in young, but not old rats	Modulation of the steady state and turnover of D2-dopamine receptors
Chronic (15 mg/kg/day i.p.) ± haloperidol (1 mg/kg/day i.p.)	↓ haloperidol-induced changes in striatal dopamine in rat brain ↓ haloperidol-induced catalepsy and apomorphine-induced stereotypies of rats	Potentiation of ↓ striatal dopamine release and haloperidol-induced release of dopamine
Chronic (16 mg/Kg/day i.p.)	↓ Effects of apomorphine in aged but not young rats ↓ Apormorphine-induced catalepsy of rats after treatment with flunarizine or haloperidol	Interaction with dopaminergic neurotransmission
Chronic (16 mg/kg, i.p.)	No effect on body weight, activity or rotation of rats under baseline conditions, but ↑ sensitivity to amphetamine (but not cocaine) in activity test ↑ sensitivity to amphetamine and cocaine in rotation test. ↓ amphetamine-induced ↑ striatal dopamine	Uridine inactive by itself, but chronic treatment modulates stimulant-induced release of dopamine

In the last column the comments presented in parenthesis are those proposed by this author and not those of the original authors
Route of administration: i.p, intraperitoneal, i.v., intravenous;↑, increase;↓, decrease; CCK, cholecystokinin.

and enhance D2 dopamine receptor turnover within the striatum of young but not adult rats (FARABEGOLI et al. 1988; Table 2). Catalepsy and exploratory locomotor activity, two behaviours associated with blockade vs activation of dopamine receptors, were affected more in young than in adult animals, suggesting uridine modulates the steady state and turnover rate of striatal D-2 dopamine receptors in young animals. Perhaps altered uridine catabolism with age accounts for the selective effects of uridine on young and not old animals.

Chronic treatment with uridine or combined with haloperidol markedly reduced dopamine release induced by an acute haloperidol challenge, haloperidol-induced catalepsy and apomorphine-induced stereotypies (AGNATI et al. 1989; Table 2), suggesting that uridine potentiates the reduction of the striatal and chronic haloperidol treatment. These effects were maximal by 7 days, reaching a plateau, whereas an acute high dose (30 mg/kg) of uridine was ineffective. Thus, uridine may mediate its effects via a change in some metabolic processes. As regards stereotypies induced by apomorphine in flunarizine or haloperidol-treated rats, uridine did not counteract the effect in young rats, while it completely antagonised the effect in older rats. Catalepsy was not observed in young rats treated with flunarizine or haloperidol, but was prominent in old rats, and was counteracted by uridine (MYERS et al. 1995; Table 2).

Further evidence for an interaction of uridine with dopaminergic neurotransmission is demonstrated by the ability of uridine to reduce rotational behaviour induced by the dopamine precursor, L-DOPA and methamphetamine in 6-hydroxydopamine-treated rats (MYERS et al. 1993; Table 2). Uridine alone did not alter rotation or its reduction by amphetamine or haloperidol, nor did it alter the biphasic response induced by chlordiazepoxide. However, uridine-treated animals with unilateral striatal lesions exhibited no rotational behaviour in the absence of drug challenge, but showed decreased rotation induced by L-DOPA, compared with controls. Uridine-treated rats also exhibited reduced rotation after repeated injections of methamphetamine in contrast to increasingly greater rotation observed in control animals (MYERS et al. 1993; Table 2). These studies show that chronic uridine administration in low doses modulates CNS dopaminergic transmission, possibly an interaction via D2-dopamine receptors, counteracts homeostatic responses aimed at increasing dopaminergic transmission, and alters drug-induced dopaminergic activity. Why uridine has such effects is still unclear, but its ability to selectively affect some neuropeptides, e.g. somatostatin in hypoglycemia and CCK in aged animals, suggests specificity within the CNS. CCK is stored with dopamine in some CNS neurons, and its ability to modulate dopaminergic receptors may be subject to modulation itself by uridine.

The effects of uridine are clearly complex; however an ability of chronically administered uridine to decrease dopamine release induced by haloperidol may indicate a potential therapeutic use in schizophrenia. In fact, early clinical results support this hypothesis, with uridine shortening the time needed for haloperidol to work, allowing a reduction of its maintenance dose, and counteracting akinetic-depression and antagonising haloperidol-induced extrapyramidal symptoms (PARIANTE 1992). Uridine possibly potentiates dopaminergic transmission at reduced levels of the amine, and therefore may be useful in treating Parkinson's disease. Pyrimidine derivatives also have anxiolytic activity in rat and mice models of anxiety (KARKISHCHENKO et al. 1990; CONNOLLY 1994) but many of these effects could arise from the known sedative-like properties of uracil and uridine described below (CONNOLLY 1994).

3. Anticonvulsant and Anxiolytic Effects

Uridine, uracil and certain derivatives have anticonvulsant and anxiolytic effects in a wide variety of behavioural paradigms (CONNOLLY 1994). These effects may arise from the fact that uridine contains within its simple ring structure the components that form the barbiturate pharmacophore. Uridine competitively inhibited [^{3}H]GABA binding to rat cerebellar membranes, frontal cortex, hippocampus and thalamus, altering receptor affinity with the number of binding sites remaining unchanged (GUARNERI et al. 1983, 1985). Uridine also produced significant reduction in bicuculline-induced seizures of rats (GUARNERI et al. 1985).

Uridine analogues like N-3-phenacyluridine interact with benzodiazepine receptors and are potent novel sedative and hypnotic agents (KIMURA et al. 1996). Uridine and its analogues modulate the $GABA_A$ receptor chloride ionophore complex (CONNOLLY 1994). Preliminary data suggest an antiepileptic effect of uridine in man (BONAVITA et al. 1975; Table 1) and an ability of pyrimidines to alter EEG activity in man and rats (BONAVITA et al. 1962, 1963).

4. Sleep and Thermoregulation

Uridine has been isolated as an endogenous sleep-inducing factor from sleep deprived rats (HONDA et al. 1984). A hypnotic action for uridine is consistent with its ability to potentiate the effects of barbiturates (CONNOLLY 1994) and to interact with GABAA and benzodiazepine receptors. The similarity in the structures of barbiturates and uracil and uridine supports a common site of action.

High doses of uridine decreased core body temperature when injected into rats or mice (PETERS et al. 1987a). The ability of uridine to reduce body temperature could serve a key physiological function, as it is consistent with a reduction in body temperature seen during sleep and the potential role of uridine as an endogenous sleep-promoting substance (HONDA et al. 1984; CONNOLLY 1994). A concomitant rise of plasma uridine with dark periods, and conversely lower plasma uridine in periods of light, due to circadian changes in uridine phosphorylase activity, may facilitate sleep.

Uridine displays a striking ability to induce fever in rabbits and humans (PETERS et al. 1987a,b) that limits the usefulness of the high doses of uridine required to alleviate 5-FU-induced toxicity. Perhaps a physiological function of uridine is to reduce the high costs of endothermy by reducing the energy demand of the brain and other tissues. Resolution of the sleep promoting and thermoregulatory actions of uridine could benefit not only pyrimidine-based treatment of cancer but also produce a method of controlling fever, and could even result in novel neuroprotectant agents aimed at lowering core temperature. Interestingly, the administration of uridine catabolites evoked an increase in body temperature, while inhibition of uridine phosphorylase partially prevented these effects (PETERS et al. 1987a,b).

5. Additional Actions of Uridine and its Derivatives

Further actions of uridine and it derivatives that warrant further study include their ability to improve a variety of pathologies caused by ischaemia (GEIGER and YAMASAKI 1956) and hypoglycaemia (BENZI et al. 1984), diabetic neuropathy (GALLAI et al. 1992), failing memory (DRAGO et al. 1990) and neurotoxicity (SCHMID-ELSAESSER et al. 1997). It is important to note that even though intracellular and tissue compartmentation of pyrimidines exists, pyrimidines are not metabolically isolated, but act in concert with other metabolites such as amino acids, purines and pyridines, and are subject to perturbations in the urea cycle.

In many of the examples presented here, patients or animals were treated for prolonged periods with pyrimidines, possibly suggesting their beneficial effects were mediated via changes in the metabolism. Even so, such studies highlight the idea that uridine and it derivatives can counteract specific pathologies and should prove useful adjuncts in their treatment Thus, metabolism is a crucial factor in normal physiology and pathological actions of

Table 3. Potential therapeutic applications for uridine and UDP-glucose

Disease/ailment	Benefits and proposed mechanism(s) of action
Anti-epileptic	Decrease in seizure activity by uridine and its derivatives. Interaction with GABAergic neurotransmission
Parkinsonism	Potentiation of CNS dopaminergic neurotransmission by uridine and its derivatives
Anxiety	Modulation of $GABA_A$-benzodiazepine binding sites by uridine and its derivatives
Liver and kidney dysfunction	Improve physiology due increase supply of pyrimidines for RNA, DNA and protein and carbohydrate production, leading to increased cell proliferation
Sleep dysfunction, e.g. insomnia and jet lag	Increased activation of $GABA_A$ receptors to potentiate benzodiazepine, barbiturate binding, by uridine and its derivatives
Cancer and viral infections	Alters cell metabolism or interferes with anti-cancer and anti-viral agents metabolism to improve its efficacy by uridine and UDPGlucose
Stroke, ischaemia and hypoxia	Increased lipid and glucose metabolism preventing cell necrosis by uridine and UDPGlucose

pyrimidines and further studies should provide a means of regulating their storage and release.

It is apparent that the relationships between pyrimidines and neurotransmission is complex and still poorly understood, but as with the plethora of actions of purines (RALEVIC and BURNSTOCK 1998), many new and exciting actions of uridine and related pyrimidines await discovery.

In conclusion, the ability of mammals to regulate tightly plasma and CSF uridine levels coupled with presence of receptors for uridine and its nucleotides makes this class of molecules particularly promising candidates for future development as novel therapeutic agents for a variety of ailments (Table 3).

References

Agnati LF, Fuxe K, Eneroth P, Zini I, Harfstrand A, Grimaldi R, Zoli M (1986) Intravenous uridine treatment antagonizes hypoglycaemia-induced reduction in brain somatostatin-like immunoreactivity. Acta Physiol Scand 126:525–531

Agnati LF, Fuxe K, Rugger M, Pich EM, Benfenati F, Volterra V, Ungerstedt U, Zini I (1989) Effects of chronic treatment with uridine on striatal dopamine release and dopamine related behaviours in the absence or the presence of chronic treatment with haloperidol. Neurochem Internat 15:107–113

Becroft DM, Phillips LI, Simmonds A (1969) Hereditary orotic aciduria: long-term therapy with uridine and a trial of uracil. J Pediatrics 75:885–891

Benzi G, Villa RF, Dossena M, Vercesi L, Gorini A, Pastoris O (1984) Cerebral endogenous substrate utilisation during the recovery period after profound hypoglycemia. J Neurosci Res 11:437–450

Berman P, Harley E (1984) Orotate uptake and metabolism by human erythrocytes. Advan Exper Med Biol 165:367–371

Bernardini R, De Luca G, Marino D (1983) Effect of glutamine and nucleosides on prolactin secretion in the rat. Acta Europaea Fertilitatis 14:341–344

Bocharova LS, Gordon RY, Arkhipov VI (1992) Uridine uptake and RNA synthesis in the brain of torpid and awakened ground squirrels. Comp Biochem Physiol B 101:189–192

Boehm S, Huck S, Illes P, (1995) UTP- and ATP-triggered transmitter release from rat sympathetic neurones via separate receptors. Br J Pharmacol 116:2341–2343

Bonavita V, Piccoli F, Savettieri G, Zito M (1975) Observations on the therapeutic usefulness of uridine in epilepsy (Italian). Acta Neurol 30:30–34

Bonavita V, Scarano E, Zito M (1962) The action of purine and pyrimidine nucleosides on the human brain as revealed by pentothal-activated EEG. Arch Int Pharmacodyn 138:152–156

Bonavita V, Bonasera N, Zito M, Scarano E (1963) Electrophysiological and neurochemical studies following injection of mono-nucleotides and their derivatives. J Neurochem 10:155–164

Calabresi P, Falcone A, St Clair MH, Wiemann MC, Chu SH, Darnowski JW (1990) Benzylacyclouridine reverses azidothymidine-induced marrow suppression without impairment of anti-human immunodeficiency virus activity. Blood 76:2210–2215

Chambers JK, Macdonald LE, Sarau HM, Ames RS, Freemann K, Foley JJ, Zhu Y, McLaughlin MM, Murdock P, McMillan L, Trill J, Swift A, Aiyar N, Taylor P, Vawter L, Naheed S, Szekeres P, Hervieu G, Scott C, Watson JM, Murphy AJ, Duzic E, Klein C, Bergsma DJ, Wilson S, Livi GP (2000) A G Protein-coupled Receptor for UDP-glucose. J Biol Chem 275:10767–10771

Codacci-Pisanelli G, Kralovanszky J, van der Wilt CL, Noordhuis P, Colofiore JR, Martin DS, Franchi F, Peters GJ (1997) Modulation of 5-fluorouracil in mice using uridine diphosphoglucose. Clin Cancer Res 3:309–315

Coltorti M (1975) Influence of UDPG, glutathione and vitamin B 12 therapy on various liver function indices in patients with liver cirrhosis. (Italian) Clinica Terapeutica 72:323–335

Connolly GP (1994) Neurobiological aspects of pyrimidines. Pathways Pyrimidines 2:37–44

Connolly GP, Abbott NJ, Demaine C, Duley JA (1997) Investigation of receptors responsive to pyrimidines. Trends Pharmacol Sci 18:413–414

Connolly GP, Demaine C, Duley JA (1998) Ecto-nucleotidases in isolated intact rat vagi, nodose ganglia, and superior cervical ganglia. Adv Exper Med Biol 431:769–776

Connolly GP, Demaine C, Duley JA (1998) Ecto-nucleotidase activity of cultured rat superior cervical ganglia. Cell Mol Biol Let 4:351

Connolly GP, Harrison PJ, Stone TW (1993) Action of purine and pyrimidine nucleotides on the rat superior cervical ganglion. Br J Pharmacol 110:1297–1304

Connolly GP, Harrison PJ (1995) Structure-activity relationship of a pyrimidine receptor in the rat isolated superior cervical ganglion. Br J Pharmacol 116:2764–2770

Connolly GP, Simmonds HA, Duley JA (1996) Pyrimidines and CNS regulation. Trends Pharmacol Sci 17:106–107

Dawson DM (1968) Enzymic conversion of uridine nucleotide to cytidine nucleotide by rat brain. J Neurochem 15:31–34

Dierich A, Wintzerith M, Sarlieve L, Mandel P (1980) Biosynthesis pathways of nucleotides in different nerve cell cultures. J Neurochem 34:1126–1129

Drago G, Casetti P, Nicosia F (1967) Uridine and cytidine in the treatment of hepatic coma. Preliminary note. (Italian) Osp Ital Chir 17:651–657

Drago F, D'Agata V, Valerio C, Spadaro F, Raffaele R, Nardo L, Grassi M, Freni V (1990) Memory deficits of aged male rats can be improved by pyrimidine nucleosides and n-acetyl-glutamine. Clin Neuropharm 13:290–296

el Kouni MH, Naguib FN, Park KS, Cha S, Darnowski JW, Soong SJ (1990) Circadian rhythm of hepatic uridine phosphorylase activity and plasma concentration of uridine in mice. Biochem Pharmac 40:2479–2485

Foresta C, Rossato M, Chiozzi P, Di Virgilio F (1996) Mechanism of human sperm activation by extracellular ATP. Am J Physiol 270:C1709–C1714

Hogans AF, Guroff G, Udenfriend S (1971) Studies on the origin of pyrimidines for biosynthesis of neural RNA in the rat. J Neurochemistry 18:1699–1710

Falcone A, Darnowski JW, Ruprecht RM, Chu SH, Brunetti I, Calabresi P (1990) Different effect of benzylacyclouridine on the toxic and therapeutic effects of azidothymidine in mice. Blood 76:2216–2221

Farabegoli C, Merlo Pich E, Cimino M, Agnati LF, Fuxe K (1988) Chronic uridine treatment reduces the level of [3H]spiperone-labelled dopamine receptors and enhances their turnover rate in striatum of young rats: relationship to dopamine-dependent behaviours. Acta Physiol Scan 132:209–216

Gallai V, Mazzzotta G, Montesi S, Sarchielli P, Del Gatto F (1992) Effects of uridine in the treatment of diabetic neuropathy: an electrophysiological study. Acta Neurol Scand 86:3–7

Geiger A, Yamasaki S (1956) Cytidine and uridine requirement of the brain. J Neurochem 1:93–100

Gerrits GP, Monnens LA, De Abreu RA, Schroder CH, Trijbels JM, Gabreels FJ (1991) Disturbances of cerebral purine and pyrimidine metabolism in young children with chronic renal failure. Nephron 58:310–314

Guarneri P, Piccoli F, Mocciaro C, Guarneri R (1983) Interaction of uridine with GABA binding sites in cerebellar membranes of the rat. Neurochem Res 8:1537–1545

Guarneri P, Guarneri R, La Bella V, Piccoli F (1985) Interaction between uridine and GABA-mediated inhibitory transmission: studies in vivo and in vitro. Epilepsia 26:666–671

Guroff G, Brodsky M (1971) Enzymes of nucleic acid metabolism in the brains of young and adult rats. J Neurochem 18:2077–2084
Goetz U, Da Prada M, Pletscher A (1971) Adenine-, guanine- and uridine-5'-phosphonucleotides in blood platelets and storage organelles of various species. J Pharmacol Exp Ther 178:210–215
Gonzalez FJ, Fernandez-Salguero P (1995) Diagnostic analysis, clinical importance and molecular basis of dihydropyrimidine dehydrogenase deficiency. Trends Pharmacol Sci 16:325–327
Harden TK, Lazarowski ER, Boucher RC (1997) Release, metabolism and interconversion of adenine and uridine nucleotides: implications for G protein-coupled P2 receptor agonist selectivity. Trends Pharmacol Sci 18:43–46
Honda K, Komoda Y, Nishida S, Nagasaki H, Higashi A, Uchizono K, Inoue S (1984) Uridine as an active component of sleep-promoting substance: its effects on nocturnal sleep in rats. Neurosci Res 1:243–252
Hussain SP, Alexander N, Duley JA, Connolly GP (1999) Abnormal intracellular signalling in skin fibroblasts from LNS patients and HGPRT deficient neuroblastoma. Cell Mol Biol Let 4:375–376
Kardos J, Kovacs I, Szarics E, Kovacs R, Skuban N, Nyitrai G, Dobolyi A, Juhasz G (1999) Uridine activates fast transmembrane Ca^{2+} ion fluxes in rat brain homogenates. Neuroreport 10:1577–1582
Karkishchenko NN, Makliakov IuS, Stradomskii BV (1990) Pyrimidine derivatives: their psychotropic properties and the molecular mechanisms of their central action (Russian). Farmakologiia i Toksikologiia 53:67–72
Keilbaugh SA, Hobbs GA, Simpson MV (1993) Anti-human immunodeficiency virus type 1 therapy and peripheral neuropathy: prevention of 2',3'-dideoxycytidine toxicity in PC12 cells, a neuronal model, by uridine and pyruvate. Mol Pharmacol 44:702–706
Kelsen DP, Martin D, O'Neil J, Schwartz G, Saltz L, Sung MT, von Borstel R, Bertino J (1997) Phase I trial of PN401, an oral prodrug of uridine, to prevent toxicity from fluorouracil in patients with advanced cancer. J Clinical Oncology 15:1511–1517
Kihara H (1967) Pseudouridinuria in mentally defective siblings. Am J Ment Defic 71:593–596
Kimura T, Yamamoto I, Watanabe K, Tateoka Y, Ho IK (1991) In vivo and in vitro metabolic studies of N3-benzyluridine which exhibits hypnotic activity in mice. Res Com Chem Path Pharmac 71:27–48
Kimura T, Kuze J, Watanabe K, Kondo S, Ho IK, Yamamoto I (1996) N3-phenacyluridine, a novel hypnotic compound, interacts with the benzodiazepine receptor. Eur J Pharmacol 311:265–269
Kirkman HN Jr (1992) Erythrocytic uridine diphosphate galactose in galactosaemia. J Inher Meta Dis 15:4–16
Knowles MR, Clarke LL, Boucher RC (1991) Activation by extracellular nucleotides of chloride secretion in the airway epithelia of patients with cystic fibrosis. New Engl J Med 325:533–538
Kovacs Z, Dobolyi A, Szikra T, Palkovits M, Juhasz G (1999) Uneven regional distribution of nucleotide metabolism in human brain. Neurobiology (Budapest) 6:315–321
Kunapuli SP, Daniel JL (1998) P2 receptor subtypes in the cardiovascular system. Biochem. J 336:513–523
Leyva A, van Groeningen CJ, Kraal I, Gall H, Peters GJ, Lankelma J, Pinedo HM (1984) Phase I and pharmacokinetic studies of high-dose uridine intended for rescue from 5-fluorouracil toxicity. Cancer Res 44:5928–5933
Mandel P, Edel-Harth S (1966) Free nucleotides in the rat brain during post-natal development. J Neurochem 13:591–595
Mandel P, Edel S, Poirel G (1966) Free nucleotides in rabbit and guinea pig brain. J Neurochem 13:885–886

Minelli A, Moroni M, Mezzasoma I (1997) Human seminal plasma soluble 5'-nucleotidase: regulatory aspects of the dephosphorylation of nucleoside 5'-monophosphates. Biochem Mol Med 61:95–101

Myers CS, Napolitano M, Fisher H, Wagner GC (1993) Uridine and stimulant-induced motor activity. Proc Soc Exp Biol Med 204:49–53

Myers CS, Wagner GC, Fisher H (1995) Uridine reduces rotation induced by L-dopa and methamphetamine in 6-OHDA-treated rats. Pharmacol Biochem Behav 52: 749–753

Ng WG, Xu YK, Kaufman FR, Donnell GN (1989) Deficit of uridine diphosphate galactose in galactosaemia. J Inher Metab Dise 12:257–266

Olivier KN, Bennett WD, Hohneker KW, Zeman KL, Edwards LJ, Boucher RC, Knowles MR (1996) Acute safety and effects on mucociliary clearance of aerosolized uridine 5'-triphosphate +/– amiloride in normal human adults. Am J Resp Critical Care Med 154:217–223

Okolicsanyi L, Orlando R, Lirussi F, Naccarato R, Dal Brun G (1980a) Pharmacometabolic activity in chronic hepatopathies: influence of treatment with UDPG. (Italian) Clinica Terapeutica 93:431–438

Okolicsanyi L, Orlando R, Busnardo F, Naccarato R, Veller-Fornasa C, Polin R (1980b) Effectiveness of UDPG in the treatment of porphyria cutanea tarda. Preliminary findings. (Italian) Clinica Terapeutica 94:687–694

Page T, Yu A, Fontanesi J, Nyhan WL (1997) Developmental disorder associated with increased cellular nucleotidase activity. Proc Nat Acad Sci 94:11601–11606

Paglia DE, Valentine WN, Nakatani M, Rauth BJ (1983) Selective accumulation of cytosol CDP-choline as an isolated erythrocyte defect in chronic hemolysis. Proc Nat Acad Sci USA 80:3081–3085

Parenti F (1992) Uridine associated with haloperidol in the treatment of schizophrenia: preliminary results. Clin Neuropharm 15:369

Persson BE, Sjoman M, Niklasson F, Ronquist G (1991) Uridine, xanthine and urate concentrations in prostatic fluid and seminal plasma of patients with prostatitis. Euro Urology 19:253–256

Peters GJ, van Groeningen CJ, Laurensse EJ, Lankelma J, Leyva A, Pinedo HM (1987a) Uridine-induced hypothermia in mice and rats in relation to plasma and tissue levels of uridine and its metabolites. Cancer Chemother Pharmacol 20:101–108

Peters GJ, van Groeningen CJ, Laurensse E, Kraal I, Leyva A, Lankelma J, Pinedo HM (1987b) Effect of pyrimidine nucleosides on body temperatures of man and rabbit in relation to pharmacokinetic data. Pharma Res 4:113–119

Peters GJ, Van Dijk J, Nadal JC, Van Groeningen CJ, Lankelma J, Pinedo HM (1987c) Diurnal variation in the therapeutic efficacy of 5-fluorouracil against murine colon cancer. In Vivo 1:113–117

Popov N, Schmidt S, Matthies H (1984) Protective effect of uridine on D-galactosamine-induced deficiency in brain uridine phosphates. Biomed Biochim Acta 43:1399–1404

Ralevic V, Burnstock G (1998) Receptors for purines and pyrimidines. Pharmacol Rev 60:413–492.

Rogers S, Segal S (1991) Modulation of rat tissue galactose-1-phosphate uridyltransferase by uridine and uridine triphosphate. Pediatr Res 30:222–226

Ronquist G, Niklasson F (1984) Uridine, xanthine, and urate contents in human seminal plasma. Arch Androl 13:63–70

Ronquist G, Stegmayr B, Niklasson F (1985) Sperm motility and interactions among seminal uridine, xanthine, urate, and ATPase in fertile and infertile men. Arch Androl 15:21–27

Scriver CR, Beaudet AL, Sly WS, Valle D (eds) (1995) The metabolic and molecular bases of inherited disease, vol 2, 7th edn. McGraw-Hill, New York, chaps 49–55

Schmid-Elsaesser R, Zausinger S, Hungerhaber E, Baethmann A, Reulen HJ (1997) Neuroprotective properties of a novel antioxidant (U-101033E) with improved

blood-brain barrier permeability in focal cerebral ischemia. Acta Neurochirurgica-Supplementum 70:176–178
Seifert R, Schultz G (1989) Involvement of pyrimidinoceptors in the regulation of cell functions by uridine and by uracil nucleotides. Trends Phramacol Sci 10:365–369
Siggins G, Gruol D, Padjen A, Forman D (1978) Action of purine and pyrimidine mononucleotides on central and cultured sympathetic neurons. In: Ryall RW, Kelly JS (eds) Iontophoresis and transmitter mechanisms in the mammalian central nervous system. Elsevier, Amsterdam, pp 453–455
Spector R (1985) Uridine transport and metabolism in the central nervous system. J Neurochem 45:1411–1418
Stacey NC, Ma HY, Duley JA, Connolly GP (2000) Abnormalities in cellular adhesion of neuroblastoma and fibroblast models of Lesch Nyhan Syndrome. Neuroscience 98:397–401
Tomlins AM, Foxall PJ, Lynch MJ, Parkinson J, Everett JR, Nicholson JK (1998) High resolution 1H NMR spectroscopic studies on dynamic biochemical processes in incubated human seminal fluid samples. Biochim Biophys Acta 1379:367–380
Traut TW (1994) Physiological concentrations of purines and pyrimidines. Mol Cell Biochem 140:1–22
Tremblay GC, Jimenez U, Crandall DE (1976) Pyrimidine biosynthesis and its regulation in the developing rat brain. J Neurochem 26:57–64
Usami M, Furuchi K, Ogino M, Kasahara H, Kanamaru T, Saitoh Y, Yokoyama H, Kano S (1996) The effect of a nucleotide-nucleoside solution on hepatic regeneration after partial hepatectomy in rats. Nutrition 12:797–803
Valentine WN, Fink K, Paglia DE, Harris SR, Adams WS (1974) Hereditary hemolytic anemia with human erythrocyte pyrimidine 5′-nucleotidase deficiency. J Clin Invest 54:866–879
Van-Groeningen CJ, Leyva A, Kraal I, Peters GJ, Pinedo HM (1986) Clinical and pharmacokinetic studies of prolonged administration of high-dose uridine intended for rescue from 5-FU toxicity. Cancer Treat Rep 70:745–750
Watabe T (1996) Strategic proposals for predicting drug-drug interactions during new drug development: based on sixteen deaths caused by interactions of the new antiviral sorivudine with 5-fluorouracil prodrugs. J Tox Sci 21:299–300
Williams, M, Jarvis MF (2000) Purinergic and pyrimidinergic receptors as potential drug targets. Biochem Pharmacol 59:1173–1185
Wu WL, So SC, Sun YP, Chung YW, Grima J, Wong PY, Yan YC, Chan HC (1998). Functional expression of P2U receptors in rat spermatogenic cells: dual modulation of a $Ca(2+)$-activated K+ channel. Biochem Biophys Res Comm 248:728–732
Xuma M, Turkington RW (1972) Hormonal regulation of uridine diphosphatase during spermatogenesis in the rat. Endocrinology 91:415–422
Zimmerman H, Braun N (1999) Ecto-nucleotidases – molecular structures, catalytic properties, and functional roles in the nervous system. Prog Brain Res 120:371–385

CHAPTER 15
Purinergic Signalling in Lower Urinary Tract

G. BURNSTOCK

A. Introduction

Atropine-resistant responses of the urinary bladder to stimulation of parasympathetic nerves was recognised for many years (LANGLEY and ANDERSON 1895; HENDERSON 1923; DALE and GADDUM 1930; HENDERSON and ROEPKE 1934, 1935; AMBACHE 1955; URSILLO and CLARK 1956; URSILLO 1961; BURNSTOCK and CAMPBELL 1963; HUKOVIC et al. 1965; CHESHER and THORP 1965; CHESHER and JAMES 1966; CARPENTER and RUBIN 1967) and it was later shown to be due to a non-cholinergic, non-adrenergic transmitter (AMBACHE and ZAR 1970; DUMSDAY 1971; SJÖSTRAND et al. 1972; ELMÉR 1975; CARPENTER 1977; KRELL et al. 1981; SIBLEY 1984; MAGGI et al. 1985; CRAGGS et al. 1986). However, it was not until 1972 that evidence was presented to support the view that the atropine-resistant component in guinea-pig bladder was purinergic, i.e. due to adenosine 5′-triphosphate (ATP) released from the purinergic nerves supplying the bladder (BURNSTOCK et al. 1972). The evidence in this paper included: mimicry of the non-adrenergic, non-cholinergic (NANC) nerve-mediated excitatory responses by ATP (Fig. 1); block of contractions both to NANC nerve stimulation and to exogenous application of ATP, but not to acetylcholine (ACh), by quinidine; and depression of NANC responses during tachyphylaxis produced by high concentrations of ATP. Direct evidence for ATP release from NANC nerves came in later papers (BURNSTOCK et al. 1978a) (Fig. 2).

Later studies have offered unequivocal support for this hypothesis, not only in guinea-pig bladder (BURNSTOCK et al. 1978a,b; LUKACSKO and KRELL 1982; MACKENZIE et al. 1982; KASAKOV and BURNSTOCK 1983; WESTFALL et al. 1983; HOURANI 1984; MACKENZIE and BURNSTOCK 1984; MELDRUM and BURNSTOCK 1985; CROWE et al. 1986; HOYLE et al. 1986, 1990; FUJII 1988; HISAYAMA et al. 1988; BRADING and MOSTWIN 1989; PETERSON and NORONHA-BLOB 1989; BRADING and WILLIAMS 1990; IACOVOU et al. 1990; KURA et al. 1992; HASHITANI and SUZUKI 1995; ZIGANSHIN et al. 1995a; BRAMICH and BRADING 1996; PATRA and WESTFALL 1996; USUNE et al. 1996; WERKSTRÖM et al. 1997), but also the bladders of many other species, including: mouse (ACEVEDO and CONTRERAS 1985, 1989; HOLT et al. 1985; SANTICIOLI et al. 1986; MOSS et al. 1989; SCHAUFELE et al. 1995; WATERMAN 1996); pig (SIBLEY 1984; FUJII 1987, 1988; MASUDA et al.

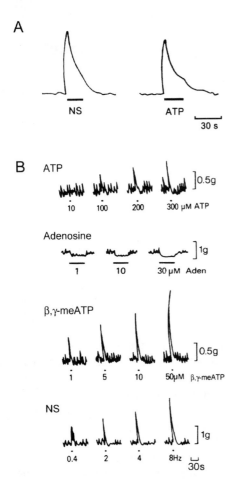

Fig. 1. A Contractile responses of the guinea-pig bladder strip to intramural nerve stimulation (NS; 2 Hz, 0.2 ms pulse duration, supramaximal voltage for 20 s) and ATP (8.5 μmol/l). Atropine (1.4 μmol/l) and guanethidine (3.4 μmol/l) were present throughout. **B** Responses of the rat isolated bladder to ATP, adenosine (Aden), β,γ-methylene ATP (β,γ-meATP) and intramural nerve stimulation (NS; 0.2 ms pulse duration, 30–50 V for 10 s). Guanethidine (3.4 μmol/l) was present throughout and atropine (1.4 μmol/l) was added to the bath 60 min before NS. Reproduced from **A** BURNSTOCK et al. (1978a) and **B** BROWN et al. (1979)

1995); hamster (PINNA et al. 1998a); marmoset and ferret (Moss and BURNSTOCK 1985); dog (TAIRA 1972); monkey (CREED et al. 1994); cat (THEOBALD 1982, 1983a, 1992; KOLEY et al. 1984; LEVIN et al. 1990); shrew (HOYLE et al. 1998); sheep (CREED et al. 1994; COTTON et al. 1996); rat (BROWN et al. 1979; BHAT et al. 1989a; BO and BURNSTOCK 1989; MAGGI 1991; PARIJA et al. 1991; IGAWA et al. 1993; BO et al. 1994; IGAWA et al. 1994; HASHIMOTO and KOKUBUN 1995; LONGHURST et al. 1995; VAUGHT et al. 1996; BOSELLI et al. 1997; TONG et al. 1997a; HANSEN et al. 1998; HEGDE et al. 1998; NUNN and NEWGREEN 1999); rabbit (DOWNIE and DEAN 1977; DEAN and DOWNIE 1978a,b; HUSTED et al. 1980a,b; LEVIN et al. 1980a, 1986a, 1990; ANDERSON 1982; LONGHURST et al. 1984; HOYLE and BURNSTOCK 1985; CROWE and BURNSTOCK 1985; FUJII 1988; CHEN and BRADING 1991; CHANCELLOR et al. 1992; KISHII et al. 1992; TAMMELA et al. 1992; ZHAO et al. 1993; ZIGANSHIN et al. 1993; YOKOTA and YAMAGUCHI

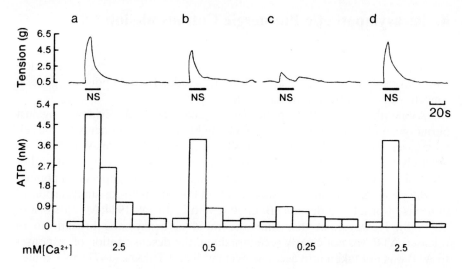

Fig. 2. Effect of changing the Ca^{2+} concentration on the release of ATP from the guinea-pig isolated bladder strip during stimulation of intramural nerves. *Upper trace*: mechanical recording of changes in tension (g) during intramural nerve stimulation (NS: 2 Hz, 0.2 ms pulse duration, supramaximal voltage for 20 s). *Lower trace*: concentration of ATP in consecutive 20-s fractions of the superfusate. The Ca^{2+} concentration in the superfusate varied as follows: *a* 2.5 mmol/l (normal Krebs); *b* 0.5 mmol/l; *c* 0.25 mmol/l; *d* 2.5 mmol/l. The successive contractions were separated by 60-min intervals as indicated by the breaks in the mechanical trace. Atropine (1.4 µmol/l) and guanethidine (3.4 µmol/l) were present throughout. The temperature of the superfusate was between 22 °C and 23 °C. Reproduced from BURNSTOCK et al. (1978a)

1996; LIU et al. 1998); and human (HUSTED et al. 1983; HOYLE et al. 1989; GHONIEM and SHOUKRY 1991; INOUE and BRADING 1991; PALEA et al. 1993, 1994, 1995; SAITO et al. 1993a,b; BO and BURNSTOCK 1995; VALERA et al. 1995; WAMMACK et al. 1995; MICHEL et al. 1996; BAYLISS et al. 1999; WU et al. 1999).

Earlier review articles, which include some aspects of purinergic signalling in the lower urinary tract, are available (BURNSTOCK 1972, 1975, 1976, 1978, 1982, 1986a,b, 1990a, 1996a; WHITE 1984; BURNSTOCK and KENNEDY 1985; BRADING 1987; CROWE and BURNSTOCK 1989; HOYLE and BURNSTOCK 1991, 1993; ANDERSSON 1993, 1997; LINCOLN and BURNSTOCK 1993; HOYLE 1994; HOYLE et al. 1994; FERGUSON and CHRISTOPHER 1996; YOSHIMURA 1999).

In this review the evidence in support of purinergic signalling in bladder and other parts of the lower urinary tract is detailed and the changes that occur in the signalling process in development, ageing, after trauma or surgery, and in various disease conditions are discussed. The possible involvement of purinergic mechano-sensory sensory transduction in pain induced by dilation of bladder and ureter will be considered.

B. Parasympathetic Purinergic Cotransmission

Following the initial proposal that ATP contributed to the contractile responses of the urinary bladder to parasympathetic nerve stimulation (BURNSTOCK 1972), much debate followed, as indeed it did about the general concept of purinergic neurotransmission.

AMBACHE et al. (1977) published a paper entitled 'Evidence against purinergic motor transmission in guinea-pig bladder' based mainly on the relative insensitivity of the bladder to ATP and the inability of ATP to match precisely the atropine-resistant neurogenic responses, extending their earlier conclusion (AMBACHE and ZAR 1970), although very early papers had pointed out the close mimicry of the responses of ATP to atropine-resistant responses in terms of onset and decline (BUCHTHAL and KAHLSON 1944; MATSUMURA et al. 1968). However, at that time the rapid ectoenzymatic breakdown of released ATP was not clearly recognised and the desensitisation of responses to ATP was not taken into account. WEETMAN and TURNER (1977) also argued against purinergic transmission on the basis of the lack of specific effects of several ATP receptor blocking agents that had been claimed to be effective on ATP responses in the guinea-pig taenia coli: "quinidine reduced the response to nerve stimulation without affecting the histamine controls, although this was probably due to a local anaesthetic effect" and "phentolamine and two experimental drugs (2,2′-prydiylisatogen, and 2,2′-methoxyphenylisatogen), that are active against ATP-induced relaxation of the guinea-pig isolated taenia, were non-specific in their blockade of contractions of the bladder to nerve stimulation". Other views expressed were that since ATP-induced contractions, but not contractions to nerve stimulation below 5Hz, were potentiated by Mg^{2+}-free Krebs, this made it unlikely that the atropine-resistant contractions of the guinea-pig urinary bladder were mediated by ATP (JOHNS 1979). The lack of effect of theophylline or dipyridamole on the excitatory junction potentials in the rabbit bladder in response to intramuscular nerve stimulation was also taken as evidence against ATP being the non-cholinergic excitatory transmitter (CREED et al. 1983), but since these agents only affect the P1 receptor-mediated actions of adenosine, but not ATP, this was clearly not a valid argument. Tetrodotoxin-resistant release of ATP was taken to indicate ATP release from muscle during transmural stimulation and argued against ATP as a neurotransmitter in the rabbit bladder (CHAUDHRY et al. 1984). Since responses to electrical field stimulation in the presence of atropine were reduced, but not abolished, following desensitisation of the ATP receptor, it was concluded that ATP was unlikely to be the sole non-cholinergic motor transmitter in the rat detrusor (LUHESHI and ZAR 1990a).

Despite these negative papers, several other laboratories confirmed and extended the evidence in favour of purinergic transmission. DEAN and DOWNIE (1978b) showed that desensitisation with ATP selectively depressed responses to ATP and to field stimulation (particularly at low frequencies), but not those in response to carbachol. BURNSTOCK et al. (1978a) extended their earlier findings: quinacrine, a fluorescent dye known to bind to high levels of ATP in granular vesicles, produced positive staining in neurones and nerve fibres in the bladder; release of ATP during stimulation of NANC excitatory nerves was demonstrated using the firefly luciferin-luciferase assay method (also reported in BURNSTOCK et al. 1978b); and sympathectomy with 6-hydroxydopamine did not affect the release of ATP in response to intramural nerve stimulation. Compared to ATP, 100-fold lower concentrations of the slowly-degradable analogue β,γ-methylene ATP (β,γ-meATP) were shown to mimic contractions of the atropine-resistant responses of the rat bladder, suggesting that the relative insensitivity of the bladder to ATP is due to its rapid degradation to adenosine 5'-monophosphate (AMP) and adenosine, which cause relaxation of the bladder (BROWN et al. 1979) (Fig. 1B). The functional effects of purinergic innervation of the rabbit urinary bladder were also reported (LEVIN et al. 1986a).

Evidence for purinergic and cholinergic components of the responses of the bladder to purine nerve stimulation in an in vivo preparation of urethane-anaesthetised guinea-pigs has been presented (PETERSON and NORONHA-BLOB 1989). In anaesthetised cats the ganglion stimulants, nicotine and dimethylphenylpiperazinium (DMPP), increased intravesicular pressure by an atropine-resistant mechanism, which was mimicked by ATP (KOLEY et al. 1984). In a more recent study of the in vivo responses of the cat bladder to pelvic nerve stimulation it was concluded that purinergic transmission plays a role in the initiation of bladder contraction and perhaps in the initiation of urine flow, in contrast to cholinergic transmission that is involved in maintenance of contractile activity and flow (THEOBALD 1995).

In the late 1970s and the 1980s, neuropeptides, particularly vasoactive intestinal peptide (VIP), became the favoured contenders for NANC transmission in a variety of preparations, including those of the lower urinary tract and penile erectile tissues (see ANDERSSON 1993; HOYLE and BURNSTOCK 1993), but in a study designed to compare the effects of substance P, VIP and its structurally-related polypeptide PHI, on the guinea-pig bladder with the effects of field stimulation and ATP (MACKENZIE and BURNSTOCK 1984), the slow sustained excitation elicited by VIP contrasted clearly with the fast transitory responses elicited by both ATP and field nerve stimulation. In a later study, MELDRUM and BURNSTOCK (1985) showed that P2 purinergic receptor desensitisation with α,β-methylene ATP (α,β-meATP) did not alter the responses to VIP while blocking NANC excitation.

In addition to evidence that ATP mimics the atropine-resistant component of parasympathetic nerve stimulation and the parallel block by ATP antagonists (see below, Sect. B.I.), there is good evidence for release of ATP during stimulation of parasympathetic nerves (BURNSTOCK et al. 1978a,b; TONG et al. 1997b).

I. Parallel Block of Responses to ATP and NANC Nerve Stimulation

It was reported that various drugs that inhibited ATP-induced contractions also inhibited the responses induced by electric field stimulation (Sjögren and Andersson 1979). NANC nerve-mediated responses of strips of guinea-pig urinary bladder were markedly reduced following desensitisation with ATP, but only slightly with guanosine 5′-triphosphate (GTP) or cytidine 5′-triphosphate (CTP) (Lukacsko and Krell 1981).

Reactive blue 2 was reported to antagonise selectively the ATP-induced relaxations of the guinea-pig distal colon (Kerr and Krantis 1979). Reactive blue 2 was also shown to inhibit the responses to ATP and to NANC nerve stimulation in both guinea-pig and rat bladders, although the authors cautioned about the use of Reactive blue 2 since it could also reduce the response to ACh after prolonged exposure, albeit in a non-competitive manner (Choo 1981). At about this time, arylazido-aminopropionyl ATP (ANAPP$_3$) was also proposed as a specific antagonist to ATP (Hogaboom et al. 1980) and was shown to inhibit contractile responses of the cat and guinea-pig bladder to both ATP and pelvic or intramural nerve stimulation (Theobald 1982, 1983a, 1986a; Westfall et al. 1983; Theobald and Hoffman 1986) (Fig. 3A). In the rabbit bladder ANAPP$_3$ blocked the atropine-resistant neurogenic response, but apparently not responses to exogenous ATP (Longhurst et al. 1984). Kasakov and Burnstock (1983) showed that the slowly degradable analogue of ATP, α,β-meATP, produced selective desensitisation of the P2 purinoceptor and that it abolished NANC excitatory responses of the guinea-pig urinary bladder (Fig. 3B). This was confirmed in later studies of both guinea-pig and rat bladder (Brading and Williams 1990; Iacovou et al. 1990; Kura et al. 1992).

Fig. 3A–D. Inhibition of contractions of urinary bladder. **A** Antagonism of urinary bladder contractions of the anaesthetised cat by atropine (2 mg/kg, i.a.) and ANAPP$_3$ (0.5 μmoles/kg, i.a.). PNS-1 and PNS-2 are phase 1 and phase 2 of contractions evoked by pelvic nerve stimulation (5 V, 5 Hz). Acetylcholine (ACh, 10 μg, i.a.), ATP (10 μmoles, i.a) and β,γ-methylene ATP (β,γ-meATP, 0.1 μmoles, i.a.) induced contractions. $p < 0.05$ was considered statistically significant ($n = 5$–10). **B** The effect of α,β-methylene ATP (α,β-meATP) on the response of isolated guinea-pig bladder strips to nerve stimulation (NS), ATP (Δ) and histamine (Hist). *Upper trace*: Control responses; lower trace, desensitisation attained by five successive applications of α,β-meATP (50 μmol/l, ▲), at 4-min intervals, completely abolished nerve-mediated and ATP-induced contractions, although histamine-induced contraction is only slightly reduced. **C,D** Concentration-dependent inhibitory effect of suramin (**C**) and PPADS (**D**) on the NANC contractile responses to electrical field stimulation (3 Hz, 0.2 ms pulse duration, 80 V for 10 s) of rat bladder strips. Control group represents 100% contraction. NANC contractions were recorded in the presence of atropine (1 μmol/l), guanethidine (1 μmol/l), phentolamine (1 μmol/l) and propranolol (1 μmol/l). Each column represents the mean ± S.E.M. of 6–8 experiments. Statistical significance (unpaired *t*-test) relative to NANC group is indicated: *$p < 0.05$, **$p < 0.01$, ***$p < 0.001$. Reproduced from **A** Theobald (1983a), **B** Kasakov and Burnstock (1983), and **C,D** Tong et al. (1997b)

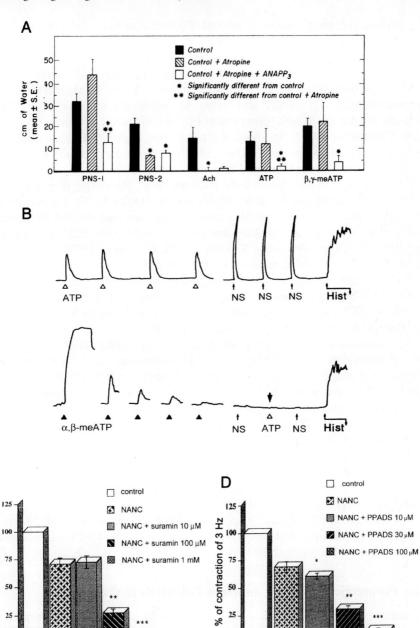

When the enantiomers of adenyl 5'-(β,γ-methylene)-diphosphonate (in preference to α,β-meATP) were used to desensitise the ATP receptors in the guinea-pig bladder, again responses to ATP and non-cholinergic nerve stimulation were inhibited, supporting the proposal for ATP as the NANC-excitatory transmitter in the bladder (HOURANI 1984). In the first study of mouse bladder, α,β-meATP was shown to abolish the response to ATP and greatly reduce the NANC component of the neurogenic response (ACEVEDO and CONTRERAS 1985). At about the same time, α,β-meATP desensitisation experiments also supported NANC excitatory transmission in the bladders of ferret and marmoset (Moss and BURNSTOCK 1985).

After suramin was shown to be a reversible P2 purinergic receptor antagonist in the mouse vas deferens (DUNN and BLAKELEY 1988), it was reported to reduce the responses to both purinergic agonists and the NANC component of neural responses in the guinea-pig, rat and shrew bladders (HOYLE et al. 1990, 1998; HASHIMOTO and KOKUBUN 1995; TONG et al. 1997b) (Fig. 3C). In a study of the effects of suramin on the responses to nerve stimulation and ATP in the bladder muscle strips from guinea-pigs, rabbits, monkeys and sheep and detrusor strips from humans, it was shown that it produced parallel inhibition in guinea-pig and rabbit, but in sheep and human tissue, where the purinergic nerve component was smaller, the effect of suramin was difficult to assess because of increase in spontaneous activity (CREED et al. 1994).

Pyridoxalphosphate-6-azophenyl-2',4'-disulphonic acid (PPADS) was introduced as a P2X antagonist in the vas deferens in 1992 (LAMBRECHT et al. 1992) and later was also shown to be effective in selectively antagonising P2X purinoceptor-mediated contractions in the rabbit urinary bladder produced by exogenous α,β-meATP and by purinergic nerve stimulation (ZIGANSHIN et al. 1993; TONG et al. 1997b) (Fig. 3D). P2X receptors in the guinea-pig bladder were shown to be more sensitive to PPADS than suramin and diadenosine tetraphosphate (Ap_4A) appeared to be acting through this P2X receptor since, like ATP responses, responses to Ap_4A were abolished after desensitisation with α,β-meATP (USUNE et al. 1996).

Reactive blue 2 reduced the postcontractile *relaxation* of the bladder neck of the male mini-pig and this was taken to suggest that P2Y purinoceptors were involved (TONG et al. 1997b).

II. Purinergic Excitatory Junction Potentials in the Bladder

Excitatory junction potentials (EJPs) elicited by stimulation of sympathetic nerves supplying the guinea-pig vas deferens were first recorded by Burnstock and Holman in the early 1960s (BURNSTOCK and HOLMAN 1960, 1961), although it was not until the 1980s that EJPs were shown to be due to the actions of neuronally released ATP (SNEDDON et al. 1982; SNEDDON and BURNSTOCK 1984).

The first recordings of EJPs (using both microelectrode and sucrose-gap methods) in smooth muscle cells of the urinary bladder in response to intramural nerve stimulation were published in 1983 (CREED et al. 1983), although the authors presented data that they interpreted as not supporting the proposal that ATP was the NANC-excitatory transmitter. Later papers, however, clearly showed that the atropine-resistant EJPs recorded in the bladders of rabbits, guinea-pigs and pigs were inhibited by desensitisation of the ATP receptor with α,β-meATP (Fig. 4) and are therefore the result of purinergic transmission (HOYLE and BURNSTOCK 1985; FUJII 1988; BRADING and MOSTWIN 1989; BRAMICH and BRADING 1996). In other studies, EJPs recorded in the guinea-pig bladder were reduced by the ATP antagonist suramin (BUCK et al. 1974; CREED et al. 1994).

In an elegant study employing the whole-cell patch clamp technique on single smooth muscle cells isolated from guinea-pig bladder, it was possible to show that ATP could closely mimic the EJP using a concentration jump technique (time constant about 10 ms) and this was taken as support for the concept that ATP is the transmitter responsible for fast neurotransmission in the bladder (INOUE and BRADING 1990). Patch-clamp studies on isolated smooth muscle cells from sheep bladder suggested that Cibacron blue is a potent activator of a Ca^{2+}-dependent outward current in addition to its action as a purinergic antagonist (COTTON et al. 1996). Using a voltage-clamp of smooth muscle cells from guinea-pig bladder, ATP, ADP, α,β-meATP and β,γ-meATP were shown to produce rises in fast inward transmembrane current, while GTP, inosine 5'-triphosphate (ITP) and AMP and adenosine failed to activate this current (MARCHENKO et al. 1987).

Analysis of the EJPs recorded in the guinea-pig bladder (BRAMICH and BRADING 1996) showed first that they varied greatly in both amplitude and time course even when recorded from cells at similar distances from the stimulating electrodes, and second that, as the strength of field stimulation was reduced, the amplitude of EJPs was decreased in two or three discreet steps, rather than gradually. Spontaneous excitatory junction potentials (SEJPs) were also recorded from most cells (Fig. 4D). The authors raise the possibility that EJPs result from the activation of two different membrane conductances and that the variation in EJP amplitude may be related to the degree of coupling between smooth muscle cells in and between muscle bundles.

III. Purinergic/Cholinergic Cotransmission

The concept of cotransmission is now well accepted (see BURNSTOCK 1976, 1990a, 1999a), including strong evidence that ATP acts as a cotransmitter with noradrenaline (NA) in the sympathetic nervous system (see BURNSTOCK 1990b, 1995). It is surprising that there is much less information about cotransmission with ATP in the parasympathetic nervous system. ATP is released from synaptic vesicles from motor nerve terminals together with ACh in the rat diaphragm and in teleost electric organs (SILINSKY and HUBBARD 1973; SILINSKY 1975; ZIMMERMANN 1978, 1982), and there is also evidence that ATP is coreleased with ACh from sympathetic nerves supplying catfish chro-

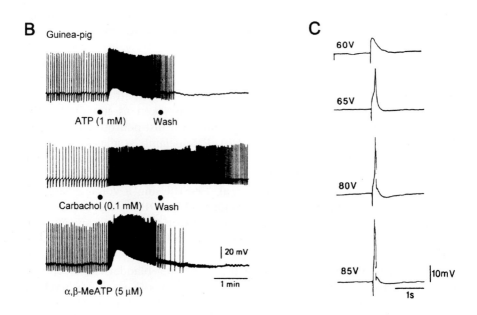

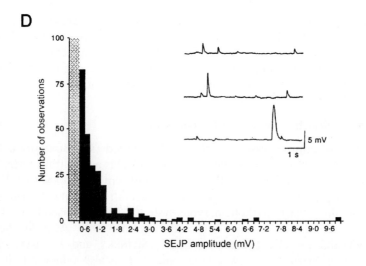

matophores (Fujii and Oshima 1986). The paucity of information about parasympathetic cotransmission is probably due to the fact that it is easier to eliminate surgically postganglionic sympathetic nerves, or chemically denervate with sympatholytics such as guanethidine or 6-hydroxydopamine, than it is to disrupt surgically or chemically postganglionic parasympathetic nerves.

Perhaps the first hint that ATP and ACh might be cotransmitters in parasympathetic nerves supplying the bladder came from an ultrastructural study of nerves supplying the smooth muscle of the bladder, where the vesicular composition of nerve profiles containing small agranular and large opaque vesicles led the authors to propose that cholinergic and NANC transmitters were colocalised (Hoyes et al. 1975). Further indirect evidence for purinergic cotransmission came from binding studies and regional studies of the responses of strips taken from five different areas of the rabbit bladder, where it was shown that the bladder body and base showed parallel sensitivity to urecholine (a muscarinic agonist), whereas the sensitivity to ATP was highest in the body and lowest in the base. On the other hand, α-adrenergic responses and receptor concentration were higher in the base than in the body of the bladder (Levin et al. 1980a). Other indirect evidence for purinergic cotransmission came from studies of purified botulinum neurotoxin type A and neuromuscular transmission in the guinea-pig bladder; both cholinergic and purinergic components of the excitatory responses to nerve stimulation were significantly reduced by botulinum toxin (MacKenzie et al. 1982). The prejunctional inhibition of both cholinergic and purinergic components of the nerve-mediated responses of the rat bladder by adenosine was taken as evidence in support of cotransmission (Parija et al. 1991).

Fig. 4. A Rat urinary bladder detrusor, sucrose-gap recording. 33°C, atropine 0.3 µmol/l. Effect of α,β-methylene ATP on excitatory junction potentials (EJPs) evoked by field stimulation (•, 0.5 Hz, 0.3 ms, 5 V, continuously) before (*left trace*) and during desensitisation with α,β-methylene ATP (α,β-meATP, 10 µmol/l) (*right trace*). At control membrane potential, EJPs are no longer visible during desensitisation with α,β-meATP. **B** Effects of ATP (1 mmol/l), carbachol (100 µmol/l) and α,β-meATP (5 µmol/l) on the membrane electrical activity recorded with microelectrodes from the guinea-pig bladder. ATP produced a large depolarisation and a greater increase in spike frequency with a faster time course than did carbachol. Applied continuously, α,β-meATP produced a large, transient depolarisation and an increase in spike frequency, the action being similar to that elicited by ATP but at a lower concentration. **C** Electrical responses elicited by increasing intensity of transmural nerve stimulation (pulse duration 0.05 ms, voltage as indicated on each trace) in a single detrusor muscle cell of guinea-pig. Membrane potential –47 V. **D** Amplitude-frequency histogram of spontaneous excitatory junction potentials (SEJPs) recorded from detrusor muscle of guinea-pig urinary bladder. Intracellular recordings from a single cell (*inset*) in the presence of nifedipine (10 µM) indicate the stable membrane potential (–45 V) is interrupted by SEJPs of variable amplitude. The *stippled column* of resultant amplitude-frequency histogram for SEJPs for this cell represents twice the background noise and the histogram has a skewed distribution with the peak disappearing into the recording noise. Reproduced from **A** Hoyle and Burnstock (1985), **B** Fujii (1988), **C** Hashitani and Suzuki (1995) and **D** Bramich and Brading (1996).

Another parasympatholytic agent, ethylcholine mustard aziridinium ion (AF64A), causes only partial loss of peripheral cholinergic nerves in the guinea-pig at sub-lethal doses, and although it causes impairment of cholinergic transmission in the intestine and the bladder, it does not significantly compromise purinergic transmission in the bladder (HOYLE et al. 1986). However, AF64A-treated animals have an increased sensitivity of the purinoceptors in the urinary bladder, possibly indicative of an early stage of degeneration of the purinergic component.

In the cat bladder, a population of nerves had been described that have a local origin in the intramural ganglia, and that contain high proportions of NANC vesicles (FEHÉR et al. 1979). As noted above, ganglionic stimulants may selectively activate non-cholinergic neurones (KOLEY et al. 1984), and the cat bladder supports a predominantly non-cholinergic, ANAPP$_3$-sensitive innervation. It seems likely from pharmacological studies that a spectrum of nerves exists, utilising different proportions of ATP and ACh, from predominantly ATP in cat and guinea-pig through to roughly 50:50 in rat and dog to predominantly ACh in human bladder.

IV. Extracellular Calcium, Calcium Channel Blockers and Potassium Channel Openers

The sources of calcium for ATP contraction of the bladder smooth muscle have been examined. In the rat bladder the responses to ATP were reversibly abolished in Ca^{2+}-free media and were never inhibited less than 45% by verapamil and diltiazem. It was concluded that, while extracellular Ca^{2+} was largely involved in the actions of ATP, some intracellular Ca^{2+} was also involved (HUDDART and BUTLER 1986; MAGGI et al. 1988; BHAT et al. 1989b). In the mouse bladder, responses to ATP and electrical field stimulation were also mainly dependent upon extracellular calcium (ACEVEDO and CONTRERAS 1989). In studies of guinea-pig detrusor and dispersed smooth muscle cells from rabbit bladder, it was concluded that stimulation of purinergic receptors opens an ion channel and allows influx of Ca^{2+}, while muscarinic receptor stimulation mobilises intracellular Ca^{2+} via hydrolysis of inositol phospholipid (IACOVOU et al. 1990; CREED et al. 1991). Experiments on cultured smooth muscle cells from rabbit bladder, performed with the fura-2 technique to measure changes in intracellular Ca^{2+}, showed that ATP produced a rapid but transient increase in $[Ca^{2+}]_I$, while ACh produced a delayed, prolonged increase (OIKE et al. 1998). Studies of isolated smooth muscle cells from guinea-pig bladder, using a whole-cell voltage-clamp technique, confirmed that purinergic receptor stimulation opens non-selective cation channels, while muscarinic stimulation triggers Ca^{2+} release from intracellular stores (NAKAYAMA 1993).

Different types of Ca^{2+} channels are present prejunctionally and postjunctionally in the urinary bladder. L-type channels appear to be predominantly present in bladder smooth muscle (KISHII et al. 1992; MARTI-CABRERA et al.

1994). Whereas ATP release from parasympathetic nerves in the bladder involves predominantly P- and Q-type calcium channels, ACh release depends primarily on N-type channels (WATERMAN 1996). However, activation of P- and Q-type Ca^{2+} channels, through phosphorylation by protein kinase C (PKC) may be involved in the enhancing effect of the PKC activator, β-phorbol-12,13-dibutyrate (β-PDBu), on the muscle contractions elicited by excitatory purinergic neurotransmission in mouse detrusor strips (LIN et al. 1998).

Since ATP was known to act by opening Ca^{2+} channels, the effect of several calcium channel blockers on NANC nerve-mediated responses of the urinary bladder were examined. Early studies using rabbit and rat urinary bladder, hinted that, while terodiline was largely anticholinergic, nifedipine, an L-type Ca^{2+} channel blocker, might be effective against the NANC-mediated component (CARPENTER and RUBIN 1967; HUSTED et al. 1980c; MAGGI et al. 1982, 1988; IRAVANI et al. 1988). Nifedipine was shown nearly to abolish the responses of the rat bladder to β,γ-meATP and the purinergic neuronal component, while a substantial proportion of the responses to ACh and the cholinergic neuronal component was resistant to nifedipine (Bo and BURNSTOCK 1990a; ZAR et al. 1990). Verapamil was more potent than diltiazem in inhibiting both ATP and NANC fast initial neurogenic responses, but ACh-mediated responses were also affected (BHAT et al. 1989a; ZHAO et al. 1993). Bay K 8644, a 1,4-dihydropyridine, which is an L-type channel activator, substantially increased the contraction of the bladder to β,γ-meATP and NANC nerve-mediated responses (Bo and BURNSTOCK 1990a). In another study of the rat bladder (MAGGI 1991), the effects of nifedipine and Bay K 8644 were confirmed and ω-conotoxin, an N-type channel blocker, was shown to reduce the purinergic component of nerve-mediated responses, although to a lesser extent than those of the cholinergic component, and had little effect on the responses to either ATP or ACh. In the guinea-pig bladder, too, ATP-evoked contractions were markedly inhibited by dihydropyridine-like Ca^{2+} antagonists such as nifedipine and nitrendipine, but not by D-600, ω-conotoxin or tetramethrin (KATSURAGI et al. 1990). Benzodiazapine (diazepam) also antagonises the responses to ATP, probably by decreasing Ca^{2+} entry (SCHAUFELE et al. 1995).

The potassium channel opener, YM934, was shown to inhibit markedly the contractile responses of the guinea-pig detrusor smooth muscle to exogenously applied α,β-meATP, but only slightly inhibited the contractions produced by carbachol; it was concluded that YM934 may hyperpolarise the smooth muscle membranes by opening ATP-sensitive potassium channels and as a result may functionally inhibit the contractile response to purinergic nerve stimulation that elicits membrane depolarisation (MASUDA et al. 1995). Other K^+ channel openers cromakalim and ZM244085 were shown to hyperpolarise and reduce contractions of bladder smooth muscle; this effect was blocked by glibenclamide, but was unaffected by apamin (FUJII et al. 1990; LI et al. 1996). In a more recent study, cromakalim was shown to affect profoundly the responses to exogenous ATP, but had little action on the responses to carbachol (BOSELLI et al. 1997). The authors concluded that cromakalim acts on purinergic transmission predominantly postjunctionally, whereas its minor action on cholinergic transmission is mainly at the prejunctional level.

V. Involvement of Prostaglandins in Purinergic Transmission

It was shown in 1974 that adenine nucleotides induce prostaglandin synthesis (NEEDLEMAN et al. 1974) and, soon after, evidence was presented that prostaglandins were responsible for the rebound contractions of the guinea-pig taenia coli that follow stimulation of purinergic inhibitory nerves (BURNSTOCK et al. 1975).

Since then, evidence has accumulated that prostaglandins are generated in bladder smooth muscle as a result of purinergic neurotransmitter activity. In the guinea-pig, rabbit and monkey isolated detrusor, prostaglandin E_2 (PGE_2) and prostaglandin $F_{2\alpha}$ ($PGF_{2\alpha}$) caused potent contractions (JOHNS and PATON 1977; JOHNS 1981). In response to electrical field stimulation, there is an initial phasic contraction followed by a secondary tonic contraction. Indomethacin, a prostaglandin synthesis inhibitor, reduced the initial phasic contraction (purinergic) and the response to ATP (JOHNS and PATON 1977; DEAN and DOWNIE 1978a; CHOO and MITCHELSON 1980). PGE_2 was later shown to be released from detrusor smooth muscle as a result of neural activity, but not from nerves (ALKONDON and GANGULY 1980), and atropine-resistant contractions were also reduced by the prostaglandin antagonist SC19220 (DOWNIE and LARSSON 1981).

ATP, but not adenosine, ACh or carbachol, evoked the release of prostaglandins from detrusor muscle (ANDERSON 1982; KASAKOV and VLASKOVSKA 1985). In the rabbit detrusor muscle, ATP evoked a biphasic contraction consisting of an initial phasic contraction followed by a delayed secondary tonic contraction. Indomethacin prevented the secondary but not the initial phasic response (HUSTED et al. 1980a,b; ANDERSSON et al. 1980; ANDERSON 1982) (Fig. 5). ADP produced only a slow tonic contraction that was almost abolished by indomethacin. Similar results were found in the human bladder (HUSTED et al. 1983). β,γ-MeATP produces an indomethacin-resistant initial phasic contraction (HUSTED et al. 1980a,b). This was explained in a later study that showed that the structural conformation of the polyphosphate chain of the ATP molecule is critical for stimulation of prostaglandin biosynthesis (BROWN and BURNSTOCK 1981). ATP can also lead to production of prostanoids in uroepithelial cells of the bladder, an effect which is enhanced in pathological conditions (PINNA et al. 2000a).

Prostaglandins may play an important role in bladder or urethral contractility in physiological or pathophysiological conditions. For example, in the rat, SC19220 reduced detrusor tone resulting in an increased bladder capacity and decrease in voiding efficiency (MAGGI et al. 1988). Similarly, treatment with indomethacin causes an increase in the residual volume of the bladder at the end of micturition (MAGGI et al. 1989a). Vesical distension in dogs causes a reflex decrease in urethral resistance, accompanied by a large increase in PGE_2

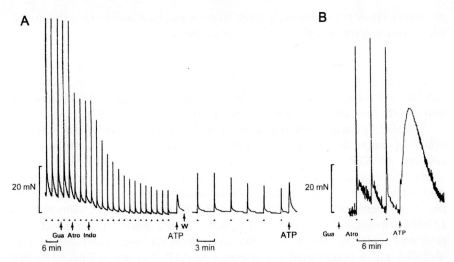

Fig. 5. A Contractile responses of the rabbit detrusor to electrical field stimulation and to ATP (100 µmol/l) in the presence of guanethidine (*Gua*, 3.4 µmol/l), atropine (*Atro*, 1.0 µmol/l) and indomethacin (*Indo*, 5 µmol/l). *W*, 3 washouts during a 20-min period with guanethidine, atropine and indomethacin still present. **B** Contractile responses of the rabbit detrusor to electrical field stimulation and to ATP (100 µmol/l) in the presence of guanethidine (3.4 µmol/l) and atropine (1.0 µmol/l). Reproduced from ANDERSSON et al. (1980)

in urethral venous outflow. The reflex inhibition and release of PGE_2 are prevented when ganglionic transmission is blocked by pentolinium (GHONEIM et al. 1976). These results imply that a physiological mechanism stimulates PGE_2 synthesis in the urethra, and that endogenous prostaglandins are important in maintaining tone in the smooth muscle, as shown by indomethacin or SC19220 causing relaxation of isolated preparations (ITO and KIMOTO 1985).

VI. Ectoenzymatic Breakdown of ATP

It has been recognised for a long time that ectoenzymes can hydrolyse ATP released from nerves and non-neuronal cells. Much is now known about the various enzymes involved and the roles of the breakdown products in different tissues (see ZIMMERMANN 1996; ZIMMERMANN, Chap. 8, this volume). BURNSTOCK (1978) proposed the concept that ATP acts on P2 receptors and that after breakdown by ectoenzymes to adenosine, it then acts via another receptor, the P1 receptor.

In an early study of adenosine triphosphatase (ATPase) activity in the bladder, it was shown that less than 0.1% of the ATP present in an in vitro bladder preparation was hydrolysed and it was therefore suggested that the low potency of ATP relative to the slowly degradable analogue β,γ-meATP could not be easily explained in terms of ectoenzymatic breakdown (LEVIN et al. 1981a). However, a relatively large 30-ml organ bath was used in this study, which may not reflect the physiological condition.

Studies of the guinea-pig bladder revealed that ATP analogues which are resistant to enzymatic degradation were more potent in eliciting a contraction via P2X receptors (WELFORD et al. 1987), although genuine differences in structure-activity relationships do exist for methylene analogues of ATP that are only slowly degraded by ectonucleotidases (CUSACK et al. 1987).

The role of enzymatic degradation was also evaluated by studying the effect of putative inhibitors of ectonucleotides on the pharmacological response of ATP in guinea-pig bladder (HOURANI and CHOWN 1989). Some inhibitors of ecto-ATPase were identified, including suramin and ethacrynic acid, which were strongly effective in enhancing the response to ATP; difluoro-dinitrobenzene was less effective, while N-ethylmaleimide, adenosine 5'-O-3-thiotriphosphate (ATPγS) and Reactive blue 2 were without effect. In a later study, ARL 67156 was identified as a potent ATPase inhibitor (CRACK et al. 1995) which potentiated the response to ATP (but not α,β-meATP) and responses to atropine-resistant responses to nerve stimulation (WESTFALL et al. 1997) (Fig. 6). Cyclopiazonic acid (CPA), an inhibitor of sarcoplasmic ATPase, potentiated the contractile response of the guinea-pig bladder to exogenous ATP and NANC excitatory nerve stimulation but this was non-specific (ZIGANSHIN et al. 1994). In the same paper, ecto-ATPase in the bladder was estimated to have a V_{max} of 0.98 nmol P_i 30 min^{-1} mg^{-1} wet tissue, with a K_m of 881 μmol/l ATP respectively; CPA (10 μmol/l) inhibited ecto-ATPase activity by about 18%. Subsequent studies (ZIGANSHIN et al. 1995b) revealed

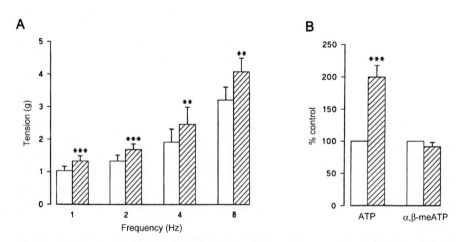

Fig. 6A,B. The effect of ARL 67156 in guinea-pig bladder strips on **A** neurogenic contractions (1–8 Hz, for 20s, $n = 8$) and **B** contractions to exogenous ATP (100 μmol/l, $n = 16$) and α,β-methylene ATP (α,β-meATP, 5 μmol/l, $n = 7$). *Open bars* represent control responses and the *hatched bars* those in the presence of ARL 67156 (100 μmol/l). **$p < 0.01$, ***$p < 0.001$. Reproduced from WESTFALL et al. (1997)

that some divalent cations, such as Cu^{2+}, Ni^{2+}, Zn^{2+} and La^{3+}, inhibit the ecto-ATPase activity in a concentration-dependent manner in the guinea-pig bladder, but not all of them potentiate contractions to ATP.

Magnesium-dependent adenosine triphosphatase (Mg^{2+}-ATPase) as well as 5'-nucleotidase and alkaline phosphatase were identified in the epithelial cells of the rat urinary bladder (ZHANG et al. 1991), perhaps functioning to degrade the ATP released from uroepithelial cells (FERGUSON et al. 1997; KNIGHT et al. 1999; HAWRANKO et al. 1999) during purinergic mechanosensory transduction (BURNSTOCK 1999b; NAMASIVAYAM et al. 1999).

VII. In Vitro and In Vivo Studies

In a whole rabbit bladder in vitro preparation exogenous ATP and electrical field stimulation in the presence of atropine produced a transient rapid rise in intravesical pressure (LEVIN and WEIN 1982) (Fig. 7A). However, these purinergic responses did not result in significant bladder emptying, suggesting that they may be complimentary, but functionally different, from those which occur in response to cholinergic transmission (LEVIN et al. 1983a, 1986a). A comparison of the purinergic responses in the whole bladder in vitro preparations of cat and rabbit revealed both qualitative and quantitative species differences (LEVIN et al. 1990). In particular, the component of purinergic NANC transmission in the bladder of the cat was considerably less than that found for the rabbit bladder. Studies of whole rabbit bladders by another group led to the conclusion that neurally released ATP is important in the initiation of micturition, but ACh is necessary for bladder emptying (CHANCELLOR et al. 1992). Another recent study, in which substances were administered intraarterially to the whole rabbit bladder preparation, showed that pretreatment with isoprenaline, a β-adrenergic agonist, significantly inhibited contractions to ACh or ATP (LIU et al. 1998). Thus in pathological conditions such as bladder-urethral dyssynergia, involving simultaneous firing of sympathetic and parasympathetic nerves, both cholinergic and purinergic bladder contractions could be suppressed while the urethra was contracted.

Pharmacological studies of the cat bladder in vivo (DE GROAT and SAUM 1976; LEVIN et al. 1986b; FLOOD et al. 1988) showed a clear atropine-resistant contraction evoked by pelvic nerve stimulation, which had a purinergic component (THEOBALD 1983a). Purinergic transmission also contributes to bladder contractions evoked by stimulation of the hypogastric (sympathetic) nerves (THEOBALD 1983b).

In another study using unanaesthetised rats and continuous cystometry, it was shown that ATP, or α,β-meATP, administered intraarterially close to the bladder, produced rapid, phasic and dose-dependent increases in bladder pressure and micturition immediately after injection (IGAWA et al. 1993); pretreatment with α,β-meATP blocked the effects of ATP. Carbachol produced sustained increases in bladder pressure and micturition. After blockade of the micturition reflex with morphine (10 μg intrathecally), ATP, α,β-meATP and

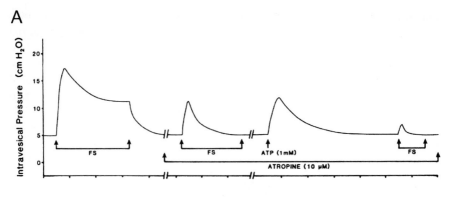
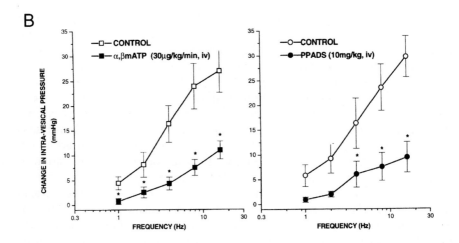
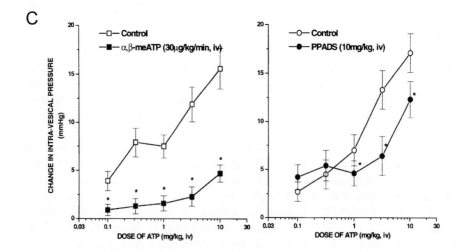

carbachol were unable to induce bladder emptying. In a later study this group examined purinergic responses in unanaesthetised rats with bladder outlet obstruction (IGAWA et al. 1994) (see Sect. L.II.). In an in vivo anaesthetised rat model, it was shown that the contractile response of the bladder to pelvic nerve stimulation consists of a phasic purinergic component that predominates at lower stimulation frequencies, followed by a tonic cholinergic component predominating at higher frequencies (NUNN and NEWGREEN 1999).

Evidence has been presented for purinergic transmission in the urinary bladder of *pithed rats* (HEGDE et al. 1998). Spinal electrical stimulation (L6-S2) evoked increases in intravesicular pressure and the major NANC component was antagonised by α,β-meATP or PPADS (Fig. 7B,C). ATP produced dose-dependent increases in intravesicular pressure. It was concluded that purinergic transmission mediated by ATP acting on P2X receptors represents a major component of excitatory innervation of the rat urinary bladder.

Implantation of chronic bladder catheters and cystometrography was used to study the micturition reflexes in *unanaesthetised* rats and it was shown that, at the spinal level, xanthine-sensitive P1 (adenosine) receptors, probably located on an excitatory interneuronal link, inhibited the volume-evoked micturition reflex (SOSNOWSKI and YAKSH 1990).

In urethane-anaesthestised rats, intrathecal administration of α,β-meATP or ATPγS induced transient bladder contractions followed by a prolonged depression of reflex bladder activity recorded under isovolumetric conditions (KIM et al. 1999). These agents also elicited bladder contractions and a secondary inhibition of reflex bladder contractions when administered intravenously. The excitatory effect of intravenous α,β-meATP was reduced by atropine or hexamethonium, a ganglionic blocking agent, indicating that this response was mediated in part by stimulation of the parasympathetic efferent pathways to the bladder and in part by a direct effect on the bladder smooth muscle. On the other hand, the excitatory effect of intravenous ATPγS was not suppressed by atropine but was completely blocked by hexamethonium, indicating that the effect was mediated by reflex activation of noncholinergic pathways to the bladder, possibly due to stimulation of bladder afferent nerves. It was concluded that excitatory and inhibitory purinergic mechanisms are present not only in the peripheral nervous system/smooth muscle of the lower

---

Fig. 7A–C. Responses of whole bladder preparations. **A** Increase in intravesicular pressure of the rabbit closed whole-bladder preparation to field stimulation (FS; 30 Hz, 1 ms pulse duration, 80 V, biphasic square wave) in the presence of atropine (10 μmol/l), ATP (1 mmol/l) in the presence of atropine, and FS in the presence of atropine and ATP. **B** Effect of α,β-methylene ATP (α,β-meATP) and PPADS on increases in intravesicular pressure evoked by segmental spinal stimulation (NANC component) in pithed rats ($n = 5$ per group; $*p < 0.05$ vs control). **C** Effect of α,β-meATP and PPADS on increases in intravesicular pressure induced by ATP in pithed rats ($n = 7$–8 per group; $*p < 0.05$ vs control). Reproduced from **A** LEVIN et al. (1986a) and **B,C** HEGDE et al. (1998)

urinary tract but also in reflex pathways in the spinal cord that control micturition.

Prolonged modulation of the parasympathetic micturition reflex was studied in anaesthetised cats, reflex discharges being recorded from a thin pelvic nerve branch to the bladder and evoked by stimulation of the remaining ipsilateral bladder pelvic nerves or urethral branches of the pudendal nerve (JIANG and LINDSTRÖM 1999). The results have led to the proposal that prolonged modulation of the micturition reflex represents physiological adaptive processes, preserving bladder function.

Recent experiments in the anaesthetised rat (ROCHA et al. 2000) have evaluated the effects on bladder function of local injection of ATP or α,β-meATP into the brain. Injection of either agent into the periaqueductal grey matter or the locus coeruleus, two brainstem areas which play an important role in the supraspinal control of micturition, led to increases in pelvic neural activity and bladder pressure and/or rate of bladder contractions. Since electrical stimulation in these same areas also activates the parasympathetic pathways to the bladder (NOTO et al. 1989), it seems likely that purinergic excitatory receptors are present in the micturition reflex circuitry in the brain.

C. Receptors to Purines and Pyrimidines

Since the potent actions of ATP in the bladder were first recognised, substantial advances have been made in identifying the ATP receptors involved. From cloning and second messenger studies in the early 1990s, it was proposed that receptors to ATP belong to two families: a P2X ion channel family and a P2Y G protein-coupled receptor family (ABBRACCHIO and BURNSTOCK 1994). Seven P2X subtypes and six P2Y subtypes are currently recognised (RALEVIC and BURNSTOCK 1998; CHESSELL et al., Chap. 3, this volume; BOARDER and WEBB, Chap. 4, this volume).

Although ATP undisputedly contracts the urinary bladder of most species (see BUCHTHAL and KAHLSON 1944; ANDERSSON 1993; HOYLE and BURNSTOCK 1993), it can also induce relaxation. Thus it is likely that multiple purinergic receptors are present in the bladder. A number of recent studies have been carried out to try to identify the receptor subtypes mediating excitation and inhibition in the lower urinary tract.

I. P2X Receptors Mediating Contraction of the Bladder

1. Agonist Potencies

An analysis of the excitatory actions of purine and pyrimidine nucleotides on the guinea pig bladder revealed the following order of potency: β,γ-meATP > ATP > GTP = CTP > ADP, while adenosine, AMP, GDP, GMP, guanosine, CDP, CMP and cytidine had no contractile activity up to 10^{-3} mol/l (LUKACSKO and

KRELL 1982). In retrospect, this potency series, although incomplete, was already suggestive of P2X receptor-mediated responses, particularly $P2X_1$ or $P2X_3$, especially since ATP responses exhibited rapid desensitisation, a property of the latter receptors (BURNSTOCK et al. 1978b). In the rabbit urinary bladder, the response to ATP is biphasic; however following desensitisation by α,β-meATP, the response became monophasic, suggesting that more than one type of excitatory receptor to ATP was present (CHEN and BRADING 1991).

Studies of the responses of the feline bladder to purines and pyrimidines revealed the following potency order: 5′-adenylimidodiphosphate (AMP-PNP) = β,γ-meATP > ATPγS = 2-methylthio ATP (2-meSATP) > ATP > UTP = CTP = GTP (THEOBALD 1992). Reactive blue 2 and Coomassie Brilliant blue G, mistakenly regarded as selective P2Y receptor antagonists at that time, both antagonised the purine-induced contractions, prompting the suggestion that multiple purine receptors were present in detrusor smooth muscle (THEOBALD 1992). In the rat and dog urinary bladder, a potency order of α,β-meATP > ATP > ADP was reported (SUZUKI and KOKUBUN 1994) and the authors concluded that three subtypes of purinoceptor might be present in rat bladder: P1 receptors (mediating relaxation), P2X receptors, and another type of P2 receptor (mediating contractions); but only a single receptor type (P2X) in dog bladder (Fig. 8A).

Contractile responses of the rat bladder induced by ATP and α,β-meATP were fast and transient, reaching a maximum in about 20s; in contrast, contractions in response to adenosine 5′-O-2-thiodiphosphate (ADPβS) and UTP were slower and sustained and were barely affected by α,β-meATP desensitisation, suggesting two different receptor subtypes (BOLEGO et al. 1995a). It has been proposed that about 20% of the neurogenic contraction of rat bladder is mediated by purinergic receptors sensitive to ADPβS (HASHIMOTO and KOKUBUN 1995) (Fig. 8B). The diadenosine polyphosphate, Ap_4A, contracts the guinea-pig bladder and since this was abolished after desensitisation with α,β-meATP and by suramin, PPADS and nifedipine, it is likely to act through P2X receptors (USUNE et al. 1996).

2. Localisation of P2X Receptors

An early study of the characteristics of [^{3}H]-ATP binding to homogenates of the rabbit urinary bladder using radioligand filtration methodology showed high affinity binding, favouring the view that ATP receptors are present in smooth muscle membranes (LEVIN et al. 1983b).

Autoradiographic studies of the distribution of [^{3}H]-α,β-meATP in the rat bladder showed a high level of labelling in the smooth muscle of the detrusor, but none in the urethra, which correlated closely with the pharmacology (Bo and BURNSTOCK 1989) (Fig. 9A,B). In a parallel study, high- and low-affinity binding sites for [^{3}H]-α,β-meATP were demonstrated in rat urinary bladder membranes and displacement experiments with unlabelled purinoceptor ligands confirmed that [^{3}H]-α,β-meATP mainly binds to P2X receptors with a

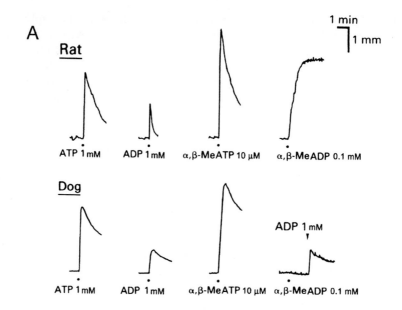

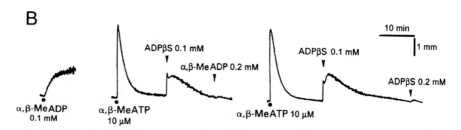

Fig. 8. A Contractile effects of various purines on rat and dog urinary bladder smooth muscle strips. Contractions were induced by ATP (1 mmol/l), ADP (1 mmol/l), α,β-methylene ATP (α,β-meATP, 10 μmol/l) and α,β-methylene ADP (α,β-meADP, 0.1 mmol/l). In each panel responses to ATP, ADP and α,β-meATP were obtained from the same strip, whereas those to α,β-meADP were obtained from a different strip. **B** Contractions of rat urinary bladder smooth muscle strips induced by α,β-meADP and ADPβS in the absence and presence of α,β-meATP (10 μmol/l) and ADPβS (0.1 mmol/l). Experiments were performed in the presence of atropine (1 μmol/l) and guanethidine (5 μmol/l). From left to right, the traces show a control contraction to 0.1 mmol/l α,β-meADP, the contraction induced by 0.2 mmol/l α,β-meADP after pretreatment with 10 μmol/l α,β-meATP and 0.1 mmol/l ADPβS, and the contraction to cumulative application of ADPβS (final concentration 0.2 mmol/l) in the presence of α,β-meATP (10 μmol/l). Reproduced from **A** Suzuki and Kokubun (1994) and **B** Hashimoto and Kokubun (1995)

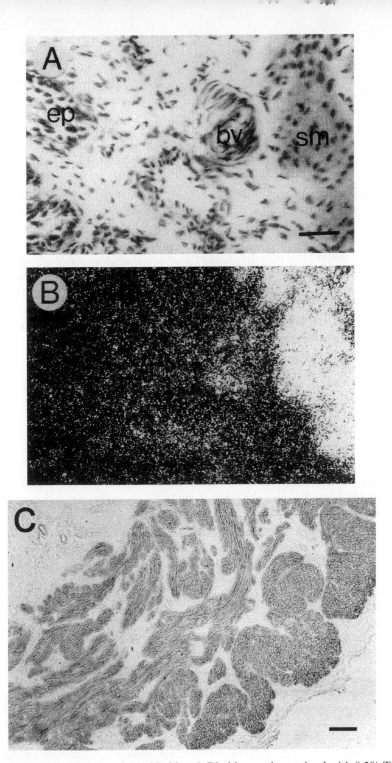

Fig. 9A–C. Sections of rat urinary bladder. **A** Bladder section stained with 0.5% Toluidine blue and viewed with bright-field optics. *bv*, blood vessel; *ep*, epithelium; *sm*, smooth muscle. *Calibration bar* = 50 μm. **B** Autoradiograph over section shown in **A** viewed with dark-field optics, showing the overall distribution of binding sites of [H^3]α,β-methylene ATP. **C** Transverse section of detrusor muscle immunostained for P2X$_1$ receptors. *Calibration bar* = 100 μm. Reproduced from **A,B** Bo and BURNSTOCK (1989) and **C** LEE et al. 2000

potency order of α,β-meATP > β,γ-meATP > suramin > ATP > ADP > 2-meSATP >> adenosine (Bo and BURNSTOCK 1990b). In later studies, autoradiographic localisation and characterisation of [^{3}H] α,β-meATP binding sites were described in the urinary bladders of guinea-pigs, rabbits, cats and humans (Bo and BURNSTOCK 1992, 1995; MICHEL et al. 1996; ZHAO et al. 1996). High affinity binding sites for [^{35}S]ATPγS in the human bladder have also been described (MICHEL et al. 1996).

Immunohistochemical studies which have been carried out with specific antibodies to the different P2X receptor subtypes showed that $P2X_1$ receptors are the dominant subtype in the membranes of the smooth muscle cells in the rat detrusor and also vascular smooth muscle in blood vessels in the bladder (LEE et al. 2000) (Fig. 9C). In another immunohistochemical study of the rat bladder, clusters of $P2X_1$ receptors were described on smooth muscle cells, some, but not all, of which were closely related to nerve varicosities (HANSEN et al. 1998).

Northern blotting and in situ hybridisation have also shown the presence of $P2X_1$ mRNA in urinary bladder (VALERA et al. 1994). Northern blot analysis also detected the expression of the human $P2X_4$ gene expression in the bladder (DHULIPALA et al. 1998), but immunohistochemistry has not detected staining for this receptor although it apparently is expressed in the bladder in late pregnancy (M. Bennett, personal communication).

3. Structure-Activity Studies

The first structure-activity studies on bladder were carried out by Noel Cusack and colleagues in collaboration with the Burnstock laboratory (BURNSTOCK et al. 1983). Neither 2-chloro-ATP nor 2-meSATP were significantly more effective than ATP itself in producing contraction of guinea-pig bladder. For each pair of enantiomers, some stereoselectivity was observed at low concentrations, but this was lost or even reversed with higher concentrations. However, it was recognised that the rapid ectoenzymatic breakdown of ATP to adenosine, which relaxes the bladder (BROWN et al. 1979), probably distorted the results. Adenylyl 5'-(β,γ-methylene)-diphosphonate (AMP-PCP) had much greater contractile effects on the guinea-pig bladder than ATP and the enantiomer of AMP-PCP, L-AMP-PCP, was even more potent, probably because L-AMP-PCP was completely resistant to degradation (CUSACK and HOURANI 1984), although it was inactive on P2Y receptors in the taenia coli (HOURANI et al. 1985) Of the phosphorothioate analogues of ATP, ADP and AMP, it was found that adenosine 5'-O-1-thiotriphosphate (ATPαS), ATPβS, Rp-ATPβS and Sp-ATPβS were much more potent than ATP in the bladder (BURNSTOCK et al. 1984). Among the 2-methylthio derivatives of AMP-PCP the potency order in the bladder was: difluoromethylene > methylene > dichloromethylene (CUSACK et al. 1987). None of the analogues was degraded by ectonucleotidases, and restoration of the electro-negativity of the triphosphate chain did not further enhance their potency. Unlike the effects of these agents on the taenia coli, the order of potency in the bladder did not reflect their order of *acidity*; this may be because some distortion of the triphosphate chain is necessary to accommodate the bulky chloro groups, whereas the difluoro analogue is sterically more similar to ATP (BLACKBURN et al. 1984). A study of the structure-activity relationships of nucleotide effects on excitatory P2 receptors in the bladder provided evidence that dephosphorylation of ATP analogues reduces pharmacological potency.

Analogues designed to alter the three main parts of the ATP molecule, i.e. the triphosphate, ribose and base, were tested on the guinea-pig bladder and it was concluded that the triphosphate group is responsible for the efficacy of the agonist, whereas the ribose and adenine moieties are associated with affinity (HOWSON et al. 1988).

Using a radioligand binding assay it was found that adenosine, adenine and xanthine had no significant effect on [^{3}H]-α,β-meATP binding to membrane fractions prepared from rat urinary bladder, while pentasodium triphosphate and disodium pyrophosphate could effectively displace the binding; these results were taken to indicate that the phosphate side-chain of ATP and its analogues is the key structure responsible for the binding to P2X receptors (Bo and BURNSTOCK 1993). A further study of the affinities of ATP derivatives for P2X purinoceptors in rat bladder was carried out with modifications of the polyphosphate chain as well as the adenine and ribose moieties (Bo et al. 1994). Replacement of the bridging oxygen in the triphosphate chain of ATP with a methylene or imido group markedly increased the affinity, modifications at N^6, N^1 and C-8 positions of the purine base reduced the affinity of ATP, attachment of an alkylthio group to the C-2 position increased affinities, while replacement of the 3'-hydroxyl group on the ribose with substituted amino or acylamino groups produced more potent P2X receptor agonists. Diadenosine polyphosphates (Ap$_n$A) were also shown to displace [^{3}H]-α,β-meATP binding with a rank order of potency Ap$_6$A > Ap$_5$A > Ap$_4$A >> Ap$_3$A >> Ap$_2$A. Suramin, PPADS and Reactive blue 2 competitively displaced the binding of [^{3}H]-α,β-meATP to P2X receptors. An extensive study of structure-activity relationships for derivatives of ATP as agonists at P2X and P2Y receptors was carried out (BURNSTOCK et al. 1994). For example, 3'-benzylamino-3'-deoxy-ATP was found to be very potent in the guinea-pig bladder, but was inactive at P2Y receptors.

II. P2Y Receptors Mediating Relaxation of the Bladder

ATP, as well as adenosine, has been shown to reduce pelvic nerve-evoked bladder contractions; however, since methylxanthines did not fully antagonise the responses (THEOBALD and DE GROAT 1989), this suggests that P2Y receptors as well as P1 (adenosine) receptors might be involved in purine inhibition. These P2Y receptors are likely to be on nerve terminals in the bladder providing prejunctional inhibition of release of excitatory neurotransmitters and both Reactive blue 2 and Coomassie brilliant blue G antagonise the inhibitory actions of ATP and analogues on nerve-mediated contractions (THEOBALD 1992) (see Sect. F.). However, these experiments did not exclude the possibility that ATP could also be acting through postjunctional P2Y receptors on bladder smooth muscle mediating direct relaxation, although usually masked by the dominating contractile actions of ATP through P2X receptors.

The first direct evidence for ATP-induced relaxation of smooth muscle came from studies of the mouse bladder (BOLAND et al. 1993). In carbachol precontracted preparations, ATP elicited an initial contraction, followed by a sustained relaxation, while on K^+ precontracted preparations, ATP caused relaxation only, which was not inhibited by 8-phenyltheophylline (8-PT). The order of potency for relaxation was: 2-meSATP > ATP > β,γ-meATP, perhaps indicative of P2Y receptors. A biphasic response of bladder strips in response to ATP was also described in the rat (BOLEGO et al. 1995b). The initial contraction was abolished after desensitisation of the P2X receptor with α,β-meATP, revealing a clear relaxation response to ATP. The evidence put forward that these relaxant responses were mediated by a P2Y receptor was first that 2-meSATP was more potent than ATP, and second that G proteins were involved, since the G protein activator, guanosine 5'-O-3-thiotriphosphate (GTPγS) significantly potentiated the relaxant responses, while the G protein blocking agent, guanosine 5'-O-2-thiodiphosphate (GDPβS) completely abolished the relaxation; these agents had no effect on the ATP-induced contractions.

In the mini-pig bladder it appears that the neurally-evoked relaxation which follows the initial cholinergic contraction of the bladder neck is mediated by P2Y receptors (TONG et al. 1997a). 8-PT did not effect the relaxations, negating P1 receptor mediation, but the P2Y antagonist Reactive blue 2 reduced the relaxations by about 80%. However, it is well known that Reactive blue 2 can produce non-specific inhibitory effects with prolonged exposure or with high concentrations, so more experiments will need to be carried out to confirm this claim. Nevertheless, supporting evidence for postjunctional purinoceptor subtypes has come from recent studies in the marmoset urinary bladder (MCMURRAY et al. 1998) where a biphasic response to ATP was demonstrated. The potency order for the relaxation phase was ATP = 2-meSATP ≥ ADP >> α,β-meATP. When the initial contraction was abolished by desensitisation with α,β-meATP, a relaxation response clearly remained, which was abolished by the G protein inactivator, GDPβS (Fig. 10). The relaxation was unaffected by 8-PT, or the nitric oxide synthase (NOS) inhibitor, L-NOARG, but was blocked by Cibacron blue which is regarded as a P2Y antagonist, at least on native receptors, and which did not affect the contractile responses to ATP. Since N-tosyl-L-phenylalanine chloromethyl ketone (TPCK), an inhibitor of cyclic AMP-dependent protein kinase A (PKA), significantly shifted the curve for the ATP-induced relaxation to the right, it was suggested that the subtype of the P2Y G protein-coupled receptor involved might be one that acts through adenylate cyclase.

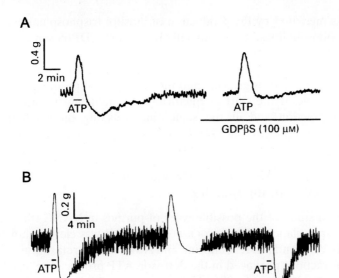

Fig. 10A,B. Response of a marmoset detrusor smooth muscle strip to ATP (1 mmol/l, 60 s application). **A** In the absence and presence of the G protein inactivator GDPβS (100 μmol/l), the strip being exposed to GDPβS for 20 min before rechallenge with ATP. **B** Before and after desensitisation of P2X receptors through prolonged exposure to α,β-methylene ATP (α,β-meATP, 10 μmol/l). Reproduced from McMurray et al. (1998)

ATP-induced relaxation was also detected in rabbit bladder smooth muscle strips precontracted with carbachol (Gupta et al. 2000). The concentration-response curve for ATP relaxations was shifted to the right by Reactive blue 2, but not by PPADS or 8-PT. The order of potency among the ATP analogues was 2-meSATP > ATP > ADP >> UTP, consistent with the $P2Y_1$ receptor pharmacological profile and this was supported by the demonstration of mRNA for $P2Y_1$ in the rabbit bladder using reverse transcription polymerase chain reaction (RT-PCR). Evidence for P2Y receptors in the rat bladder has also been provided using RT-PCR, Northern blotting and in situ hybridisation (Obara et al. 1998). The in situ hybridisation technique showed the presence of $P2Y_1$ mRNA in detrusor smooth muscle and blood vessels in the bladder, but no positive staining was seen in urethral smooth muscle.

P2Y receptors might also mediate excitatory responses in the bladder because UTP contracts the rat bladder (Bolego et al. 1995a). According to current thinking UTP would be acting through a $P2Y_2$ or $P2Y_4$ receptor or may be acting through an as yet unidentified P2X receptor subtype. It has been claimed (Naramatsu et al. 1997) that, in addition to a P2X receptor (probably the $P2X_1$ subtype), there are ADPβS-sensitive receptors in the rat bladder. Since these receptors mainly depend on Ca^{2+} released from intracellular stores and this is mediated by the production of inositol trisphosphate via the activation of phospholipase C, it seems likely that the ADPβS-sensitive receptor might be a P2Y receptor.

In summary, the evidence for the presence of P2Y receptors on smooth muscle of the bladder is growing and the concept that purinergic innervation may play a role at the start of micturition by inducing the initial detrusor muscle contraction and at the same time relaxing the bladder neck, is attractive.

III. P1 Receptors Mediating Relaxation and Contraction of the Bladder

In the initial study of the possible roles of purines in NANC transmission in the guinea-pig bladder (BURNSTOCK et al. 1972), it was clearly shown that, in contrast to ATP and ADP, adenosine and AMP caused relaxation. In contrast to the contractions produced in the detrusor, ATP produced relaxation of the bladder neck (HILLS et al. 1984). However, since ATP is rapidly broken down to adenosine, and ATP and adenosine are equipotent in this tissue, it seems likely that ATP exerts its action via a P1 receptor. In support of this view, α,β-meATP, which is slowly degradable, is without effect on the bladder neck and both ATP and adenosine concentration curves are shifted to the right by the P1 receptor antagonist 8-PT. However, 8-PT failed to modify nerve-mediated relaxations in the bladder neck (HILLS et al. 1984; KLARSKOV 1987a), so it was interesting that the NANC responses were later identified as nitrergic (PERSSON and ANDERSSON 1992; THORNBURY et al. 1992).

In the rat bladder, adenosine and 5'-N-ethylcarboxamidoadenosine (NECA) inhibited the contractions induced by carbachol. Since NECA was much more potent than CPA and adenosine, the P1 receptor subtype involved was likely to be A_2 (NICHOLLS et al. 1992a).

It has been claimed recently that P1 (A_1) receptors mediate 2-chloroadenosine (2 CADO) *contractions* in cat detrusor muscle and that the contraction depends on a pertussis toxin-sensitive G_{i3} protein, PLC-β_3 and the release of intracellular Ca^{2+} (YANG et al. 2000).

D. Sympathetic Cotransmission

Sympathetic innervation of the detrusor has in general been reported to be sparse, although the trigone region is relatively densely innervated by sympathetic nerves (see ALM and ELMÉR 1975; HOYLE and BURNSTOCK 1993). Sympathetic nerve fibres reach the bladder largely in the hypogastric nerve, although some parasympathetic fibres may also be present in some species (see LINCOLN and BURNSTOCK 1993). Stimulation of the hypogastric nerve may cause an increase or decrease in pressure in the urinary bladder, but always excites the urethra (see DE GROAT and THEOBALD 1976; CREED 1979; IMAGAWA et al. 1989). Physiologically inhibitory sympathetic transmission in the detrusor is important during the filling phase of the voiding cycle (DE SY et al. 1974; LABADIA et al. 1988; MAGGI et al. 1989a). In addition to transmitter released from sympathetic nerves acting directly on smooth muscle, they may act prejunctionally on parasympathetic nerve terminals to inhibit both cholinergic and non-cholinergic (purinergic) excitation, which are invoked in the voiding phase of the micturition cycle.

Although purinergic transmission predominantly originates from postganglionic parasympathetic or intramural nerves, in the cat at least, ATP may also be released from the hypogastric nerve. This nerve is predominantly sympathetic, but may also contain parasympathetic elements (see LINCOLN and BURNSTOCK 1993). When the hypogastric nerve is stimulated in the cat it causes the bladder to contract; this contraction is reduced by $ANAPP_3$ (THEOBALD 1982, 1983a; THEOBALD and HOFFMAN 1986), implying that ATP is being released. Further, 6-hydroxydopamine, which destroys sympathetic nerves, prevents this contractile response, indicating that the ATP is released from sympathetic nerves (THEOBALD 1983b). Guanethidine, in a dose which blocked the bladder relaxation induced by hypogastric nerve stimulation and mediated by noradrenaline acting on β-adrenergic receptors (DE GROAT and SAUM 1972), did not affect hypogastric nerve-mediated excitation (THEOBALD 1983b). However, in guanethidine-treated animals, $ANAPP_3$ blocked the excitation. These findings suggest that ATP may be released from the hypogastric nerve.

Nicotinic-induced contractions of the guinea-pig bladder in the presence of atropine, were abolished by desensitisation of the P2X receptor with α,β-meATP (HISAYAMA et al. 1988). Several possible mechanisms might be involved, namely that nicotine might produce a contraction by activating nicotinic receptors on:

1. Parasympathetic nerve terminals coreleasing ACh and ATP
2. Sympathetic nerve terminals coreleasing NA and ATP
3. Intramural bladder neurones that corelease ATP with peptides

One of the basic rules about neuromuscular cotransmission appears to be that the cotransmitters released act synergistically (see BURNSTOCK 1990a). There is evidence that NA and ATP released as cotransmitters from sympathetic nerves act synergistically (HOLCK and MARKS 1978), but there do not appear to be any reports of synergistic cotransmission in the urinary bladder involving either parasympathetic (ACh and ATP) or sympathetic (NA and ATP) nerves.

By analogy with most other blood vessels, it is likely that ATP is a cotransmitter with NA in perivascular sympathetic nerves supplying blood vessels in the bladder (see BURNSTOCK 1990b).

E. Intramural Bladder Neurones and Pelvic Ganglia

Intramural ganglia have been described in the bladder of several mammalian species, including humans (ALM 1978; GILPIN et al. 1983; CROWE et al. 1986, 1988; BURNSTOCK et al. 1987; PITTAM et al. 1987; CROWE and BURNSTOCK 1989). Quinacrine, a fluorescent dye, that selectively labels high levels of ATP bound to peptides in granular vesicles, stained a subpopulation of neurones in ganglia in the guinea-pig bladder (BURNSTOCK et al. 1978a; CROWE et al. 1986) (Fig. 11A). Subpopulations of neurones in bladder ganglia also stained positively for VIP, somatostatin, substance P, 6-hydroxytryptamine and acetylcholinesterase. Thus intramural ganglia, perhaps largely parasympathetic postganglionic neurones contain and probably release ATP. They also respond to microapplication of ATP (BURNSTOCK et al. 1987) (Fig. 11B,C). No intramural neurones have been observed in the rat bladder, but several weeks after unilateral pelvic ganglion destruction, intramural neurones were consistently observed along the remnants of nerves in the originally denervated half of the bladder (UVELIUS and GABELLA 1995).

Parasympathetic ganglia on the surface of the cat urinary bladder have provided useful preparations for examining synaptic modulatory mechanisms (see DE GROAT and BOOTH 1993). These ganglia contain several types of principal ganglion cells (coexpressing various neuropeptides, ACh, NA, ATP and nitric oxide) as well as small intensely fluorescent cells (SIF cells). They receive an innervation from both parasympathetic and sympathetic preganglionic axons. Parasympathetic preganglionic axons, which arise in the sacral segments of the spinal cord and travel in the pelvic nerve, represent the principal excitatory pathway to the cholinergic-purinergic ganglion cells (DE GROAT 1975; DE GROAT et al. 1979), which in turn provide an excitatory input to the detrusor smooth muscle. The sympathetic innervation originates in the lumbar spinal cord and passes to the bladder via the hypogastric nerves and the sympathetic chain. The sympathetic system exerts an inhibitory control over activity of the detrusor muscle and an excitatory input to the trigone and urethra.

Various purinergic agonists including ATP, α,β-meATP, ADP, AMP, adenosine and 2 CADO administered intraarterially depress cholinergic transmission and depress the bladder contractions elicited by stimulation of preganglionic axons in the pelvic nerve (DE GROAT and BOOTH 1980; THEOBALD and DE GROAT 1989). High doses of ATP also produce postganglionic firing in unstimulated, decentralised ganglia, indicating a direct excitatory effect of ATP on bladder ganglion cells. Other nucleotides and related substances such as cyclic AMP, dibutyryl cyclic AMP, adenosine, inosine and ITP have weak or no effects on transmission. ATP, ADP, AMP and adenosine are equipotent in depressing transmission, whereas 2-ClAdo, an agent that is more resistant to cellular uptake and metabolism, is ten times more potent than adenosine. This indicates that metabolism could have a significant influence on the effectiveness of purinergic agents. This is also indicated by the

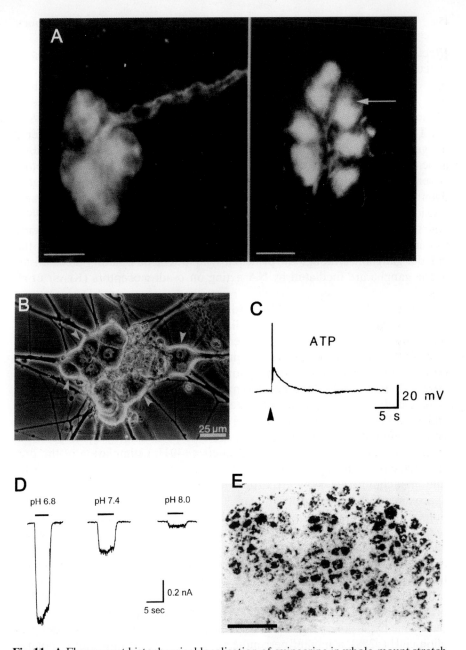

Fig. 11. A Fluorescent histochemical localisation of quinacrine in whole-mount stretch preparations of adult rabbit urinary bladder. *Left-hand panel*: quinacrine-positive ganglion cells associated with a nerve fibre bundle. *Right-hand panel*: a ganglion cell containing at least 6 fluorescent nerve cells. The nuclei (*arrow*) are non-fluorescent. *Calibration bars* = 50 µm. **B** Phase contrast micrograph of a group of intramural bladder neurones with examples of neuronal somata (*arrowheads*). **C** Depolarising and firing in response to 20-ms microapplication of 1 mmol/l ATP in an intramural bladder neurone. **D** ATP-induced currents recorded from a rat major pelvic ganglion neurone at different pH. The *horizontal bars* indicate the application time of ATP (30 µmol/l). **E** Detection of $P2X_2$ receptor mRNA in the rat major pelvic ganglion using in situ hybridisation. Distinct labelling was observed with the $P2X_2$ probe, although there was great variation in the intensity from cell to cell. *Calibration bar* = 100 µm. Reproduced from **A** Crowe and Burnstock (1985), **B,C** Burnstock et al. (1987), and **D,E** Zhong et al. (1998)

effect of dipyridamole to enhance and prolong the inhibitory responses to injected purinergic agents. Dipyridamole, which slows the cellular uptake of adenosine, enhances the inhibitory actions of AMP and adenosine as well as those of ATP and ADP, suggesting that the latter agents can be converted to adenosine.

Theophylline and caffeine block the inhibitory effects of purinergic agents on ganglionic transmission and on neurally-evoked bladder contractions, indicating that the inhibition is mediated by P1 receptors. The P1 receptors appear to be located presynaptically as well as postsynaptically on the ganglion cells. Since the sympathetic input has a modulatory effect on transmission in bladder ganglia (DE GROAT and SAUM 1972; KUO et al. 1983, 1984), and since ATP can be released as a cotransmitter from sympathetic nerve terminals, sympathetic nerves may be a source of ATP released within the bladder ganglion, although the principal sympathetic modulatory mechanisms in the ganglia are mediated by NA acting on α-adrenoceptors (KEAST et al. 1990).

Since ATP can be released from adrenergic and cholinergic nerves, studies were conducted on cat bladder ganglia to determine whether endogenously released substances might elicit purinergic inhibition. Extracellular recordings in situ did not detect a theophylline-sensitive (purinergic) component in either the inhibition of ganglionic transmission elicited by stimulation of sympathetic nerves (hypogastric) or the heterosynaptic inhibition elicited by stimulation of preganglionic axons in the pelvic nerves (DE GROAT and KAWATANI 1989a). On the other hand, intracellular recordings from isolated bladder ganglia in vitro identified a non-cholinergic, slow hyperpolarising synaptic potential elicited by high intensity and high frequency (40 Hz) stimulation of the preganglionic nerve trunk (AKASU et al. 1984; SHINNICK-GALLAGHER et al. 1986). The non-cholinergic slow hyperpolarisation, which has an amplitude of approximately 5 mV and a duration of 30 s, is increased in amplitude and duration by dipyridamole, an agent which blocks the uptake of adenosine, and is reduced in amplitude by adenosine deaminase, an enzyme that metabolises adenosine. Caffeine, a P1 receptor antagonist, also blocks the synaptic potential. The slow hyperpolarising synaptic potential is mimicked by the administration of exogenous purinergic agonists (500 nmol/l–1 mmol/l), the relative order of potency being 2 CADO >> AMP > adenosine > ADP > ATP. This order of potency is consistent with a response mediated by a P1 receptor.

While hyperpolarising responses are detected in virtually all bladder ganglion cells (92%), a smaller percentage of cells (52%) exhibit fast depolarising responses to ATP and other purinergic agonists (SHINNICK-GALLAGHER et al. 1986). In some cells the fast depolarising response is followed by a more prolonged slow hyperpolarisation lasting 1–1.5 min. The ATP depolarisation is associated with a decrease in membrane resistance, reverses polarity at –7 mV and is dependent on the concentration of Na^+ and not K^+ ions. The relative order of potency among purinergic agents to produce the fast depolarisation is ATP > ADP >> AMP > adenosine. This depolarising action of ATP no doubt mediates the ganglionic excitatory effects of ATP noted during in situ experiments (THEOBALD and DE GROAT 1989).

The precise physiological roles of purinergic agents in the control of transmission in bladder ganglia need resolution. The demonstration of purinergic slow hyperpolarising potentials following stimulation of preganglionic nerves indicates that purinergic agents can be released in ganglia during neural activity. However, since ATP is present in bladder postganglionic neurones as well as in cholinergic and adrenergic nerve terminals, there are various possible sources of purinergic transmitter. In addition, although adenosine has been proposed as the inhibitory transmitter (AKASU et al. 1984), it is possible that the extracellular catabolism of ATP to adenosine could be important in the mediation of the slow hyperpolarising responses. Whether this catabolism could occur within the latency for evoking the hyperpolarising potentials is not known. It is also important to note that preganglionic nerves contain cholinergic and adrenergic postganglionic axons as well as preganglionic axons and therefore various neural pathways could be involved in eliciting the slow hyperpolarising potential. Adenosine deaminase has been identified in sacral preganglionic neurones in the rat (SENBA et al. 1987). This has prompted the speculation that preganglionic pathways may be purinergic as well as cholinergic. It is not known whether a similar situation exists in the cat.

Single electrode voltage-clamp techniques in *rabbit* vesical parasympathetic ganglion cells (NISHIMURA and TOKIMASA 1996) showed that ATP and ADP, but not AMP or adenosine, caused an inward current associated with increased conductance. Suramin and Reactive blue 2, but not hexamethonium, reversibly depressed the actions of ATP and ADP, suggesting that ATP activates cation channels through P2X receptors in rabbit parasympathetic neurones. Application of ATP also modulates the amplitude of nicotinic fast EPSPs in the rabbit vesical parasympathetic ganglia (NISHIMURA and AKASU 1994).

P2X receptors have been demonstrated recently in pelvic ganglion neurones of rat (ZHONG et al. 1998) and guinea-pig (ZHONG et al. 2000). In the rat, evidence from the pharmacological characteristics of pelvic ganglion neurones in response to P2 agonists and antagonists recorded with the whole cell voltage-clamp technique combined with in situ hybridisation and immunohistochemistry led to the conclusion that $P2X_2$ receptors are the predominant P2X subtype present in about 39% of the neurones (Fig. 11D,E). In contrast, in the guinea-pig, at least three distinct P2X receptors were shown to be present in different subpopulations of neurones in the pelvic ganglion, probably $P2X_2$ and $P2X_3$ homomultimers in 5% and 70% of the neurones respectively, and about 25% with heteromultimeric $P2X_{2/3}$ receptors, but the possibility that an unidentified P2X receptor subtype is also present was not discounted.

F. Neuromodulation in the Bladder

ATP released as a cotransmitter at various sites including sympathetic, parasympathetic, sensory and motor nerve terminals, at synapses in autonomic ganglia and in the central nervous system can be broken down by ectoenzymes to adenosine that then acts on prejunctional P1 receptors to modulate the release of neurotransmitters (see DE MEY et al. 1979; RIBEIRO 1979; MCDONALD and WHITE 1984; SNYDER 1985; KATSURAGI et al. 1986; CUSACK et al. 1988; SPERLÁGH and VIZI 1991; ACEVEDO et al. 1992; VON KÜGELGEN 1994; FUDER and MUSCHOLL 1995; MACDERMOTT et al. 1999). P1 receptors on the nerve terminals on the bladder are of the A_1 subtype, while postjunctional smooth muscle receptors are of the A_2 subtype (BURNSTOCK 1990a; ACEVEDO et al. 1992). At some sites, especially where the junctional cleft is narrow, ATP itself acts on prejunctional P2 receptors to modulate transmitter release (WIKLUND and GUSTAFSSON 1986; VON KÜGELGEN et al. 1989; SHINOZUKA et al. 1990; FUDER and MUTH 1993). This has also been described in rat bladder (KING et al. 1997). ATP can also act as a postjunctional modulator of cholinergic (nicotinic) transmission to skeletal muscle (HENNING 1997) and of sympathetic responses in vas deferens (HOLCK and MARKS 1978).

In the rat urinary bladder, ATP, adenosine and β,γ-meATP were all shown to produce a dose-dependent and reversible inhibition of the atropine-resistant contractile responses to transmural nerve stimulation (DAHLÉN and HEDQVIST 1980), suggesting that both P1 and P2 receptors are present on terminals of parasympathetic nerves in this bladder preparation. A study of natural products from *Nauclea latifolia*, a tree that grows in the northern part of Nigeria, showed that the *leaf* extract was very potent in potentiating purinergic neurotransmission and ATP-induced contractions in rat bladder, while the *root* extract depressed purinergic contraction by a direct action in smooth muscle, since it did not modify ATP-induced contractions (UDOH 1995).

In the mouse bladder, 5-hydroxytryptamine (5-HT), perhaps released from circulating platelets, was shown to potentiate strongly the predominantly purinergic parasympathetic nerve mediated responses, probably via 5-HT_{1B} receptors (HOLT et al. 1985). This was later confirmed in guinea-pig bladder, where the effects of 5-HT were shown to be mediated by 5-HT_{2A} and 5-HT_4 receptors (MESSORI et al. 1995) (Fig. 12A) and also in bladder of pigs (BUSHFIELD et al. 1996), humans (CORSI et al. 1991) and rabbits (BARRAS et al. 1996). In contrast, γ-aminobutyric acid (GABA), acting through GABA_B prejunctional receptors, inhibited nerve-mediated contractions in mouse bladder (SANTICIOLI et al. 1986). Morphine, however, did not alter the responses to nerve stimulation or exogenously applied ATP in the mouse bladder. Cannabinoid CB_1 receptors reduced nerve-mediated contractions, but did not affect responses to ACh or α,β-meATP, providing evidence for prejunctional modulation of release of transmitter from parasympathetic nerves (PERTWEE and FERNANDO 1996).

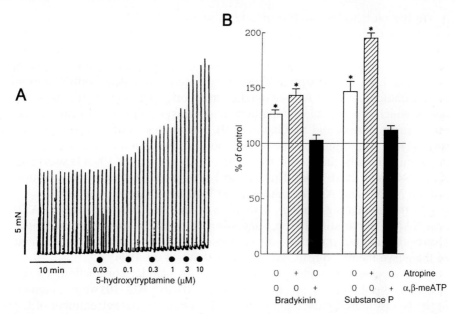

Fig. 12. A Trace of guinea-pig detrusor strips illustrating the potentiating effect of 5-hydroxytryptamine (0.03–10 μmol/l) on the excitatory neuromuscular transmission elicited by repetitive trains of electrical pulses. **B** Effect of bradykinin (10 μmol/l) and substance P (1 μmol/l) on neurogenic responses (4 Hz, for 1 min) of guinea-pig urinary bladder strips. *Open bars* show effects of bradykinin or substance P in the absence of other drugs. *Hatched bars* show effects on the purinergic component of neurogenic contraction (after treatment with atropine to block cholinergic component). *Solid bars* show effects on the cholinergic component of neurogenic contraction (after treatment with α,β-meATP to block purinergic component). Note that both bradykinin or substance P potentiate intact neurogenic response and purinergic component but not cholinergic component. Each *bar* represents the mean ± S.E.M. of 9–16 experiments; * indicates significant difference from control. Reproduced from **A** MESSORI et al. (1995) and **B** PATRA and WESTFALL (1996)

In the rat bladder, bradykinin and substance P have been shown to facilitate the purinergic component of parasympathetic nerve responses and the responses to exogenous ATP, implying that in this case the mechanism of action is at the postjunctional site (Acevedo et al. 1990; Patra and Westfall 1996) (Fig. 12B). Neuropeptide Y also potentiated α,β-meATP contractions in the rat bladder, but not ACh-evoked contractions, suggesting that neuropeptide Y, which is present in sympathetic and parasympathetic pathways to the rat detrusor (Keast and de Groat 1989), contributes to transmission in two ways: (1) by promoting non-cholinergic motor transmission, and (2) by inhibiting prejunctionally cholinergic transmission; (Zoubek et al. 1993; Iravani and Zar 1994; Tran et al. 1994). Endothelin-1 produced long lasting potentiation of both NANC and ATP responses in rat bladder (Donoso et al. 1994).

In the guinea-pig bladder, histamine potentiated the responses to ATP and NANC nerve stimulation, suggesting an action at postjunctional sites. However, it did not potentiate the response to ACh or cholinergic component of the nerve-mediated response (Patra and Westfall 1994).

The mechanisms underlying stimulation of bladder contractions by the selective neurokinin NK_2 receptor agonist [β-Ala^8]NKA(4–10) were examined in the anaesthetised guinea-pig (Ahluwalia et al. 1998). Pretreatment of the animals with both atropine and α,β-meATP or by ganglion blockers led to complete blockade of NK_2 receptor-induced contractions. These results suggest that stimulation of NK_2 receptors located on capsaicin-sensitive sensory nerves (where NK_2 receptors have been demonstrated autoradiographically) leads to bladder contractions via both cholinergic and purinergic parasympathetic motor nerves. Substance P and bradykinin both potentiate the neurogenic responses of the guinea-pig bladder by influencing the purinergic component of the excitatory motor innervation, apparently at a postjunctional site (Patra and Westfall 1996).

G. Afferent Pathways in Bladder and Purinergic Mechanosensory Transduction

Most of the afferent supply of the bladder and urethra originates in dorsal root ganglia (DRG) at the lumbosacral region of the spinal cord and passes peripherally through the pelvic nerve (de Groat 1986; Nance et al. 1988). Although smaller in number, afferents also project to the urogenital tract through sympathetic nerves (hypogastric) from DRG at the thoracolumbar level (see Morgan et al. 1986; Jancsó and Maggi 1987; Yoshimura 1999). In addition, the afferent supply of the external urethral sphincter travels by the pudendal nerve to the sacral region of the spinal cord (see Hoyle et al. 1994). The central projections of pelvic and pudendal afferents overlap within the spinal cord, allowing integration of somatic and parasympathetic motor activity (de Groat et al. 1993). Activity of afferents from one region of the pelvic viscera can influence the efferent output to another region. Thus, stimulation

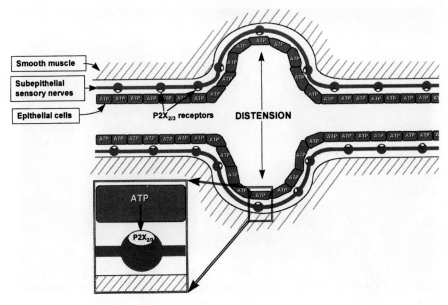

Fig. 13. Schematic of hypothesis for purinergic mechanosensory transduction in tubes (e.g. ureter, vagina, salivary and bile duct, gut) and sacs (e.g. urinary and gall bladders, and lung). It is proposed that distension leads to release of ATP from epithelium lining the tube or sac, which then acts on $P2X_{2/3}$ receptors on subepithelial sensory nerves to convey sensory/nociceptive signals to the CNS. Reproduced from BURNSTOCK (1999b)

of anal, rectal or vaginal afferents can inhibit micturition (CONTE et al. 1989). It is also now becoming clear that sensory afferents can modulate efferent activity in peripheral autonomic ganglia as well as in the central nervous system. Afferent neurones can release neurotransmitters from peripheral terminals in the viscera as well as from central terminals in the spinal cord. In the bladder, transmitter release from sensory nerve endings may also play an efferent role by direct postjunctional effects on the detrusor muscle (DE GROAT 1987; MAGGI 1993).

Recently, BURNSTOCK (1999b) put forward a hypothesis, indicating that distension of epithelial cells lining the tubes (including ureter) and sacs (including urinary bladder) in the body leads to release of ATP which then acts on $P2X_3$ receptors on suburothelial sensory nerves to modulate afferent firing that can lead to bladder voiding and pain (see Fig. 13). The evidence in support of this hypothesis from earlier as well as later papers follows.

I. Evidence for a Suburothelial Sensory Nerve Plexus with $P2X_3$ Receptors

An extensive plexus of suburothelial sensory nerve fibres has been described in the bladder of several mammalian species (HÖKFELT et al. 1975; ALM et al.

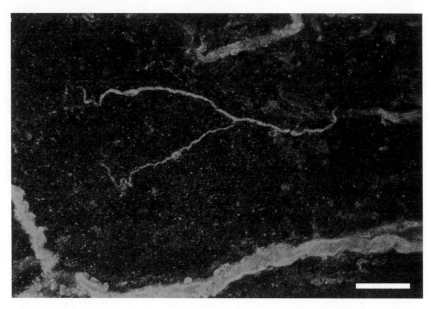

Fig. 14. P2X$_3$ receptor immunoreactivity in the mouse bladder. Immunostaining is seen on small suburoepithelial nerve fibres and on larger nerve bundles. *Calibration bar* = 50 μm. Reproduced from COCKAYNE et al. (2000)

1978; ELBADAWI 1982; FAHRENKRUG and HANNIBAL 1998; GABELLA and DAVIS 1998), including humans (GU et al. 1982; GOSLING et al. 1983; ISLAM et al. 1983; DIXON and GILPIN 1987). These authors described a suburoepithelial plexus of sensory nerves without neurolemminal covering immediately beneath the basal lamina or, on occasion, penetrating the basal lamina to end in close association with the membranes of urothelial cells (COCKAYNE et al. 2000; VLASKOVSKA et al. 2000) (Fig. 14). P2X$_3$ receptors have been localised immunohistochemically on nerves in the bladder and in the subepithelial afferent plexus of the ureter of the rat (LEE et al. 2000).

II. Afferent Nerve Activity During Bladder Distension and Evidence of Involvement of Purines

Lumbosacral afferent neurones which monitor the volume of the bladder or the amplitude of bladder contractions have myelinated (Aδ) and unmyelinated (C) axons (JÄNIG and MORRISON 1986; HÄBLER et al. 1990, 1993; DE GROAT et al. 1999; MORRISON 1999). Sensing bladder volume is of particular relevance during urine storage, whereas afferent discharges that occur during a bladder contraction have an important reflex function and appear to reinforce the central drive that maintains voiding. Aδ- and C-fibre mechanosensitive afferent axons which respond to both bladder distension and contraction, i.e.

tension receptors, have been identified in the pelvic and hypogastric nerves of cats and rats (MORRISON 1999). Afferents that respond only to bladder filling have been identified in the rat. These afferents have C-fibre axons and appear to function as volume receptors that are sensitive to stretching of the mucosa. Afferents that are only activated by very high bladder pressures or that are unresponsive to mechanical stimulation (i.e. silent C-fibres), but responsive to chemical noxious stimulation, have also been identified. In the cat, the majority of bladder C-fibre afferents are mechanoinsensitive and chemosensitive (HÄBLER et al. 1990). In the rat, single unit recordings from S_1 dorsal root afferent fibres innervating the bladder revealed that 80% of the units were low threshold mechanoreceptors and a smaller proportion (20%) were high threshold receptors. About half of the high threshold units were Aδ-fibres and the other half C-fibres (SENGUPTA and GEBHART 1994). Most of the fibres responding to noxious urinary bladder distension were C-fibres, but some Aδ-fibres also responded. In a recent study in the Burnstock laboratory, distension of the mouse bladder evoked activity in sensory fibres in the pelvic nerve, concomitant with urothelial release of ATP; ATP and α,β-meATP evoked comparable activity in the low threshold afferent nerve fibres (COCKAYNE et al. 2000; VLASKOVSKA et al. 2000).

Activity of mechanosensitive as well as chemosensitive bladder afferents may be modulated by neurotransmitters such as neurokinins, nitric oxide (NO), ATP, prostaglandins, as well as by capsaicin or by the chemical environment in the bladder, e.g. the pH, osmolality and K^+ concentration of the urine (see review, DE GROAT et al. 1999). The mechanisms underlying afferent modulation may involve the release of endogenous neurokinins (neurokinin A or substance P) from afferent nerves which then sensitise afferents by acting on autoreceptor NK_2 binding sites on the sensory nerve endings (NIMMO et al. 1991; MORRISON 1999). In addition, release of ATP or NO from uroepithelium or afferent nerves may modulate the sensitivity of mucosal mechanoreceptors or chemoreceptors.

The contribution of neurokinin-containing C-fibre afferents to bladder function is well established (MAGGI 1993). Long-term treatment of rat urinary bladder with capsaicin, a specific C-fibre afferent neurotoxin, destroys substance P-containing nerves and markedly alters the sensory threshold of micturition, causing urine retention (SHARKEY et al. 1983; SANTICIOLI et al. 1985), although there are marked species-related differences in the functions mediated by capsaicin-sensitive nerves (MAGGI et al. 1987a). Evidence has been presented that NK_2 receptors located on sensory nerves in the bladder wall participate in the tachykinin-induced activation of reflex micturition (MAGGI et al. 1987b). This reflex was largely impaired in rats chronically pretreated with capsaicin implicating the involvement of capsaicin-sensitive afferent nerves.

The contractile response of the bladder was examined after sensory denervation by capsaicin treatment of newborn rats (ZIGANSHIN et al. 1995b). It was concluded that sensory nerves have a trophic influence on the development of parasympathetic nerves; early removal of sensory nerves resulted in an increase in the cholinergic and to a lesser extent the purinergic component.

Many years ago BARRINGTON (1931) reported that distension of the bladder or active contractions of its wall induced a reflex relaxation of the urethra and this has been confirmed in many studies since that time (see LINCOLN and BURNSTOCK 1993; HOYLE et al. 1994).

III. Evidence for ATP Release from Urothelial Cells During Distension

Recent studies from several laboratories (FERGUSON et al. 1997; BIRDER et al. 1998a,b, 1999a,b; FERGUSON 1999; KNIGHT et al. 1999; BUFFINGTON et al. 2000) indicate that urothelial cells have neuronal-like properties, including the expression of receptors for capsaicin, NA, ACh, substance P and calcitonin gene-related peptide, and the ability to release transmitters such as NO and ATP. In some respects urothelial cells are similar to endothelial cells which line the luminal surface of blood vessels. ATP has been shown to be rapidly released from freshly dissociated or cultured endothelial cells exposed to sheer stress (see BODIN et al. 1991, 1992; BODIN and BURNSTOCK 1995) and from different vascular beds in response to increased perfusion flow rate (MILNER et al. 1990a,b; BODIN et al. 1991, 1992; BODIN and BURNSTOCK 1995). Flow-induced release of uridine nucleotides as well as purine nucleotides has been demonstrated from cultured endothelial cells of rabbit aorta (SAÏAG et al. 1995). The ATP release is involved in both short- and long-term signalling in the vessel wall: it acts on P2Y endothelial receptors leading to release of NO and vasodilatation, and it is involved in the proliferation and growth of cells involved in restenosis following angioplasty and in atherosclerosis and hypertension (see ABBRACCHIO and BURNSTOCK 1998). ATP has also been shown to be stored at high levels in marginal cells of the stria vascularis in the inner ear and released by mechanical stress leading to mechanosensory transduction in this system (WHITE et al. 1995); ATP release has also been demonstrated from the organ of Corti (WANGEMANN 1996). ATP released from cells by mechanical distension may also be involved in embryonic development (see BURNSTOCK 1996b) in bone remodelling (release from osteoblasts) (BOWLER et al. 1998) and cystic fibrosis (WATT et al. 1998).

In the rabbit bladder there is evidence for release of ATP from uroepithelial cells by hydrostatic pressure changes (FERGUSON et al. 1997) and it was later shown that the ATP is released from the inner basolateral surface of urothelium close to the sensory nerve plexus, rather than from the luminal surface (FERGUSON 1999). After removal of the urothelium there was very little release of ATP. On the other hand, in the rat, distension of the bladder in vivo released ATP from the luminal surface of the urothelium (HAWRANKO et al. 1999). Similarly, distension of the guinea-pig ureter leads to substantial release of ATP from urothelial cells into the lumen, without damaging these cells (KNIGHT et al. 1999). This was taken a step further when multiunit recordings were taken from mechanosensitive pelvic nerve afferents in response to bladder distension. It was shown that suramin, a P2 antagonist, or prolonged exposure to α,β-meATP that would desensitise the $P2X_3$ receptor, both produced a significant reduction in afferent nerve activity (NAMASIVAYAM et al. 1999). Recent experiments from the Burnstock laboratory demonstrated a significant reduction in afferent nerve activity in the distended bladder of $P2X_3$ knockout mice (COCKAYNE et al.; VLASKOVSKA et al. 2000). These results indicate that purinergic receptors are involved in mechanosensory signalling in the bladder.

Discovering precisely how ATP is released from a variety of cell types, including endothelial and urothelial cells, odontoblasts and osteoblasts, in response to mechanical stimuli is an exciting challenge. It appears likely that this release involves a special ATP transport mechanism as distinct from exocytotic release from nerves. There is considerable current interest in the possibility that mechanically stimulated ATP transport involves ATP-binding cassette (ABC) proteins, sulphonylurea receptors (SUR) and/or cystic fibrosis transmembrane conductance regulator (CFTR) channel proteins (GUIDOTTI 1996; CANTIELLO et al. 1998; HAMADA et al. 1998; SPRAGUE et al. 1998; BEIGI et al. 1999; ROMAN and FITZ 1999). It is interesting in this respect that glibenclamide, a drug that blocks ATP-sensitive K^+ channels which contain the SUR subunit, was reported to block flow-induced release of ATP from endothelial cells of the rat pulmonary vascular bed (HASSÉSSIAN et al. 1993). Clearly, discovery of ATP transporters and of agents that can enhance or inhibit release of ATP to mechanical stimulation would have significant therapeutic potential.

IV. Evidence for Purinergic Involvement in Bladder Nociception

Electrophysiological recordings from single primary afferent fibres in the sacral dorsal roots supplying the cat bladder showed that, with increase in intravesical pressure, mainly thin myelinated, low threshold mechanoreceptors were excited; some unmyelinated visceral afferents responded to a mechanical stimulus, but all had high, perhaps noxious thresholds (HÄBLER et al. 1990, 1993). Of the many unmyelinated afferents without appreciable mechanosensitivity, an entirely new subpopulation was activated by chemical irritants during acute inflammation (induced by intravesicle injections of mustard or turpentine oil). It was suggested that this novel population of sensory neurones might contribute to the pathogenesis of visceral pain states and reflex disturbances of bladder motility.

A study of the rabbit bladder, using local application of capsaicin, led the authors to propose that small diameter sensory neurones in the bladder wall may have a role in the transmission of the sensation of pain and in the triggering of inflammatory reactions rather than forming the afferent limb of the micturition reflex (HARRISON et al. 1990a).

Unmyelinated (C-fibre) sensory afferents from the urinary bladder in the hypogastric and pelvic nerves encode noxious events in different ways:

hypogastric fibres are low threshold mechanosensory units that signal intravesicle pressure from innocuous to noxious on an intensity-coding basis (BAHNS et al. 1986), while C-fibre afferents in the pelvic nerve are high threshold units that exclusively signal strong intravesicular pressure in the range usually associated with bladder pain (HÄBLER et al. 1990). Bladder Aδ afferents, whether coursing in hypogastric or pelvic nerves, are intensity-encoding units (BAHNS et al. 1987). In the rat bladder, the ratio of low-threshold to high-threshold pelvic nerve afferents was 4:1 (SENGUPTA and GEBHART 1995). Activation of the proto-oncogene, c-*fos*, in the spinal cord has been used as an indirect method to study afferent activity following noxious and nonnoxious stimulation of the rat urinary bladder (BIRDER and DE GROAT 1993; CRUZ et al. 1994). *Fos*-positive spinal neurones were detected in the intermediolateral grey matter and dorsal commissure following both mechanical and noxious stimulation of the bladder. Pretreatment with capsaicin which desensitises C-fibre bladder afferents markedly reduced the *Fos* expression evoked by noxious stimulation (BIRDER and DE GROAT 1998). However, it should be noted that receptors for capsaicin (i.e. VR_1) are present not only on small diameter afferent nerves but also on urothelial cells in rat, cat and human (BIRDER et al. 1998a,b, 1999a,b; BUFFINGTON et al. 2000). Thus, capsaicin could alter sensory mechanisms in the bladder by direct actions on afferent nerves or the urothelium.

There were early hints about the actions of ATP on nociceptive sensory nerves (BLEEHEN and KEELE 1977; BLEEHEN 1978; COUTTS et al. 1981; BURNSTOCK 1981; KRISHTAL et al. 1988; BOUVIER et al. 1991). However, more recently, cloning and characterisation of an extracellular receptor for ATP has provided direct evidence for $P2X_3$ receptor homomultimers and $P2X_{2/3}$ receptor heteromultimers on nociceptive sensory neurones (CHEN et al. 1995; LEWIS et al. 1995; BURNSTOCK and WOOD 1996). These receptors have been localised on subpopulations of nerve cell bodies in DRG, trigeminal and nodose ganglia and on their central and peripheral extensions with in situ hybridisation and immunohistochemical methods (VULCHANOVA et al. 1996, 1997; BRADBURY et al. 1998; LLEWELLYN-SMITH and BURNSTOCK 1998). Evidence in support of this hypothesis is beginning to appear from application of ATP and purinoceptor antagonists to in vivo pain models (e.g. BLAND-WARD and HUMPHREY 1997; TREZISE and HUMPHREY 1997; DOWD et al. 1998; see BURNSTOCK 2000; BURNSTOCK et al. 2000).

The possible sources of the ATP acting on $P2X_{2/3}$ receptors was discussed by BURNSTOCK (1996c) in relation to pain associated with causalgia, reflex sympathetic dystrophy, cancer and vascular pain such as migraine, angina and ischaemia, where it was suggested that endothelial cells in the microcirculation might provide the source of the ATP acting on perivascular sensory nerve terminals.

It is now proposed that the pain caused by distension of the bladder works through a purinergic mechanosensory transduction mechanism (BURNSTOCK 1999b). There is already supportive evidence for this concept in the bladder (see FERGUSON et al. 1997; MORRISON et al. 1998; NAMASIVAYAM et al. 1999) and in the ureter (HOYES and BARBER 1976; SIKRI et al. 1981; AMANN et al. 1988; CERVERO and SANN 1989).

Nociception exerts a naloxone-resistant suppression of the volume-evoked micturition reflex which involves inhibition of release of transmitters from motor nerves in the rat bladder, and perhaps also afferent fibres; nociception did not affect the contraction of the bladder to ACh and ATP (GIULIANI et al. 1998).

H. Perinatal Development of Purinergic Signalling in Urinary Bladder

ATP and ACh are cotransmitters in parasympathetic nerves supplying the adult bladder (see above). In an early study of the responses of the rabbit urinary bladder to autonomic neurotransmitters during development, receptors to ATP and ACh were recognised in the newborn animals, while adrenoceptors were poorly expressed at this stage (LEVIN et al. 1981b). In a later study, newborn rabbit bladders were shown to generate much greater tension in response to ATP than in adult tissue and then decline, while the response to cholinergic agonists did not decline (KEATING et al. 1990; ZDERIC et al. 1990; SNEDDON and McLEES 1992) (Fig. 15). CROWE and BURNSTOCK (1985) carried out a histochemical study using markers for cholinergic, adrenergic and purinergic transmission during perinatal development of rabbit bladder. Acetylcholinesterase-positive nerve fibres and ganglion cells and quinacrine-positive ganglion cells were both present on day 23 of gestation, while quinacrine-positive varicose nerve fibre were first seen on day 24. At fetal day 26, large numbers of ganglia (25–38), each containing 30–40 quinacrine-positive neurones, were seen in the detrusor wall. In contrast, only 5–12 ganglia contained 3–12 acetylcholinesterase-positive nerve cell bodies at the same fetal age. No catecholamine-containing nerve cell bodies were seen at any fetal age or in the adult and catecholamine-containing nerve fibres were not detected until 28 days of gestation. In adult bladder there was a reduction of 25–30% in the number of quinacrine-positive cell bodies within the ganglia when compared with 1-day-old bladders, although there was an increase of about 50% in nerve fibres.

The postnatal development of purinoceptors in rat urinary bladder has also been examined. Neurogenic contractions of bladders from newborn rats were atropine-sensitive in the whole range of frequencies studied. During the first 2 weeks, the atropine-resistant component of these contractions increased progressively to reach adult-like conditions, i.e. atropine-resistant contractions consisted of over 90% of contractions at 0.1 Hz and about 60% at 1–20 Hz (MAGGI et al. 1984). Responses to adenosine (inhibitory) and ATP (excitatory) mediated by P1 and P2X receptors, respectively, were present as early as postnatal day 2, the earliest day studied (NICHOLLS et al. 1990). Adenosine was more

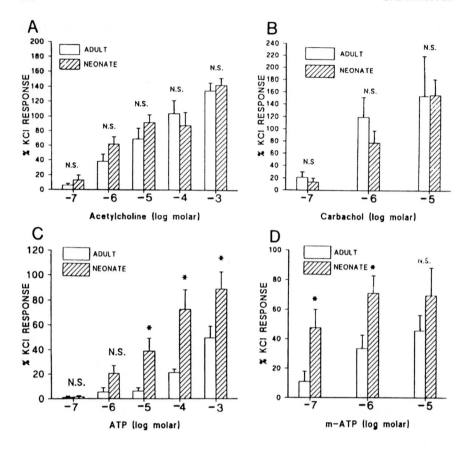

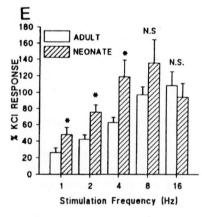

Fig. 15. Contractile responses of isolated strips of urinary bladder of adult and neonate (2- to 6-day-old) rabbits to: **A** acetylcholine, **B** carbachol, **C** ATP, **D** α,β-methylene ATP (m-ATP), and **E** nerve stimulation. Note that in **A** and **B** there was no difference between the response in the adult and the neonate to acetylcholine or carbachol, but **C** and **D** show that both ATP and α,β-methylene ATP were significantly more potent in the neonatal than the adult tissue, and **E** shows that responses to nerve stimulation were also greater in neonatal tissue. *Bars* represent the mean response ± S.E.M. for 5–7 experiments. N.S. indicates no significant difference, and * indicates $p < 0.05$ (Student's t-test for unpaired data). Reproduced from SNEDDON and MCLEES (1992)

potent in the neonate than in the adult, while the potency of ATP initially increased with age, but then declined, being highest between postnatal days 10 and 25. In vivo evidence for the functional roles of cholinergic and purinergic components of parasympathetic cotransmission for micturition contractions in normal unanaesthetised rats has been presented (IGAWA et al. 1993).

The rate and pattern of breakdown of ATP and adenosine by ectoenzymes in the rat urinary bladder was shown to be identical in neonates and adults, indicating that the marked differences in potency to ATP and adenosine during development is likely to be due to changes in receptor number and/or agonist affinity or efficacy (NICHOLLS et al. 1992b).

Using various markers for sensory and motor nerves, it was concluded that both nerve types were present at birth and that sensory (calcitonin gene-related peptide-positive and substance P-positive) nerve fibres approached adult levels at the end of the second week, shortly before the micturition reflex was fully developed (SANN et al. 1997).

In a recent review of postnatal development in several rat visceral smooth muscle preparations, it was concluded that in most organs, in contrast to vas deferens, purinergic mechanisms were more important in the neonate than in the adult (HOURANI 1999).

Few reports were found describing changes in purinergic signalling in the ageing bladder, although a comparison of contractions in detrusor muscle strips from unobstructed bladders of young and aged rats showed that, with age, there is an *increased* sensitivity to ATP as well as NA, but with no change in response to ACh and KCl (FERGUSON and CHRISTOPHER 1996). A reduction in acetylcholinesterase-positive nerve fibres in the human bladder with increasing age has been reported (GILPIN et al. 1983) and a decreased fluorescence intensity for catecholamines in neurones in the hypogastric ganglion which supplies sympathetic fibres to the bladder has also been shown (PARTANEN et al. 1980).

I. Urethra

In many species, including rabbit, cat, pig and humans, there is a non-adrenergic, non-cholinergic inhibitory transmission to the urethra (McGUIRE and HERLIHY 1978; ANDERSSON et al. 1983; HILLS et al. 1984; ITO and KIMOTO 1985; KLARSKOV 1987a; SJÖGREN et al. 1988; MATTIASSON et al. 1990). Amongst compounds which cause relaxation, the putative transmitter was shown not to be VIP, ATP, 5-HT or adenosine because blockade of these responses by pharmacological manipulation did not produce a parallel effect on the neurogenic response (HILLS et al. 1984; KLARSKOV 1987b; HASHIMOTO et al. 1992; WERKSTRÖM et al. 1997). However, the principal NANC inhibitory transmitter is now clearly established, namely NO (ANDERSSON et al. 1991; DOKITA et al. 1991; GARCIA-PASCUAL et al. 1991; THORNBURY et al. 1992; HASHIMOTO et al. 1993; LEE et al. 1994; TAKEDA and LEPOR 1995; PINNA et al. 1996).

When the tone of the urethra is raised, ATP causes relaxation, but if the tone of the urethra is low, high concentrations of ATP can cause contraction (Persson 1976; Creed and Tulloch 1978; Callahan and Creed 1981). Bursts of spikes in the urethra were initiated by NA or ACh, but inhibited by ATP (Callahan and Creed 1981), perhaps after breaking down to adenosine. In a study of adrenergic, cholinergic and NANC nerve-mediated *contractions* of the female rabbit bladder neck and proximal, medial and distal urethra, the residual contractile responses after combined treatment with prazosin and atropine were insensitive to P2X receptor desensitisation with α,β-meATP (Deplanne et al. 1998).

Sympathetic (hypogastric) nerve stimulation produced a contraction of the urethra, which was significantly reduced by quinidine (Creed 1979), suggestive of sympathetic purinergic cotransmission. The responses to other sympathetic efferent pathways projecting to the urethra (Kihara and de Groat 1997) have not been examined.

While NO was recognised as the principal transmitter involved in NANC relaxation of the hamster proximal urethra, experiments suggested that another inhibitory transmitter might be involved (Pinna et al. 1996). In the precontracted proximal urethra, NANC nerve stimulation and exogenous ATP were also shown to produce relaxations, which were attenuated by suramin and Reactive blue 2, and to a lesser extent by 8-PT, but not by PPADS. ATP-induced relaxations were also reduced by indomethacin and were urothelium- and NO-independent since they were not affected by removal of the urothelium or by the NOS inhibitor, L-NAME (Pinna et al. 1998a). Thus, P2Y as well as P1 receptors appear to mediate the relaxing effect of ATP released from a NANC nerve pathway which has a subordinate role to the major nitrergic pathway. In a study of the roles of purines in neurally mediated urethral relaxation in male rabbits, NANC relaxations were shown to be reduced by suramin as well as by the NOS inhibitor L-NOARG, and in superfusion experiments electrical field stimulation markedly increased the outflow of ATP into the superfusate. It was suggested that P2Y receptors exist in male rabbit urethra and that ATP and related compounds may play a role in NANC transmission (Ohnishi et al. 1997).

In a recent microelectrode study, transmural stimulation of longitudinal smooth muscle strips from guinea-pig urethra evoked EJPs and triggered slow waves that were abolished by α,β-meATP as well as by tetrodotoxin (Hashitani and Edwards 1999). The authors concluded that stimulation of purinoceptors by neurally released ATP initiates EJPs in the guinea-pig urethra and also causes the release of Ca^{2+} from intracellular stores to evoke slow waves.

ATP may also be released from sensory nerve fibres supplying the urethra during axon reflex activity (see Burnstock 1993).

J. Ureter

The innervation of the ureter is sparse, perhaps because peristaltic activity is myogenic rather than neurogenic as in the gut. The dominant nerve components are sensory nerves, largely confined to a suburothelial plexus (HOYES 1984).

Application of capsaicin or electrical field stimulation of the rat ureter produced a transient inhibition of neurokinin-activated unilateral motility in control preparations, but not after destruction of the C-fibres by capsaicin pretreatment. It was concluded that capsaicin-sensitive neurokinin-containing inhibitory innervation exists in the rat ureter (MAGGI et al. 1986), perhaps involved in reflex or axon-reflex activity. These afferents might play a role during vesicoureteral reflux which could distend the ureter and activate reflexes that modulate urine delivery to the bladder (THEOBALD 1986b). When detrusor pressure is increased in the cat, the ureteral peristaltic frequency decreases. Pelvic nerve stimulation produced a modest, transient decrease in ureteral peristaltic frequency, perhaps associated with the bladder contractions produced in response to the stimulus, while hypogastric nerve stimulation produced different responses depending on the detrusor pressure. ATP, β,γ-meATP and adenosine produced transient, but definite decreases in ureteral peristaltic frequency; these purines also decreased the spontaneous firing of the renal nerve. Theophylline failed to block the effects of ATP, suggesting that P2 as well as P1 receptors are involved. It was later proposed that calcitonin gene-related peptide was an inhibitory transmitter which, when released from capsaicin-sensitive sensory nerve fibres, also participated in the control of ureteral motility (HUA 1986; MAGGI et al. 1987c).

A recent paper reports that adenosine relaxes the pig intravesicle ureter, via A_{2B} receptors on the smooth muscle and may modulate the ureteral NANC excitatory transmission through a postjunctional mechanism (HERNÁNDEZ et al. 1999).

The electrical activity in mechanosensitive C-fibre afferent units was recorded in small ureteric branches of the hypogastric nerve during various mechanical stimuli, including probing of the ureter with small glass probes, insertion of an intraluminal glass bead to mimic kidney stones and distension of the ureter using hydrostatic pressure (SANN and CERVERO 1988; CERVERO and SANN 1989). It was suggested that some of these afferent fibres might be involved in the signalling of nociceptive events. In the later paper from this group, they distinguished two classes of mechanosensitive afferent fibres in the guinea-pig ureter: U1 units monitoring normal peristalsis and U2 units perhaps involved in the signalling of noxious events. A third class of mechanosensitive units was identified in the chicken ureter (HAMMER et al. 1993). Units responding to peristaltic movements of the ureter have also been reported by another group, who suggested that one of the functions of ureteric afferents might be the monitoring of peristaltic rhythms (JÄNIG and MORRISON 1986).

Evidence has been presented that spinothalamic tract neurones mediate nociceptive responses to ureteral occlusion (AMMONS 1989). Recordings from dorsal horn neurones in the spinal cord (T_{12}-L_1) in anaesthetised rats led to the conclusion that they receive both noxious and innocuous ureter stimulation mainly from high-threshold afferents and their response properties correlate well with ureteric pain sensations in humans (LAIRD et al. 1996). In a later paper from this group, spinal neurone recording after implantation of an experimental ureteric stone led to the conclusion that the presence of ureteric stone evokes excitability changes in spinal neurones (enhanced background activity, a greater number of ureter-driven cells, decreased threshold of convergent somatic receptor fields) which likely account for the referred hyperalgesia seen in rats with calculosis (ROZA et al. 1998). The possibility that ATP released from urothelial cells during distension acting on $P2X_3$ receptors on suburothelial sensory nerves is involved in the mechanosensory transduction mechanism involved in ureteric pain is discussed in Sect. G.I.

K. Plasticity of Purinergic Signalling in Bladder, Urethra and Ureter

I. Changes Occurring During Pregnancy or Hormone Therapies

Incontinence is a common problem in adult women (TURAN et al. 1996). The first symptoms of urinary incontinence can arise after the first pregnancy and the risk of incontinence increases with multiple deliveries (RYHAMMER et al. 1995). However, the sensitivity of the rat detrusor muscle to ATP was not modified by multiple pregnancies, while there was increased sensitivity to adrenergic and cholinergic stimulation (GRANDADAM et al. 1999). However, other studies reported that the responses to adrenergic and cholinergic stimulation were reduced (LEVIN et al. 1991a; TONG et al. 1992), and the responses to ATP increased during pregnancy in both rat and rabbit bladders (TONG et al. 1995) (Fig. 16A). These latter authors concluded that suppressed bladder contractility during pregnancy, due to a reduction in cholinergic and less importantly α-adrenergic function, is associated with decreased muscarinic receptor density, while the affinity of purinergic receptors for ATP is increased.

An old concept is that incontinence during pregnancy is related to hormonal factors (STANTON et al. 1980; IOSIF et al. 1981). Oestrogen has been used for the treatment of urinary stress incontinence in women (SALMON et al. 1941; BROWN 1977). Oestrogen is known to have a profound influence on the function of smooth muscle (BATRA 1980) and receptors for oestrogen have been identified in both rabbit myometrium (BATRA 1979) and human female lower urinary tract (IOSIF et al. 1981). When ovariectomized rabbits were injected i.v. with [^{3}H]-oestradiol, high affinity binding sites were clearly demonstrated in the female urethra and urinary bladder (IOSIF et al. 1981). The

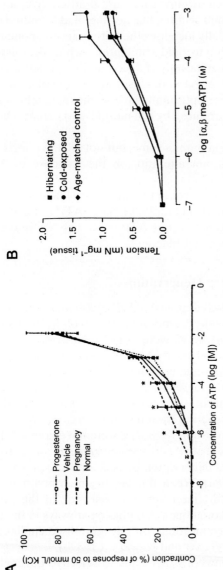

Fig. 16. A ATP concentration-response curves of urinary bladder muscle strips from normal female rats, 2-week pregnant rats, rats treated daily with progesterone 5 mg/kg i.m. or rats treated with vehicle for 2 weeks ($n = 8$ for each group). The contractile response of the muscle strips was expressed as a percentage of the response to 50 mmol/l KCl. The contractile responses of the pregnant group at 10^{-6} to 10^{-4} mol/l ATP were significantly higher than those of the other three groups (*$p < 0.05$). **B** Non-cumulative concentration-response curves to α,β-methylene ATP (α,β-meATP) in longitudinal smooth muscle strips of urinary bladder from age-matched control hamsters, cold-exposed hamsters and hibernating hamsters. In cold-exposed and hibernating hamsters α,β-meATP elicited decreased contractions (ANOVA, $p < 0.05$; $n = 6$). Reproduced from **A** Tong et al. (1995) and **B** Pinna et al. (1998b)

amplitude of NANC transmission in detrusor strips from mature female rats was increased in oestrogen-treated, but not ovariectomized animals (Eika et al. 1988).

Chronic treatment with oestrogen induced a marked increase in the responses to purinergic (as well as muscarinic and α-adrenergic) agonists in the rabbit bladder body and mid-section, but not the bladder base (Levin et al. 1980b). Pregnancy substantially increases the purinergic components of the response of the rabbit bladder to field stimulation, while the response of bladder to bethanechol was significantly reduced and was associated with a 50% decrease in muscarinic receptor desensitisation (Levin et al. 1991b). Oestradiol and the oestrogen receptor antagonist, tamoxifen, inhibit contractions of rabbit detrusor strips produced by α,β-meATP and bethanechol (Ratz et al. 1999).

Progesterone administration mimics some, but not all, the effects of pregnancy; for example, no significant alteration in the response to ATP was observed (Tong et al. 1995).

Degeneration of adrenergic nerves in the rat urinary bladder during pregnancy has been described (Qayyum et al. 1989). Since ATP is a cotransmitter in sympathetic nerves, it is likely that less ATP as well as NA is available in pregnant compared to non-pregnant bladders.

II. Changes Due to Selective Denervation

When the rat urinary bladder was deprived of half of its innervation by removing the pelvic ganglion on one side, the motor responses of the bladder to stimulation of the remaining pelvic nerve were larger than those of the control bladder at 1 week, 1 month and especially 2 months postoperatively (Ekström and Elmér 1980). Further experiments with atropine and eserine led to the conclusion that the increased responses 1 week postoperatively are mainly due to sensitisation of the muscarinic receptors, while those observed at later stages are due to collateral sprouting from the cholinergic nerve fibres in the intact pelvic nerve. Atropine-resistant responses were not examined. In a later study, development of supersensitivity to methacholine in rat detrusor following either parasympathetic denervation or decentralisation was reported (Ekström and Malmberg 1984). When the sacral parasympathetic preganglionic pathways were surgically interrupted on one side of the cat urinary bladder, it was claimed that cholinergic sympathetic pathways in the hypogastric nerve make sympathetic connections with decentralised cholinergic ganglion cells in the bladder (de Groat and Kawatani 1989b).

Capsaicin treatment of newborn rats leads to selective degeneration of sensory nerve fibres (see Holzer 1991). In a study of rat bladder in 3-month-old rats treated at birth with capsaicin, contractions evoked by electrical field stimulation were significantly larger than those of control (vehicle-treated) animals, an effect which preferentially involves the cholinergic component of the response, although there was some increase, too, in the purinergic component (ZIGANSHIN et al. 1995b). However, since contractions in response to exogenous carbachol or ATP were not significantly different, this suggested that the changes involve prejunctional mechanisms, probably a trophic increase in parasympathetic innervation. Capsaicin treatment, causing selective sensory denervation of the rat ureter, leads to increased sympathetic innervation (SANN et al. 1995), presumably leading to increase in release of both NA and ATP.

After bilateral sympathectomy by cutting the hypogastric nerves distal to the hypogastric ganglia, the adrenergic nerve supply to the bladder did not differ from normal bladder either at 10 days or 6–9 weeks after denervation; in contrast, 10 days after total unilateral postganglionic denervation by removal of the left pelvic ganglion, few adrenergic nerve fibres were seen in the left half of the bladder (ALM and ELMÉR 1979). However, 6–9 weeks after pelvic ganglion removal, the adrenergic innervation had reappeared, although the origin of the regenerating fibres was not resolved. Studies on the vascular system show that P1 receptor agonists prevent the trophic changes caused by sympathetic denervation, which was taken to be consistent with an involvement of purines in the trophic effects of sympathetic innervation (ALBINO-TEIXEIRA et al. 1990). In spontaneously hypertensive rats, where there is increased sympathetic nerve activity, there is hyperactive bladder voiding that appears to be associated with higher secretion of nerve growth factor by bladder smooth muscle and hyperinnervation (CLEMOW et al. 1998; SPITSBERGEN et al. 1998).

Over-distension of the bladder is caused by urinary retention, but it has also been used as a method for treating unstable bladder or interstitial cystitis (DUNN et al. 1974; JÖRGENSEN et al. 1985). However, micturition problems are often encountered after long-term over-distension (TAMMELA et al. 1986); for example, distension of the rat urinary bladder for 3h led to depletion of catecholamines which was complete after 2 days, although partially recovered after 5–7 days (TAMMELA et al. 1990). The urinary bladder of the rat, deprived of its motor innervation, increases several-fold in weight in response to distension (EKSTRÖM and UVELIUS 1981). This increase in weight is due to both hyperplasia and hypertrophy of the smooth muscle (EKSTRÖM et al. 1984). Since it is now known that distension of the bladder leads to substantial release of ATP from urothelial cells (see Sect. G.II.) and ATP is known to have trophic effects (ABBRACCHIO and BURNSTOCK 1998), it is possible that purines participate in the trophic changes that occur in the bladder.

Damage to the spinal cord rostral to the lumbosacral level can induce marked changes in the neural control of the lower urinary tract; following spinal cord injury that interrupts the normal supraspinal pathway regulating micturition, the urinary bladder is initially areflexic, but over the course of several weeks becomes hyperreflexic and hypertrophic (DE GROAT et al. 1990; KRUSE et al. 1993). Little is known about the mechanisms underlying these changes (DE GROAT et al. 1990), although it has been shown recently that chronic spinal injury enhances the electrical excitability of bladder afferent neurones by increasing the expression of low-threshold tetrodotoxin-sensitive Na⁺ channels (YOSHIMURA and DE GROAT 1997).

III. Bladder Grafts

The gastrointestinal tract has been the chief source of material for bladder augmentation and substitution despite complications such as malignancy, electrolyte abnormalities, infection, obstruction, the inherent need for catheterisation, mucus production and perforation (see MITCHELL et al. 1987). A small intestine submucosal (SIS) preparation has been developed more recently and used as a bladder patch in rats that produced both smooth muscle and urothelial cell regeneration (KROPP et al. 1995; KROPP 1998). A further study indicated that small intestine submucosal-regenerated bladder exhibits contractile activity, expresses muscarinic, purinergic and β-adrenergic receptors and exhibits functional cholinergic and purinergic transmission (VAUGHT et al. 1996).

Another approach has been to use autologous cultured urothelium for bladder reconstruction (ATALA et al. 1992; HUTTON et al. 1993). Collagen-based and biodegradable materials have also been shown to have regenerative and functional capacities and a bladder acellular matrix graft (BAMG) has been claimed to be successful for augmentation cytoplasty in the rat model leading to structural and functional regeneration of detrusor smooth muscle (PROBST et al. 1997) including contractile activity to electrical field stimulation showing responses to muscarinic, purinergic and adrenergic agonists (PIECHOTA et al. 1998).

IV. Hibernation

Purinergic and cholinergic components of parasympathetic neurotransmission were investigated in hibernating hamsters (PINNA et al. 1998b). Perhaps surprisingly, 4 weeks of hibernation significantly increased both cholinergic and purinergic neurogenic responses of the hamster urinary bladder. This appears to be due to an increase in postjunctional responses to ACh, while there was a decrease in the postjunctional response to ATP (Fig. 16B).

L. Purinergic Signalling in the Human Bladder in Health and Disease

I. Healthy Bladder

Although the presence of an atropine-resistant component of parasympathetic nerve-mediated contraction in experimental animals was well established many years ago and later shown to be purinergic (see above), the presence of an atropine-resistant nerve component in the human bladder has been controversial, although most authors did find a small component, usually less than

5% in healthy bladder (HINDMARSH et al. 1977; ANDERSSON and SJÖGREN 1982; SJÖGREN et al. 1982; NERGÅRDH and KINN 1983; MAGGI et al. 1984; SIBLEY 1984; KINDER and MUNDY 1985; SPEAKMAN et al. 1988; LUHESHI and ZAR 1990b; RUGGIERI et al. 1990; GHONIEM and SHOUKRY 1991; CREED et al. 1994; TAGLIANI et al. 1997; BAYLISS et al. 1999). One early report suggested that the NANC component in the human female bladder was greater than in the male bladder, amounting to about 50% of the nerve-mediated contractile response of the bladder (COWAN and DANIEL 1983). In a paper concerned with anticholinergic drugs, it was claimed that terodiline and propiverine significantly inhibited the atropine-resistant contractions in the human bladder (WADA et al. 1995).

Atropine-resistant responses of the human bladder were significantly reduced by ATP (possibly mediated by postjunctional desensitisation and/or prejunctional inhibition) and indomethacin and were abolished by nifedipine (HUSTED et al. 1983). One recent paper claimed that the atropine-resistant component of excitatory transmission in the human bladder was not mediated by neural release of ATP in spite of the presence of P2 receptors in the effector cells (TAGLIANI et al. 1997).

A NANC nerve-mediated *relaxation* following the initial excitation was identified in human detrusor muscle (KLARSKOV 1987b; and see Sect. B.). Transmural stimulation of muscle strips from the human trigone revealed a NANC response which represented 40% of maximal contractions at 5Hz; NANC relaxation responses were also identified in trigone (SPEAKMAN et al. 1988) and in detrusor where they might be due to NO (JAMES et al. 1993). Responses of human bladder strips to NANC nerve stimulation and ATP or P^1P^6-diadenosine hexaphosphate were blocked following desensitisation of $P2X_1$ receptors with α,β-meATP (HOYLE et al. 1989) (Fig. 17).

There is clear evidence for the presence of P2X purinoceptors in the human bladder from pharmacological studies where ATP produces contractions (HUSTED et al. 1983; HOYLE et al. 1989; INOUE and BRADING 1991). ATP, α,β-meATP, and P^1P^6-diadenosine hexaphosphate caused concentration-dependent contractions of human detrusor muscle strips (HOYLE et al. 1989). ATP elicits large inward currents (INOUE and BRADING 1991) and increases in intracellular Ca^{2+} (WU et al. 1995; BAYLISS et al. 1999) in dispersed human bladder smooth muscle cells (Fig. 18A,B). In a study of P2 receptor subtypes in human bladder strips the agonist rank order of potency was: α,β-meATP = ADPβS > 2-meSATP > ATP >> UTP. In addition, it was reported that responses to α,β-meATP and ADPβS were additive and that the P2 antagonist *p*-chloromercuribenzene sulphonic acid (pCMBS) (WIKLUND and GUSTAFSSON 1988) antagonised ADPβS-induced contractions, but was inactive against α,β-meATP, while Reactive blue 2 had no effect against ADPβS contractions (PALEA et al. 1994). The authors concluded that the human detrusor muscle contains two contractile purinoceptor subtypes: one is activated by α,β-meATP and is probably a P2X receptor; the other receptor is activated by ADPβS and appears to be different from those which are included in the current classification system (Fig. 18C). In a later paper from this group (PALEA

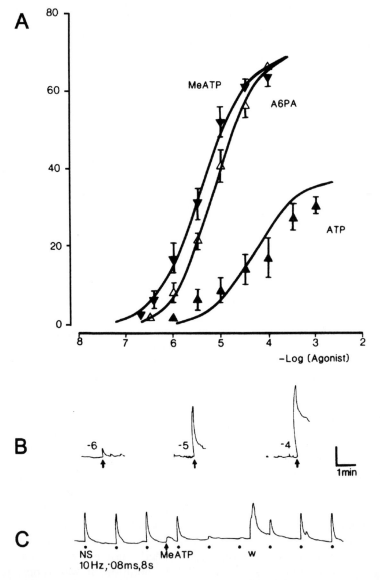

Fig. 17A–C. Responses to ATP, α,β-methylene ATP (MeATP) and P_1,P_6-diadenosine 5'-hexaphosphate (A6PA) in isolated human urinary bladder detrusor muscle. **A** Concentration–response relationships. The response curve relates contractions due to the agonists to the standard contraction to KCl (150 mmol/l). Points show mean ± S.E.M. unless occluded by symbol. Curves are fitted following probit transformation and horizontal averaging. **B** Examples of contractions evoked by A6PA, log molar concentrations applied as indicated by *arrows*. *Scale bar* represents 100 mg. **C** Electrical field stimulation of the intramural nerves (NS, ●) evoked contractions. MeATP (0.3 μmol/l) caused a small contraction which faded and blocked neurogenic contractions. Following washout of MeATP (W), the neurogenic responses returned. Record obtained in the presence of atropine (0.3 μmol/l). *Scale bar* represents 50 mg. Reproduced from HOYLE et al. (1989)

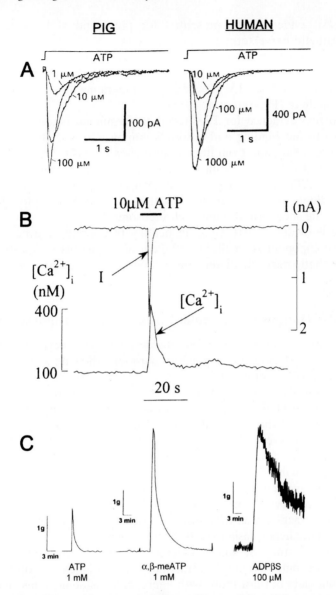

Fig. 18. A ATP activates large inward currents in pig (*left panel*) and human (*right panel*) detrusor muscle. The currents recorded at three different concentrations of ATP are superimposed and labelled. ATP was applied by a concentration-jump method and was present throughout each record. Note the rapid activation and inactivation of the inward current, and the difference in the current scale on the two panels. **B** Human detrusor muscle cell: a simultaneous measurement of membrane current (I) and intracellular Ca^{2+} concentration ($[Ca^{2+}]_i$) in response to brief exposure to $10\,\mu mol/l$ ATP. **C** Contractile effect of ATP (1 mmol/l), α,β-methylene ATP (α,β-meATP, 1 mmol/l) and ADPβS ($100\,\mu mol/l$) in three different human detrusor muscle strips. Note the increase in spontaneous motility induced by prolonged contact with ADPβS. Reproduced from **A** INOUE and BRADING (1991), **B** BAYLISS et al. (1999), and **C** PALEA et al. (1994)

et al. 1995), evidence was presented for prejunctional P2 receptors on parasympathetic nerve terminals as well as two postjunctional P2 receptor subtypes, one of which was insensitive to suramin. ATP-induced contractions were reduced about 30% by indomethacin, indicating involvement of prostaglandins (see Sect. B.V.), by 48% after nifedipine and were abolished in Ca^{2+}-free medium (HUSTED et al. 1983).

Supporting evidence for P2X receptors in human bladder comes from radioactive ligand binding and autoradiography (Bo and BURNSTOCK 1995; MICHEL et al. 1996), and from immunohistochemistry (O'REILLY et al. 2000). Additionally, a cDNA encoding an ion channel (hP2X) receptor gated to extracellular ATP was isolated from human urinary bladder (VALERA et al. 1995). By fluorescent in situ hybridisation, the hP2X receptor gene was mapped to the short arm of human chromosome 17.

There is evidence that, while in healthy bladder the purinergic nerve-mediated component is small, in pathological conditions up to 40% of the parasympathetic nerve-mediated contractions are purinergic (see Sects. L.II. and L.III.).

II. Bladder Outflow Obstruction: Unstable, Hypertrophic Bladders

In cystometrically verified unstable bladder, a varying degree of atropine-resistance was described, with some preparations showing a 50% resistance to atropine (SJÖGREN et al. 1982).

Atropine-resistant nerve-mediated contractions have been demonstrated in hypertrophied bladders, secondary to benign prostatic hyperplasia and it was suggested that the NANC component might be related to the hyperactivity observed in these bladders (SJÖGREN et al. 1982; RUGGIERI et al. 1990). Increased atropine-resistant nerve-mediated responses have also been shown in the bladders of myelodysplastic children (GHONIEM and SHOUKRY 1991). This finding was supported in a later study where, in contrast to control bladders, a NANC excitatory response amounting to about 25% of the total nerve-mediated contraction was described in strips from bladders obstructed by benign prostatic hyperplasia (SMITH and CHAPPLE 1994).

In a recent study of human detrusor muscle, it was reported that the purinergic atropine-resistant contraction was prominent in obstructed or unstable bladders but not those with neurogenic instability. This change was not caused by a differential sensitivity of the muscle to ATP or cholinergic agonists (BAYLISS et al. 1999). In a follow-up paper by the same group (Wu et al. 1999), it was confirmed that the generation of purinergic contractions in detrusor strips from unstable bladders was not due to altered sensitivity of detrusor muscle to ATP. The possibility that the increase in purinergic transmission is due to increased neural release of ATP, reduction in ecto-ATPase activity or to changes in gap junctions between muscle cells has been raised (BAYLISS et al. 1999; FRY and WU 2000). On the other hand, an earlier study (HUSTED et al. 1983) showed that preparations obtained from hypertrophic human bladders were more sensitive to ATP than macroscopically normal preparations.

In animal models, the results of outlet obstruction were also variable. For example, in a rabbit model of bladder outflow obstruction there appeared to be a reduction of both atropine-sensitive and atropine-resistant responses, suggesting nerve damage (HARRISON et al. 1990b). However, when the contribution of cholinergic and purinergic neurotransmission to micturition contractions and bladder hyperactivity was investigated by continuous cystometry in unanaesthetised rats with outlet obstruction (IGAWA et al. 1994), it was concluded that both cholinergic and purinergic transmission are important for pressure generation and emptying of the bladder.

The effects of purinergic receptor agonists were examined on hypertrophied smooth muscle of rat bladder, induced by partial ligation of the urethra giving an increase in bladder weight from 65 mg to 300 mg (SJUVE et al. 1995). The force of contraction produced by ATP and α,β-meATP was significantly *lower* than in controls, and the rate of contraction slower (Fig. 19A).

The sympathetic innervation of the bladder neck appears to be diminished in patients with bladder outlet obstruction (PARK et al. 1986), perhaps indicating in this condition a reduced role for sympathetic nerve-released ATP as well as NA.

An increase in density of subepithelial sensory nerves has been described in the bladder wall of women with idiopathic detrusor instability; the authors speculated that this may serve to increase the appreciation of bladder filling, giving rise to the frequency and urgency of micturition which are characteristic of patients with detrusor instability (MOORE et al. 1992).

III. Neurogenic Hyperreflexive Bladder

Studies have been conducted on isolated bladder strips from patients with neurogenic bladder who underwent ileocystoplasty in order to resolve intractable incontinence and/or vesicoureteric reflux due to low compliance or severe detrusor uninhibited contractions (SAITO et al. 1993a) (Fig. 19B). Atropine-resistant responses to field stimulation of neurogenic bladder strips were about 30% compared with 4% from control bladder strips. In a later paper, this group showed that neurogenic bladders are hyper responsive to ATP (SAITO et al. 1993b). In another study of muscle taken from neurogenic bladders, a NANC component of 40% was identified, which was regarded as purinergic since it was blocked by suramin (WAMMACK et al. 1995).

In paraplegic patients with suprasacral lesions, the management of urinary incontinence resulting from hyperreflexic detrusor contraction is a frequent problem. A study of changes in cholinergic and purinergic transmission was carried out in detrusor muscle strips taken from chronic spinal rabbits (spinal cord transected at thoracic level T_9–T_{10}) with detrusor hyperreflexia and detrusor sphincter dyssynergia (YOKOTA and YAMAGUCHI 1996). The results showed that the relative amplitudes of the cholinergic and purinergic contractions

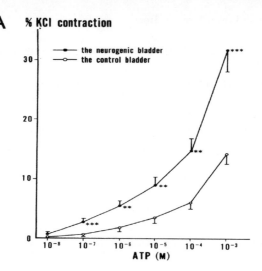
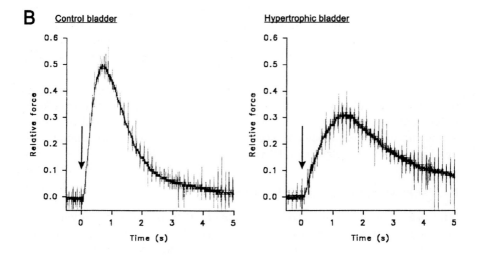
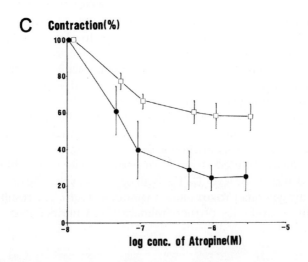

shifted from a control ratio of 40:60 to 75:25 in the pathologic detrusor, indicating a shift to cholinergic parasympathetic dominance in neurogenic bladders affected by detrusor hyperreflexia and sphincter dyssynergia after spinal chord injury (Fig. 19C). There were no differences in the dose-response curves for ACh and ATP between normal and pathologic detrusors.

The induced synthesis of prostaglandins may become important in pathological conditions. For example, in patients undergoing retropubic prostatectomy, the detrusor has a larger non-cholinergic excitatory component than in patients undergoing cystourethectomy (HUSTED et al. 1983). In the former group, indomethacin causes a significant reduction of the response to non-cholinergic nerve stimulation, whereas in the latter group, indomethacin has no such effect. In patients with chronic neurogenic vesical dysfunction, the sensitivity of the bladder to intravenous infusion of an analogue of $PGF_{2\alpha}$ is dramatically greater than in control patients (VAIDYANATHAN et al. 1982). The significance of this is unknown, but in view of the relationship of ATP with prostaglandin synthesis, it may be related to a degeneration of parasympathetic nerves resulting in supersensitivity to effectors. That is to say, a loss of purinergic transmission might have led to an increase in sensitivity of its effector mechanisms, one of which is prostaglandin activity (HOYLE and BURNSTOCK 1993).

Reflex sympathetic dystrophy is a disabling syndrome characterised by severe pain with autonomic disturbances, including urological problems (GALLOWAY et al. 1991; SCHWARTZMAN 1993; CHANCELLOR et al. 1996). Since hyperactivity of sympathetic nerves is usually implicated in reflex sympathetic dystrophy, more ATP would be released as a cotransmitter to target both $P2X_1$ receptors in smooth muscle mediating bladder contractions and $P2X_3$ receptors on the terminals of sensory nerve fibres mediating bladder reflexes and nociception.

IV. Multiple Sclerosis (MS)

MS patients often have peripheral symptoms, including bladder dysfunction (MILLER et al. 1965; BLAIVAS et al. 1979, 1984) and it has been claimed that peripheral nerve damage occurs in the MS bladder (GU et al. 1984). Mice

◄───────────────────────────

Fig. 19. A In vitro dose-response curve to ATP in human neurogenic bladder muscle strips ($n = 28$) and control bladder strips ($n = 16$). **$p < 0.01$, ***$p < 0.001$ (Wilcoxon's test). **B** Force transient responses of rat bladder muscle strips after photolytic release of ATP from caged-ATP. A control and a hypertrophic muscle preparation were held in Krebs solution containing $100 \mu mol/l$ caged-ATP. The light flash is indicated by an *arrow*. Force values are normalised to the active force during KCl stimulation. **C** Effect of atropine on electrical field stimulation of normal rabbit detrusor (□, $n = 10$) and pathologic detrusor from chronic spinal rabbit (●, $n = 12$) (stimulation parameters: supramaximal voltage, pulse duration 0.5 ms, 20 Hz for 20 s). Reproduced from **A** SJUVE et al. (1995), **B** SAITO et al. (1993a), and **C** YOKOTA and YAMAGUCHI (1996).

infected with the Semliki Forest Virus have been proposed as a model for the demyelinating disease, MS (Webb et al. 1978). This model was used to study purinergic and cholinergic neurotransmission in the mouse bladder (Moss et al. 1989). A selective change in purinergic transmission occurred in infected mice, while cholinergic transmission remained unchanged. There was a significant increase in the contractile responses to β,γ-meATP and in the purinergic (atropine-resistant) component of nerve-mediated contractions. The question was raised as to whether the increase in purinergic signalling is secondary to the bladder hypertrophy that occurs in this model or whether it is a primary event.

Bladder hyperactivity and incontinence in MS patients seems to be mediated in part by the emergence of involuntary bladder contractions induced by C-fibre bladder afferents. Desensitisation of the C-fibre afferents by intravesicle administration of afferent neurotoxins (capsaicin or resiniferatoxin) increases bladder capacity and reduces the number of incontinence episodes (Fowler et al. 1992, 1994; Chancellor and de Groat 1999). A role for purinergic mechanisms in the activation of bladder C-fibre afferents in MS patients has yet to be established.

V. Post-Irradiation Bladder Dysfunction

Most interpretations of late irradiation injury of the urinary bladder have focussed on urothelial damage and fibrosis (see Gowing 1960; Stewart 1986). However, such mechanisms would not account for the development of detrusor instability in up to one-third of patients (Parkin et al. 1988). In a study of rat detrusor strips taken 6 months after bladder X-irradiation at doses of 15 Gy and 25 Gy, there was an *increase* in sensitivity to the purinergic agonist α,β-meATP, but no changes in the sensitivity to ACh or NA (Vale et al. 1994). This led the authors to suggest that purinergic hypersensitivity in irradiated bladder, coupled with ultrastructural evidence of neural injury, leads to denervation supersensitivity that may contribute to the pathophysiology of post-irradiation bladder dysfunction.

VI. Ischaemic Bladder

Rabbit detrusor contractions elicited by nerve stimulation were more sensitive to bilateral ischaemia (3, 6 *or* 18h duration) than were contractions to carbachol and ATP, indicating ischaemic damage to nerves (Bratslavsky et al. 1999). This prejunctional change is consistent with previous studies of bladder outlet obstruction and ischaemia (Sibley 1984; Harrison et al. 1990a).

VII. Chronic Alcohol Consumption and Bladder Function

Amongst the many adverse effects of chronic alcohol consumption, autonomic neuropathies, affecting both sympathetic and parasympathetic systems, are very common (JOHNSON et al. 1986). In a study of the effects of chronic (12 weeks duration) ethanol consumption in rats on bladder activity, it was found that neurally evoked contractions and contractile responses to both carbachol and β,γ-meATP were potentiated. Cholinergic responses were more sensitive to ethanol than the purinergic responses which showed limited potentiation at higher stimulation frequency and concentrations (KNIGHT et al. 1995).

Ethanol has been shown to alter the neuronal P2X receptor so that the ATP concentration-response curve is shifted to the right, which involves an allosteric action to decrease agonist affinity (LI et al. 1998).

VIII. Vitamin E Deficiency

Vitamin E (α-tocopherol) is essential for normal neuronal physiology and its deficiency results in neuropathic changes (see MULLER and GOSS-SAMPSON 1990). The effects of vitamin E deficiency were studied on neuromuscular transmission in the caecum, vas deferens and urinary bladder of the rat (HOYLE et al. 1995). While both pre- and postjunctional dysfunction were produced in the caecum, no changes in sympathetic neuromuscular transmission were observed in the vas deferens or in parasympathetic neuromuscular transmission in the bladder.

IX. Interstitial Cystitis

Interstitial cystitis is a chronic bladder disorder of unknown aetiology that mainly affects middle-aged women; the most frequent symptoms are frequency, pollakiuria, dysuria and chronic pelvic pain in the absence of any urinary tract infection (HOLM-BENTZEN and LOSE 1987; KOZIOL et al. 1993).

In detrusor strips taken from patients with interstitial cystitis, the atropine-resistant contractile component was about 43% of the total responses, while this component was not observed in controls (PALEA et al. 1993). The NANC component in the neurogenic bladder was abolished following desensitisation with α,β-meATP and the detrusor muscle showed increased sensitivity to the agonist actions of α,β-meATP, in contrast to decreased sensitivity to ACh and histamine (Fig. 20).

Cats suffer a naturally occurring chronic idiopathic cystitis, termed feline interstitial cystitis (FIC), with features similar to human interstitial cystitis (LAVELLE et al. 2000). Intracellular Ca^{2+} measurements in cultured urothelial cells revealed that purinergic responses of the urothelium are changed in FIC (BUFFINGTON et al. 2000). Urothelial cells from normal cats showed increased intracellular Ca^{2+} levels in response to 2-meSATP but not to α,β-meATP, suggesting the presence of P2Y but not P2X receptors in normal tissue. However, urothelial cells from FIC cats responded to 2-meSATP and to α,β-meATP, indicating the increased expression of P2X receptors in the urothelium of animals with the disorder. NO release was also altered in FIC (ERICKSON et al. 1998). Capsaicin-induced release of NO was reduced in the mucosal strips

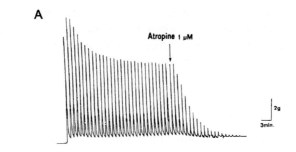
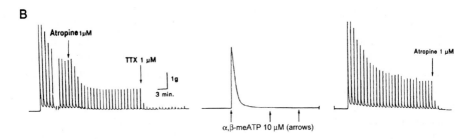
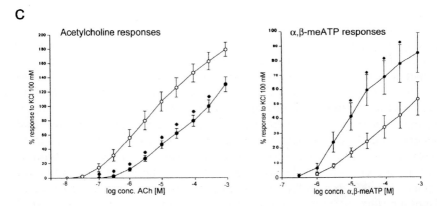

Fig. 20A–C. Detrusor contractile responses in interstitial cystitis. **A** Control human detrusor muscle strip (man, 64 years old, carcinoma of bladder), showing contractile response to electrical field stimulation (50 V, 20 Hz, 0.1 ms, trains of 5 s every 60 s) and the effect of atropine 1 µmol/l. **B** Human detrusor muscle strip from patient with interstitial cystitis. Basal contractile responses to electrical field stimulation (as in **A**) and effect of atropine (1 µmol/l) and tetrodotoxin (TTX, 1 µmol/l) before (*left trace*) and after (*right trace*) P2X receptor desensitisation with doses of α,β-methylene ATP (α,β-meATP, 10 µmol/l; *centre trace*) added cumulatively in a quiescent bladder (*arrows*). **C** Contractile effects of acetylcholine (ACh) and α,β-meATP in human detrusor strips taken from patients affected by carcinoma (controls, ○) and by interstitial cystitis (●) Each value is mean ± S.E.M. of 8 (ACh data) or 5 (α,β-meATP data) experiments. *$p < 0.05$ vs control (Student's *t*-test for unpaired data). Reproduced from PALEA et al. (1993)

from FIC cats compared to controls, whereas basal NO release mediated by inducible NOS was increased in FIC cats. Increased expression of c-*jun* and NK_2 receptors was also noted in bladder afferent neurones in FIC cats. These results indicate chemical signalling in the urothelium and in bladder sensory nerves is altered in chronic cystitis in cats.

A model for interstitial cystitis has been developed in rats by administering cyclophosphamide, an anticancer drug which is metabolised in the body to acrolein, a chemical irritant that is excreted in the urine. Rats treated with cyclophosphamide develop characteristic behavioural signs associated with bladder pain in parallel with the development of bladder lesions and increased expression of immediate early gene-encoded proteins c-*fos* and Krox-24 in the spinal cord (LANTÉRI-MINET et al. 1995; VIZZARD et al. 1996). In bladder afferent neurones, a marked increase in the excitability (YOSHIMURA and DE GROAT 1999) and in the expression of neuronal NOS (VIZZARD et al. 1996) has also been detected after cyclophosphamide treatment, indicating a change in the electrical and chemical properties of bladder afferents following chronic inflammation. It would be interesting to see if any changes in motor and/or sensory purinergic signalling occur in this model. Another model that shows many of the characteristics of interstitial cystitis has recently been proposed, where hypersensitivity inflammation of the bladder in vivo is induced by local application of ovalbumen in ovalbumen-sensitive female rats (AHLUWALIA et al. 1998).

X. Diabetes

Disturbances in micturition and damage to autonomic nerves supplying the human urinary bladder in diabetes has been known for many years (see BARTLEY et al. 1966; FAERMAN et al. 1973; ANDERSEN and BRADLEY 1976; BRADLEY 1980; ELLENBERG 1980; FRIMODT-MOLLER 1980; MASTRI 1980; FRIEDLAND and PERKASH 1983; BARKAI and SZABO 1993). The major clinical feature of diabetic bladder dysfunction is a gradual loss of bladder sensation and motor function, resulting in a large bladder and chronic residual urine volume. Similar changes have been identified in the streptozotocin-induced diabetic (STZ) rat model (see LINCOLN et al. 1984a; KOLTA et al. 1985; ANDERSSON et al. 1988; KUDLACZ et al. 1988; LONGHURST et al. 1990; STEERS et al. 1994), in the diabetic Chinese hamster (DAIL et al. 1977), and in most reports of the alloxan-induced diabetic rat model (TOMLINSON and YUSOF 1983; UVELIUS 1986; PARO and PROSDOCIMI 1987; PARO et al. 1990; PROSDOCIMI and PARO 1990). Functional abnormalities associated with progressive axonopathy of afferent myelinated sensory nerve fibres and later of unmyelinated efferent preganglionic fibres were also described in the spontaneously diabetic BB rat (PARO et al. 1989). In urethral rings from STZ diabetic rats, the contractile responses to field stimulation, ACh and NA were unchanged compared to controls (MAGGI et al. 1989b). It has been claimed that in the rat urinary bladder, STZ diabetes causes impairment of capsaicin-sensitive sensory fibres even at

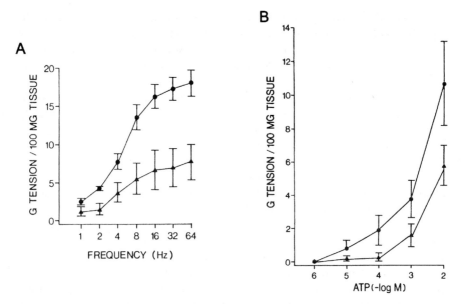

Fig. 21A,B. Responses of bladder body strips from streptozotocin-diabetic rats. **A** Responses to nerve stimulation (supramaximal voltage, 0.5 ms pulse duration for 2s every 2 min) of bladder strips from 2-month streptozotocin-diabetic (▲, $n = 5$) and control rats (●, $n = 5$). **B** Responses to ATP of bladder strips from 2-month streptozotocin-diabetic (▲, $n = 5$) and control rats (●, $n = 5$). For both graphs, each point represents the mean ± S.E.M. and data are expressed as grams tension per 100 mg tissue. Reproduced from LONGHURST and BELIS (1986)

4 weeks, but not of the cholinergic system, and also stimulates the release of epithelial contracting factors, and further that epithelium removal impairs ACh-induced contractions in diabetic bladder, but not in controls (PINNA et al. 1994).

An early study of nerve-mediated contractions of the diabetic rat bladder showed reduced responses, but only a tendency for reduced responses to ATP, ACh and KCl (LONGHURST and BELIS 1986) (Fig. 21). Although there was some indication of neuropathy of motor nerves in the STZ diabetic bladder (LINCOLN et al. 1984b), no sign of damage to capsaicin-sensitive sensory nerves in the bladder was observed, at least in 7–9 week diabetic animals (SANTICIOLI et al. 1987). A transient increase in sensitivity of the 6- and 12-week STZ rat whole bladder preparations to α,β-meATP was reported (Moss et al. 1987); the biphasic response of the bladder to α,β-meATP was not changed significantly in earlier 4–5-week STZ rats (KUDLACZ et al. 1989). In another study of bladder strips taken from STZ rats, it was shown that atropine-resistant (purinergic) responses to field stimulation were reduced and it was concluded that this was probably the result of a reduction in release of the NANC transmitter (LUHESHI and ZAR 1990c). The same authors found a potentiation of cholinergic transmission in the STZ bladder, and suggested this was due to enhanced release of ACh (LUHESHI and ZAR 1991).

The calcium channel blocker, nifedipine, has been shown to block the purinergic component of the parasympathetic contractile responses of the bladder (Bo and BURNSTOCK 1990a). No significant differences were found in the sensitivity of bladder strips from control and STZ diabetic rats to antagonism by nifedipine (LONGHURST et al. 1992). ATP significantly increases the endogenous release of PGE_2 and $PGF_{2\alpha}$ from the urothelium of 4-week STZ diabetic rat bladder and it was proposed that P2X receptors are present on urothelial cells as well as smooth muscle (PINNA et al. 2000a). Increased synthesis of prostanoids during epithelial irritation may produce hyperactivity or spasm of the detrusor muscle (HOYLE 1994).

In contrast to normal mouse bladder, the urinary bladder of STZ mice had weaker neurogenic contractile responses to electrical field stimulation (LIU and LIN-SHIAU 1996). Nerve-mediated contractions in diabetic bladders were also less sensitive to the depressant actions of uranyl nitrate, which was shown in the same study to be selective for the non-cholinergic contractile component. Since high Ca^{2+} or calmodulin inhibitors antagonised the suppressant effect of uranyl nitrate, it was postulated that Ca^{2+} regulation of ATP release might be impaired in the diabetic state.

Bladders from 8-week STZ diabetic rats showed enhanced relaxant responses to ATP and adenosine, as well as increased contractile responses to ATP (GÜR and KARAHAN 1997). Enhanced responses to ATP, but not to ACh or KCl, were also reported in the 4-week STZ diabetic bladder, and it was also shown that the responses of detrusor muscle from diabetic ovariectomized rats were decreased, although partially recovered to control values by oestrogen treatment (PINNA et al. 2000b).

In summary, in the hands of some workers, diabetes appears to reduce neurogenic contractions of the bladder, largely as a consequence of reduced release of ATP; other workers report little change in neurogenic responses, but with increased contractile and relaxant responses to exogenous ATP.

M. Concluding Comments

Despite earlier scepticism, there is now abundant evidence for purinergic signalling in the lower urinary tract, in particular parasympathetic purinergic contractions of urinary bladder (via $P2X_1$ receptors) and $P2X_3$ receptors on sensory nerve terminals involved in the micturition reflex and pain. Purinergic synaptic transmission occurs in pelvic ganglia.

Of special interest is the evidence for enhanced purinergic signalling in pathological conditions such as obstructive bladder and interstitial cystitis encouraging the future development of purinergic therapeutic drugs.

Acknowledgments. I appreciate the very valuable suggestions of Chet de Groat for improvements to this manuscript and the highly skilful help of Roy Jordan and the patience and perseverance of Annie Evans in its preparation. The valuable support of Roche Bioscience (Palo Alto, Calif.) is also acknowledged.

References

Abbracchio MP, Burnstock G (1994) Purinoceptors: are there families of P_{2X} and P_{2Y} purinoceptors? Pharmacol Ther 64:445–475
Abbracchio MP, Burnstock G (1998) Purinergic signalling: pathophysiological roles. Jpn J Pharmacol 78:113–145
Acevedo CG, Contreras E (1985) Possible involvement of adenine nucleotides in the neurotransmission of the mouse urinary bladder. Comp Biochem Physiol C 82:357–361
Acevedo CG, Contreras E (1989) Effect of extracellular calcium and calcium channel antagonists on ATP and field stimulation induced contractions of the mouse urinary bladder. Gen Pharmacol 20:811–815
Acevedo CG, Lewin J, Contreras E, Huidobro-Toro JP (1990) Bradykinin facilitates the purinergic motor component of the rat bladder neurotransmission. Neurosci Lett 113:227–232
Acevedo CG, Contreras E, Escalona J, Lewin J, Huidobro-Toro JP (1992) Pharmacological characterization of adenosine A_1 and A_2 receptors in the bladder: evidence for a modulatory adenosine tone regulating non-adrenergic non-cholinergic neurotransmission. Br J Pharmacol 107:120–126
Ahluwalia A, Giuliani S, Scotland R, Maggi CA (1998) Ovalbumin-induced neurogenic inflammation in the bladder of sensitized rats. Br J Pharmacol 124:190–196
Akasu T, Shinnick-Gallagher P, Gallagher JP (1984) Adenosine mediates a slow hyperpolarizing synaptic potential in autonomic neurones. Nature 311:62–65
Albino-Teixeira A, Azevedo I, Branco D, Osswald W (1990) Purine agonists prevent trophic changes caused by sympathetic denervation. Eur J Pharmacol 179:141–149
Alkondon M, Ganguly DK (1980) Release of prostaglandin E from the isolated urinary bladder of the guinea-pig. Br J Pharmacol 69:573–577
Alm P (1978) Cholinergic innervation of the human urethra and urinary bladder: A histochemical study & review of methodology. Acta Pharmacol Toxicol 43:56–62
Alm P, Elmér M (1975) Adrenergic and cholinergic innervation of the rat urinary bladder. Acta Physiol Scand 94:36–45
Alm P, Elmér M (1979) Adrenergic reinnervation of the denervated rat urinary bladder. Experientia 35:1387–1388
Alm P, Alumets J, Brodin E, Hakanson R, Nilsson G, Sjöberg NO, Sundler F (1978) Peptidergic (substance P) nerves in the genito-urinary tract. Neuroscience 3:419–425
Amann R, Dray A, Hankins MW (1988) Stimulation of afferent fibres of the guinea-pig ureter evokes potentials in inferior mesenteric ganglion neurones. J Physiol 402:543–553
Ambache N (1955) The use and limitations of atropine for pharmacological studies on autonomic effectors. Pharmacol Rev 7:467–494
Ambache N, Zar MA (1970) Non-cholinergic transmission by post-ganglionic motor neurones in the mammalian bladder. J Physiol 210:761–783
Ambache N, Killick SW, Woodley JP (1977) Evidence against purinergic motor transmission in guinea-pig urinary bladder. Br J Pharmacol 61:464P
Ammons WS (1989) Primate spinothalamic cell response to ureteral occlusion. Brain Res 496:124–130
Andersen JT, Bradley WE (1976) Abnormalities of bladder innervation in diabetes mellitus. Urology 7:442–448
Anderson GF (1982) Evidence for a prostaglandin link in the purinergic activation of rabbit bladder smooth muscle. J Pharmacol Exp Ther 220:347–352

Andersson K-E (1993) Pharmacology of lower urinary tract smooth muscles and penile erectile tissues. Pharmacol Rev 45:253–308

Andersson K-E (1997) Emptying against outflow obstruction–pharmacological aspects. Scand J Urol Nephrol Suppl 184:77–84

Andersson K-E, Sjögren C (1982) Aspects on the physiology and pharmacology of the bladder and urethra. Prog Neurobiol 19:71–89

Andersson K-E, Husted S, Sjögren C (1980) Contribution of prostaglandins to the adenosine triphosphate-induced contraction of rabbit urinary bladder. Br J Pharmacol 70:443–452

Andersson K-E, Mattiasson A, Sjögren C (1983) Electrically induced relaxation of the noradrenaline contracted isolated urethra from rabbit and man. J Urol 129:210–214

Andersson K-E, Garcia-Pascual A, Forman A, Tøttrup A (1991) Non-adrenergic, non-cholinergic nerve-mediated relaxation of rabbit urethra is caused by nitric oxide. Acta Physiol Scand 141:133–134

Andersson PO, Malmgren A, Uvelius B (1988) Cystometrical and in vitro evaluation of urinary bladder function in rats with streptozotocin-induced diabetes. J Urol 139:1359–1362

Atala A, Vacanti JP, Peters CA, Mandell J, Retik AB, Freeman MR (1992) Formation of urothelial structures in vivo from dissociated cells attached to biodegradable polymer scaffolds in vitro. J Urol 148:658–662

Bahns E, Ernsberger U, Jänig W, Nelke A (1986) Functional characteristics of lumbar visceral afferent fibres from the urinary bladder and the urethra in the cat. Pflugers Arch 407:510–518

Bahns E, Halsband U, Jänig W (1987) Responses of sacral visceral afferents from the lower urinary tract, colon and anus to mechanical stimulation. Pflugers Arch 410: 296–303

Barkai L, Szabo L (1993) Urinary bladder dysfunction in diabetic children with and without subclinical cardiovascular autonomic neuropathy. Eur J Pediatr 152: 190–192

Barras M, Van der Graaf PH, Angel I (1996) Characterization of the 5-HT receptor potentiating neurotransmission in rabbit bladder. Eur J Pharmacol 318:425–428

Barrington FJF (1931) The component reflexes of micturition in the cat. Parts I and II. Brain 54:177–188

Bartley O, Brolin I, Fagerberg SE, Wilhelmsen L (1966) Neurogenic disorders of the bladder in diabetes mellitus. A clinical-roentgenological investigation. Acta Med Scand 180:187–198

Batra S (1979) Subcellular distribution and cytosolic receptors of progesterone and of estradiol-17β in the rabbit myometrium: effect of progesterone treatment. Biol Reprod 21:483–489

Batra S (1980) Estrogen and smooth muscle function. Trends Pharmacol Sci 1:388

Bayliss M, Wu C, Newgreen D, Mundy AR, Fry CH (1999) A quantitative study of atropine-resistant contractile responses in human detrusor smooth muscle, from stable, unstable and obstructed bladders. J Urol 162:1833–1839

Beigi R, Kobatake E, Aizawa M, Dubyak GR (1999) Detection of local ATP release from activated platelets using cell surface-attached firefly luciferase. Am J Physiol 276:C267–C278

Bhat MB, Mishra SK, Raviprakash V (1989a) Differential susceptibility of cholinergic and noncholinergic neurogenic responses to calcium channel blockers and low Ca^{2+} medium in rat urinary bladder. Br J Pharmacol 96:837–842

Bhat MB, Mishra SK, Raviprakash V (1989b) Sources of calcium for ATP-induced contractions in rat urinary bladder smooth muscle. Eur J Pharmacol 164:163–166

Birder LA, de Groat WC (1993) Induction of c-*fos* expression in spinal neurons by nociceptive and non-nociceptive stimulation of the lower urinary tract. Am J Physiol 265:R326–R333

Birder LA, de Groat WC (1998) Contribution of C-fiber afferent nerves and autonomic pathways in the urinary bladder to spinal c-fos expression induced by bladder irritation. Somatosens Mot Res 15:5–12

Birder LA, Apodaca G, de Groat WC, Kanai AJ (1998a) Adrenergic- and capsaicin-evoked nitric oxide release from urothelium and afferent nerves in urinary bladder. Am J Physiol 275:F226–F229

Birder LA, de Groat WC, Truschel ST, Apodaca G (1998b) Norepinephrine, acetylcholine and capsaicin evoked nitric oxide release in the urinary bladder epithelium, measured using a porphyrinic microsensor. Soc Neurosci Abst 24:1619

Birder LA, Apodaca G, Truschel ST, Kanai AJ, de Groat WC (1999a) Nitric oxide is released from urinary bladder epithelial cells and modifies epithelial function. FASEB J 13:A728

Birder LA, de Groat WC, Truschel ST, Apodaca G (1999b) Effect of chronic spinal injury on capsaicin-sensitive (CAPS) afferent innervation of the urinary bladder and on nitric oxide (NO) release from the urothelium. Soc Neurosci Abst 25:401

Blackburn GM, Kent DE, Kolkmann F (1984) The synthesis and metal binding characteristics of novel isopolar phosphonate analogues of nucleotides. J Chem Soc Perkin Trans 11:1119–1125

Blaivas JG, Bhimani G, Labib KB (1979) Vesicourethral dysfunction in multiple sclerosis. J Urol 122:342–347

Blaivas JG, Holland NJ, Giesser B, LaRocca N, Madonna M, Scheinberg L (1984) Multiple sclerosis bladder. Studies and care. Ann N Y Acad Sci 436:328–346

Bland-Ward PA, Humphrey PPA (1997) Acute nociception mediated by hindpaw P2X receptor activation in the rat. Br J Pharmacol 122:365–371

Bleehen T (1978) The effects of adenine nucleotides on cutaneous afferent nerve activity. Br J Pharmacol 62:573–577

Bleehen T, Keele CA (1977) Observations on the algogenic actions of adenosine compounds on human blister base preparation. Pain 3:367–377

Bo X, Burnstock G (1989) [^{3}H]-α,β-Methylene ATP, a radioligand labelling P_2-purinoceptors. J Auton Nerv Syst 28:85–88

Bo X, Burnstock G (1990a) The effects of Bay K8644 and nifedipine on the responses of rat urinary bladder to electrical field stimulation, β,γ-methylene ATP and acetylcholine. Br J Pharmacol 101:494–498

Bo X, Burnstock G (1990b) High- and low-affinity binding sites for [^{3}H]-α,β-methylene ATP in rat urinary bladder membranes. Br J Pharmacol 101:291–296

Bo X, Burnstock G (1992) Species differences in localization of [^{3}H]α,β-methylene ATP binding sites in urinary bladder and urethra of rat, guinea-pig and rabbit. Eur J Pharmacol 216:59–66

Bo X, Burnstock G (1993) Triphosphate, the key structure of the ATP molecule responsible for interaction with P_{2X} purinoceptors. Gen Pharmacol 24:637–640

Bo X, Burnstock G (1995) Characterization and autoradiographic localization of [^{3}H]α,β-methylene adenosine 5'-triphosphate binding sites in human urinary bladder. Br J Urol 76:297–302

Bo X, Fischer B, Maillard M, Jacobson KA, Burnstock G (1994) Comparative studies on the affinities of ATP derivatives for P_{2X}-purinoceptors in rat urinary bladder. Br J Pharmacol 112:1151–1159

Bodin P, Burnstock G (1995) Synergistic effect of acute hypoxia on flow-induced release of ATP from cultured endothelial cells. Experientia 51:256–259

Bodin P, Bailey DJ, Burnstock G (1991) Increased flow-induced ATP release from isolated vascular endothelial but not smooth muscle cells. Br J Pharmacol 103:1203–1205

Bodin P, Milner P, Winter R, Burnstock G (1992) Chronic hypoxia changes the ratio of endothelin to ATP release from rat aortic endothelial cells exposed to high flow. Proc R Soc Lond B Biol Sci 247:131–135

Boland B, Himpens B, Paques C, Casteels R, Gillis JM (1993) ATP induced-relaxation in the mouse bladder smooth muscle. Br J Pharmacol 108:749–753

Bolego C, Abbracchio MP, Cattabeni F, Ruzza R, Puglisi L (1995a) Effects of ADPβS and UTP on the rat urinary bladder smooth muscle. Res Commun Mol Pathol Pharmacol 87:75–76

Bolego C, Pinna C, Abbracchio MP, Cattabeni F, Puglisi L (1995b) The biphasic response of rat vesical smooth muscle to ATP. Br J Pharmacol 114:1557–1562

Boselli C, Bianchi L, Grana E (1997) Effect of cromakalim on the purinergic and cholinergic transmission in the rat detrusor muscle. Eur J Pharmacol 335:23–30

Bouvier MM, Evans ML, Benham CD (1991) Calcium influx induced by stimulation of ATP receptors on neurons cultured from rat dorsal root ganglia. Eur J Neurosci 3:285–291

Bowler WB, Tattersall JA, Hussein R, Dixon CJ, Cobbold PH, Gallagher JA (1998) Release of ATP by osteoblasts; modulation by fluid shear forces. Bone 22:3S

Bradbury EJ, Burnstock G, McMahon SB (1998) The expression of P2X$_3$ purinoceptors in sensory neurons: effects of axotomy and glial-derived neurotrophic factor. Mol Cell Neurosci 12:256–268

Brading AF (1987) Physiology of bladder smooth muscle. In: Torrens M, Morrison JFB (eds) The Physiology of the Lower Urinary Tract. Springer-Verlag, London, pp 161–191

Brading AF, Mostwin JL (1989) Electrical and mechanical responses of guinea-pig bladder muscle to nerve stimulation. Br J Pharmacol 98:1083–1090

Brading AF, Williams JH (1990) Contractile responses of smooth muscle strips from rat and guinea-pig urinary bladder to transmural stimulation: effects of atropine and α,β-methylene ATP. Br J Pharmacol 99:493–498

Bradley WE (1980) Diagnosis of urinary bladder dysfunction in diabetes mellitus. Ann Intern Med 92:323–326

Bramich NJ, Brading AF (1996) Electrical properties of smooth muscle in the guinea-pig urinary bladder. J Physiol 492:185–198

Bratslavsky G, Whitbeck C, Horan P, Levin RM (1999) Effects of in vivo ischemia on contractile responses of rabbit bladder to field stimulation, carbachol, ATP and KCl. Pharmacology 59:221–226

Brown AD (1977) Postmenopausal urinary problems. Clin Obstet Gynaecol 4:181–206

Brown CM, Burnstock G (1981) The structural conformation of the polyphosphate chain of the ATP molecule is critical for its promotion of prostaglandin biosynthesis. Eur J Pharmacol 69:81–86

Brown C, Burnstock G, Cocks T (1979) Effects of adenosine 5'-triphosphate (ATP) and β,γ-methylene ATP on the rat urinary bladder. Br J Pharmacol 65:97–102

Buchthal F, Kahlson G (1944) The motor effect of adenosine triphosphate and allied phosphorus compounds on smooth mammalian muscle. Acta Physiol Scand 8:325–334

Buck AC, McRae C, Reed PI, Chisholm GD (1974) The diabetic bladder. Proc Roy Soc Med 67:81–83

Buffington CA, Kiss S, Kanai AJ, Dineley K, Roppolo JR, Reynolds IR, de Groat WC, Birder LA (2000) Alterations in urothelium and bladder afferents in feline interstitial cystitis. Soc Neurosci Abst (In press)

Burnstock G (1972) Purinergic nerves. Pharmacol Rev 24:509–581

Burnstock G (1975) Comparative studies of purinergic nerves. J Exp Zool 194:103–133

Burnstock G (1976) Do some nerve cells release more than one transmitter? Neuroscience 1:239–248

Burnstock G (1978) A basis for distinguishing two types of purinergic receptor. In: Straub RW, Bolis L (eds) Cell Membrane Receptors for Drugs and Hormones: A Multidisciplinary Approach. Raven Press, New York, pp 107–118

Burnstock G (1981) Pathophysiology of migraine: a new hypothesis. Lancet i:1397–1399

Burnstock G (1982) The co-transmitter hypothesis, with special reference to the storage and release of ATP with noradrenaline and acetylcholine. In: Cuello AC (ed) Co-transmission. Macmillan Press, London, pp 151–163

Burnstock G (1986a) Non-adrenergic neurotransmitters in relation to sympathetic nervous control of the lower urinary tract. Clin Sci 70 (Suppl. 14):15s–20s

Burnstock G (1986b) Purinergic receptors. In: Magistretti JH, Morrison JH, Reisine TD (eds) Discussions in Neurosciences, Vol. III, No. 3: Transduction of Neuronal Signals. FESN, Geneva, pp 48–55

Burnstock G (1990a) Co-transmission. The Fifth Heymans Lecture – Ghent, February 17, 1990. Arch Int Pharmacodyn Ther 304:7–33

Burnstock G (1990b) Noradrenaline and ATP as cotransmitters in sympathetic nerves. Neurochem Int 17:357–368

Burnstock G (1993) Introduction: Changing face of autonomic and sensory nerves in the circulation. Vascular Innervation and Receptor Mechanisms: New Perspectives Eds. L. Edvinsson and R. Uddman. Academic Press Inc, USA, pp 1–22

Burnstock G (1995) Noradrenaline and ATP: cotransmitters and neuromodulators. J Physiol Pharmacol 46:365–384

Burnstock G (1996a) Cotransmission with particular emphasis on the involvement of ATP. In: Fuxe K, Hökfelt T, Olson L, Ottoson D, Dahlström A, Björklund A (eds) Molecular Mechanisms of Neuronal Communication. A Tribute to Nils-Ake Hillarp. Pergamon Press, Oxford, pp 67–87

Burnstock G (1996b) Purinoceptors: Ontogeny and Phylogeny. Drug Dev Res 39:204–242

Burnstock G (1996c) A unifying purinergic hypothesis for the initiation of pain. Lancet 347:1604–1605

Burnstock G (1999a) Purinergic cotransmission. Brain Res Bull 50:355–357

Burnstock G (1999b) Release of vasoactive substances from endothelial cells by shear stress and purinergic mechanosensory transduction. J Anat 194:335–342

Burnstock G (2000) P2X receptors in sensory neurones. Br J Anaesth 84:476–488

Burnstock G, Campbell G (1963) Comparative physiology of the vertebrate autonomic nervous system. II. Innervation of the urinary bladder of the ringtail possum (*Pseudocheirus peregrinus*). J Exp Biol 40:421–436

Burnstock G, Holman ME (1960) Autonomic nerve-smooth muscle transmission. Nature 187:951–952

Burnstock G, Holman ME (1961) The transmission of excitation from autonomic nerve to smooth muscle. J Physiol 155:115–133

Burnstock G, Kennedy C (1985) Is there a basis for distinguishing two types of P_2-purinoceptor? Gen Pharmacol 16:433–440

Burnstock G, Wood JN (1996) Purinergic receptors: their role in nociception and primary afferent neurotransmission. Curr Opin Neurobiol 6:526–532

Burnstock G, Dumsday B, Smythe A (1972) Atropine resistant excitation of the urinary bladder: the possibility of transmission via nerves releasing a purine nucleotide. Br J Pharmacol 44:451–461

Burnstock G, Cocks T, Paddle B, Staszewska-Barczak J (1975) Evidence that prostaglandin is responsible for the 'rebound contraction' following stimulation of non-adrenergic, non-cholinergic ('purinergic') inhibitory nerves. Eur J Pharmacol 31:360–362

Burnstock G, Cocks T, Crowe R, Kasakov L (1978a) Purinergic innervation of the guinea-pig urinary bladder. Br J Pharmacol 63:125–138

Burnstock G, Cocks T, Kasakov L, Wong HK (1978b) Direct evidence for ATP release from non-adrenergic, non-cholinergic ("purinergic") nerves in the guinea-pig taenia coli and bladder. Eur J Pharmacol 49:145–149

Burnstock G, Cusack NJ, Hills JM, MacKenzie I, Meghji P (1983) Studies on the stereoselectivity of the P_2-purinoceptor. Br J Pharmacol 79:907–913

Burnstock G, Cusack NJ, Meldrum LA (1984) Effects of phosphorothioate analogues of ATP, ADP and AMP on guinea-pig taenia coli and urinary bladder. Br J Pharmacol 82:369–374

Burnstock G, Allen TGJ, Hassall CJS, Pittam BS (1987) Properties of intramural neurones cultured from the heart and bladder. In: Heym C (ed) Histochemistry and

Cell Biology of Autonomic Neurons and Paraganglia. Exp. Brain Res. Ser. 16. Springer Verlag, Heidelberg, pp 323-328

Burnstock G, Fischer B, Hoyle CHV, Maillard M, Ziganshin AU, Brizzolara AL, von Isakovics A, Boyer JL, Harden TK, Jacobson KA (1994) Structure activity relationships for derivatives of adenosine 5'-triphosphate as agonists at P_2 purinoceptors: heterogeneity within P_{2X} and P_{2Y} subtypes. Drug Dev Res 31:206–219

Burnstock G, McMahon SB, Humphrey PPA, Hamilton SG (2000) ATP (P2X) receptors and pain. In: Devor M, Rowbottom MC, Wiesenfeld-Hallin Z (eds) Proceedings of the 9th World Congress on Pain. Progress in Pain Research and Management, Vol 17. IASP Press, Seattle, pp 63–76

Bushfield M, Kenny BA, Parker N (1996) Facilitation by 5-HT of ATP-mediated electrically-stimulated contractions in the pig urinary bladder. Br J Pharmacol 117:201P

Callahan SM, Creed KE (1981) Electrical and mechanical activity of the isolated lower urinary tract of the guinea-pig. Br J Pharmacol 74:353–358

Cantiello HF, Jackson GR Jr, Grosman CF, Prat AG, Borkan SC, Wang Y, Reisin IL, O'Riordan CR, Ausiello DA (1998) Electrodiffusional ATP movement through the cystic fibrosis transmembrane conductance regulator. Am J Physiol 274:C799–C809

Carpenter FG (1977) Atropine resistance and muscarinic receptors in the rat urinary bladder. Br J Pharmacol 59:43–49

Carpenter FG, Rubin RM (1967) The motor innervation of the rat urinary bladder. J Physiol 192:609–617

Cervero F, Sann H (1989) Mechanically evoked responses of afferent fibres innervating the guinea-pig's ureter: an *in vitro* study. J Physiol 412:245–266

Chancellor MB, de Groat WC (1999) Intravesical capsaicin and resiniferatoxin therapy: spicing up the ways to treat the overactive bladder. J Urol 162:3–11

Chancellor MB, Kaplan SA, Blaivas JG (1992) The cholinergic and purinergic components of detrusor contractility in a whole rabbit bladder model. J Urol 148:906–909

Chancellor MB, Shenot PJ, Rivas DA, Mandel S, Schwartzman RJ (1996) Urological symptomatology in patients with reflex sympathetic dystrophy. J Urol 155:634–637

Chaudhry A, Downie JW, White TD (1984) Tetrodotoxin-resistant release of ATP from superfused rabbit detrusor muscle during electrical field stimulation in the presence of luciferin-luciferase. Can J Physiol Pharmacol 62:153–156

Chen CC, Akopian AN, Sivilotti L, Colquhoun D, Burnstock G, Wood JN (1995) A P2X purinoceptor expressed by a subset of sensory neurons. Nature 377:428–431

Chen H-I, Brading AF (1991) The mechanism of action of putative non-adrenergic, non-cholinergic transmitters on the rabbit urinary bladder. J Auton Nerv Syst 33:178–179

Chesher GB, James B (1966) The "nicotinic" and "muscarinic" receptors of the urinary bladder of the guinea-pig. J Pharm Pharmacol 18:417–423

Chesher GB, Thorp RH (1965) The atropine-resistance of the response to intrinsic nerve stimulation of the guinea-pig bladder. Br J Pharmacol 25:288–294

Choo LK (1981) The effect of reactive blue, an antagonist of ATP, on the isolated urinary bladders of guinea-pig and rat. J Pharm Pharmacol 33:248–250

Choo LK, Mitchelson F (1980) The effect of indomethacin and adenosine 5'-triphosphate on the excitatory innervation of the rate urinary bladder. Can J Physiol Pharmacol 58:1042–1048

Clemow DB, Steers WD, McCarty R, Tuttle JB (1998) Altered regulation of bladder nerve growth factor and neurally mediated hyperactive voiding. Am J Physiol 275:R1279–R1286

Cockayne DA, Hamilton SG, Zhu Q-M, Dunn PM, Zhong Y, Novakovic S, Malmberg AB, Cain G, Berson K, Kassotakis L, Hedley L, Lachnit WG, Burnstock G, McMahon SB, Ford APDW (2000) Urinary bladder hyporeflexia and reduced pain-related behaviour in $P2X_3$-deficient mice. Nature 407:1011–1015

Conte B, Maggi CA, Meli A (1989) Vesico-inhibitory responses and capsaicin-sensitive afferents in rats. Naunyn Schmiedebergs Arch Pharmacol 339:178–183

Corsi M, Pietra C, Toson G, Trist D, Tuccitto G, Artibani W (1991) Pharmacological analysis of 5-hydroxytryptamine effects on electrically stimulated human isolated urinary bladder. Br J Pharmacol 104:719–725

Cotton KD, Hollywood MA, Thornbury KD, McHale NG (1996) Effect of purinergic blockers on outward current in isolated smooth muscle cells of the sheep bladder. Am J Physiol 270:C969–C973

Coutts AA, Jorizzo JL, Eady RAJ, Greaves MW, Burnstock G (1981) Adenosine triphosphate-evoked vascular changes in human skin: mechanism of action. Eur J Pharmacol 76:391–401

Cowan WD, Daniel EE (1983) Human female bladder and its noncholinergic contractile function. Can J Physiol Pharmacol 61:1236–1246

Crack BE, Pollard CE, Beukers MW, Roberts SM, Hunt SF, Ingall AH, McKechnie KC, Ijzerman AP, Leff P (1995) Pharmacological and biochemical analysis of FPL 67156, a novel, selective inhibitor of ecto-ATPase. Br J Pharmacol 114:475–481

Craggs MD, Rushton DN, Stephenson JD (1986) A putative non-cholinergic mechanism in urinary bladders of New but not Old World primates. J Urol 136:1348–1350

Creed KE (1979) The role of the hypogastric nerve in bladder and urethral activity of the dog. Br J Pharmacol 65:367–375

Creed KE, Tulloch AG (1978) The effect of pelvic nerve stimulation and some drugs on the urethra and bladder of the dog. Br J Urol 50:398–405

Creed KE, Ishikawa S, Ito Y (1983) Electrical and mechanical activity recorded from rabbit urinary bladder in response to nerve stimulation. J Physiol 338:149–164

Creed KE, Ito Y, Katsuyama H (1991) Neurotransmission in the urinary bladder of rabbits and guinea pigs. Am J Physiol 261:C271–C277

Creed KE, Callahan SM, Ito Y (1994) Excitatory neurotransmission in the mammalian bladder and the effects of suramin. Br J Urol 74:736–743

Crowe R, Burnstock G (1985) Perinatal development of adrenergic, cholinergic and non-adrenergic, non-cholinergic nerves and SIF cells in the rabbit urinary bladder. Int J Dev Neurosci 3:89–101

Crowe R, Burnstock G (1989) A histochemical and immunohistochemical study of the autonomic innervation of the lower urinary tract of the female pig. Is the pig a good model for the human bladder and urethra. J Urol 141:414–422

Crowe R, Haven AJ, Burnstock G (1986) Intramural neurons of the guinea-pig urinary bladder: histochemical localization of putative neurotransmitters in cultures and newborn animals. J Auton Nerv Syst 15:319–339

Crowe R, Burnstock G, Light JK (1988) Intramural ganglia in the human urethra. J Urol 140:183–187

Cruz F, Avelino A, Lima D, Coimbra A (1994) Activation of the c-*fos* proto-oncogene in the spinal cord following noxious stimulation of the urinary bladder. Somatosens Mot Res 11:319–325

Cusack NJ, Hourani SMO (1984) Some pharmacological and biochemical interactions of the enantiomers of adenylyl 5'-(β,γ-methylene)-diphosphonate with the guinea-pig urinary bladder. Br J Pharmacol 82:155–159

Cusack NJ, Hourani SMO, Loizou GD, Welford LA (1987) Pharmacological effects of isopolar phosphonate analogues of ATP on P2-purinoceptors in guinea-pig taenia coli and urinary bladder. Br J Pharmacol 90:791–795

Cusack NJ, Hourani SMO, Welford LA (1988) The role of ectonucleotidases in pharmacological responses to nucleotide analogues. In: Paton DM (ed) Adenosine and Adenine Nucleotides. Taylor & Francis, London, pp 93–100

Dahlén SE, Hedqvist P (1980) ATP, β-γ-methylene-ATP, and adenosine inhibit noncholinergic non-adrenergic transmission in rat urinary bladder. Acta Physiol Scand 109:137–142

Dail WG, Evan AP, Gerritsen GC, Dulin WE (1977) Abnormalities in pelvic visceral nerves. A basis for neurogenic bladder in the diabetic Chinese hamster. Invest Urol 15:161–166

Dale HH, Gaddum JH (1930) Reactions of denervated voluntary muscle, and their bearing on the mode of action of parasympathetic and related nerves. J Physiol 70:109–144
de Groat WC (1975) Nervous control of the urinary bladder of the cat. Brain Res 87:201–211
de Groat WC (1987) Neuropeptides in pelvic afferent pathways. Experientia 43:801–813
de Groat WC (1986) Spinal cord projections and neuropeptides in visceral afferent neurons. In: Cervero F, Morrison JFB (eds) Visceral Sensation, Progress in Brain Reseach, Vol. 67. Elservier Science B.V., Amsterdam, pp 165–188
de Groat WC, Booth AM (1980) Inhibition and facilitation in parasympathetic ganglia of the urinary bladder. Fed Proc 39:2990–2996
de Groat WC, Booth AM (1993) Synaptic transmission in pelvic ganglia. In: Maggi C (ed) The Autonomic Nervous System, Vol. 3. Nervous Control of the Urogenital System. Harwood Academic Publishers, Chur, Switzerland, pp 291–348
de Groat WC, Kawatani M (1989a) Enkephalinergic inhibition in parasympathetic ganglia of the urinary bladder of the cat. J Physiol 413:13–29
de Groat WC, Kawatani M (1989b) Reorganization of sympathetic preganglionic connections in cat bladder ganglia following parasympathetic denervation. J Physiol 409:431–449
de Groat WC, Saum WR (1972) Sympathetic inhibition of the urinary bladder and of pelvic ganglionic transmission in the cat. J Physiol 220:297–314
de Groat WC, Saum WR (1976) Synaptic transmission in parasympathetic ganglia in the urinary bladder of the cat. J Physiol 256:137–158
de Groat WC, Theobald RJ (1976) Reflex activation of sympathetic pathways to vesical smooth muscle and parasympathetic ganglia by electrical stimulation of vesical afferents. J Physiol 259:223–237
de Groat WC, Booth AM, Krier J, Milne RJ, Morgan C, Nadelhaft I (1979) Neural control of the urinary bladder and large intestine. In: Brooks CM, Koizumi K, Sato A (eds) Integrative Functions of the Autonomic Nervous System. Elsevier/North Holland Biomedical Press, Amaterdam, pp 50–67
de Groat WC, Kawatani M, Hisamitsu T, Cheng C-L, Ma C-P, Thor K, Steers W, Roppolo JR (1990) Mechanisms underlying the recovery of urinary bladder function following spinal cord injury. J Auton Nerv Syst 30:S71–S77
de Groat WC, Booth WC, Yoshimura M (1993) Neurophysiology of micturition and its modification in animal models of human disease. In: Maggi CA (ed) The Autonomic Nervous System. Volume 3. Nervous Control of the Urogenital System. Harwood Academic Publishers, Chur,Switzerland, pp 227–290
de Groat WC, Downie JW, Levin RM, Long Lin AT, Morrison JFB, Nishizawa O, Steers W, Thor KB (1999) Basic neurophysiology and pharmacology. 1st International Consultation on Incontinence. World Health Organization, New York, p 105
De Mey J, Burnstock G, Vanhoutte PM (1979) Modulation of the evoked release of noradrenaline in canine saphenous vein via presynaptic receptors for adenosine but not ATP. Eur J Pharmacol 55:401–405
De Sy W, Lacroix E, Leusen I (1974) An analysis of the urinary bladder response to hypogastric nerve stimulation in the cat. Invest Urol 11:508–516
Dean DM, Downie JW (1978a) Interaction of prostaglandins and adenosine 5'-triphosphate in the noncholinergic neurotransmission in rabbit detrusor. Prostaglandins 16:245–251
Dean DM, Downie JW (1978b) Contribution of adrenergic and 'purinergic' neurotransmission to contraction in rabbit detrusor. J Pharmacol Exp Ther 207:431–445
Deplanne V, Palea S, Angel I (1998) The adrenergic, cholinergic and NANC nerve-mediated contractions of the female rabbit bladder neck and proximal, medial and distal urethra. Br J Pharmacol 123:1517–1524
Dhulipala PD, Wang YX, Kotlikoff MI (1998) The human $P2X_4$ receptor gene is alternatively spliced. Gene 207:259–266

Dixon JS, Gilpin CJ (1987) Presumptive sensory axons of the human urinary bladder: a fine structural study. J Anat 151:199–207

Dokita S, Morgan WR, Wheeler MA, Yoshida M, Latifpour J, Weiss RM (1991) N^G-nitro-L-arginine inhibits non-adrenergic, non-cholinergic relaxation in rabbit urethral smooth muscle. Life Sci 48:2429–2436

Donoso MV, Salas C, Sepulveda G, Lewin J, Fournier A, Huidobro-Toro JP (1994) Involvement of ET_A receptors in the facilitation by endothelin-1 of non-adrenergic non-cholinergic transmission in the rat urinary bladder. Br J Pharmacol 111:473–482

Dowd E, McQueen DS, Chessell IP, Humphrey PPA (1998) P2X receptor-mediated excitation of nociceptive afferents in the normal and arthritic rat knee joint. Br J Pharmacol 125:341–346

Downie JW, Dean DM (1977) The contribution of cholinergic postganglionic neurotransmission to contractions of rabbit detrusor. J Pharmacol Exp Ther 203:417–425

Downie JW, Larsson C (1981) Prostaglandin involvement in contractions evoked in rabbit detrusor by field stimulation and by adenosine 5'-triphosphate. Can J Physiol Pharmacol 59:253–260

Dumsday B (1971) Atropine-resistance of the urinary bladder innervation. J Pharm Pharmacol 23:222–225

Dunn M, Smith JC, Ardran GM (1974) Prolonged bladder distension as a treatment of urgency and urge incontinence of urine. Br J Urol 46:645–652

Dunn PM, Blakeley AGH (1988) Suramin: a reversible P_2-purinceptor antagonist in the mouse vas deferens. Br J Pharmacol 93:243–245

Eika B, Salling LN, Loft L, Laurberg S, Lundbeck F (1988) Effect of estrogen on NANC transmission in bladder of mature female rats. Neurourol Urodyn 7:201–203

Ekström J, Elmér M (1980) Compensatory increase of responses to nerve stimulation of the partially denervated rat urinary bladder. Acta Physiol Scand 110:21–29

Ekström J, Malmberg L (1984) Development of supersensitivity to methacholine in the rat detrusor following either parasympathetic denervation or decentralization. Acta Physiol Scand 122:175–179

Ekström J, Uvelius B (1981) Length-tension relations of smooth muscle from normal and denervated rat urinary bladders. Acta Physiol Scand 112:443–447

Ekström J, Henningsson AC, Henningsson S, Malmberg L (1984) Hyperplasia and hypertrophia in the denervated and distended rat urinary bladder. Acta Physiol Scand 122:45–48

Elbadawi A (1982) Neuromorphological basis of vesicourethral function. I. Histochemistry, ultrastructure and function of intrinsic nerves of the bladder and urethra. Neurourol Urodyn 1:3–50

Ellenberg M (1980) Development of urinary bladder dysfunction in diabetes mellitus. Ann Intern Med 92:321–323

Elmér M (1975) Atropine sensitivity of the rat urinary bladder during nerve degeneration. Acta Physiol Scand 93:202–205

Erickson K, Buffington CA, Kanai AJ, de Groat WC, Bullo A, D'Alatri L, Edwards D, Birder LA (1998) Alterations in nitric oxide (NO) production of NK2 immunoreactivity in urinary bladder from cats with feline interstitial cystitis. Soc Neurosci Abst 24:1619

Faerman I, Glocer L, Celener D, Jadzinsky M, Fox D, Maler M, Alvarez E (1973) Autonomic nervous system and diabetes. Histological and histochemical study of the autonomic nerve fibers of the urinary bladder in diabetic patients. Diabetes 22:225–237

Fahrenkrug J, Hannibal J (1998) Pituitary adenylate cyclase activating polypeptide immunoreactivity in capsaicin-sensitive nerve fibres supplying the rat urinary tract. Neuroscience 83:1261–1272

Fehér E, Csanyi K, Vajda J (1979) Ultrastructure of the nerve cells and fibres in the urinary bladder wall of the cat. Acta Anat (Basel) 103:109–118

Ferguson D, Christopher N (1996) Urinary bladder function and drug development. Trends Pharmacol Sci 17:161–165

Ferguson DR (1999) Urothelial function. BJU Int 84:235–242
Ferguson DR, Kennedy I, Burton TJ (1997) ATP is released from rabbit urinary bladder epithelial cells by hydrostatic pressure changes – a possible sensory mechanism? J Physiol 505:503–511
Flood HD, Downie JW, Awad SA (1988) Influence of filling rates and sympathectomy on bladder compliance in the chloralose-anaesthetized cat. Neurourol Urodyn 7:377–384
Fowler CJ, Jewkes D, McDonald WI, Lynn B, de Groat WC (1992) Intravesical capsaicin for neurogenic bladder dysfunction. Lancet 339:1239
Fowler CJ, Beck RO, Gerrard S, Betts CD, Fowler CG (1994) Intravesical capsaicin for treatment of detrusor hyperreflexia. J Neurol Neurosurg Psychiatry 57:169–173
Friedland GW, Perkash I (1983) Neuromuscular dysfunction of the bladder and urethra. Semin Roentgenol 18:255–266
Frimodt-Moller C (1980) Diabetic cystopathy: epidemiology and related disorders. Ann Intern Med 92:318–321
Fry CH, Wu C (2000) Determinants of mechanical activity in detrusor smooth muscle. J Physiol 523:61P
Fuder H, Muscholl E (1995) Heteroreceptor-mediated modulation of noradrenaline and acetylcholine release from peripheral nerves. Rev Physiol Biochem Pharmacol 126:265–412
Fuder H, Muth U (1993) ATP and endogenous agonists inhibit evoked [^{3}H]-noradrenaline release in rat iris via A_1 and P_{2y}-like purinoceptors. Naunyn Schmiedebergs Arch Pharmacol 348:352–357
Fujii K (1987) Electrophysiological evidence that adenosine triphosphate (ATP) is a cotransmitter with acetylcholine (ACh) in isolated guinea-pig, rabbit and pig urinary bladder. Proc Physiol Soc 394:26P
Fujii K (1988) Evidence for adenosine triphosphate as an excitatory transmitter in guinea-pig, rabbit and pig urinary bladder. J Physiol 404:39–52
Fujii K, Foster CD, Brading AF, Parekh AB (1990) Potassium channel blockers and the effects of cromakalim on the smooth muscle of the guinea-pig bladder. Br J Pharmacol 99:779–785
Fujii R, Oshima N (1986) Control of chromatophore movements in teleost fish. Zool Sci 3:13–47
Gabella G, Davis C (1998) Distribution of afferent axons in the bladder of rats. J Neurocytol 27:141–155
Galloway NT, Gabale DR, Irwin PP (1991) Interstitial cystitis or reflex sympathetic dystrophy of the bladder? Semin Urol 9:148–153
Garcia-Pascual A, Costa G, Garcia-Sacristan A, Andersson K-E (1991) Relaxation of sheep urethral muscle induced by electrical stimulation of nerves: involvement of nitric oxide. Acta Physiol Scand 141:531–539
Ghoneim MA, Fretin JA, Gagnon DJ, Lebel E, Van Lier J, Arsenault A, Susset JG (1976) The influence of vesical distension on the urethral resistance to flow: a possible role for prostaglandins? J Urol 116:739–743
Ghoniem GM, Shoukry MS (1991) Atropine resistance phenomenon in human bladders of myelodysplastic children. Neurourol Urodyn 10:304
Gilpin CJ, Dixon JS, Gilpin SA, Gosling JA (1983) The fine structure of autonomic neurons in the wall of the human urinary bladder. J Anat 137:705–713
Giuliani S, Lecci A, Tramontana M, Maggi CA (1998) The inhibitory effect of nociceptin on the micturition reflex in anaesthetized rats. Br J Pharmacol 124:1566–1572
Gosling JA, Dixon JS, Humpherson JA (1983) Functional Anatomy of the Urinary Tract: an Integrated Text and Colur Atlas. Churchill Livingstone, Edinburgh
Gowing NFC (1960) Pathological changes in the bladder following irradiation. Br J Radiol 33:484–487
Grandadam F, Lluel P, Palea S, Martin DJ (1999) Pharmacological and urodynamic changes in rat urinary bladder function after multiple pregnancies. BJU Int 84:861–866

Gu J, Islam KN, Restorick J, Adrian TE, Bloom SR, Polak JM (1982) Peptidergic innervation of the human urinary tract. Journal of Pathology 138:89–90

Gu J, Polak JM, Deane A, Cocchia D, Michetti F (1984) Increase of S-100 immunoreactivity in the urinary bladder from patients with multiple sclerosis, an indication of peripheral neuronal lesion. Am J Clin Pathol 82:649–654

Guidotti G (1996) ATP transport and ABC proteins. Chem Biol 3:703–706

Gupta S, Gomes CM, Dhulipala PDK, Wein AJ, Kotlikoff MI (2000) The expression of functional $P2Y_1$ receptors in rabbit bladder smooth muscle. J Urol (In Press)

Gür S, Karahan ST (1997) Effects of adenosine 5′-triphosphate, adenosine and acetylcholine in urinary bladder and colon muscles from streptozotocin diabetic rats. Arzneimittelforschung 47:1226–1229

Häbler HJ, Jänig W, Koltzenburg M (1990) Activation of unmyelinated afferent fibres by mechanical stimuli and inflammation of the urinary bladder in the cat. J Physiol 425:545–562

Häbler HJ, Jänig W, Koltzenburg M (1993) Myelinated primary afferents of the sacral spinal cord responding to slow filling and distension of the cat urinary bladder. J Physiol 463:449–460

Hamada K, Takuwa N, Yokoyama K, Takuwa Y (1998) Stretch activates Jun N-terminal kinase/stress-activated protein kinase in vascular smooth muscle cells through mechanisms involving autocrine ATP stimulation of purinoceptors. J Biol Chem 273:6334–6340

Hammer K, Sann H, Pierau F-K (1993) Functional properties of mechanosensitive units from the chicken ureter in vitro. Pflugers Arch 425:353–361

Hansen MA, Balcar VJ, Barden JA, Bennett MR (1998) The distribution of single P_{2X1}-receptor clusters on smooth muscle cells in relation to nerve varicosities in the rat urinary bladder. J Neurocytol 27:529–539

Harrison SC, Ferguson DR, Doyle PT (1990a) Effect of bladder outflow obstruction on the innervation of the rabbit urinary bladder. Br J Urol 66:372–379

Harrison SC, Ferguson DR, Hanley MR (1990b) Effect of capsaicin on the rabbit urinary bladder. What is the function of sensory nerves that contain substance P? Br J Urol 66:155–161

Hashimoto M, Kokubun S (1995) Contribution of P_2-purinoceptors to neurogenic contraction of rat urinary bladder smooth muscle. Br J Pharmacol 115:636–640

Hashimoto S, Kigoshi S, Muramatsu I (1992) Neurogenic responses of urethra isolated from the dog. Eur J Pharmacol 213:117–123

Hashimoto S, Kigoshi S, Muramatsu I (1993) Nitric oxide-dependent and -independent neurogenic relaxation of isolated dog urethra. Eur J Pharmacol 231:209–214

Hashitani H, Edwards FR (1999) Spontaneous and neurally activated depolarizations in smooth muscle cells of the guinea-pig urethra. J Physiol 514:459–470

Hashitani H, Suzuki H (1995) Electrical and mechanical responses produced by nerve stimulation in detrusor smooth muscle of the guinea-pig. Eur J Pharmacol 284: 177–183

Hasséssian H, Bodin P, Burnstock G (1993) Blockade by glibenclamide of the flow-induced endothelial release of ATP that contributes to vasodilatation in the pulmonary vascular bed of the rat. Br J Pharmacol 109:466–472

Hawranko AA, Barrick S, Birder LA, de Groat WC (1999) Effects of capsaicin and cyclophosphamide on the distension-dependent release of ATP into the lumenof the lower urinary tractspinal and peripheral administration of purinergic agonists on the micturition reflex in the rat. Soc Neurosci Abst 25:1171

Hegde SS, Mandel DA, Wilford MR, Briaud S, Ford APDW, Eglen RM (1998) Evidence for purinergic neurotransmission in the urinary bladder of pithed rats. Eur J Pharmacol 349:75–82

Henderson VE (1923) The action of atropine on intestine and urinary bladder. Arch Int Pharmacodyn Ther 27:205–211

Henderson VE, Roepke MH (1934) The role of acetylcholine in bladder contractile mechanisms and in parasympathetic ganglia. J Pharmacol Exp Ther 51:97–111

Henderson VE, Roepke MH (1935) The urinary bladder mechanisms. J Pharmacol Exp Ther 54:408–414

Henning RH (1997) Purinoceptors in neuromuscular transmission. Pharmacol Ther 74:115–128

Hernández M, Barahona MV, Bustamante S, García-Sacristín A, Orensanz LM (1999) A_{2B} adenosine receptors mediate relaxation of the pig intravesical ureter: adenosine modulation of non adrenergic non cholinergic excitatory neurotransmission. Br J Pharmacol 126:969–978

Hills J, Meldrum LA, Klarskov P, Burnstock G (1984) A novel non-adrenergic, non-cholinergic nerve-mediated relaxation of the pig bladder neck: an examination of possible neurotransmitter candidates. Eur J Pharmacol 99:287–293

Hindmarsh JR, Idowu OA, Yeates WK, Zar MA (1977) Pharmacology of electrically evoked contractions of human bladder. Br J Pharmacol 61:115P

Hisayama T, Shinkai M, Takayanagi I, Toyoda T (1988) Mechanism of action of nicotine in isolated urinary bladder of guinea- pig. Br J Pharmacol 95:465–472

Hogaboom GK, O'Donnell JP, Fedan JS (1980) Purinergic receptors: photoaffinity analog of adenosine triphosphate is a specific adenosine triphosphate antagonist. Science 208:1273–1276

Holck MI, Marks BH (1978) Purine nucleoside and nucleotide interactions on normal and subsensitive alpha adrenoreceptor responsiveness in guinea-pig vas deferens. J Pharmacol Exp Ther 205:104–117

Holm-Bentzen M, Lose G (1987) Pathology and pathogenesis of interstitial cystitis. Urology 29:8–13

Holt SE, Cooper M, Wyllie JH (1985) Evidence for purinergic transmission in mouse bladder and for modulation of responses to electrical stimulation by 5- hydroxytryptamine. Eur J Pharmacol 116:105–111

Holzer P (1991) Capsaicin: cellular targets, mechanisms of action, and selectivity for thin sensory neurons. Pharmacol Rev 43:143–201

Hourani SMO (1984) Desensitization of the guinea-pig urinary bladder by the enantiomers of adenylyl 5'-(β,γ-methylene)-diphosphonate and by substance P. Br J Pharmacol 82:161–164

Hourani SMO (1999) Postnatal development of purinoceptors in rat visceral smooth muscle preparations. Gen Pharmacol 32:3–7

Hourani SM, Chown JA (1989) The effects of some possible inhibitors of ectonucleotidases on the breakdown and pharmacological effects of ATP in the guinea-pig urinary bladder. Gen Pharmacol 20:413–416

Hourani SMO, Welford LA, Cusack NJ (1985) L-AMP-PCP, an ATP receptor agonist in guinea-pig bladder, is inactive on taenia coli. Eur J Pharmacol 108:197–200

Howson W, Taylor EM, Parsons ME, Novelli R, Wilczynska MA, Harris DT (1988) Synthesis and biological evaluation of ATP analogues acting at putative purinergic P_{2X}-receptors (on guinea pig bladder). Eur J Med Chem 23:433–439

Hoyes AD (1984) Fine structure and response to capsaicin of primary afferent nociceptive axons in the rat and guinea-pig ureter. In: Hamann W, Iggo A (eds) Sensory Receptor Mechanisms. World Scientific Publ., Singapore, pp 25–34

Hoyes AD, Barber P (1976) Parameters of fixation of the putative pain afferents in the ureter: preservation of the dense cores of the large vesicles in the axonal terminals. J Anat 122:113–120

Hoyes AD, Barber P, Martin BG (1975) Comparative ultrastructure of the nerves innervating the muscle of the body of the bladder. Cell Tissue Res 164:133–144

Hoyle CH (1994) Non-adrenergic, non-cholinergic control of the urinary bladder. World J Urol 12:233–244

Hoyle CHV, Burnstock G (1985) Atropine-resistant excitatory junction potentials in rabbit bladder are blocked by α,β-methylene ATP. Eur J Pharmacol 114:239–240

Hoyle CHV, Burnstock G (1991) ATP receptors and their physiological roles. In: Stone TW (ed) Adenosine in the Nervous System. Academic Press, London, pp 43–76

Hoyle CHV, Burnstock G (1993) Postganglionic efferent transmission to the bladder and urethra. In: Maggi C (ed) The Autonomic Nervous System, Vol. 3. Nervous Control of the Urogenital System. Harwood Academic Publishers, Switzerland, pp 349–383

Hoyle CHV, Moss HE, Burnstock G (1986) Ethylcholine mustard aziridinium ion (AF64A) impairs cholinergic neuromuscular transmission in the guinea-pig ileum and urinary bladder, and cholinergic neuromodulation in the enteric nervous system of the guinea-pig distal colon. Gen Pharmacol 17:543–548

Hoyle CHV, Chapple C, Burnstock G (1989) Isolated human bladder: evidence for an adenine dinucleotide acting on P_{2X}-purinoceptors and for purinergic transmission. Eur J Pharmacol 174:115–118

Hoyle CHV, Knight GE, Burnstock G (1990) Suramin antagonizes responses to P_2-purinoceptor agonists and purinergic nerve stimulation in the guinea-pig urinary bladder. Br J Pharmacol 99:617–621

Hoyle CHV, Lincoln J, Burnstock G (1994) Neural control of pelvic organs. In: Rushton DN (ed) Handbook of Neuro-Urology. Marcel Dekker, Inc., New York, pp 1–54

Hoyle CHV, Ralevic V, Lincoln J, Knight GE, Goss-Sampson MA, Milla PJ, Burnstock G (1995) Effects of vitamin E deficiency on autonomic neuroeffector mechanisms in the rat caecum, vas deferens and urinary bladder. J Physiol 487:773–786

Hoyle CH, Chakrabarti G, Pendleton NP, Andrews PL (1998) Neuromuscular transmission and innervation in the urinary bladder of the insectivore *Suncus murinus*. J Auton Nerv Syst 69:31–38

Hökfelt T, Kellerth JO, Nilsson G, Pernow B (1975) Experimental immunohistochemical studies on the localization and distribution of substance P in cat primary sensory neurons. Brain Res 100:235–252

Hua X-Y (1986) Tachykinins and calcitonin gene-related peptide in relation to peripheral functions of capsaicin-sensitive sensory neurons. Acta Physiol Scand Suppl 551:1–45

Huddart H, Butler DJ (1986) Field stimulation responses of rat urinary bladder detrusor smooth-muscle. Dependence upon slow calcium channel activity determined by K^+ depolarization and calcium antagonists. Gen Pharmacol 17:695–703

Hukovic S, Rand MJ, Vanov S (1965) Observations on an isolated, innervated preparation of rat urinary bladder. Br J Pharmacol 24:178–188

Husted S, Sjögren C, Andersson K-E (1980a) Mechanisms of the responses to non-cholinergic, non-adrenergic nerve stimulation and to ATP in isolated rabbit urinary bladder: evidence for ADP evoked prostaglandin release. Acta Pharmacol Toxicol 47:84–92

Husted S, Sjögren C, Andersson K-E (1980b) Role of prostaglandins in the responses of rabbit detrusor to non-cholinergic, non-adrenergic nerve stimulation and to ATP. Arch Int Pharmacodyn Ther 246:84–97

Husted S, Andersson K-E, Sommer L, Østergaard JR (1980c) Anticholinergic and calcium antagonistic effects of terodiline in rabbit urinary bladder. Acta Pharmacol Toxicol 46 Suppl 1:20–30

Husted S, Sjögren C, Andersson K-E (1983) Direct effects of adenosine and adenine nucleotides on isolated human urinary bladder and their influence on electrically induced contractions. J Urol 130:392–398

Hutton KA, Trejdosiewicz LK, Thomas DF, Southgate J (1993) Urothelial tissue culture for bladder reconstruction: an experimental study. J Urol 150:721–725

Iacovou JW, Hill SJ, Birmingham AT (1990) Agonist-induced contraction and accumulation of inositol phosphates in the guinea-pig detrusor: evidence that muscarinic and purinergic receptors raise intracellular calcium by different mechanisms. J Urol 144:775–779

Igawa Y, Mattiasson A, Andersson K-E (1993) Functional importance of cholinergic and purinergic neurotransmission for micturition contraction in the normal, unanaesthetized rat. Br J Pharmacol 109:473–479

Igawa Y, Mattiasson A, Andersson K-E (1994) Micturition and premicturition contractions in unanesthetized rats with bladder outlet obstruction. J Urol 151: 244–249

Imagawa J, Akima M, Sakai K (1989) Functional evaluation of sympathetically mediated responses in in vivo lower urinary tract of dogs. J Pharmacol Methods 22: 103–111

Inoue R, Brading AF (1990) The properties of the ATP-induced depolarization and current in single cells isolated from the guinea-pig urinary bladder. Br J Pharmacol 100:619–625

Inoue R, Brading AF (1991) Human, pig and guinea-pig bladder smooth muscle cells generate similar inward currents in response to purinoceptor activation. Br J Pharmacol 103:1840–1841

Iosif CS, Batra S, Ek A, Astedt B (1981) Estrogen receptors in the human female lower uninary tract. Am J Obstet Gynecol 141:817–820

Iravani MM, Zar MA (1994) Neuropeptide Y in rat detrusor and its effect on nerve-mediated and acetylcholine-evoked contractions. Br J Pharmacol 113: 95–102

Iravani MM, Luheshi GN, Zar MA (1988) Inhibition of non-cholinergic motor transmission in isolated rat bladder by nifedipine. J Physiol 410:61P

Islam KN, Gu J, McGregor GP, Shuttleworth KED, Bloom SR, Polak JM (1983) Morphological evidence of physiological functioning peptide-containing nerves in the human urinary bladder. Proceedings of the 13th Annual Meeting of the International Continence Society, Aachen: 247–249

Ito Y, Kimoto Y (1985) The neural and non-neural mechanisms involved in urethral activity in rabbits. J Physiol 367:57–72

James MJ, Birmingham AT, Hill SJ (1993) Partial mediation by nitric oxide of the relaxation of human isolated detrusor strips in response to electrical field stimulation. Br J Clin Pharmacol 35:366–372

Jancsó G, Maggi CA (1987) Distribution of capsaicin-sensitive urinary bladder afferents in the rat spinal cord. Brain Res 418:371–376

Jänig W, Morrison JFB (1986) Functional properties of spinal visceral afferents supplying abdominal and pelvic organs, with special emphasis on visceral nociception. In: Cervero F, Morrison JFB (eds) Visceral Sensation, Progress in Brain Reseach, Vol. 67. Elservier Science B.V., Amsterdam, pp 87–114

Jiang CH, Lindström S (1999) Prolonged enhancement of the micturition reflex in the cat by repetitive stimulation of bladder afferents. J Physiol 517:599–605

Johns A (1979) The effect of magnesium-free solutions on the responses of the guinea pig urinary bladder to adenosine 5′-triphosphate and nerve stimulation. Can J Physiol Pharmacol 57:1320–1323

Johns A (1981) The effect of indomethacin and substance P on the guinea pig urinary bladder. Life Sci 29:1803–1809

Johns A, Paton DM (1977) Effect of indomethacin on atropine-resistant transmission in rabbit and monkey urinary bladder: evidence for involvement of prostaglandins in transmission. Prostaglandins 13:245–254

Johnson RH, Eisenhofer G, Lambie DG (1986) The effects of acute and chronic ingestion of ethanol on the autonomic nervous system. Drug Alcohol Depend 18: 319–328

Jörgensen L, Mortensen SO, Colstrup H, Andersen JT (1985) Bladder distension in the management of detrusor instability. Scand J Urol Nephrol 19:101–104

Kasakov L, Burnstock G (1983) The use of the slowly degradable analog, α,β-methylene ATP, to produce desensitization of the P_2-purinoceptor: effect on non-adrenergic, non-cholinergic responses of the guinea-pig urinary bladder. Eur J Pharmacol 86:291–294

Kasakov LN, Vlaskovska MV (1985) Profile of prostaglandins generated in the detrusor muscle of rat urinary bladder: effects of adenosine triphosphate and adenosine. Eur J Pharmacol 113:431–436

Katsuragi T, Kuratomi L, Furukawa T (1986) Clonidine-evoked selective P_1-purinoceptor antagonism of contraction of guinea-pig urinary bladder. Eur J Pharmacol 121:119–122

Katsuragi T, Usune S, Furukawa T (1990) Antagonism by nifedipine of contraction and Ca^{2+}-influx evoked by ATP in guinea-pig urinary bladder. Br J Pharmacol 100:370–374

Keast JR, de Groat WC (1989) Immunohistochemical characterization of pelvic neurons which project to the bladder, colon, or penis in rats. J Comp Neurol 288:387–400

Keast JR, Kawatani M, de Groat WC (1990) Sympathetic modulation of cholinergic transmission in cat vesical ganglia is mediated by α_1- and α_2-adrenoceptors. Am J Physiol 258:R44–R50

Keating MA, Duckett JW, Snyder HM, Wein AJ, Potter L, Levin RM (1990) Ontogeny of bladder function in the rabbit. J Urol 144:766–769

Kerr DIB, Krantis A (1979) A new class of ATP antagonist. Proc Aust Physiol Pharmacol Soc 10:156P

Kim D-Y, Hawranko AA, Fraser MO, Yoshiyama M, Chancellor MB, Cheng C-L, de Groat WC (1999) The effects of spinal and peripheral administration of purinergic agonists on the micturition reflex in the rat. Soc Neurosci Abst 25:1170

Kihara K, de Groat WC (1997) Sympathetic efferent pathways projecting to the bladder neck and proximal urethra in the rat. J Auton Nerv Syst 62:134–142

Kinder RB, Mundy AR (1985) Atropine blockade of nerve-mediated stimulation of the human detrusor. Br J Urol 57:418–421

King JA, Huddart H, Staff WG (1997) Purinergic modulation of rat urinary bladder detrusor smooth muscle. Gen Pharmacol 29:597–604

Kishii K-I, Hisayama T, Takayanagi I (1992) Comparison of contractile mechanisms by carbachol and ATP in detrusor strips of rabbit urinary bladder. Jpn J Pharmacol 58:219–229

Klarskov P (1987a) Non-cholinergic, non-adrenergic inhibitory nerve responses of bladder outlet smooth muscle in vitro. Br J Urol 60:337–342

Klarskov P (1987b) Non-cholinergic, non-adrenergic nerve-mediated relaxation of pig and human detrusor muscle in vitro. Br J Urol 59:414–419

Knight GE, Brizzolara AL, Soediono P, Karoon P, Burnstock G (1995) Chronic ethanol consumption affects cholinoceptor- and purinoceptor- mediated contractions of the isolated rat bladder. Alcohol 12:183–188

Knight GE, Bodin P, de Groat WC, Burnstock G (1999) Distension of the guinea pig ureter releases ATP from the epithelium. Soc Neurosci Abst 25:1171

Koley B, Koley J, Saha JK (1984) The effects of nicotine on spontaneous contractions of cat urinary bladder in situ. Br J Pharmacol 83:347–355

Kolta MG, Wallace LJ, Gerald MC (1985) Streptozocin-induced diabetes affects rat urinary bladder response to autonomic agents. Diabetes 34:917–921

Koziol JA, Clark DC, Gittes RF, Tan EM (1993) The natural history of interstitial cystitis: a survey of 374 patients. J Urol 149:465–469

Krell RD, McCoy JL, Ridley PT (1981) Pharmacological characterization of the excitatory innervation to the guinea-pig urinary bladder in vitro: evidence for both cholinergic and non-adrenergic-non-cholinergic neurotransmission. Br J Pharmacol 74:15–22

Krishtal OA, Marchenko SM, Obukhov AG (1988) Cationic channels activated by extracellular ATP in rat sensory neurons. Neuroscience 27:995–1000

Kropp BP (1998) Small-intestinal submucosa for bladder augmentation: a review of preclinical studies. World J Urol 16:262–267

Kropp BP, Eppley BL, Prevel CD, Rippy MK, Harruff RC, Badylak SF, Adams MC, Rink RC, Keating MA (1995) Experimental assessment of small intestinal submucosa as a bladder wall substitute. Urology 46:396–400

Kruse MN, Belton AL, de Groat WC (1993) Changes in bladder and external urethral sphincter function after spinal cord injury in the rat. Am J Physiol 264:R1157–R1163

Kudlacz EM, Chun AL, Skau KA, Gerald MC, Wallace LJ (1988) Diabetes and diuretic-induced alterations in function of rat urinary bladder. Diabetes 37:949–955

Kudlacz EM, Gerald MC, Wallace LJ (1989) Effects of diabetes and diuresis on contraction and relaxation mechanisms in rat urinary bladder. Diabetes 38:278–284

Kuo DC, Hisamitsu T, de Groat WC (1983) The function of efferent projections from the lumbosacral sympathetic chain to the urinary bladder in the cat. Soc Neurosci Abst 9:610

Kuo DC, Hisamitsu T, de Groat WC (1984) A sympathetic projection from sacral paravertebral ganglia to the pelvic nerve and to postganglionic nerves on the surface of the urinary bladder and large intestine of the cat. J Comp Neurol 226:76–86

Kura H, Obara K, Yabu H (1992) Contractile responses to electrical field stimulation and ATP in guinea- pig urinary bladder. Comp Biochem Physiol C 102:193–197

Labadia A, Rivera L, Costa G, Garcia-Sacristan A (1988) Influence of the autonomic nervous system in the horse urinary bladder. Res Vet Sci 44:282–285

Laird JM, Roza C, Cervero F (1996) Spinal dorsal horn neurons responding to noxious distension of the ureter in anesthetized rats. J Neurophysiol 76:3239–3248

Lambrecht G, Friebe T, Grimm U, Windscheif U, Bungardt E, Hildebrandt C, Bäumert HG, Spatz-Kümbel G, Mutschler E (1992) PPADS, a novel functionally selective antagonist of P_2 purinoceptor-mediated responses. Eur J Pharmacol 217:217–219

Langley KN, Anderson HK (1895) The innervation of the pelvic and adjoining viscera. Part II. The bladder. J Physiol 19:71–84

Lantéri-Minet M, Bon K, de Pommery J, Michiels JF, Menétrey D (1995) Cyclophosphamide cystitis as a model of visceral pain in rats: model elaboration and spinal structures involved as revealed by the expression of c-Fos and Krox-24 proteins. Exp Brain Res 105:220–232

Lavelle JP, Meyers SA, Ruiz WG, Buffington CA, Zeidel ML, Apodaca G (2000) Urothelial pathophysiological changes in feline interstitial cystitis: a human model. Am J Physiol 278:F540–F553

Lee HY, Bardini M, Burnstock G (2000) Distribution of P2X receptors in the urinary bladder and ureter of the rat. J Urol 163:2002–2007

Lee JG, Wein AJ, Levin RM (1994) Comparative pharmacology of the male and female rabbit bladder neck and urethra: involvement of nitric oxide. Pharmacology 48:250–259

Levin RM, Wein AJ (1982) Response of the in vitro whole bladder (rabbit) preparation to autonomic agonists. J Urol 128:1087–1090

Levin RM, Shofer FS, Wein AJ (1980a) Cholinergic, adrenergic and purinergic response of sequential strips of rabbit urinary bladder. J Pharmacol Exp Ther 212:536–540

Levin RM, Shofer FS, Wein AJ (1980b) Estrogen-induced alterations in the autonomic responses of the rabbit urinary bladder. J Pharmacol Exp Ther 215:614–618

Levin RM, Jacoby R, Wein AJ (1981a) Effect of adenosine triphosphate on contractility and adenosine triphosphatase activity of the rabbit urinary bladder. Mol Pharmacol 19:525–528

Levin RM, Malkowicz SB, Jacobowitz D, Wein AJ (1981b) The ontogeny of the autonomic innervation and contractile response of the rabbit urinary bladder. J Pharmacol Exp Ther 219:250–257

Levin RM, Brendler K, Wein AJ (1983a) Comparative pharmacological response of an in vitro whole bladder preparation (rabbit) with response of isolated smooth muscle strips. J Urol 130:377–381

Levin RM, Jacoby R, Wein AJ (1983b) High-affinity, divalent ion-specific binding of ^{3}H-ATP to homogenate derived from rabbit urinary bladder. Comparison with divalent-ion ATPase activity. Mol Pharmacol 23:1–7

Levin RM, Ruggieri MR, Wein AJ (1986a) Functional effects of the purinergic innervation of the rabbit urinary bladder. J Pharmacol Exp Ther 236:452–457

Levin RM, Ruggieri MR, Velagapudi S, Gordon D, Altman B, Wein AJ (1986b) Relevance of spontaneous activity to urinary bladder function: an in vitro and in vivo study. J Urol 136:517–521

Levin RM, Longhurst PA, Kato K, McGuire EJ, Elbadawi A, Wein AJ (1990) Comparative physiology and pharmacology of the cat and rabbit urinary bladder. J Urol 143:848–852

Levin RM, Tong Y-C, Wein AJ (1991a) Effect of pregnancy on the autonomic response of the rabbit urinary bladder. Neurourol Urodyn 10:313

Levin RM, Zderic SA, Ewalt DH, Duckett JW, Wein AJ (1991b) Effects of pregnancy on muscarinic receptor density and function in the rabbit urinary bladder. Pharmacology 43:69–77

Lewis C, Neidhart S, Holy C, North RA, Surprenant A (1995) Coexpression of $P2X_2$ and $P2X_3$ receptor subunits can account for ATP-gated currents in sensory neurons. Nature 377:432–435

Li C, Peoples RW, Weight FF (1998) Ethanol-induced inhibition of a neuronal P2X purinoceptor by an allosteric mechanism. Br J Pharmacol 123:1–3

Li JH, Yasay GD, Kau ST, Ohnmacht CJ, Trainor DA, Bonev AD, Heppner TJ, Nelson MT (1996) Studies of the K_{ATP} channel opening activity of the new dihydropyridine compound 9-(3-cyanophenyl)-3,4,6,7,9,10-hexahydro-1,8-(2H,5H)-acridinedione in bladder detrusor in vitro. Arzneimittelforschung 46:525–530

Lin MJ, Liu S-H, Lin-Shiau S-Y (1998) Phorbol ester-induced contractions of mouse detrusor muscle are inhibited by nifedipine. Naunyn Schmiedebergs Arch Pharmacol 357:553–557

Lincoln J, Burnstock G (1993) Autonomic innervation of the urinary bladder and urethra. In: Maggi C (ed) The Autonomic Nervous System, Vol. 3. Nervous Control of the Urogenital System. Harwood Academic Publishers, Switzerland, pp 33–69

Lincoln J, Haven AJ, Sawyer M, Burnstock G (1984a) The smooth muscle of rat bladder in the early stages of streptozotocin-induced diabetes. Br J Urol 56:24–30

Lincoln J, Crockett M, Haven AJ, Burnstock G (1984b) Rat bladder in the early stages of streptozotocin-induced diabetes: adrenergic and cholinergic innervation. Diabetologia 26:81–87

Liu S-H, Lin-Shiau S-Y (1996) The effects of uranyl ions on neuromuscular transmission in the urinary bladder of the normal and streptozotocin-diabetic mouse. Naunyn Schmiedebergs Arch Pharmacol 354:773–778

Liu SP, Horan P, Levin RM (1998) Effects of atropine, isoproterenol and propranolol on the rabbit bladder contraction induced by intra-arterial administration of acetylcholine and ATP. J Urol 160:1863–1866

Llewellyn-Smith IJ, Burnstock G (1998) Ultrastructural localization of $P2X_3$ receptors in rat sensory neurons. Neuroreport 9:2245–2250

Longhurst PA, Belis JA (1986) Abnormalities of rat bladder contractility in streptozotocin-induced diabetes mellitus. J Pharmacol Exp Ther 238:773–777

Longhurst PA, Belis JA, O'Donnell JP, Galie JR, Westfall DP (1984) A study of the atropine-resistant component of the neurogenic response of the rabbit urinary bladder. Eur J Pharmacol 99:295–302

Longhurst PA, Kang J, Wein AJ, Levin RM (1990) The influence of intravesical volume upon contractile responses of the whole bladder preparation from streptozotocin-diabetic rats. Gen Pharmacol 21:687–692

Longhurst PA, Brotcke TP, Leggett RE, Levin RM (1992) The influence of streptozotocin-induced diabetes mellitus on the sensitivity of rat urinary bladder body and base strips to changes in extracellular calcium. Gen Pharmacol 23:83–88

Longhurst PA, Leggett RE, Briscoe JA (1995) Influence of strip size and location on contractile responses of rat urinary bladder body strips. Gen Pharmacol 26:1519–1527

Luheshi G, Zar A (1990a) Purinoceptor desensitization impairs but does not abolish the non-cholinergic motor transmission in rat isolated urinary bladder. Eur J Pharmacol 185:203–208

Luheshi GN, Zar MA (1990b) Presence of non-cholinergic motor transmission in human isolated bladder. J Pharm Pharmacol 42:223–224

Luheshi GN, Zar MA (1990c) Inhibitory effect of streptozotocin-induced diabetes on non-cholinergic motor transmission in rat detrusor and its prevention by sorbinil. Br J Pharmacol 101:411–417

Luheshi GN, Zar MA (1991) The effect of streptozotocin-induced diabetes on cholinergic motor transmission in the rat urinary bladder. Br J Pharmacol 103:1657–1662

Lukacsko P, Krell RD (1981) The effects of nucleotides on the response of the isolated guinea pig urinary bladder to nonadrenergic, noncholinergic nerve stimulation. Can J Physiol Pharmacol 59:1199–1201

Lukacsko P, Krell RD (1982) Response of the guinea-pig urinary bladder to purine and pyrimidine nucleotides. Eur J Pharmacol 80:401–406

MacDermott AB, Role LW, Siegelbaum SA (1999) Presynaptic ionotropic receptors and the control of transmitter release. Annu Rev Neurosci 22:443–485

MacKenzie I, Burnstock G (1984) Neuropeptide action on the guinea-pig bladder; a comparison with the effects of field stimulation and ATP. Eur J Pharmacol 105:85–94

MacKenzie I, Burnstock G, Dolly JO (1982) The effects of purified botulinum neurotoxin type A on cholinergic, adrenergic and non-adrenergic, atropine-resistant autonomic neuromuscular transmission. Neuroscience 7:997–1006

Maggi CA (1991) Omega conotoxin and prejunctional modulation of the biphasic response of the rat isolated urinary bladder to single pulse electrical field stimulation. J Auton Pharmacol 11:295–304

Maggi CA, Grimaldi G, Meli A (1982) The effects of nifedipine and verapamil on spontaneous and carbachol-stimulated contractions of rat urinary bladder "in vivo". Arch Int Pharmacodyn Ther 257:288–294

Maggi CA, Santicioli P, Meli A (1984) Postnatal development of myogenic contractile activity and excitatory innervation of rat urinary bladder. Am J Physiol 247:R972–R978

Maggi CA (1993) The dual sensory and efferent function of capsaicin-sensitive primary sensory nerves in the bladder and urethra. In: Maggi C (ed) The Autonomic Nervous System, Vol. 3. Nervous Control of the Urogenital System. Harwood Academic Publishers, Chur, Switzerland, pp 383–422

Maggi CA, Santicioli P, Meli A (1985) Pharmacological evidence for the existence of two components in the twitch response to field stimulation of detrusor strips from the rat urinary bladder. J Auton Pharmacol 5:221–229

Maggi CA, Santicioli P, Giuliani S, Abelli L, Meli A (1986) The motor effect of the capsaicin-sensitive inhibitory innervation of the rat ureter. Eur J Pharmacol 126:333–336

Maggi CA, Giuliani S, Santicioli P, Abelli L, Geppetti P, Somma V, Renzi D, Meli A (1987a) Species-related variations in the effects of capsaicin on urinary bladder functions: relation to bladder content of substance P-like immunoreactivity. Naunyn Schmiedebergs Arch Pharmacol 336:546–555

Maggi CA, Giuliani S, Santicioli P, Abelli L, Regoli D, Meli A (1987b) Further studies on the mechanisms of the tachykinin-induced activation of micturition reflex in rats: evidence for the involvement of the capsaicin-sensitive bladder mechanoreceptors. Eur J Pharmacol 136:189–205

Maggi CA, Giuliani S, Santicioli P, Abelli L, Meli A (1987c) Visceromotor responses to calcitonin gene-related peptide (CGRP) in the rat lower urinary tract: evidence for a transmitter role in the capsaicin-sensitive nerves of the ureter. Eur J Pharmacol 143:73–82

Maggi CA, Manzini S, Parlani M, Conte B, Giuliani S, Meli A (1988) The effect of nifedipine on spontaneous, drug-induced and reflexly-activated contractions of the rat urinary bladder: evidence for the participation of an intracellular calcium store to micturition contraction. Gen Pharmacol 19:73–81

Maggi CA, Conte B, Furio M, Santicioli P, Giuliani S, Meli A (1989a) Further studies on mechanisms regulating the voiding cycle of the rat urinary bladder. Gen Pharmacol 20:833–838

Maggi CA, Santicioli P, Manzini S, Conti S, Giuliani S, Patacchini R, Meli A (1989b) Functional studies on the cholinergic and sympathetic innervation of the rat proximal urethra: effect of pelvic ganglionectomy or experimental diabetes. J Auton Pharmacol 9:231–241

Marchenko SM, Volkova TM, Fedorov OI (1987) ATP-activated ion conductance in isolated smooth muscle cells of the urinary bladder of the guinea pig. Neirofiziologiia 19:95–100

Marti-Cabrera M, Llopis P, Abengochea A, Ortiz JL, Climent VJ, Cortijo J, Morcillo EJ (1994) Effects of Ca^{2+} channel antagonists and benzodiazepine receptor ligands in normal and skinned rat urinary bladder. Eur J Pharmacol 255:157–165

Mastri AR (1980) Neuropathology of diabetic neurogenic bladder. Ann Intern Med 92:316–318

Masuda N, Uchida W, Shirai Y, Shibasaki K, Goto K, Takenaka T (1995) Effect of the potassium channel opener YM934 on the contractile response to electrical field stimulation in pig detrusor smooth muscle. J Urol 154:1914–1920

Matsumura S, Taira N, Hashimoto K (1968) The pharmacological behaviour of the urinary bladder and its vasculature of the dog. Tohoku J Exp Med 96:247–258

Mattiasson A, Andersson KE, Andersson PO, Larsson B, Sjogren C, Uvelius B (1990) Nerve-mediated functions in the circular and longitudinal muscle layers of the proximal female rabbit urethra. J Urol 143:155–160

McDonald WF, White TD (1984) Adenosine released from synaptosomes is derived from the extracelluar dephosphonylation of released ATP. Prog Neuropsychopharmacol Biol Psychiatr 8:487–494

McGuire EJ, Herlihy E (1978) Bladder and urethral responses to isolated sacral motor root stimulation. Invest Urol 16:219–223

McMurray G, Dass N, Brading AF (1998) Purinoceptor subtypes mediating contraction and relaxation of marmoset urinary bladder smooth muscle. Br J Pharmacol 123:1579–1586

Meldrum LA, Burnstock G (1985) Evidence against VIP or substance P being the transmitter in non-cholinergic excitatory nerves supplying the guinea-pig bladder. J Pharm Pharmacol 37:432–434

Messori E, Rizzi CA, Candura SM, Lucchelli A, Balestra B, Tonini M (1995) 5-Hydroxytryptamine receptors that facilitate excitatory neuromuscular transmission in the guinea-pig isolated detrusor muscle. Br J Pharmacol 115:677–683

Michel AD, Lundström K, Buell GN, Surprenant A, Valera S, Humphrey PP (1996) The binding characteristics of a human bladder recombinant P_{2X} purinoceptor, labelled with $[^3H]$-$\alpha\beta$meATP, $[^{35}S]$-ATPγS or $[^{33}P]$-ATP. Br J Pharmacol 117:1254–1260

Miller H, Simpson CA, Yeates WK (1965) Bladder dysfunction in multiple sclerosis. Br Med J 1:1265–1269

Milner P, Kirkpatrick KA, Ralevic V, Toothill V, Pearson JD, Burnstock G (1990a) Endothelial cells cultured from human umbilical vein release ATP, substance P and acetylcholine in response to increased flow. Proc R Soc Lond B Biol Sci 241:245–248

Milner P, Bodin P, Loesch A, Burnstock G (1990b) Rapid release of endothelin and ATP from isolated aortic endothelial cells exposed to increased flow. Biochem Biophys Res Commun 170:649–656

Mitchell ME, Gonzales R, Cabral BH, Bauer SB, Gearhart JP, Filmer RB (1987) Bladder augmentation problems in neurovesical dysfunction. Dialogues Ped Urol 10:1

Moore KH, Gilpin SA, Dixon JS, Richmond DH, Sutherst JR (1992) Increase in presumptive sensory nerves of the urinary bladder in idiopathic detrusor instability. Br J Urol 70:370–372

Morgan C, de Groat WC, Nadelhaft I (1986) The spinal distribution of sympathetic preganglionic and visceral primary afferent neurons that send axons into the hypogastric nerves of the cat. J Comp Neurol 243:23–40

Morrison JFB (1999) The activation of bladder wall afferent nerves. Exp Physiol 84:131–136

Morrison JFB, Namasivayam S, Eardley I (1998) ATP may be a natural modulator of the sensitivity of bladder mechanoreceptors during slow distention. Proc 1st Int Consultation on Incontinence: 84

Moss HE, Burnstock G (1985) A comparative study of electrical field stimulation of the guinea-pig, ferret and marmoset urinary bladder. Eur J Pharmacol 114:311–316

Moss HE, Lincoln J, Burnstock G (1987) A study of bladder dysfunction during streptozotocin-induced diabetes in the rat using an in vitro whole bladder preparation. J Urol 138:1279–1284

Moss HE, Tansey EM, Burnstock G (1989) Abnormalities of responses to autonomic stimulation in the mouse urinary bladder associated with Semliki Forest virus-induced demyelination. J Urol 142:850–854

Muller DP, Goss-Sampson MA (1990) Neurochemical, neurophysiological, and neuropathological studies in vitamin E deficiency. Crit Rev Neurobiol 5:239–263

Nakayama S (1993) Effects of excitatory neurotransmitters on Ca^{2+} channel current in smooth muscle cells isolated from guinea-pig urinary bladder. Br J Pharmacol 110:317–325

Namasivayam S, Eardley I, Morrison JFB (1999) Purinergic sensory neurotransmission in the urinary bladder: an in vitro study in the rat. BJU Int 84:854–860

Nance DM, Burns J, Klein CM, Burden HW (1988) Afferent fibers in the reproductive system and pelvic viscera of female rats: anterograde tracing and immunocytochemical studies. Brain Res Bull 21:701–709

Naramatsu M, Yamashita T, Kokubun S (1997) The signalling pathway which causes contraction via P2-purinoceptors in rat urinary bladder smooth muscle. Br J Pharmacol 122:558–562

Needleman P, Minkes MS, Douglas JR (1974) Stimulation of prostaglandin biosynthesis by adenine nucleotides. Profile of prostaglandin release by perfused organs. Circ Res 34:455–460

Nergårdh A, Kinn AC (1983) Neurotransmission in activation of the contractile response in the human urinary bladder. Scand J Urol Nephrol 17:153–157

Nicholls J, Hourani SM, Kitchen I (1990) The ontogeny of purinoceptors in rat urinary bladder and duodenum. Br J Pharmacol 100:874–878

Nicholls J, Hourani SMO, Kitchen I (1992a) Characterization of P_1-purinoceptors on rat duodenum and urinary bladder. Br J Pharmacol 105:639–642

Nicholls J, Hourani SMO, Kitchen I (1992b) Degradation of extracellular adenosine and ATP by adult and neonate rat duodenum and urinary bladder. Pharmacol Commun 2:203–210

Nimmo AJ, Andersson PO, Morrison JFB (1991) Reactive changes in neurokinin receptor density following selective denervations and outlet obstruction of rat bladder. J Physiol (Lond) 446:524P

Nishimura T, Akasu T (1994) Endogenous ATP modulates nicotinic transmission through presynaptic P2 receptors in rabbit parasympathetic ganglia. Neurosci Res 19:S31

Nishimura T, Tokimasa T (1996) Purinergic cation channels in neurons of rabbit vesical parasympathetic ganglia. Neurosci Lett 212:215–217

Noto H, Roppolo JR, Steers WD, de Groat WC (1989) Excitatory and inhibitory influences on bladder activity elicited by electrical stimulation in the pontine micturition center in the rat. Brain Res 492:99–115

Nunn PA, Newgreen DT (1999) An investigation into the bladder responsesinduced via pelvic nerve stimulation in the anaesthetized rat. Br J Pharmacol 126:227P

O'Reilly B, Knight GE, Popert R, Burnstock G, McMahon SB (2000) Distribution and function of P2X receptors in the unstable bladder. BJU Int (In press)

Obara K, Lepor H, Walden PD (1998) Localization of P_{2Y1} purinoceptor transcripts in the rat penis and urinary bladder. J Urol 160:587–591

Ohnishi N, Park YC, Kurita T, Kajimoto N (1997) Role of ATP and related purine compounds on urethral relaxation in male rabbits. Int J Urol 4:191–197

Oike M, Creed KE, Onoue H, Tanaka H, Ito Y (1998) Increase in calcium in smooth muscle cells of the rabbit bladder induced by acetylcholine and ATP. J Auton Nerv Syst 69:141–147

Palea S, Artibani W, Ostardo E, Trist DG, Pietra C (1993) Evidence for purinergic neurotransmission in human urinary bladder affected by interstitial cystitis. J Urol 150:2007–2012

Palea S, Corsi M, Pietra C, Artibani W, Calpista A, Gaviraghi G, Trist DG (1994) ADPβS induces contraction of the human isolated urinary bladder through a purinoceptor subtype different from P_{2X} and P_{2Y}. J Pharmacol Exp Ther 269:193–197

Palea S, Pietra C, Trist DG, Artibani W, Calpista A, Corsi M (1995) Evidence for the presence of both pre- and postjunctional P_2-purinoceptor subtypes in human isolated urinary bladder. Br J Pharmacol 114:35–40

Parija SC, Raviprakash V, Mishra SK (1991) Adenosine- and α,β-methylene ATP-induced differential inhibition of cholinergic and non-cholinergic neurogenic responses in rat urinary bladder. Br J Pharmacol 102:396–400

Park Y-C, Sugiyama T, Kaneko S, Kurita T (1986) Sympathetic contribution to bladder outlet obstructions: quantitative analysis of tissue catecholamine content. Neurourol Urodyn 5:573–577

Parkin DE, Davis JA, Symonds RP (1988) Urodynamic findings following radiotherapy for cervical carcinoma. Br J Urol 61:213–217

Paro M, Prosdocimi M (1987) Experimental diabetes in the rat: alterations in the vesical function. J Auton Nerv Syst 21:59–66

Paro M, Prosdocimi M, Zhang WX, Sutherland G, Sima AA (1989) Autonomic neuropathy in BB rats and alterations in bladder function. Diabetes 38:1023–1030

Paro M, Italiano G, Travagli RA, Petrelli L, Zanoni R, Prosdocimi M, Fiori MG (1990) Cystometric changes in alloxan diabetic rats: evidence for functional and structural correlates of diabetic autonomic neuropathy. J Auton Nerv Syst 30:1–11

Partanen M, Santer RM, Hervonen A (1980) The effect of ageing on the histochemically demonstrable catecholamines in the hypogastric (main pelvic) ganglion of the rat. Histochem J 12:527–535

Patra PB, Westfall DP (1994) Potentiation of purinergic neurotransmission in guinea pig urinary bladder by histamine. J Urol 151:787–790

Patra PB, Westfall DP (1996) Potentiation by bradykinin and substance P of purinergic neurotransmission in urinary bladder. J Urol 156:532–535

Persson CGA (1976) Inhbitory effect at the bladder-urethral junction. Acta Physiol Scand 97:139–141

Persson K, Andersson K-E (1992) Nitric oxide and relaxation of pig lower urinary tract. Br J Pharmacol 106:416–422

Pertwee RG, Fernando SR (1996) Evidence for the presence of cannabinoid CB_1 receptors in mouse urinary bladder. Br J Pharmacol 118:2053–2058

Peterson JS, Noronha-Blob L (1989) Effects of selective cholinergic antagonists and α,β-methylene ATP on guinea-pig urinary bladder contractions in vivo following pelvic nerve stimulation. J Auton Pharmacol 9:303–313

Piechota HJ, Dahms SE, Nunes LS, Dahiya R, Lue TF, Tanagho EA (1998) In vitro functional properties of the rat bladder regenerated by the bladder acellular matrix graft. J Urol 159:1717–1724

Pinna C, Bolego C, Puglisi L (1994) Effect of substance P and capsaicin on urinary bladder of diabetic rats and the role of the epithelium. Eur J Pharmacol 271:151–158

Pinna C, Ventura S, Puglisi L, Burnstock G (1996) A pharmacological and histochemical study of hamster urethra and the role of urothelium. Br J Pharmacol 119:655–662

Pinna C, Puglisi L, Burnstock G (1998a) ATP and vasoactive intestinal polypeptide relaxant responses in hamster isolated proximal urethra. Br J Pharmacol 124: 1069–1074

Pinna C, Knight G, Puglisi L, Burnstock G (1998b) Neurogenic and non-neurogenic responses in the urinary bladder of hibernating hamster. Br J Pharmacol 123: 1281–1287

Pinna C, Zanardo R, Puglisi L (2000a) Prostaglandin-release impairment in the bladder epithelium of streptozotocin-induced diabetic rats. Eur J Pharmacol 388:267–273

Pinna C, Zanardo R, Cignarella A, Bolego C, Eberini I, Nardi F, Zancan V, Puglisi L (2000b) Diabetes influences the effet of 17β-estradiol on mechanical responses of rat urethra and detrusor strips. Life Sci 6:617–627

Pittam BS, Burnstock G, Purves RD (1987) Urinary bladder intramural neurones: an electrophysiological study utilizing a tissue culture preparation. Brain Res 403: 267–278

Probst M, Dahiya R, Carrier S, Tanagho EA (1997) Reproduction of functional smooth muscle tissue and partial bladder replacement. Br J Urol 79:505–515

Prosdocimi M, Paro M (1990) Urinary bladder innervation in experimental diabetes. J Auton Nerv Syst 30 Suppl: S123–S127

Qayyum MA, Fatani JA, Abbas MO (1989) Degeneration of adrenergic nerves in the urinary bladder during pregnancy. Acta Anat (Basel) 136:303–305

Ralevic V, Burnstock G (1998) Receptors for purines and pyrimidines. Pharmacol Rev 50:413–492

Ratz PH, McCammon KA, Altstatt D, Blackmore PF, Shenfeld OZ, Schlossberg SM (1999) Differential effects of sex hormones and phytoestrogens on peak and steady state contractions in isolated rabbit detrusor. J Urol 162:1821–1828

Ribeiro JA (1979) Purinergic modulation of transmitter release. J Theor Biol 80:259–270

Rocha I, Burnstock G, Spyer KM (2000) Effect on urinary bladder function and blood pressure of the activation of purine receptors in brainstem areas. Autonom Neurosci (In press)

Roman RM, Fitz JG (1999) Emerging roles of purinergic signaling in gastrointestinal epithelial secretion and hepatobiliary function. Gastroenterology 116:964–979

Roza C, Laird JM, Cervero F (1998) Spinal mechanisms underlying persistent pain and referred hyperalgesia in rats with an experimental ureteric stone. J Neurophysiol 79:1603–1612

Ruggieri MR, Whitmore KE, Levin RM (1990) Bladder purinergic receptors. J Urol 144:176–181

Ryhammer AM, Bek KM, Laurberg S (1995) Multiple vaginal deliveries increase the risk of permanent incontinence of flatus urine in normal premenopausal women. Dis Colon Rectum 38:1206–1209

Saïag B, Bodin P, Shacoori V, Catheline M, Rault B, Burnstock G (1995) Uptake and flow-induced release of uridine nucleotides from isolated vascular endothelial cells. Endothelium 2:279–285

Saito M, Kondo A, Kato T, Levin RM (1993a) Response of isolated human neurogenic detrusor smooth muscle to intramural nerve stimulation. Br J Urol 72:723–727

Saito M, Kondo A, Kato T, Hasegawa S, Miyake K (1993b) Response of the human neurogenic bladder to KCl, carbachol, ATP and $CaCl_2$. Br J Urol 72:298–302

Salmon UJ, Walter RI, Geist SH (1941) The use of estrogens in the treatment of dysuria and incontinence in postmenopausal women. Am J Obstet Gynecol 42:845

Sann H, Cervero F (1988) Afferent innervation of the guinea-pig's ureter. Agents Actions 25:243–245

Sann H, Jancsó G, Ambrus A, Pierau F-K (1995) Capsaicin treatment induces selective sensory degeneration and increased sympathetic innervation in the rat ureter. Neuroscience 67:953–966

Sann H, Walb G, Pierau FK (1997) Postnatal development of the autonomic and sensory innervation of the musculature in the rat urinary bladder. Neurosci Lett 236:29–32

Santicioli P, Maggi CA, Meli A (1985) The effect of capsaicin pretreatment on the cystometrograms of urethane anesthetized rats. J Urol 133:700–703
Santicioli P, Maggi CA, Meli A (1986) The postganglionic excitatory innervation of the mouse urinary bladder and its modulation by prejunctional $GABA_B$ receptors. J Auton Pharmacol 6:53–66
Santicioli P, Gamse R, Maggi CA, Meli A (1987) Cystometric changes in the early phase of streptozotocin-induced diabetes in rats: evidence for sensory changes not correlated to diabetic neuropathy. Naunyn Schmiedebergs Arch Pharmacol 335:580–587
Schaufele P, Schumacher E, Acevedo CG, Contreras E (1995) Diazepam, adenosine analogues and calcium channel antagonists inhibit the contractile activity of the mouse urinary bladder. Arch Int Pharmacodyn Ther 329:454–466
Schwartzman RJ (1993) Reflex sympathetic dystrophy. Curr Opin Neurol Neurosurg 6:531–536
Senba E, Daddona PE, Nagy JI (1987) Development of adenosine deaminase-immunoreactive neurons in the rat brain. Brain Res 428:59–71
Sengupta JN, Gebhart GF (1994) Mechanosensitive properties of pelvic nerve afferent fibers innervating the urinary bladder of the rat. J Neurophysiol 72:2420–2430
Sengupta JN, Gebhart GF (1995) Mechanosensitive afferent fibers in the gastrointestinal and lower urinary tracts. In: Gebhart GF (ed) Visceral Pain. IASP Press, Seattle, pp 75–98
Sharkey KA, Williams RG, Schultzberg M, Dockray GJ (1983) Sensory substance P-innervation of the urinary bladder: possible site of action of capsaicin in causing urine retention in rats. Neuroscience 10:861–868
Shinnick-Gallagher P, Gallagher JA, Griffith WH (1986) Inhibition in parasympathetic ganglia. In: Karczmar AG, Koketsu K, Nishi S (eds) Autonomic and Enteric Ganglia. Plenum Publishing Co., New York, pp 335–367
Shinozuka K, Bjur RA, Westfall DP (1990) Effects of α,β-methylene ATP on the prejunctional purinoceptors of the sympathetic nerves of the rat caudal artery. J Pharmacol Exp Ther 254:900–904
Sibley GNA (1984) A comparison of spontaneous and nerve-mediated activity in bladder muscle from man, pig and rabbit. J Physiol 354:431–443
Sikri KL, Hoyes AD, Barber P, Jagessar M (1981) Substance P-like immunoreactivity in the intramural nerve plexuses of the guinea-pig ureter: a light and electron microscopical study. J Anat 133:425–442
Silinsky EM (1975) On the association between transmitter secretion and the release of adenine nucleotides from mammalian motor nerve terminals. J Physiol 247:145–162
Silinsky EM, Hubbard JI (1973) Release of ATP from motor nerve terminals. Nature 243:404–405
Sjögren C, Andersson K-E (1979) Effects of cholinoceptor blocking drugs, adrenoceptor stimulants, and calcium antagonists on the transmurally stimulated guinea-pig urinary bladder in vitro and in vivo. Acta Pharmacol Toxicol 44:228–234
Sjögren C, Andersson K-E, Husted S, Mattiasson A, Moller-Madsen B (1982) Atropine resistance of transmurally stimulated isolated human bladder muscle. J Urol 128:1368–1371
Sjögren C, Andersson K-E, Andersson PO, Mattiasson A, Uvelius B (1988) Different effects of neuropeptide Y on electrically induced contractions in the longitudinal and circular smooth muscle layers of the female rabbit urethra. Acta Physiol Scand 133:177–181
Sjöstrand SE, Sjögren C, Schmiterlöw CG (1972) Responses of the rabbit and cat urinary bladders in situ to drugs and to nerve stimulation. Acta Pharmacol Toxicol 31:241–254
Sjuve R, Ingvarson T, Arner A, Uvelius B (1995) Effects of purinoceptor agonists on smooth muscle from hypertrophied rat urinary bladder. Eur J Pharmacol 276:137–144

Smith DJ, Chapple CR (1994) In vitro response of human bladder smooth muscle in unstable obstructed male bladders: a study of pathophysiological causes? Neurourol Urodyn 13:414–415

Sneddon P, Burnstock G (1984) Inhibition of excitatory junction potentials in guinea-pig vas deferens by α,β-methylene-ATP: further evidence for ATP and noradrenaline as cotransmitters. Eur J Pharmacol 100:85–90

Sneddon P, McLees A (1992) Purinergic and cholinergic contractions in adult and neonatal rabbit bladder. Eur J Pharmacol 214:7–12

Sneddon P, Westfall DP, Fedan JS (1982) Cotransmitters in the motor nerves of the guinea pig vas deferens: electrophysiological evidence. Science 218:693–695

Snyder SH (1985) Adenosine as a neuromodulator. Annu Rev Neurosci 8:103–124

Sosnowski M, Yaksh TL (1990) The role of spinal and brainstem adenosine receptors in the modulation of the volume-evoked micturition reflex in the unanesthetized rat. Brain Res 515:207–213

Speakman MJ, Walmsley D, Brading AF (1988) An in vitro pharmacological study of the human trigone – a site of non-adrenergic, non-cholinergic neurotransmission. Br J Urol 61:304–309

Sperlágh B, Vizi ES (1991) Effect of presynaptic P_2 receptor stimulation on transmitter release. J Neurochem 56:1466–1470

Spitsbergen JM, Clemow DB, McCarty R, Steers WD, Tuttle JB (1998) Neurally mediated hyperactive voiding in spontaneously hypertensive rats. Brain Res 790:151–159

Sprague RS, Ellsworth ML, Stephenson AH, Kleinhenz ME, Lonigro AJ (1998) Deformation-induced ATP release from red blood cells requires CFTR activity. Am J Physiol 275:H1726–H1732

Stanton SL, Kerr-Wilson R, Harris VG (1980) The incidence of urological symptoms in normal pregnancy. Br J Obstet Gynaecol 87:897–900

Steers WD, Mackway-Gerardi AM, Ciambotti J, de Groat WC (1994) Alterations in neural pathways to the urinary bladder of the rat in response to streptozotocin-induced diabetes. J Auton Nerv Syst 47:83–94

Stewart FA (1986) Mechanism of bladder damage and repair after treatment with radiation and cytostatic drugs. Br J Cancer 53:280–291

Suzuki H, Kokubun S (1994) Subtypes of purinoceptors in rat and dog urinary bladder smooth muscles. Br J Pharmacol 112:117–122

Tagliani M, Candura SM, Di Nucci A, Franceschetti GP, D'Agostino G, Ricotti P, Fiori E, Tonini M (1997) A re-appraisal of the nature of the atropine-resistant contraction to electrical field stimulation in the human isolated detrusor muscle. Naunyn Schmiedebergs Arch Pharmacol 356:750–755

Taira N (1972) The autonomic pharmacology of the bladder. Annu Rev Pharmacol 12:197–208

Takeda M, Lepor H (1995) Nitric oxide synthase in dog urethra: a histochemical and pharmacological analysis. Br J Pharmacol 116:2517–2523

Tammela T, Kontturi M, Lukkarinen O (1986) Postoperative urinary retention. II. Micturition problems after the first catheterization. Scand J Urol Nephrol 20:257–260

Tammela T, Lasanen L, Waris T (1990) Effect of distension on adrenergic innervation of the rat urinary bladder. Urol Res 18:345–348

Tammela TL, Wein AJ, Levin RM (1992) Effect of tetrodotoxin on the phasic and tonic responses of isolated rabbit urinary bladder smooth muscle to field stimulation. J Urol 148:1937–1940

Theobald RJ Jr (1982) Arylazido aminopropionyl ATP ($ANAPP_3$) antagonism of cat urinary bladder contractions. J Auton Pharmacol 2:175–179

Theobald RJ Jr (1983a) The effect of arylazido aminopropionyl ATP on atropine resistant contractions of the cat urinary bladder. Life Sci 32:2479–2484

Theobald RJ Jr (1983b) Evidence against purinergic nerve fibres in the hypogastric nerves of the cat. J Auton Pharmacol 3:235–239

Theobald RJ Jr (1986a) The effect of arylazido aminopropionyl ATP ($ANAPP_3$) on inhibition of pelvic nerve evoked contractions of the cat urinary bladder. Eur J Pharmacol 120:351–354

Theobald RJ Jr (1986b) Changes in ureteral peristaltic activity induced by various stimuli. Neurourol Urodyn 5:11–22

Theobald RJ Jr (1992) Subclasses of purinoceptors in feline bladder. Eur J Pharmacol 229:125–130

Theobald RJ Jr (1995) Purinergic and cholinergic components of bladder contractility and flow. Life Sci 56:445–454

Theobald RJ Jr, de Groat WC (1989) The effects of purine nucleotides on transmission in vesical parasympathetic ganglia of the cat. J Auton Pharmacol 9:167–182

Theobald RJ Jr, Hoffman V (1986) Long-lasting blockade of P_2-receptors of the urinary bladder in vivo following photolysis of arylazido aminopropionyl ATP, a photoaffinity label. Life Sci 38:1591–1595

Thornbury KD, Hollywood MA, McHale NG (1992) Mediation by nitric oxide of neurogenic relaxation of the urinary bladder neck muscle in sheep. J Physiol 451:133–144

Tomlinson DR, Yusof AP (1983) Autonomic neuropathy in the alloxan-diabetic rat. J Auton Pharmacol 3:257–263

Theobald RJ Jr (1986b) Changes in ureteral peristaltic activity induced by various stimuli. Neurourol Urodyn 5:11–22

Tong Y-C, Hung Y-C, Lin JSN, Hsu C-T, Cheng J-T (1995) Effects of pregnancy and progesterone on autonomic function in the rat urinary bladder. Pharmacology 50:192–200

Tong YC, Hung Y-C, Cheng JT (1997a) Evidence of P_{2Y}-purinoceptor mediated bladder neck smooth muscle post-contractile relaxation in the male mini-pig. Neurosci Lett 225:181–184

Tong YC, Hung Y-C, Shinozuka K, Kunitomo M, Cheng JT (1997b) Evidence of adenosine 5'-triphosphate release from nerve and P2X-purinoceptor mediated contraction during electrical stimulation of rat urinary bladder smooth muscle. J Urol 158:1973–1977

Tran LV, Somogyi GT, de Groat WC (1994) Inhibitory effect of neuropeptide Y on adrenergic and cholinergic transmission in rat urinary bladder and urethra. Am J Physiol 266:R1411–R1417

Trezise DJ, Humphrey PPA (1997) Activation of cutaneous afferent neurones by ATP. In: Olesen J, Moscowitz MA (eds) Experimental Headache Models, Frontiers in Headache Research, Vol 6. Raven Press, New York, pp 111–116

Turan C, Zorlu CG, Ekin M, Hancerliogullari N, Saracoglu F (1996) Urinary incontinence in women of reproductive age. Gynecol Obstet Invest 41:132–134

Udoh FV (1995) Effects of leaf and root extracts of *Nauclea latifolia* on purinergic neurotransmission in the rat bladder. Phytother Res 9:239–243

Ursillo RC (1961) Investigation of certain aspects of atropine-resistant nerve effects. J Pharmacol Exp Ther 131:231–236

Ursillo RC, Clark BB (1956) The action of atropine on the urinary bladder of the dog and on the isolated nerve-bladder strip preparation of the rabbit. J Pharmacol Exp Ther 118:338–347

Usune S, Katsuragi T, Furukawa T (1996) Effects of PPADS and suramin on contractions and cytoplasmic Ca^{2+} changes evoked by AP_4A, ATP and α,β-methylene ATP in guinea-pig urinary bladder. Br J Pharmacol 117:698–702

Uvelius B (1986) Detrusor smooth muscle in rats with alloxan-induced diabetes. J Urol 136:949–952

Uvelius B, Gabella G (1995) Intramural neurones appear in the urinary bladder wall following excision of the pelvic ganglion in the rat. Neuroreport 6:2213–2216

Vaidyanathan S, Rao MS, Mapa MK, Rao K, Sharma PL, Chary KS (1982) Detrusor supersensitivity to 15(S),15-methyl prostaglandin $F_2\alpha$ in chronic neurogenic vesical dysfunction. Indian J Med Res 75:839–845

Vale JA, Liu K, Whitfield HN, Trott KR (1994) Post-irradiation bladder dysfunction: muscle strip findings. Urol Res 22:51–55

Valera S, Hussy N, Evans RJ, Adani N, North RA, Surprenant A, Buell G (1994) A new class of ligand-gated ion channel defined by P_{2X} receptor for extra-cellular ATP. Nature 371:516–519

Valera S, Talabot F, Evans RJ, Gos A, Antonarakis SE, Morris MA, Buell GN (1995) Characterization and chromosomal localization of a human P_{2X} receptor from the urinary bladder. Receptors Channels 3:283–289

Vaught JD, Kropp BP, Sawyer BD, Rippy MK, Badylak SF, Shannon HE, Thor KB (1996) Detrusor regeneration in the rat using porcine small intestinal submucosal grafts: functional innervation and receptor expression. J Urol 155:374–378

Vizzard MA, Erdman SL, de Groat WC (1996) Increased expression of neuronal nitric oxide synthase in bladder afferent pathways following chronic bladder irritation. J Comp Neurol 370:191–202

Vlaskovska MV, Kasakov L, Rong W, Bodin P, Bardini M, Koch B, Cockayne D, Ford A, Burnstock G (2000) P2X$_3$ knockout mice reveal a major sensory role for urothelial released ATP. Soc Neurosci Abst (In Press)

von Kügelgen I (1994) Purinoceptors modulating the release of noradrenaline. J Auton Pharmacol 14:11–12

von Kügelgen I, Schöffel E, Starke K (1989) Inhibition by nucleotides acting at presynaptic P_2-receptors of sympathetic neuro-effector transmission in the mouse isolated vas deferens. Naunyn Schmiedebergs Arch Pharmacol 340:522–532

Vulchanova L, Arvidsson U, Riedl M, Wang J, Buell G, Surprenant A, North RA, Elde R (1996) Differential distribution of two ATP-gated channels (P2X receptors) determined by imunocytochemistry. Proc Natl Acad Sci USA 93:8063–8067

Vulchanova L, Arvidsson U, Riedl M, Wang J, Buell G, Surprenant A, North RA, Elde R (1997) Imunocytochemical study of the P2X$_2$ and P2X$_3$ receptor subunits in rat and monkey sensory neurons and their central terminals. Neuropharmacology 36:1229–1242

Wada Y, Yoshida M, Kitani K, Kikukawa H, Ichinose A, Takahashi W, Gotoh S, Inadome A, Machida J, Ueda S (1995) Comparison of the effects of various anticholinergic drugs on human isolated urinary bladder. Arch Int Pharmacodyn Ther 330:76–89

Wammack R, Weihe E, Dienes HP, Hohenfeller R (1995) Die neurogene blase in vitro. Akt Urol 26:16–18

Wangemann P (1996) Ca^{2+}-dependent release of ATP from the organ of Corti measured with a luciferin-lucifrease bioluminescence assay. Auditory Neurosci 2:187–192

Waterman SA (1996) Multiple subtypes of voltage-gated calcium channel mediate transmitter release from parasympathetic neurons in the mouse bladder. J Neurosci 16:4155–4161

Watt WC, Lazarowski ER, Boucher RC (1998) Cystic fibrosis transmembrane regulator-independent release of ATP. Its implications for the regulation of P2Y$_2$ receptors in airway epithelia. J Biol Chem 273:14053–14058

Webb HE, Chew-Lim M, Jagelman S, Oaten SW, Pathak A, Suckling AJ, MacKenzie A (1978) Semliki forest virus infections in mice as a model for studying acute and chronic CNS virus infections in man. In: Clifford Rose F (ed) Clinical Neuroimmunology. Blackwell, Oxford, pp 369–390

Weetman DF, Turner N (1977) The effects of ATP-receptor blocking agents on the response to the guinea-pig isolated bladder preparation to hyoscine-resistant nerve stimulation. Arch Int Pharmacodyn Ther 228:10–14

Welford LA, Cusack NJ, Hourani SMO (1987) The structure-activity relationships of ectonucleotidases and of excitatory P_2-purinoceptors: evidence that dephosphorylation of ATP analogues reduces pharmacological potency. Eur J Pharmacol 141:123–130

Werkström V, Persson K, Andersson K-E (1997) NANC transmitters in the female pig urethra – localization and modulation of release via α_2-adrenoceptors and potassium channels. Br J Pharmacol 121:1605–1612

Westfall DP, Fedan JS, Colby J, Hogaboom GK, O'Donnell JP (1983) Evidence for a contribution by purines to the neurogenic response of the guinea-pig urinary bladder. Eur J Pharmacol 87:415–422

Westfall TD, Kennedy C, Sneddon P (1997) The ecto-ATPase inhibitor ARL 67156 enhances parasympathetic neurotransmission in the guinea-pig urinary bladder. Eur J Pharmacol 329:169–173

White PN, Thorne PR, Housley GD, Mockett B, Billett TE, Burnstock G (1995) Quinacrine staining of marginal cells in the stria vascularis of the guinea-pig cochlea: a possible source of extracellular ATP? Hear Res 90:97–105

White TD (1984) Characteristics of neuronal release of ATP. Prog Neuropsychopharmacol Biol Psychiatry 8:487–493

Wiklund NP, Gustafsson LE (1986) Neuromodulation by adenine nucleotides, as indicated by experiments with inhibitors of nucleotide inactivation. Acta Physiol Scand 126:217–223

Wiklund NP, Gustafsson LE (1988) Indications for P_2-purinoceptor subtypes in guinea pig smooth muscle. Eur J Pharmacol 148:361–370

Wu C, Wallis WRJ, Fry CH (1995) Purinergic activation induces a transient rise of intracellular Ca^{2+} in isolated human detrusor myocytes. J Physiol 489:136P–137P

Wu C, Bayliss M, Newgreen D, Mundy AR, Fry CH (1999) A comparison of the mode of action of ATP and carbachol on isolated human detrusor smooth muscle. J Urol 162:1840–1847

Yang SJ, An JY, Shim JO, Park CH, Huh IH, Sohn UD (2000) The mechanism of contraction by 2-chloroadenosine in cat detrusor muscle cells. J Urol 163:652–658

Yokota T, Yamaguchi O (1996) Changes in cholinergic and purinergic neurotransmission in pathologic bladder of chronic spinal rabbit. J Urol 156:1862–1866

Yoshimura N (1999) Bladder afferent pathway and spinal cord injury: possible mechanisms inducing hyperreflexia of the urinary bladder. Prog Neurobiol 57:583–606

Yoshimura N, de Groat WC (1997) Plasticity of Na^+ channels in afferent neurones innervating rat urinary bladder following spinal cord injury. J Physiol 503:269–276

Yoshimura N, de Groat WC (1999) Increased excitability of afferent neurons innervating rat urinary bladder after chronic bladder inflammation. J Neurosci 19:4644–4653

Zar MA, Iravani MM, Luheshi GN (1990) Effect of nifedipine on the contractile responses of the isolated rat bladder. J Urol 143:835–839

Zderic SA, Duckett JW, Wein AJ, Snyder HM, III, Levin RM (1990) Development factors in the contractile response of the rabbit bladder to both autonomic and non-autonomic agents. Pharmacology 41:119–123

Zhang SX, Kobayashi T, Okada T, Garcia del Saz E, Seguchi H (1991) Alkaline phosphatase, 5'-nucleotidase and magnesium-dependent adenosine triphosphatase activities in the transitional epithelium of the rat urinary bladder. Histol Histopathol 6:309–315

Zhao M, Bo X, Neely CF, Burnstock G (1996) Characterization and autoradiographic localization of [^{3}H] α,β-methylene ATP binding sites in cat urinary bladder. Gen Pharmacol 27:509–512

Zhao Y, Wein AJ, Levin RM (1993) Role of calcium in mediating the biphasic contraction of the rabbit urinary bladder. Gen Pharmacol 24:727–731

Zhong Y, Dunn PM, Burnstock G (2000) Multiple P2X receptors on guinea pig pelvic ganglion neurons exhibit novel pharmacological properties. Br J Pharmacol (In Press)

Zhong Y, Dunn PM, Xiang Z, Bo X, Burnstock G (1998) Pharmacological and molecular characterization of P2X purinoceptors in rat pelvic ganglion neurons. Br J Pharmacol 125:771–781

Ziganshin AU, Hoyle CHV, Bo X, Lambrecht G, Mutschler E, Bäumert HG, Burnstock G (1993) PPADS selectively antagonizes P_{2X}-purinoceptor-mediated responses in the rabbit urinary bladder. Br J Pharmacol 110:1491–1495

Ziganshin AU, Hoyle CHV, Ziganshina LE, Burnstock G (1994) Effects of cyclopiazonic acid on contractility and ecto-ATPase activity in guinea-pig urinary bladder and vas deferens. Br J Pharmacol 113:669–674

Ziganshin AU, Ziganshina LE, Hoyle CHV, Burnstock G (1995a) Effects of divalent cations and La^{3+} on contractility and ecto-ATPase activity in the guinea-pig urinary bladder. Br J Pharmacol 114:632–639

Ziganshin AU, Ralevic V, Burnstock G (1995b) Contractility of urinary bladder and vas deferens after sensory denervation by capsaicin treatment of newborn rats. Br J Pharmacol 114:166–170

Zimmermann H (1978) Turnover of adenine nucleotides in cholinergic synaptic vesicles of the Torpedo electric organ. Neuroscience 3:827–836

Zimmermann H (1982) Co-existence of adenosine 5'-triphosphate and acetylcholine in the electromotor synapse. In: Cuello AC (ed) Co-transmission. MacMillan Press, London, pp 243–259

Zimmermann H (1996) Biochemistry, localization and functional roles of ecto-nucleotidases in the nervous system. Prog Neurobiol 49:589–618

Zoubek J, Somogyi GT, de Groat WC (1993) A comparison of inhibitory effects of neuropeptide Y on rat urinary bladder, urethra, and vas deferens. Am J Physiol 265:R537–R543

Subject Index

A_1 receptor   19, 108, 113, 251–253, 374, 378
– agonists   133–136
– allosteric modulators   29
– antagonists   140–142
– bronchoconstriction   21
– Ca^{2+} currents   28
– G-proteins   27, 31
– G_α subunits   27
– inositol phosphate production   28
– interaction with other receptors   27, 28
– lipolysis   21
– nociception   21
– PD 81 723   29
– renin release   21
– tissue distribution   20
A_{2A} receptor   19, 20, 23, 251–253
– A_2-like-receptor   253, 254
– acetylcholine release, effect on   255
– agonists   136–138
– antagonists   142–144
– atypical   254, 255
– calcium channels, activation   255
– dopamine, interaction   256, 264
– functional roles   21
– GABA-receptor, interaction   256
– interaction with A_1 255
– tissue distribution   21
A_{2B} receptor   19, 20, 251, 253, 323
– adenylate cyclase   30
– antagonists   145
– functional roles   21
– phospholipase C   30
– tissue distribution   22
A_3 receptor   19, 20, 251, 253, 264, 376
– antagonists   145–147
– functional role   23
– tissue distribution   23
A3P5PS, structure   159
ABC proteins   463
ABT-702   5, 147

acetylcholine   22, 115
aciduria, chronic orotic   407
ADA immunodeficiency syndrome   309, 315
ADAC   134, 135
– structure   134
addiction   260, 261
adenosine
– adenosine 5′-(β,γ-imido)triphosphate   318
– adenosine 5′-O-(3-thiotrisphosphate)   71
– adenosine collateral hypothesis   307
– analgesic effects   385, 388–392
– – sites of action   391
– binding proteins   32–34
– central sensitization   388
– effect on neurotransmitter release   253, 255, 378
– extracellular concentration in brain   268–271
– functional role   264–267
– half life in CSF   387
– intraoperative use   390
– levels, in brain   182
– phosphorylation   271
– protective effect
– – epilepsy   267
– – heart   308
– – in inflammation   376, 382
– release   181–183, 188
– transporters   268, 269
adenosine deaminase   105, 114, 182, 271, 350, 455
adenosine kinase   5, 181, 268, 271, 376
– inhibitors   147
adenotin
– -1   20, 33, 34
– -2   20, 34
adenyl 5′-(β,γ-methylene)-diphosphonate   430
adenylate charge   4

adenylate cyclase 21, 30
adenylate kinase 181
ADP 181, 346
– α,β-meADP 154, 227, 297
– – structure 147
– ADPβS 317, 446, 449
– – structure 150, 297
– hydrolysis, inhibitors 219
– ecto-ribosylation 234, 235
– 2-meSADP 151
– – structure 150
– 2-methylthioADP 71
α_2-adrenoceptors 188
– agonists 190
AIT-082 314
alcohol 482
allodynia 379
Alzheimer's disease 311, 312
aminophylline 318
AMP 181
– AMP 579, structure 132
– deaminase inhibitor, structure 147
AMPA/kainate 190
analgesia 133, 257, 258
ANAPP$_3$ 292, 434, 451
angiogenesis 97, 306, 307, 341, 350
angiotensin-II 193
anticonvulsant activity 132
antiviral therapy 408
anxiolysis 257
Ap3A, structure 150
Ap4A 9, 148, 181, 276, 430
– structure 150
Ap5A 9, 181
– structure 151
apamin 104, 435
APEC 137, 138
– structure 136
APNEA, structure 132
apoptosis 95, 100, 138, 265, 305, 308, 309, 312, 314, 315, 326
– purine release in 199
apyrase 3, 266
– conserved regions 211
l-arginine transport 106
ARL 66,096, structure 159, 162
ARL 67156 298, 438
ARL 69931MX 162
arousal 258, 259
arthritis 230
ASA 307
astrocytes 310–312
atherosclerosis 308
ATP
– $\alpha\beta$meATP 48, 49, 51, 55, 58, 95, 153, 154, 292, 299, 310, 311, 322, 325, 345, 346, 347, 351, 377, 428, 430, 435, 439, 441, 458, 463, 468, 479, 486
– – structure 149
– 2-(4-Aminophenylethylthio)ATP, structure 151
– ATP/ADP ratio 214
– ATPαS 446
– – structure 150
– ATP-β-F 218
– ATP-β-S 218
– ATP-γ-F 218
– ATPγS 104, 106, 155, 441
– ATPγS, structure 151
– $\beta\gamma$meATP 154, 311, 427, 436, 437, 443, 448, 456, 482
– – structure 149
– $\beta\gamma$me-L-ATP, structure 149
– 3'-Benzylamino 3'doxy-ATP 154, 155
– – structure 151
– 8-bromo-ATP 218
– 2-chloro-ATP 218
– 2-(7-Cyanohexylthio)-ATP 153
– extracellular, sources 184
– 2-HexylthioATP, structure 151
– hydrolysis, inhibitors 219
– 2meSATP 49, 97, 152, 153, 306, 311, 323, 345, 347, 351, 443, 448
– – structure 149
– 2meS-β,γ-meATP, structure 150
– 2MeS-L-ATP, structure 150
– 2-methylthioATP 71
– – structure 158
– release 292, 295, 379
– – from astrocytes 310
– – in kidney 316, 317
– – involvement of calcium channels 297
– – urothelial cells 462
– response in tissues 273
– releasable stores 179–181, 184, 196
– release 183–193
– – tetrodotoxin-insensitive 190
– [^{35}S]ATP 57
– structure 149
– synaptic response 275
ATPase 437
– ecto 210, 354
– F-type 209
– P-type 209
– V-type 209
ATPase, ecto 354
autotaxin 225, 230
avoidance learning 225

basilen blue 158
BG 9719 142

Subject Index

BIIP20, structure 141
bisphosphate analogues 72
bladder
– dysfunction 298
– grafts 474
– hyperactive 473
– hyperreflexive 479
bladder outflow obstruction 478
blood flow, coronary 1
bone 104, 105
bone cells 319, 320
bradycardia 133, 142
bradykinin 194
Brilliant Blue G 160
bronchoconstriction 22
– A_1 receptor 21
BW-A 1433, structure 140
BzATP 49, 50, 55, 155, 156, 161
– structure 151

Cl-IB-MECA 7
2-CADO 2, 32, 257, 269, 314, 317, 358, 452
caffeine 2, 7, 130, 257, 258, 260, 392, 454
– structure 140
calcification 230
calculosis 470
cAMP efflux 270
cancer 315, 326, 407–409, 416
capacitation 89, 320
cardiovascular system 95–97, 109–111
catalepsy 264, 414
CB-MECA, structure 139
CCPA 133
– structure 134
CD38 234
CD39 211, 224
CD39L3 211
CD73 231
CDP 214
central nervous system 101, 102, 106–108
– role of purines and pyrimidines 251–276
CFTR (cystic fibrosis transmembrane conductant regulator) 141
CGRP (calcitonin gene-related peptide) 90
CGS 15943, structure 140
CGS 21680 7, 107, 130, 136–138, 260, 264, 375
– structure 136
CHA 7, 107
– structure 134
chemotaxis, inhibition 23
chick embryo 91–98
chloride secretion 22

2-chloroadenosine (2-CADO) structure 132
cholecystokinin CCK_1 receptor 29
chondrocytes 98
chronic bronchitis 403
CI-936 264
– structure 132
cibacron blue 52, 155, 158, 218, 431
Cl-IB-MECA 139, 145
CNS modulation 410–412
cochlear 351–359
coma, hepatic 405
condo red 218
ω-conotoxin 193
Coomassie Brilliant blue 218, 443, 447
COPD (chronic obstructive pulmonary disease) 403
coronary artery disease 385
cotransmission
– by ATP 3, 289, 296, 299, 426, 427, 431, 433, 452
– sympathetic 450, 451
COX-1 376
COX-2 311, 326, 376
CP 66713, structure 140
CPA 7, 133, 260, 264
– structure 134
CPT 7
CPX 7
CSC 143, 144, 254, 350
– structure 143
CTP 214, 428
– structure 151
CVT 124 142
– structure 141
CVT-510, structure 134
cystic fibrosis 403, 462, 463
– uridine nucleotides 406
cystitis, interstitial 298, 483
cytokines 305, 311, 376

dendogram analysis 73
dense cored vesicles, large / small 180
development, purinergic signalling 89–116
– bladder 465–468
– embryonal development 89–106
– postnatal development 106–116
developmental delay, pervasive 407
diabetes 230, 485
diadenosine polyphosphate (see also Ap3A; Ap4A; Ap5A) 230, 231
– physiological effects 276
DIDS, structure 159
dihydropyridine 146
diphosphokinase, ecto-nucleoside 233
dipyridamole 307, 314, 315, 454

diseases (see syndromes)
divalent cations   56
DMPX   358, 375
L-DOPA   143
dopamine   190
dopamine system, interaction with adenosine   263
dorsal horn neurons   373, 374
– actions by purines and pyrimidines on 376–380
DPCPX   102, 140, 141, 264, 267, 358
– structure   141
DPMA   137
– structure   136
drug addiction   257
Duchenne muscular dystrophy   309
duodenum, contraction / relaxation   113
dipyridamole   182

ectoapyrase   199
ectoenzymes   105
– ecto-5′-nucleotidase   3, 210, 231, 232, 266
EF-hand   227, 228
EHNA   343
energetic condition   195
energy
– demand   1
– deprivation, release of purines   195, 196
– energy supply / demand balance   5, 19
enkephalin   269
E-NPP   209
– catalytic properties   226, 227
– cell and tissue distribution   228, 229
– functional roles   229, 230
– types and structural properties   225, 226
enprofylline   145
E-NTPDase   148, 154, 157, 209, 211–225
– apyrase conserved regions   216, 217
– general structural properties   212–214
– inhibitors   217–219
– physiological and pathological implications   223–225
– regulation of expression   223
– subtypes   211
– tissue distribution   219–223
– vertebrate isoforms   211, 212
epilepsy   407, 416
– adenosine levels   267
epileptic episode   2
epileptic seizures   265
ERK   321, 323
ethanol   261–263

Evans blue   232
excitatory transmission
– by ATP   374, 376, 442
– by pyrimidines   410
– by UTP   449
excitotoxicity, kainate-induced   137
exocytosis, tetrodotoxin sensitive   183

fertilisation   89
fibroblasts   318, 319
FK 453, structure   141
FMRFamide peptide-gated sodium channels   47
FPL 67156   217
FR 113452, structure   141
frog embryo   89–91
5′FSBA   219

G protein-coupled receptors (GPCR)   66
GABA   22, 190, 252, 253, 345
galactosaemia   407
ganglia   97, 98
gastrointestinal tract   104
gastrulation   90
GDP   214
GFAP   311
GIRK   252, 274
glaucoma   359
glial cells   94, 95
gliosis   310, 311
glucose transport   351
glutamate receptor
– activation   182
– effects of extracellular adenosine concentration   269
glycosylation
– asparagine-linked   77
– N-linked   24
GP 3269   5
G-proteins   26, 27, 30
GR 79236   135
– structure   134
GTP   214, 428, 431
guanine nucleotides   181
guanosine   100, 313
gustatory system   341

H_1 histamine receptor   29
H-9   34
HB6   211
hearing loss   359
heart   109
– congestive failure   142
– ischemic disease   384
HE-NECA   130, 138

– structure 136
hepatic coma 405
hibernation 474
His-251 in TM 6 24
His-278 in TM 7 24
histamine
– H_2 194
– plasma level 139
– release 23
homeostatic modulator 1
HT-AMP 149, 156, 157
hyperalgesia 379
hyperemia 349
– reactive 1
hypertension 308
hypoglycemia, release of purines 195, 196, 199
hypotension 133, 139
hypotensive agent 1
hypotonia, cellular, rellease of purines 198, 199
hypoxanthine 100, 349
hypoxia 19, 110, 111, 265, 266
– release of purines 195, 196, 199
– cardial 21

I-AB-MECA 133
– structure 132
I-ABOPX, structure 140
IB-MECA 130, 138, 139
– structure 139
IB-MECA 7
IDP 214
IES 218
IL-1β 55, 315
immune cells 314–316
incontinence 470
inflammation 311, 312
– release of purine 196–199
inhibitory transmission
– adenosine 454
– in urinary bladder 455
inosine 32, 100, 182
inspiratory neurons 107
insulin
– resistance 230
– secretion 116, 153, 317
integrins 77
ion channels, regulation 81
Ip5I 9, 148, 162
ischemia 136, 138, 140, 182, 197, 224, 225, 265, 311, 317, 326, 349, 416
ischemic preconditioning 266, 308
isolectin B4 51
ITP 214, 431, 452
– structure 151

jet lag 416

K-252a 34
K_{ATP} channel 382, 393
KF 17837 254, 264
– structure 143
KFM 19 142
– structure 141
kidney
– cells 316, 317
– dysfunction 416
KN-62 9, 55, 148, 162, 163, 346
– structure 159
knockout mice 10, 258, 320
– P2X$_3$ 463
KT 5720 34
KT 5723 34
KW 3902, structure 141
KW 6002 144, 264
– structure 143

L-249313 145
– structure 146
L-AMP-PCP 154
LALP70 211
lead poisoning 406
Lesch-Nyhan syndrome 404
lipolysis 21
– A_1 receptor 21
liver
– dysfunction 416
– failure 405
locomotor activity / effects 257, 264
– depression 23
– spontaneous 21
LPS 196
LTP 185, 187, 254, 273
luciferin-luciferase assay 427
lung 105
– surfactant secretion 111

MAPK 106, 321, 324
– cascades 82
mechanical stress, ATP release 462
mechanosensory signalling, bladder 463
memory deficit 416
Meniere's disease 359
Merkel cells 104
metabolite, retaliatory 1, 195
metrifudil 130–133
– structure 132
micturition 473, 479
mitochondria 10
– function 269
– purinergic receptors 9, 10
molecular modeling 129

motor impairment 386
MRE 3008-F20 7
– structure 146
MRS-1191 146
MRS 1220 146
MRS 1754 145
MRS 2142 161
MRS 2160 161
MRS 2179 162
– structure 159
MRS 2191, structure 159
MRS 2192 161
MRS 2209 162
MRS 2220 161
multiple sclerosis 481, 482
muscarinic receptors 194
muscular dystrophy, Duchenne 309
myelination, active 230
myokinase, ecto- 233
myotube depolarisation 95

N-0861, structure 141
Na$^+$/H$^+$ exchanger regulatory factor 77–79
NAD$^+$ hydrolysis, extracellular 233, 234
NAD$^+$-glycohydrolase 234
L-NAME 196, 358
NANC (non-adrenergic, non-cholinergic transmission) 2, 104, 423, 427, 438, 439, 441, 450, 458, 467, 469
NBI (*see also* nitrobenzylthioinosine) 181
NECA 26, 32, 107, 109, 130, 138, 142, 314, 318, 324, 450
– [^{3}H]NECA 343
– N^6-benzyl-NECA 376
– structure 132
neurite outgrowth 32
neurogenesis 313
neuroleptics 21
neuronal activity 19
neuropathy, peripheral nerve , diabetic-induced 407
neuropeptide Y 193, 289
neuroprotection 343
– role of A$_1$ receptors 265–267
neurotransmission
– retinal 342–344
– somatosensory 371–383
neurotransmitter
– criteria to define 183
– sympathetic 290–292
neutrophil activation 21
NF 023 9, 157
– structure 158
NF 279 157
– structure 158

nicotine 113
nicotinic channels, sensitivity to ATP 90
nicotinic cholinergic receptor construct 10
nicotinic receptors 194
nitrobenzylthioinosine (*see also* NBI) 314
7-nitroindazole 196
NMDA 190
– receptor, blokade by suramin 381
NNC 21-0136 136
– structure 134
NNC 21-0238 140
– structure 139
NNC 53-0055, structure 139
NO (nitric oxid) 182, 289, 353, 448
– interaction with adenosine 257
– interaction with ATP 452, 461, 468
– nitric oxide synthase 113, 196
nociception 21, 110, 340
– in bladder 463
– clinical aspects 383–393
– roles of purines 371–393
NPP 226
– E-NPP (*see there*)
NTPDase 157
– catalytic properties 214–216
– E-NTPDase (*see there*)
– *figure* 213
– NTPDase-1-deficient mice 224
– subtypes 211, 212
nucleoside transport 195
nucleotidase
– ecto-5′-nucleotidase 3, 210, 231, 232, 266
– ecto-nucleotidases 209–236, 266, 438
– 5′-nucleotidase 95, 107, 195, 408
– – in ocular dominance 101
– – redistribution 101
– soluble 3
nucleotide diphosphokinase 215
nucleotide metabolism 210, 298, 310, 343, 379, 437
– extracellular 270, 271

oestrogens, interaction with purines 472
6-OHDA 184, 188, 190
olfactory system 341
oligonucleotides, antisense 327
one letter amino acid code 66
OPRT (oritidine phosphoribosyl transferase) deficiency 404
oxidative phosphorylation, mitochondrial 179
oxygen species, reactive 312

Subject Index

P1 receptor   6–8, 10
- agonists, non-selective   130–133
- antagonists   140–147
- - recognition   24
- in bone   319, 320
- cancer   326
- in cardiovascular diseases   306, 307
- in cell growth   305–307, 313, 319, 350, 357
- distribution   310, 347
- in endocrine cells   317, 318
- expression   343
- G-proteins   26
- glycosilation   24
- growth factors, interaction   317, 318
- in hearing   351–359
- in immune diseases   314–316
- ligands   130–147
- localization and function   19–23
- modulators   147
- nervous tissue distribution   251, 252
- in nociception   375, 380
- olfactory transduction   341
- palmitoylation   24
- in retina   344
- in sensory system   339–359
- Ser-277   26
- signalling   313, 316, 320–325, 343, 350, 351, 355
- signal transduction   26–32
- species differences   24
- structure   23–26
- TM 1-4   26
- TM 5   26
- in urinary bladder   450
- in vision   341–350

P2X receptor   8, 9, 10, 47, 48, 372
- agonists   149–157
- allosteric regulation   57
- antagonists   157–163
- in auditory transmission   356
- binding assay   148
- in bone   319, 320
- calcium influx   293
- cancer   326
- in cell growth   305–307, 313, 319, 350, 357
- desensitisation   53, 148, 155, 295, 298, 428, 430, 443, 468
- distribution   272, 273, 310, 347, 373, 443, 448
- in endocrine cells   317, 318
- in exocrine cells   316
- functional roles   296
- growth factors, interaction   317, 318
- in hearing   351–359

- heteropolymeric   48, 57, 58, 455
- - in nociceptive sensory neurones   464
- - in spinal cord   377
- homomeric, effects of ions   56, 57
- in immune diseases   314–316
- ligands   147–163
- localization   356
- multimers   148
- in nociception   375, 380
- olfactory transduction   341
- $P2X_1$ 48, 49, 110, 115, 155, 161, 275, 292, 487
- $P2X_2$ 50, 58, 158, 161, 272, 274, 275, 340, 351, 354, 374, 455
- - subunit   345
- $P2X_3$ 50, 51, 58, 102, 161, 340, 373, 380, 455, 459, 460, 463, 487
- $P2X_4$ 51–53, 58, 96, 111, 158, 272, 374, 377, 446
- $P2X_5$ 53, 58, 272, 275
- $P2X_6$ 53, 58, 272, 345, 374, 377
- $P2X_7$ 47, 53–55, 105, 108, 155, 161, 198, 315, 320, 346
- - pore formation   54
- - receptor   190
- pharmacological profile   49, 443, 446
- in retina   344
- in sensory system   339–359
- signalling   313, 316, 320–325, 350, 351, 355, 434, 464
- structure   49, 357
- structure activity studies   446
- in vision   341–350

P2Y receptor   9, 10, 65
- agonists   70, 71, 149–157
- antagonists   71, 72, 157–163
- asparagine-linked glycosilation   77
- associations
- - integrins   77
- - Na^+/H^+ exchanger regulatory factor   77–79
- ATP as an antagonist   69
- in auditory transmission   356
- in bone   319, 320
- cancer   326
- cell signalling   80
- coupling to phospholipase C   80
- cyclic AMP, regulation   80, 81
- dendogram   73
- desensitisation   79
- distribution   272, 273, 310, 347, 373
- in endocrine cells   317, 318
- in exocrine cells   316
- G protein-coupled receptors (GPCRs)   66
- growth factors, interaction   317, 318

- in hearing 351–359
- in immune diseases 314–316
- ion channels, regulation 81
- ligands 147–163
- localization 356
- mitogen activated protein kinases (MAP kinases), stimulation 81, 82
- mutations 75
- in nociception 375, 380
- nucleotide selectivity 69
- olfactory transduction 341
- $P2Y_1$ 67, 69, 70, 72, 106, 157, 158, 161, 272, 325, 340, 349, 449
- $P2Y_2$ 67, 106, 109, 111, 156–158, 161, 273, 274, 313, 323, 325, 345, 349, 449
- $P2Y_3$ 72
- $P2Y_4$ 67, 70, 72, 109, 156–158, 161, 273, 313, 325, 355, 449
- $P2Y_6$ 67, 72, 109, 156
- $P2Y_7$ 67
- $P2Y_8$ 89
- $P2Y_{11}$ 67, 72
- pharmacological profile 448
- phospholipase C independent Ca^{2+} mobilisation 80
- phosphorylation sites 77
- in retina 344
- in sensory system 339–359
- sequences 68, 72
- signalling 66, 313, 316, 320–325, 344, 350, 351, 355, 434, 448, 449, 464
- structure 74
- tyrosine kinase, stimulation 81, 82
- in urinary bladder 447
- in vision 341–350
P3 receptor 274
$p21^{ras}$ 32
p38 324
pain 2, 258, 378
- neurogenic 386
- roles of purines 371–393
- sympathetically maintained 372
- threshold 387
pancreatic islets 115
Parkinson's disease 143, 144, 263, 416
PD 81,273 265
PD 81 723 29, 147
permeabilisation, receptor-mediated 53
8-phenyltheophylline 32
phosphatase, ecto-alkaline 209, 232, 233
phosphorylation, ecto- 234, 235
PIA 343
plasma half life, adenosine 1
platelet aggregation 4, 307
PN401 409
porphyria cutanea 407
PPADS 49, 50, 51, 55, 72, 92, 160, 161, 196, 218, 226, 231, 232, 275, 292, 293, 296, 311, 345, 353, 430, 441, 449, 468
- iso-PPADS 161
- – structure 159
- structure 159
propentofylline 312
prostaglandins, release by ATP 436, 487
protein kinase, mitogen-activated (see also MAPK) 32
proton gradient 181
2,2′-pyridylisatogen 426
pseudouridinuria 405
8-PT 448, 450, 468
purine
- availability 3–6
- extracellular level 195
- interaction with other transmitters 456, 458
- micturition 442
- peripheral nervous system 289
- release 3
- – co-release, other transmitters 191, 193
- – effect of pre- or post-synaptic receptors 194, 195
- – regulation 179–200
- – retrograde 191
- trophic roles 305–327
purinergic cascade 3
purinergic signalling in development 89–116
- bladder 465–468
- brain 313
- chick embryos 91
- – cardiovascular system 95–97
- – chondrocytes 98
- – ganglia 97, 98
- – glial cells 94, 95
- – retina 91–93
- – skeletal muscle 95
- frog embryos 89–91
- human embryos 106
- mammalian embryos 98–105
- – bone 104, 105
- – cardiorespiratory system 102, 103
- – central nervous system 101, 102
- – ectoenzymes 105
- – gastrointestinal tract 104
- – lung 105
- – skeletal muscle 103, 104
- – skin 104
- postnatal development
- – airways 111
- – cardiovascular system 109–111
- – central nervous system 106–108
- – gastrointestinal tract 111–114
- – pancreatic islets 115

Subject Index

– – salivary glands 115
– – urinary bladder 115
– – vas deferens 114, 115
purinergic signalling, in urinary bladder 426, 428, 444, 447, 474–478
pyrimidines
– metabolism 404
– – inborn errors 411
– reproductive function 408
– synthesis and salvage 403–405
– trophic roles 305–327

quinidine 423

R-PIA 32, 96, 111, 133, 386
– structure 134
radioligand binding assay, P2 receptor 8
Reactive blue 2 (RB2) 92, 157–160, 311, 428, 443, 447, 448, 455, 468
– structure 158
receptors
– A_1 (see there)
– A_{2A} (see there)
– A_{2B} (see there)
– A_3 (see there)
– cholecystokinin CCK_1 receptor 29
– G protein-coupled receptors 66
– glutamate receptor 182, 269
– H_1 histamine receptor 29
– heteropolymeric 48
– leukotriene receptor 67
– mitochondrial purinergic 9, 10
– nicotinic cholinergic receptor construct 10
– nomenclature 6–8
– P1 (see there)
– P2X (see there)
– P2Y (see there)
– P3 274
– Substance P receptors 194
REM sleep 259
renal failure, chronic 405
renin release 21
– A_1 receptor 21
reperfusion, post-ischemic 349
reproduction 320–325, 408
retina 91–93
retinopathy 350, 359
RG 14202, structure 134
RGD-tripeptide 227, 228

SAH (S-adenosyl-homocysteine) 180, 181, 268
salivary cells 316
salivary glands 115
SCH 58261 7, 33, 137, 142, 143
– structure 143
schizophrenia 263, 407
SDZ WAG 994 135, 136
– structure 134
secretory cells 316–318
sedative actions 2
seizures, pentylenetetrazole-induced 138
Ser-277 26
serotonin 190
signal of life 2
signal transduction, adenosine receptors 26–32
site and event specific 5
skeletal muscle 95, 103, 104
skin 104
sleep 257–259
sodium channels, FMRFamide peptide-gated 47
sodium excretion 142
soluflazine 182
somatomedin B-like domain 227, 228
sperm movement 89
spinal cord damage 473, 479
status epilepticus 225
stroke 2, 416
Substance P receptors 194
8-p-sulfophenyltheophylline, structure 140
suramin 9, 27, 49, 50, 71, 90, 92, 104, 148, 157, 196, 219, 231, 232, 292, 293, 296, 430, 455, 478
– blokade of NMDA receptor 381
– structure 158
synaptic response, ATP 275
synaptic signaling, UTP 5
synaptic transmission, inhibition 253, 266, 358, 380
syndromes / diseases
– Alzheimer's disease 311, 312
– Lesch-Nyhan syndrome 404
– Meniere's disease 359
– Parkinson's disease 143, 144, 263, 416

T-cell, human, inhibition of activation 21
tetanus toxin 193
THA 219
theophylline 7, 140, 257, 258, 266, 454
thermoregulation 260, 415, 416
thrombosis 223, 224
TM 1-4 26
TM 5 26
TNF-α 376
– production 140
TNP-ATP 9, 49, 58, 162

transgenic mice   327
transmission, non-adrenergic, non-cholinergic (NANC)   2, 104, 423, 427, 438, 439, 441, 450, 458, 467, 469
trauma   311
trimeric combinations   9
trimers   47
trypan blue   90, 218
tubercidin   314
tubes and sacs hypothesis   459
d-tubocurarine   155
tumor rejection antigen 1   34

UDP   156, 214
UDPase   211
ureter   469
urethra   467, 468
uridine   403
– antiepileptic effect   415
– barbiturate pharmacophore   415
– brain   404
– dopamine interaction   414
– functional effects   410, 412, 413
– plasma / CSF   405
– salvage   411
– sedative property   415
– sleep inducing factor   415
– thermoregulation   415, 416
urinary bladder   115
urinary retention   473
urinary tract, lower, purinergic signalling   423–487
urine flow rate   142
UTP   154, 156, 181, 214
– structure   149

– synaptic signaling   5
– UTPγS   156
– – structure   149

vanilloid-1   51
vas deferens   114, 115
vascular reperfusion   223, 224
vascular system   109–111
vasectomy   408
vasodilatation   22
– coronary   21
VDCCs   252
VEGF (vascular endothelial growth factor)   306, 341
ventilatory depression   111
veratridine   184
vestibular system   351–359
vitamin E deficiency   483
VTA (ventral tegmental area)   273

Walker ATP binding motifs   216
wound healing   316, 318, 326
WRC-0342   30
WRC-0571   142
– structure   141

XAC, structure   140
XAMR0721   157
– structure   158
xanthine-insensitive effects   33

ZM 241385   144
– structure   143